ACTIVE AND PASSIVE
THIN FILM DEVICES

ACTIVE AND PASSIVE THIN FILM DEVICES

Edited by
T. J. Coutts

*Industrial Science Research Centre
Cranfield Institute of Technology
Cranfield, Bedford
England*

1978

ACADEMIC PRESS
London · New York · San Francisco

A Subsidiary of Harcourt Brace Jovanovich, Publishers

ACADEMIC PRESS INC. (LONDON) LTD.
24/28 Oval Road
London NW1

United States Edition published by
ACADEMIC PRESS INC.
111 Fifth Avenue
New York, New York 10003

Copyright © 1978 by
ACADEMIC PRESS INC. (LONDON) LTD.

All Rights Reserved

No part of this book may be reproduced in any form by photostat, microfilm, or any other means, without written permission from the publishers

Library of Congress Catalog Card Number: 77-75371
ISBN: 0-12-193850-6

Printed in Great Britain by
Thomson Litho Ltd, East Kilbride, Scotland.

Foreword

During the past seventy years, a new branch of learning has emerged, stimulated by the need for new and improved optical and electronic devices in industrial and military applications, and also by the exciting possibilities opened up for basic research into the physics of the solid state and the physics and chemistry of interfaces.

This new branch of science has become known as " thin solid films," and the field has developed so rapidly that more than 25 000 relevant scientific and technical papers have been published, the majority of these since 1940. Most universities, polytechnics and centres of advanced learning throughout the world, in physics, material science or engineering, have conducted research on some aspect of thin film science or technology.

The attractions of the field arise from the wide variety of methods available for depositing thin films, each of which introduces its own characteristic properties. These methods, in general, enable extremely thin layers of very pure materials to be deposited on crystalline or amorphous substrates under precisely controlled conditions, and the possibilities for basic research are manifold. As in most fields, the technology has advanced faster than the science. Publication of the results of both the basic research and the applications is spread over many different journals and reported in many different languages. Thus, there is a need, at appropriate points in time, to take stock of the state of the art, both in terms of the available technology and the basic understanding of thin film phenomena. The global situation regarding publication in this field is quite astonishing, there is evidence of duplication and "re-discovery" occurring, and quite obviously there are many authors who have neglected to read some of the authoritative early papers. The need for a comprehensive review is, therefore, urgent.

Thus, the publication of the present book is timely. It is intended to review the present state of knowledge concerning the applications of thin films in the fields of optics and electronics. It has been prepared by a group of people who are not only expert in their own particular speciality, but who also have a sufficient knowledge of the field as a whole to be aware of developments occurring across a broad front.

The material of this volume has been prepared with a number of aims in mind. Firstly, to introduce the new reader to the subject, secondly to take the development of each field right up to the present day level of application so that the expert in the field will also be able to benefit, and finally to provide the environment for cross-fertilization to occur through a careful subdivision of the subject matter into suitable chapters.

This is a very comprehensive work in the fields that it covers, and its publication represents yet another stepping stone in the technological progress of the subject. Let us hope that its appearance will ensure that new workers in the field will be so enlightened that they will not feel the need to repeat earlier work and will no longer remain unaware of the existence of a massive body of prior art.

April 1978
G. Siddall
The University of Strathclyde,
Glasgow
Editor of the International Journal of
Thin Solid Films

Contents

	Foreword	
	G. Siddall	v
	List of Symbols	ix
Chapter 1	**Practical Applications of Thin Films**	
	T. J. Coutts	1
Chapter 2	**Preparation Methods for Thin Film Devices**	
	D. S. Campbell	23
Chapter 3	**Electrical Properties and Applications of Thin Metal and Alloy Films**	
	T. J. Coutts	57
Chapter 4	**Thin Film Dielectrics**	
	D. S. Campbell	113
Chapter 5	**Electrical Properties and Applications of Metal/Dielectric Films**	
	T. J. Coutts	163
Chapter 6	**Thin Film Transistors**	
	J. C. Anderson	207
Chapter 7	**Amorphous Semiconducting Films**	
	J. R. Bosnell	245
Chapter 8	**Thin Film Optical Devices**	
	H. A. M. Macleod	321
Chapter 9	**Thin Film Photoconductors**	
	V. Vincent	429
Chapter 10	**Thin Film Solar Cells**	
	R. Hill	487

Chapter 11	**Magnetic Thin Films and Devices**	
	B. K. Middleton	603
Chapter 12	**Pyroelectric and Ferroelectric Thin Film Devices**	
	J. C. Burfoot.	697
Chapter 13	**Superconducting Thin Film Devices**	
	Gordon B. Donaldson.	743
	Subject Index	845

Symbols

Chapter 3

Symbol	Definition
f_0	unperturbed probability distribution function (Fermi distribution)
f	perturbed probability distribution function
x, y, z	Cartesian co-ordinates
t	time
$p_{x,y,z}$	components of momentum
X, Y, Z	components of forces
m	mass of electron
e	charge of electron
$V_{x,y,z}$	components of velocity
$\vec{F}$	vector force
$\vec{V}$	vector velocity
τ	relaxation time (also $\tau_{phonons}$, $\tau_{imp.}$, $\tau_{defects}$)
E	electric field (and kinetic energy of electrons)
σ	conductivity
j	current density
$k_{x,y,z}$	electron wave numbers
N_{eff}	effective number of free electrons
$\tau(k_{max})$	relaxation time of electrons with Fermi energy
h	Planck's constant ($\hbar = h/2\pi$)
ν	frequency of lattice vibration (and Poisson's ratio)
ν_{max}	cut-off frequency of vibrational spectrum
k	Boltzmann's constant (and $k = d/\lambda_0$)
T	absolute temperature
Θ	Debye temperature ($h\nu_{max} = \Theta$)
λ	mean free path of conduction electrons
V_F	Fermi velocity
α	temperature coefficient of resistance $\left(\text{and } \alpha = \dfrac{\lambda_0}{r^1} \cdot \dfrac{R}{1-R}\right)$

Symbol	Definition
$V(r)$	potential at distance r from impurity
k_F	wave number of Fermi electrons
Z	effective charge (or valence difference)
$\sigma(\theta)$	effective scattering cross section
θ	scattering angle from impurity atom
Λ	mean free path for impurity scattering
N_i	density of impurity atoms
ρ	resistivity (also ρ_{phonons}, $\rho_{\text{imp.}}$, ρ_{defects})
ϕ	arbitrary function of V
d	film thickness
λ_0	mean free path in bulk impure material
λ_b	mean free path in pure bulk material
ρ_0, ρ_b	resistivity of impure and pure bulk materials
σ_0, σ_b	conductivity of impure and pure bulk materials
α_0, α_b	T.C.R. of impure and pure bulk materials
p	specular reflection coefficient
ρ_g	resistivity inclusive of grain boundary scattering
r^1	radius of grain
R	grain boundary reflection coefficient
F_0	characteristic function for annealing of combined defects
R_i	residual resistance
E_0	characteristic energy for annealing of combined defects
ρ_{300}	resistivity at 300K
$\rho_{4.2}$	resistivity at 4.2K
S_0	thermoelectric power of bulk material
S	thermoelectric power of thin film
E_F	Fermi energy
A	area of Fermi surface
β	$= d/r^*$
r^*	radius of electron helix in magnetic field
I_s	sputtering current
P_{N_2}	partial pressure of nitrogen
γ	strain sensitivity of strain gauge

Chapter 5

Symbol	Definition
e	electronic charge
r	island radius and rate of tunnelling
n	number of charged islands
N	total number of islands
δE	activation energy

k	Boltzmann's constant
T	absolute temperature
ε_0	permittivity of free space
ε_r	relative permittivity
D	transmission coefficient (and dimensionality of lattice)
$P_{i,j,h}$	tunnelling probabilities
$f_{i,j,h}$	Fermi functions
E	potential energy
V	potential difference
s	island separation
τ	average time per transition
v	average electron velocity
σ	conductivity
μ	electron mobility
F	electric field
E_F	Fermi energy
E_x	electron energy in x direction
E_r	electron energy in y, z plane
J	current density
m	electron mass
h	Planck's constant
$\phi(x)$	height of potential barrier as a function of x
x	position between islands (and volume fraction of metal)
$\bar{\phi}$	average height of potential barrier
a	island diameter
$\beta(s)$ or $g(s)$	distribution of island separations
σ_L	low field conductivity
χ_0	wave function decay constant related to potential barrier
$\rho(E)$	density of electronic states function
G	volume of islands
ν	rate of electron impingement of potential barrier or frequency
γ	geometrical factor and gauge factor
S_0	average island separation
a_0	average island diameter
F_0	constant related to field induced tunnelling
F^*	normalised field
T^*	normalised temperature
σ^*	normalised conductivity
σ_H	high field conductivity
s'	nearest neighbour separation
α	temperature coefficient of resistance (also power exchange parameter and wave function delay constant)

ε	strain
W_ν	radiated energy at frequency ν
T_{el}	electron temperature
I	total current through film
V_c	critical volume fraction
A_c	critical area fraction
p_c	critical percolation probability
f	packing fraction
β	constant in percolation conductivity
α_T	temperature coefficient of tunnelling current
Z	co-ordination number
N_c	critical co-ordination number
R	$2r+s$
$f(R)$	distribution of centre to centre separations
E_s	energy range over which hopping occurs
$g(E_s)$	distribution of hopping energies
E_u	lower limit of E_s
E_v	upper limit of E_s

Chapter 6

Symbol	Definition
W	width of device
L	length of device
d	thickness of dielectric
h	thickness of semiconductor
V_g	gate voltage
$V(x)$	voltage as a function of distance, x, from source
V_d	source-drain voltage
I_d	source-drain current
J_D	drain current density
$G_s(x)$	sheet conductivity of the semiconductor
q	charge on the electron
μ	electron mobility
n_0	number of carriers per unit area in the semiconductor film for zero gate voltage (bulk carrier density × film thickness)
C_g	gate capacitance per unit area
μ_c	semiconductor bulk electron conductivity mobility
V_0	pinch-off voltage
I_{ds}	saturation source-drain current

Symbols

λ	space charge depth
N_d	donor density
E	energy
ϕ_0	equilibrium height of potential barrier at grain boundary
$\bar{c}$	mean thermal velocity of carriers
n	carrier density in the bulk crystallite
m_i^*	inertial effective mass
V_s	surface potential
N_L	number of barriers per unit length of film
μ_g	grain boundary limited electron mobility
μ_b	total mobility of electrons
n_{c0}	carrier density in bulk crystallite per unit area
n_{b0}	carrier density at the top of the barrier per unit area
R_H	Hall coefficient
μ_H	Hall electron mobility
Δn	sheet charge density induced by the gate voltage
Δn_{bt}	amount of induced charge which is trapped at the intergrain boundaries
θ_b	barrier trapping parameter
μ_{FE}	field effect mobility
t	time in seconds
σ	conductivity
N_{ss}	surface slow-state density per unit area
X_m	maximum depth to which charge penetrates into the insulator in the measurement time, T_m
Δv_s	change in surface potential
L_t	Debye length in the insulator
N_t	trap density per unit volume at an energy E_t in the insulator band gap
ε_r	relative permittivity of dielectric
T_m	measurement time in seconds
K_0	wave function decay constant
Σ_0	capture cross-section of an insulator trap at the interface
$\bar{v}$	mean electron velocity in the X direction
n_s	surface density of electrons
p_s	surface density of holes
n_1	equilibrium carrier density of electrons
p_1	equilibrium carrier density of holes
ψ_0	energy barrier at the interface
ε_x	x-directed kinetic energy of the electron
E_t	trap energy
v_s	reduced surface potential in units of kT

v_{fb}	flat-band voltage
g_m	transconductance
V_b	pinch-off voltage in barrier theory
ϕ_B	intercrystalline potential barrier
θ_c	bulk trapping parameter
e_n	spot noise voltage ($V\sqrt{\text{Hz}}$)
g	gate electrode width
$V_{g\max}$	maximum allowable gate voltage
R_{on}	"on"-resistance (corresponding to I_{ds})
R_{off}	"off"-resistance (for $V_g = 0$)

Chapter 8

Symbol	Definition
a	diameter of cones in "moth's eye coating"
A	absorptance
$\mathscr{A}$	a resultant amplitude
B	component of electric field at multilayer front surface (equations 11–13)
$\mathscr{B}$	a resultant amplitude
c	velocity of light *in vacuo*
C	component of H/Y_0 at multilayer front surface (equations 11–13)
d	geometrical thickness
D	optical thickness nd
E	electric vector
E	magnitude of electric field
$\mathscr{E}$	vector amplitude of **E**
F	function used in analysis of Fabry Perot interferometer (equation 21)
$\mathscr{F}$	finesse (equation 24)
$\hat{\mathbf{g}}$	unit vector along diffraction grating perpendicular to grooves
g	$\dfrac{\lambda_0}{\lambda} = \dfrac{v}{v_0}$ where λ_0 and v_0 are reference wavelength and wavenumber respectively
G	$\dfrac{2\pi}{\Lambda}$
G	$G\hat{\mathbf{g}}$
H	magnetic vector
H	magnitude of **H**

Symbols

$\mathscr{H}$	vector amplitude of H
H	quarter wave of high index
I	vector intensity of wave
I	scalar intensity
j	mode number for guided waves
k, **K**, **K**′	wave vectors
l	width of film guide
L	quarter wave of low index
m	an integer, usually an order number
M	quarter wave of intermediate index
n	real part of refractive index, usually referred to as simply refractive index
n_e	equivalent index
n^*	effective index for tilted filters
N	complex refractive index, $N = n - i\kappa$
p	an integer
P	Poynting vector
q	an integer
Q	reciprocal lattice vector
r	position vector
R	reflectance
s	an integer
$\hat{\mathbf{s}}$	unit vector in direction of propagation
t	time
T	transmittance
U	scattering parameter
V	half width of filter in terms of wavenumber
W	thickness of thin film guide
x	coupling length in prism coupler
Y	optical admittance of medium
Y_0	optical admittance of free space
β	propagation constant of mode in thin film guide
γ	decay constant for evanescent wave
δ	phase thickness, $\dfrac{2\pi N d \cos\theta}{\lambda}$
g	index referred to particular angle of incidence
φ	2φ is the phase shift between successive beams
Φ	2Φ is the phase shift on reflection at film guide boundary
κ	extinction coefficient
λ	wavelength
λ_p	peak wavelength of filter
$\Delta\lambda_H$	filter halfwidth expressed in terms of wavelength

xvi   Symbols

Λ	groove spacing in diffraction grating
ν	wavenumber
ω	angular frequency
ζ	ratio of optical thickness in symmetrical period, $2D_A/D_B$
ψ	phase shift on reflection
Ψ	semiangle of cone illuminating a filter
τ	amplitude transmission coefficient
θ	angle of incidence
Θ	angle of tilt of interference filter
ξ	equivalent phase thickness
ρ	amplitude reflection coefficient

Chapter 9

Symbol	Definition
I	light intensity
I_0	initial light intensity
ε	extinction coefficient or absorption coefficient
C	molar concentration
l	sample dimensions using length, width and
w	thickness in a frame of reference of length
t	representing electrode separation
n, p	concentration of free photogenerated electrons, holes
n_0, p_0	concentration of "dark" electrons, holes
τ	free carrier lifetime (sometimes defined further by suffix p or n)
N	concentration of recombination centres
v	thermal carrier velocity
S	capture cross section (specified by suffix p, n)
f	number of electron-hole pairs generated per unit volume per second
F	light flux
E_g	band gap
E_1 or E_t	trap depth
E_c, E_v	conduction band, valence band level
E_f	Fermi level
E_{fn}, E_{fp}	quasi Fermi level for electrons, holes
D_n, D_p	demarcation level for electrons, holes
ϕ	barrier height for activated photoconductive process
N_c	effective density of states in the conduction band
$N_c(E)$	density of trap states (electronic)
N_r, N_t	number of recombination, trap states (electronic)

P_r	number of recombination centres at the Fermi level
n_{b0}	number of carriers at the grain boundary
σ	conductivity
σ_0	pre-exponential conductivity factor
h	Planck's constant
ν	frequency of light
ρ	resistivity
T	absolute temperature
T_1	formal parameter
e	electronic charge
Q	charge
c	capacitance
κ	dielectric constant
V	applied voltage
J	current
T_r	transit time of carrier
τ_0	response time of sample
τ_{rel}	dielectric relaxation time
μ	mobility (specified by suffix): no suffix represents majority carrier drift mobility
d	penetration depth of light
$\theta = \dfrac{n_t}{n}$	the fraction of trapped to free charge (electronic)
G	gain
γ	the gamma or constant in the vidicon
$\mathscr{R}$	responsivity ⎫
$\mathscr{D}$	detectivity ⎬ in **IR** detectors
D, D^+, D^*	the donor molecule in respectively its ground, ionised and excited singlet states
A, A^-, A^*	the acceptor molecule in respectively its ground, ionised and excited singlet states

Chapter 10

Symbol	Definition
A	diode factor or ideallity factor
A^*	modified Richardson constant
a_{OR}, b_{OR}, c_{OR}	orthorhombic unit cell parameters for Cu_2S
$a_{mc}, b_{mc}, c_{mc}, \beta_{mc}$	monoclinic unit cell parameters for Cu_2S
c	velocity of light *in vacuo*
C	capacitance of a junction
C_i, C_s	interstitial, substitutional, cuprous ion concentrations
C_V'	cadmium vacancy concentration

C_{b++}	interstitial cupric ion concentrations
D_n, D_p	electron, hole, diffusion coefficients
D_i, D_s	interstitial, substitutional, cuprous ion diffusion coefficients
E_a	activation energy for cell degradation
E_g	semiconductor energy-gap
E_p	phonon energy
$\Delta E_{CA}, \Delta E_{CG}$	conduction band discontinuity for abrupt graded heterojunction
e	charge on electron
$F.F.$	fill-factor of current/voltage characteristics
$F(\lambda), G(\nu)$	spectral distribution functions of sunlight
f	oscillator strength
ΔG	change in Gibb's free energy
h	Planck's constant
I	light intensity
I_0	incident light intensity
J	electrical current density
J_0	reverse saturation current density
J_{00}	pre-exponential factor
J_L	photogenerated current density
J_T	current per unit area of cell available at terminals
J_d	current per unit area of cell through diode in equivalent circuit
J_{SC}	short-circuit current density
J_m	current per unit area of cell at maximum output power
k	Boltzmann's constant
K	electron wavevector
L_n, L_p	electron, hole, diffusion lengths
m_0	free electron mass
m_e, m_h	electron, hole, effective mass
N	refractive index $= n - ik$
n_0, p_0	thermal equilibrium electron, hole, density
n_p, p_n	minority electron, hole, density
$\Delta n, \Delta p$	photogenerated electron, hole, density
N_d, N_a	donor, acceptor, concentrations
N_G	number of grains in a current path
n_G, n_B	total electron density in grains, between grains
q	quantum efficiency of photogeneration
R_G	generation rate of electrons and holes
$\mathscr{R}(\nu)$	reflection coefficient
R_S, R_{SH}	series, shunt, resistance in equivalent circuit

S_n, S_p	surface recombination velocities for electrons, holes
S_{Th}	thermal stress in a thin film
ΔS	change in entropy
T	temperature
V	applied voltage
V_d	voltage across diode in equivalent circuit
V_D	diffusion voltage or built-in voltage
V_m	voltage output of cell at maximum power
V_T	voltage across terminals of solar cell
V_{OC}	open-circuit voltage
v_e, v_h	drift velocities of electrons, holes
$\langle v_{th} \rangle$	mean thermal velocity of electrons
W	width of depletion layer
x	distance from a surface
x	stoichiometry parameter in Cu_xS
Y_F	Young's modulus of a thin film
α	optical absorption coefficient
$\Delta \alpha_T$	difference in thermal expansion coefficients of a film and its substrate
ϵ	electrical permittivity
ε	electric field strength
η	power conversion efficiency of a solar cell
η_u	ultimate (zero loss) power conversion efficiency
λ	$= \dfrac{e}{kT} \quad \lambda' = \dfrac{e}{AkT}$
μ_e, μ_h	electron, hole, mobilities
ν	frequency of electromagnetic radiation
σ	electrical conductivity
τ_n, τ_p	recombination lifetimes of minority electrons, holes
χ	electron affinity
Φ	photoelectric threshold
ϕ_B	Schottky barrier height

Chapter 11

Symbol	Definition
a	transition width parameter
a_1, a_2, a_3	transition width parameters at record, relaxation, and remagnetisation stages
a_d, a_f	transition width parameters after self-demagnetisation and remagnetisation
d	head to tape separation

Symbols

	film thickness
$e, e(\bar{x})$	output voltage
E, E_k	total and anisotropy energy densities
g	half of head gap width
h	height of sensitive element of magnetoresistive head
H, H_t	total applied magnetic field
H_d	self-demagnetising field
H_{dep}	field applied during film deposition
H_k, H_k'	anisotropy and effective anisotropy fields
H_x, H_y, H_z	components of applied magnetic field
$H_w, H_w', H_c, H_r, H_{cr}$	wall motion, wall motion, coercive rotational and creep threshold fields
H_n	nucleation field
H_1, H_2	initial and final remagnetising field
H_k, H_0	fields produced under and along the surface of a recording head
H_0, H_1, H_2, H_3	critical fields for bubble domain stability
	current
$I_d, I_{dd}, I_{dt}, I_{wt}$	digit, digit threshold, digit disturb threshold, and word threshold currents
K, K_u, K_1	uniaxial, induced uniaxial, and first order uniaxial anisotropy constants
l	characteristic length of bubble materials
$M, M(H)$	magnetisation, and magnetisation as a function of applied magnetic field
M_1, M_2	initial and final values of magnetisation on the remagnetisation curve
M_x, M_y	components of magnetisation
M_r, M_s	remanent and saturation magnetisation
N_a, N_b	demagnetising factors of prolate spheroid
p_{50}	pulse width at 50% of peak amplitude
q	quality factor of bubble materials
R	bubble radius
s	squareness M_r/M_s
t	time
y	head to tape spacing during record
γ	wall energy
θ	angle between magnetisation vector and easy axis
ϕ	angle between applied field vector and easy axis
$\rho, \rho_0, \Delta\rho$	total, isotropic and magneto resistivities
χ	slope of major hysteresis loop at $H = 0$
τ	switching time

Chapter 12

Symbol	Definition
i	current
V	voltage
P	electric polarisation
P_s	spontaneous electric polarisation
p	pyroelectric coefficient
T	temperature
σ'	electric conductivity
ε	permittivity (real part)
ε''	permittivity (imaginary part)
T_c	ferroelectric transition temperature
n	refractive index
c	polar axis
ω_0	angular frequency of soft mode
q, q', q_1	charge per unit area
ρ'	resistivity
i_p	pyroelectric current
r	rate of rise of temperature
A	area (of electrode or specimen or hotpatch)
E	electric field
Δ	$\equiv T_1 - T_2$
W	radiant energy in beam (maximum value, $\hat{W}$)
C_T	thermal capacity
C	electric capacity
R_T	thermal resistance
R	electric resistance
c	specific heat
d	thickness of specimen
ρ	density
$\mathcal{R}_i$	current responsivity
$\mathcal{R}_v$	voltage responsivity
V_p	pyroelectric voltage output
V_n	noise voltage output
θ	temperature excess over ambient
T_a	ambient temperature
τ	thermal time-constant or detector response time
τ_e	electric time-constant
ω	angular frequency of mode
t	time
e	fraction of incident energy absorbed

e	emissivity
P_n	noise equivalent power (NEP)
B	noise bandwidth
D	detectivity
D^*	area-normalised detectivity
σ	Stefan's constant
W_n	noise power
W_R	radiative noise power
k	Boltzmann's constant or propagation constant of optical wave
ΔV	voltage fluctuation
Δi	current fluctuation
K	thermal conductivity
F	figure of merit
$\tan \delta$	loss factor
x	depth into $\left(\dfrac{\varepsilon''}{\varepsilon'}\right)$ pyroelectric specimen
λ	radiation wavelength
α	temperature coefficient of resistance
s	grain size
Λ	lattice vibration mode wavelength
T_{CW}	Curie–Weiss temperature
x_c	absorption length for radiant heat beam
d_{31}	one of the piezoelectric coefficients
Δn	birefringence
r_{33}	an electro-optic coefficient
ΔT	extent of thermal hysteresis

Chapter 13

Symbol	Definition
c	free space velocity of light
d	film thickness
e	electronic charge
f	force on a flux vortex
$f(E)$	Fermi function at energy E
$g(E)$	density of quasiparticle states of energy E
$g_N(\varepsilon)$	density of normal electron states of energy ε (measured from Fermi level)
$h, \hbar$	Planck's constant and reduced constant
$\mathbf{j}(\mathbf{r})$	current density at point $\mathbf{r}$
j_c	critical current density

$\mathbf{k}, k$	wave-vector, wave-number
k_B	Boltzmann's constant
l	mean free path in thin film
n	demagnetisation factor
n	integer
$\mathbf{p}$	canonical momentum
$\mathbf{r}$	position vector
t	tunnel barrier thickness
t	polycrystalline grain size
t	time
v	vortex flow velocity
v_F	Fermi velocity
$\mathbf{A}$	magnetic vector potential
A, A_1, A_2	bolometer area, transformer loop area
$\mathbf{B}$	magnetic flux density
C	capacitance
C	bolometer thermal capacity
D^*	bolometer specific detectivity
$\mathbf{E}$	electric field
E_k	energy of quasiparticle of wave-number k
F	bulk force on vortices
F_d, F_p	bulk drag, pinning, force on vortices
F_H	magnetic free energy
H_a	applied field
$H_c(T)$	bulk thermodynamic critical field at temperature T
$H_c(T)$	thermodynamic critical field in thin film at temperature T
H_{C1}, H_{C2}, H_{C3}	critical fields of type II superconductor
$H_\perp$	lower perpendicular critical field of type I superconductor
$H_{\perp\perp}$	lower perpendicular critical field of type II superconductor
H_{C2}	upper critical field of type II thin film superconductor
I_c	critical current
I_{cc}, I_{gc}	cryotron critical control and critical gate currents
I_B	cryotron bias current
I_N	noise current
J_t	transport current density
K	junction normal conductance
K	system noise figure
L, L_A, L_B, L'	inductances
M	mutual inductance

M	magnetisation
$N(E)$	number of occupied quasiparticle states of energy E
Q	quality factor of cavity
R	electrical resistance
T	temperature
T_c	critical temperature
$V_{kk'}, V$	electron–phonon–electron interaction parameter
V, V_p, V_s	voltage, primary and secondary voltage
W	condensation energy
W, W_g, W_c	film width, cryotron gate and cryotron control film widths
β_c	Stewart–McCumber hysteresis parameter
γ	electron density
γ	gauge invariant phase difference
ε_F	Fermi energy
ε_k	excitation energy, measured from Fermi level for electron of wave-vector k
κ	Ginsburg–Landau parameter
$\lambda(T)$	penetration depth at temperature T
λ_L	London penetration depth
λ_J	Josephson penetration depth
ν	frequency of Josephson oscillations
ξ	coherence length
ξ_0	coherence length in pure bulk material
τ	switching time constant
$\phi, \phi(\mathbf{r})$	phase of superconducting wave function (at point r)
$\chi, \chi(\mathbf{r})$	superconducting wave function
ω	angular oscillation frequency
ω_D	Debye frequency (angular)
ω_J	Josephson plasma frequency (angular)
$\Delta, \Delta(T)$	superconducting energy gap parameter, at temperature T
Δ_k	energy gap parameter for quasiparticles of wave-number k
Δf	bandwidth
ΔI_c	change in critical current
ΔT_c	change in critical temperature
$\Delta \nu_J$	bandwidth of Josephson frequency
Φ	flux
Φ_a	applied flux
Φ_c	critical flux
$\Phi_{DC, RF}$	D.C., R.F. bias flux
Φ_N	noise flux
Φ_o	flux quantum $hc/2e$

Chapter 1

Practical Applications of Thin Films

T. J. Coutts
Industrial Science Research Centre,
Cranfield Institute of Technology,
Cranfield, Bedford, England

1. Introduction 1
2. Electrical Properties and Devices 3
   2.1 Continuous Metal and Alloy Films 3
   2.2 Dielectric Thin Films 5
   2.3 Mixed Metal/Dielectric Thin Films 6
   2.4 Thin Film Transistors 7
   2.5 Amorphous Films 8
   2.6 Ferroelectric and Pyroelectric Films 9
3. Magnetic Properties and Devices 12
4. Optical Properties and Devices 14
5. Electro-optic Properties and Devices 15
   5.1 Thin Film Photoconductors 15
   5.2 Thin Film Solar Cells 17
6. Superconducting Properties and Devices 19
   References 21

1. Introduction

Although the study of thin film phenomena dates back well over a century, it is really only over the last two decades that they have been used to a significant extent in practical situations. This may seem surprising in view of their widespread application in many fields today but there are several reasons which account for their not being exploited earlier.

Firstly, in most cases, it has only been in comparatively recent years that there has been an advantage or a need to exploit the unusual properties of thin films. In electronic circuitry, for example, there would have been little advantage in using thin film resistors in conjunction with valve circuitry and indeed conventional carbon or wire-wound resistors were used exclusively until such time as the requirements of microminiaturisation made the use of thin, and thick, films virtually imperative. Equally, the development of computer technology led to a requirement for very high density storage techniques and it is this which has stimulated most of the research on the magnetic properties of thin films. Of course, necessity is not inevitably the mother of invention and many thin film devices have been developed which have found themselves looking for an application or, perhaps more importantly, a market. In general these devices have resulted from research into the physical properties of thin films. Research in this topic has shown that it is possible to obtain information of fundamental importance to solid state physics and their use in this connection is fully discussed by Anderson[1]. The thin film photovoltaic cell is a classic example of basic research preceeding an application. The photovoltaic effect in thin films was discovered by Reynolds[2] in 1954 but it was not until a considerable time after the start of the U.S.A. space programme that the full potential of thin film cells as satellite power supplies was appreciated.

Secondly, as well as generating ideas for new devices, fundamental research has led to a dramatic improvement in our understanding of thin films and surfaces and this in turn has resulted in a greater ability to fabricate devices with predictable, controllable and reproducible properties.

The cleanliness and nature of the substrate, the deposition conditions, post deposition heat treatment and passivation are vital process variables in thin film fabrication and in many respects it is only over comparatively recent years that their importance has been appreciated. Therefore, prior to this improvement in our understanding of thin films, it had not really been possible to apply them to real devices.

Thirdly, although there is a very wide range of thin film deposition techniques (as discussed by Campbell in Chapter 2) in general devices are fabricated using vacuum technology. Again more recent improvements in this field mean that we can now achieve quite high vacua ($< 10^{-6}$ torr) as a matter of routine even in a production environment, using relatively unskilled personnel to operate plant and processes which lend themselves particularly well to automation. This has the obvious effect of reducing device costs not only because of higher yields brought about by the use of high vacua and clean fabrication conditions but also because of the relatively low labour cost involved. In addition, the equipment used requires very little maintenance and often has a lifetime considerably greater than the five year periods over which industrial capital costs are generally amortised. The upshot of these factors is, of course, the simple and essential point that the devices can now be made economically. Although much of the finance for early thin film research originated from space or defence programmes to which the device cost was less important than its lightweight and other advantages, the major applications of thin film technology are not now exclusively in these areas but rather often lie in the domestic sector in which low cost is an essential.

The physical properties of films may be divided into two fundamental categories: those which are inherent to the film because of the fact that at least one of its dimensions is very small, and those which arise as a consequence of the particular method of preparation. In a metallic film, for example, the inherent property is the increase in the resistivity of the film with decreasing thickness. It turns out that in practical devices this is not really an effect which can be exploited because metallic films which are used as, say, conductors in microminiature circits, are generally much too thick to exhibit a noticeable size effect. On the other hand thin metal films also usually have a higher resistivity than the bulk value because they consist of much smaller crystallites and this is a property which results from the method of preparation. In some devices it is the inherent or preparational property which is exploited, whilst in others the important factor is simply the ability to prepare a device in thin film form. In the latter case this may be because of its small physical size, low mass or some other property which results in an advantage in performance, lifetime, cost or any other relevant factor. Where appropriate the chapters in this book have been organised according to a common pattern. This consists of a discussion of the basic physical property, the identification of the thin film properties in terms of inherent or preparational effects, and the identification of the various applications which exploit either of these effects or simply utilise the small dimensions obtainable in thin films. A number of the devices discussed are already well established commercially, some require further research effort whilst others are much more speculative and are perhaps never likely to be used in practice. In some chapters the authors review recent literature published on their topic and discuss several devices whilst in others the subject matter is more specifically concerned with the research activities of the author himself as related to a single device. Methods of fabrication are not discussed in detail by individual authors but a summary of these is given by Campbell in Chapter 2. We shall now discuss in brief the principal physical properties upon which the devices discussed in this book are based.

2. Electrical Properties and Devices

2.1 Continuous Metal and Alloy Films

The topic of electrical conduction in thin, but continuous, metal and alloy films is an important aspect of passive thin film devices and it is discussed by

Coutts in Chapter 3. The conductivity of pure bulk metals can be derived from Fermi-Dirac statistics and the Boltzmann transport equation. Deviation from perfect lattice periodicity arises from several causes including thermal vibrations of the lattice, impurity atoms and defects. Each of these contribute to the overall resistivity of both metals and alloys. Further, the transition metal alloys have a much higher resistivity than do other metal alloys because of the overlap of partially filled and partially empty energy bands. This feature also gives rise to a relatively low temperature coefficient of resistance so that useful resistors can be made of these materials even in their bulk form.

In the case of thin metal or alloy films the boundaries of the films make an additional contribution to the resistivity due to the well known phenomenon of surface scattering. This is actually an inherent property of thin metal films and its effects on the resistivity have been formulated by Fuchs[3]. The effect can be readily understood conceptually when it is appreciated that the typical wavelength of conduction electrons is less than the surface roughness of the films. Thus the conduction electrons tend to be diffusely scattered by the surface and lose their momentum in the direction of the applied electric field. This, like defects or impurity atoms, constitutes an additional source of resistivity. However the contribution to the overall resistivity only becomes significant when the thickness of the films is reduced to a value comparable with the conduction electron mean free path, so that an appreciable number of electrons actually do suffer collisions with the surface. In addition to the inherent property, thin metal and alloy films generally consist of grains which are of about the same size or less than the mean free path and in practice collisions between the conduction electrons and the grain boundaries often contribute more to the film resistivity than does surface scattering. The effect of grain boundaries has been formulated by Mayadas and Shatzkes[4]. In practice it is neither the phenomenon of surface nor grain boundary scattering which is exploited but rather the fact that the resistor or conductor tracks generally occupy only a very small part of the substrate area.

Even the conductors in microcircuitry are not formed from simple single component metals whilst the resistors are generally based on more complicated materials such as tantalum or its nitrides, or nichrome. Many of the problems of resistor stability have now been overcome and this is at least in part due to an appreciation of the important role of the substrate. A review of this aspect has been given by Duckworth[5]. So far as complete circuits are concerned, particularly with regard to silicon integrated devices, the problems of forming ohmic contacts to semiconductors are still difficult to describe theoretically although satisfactory film contact systems have been developed on the basis of trial and error.

Thin metal films can also be used in the construction of inductors, capacitors and strain gauges although, in terms of numbers of devices, these applications are less important than their use as conductors and resistors.

2.2 Dielectric Thin Films[6]

With dielectric thin films, as with several of the other types discussed in this book, the interest is both of a practical and a fundamental nature. In order to make either useful devices (capacitors, for example) or a meaningful fundamental investigation, the films must possess physical continuity, i.e. pinholes etc. must be avoided; mechanical stability, i.e. the films must not be in a highly stressed state or cracking as a short circuit can result; and homogeneity across their surface, since many of the properties of a dielectric depend sensitively on film thickness. These may be desirable features in all thin film devices but in this case they are particularly important since the dielectric often forms the central layer of a metal-insulator-metal sandwich and if any of the above qualities are lacking it is impossible either to make useful devices or fundamental studies of the electrical properties of the film.

Since one of the principal applications of dielectrics is in thin film capacitors, it is desirable to use materials with as high a value of permittivity as is practically realisable. The range of materials used is wide and encompasses plastics, oxides and ferroelectrics, all of which can be deposited by conventional techniques. As is the case with thin film resistors it is highly desirable that the capacitance of devices changes very little with temperature and this implies that the temperature coefficient of permittivity should be as low as possible. It turns out in practice that even for low-loss materials there is a minimum temperature coefficient of capacitance which can be achieved and this is entirely an intrinsic property of the materials of which the film is composed. In other films, there is a significant dielectric loss (usually referred to as tan δ) and this is a function of the series and parallel resistances associated with the capacitor so that it is an important quantity when considering A.C. conduction.

So far as D.C. conduction in dielectrics is concerned there are a number of mechanisms which can occur[7] and these depend very much on the level of applied voltage and the thickness of the film. At low fields, hopping conduction between localised states in the forbidden gap of the dielectric occurs and, at sufficiently low temperatures, the now well established "$T^{-\frac{1}{4}}$ law" appears to apply. At higher fields, the logarithm of the current is often found to be exponentially dependent on the square root of the applied field or, if there is a high density of impurity sites and traps, linearly on the applied field. For very thin films, in which the thickness of the dielectric is less than 50 Å, direct tunnelling of electrons between the metal electrodes can occur and, as opposed to the mechanisms mentioned immediately above, this is an effect which depends relatively weakly on temperature.

It is also now well established that many of the observed electrical characteristics of dielectrics are a function of the contacts. In much the same way as with semiconductors it is possible to form ohmic, blocking (rectifying) or injecting contacts. Care has to be taken to ensure an ohmic contact before interpreting any current/voltage data since in the case of both blocking and injecting contacts the width of the depletion or accumulation layers can be as great as the film thickness. Thus the conduction process can be limited by the electrodes and not by the dielectric itself. In some cases space charge limited current can be observed in which the current is found to be proportional to the square of the applied field and this is normally to be associated with an injecting contact. Of the electrical characteristics of thin dielectric films, possibly the most important from the point of view of applications is that the breakdown fields are generally considerably in excess of those found in bulk materials. It appears, perhaps fortuituously, the very thinness of the films actually precludes several mechanisms of breakdown which can occur in bulk dielectrics. In practice, the actual process of breakdown depends on several factors including the impedance of the voltage source but, so far as capacitor based devices are concerned, there always exists the possibility of employing the now, well-known self-healing effect in which short circuits due to melted electrodes can be removed by application of a high current.

The applications of thin dielectric films are now, of course, very widespread but are limited in numbers to capacitors and related devices such as temperature and pressure sensors (which rely for their operation on the measurement of a change of capacitance) to insulating cross-overs between conductors in thin film circuits, or to insulators between the source and drain of field effect transistors and other similar devices. Equally, but perhaps less importantly from the device standpoint, certain dielectrics also exhibit threshold and memory effects similar to the amorphous films discussed in Chapter 7 and a highly negative temperature coefficient of resistivity which possibly offers a useful thermistor application.

2.3 Mixed Metal/Dielectric Thin Films

Bulk specimens consisting of a mixture of conducting and insulating phases have been used and studied for many years; the earliest theoretical attempt to predict the variation of the resistivity of the sample with the concentration of conducting component dating back to the "effective medium theory" of Maxwell[8] published in 1881. More advanced theories based on the concept of a critical concentration are now available and these appear to explain the electrical behaviour of bulk mixed-phase systems quite well. Surprisingly, perhaps, mixed phase thin films, which are often known as cermets, also appear to be fairly well described by percolation theory, at least when the concentration of the conducting species is greater than some critical value. The critical concentration is unfortunately not well defined even for particular cermet combinations and appears to depend on several parameters. However a critical value always appears to exist and at higher concentrations the film consists of a continuous network of metal capillaries whilst at concentrations less than the critical value the network is not sufficiently well connected to present a continuous metallic path. In this case electrical conduction occurs by quantum mechanical tunnelling between discrete particles spaced by only a few tens of Angstroms. Transfer of electrons between islands requires a thermally supplied activation energy to overcome the electrostatic forces involved and this gives rise to a negative temperature coefficient of resistance; rather like a semiconductor. This can be regarded as the inherent property of the three dimensional cermet or two dimensional discontinuous film although it arises from the small size of the islands which in turn results from the nucleation processes in evidence when the film is being deposited.

From the practical point of view thin film cermets offer a very positive potential advantage as resistor materials over the more conventional compounds such as tantalum nitride or nichrome. Since they always consist of at least two phases, charge transfer can always occur by at least two paths. If, for example, as was described above, these paths happen to be metallic capillaries and discrete islands, then the T.C.R. of the film may be near zero since the T.C.R.'s of the two separate paths are opposite in sign and, under the right circumstances, can cancel. Since the film resistivity is generally high when this condition is realised, the cermet offers the possibility of achieving physically small but high value resistors which are not affected by temperature changes. In practice there are still problems related to stability and further research needs to be undertaken before the cermet can compete with more conventional thin film resistors.

2.4 Thin Film Transistors

The thin film transistor is a device which relies for its operation on a field effect whereby an external field is applied via an electrode known as the "gate" and causes modulation of the conductivity of a semiconductor between two electrodes called the "source" and the "drain". Studies of bulk field effect devices date back to 1930 although a successful silicon field effect transistor (F.E.T.) was not developed until 1963[9]. The problems of achieving a successful device hinged on the difficulty of obtaining satisfactory modulation. This depends very sensitively on the nature of the semiconductor used and although intuitively one might assume that a high mobility material should be used, in fact the first succesful thin film device was based on cadmium sulphide which has a wide band gap and a low mobility. The reasons for this are to be associated with a number of factors including interface, geometrical and structural effects and in Chapter 6 Anderson presents a theoretical derivation of the current/voltage characteristics of polycrystalline thin film transistors, together with an explanation of the choice of materials.

The principal advantage of the device is that it can be made very small and in high densities on a substrate. The ratio of its off/on resistance is about $10^5:1$ so that it is therefore suitable for use as a switch at each picture element of an x–y flat panel display. Indeed a flat T.V. screen based on an electroluminescent panel and thin film transistor switches was reported in 1975[10] and it seems probable that the device will become widely used not only in this type of panel but in the whole range of display panels presently under development.

2.5 Amorphous Films[11]

The high level of interest in amorphous materials over the last decade or so appears to stem from both a direct and fundamental research interest, and from a device interest. In the first case the interest arose because the use of quantum theory had provided an understanding of the crystalline state and it was the next logical step for physicists to ask what would happen to the physical properties of materials in the absence of long range crystalline order. In the second case, amorphous selenium has long been used in xerography whilst other amorphous oxides have been used in integrated circuitry and these applications were at least partly responsible for initiating the search for new devices.

A great deal of recent device research has been concerned with amorphous switches which were being offered commercially long before there was an understanding of the electronic processes involved and so, as is usually the case in device research, there has been a very close liaison between the fundamental and the more applied researchers.

In several areas of application, amorphous and crystalline devices are in direct competition and, from the point of view of the user, the former class offers at least two potential advantages. Firstly, the area of crystalline devices is limited by the size to which crystals may be grown whereas amorphous devices may be made in relatively much larger areas. Secondly, it is possible using rapid quenching techniques, to produce completely new materials in amorphous form which do not exist in the crystalline state. Whether or not such materials are exploited in practice depends, inevitably, upon the economics of the situation. The development of the thin film amorphous device hinges on the second of the above advantages. Because very rapid rates of quenching can be achieved using thin film techniques it is possible to extend the range of materials which can be produced in amorphous form. Thus, yet again, it is not really a fundamental thin film property which is exploited in actual devices but rather one which results from the method of preparation.

The most widespread application of amorphous films is as switching elements although Bosnell, in Chapter 7, discusses several other uses which do not rely upon the switching characteristics of these materials. There are two forms of switching. The first, the threshold switch, changes from a low to a high resistance state upon application of a voltage greater than some critical value provided that a holding voltage is maintained after switching. The second, the memory device, does not require a holding voltage to be applied after switching has occurred. The bulk of the device research has been concerned with developing switches which exhibit consistent characteristics over many switching operations whilst the fundamental research has concentrated more on the mechanisms responsible for the change of state. The memory state is quite well understood and is generally associated with the occurrence of crystallisation but there is still some controversy over the nature of the threshold device. The theoretical models invoked fall into three categories. Firstly, those which rely upon purely electronic processes, secondly, those which rely upon thermally or electrically initiated processes and thirdly, those which invoke microstructural changes. It is not appropriate to give a discussion here of the arguments presented in support of the various models but they are compared in detail in Chapter 7.

As well as the commercial possibilities of amorphous films as switches, they may yet find application as thin film waveguides (in which acousto-optical properties are exploited) as infrared detectors (although their detectivity is presently about one order less than the pyroelectric detectors discussed in Chapter 12), or in photography (in which use may be made of the change in optical transmission with intensity).

2.6 Ferroelectric and Pyroelectric Films[12]

Ferroelectric materials are that subgroup of pyroelectrics in which the spontaneous electric moment (P_s) which does not require the application of an electric field to be maintained. Pyroelectricity is due to the variation of P_s with temperature and the pyroelectric coefficient is defined as:

$$P = -\frac{dP_s}{dT}$$

Ferroelectric materials are that subgroup of pyroelectrics in which the direction of P_s can be reversed by the application of an external electric field in just the same way as the domains of a ferromagnetic material can be rotated by an external magnetic field. Thus, the presence of ferroelectricity is not a prerequisite of a pyroelectric device and it is not necessary that the latter exhibits hysteresis loops. In ideal circumstances a maximum in the permittivity of a material as a function of temperature can be taken as indicative of a spontaneous electric moment (the maximum occurring at the critical temperature T_c) but in thin films the maximum is generally reduced to a broad peak. This is due to the fact that T_c is influenced by stress and since this varies from grain to grain, there will be a spread in the values of T_c.

The pyroelectric coefficient can be assessed by heating the specimen at a known rate and measuring the resultant current, i.e.

$$p = -\frac{dP_s}{dT} = -\frac{dP_s}{dt} \cdot \frac{dt}{dT} \quad \therefore \quad p = -\frac{i_p/A}{r}$$

where i_p is the pyroelectric current measured between the electrodes of area A whilst the temperature is rising at a rate r. For ferroelectric materials which are pyroelectric, the coefficient p can be estimated by measuring $P_s(E)$ rather than $P_s(T)$.

One of the principal applications of pyroelectric materials is in the field of thermal detection and the responsivity of a given detector material can be defined as the ratio of the pyroelectric current produced by heating the detector to the thermal power absorbed, w, which causes the temperature to rise. Since measurement of the effect is usually carried out in conjunction with a high impedance amplifier, the voltage, rather than the current, responsivity is usually quoted viz:

$$R = \frac{R i_p}{w}$$

where R is the impedance of the measuring circuit. Thus, we have:

$$R = \frac{pAR}{C_T}$$

where C_T is the heat capacity of the detector which is given by:

$$C_T = cAd\rho$$

In the above expression for the heat capacity, c is the specific heat of the detector material, d and ρ being its thickness and density respectively. Therefore, the voltage responsivity can be expressed as:

$$R = \frac{pR}{cd\rho}$$

This suggests that in the ideal case the voltage responsivity should be inversely proportional to the thickness of the detector and it is this feature which has promoted most of the research into thin film pyroelectric detectors. Yet again, the important property results from the goemetry of the device rather than being associated with a fundamental physical property of the device. As Burfoot points out in Chapter 12, many of the advantages of a thin film detector can also be realised in very thin bulk specimens. In the above chapter the performance of pyroelectric detectors is compared with that of other forms of thermal detector and although improvements in the responsivity can result in thin film detectors it is not uncommon to find that noise problems, which ultimately limit the sensitivity of the device, increase. Nevertheless, thin film pyroelectric elements are now being used as the basis of thermal imaging systems for military or anti-intruder applications in the form of two dimensional arrays of detectors. The latter can be used like a vidicon with an electron beam discharging the elements.

In general ferroelectric materials exhibit unusually high values for physical properties such as permittivity and birefringence and it is this which is exploited in associated devices. Since these materials exhibit a large piezoelectric effect their major application is as transducers in which an output signal is produced across the thickness of the sensor (due to a change in charge separation) when a longitudinal stress is applied. This has been exploited in polyvinyl difluoride films which have been used as diaphragms in microphones. These tend to result in better intelligibility and signal to noise ratio than the conventional carbon microphone.

The high permittivity of ferroelectrics can be used in the production of high value capacitors and this is discussed more fully in Chapter 4 by Campbell. Because ferroelectrics exhibit a high degree of birefringence they can be used in conjunction with crossed or uncrossed polarisers as the basis of electrically activated light gates. The application of an electric field rotates the domains of the ferroelectric and this influences the way in which they rotate the plane of polarisation of incident plane-polarised light. With the domains aligned in one direction the device can transmit plane polarised light but when their direction is reversed no light is transmitted. This can be used as the basis of an optical memory.

Ferroelectrics can also be used in conjunction with photoconductors to form what has come to be known as the ferroelectric picture device or FERPIC. This involves applying an electric field, via transparent electrodes, across a ferroelectric/photoconductor sandwich. When the photoconductor is illuminated its resistance falls and all the field appears across the ferroelectric element which thus changes its optical state. This system can be used to store images and it also has been suggested as the basis of the flat screen T.V. Further applications discussed in Chapter 12 include a variation of a MOST transistor in which the conductivity of the semiconductor is modulated using electrical effects in the ferroelectric film (which replaces the oxide) and an amplitude modulator of light using sputtered light guides (discussed more fully in Chapter 8).

3. Magnetic Properties and Devices (Soohoo[13])

Thin magnetic films have been under study since the mid 1950's largely as a result of the suggestion that they had potential for application to digital data storage devices. Magnetic thin films are usually prepared in such a way that they exhibit a uniaxial anisotropy, i.e. they have lowest energy when their magnetisation direction coincides with the so-called easy axes of magnetisation, which are usually arranged to be in the film plane. Consequently when the magnetisation of a film assumes the lowest energy configuration it will lie along one or other sense of the easy axis. Clearly there are two antiparallel magnetisation directions corresponding to energy minima and also a maximum can be experienced if the magnetisation of the entire film is rotated coherently from one sense to the other. The magnitude of the applied field necessary to cause the latter change can easily be calculated and shown to be but a few oersteds in films of 80:20 NiFe Permalloy about 1000 Å thick. Therefore the sense of magnetisation of a film can be selected by the application of a magnetic field and by taking one of these senses to represent a digital one and the other a digital nought, in principle, a digital store can be realised.

Uniaxial anisotropy is the property of the films which makes this type of behaviour possible. The causes of uniaxial anisotropy are complex but, in magnetic films, result from the application of a magnetic field parallel to the plane of the film during its deposition. The resulting easy axis is usually parallel to the direction of the applied field and is termed induced anisotropy. Many aspects of the behaviour of a uniform thin film with uniaxial anisotropy can be predicted by simple theory but it is the divergence of real films from ideal uniformity that causes the complex behaviour actually observed. It is now known that local regions of films, possibly on the scale of individual crystallites, are responsible for local anisotropies differing in magnitude and direction from the mean of the entire film. This dispersion of anistropy gives rise to local variations of the magnetisation which are known as ripple. This effect means that the process of reversal is far more complicated than that of coherent rotation already mentioned. It is divergences from the ideal that have conspired to make the development of such devices as the planar thin film random access store and the plated wire thin film store far more difficult to engineer than originally anticipated. However plated wire stores did reach the market place but never in such numbers as to threaten the dominating position of the core store, which is now being eroded by the availability of semiconducting stores. Nevertheless Permalloy films are currently finding application in other areas of interest.

Large capacity memory systems employ magnetic recording of data onto tapes, discs, or occasionally drums. The principle of the recording process is the same, in broad terms, in all these devices although their geometries differ but for the highest packing densities the storage media should be thin and have high coercivity. The preparation of such films, usually of cobalt or cobalt alloys, is critical and their properties can often be related to the local anisotropies although uniaxial anistropy of the entire films is not a requirement. Although limited application of films to commercial disc and drum storage devices has been achieved it may turn out to be other than in the storage medium itself that films have their main application. There is currently considerable effort being channelled into the development of thin film record and replay heads suitable for high track density applications and these employ soft magnetic films of the type already described (i.e. Permalloy).

More recent activity has involved devices which have been labelled, in this work, "domain storage devices". Contrary to the situation in tape and disc machines, where information is stored by the selection of the sense of the magnetisation at a particular location in a moving medium, these devices involve stationary media in which the domains are moved through the medium to the required locations. In the simplest form the memories may be shift registers but for increased capacity complex storage loop systems wherein domains are continuously circulating have been developed.

Considerable effort has been devoted to "bubble domain" devices which require uniaxially anisotropic materials with easy axes of magnetisation parallel to the film normal. If a magnetic field is applied in the same direction and is of sufficient magnitude to just saturate the film and an oppositely magnetised domain is introduced it will be found to be cylindrically symmetric about the field direction. These cylindrical areas are known as bubble domains and they can be easily shifted using local field gradients produced by current-carrying conductors or by carefully designed Permalloy arrays overlaying the film. The anisotropy of the storage medium in some cases is induced during the growth process, i.e. growth induced, while in others it may be related to the symmetry of the parent crystal, in which case it is known as magnetocrystalline anisotropy. These and the other properties of various complicated media being studied and the design features of storage devices are discussed by Middleton in Chapter 11. Other domain devices are under development and these utilise uniaxial Permalloy films which have properties only slightly different to those used in the thin film store. In these the magnetisation components are in the film plane and the reverse domains are not as regularly shaped. Despite the different orientations of the magnetisation in the various devices the underlying principles of the shift register and storage loop design are very much the same.

Another area of activity involving magnetic films is that of optical beam storage. In this technique a pulsed beam of light is focussed onto the selected storage area of a film causing local heating as the light is absorbed.

If the temperature rise is sufficient, the local film properties are different to those of the film as a whole. Consequently when a magnetic field is applied, the magnitude of which is insufficient to affect the unheated parts of the film, the magnetisation within the heated spot will revert to the same sense as that of the applied field. Hence writing and erasure of information can be achieved. The magnetic properties required for such an application are a combination of the type required for domain stability, as in bubble domains, accompanied by suitable temperature-dependent properties. However, in spite of the technical feasibility the cost of the associated optical technology makes the economic viability of optical beam stores using thin magnetic film media unlikely in the near future.

4. Optical Properties and Devices (Macleod[14])

Thin film optical coatings and filters are an indispensable branch of modern optics. Antireflection coatings are vital for the operation of most modern optical instruments and represent a very large slice of the total market for thin film optics. Dichroic beam splitters are an essential part of colour television systems whilst narrow band filters are used in a wide range of applications from astronomy to chemical analysis. Lasers could not function without the low-loss thin-film laser mirror coatings and the laser fusion programme will depend for its success on the development of coatings which are capable of handling the large powers involved. Heat reflecting and cold mirrors are used in a wide range of optical instruments, notably slide projectors and lamps for use in operating theatres, and similar coatings are likely to be of increasing use in solar energy applications. Thin film coatings have also been suggested as aids to making documents, credit cards and so on more difficult to counterfeit. The appearance of thin film coatings is often startlingly beautiful and they have been much used in jewellery.

These coatings are characterised by propagation across the film and the interference effects which are responsible for the behaviour of the various devices mentioned above are a result of the film thicknesses being comparable to the wavelength of light. That is, it is a geometrical property rather than a fundamental film property which is exploited. Films of a higher refractive index than their surroundings can also support propagation parallel to their boundaries in which the energy of the wave is retained by the film in much the same way as a wave in a conventional waveguide and the propagation can be described by a series of normal modes. The propagation of these normal modes in films can be looked on as a two dimensional representation of normal three dimensional optics and equivalents of components such as lenses, reflectors, and beam splitters can be made.

There are possible applications for two dimensional optics in areas such as computing and image processing. Additionally, optical communication using fibre waveguides is a rapidly expanding and important field and optical transmitters and signal processors and receivers containing the optical equivalents of modulators, oscillators, switches, mixers and so on are currently being developed. These systems are most likely to take the form of thin-film integrated optical devices in which the light beams are contained and guided by thin film guides and where the manipulation and generation of the signals is carried out by thin film devices[15].

Optical thin films is therefore an enormous field which can be divided into main sections, the first dealing with the more traditional optical filters and coatings and the second with the relatively new and rapidly expanding field of thin film guides. Macleod reviews the topics in Chapter 8 but in a book of this size it is impossible to cover the entire field in detail and so the account is a fairly general one which attempts to pick out some of the principal features of a number of the more important devices. The emphasis is more on how the devices perform rather than on the techniques of manufacture.

5. Electro-optic Properties and Devices

5.1 Thin Film Photoconductors (Rose[16])

Present usage of thin film photoconductors ranges from light meters, for photographic use, light beam detectors for burglar alarms or servo systems for door-opening mechanisms or machine tool control, to vidicon T.V. camera tubes and photocopying processes. These applications are all based on the same essential feature of photoconductivity; that is, the interaction of light (or other radiation) causes a change in the conductivity of the material. Bulk photoconductors respond to radiation in much the same way as do thin films and the thickness of the device does not affect its inherent property. However, the majority of photoconductors generally exhibit strong absorption over a limited portion of the spectrum and so utilise only a very thin surface layer, of the order of microns, of the material. Thus for economic and practical reasons, it is pointless to fabricate thicker films than are necessary and it is usual to use vacuum-deposited thin films rather than bulk material. In general however, the properties of thin film photoconductors are less than ideal and this can be ascribed to the presence of traps in the films. Traps can arise from impurities or from lattice irregularities and since thin films generally tend to have a greater concentration of both of these defects it is to be expected that they will also contain a greater density of traps.

A study of the properties of individual photoconductors illustrates the features of practical importance as regards device requirements. Selenium is a particularly useful and extensively studied photoconductor and is fairly typical of those materials exhibiting high dark resistivity. Its photosensitivity is greatest in the visible part of the spectrum and, in addition, its conductivity changes reproducibly from light to dark conditions, even after many operations, so that its extensive use in xerography is not surprising. By treating the material as a capacitative insulator, charge can be induced on the photoconductor surface. On exposure to white light charge is conducted away from the surface of the illuminated areas so that an electrical image of the pattern is generated. This can be reproduced by dusting the photoconductor with dielectric particles which adhere to the charged areas. The particles can then be transferred to paper by bringing the latter into intimate contact with the photoconductor; the impression being made permanent by fusing the particles onto the paper. Selenium has remained virtually unchallenged as the photoconductor in xerography but the recent emergence of new organic materials is now posing a serious threat. These combine the vital properties of high dark resistance, good response in the visible part of the spectrum and speed, with a low optical density. Their principal disadvantage is fatigue due to deep traps leading eventually to unwanted image retention but once this problem has been overcome these materials will offer a formidable challenge to selenium. Electrofax is another version of electrophotography in which the plate is coated with zinc oxide and the image fused onto this.

Probably the second most studied photoconductor is cadmium sulphide, the main use of which being in radiation detectors. In these devices a voltage is applied across the photoconductor and the current flowing is directly proportional to the intensity of the incident radiation. In this application, the high gain factor (effectively the number of times the same photo-generated charge can pass between the electrodes) of cadmium sulphide is particularly advantageous. This mode of use necessitates a trade-off between response time and gain due to their opposing dependencies on free carrier lifetime. In certain applications, e.g. light meters, this limitation is not a problem but in others, such as light sensitive switches, the trade-off is important.

Antimony trisulphide is a photoconductor which has found particular use in vidicons. The device itself is limited to a gain of unity and demands a response time of tens of milliseconds so there is no virtue in using a photoconductor which can outmatch this specification. The material also has other relevant properties such as a similar spectral response to that of the human eye, a high dark resistivity and good sensitivity (the ratio of current output to optical power input). The vidicon itself operates on similar principles to xerophotography. Light from the object to be projected falls on one side of the photoconductor area of the camera tube whilst the opposite side is charged by a scanning electron beam. The charge is conducted away from the illuminated areas and when the beam rescans those areas it supplies further electrons to restore the charge. By so doing a change in signal on the electron gun is registered which thus provides an electrical record of the scene. Lead oxide is another material often used in vidicons or in this case more appropriately called plumbicons.

The materials mentioned here, and indeed even those discussed in Chapter 9, represent only a small selection of those presently being used and researched but there are of course many others which have not yet been fully investigated and whose potentials have yet to be exploited. An example of these is the vast class of mixed photoconductor systems. In particular the full range of organic compounds with potential applications is only just beginning to be appreciated. Photoconductors are now beginning to be used in biology and in developing prosthetics and it is also possible to visualise their ultimate application in advanced computers. Because of the present energy shortage it is also tempting to speculate about the possible use of photoconductors in energy generation systems, on the basis that if plants can utilise sunlight to produce food via photosynthesis then perhaps man can learn to exploit photoconductivity in energy conversion.

5.2   Thin Film Solar Cells (Wolf[17])

The historical development of solar cells can be traced from the original observation in 1839[18] of the effect of light on electrolytic cells to the discovery of the photovoltaic effect in selenium, copper oxide and other materials, to the present day when the devices offer a very real possibility of solar energy conversion for terrestrial application. The principle of operation is relatively straightforward, at least in the case of single crystal devices, and is based on the fact that under the right conditions light absorbed near the junction of a p–n diode can create electron–hole pairs which can be separated by the internal electrostatic field. This charge separation constitutes a voltage and, when an external circuit is completed, current can flow and power can be delivered to an appropriate load. The current/voltage characteristics of single crystal devices can be accounted for by simple p–n junction theory and even deviations from ideal behaviour, such as the existence of finite series and shunt resistances, can be adequately formulated. This is not always the case in thin film devices however and for many of the devices studied today there is continuing debate regarding energy band diagrams, the effect of composition, structure etc. Without a thorough understanding of the operation of thin film cells it is unlikely that their full potential can be realised.

The thickness of thin film cells is typically 20 µm and the semiconductors are very often typified by a polycrystalline structure of columnar grains extending through the thickness of the film. Their comparatively great thickness precludes any inherent property, such as a thickness-dependent resistivity, but their columnar structure is a function of the manner in which they are deposited and grow. In practical devices, the latter property is important insofar as it can modify cell performance but it is certainly not a feature which is exploited. The prospective importance of the thin film cell (if something 20 μm thick can be called thin) is that it may be possible to prepare very large areas at relatively low cost.

To date, primarily for reasons of cost, solar cells have only been used to any extent as a power source in spacecraft. In particular, the single crystal silicon homojunction cell has been extensively used and, because of the highly advanced status of silicon technology (due to its widespread use in microelectronics) it is now possible to manufacture cells with conversion efficiencies of up to, or even greater than, 15%. However, silicon is not, from the standpoint of the solid state physicist, an ideal material from which to fabricate a solar cell, and to achieve adequate light absorption it is necessary to use relatively thick wafers of about 100 μm. Since not all the incident light is absorbed in the immediate vicinity of the interface between the p- and n-regions, it is essential that the silicon contains very few structural defects at which carrier recombination can occur thereby reducing the efficiency of the device. To achieve the consequent material perfection requires the use of elaborate crystal-growing techniques which increase the cost of the power generated. Although the thin film cell, and, in particular, the cadmium sulphide/cuprous sulphide heterojunction, was originally studied in connection with space application because it had a potentially superior power/mass ratio to that of silicon and could be deposited on flexible substrates (useful for deployable arrays), the emphasis today is more on its potential as a generator of electricity for terrestrial application.

In order to make the use of photovoltaic power feasible for large scale terrestrial generation, it is necessary to achieve compatibility between production costs, material costs and the size of the arrays necessary to generate useful amounts of power. Although efforts are being made to produce low cost, but high quality, single-crystal silicon wafers, there are also a number of projects aimed at developing useful thin film polycrystalline or even amorphous silicon devices which, although probably less efficient, may produce power at a lower cost. This in fact, is the principal objective of most research on the thin film cell. Because films inevitably have inferior material qualities to single crystals, their efficiency seems certain to be inferior also although this would be unimportant if the cost of the power was less.

The CdS/Cu_2S combination is by far the most extensively studied thin film cell and although it is normally fabricated by depositing the CdS by evaporation and the Cu_2S by dip conversion, other techniques are being examined including spraying, painting and anodic sulphurisation. The latter are all potentially more suitable for the production of very large areas than vacuum deposition.

Cadmium sulphide is used in other heterojunctions in conjunction with indium phosphide, cadmium telluride and copper indium selenide, all of which show considerable promise. Hill reviews all the above and several other combinations in Chapter 10 and pays particular attention to the structure required of the materials together with the importance of parameters such as the diffusion length of minority carriers and surface recombination velocity. Finally, Hill discusses the present day status of the thin film cell as a generator of terrestrial electricity and points out the obstacles to developing commercially viable systems.

6. Superconducting Properties and Devices
(Kammerlingh Onnes[18])

Although known since 1911, superconductivity is a phenomenon which has only recently begun to yield devices with any significant potential for use outside the low temperature physicist's laboratory. The essential properties upon which devices are based are zero resistance, critical temperature and critical field. Just as important, however, is flux (or more generally, fluxoid) quantisation: this is the phenomenon that any magnetic flux which threads a hole in a superconductor does so in integral multiples of the flux quantum, $\phi_0 = hc/2e = 2.07 \times 10^{-7}$ gauss-cm^2. These properties reflect a condensation of the conduction electrons in a normal metal into a state which is highly ordered in momentum space and two consequences follow. First, "normal electrons" (or excitations from the condensed superfluid state) can only be created by injecting enough energy to surmount an energy gap Δ which is typically a few milli-electron volts per excited electron. Secondly, the condensation effectively leads to a single macroscopic electronic wave function in the metal, whose phase, although dependent on the local value of the magnetic vector potential, is coherent over macroscopic distances. This extensive phase property leads to the various quantum phenomena associated with the name of Josephson, making possible (in principle at least) first, switching processes involving picosecond times and power dissipation of the order of 10^{-17} joules, secondly, voltage to GHz frequency converters obeying the relation $v = (2e/h)V$ and lastly, magnetic flux detectors sensitive to small fractions of a flux quantum.

In many of their properties, particularly critical temperatures, superconducting thin films do not differ greatly from bulk samples of the same material. However, the magnetic behaviour is in general much more complex, and in particular it is important to realise that it may often be necessary to bear in mind the large demagnetising factor of a superconducting thin film in a perpendicular magnetic field. This can be so large as to reduce the field at which flux penetration first occurs by two or more orders of magnitude.

In Chapter 13 Donaldson reviews a number of devices including those based on "simple" properties such as zero resistance, critical temperature and critical field. Amongst these is the D.C. transformer, which employs magnetic field coupling between films separated by a layer of insulator. This is in an early stage of development as yet and does not rival more conventional transformers, although 80% power transfer efficiency has been achieved. On the other hand, the superconducting transition-edge bolometer is already the best available detector of far infrared radiation, with a noise equivalent power of 2×10^{-15} W-Hz$^{-\frac{1}{2}}$ (for comparative performance figures for other devices, see Chapter 12). It is clearly ripe for balloon and satellite-borne operation, in, for example, deep space infrared background radiation measurements.

The cryotron, in which a lot of computer industry money was invested in the 1960's, was originally studied because of its promise as a thin-film flip-flop storage element based on the superconductive resistive transition. However competing technologies proved superior on the grounds of speed, but now a successor, the Josephson tunnelling cryotron, which seems to have no rivals whatever in terms of power or speed, appears to be pointing the way back to superconducting computer systems. As storage elements, Josephson devices consume no stand-by power, and can be interrogated in a few tens of picoseconds with a dissipation of only a few hundred electron volts *per element*! Logical structures such as shift registers now have cycle rates above 1 GHz, while cell sizes down to 50 μm^2 have been achieved, so that since cooling requirements are so minimal, very high element-density should be possible.

The acronym SQUIDs has been coined for devices based on superconducting quantum interference effects. These devices are essentially magnetic flex detectors based on the Josephson effect, but which have a wide range of instrumentation applications. Fields of less than 10^{-11} gauss can now be resolved in a 1 Hz bandwidth, and there are SQUID users in geophysics, physiology and biochemistry, not to speak of communications and other fields closer to "basic" physics. So far thin films have not played much part in commercial SQUID development, but one of the most successful laboratory devices has been the thin film D.C. SQUID[19] and this seems likely to become dominant before long.

Superconductivity also has a role in detection and mixing, not only of electromagnetic radiation but also of ultrasound, at frequencies up to infrared. Applications of energy-gap related tunnelling to Raman spectroscopy and to phonon spectroscopy are also relevant, but once again it is the Josephson effect which shows the greatest promise. Large arrays of thin film Josephson elements may be operated as very efficient low-noise mixers, and as detectors and parametric amplifiers. Single element devices have already been used in such work: in particular they have led to a new determination (accurate to 0.1 ppm) of the fundamental quantity h/e, and in turn to a

Josephon voltage reference system which is calibrated against a standard of frequency, and which replaces the more traditional chemical cell. Thus it is clear that this is a rapidly developing field and it seems certain that many of the devices discussed by Donaldson in Chapter 13 will become commercially and technically viable in the near future. In addition, it is to be anticipated that other devices will appear possibly even before this book is published.

References

1. J. C. Anderson, in "The Use of Thin Film in Physical Investigations", ed., J. C. Anderson, Academic Press, New York, (1966).
2. D. C. Reynolds, *Phys. Rev.*, **96**, 533, (1954).
3. K. Fuchs, *Proc. Camb. Phil. Soc.*, **34**, 100, (1938).
4. A. F. Mayadas and M. Shatzkes, *Phys. Rev. B.*, **1**, 1382, (1970).
5. R. G. Duckworth, *Thin Solid Films*, **26**, 77, (1975).
6. S. J. Fonash, in "Physics of Non-Metallic Thin Films", eds., C. H. S. Dupuy and A. Cachard, Plenum Press, New York, (1976).
7. A. K. Jonscher and R. M. Hill, in "Physics of Thin Films", Vol. 8, eds., G. Hass, M. H. Francombe and R. W. Hoffman, Academic Press, New York, (1975).
8. J. C. Maxwell, "A Treatise on Electricity and Magnetism", Vol. 1, 2nd edn., Clarendon Press, Oxford, (1881).
9. J. E. Lilienfeld, U. S. Patent 1900018, (1935).
10. A. G. Fischer, *Thin Solid Films*, **36**, 469, (1976).
11. N. F. Mott and W. D. Twose, *Advan. Phys.*, **10**, 107, (1961).
12. E. H. Putley and D. H. Martin, in "Spectroscopic Techniques", ed. D. H. Martin, North Holland Publishing Co., Amsterdam, (1967).
13. R. F. Soohoo, "Magnetic Thin Films", Harper and Row, London, (1965).
14. H. A. M. Macleod, "Thin Film Optical Filters", Adam Hilger, London, New York, (1969).
15. T. Tamir, "Integrated Optics", ed., T. Tamir, Springer-Verlag, Heidelberg, (1975).
16. A. Rose, "Concepts in Photoconductivity and Allied Problems", Wiley, New York, (1963).
17. M. Wolf, in "International Conference on Photovoltaic Power Generation", ed., H. R. Lösch, DGLR, Hamburg, (1974).
18. E. Becquerel, *Compt. Rend.*, **9**, 561, (1834).
19. H. Kammerlingh Onnes, *Leiden Commun.*, **122b**, 124c, (1911).
20. R. P. Gifford, R. A. Webb and J. C. Wheatley, *J. Low Temp. Phys.*, **6**, 533, (1972).

Chapter 2

Preparation Methods for Thin Film Devices

D. S. Campbell
Department of Electronic and Electrical Engineering, Loughborough University of Technology, Loughborough, England

1. Introduction 24
2. Physical Deposition Techniques 25
    2.1 Evaporation 25
        2.1.1 Introduction. 25
        2.1.2 Basic principles. 26
        2.1.3 Types of evaporation sources 27
    2.2 Sputtering 32
        2.2.1 Introduction. 32
        2.2.2 Basic principles. 33
        2.2.3 Types of sputtering systems 35
3. Chemical Deposition Techniques 41
    3.1 Thermal Growth 41
    3.2 Anodisation 41
        3.2.1 Introduction. 41
        3.2.2 Basic principles. 42
        3.2.3 Practical aspects 44
        3.2.4 Gaseous anodisation 44
    3.3 Vapour Phase Growth. 44
        3.3.1 Introduction. 44
        3.3.2 Types of vapour phase reactions 44
    3.4 Electroplating 46
        3.4.1 Introduction. 46
        3.4.2 Basic principles. 47
        3.4.3 Practical aspects 49
    3.5 Chemical Reduction Plating. 49
        3.5.1 Introduction. 49

3.6	Solution Deposition		51
	3.6.1	Oxide films	51
	3.6.2	Built up films (Langmuir films)	51
	3.6.3	Liquid phase epitaxy	51
	3.6.4	Growth of polymer films	52
4.	Summary	53
	References	54

1. Introduction

Underlying the performance and economics of thin film components are the manufacturing techniques that are used to produce the devices. It is possible to classify these techniques in a variety of ways[1,2], and this chapter aims to discuss these classifications with respect to devices.

One classification is in terms of the groupings (1) chemical methods and (2) physical methods. Physical methods cover deposition techniques which depend on the evaporation or ejection of material from a source, i.e. evaporation or sputtering, whereas chemical methods depend on a specific chemical reaction. Thus, chemical reactions may depend on the electrical separation of ions as in electro-plating and anodisation, or may depend on thermal effects as in vapour phase deposition and thermal growth. However, in all these cases a definite chemical reaction is required to obtain the final film.

When one seeks to classify deposition of films by chemical methods, one finds that it is possible to further sub-divide the methods that are available, into two more classes. The first of these classes is concerned with the chemical formation of the film from the medium, and typical methods involved are electro-plating, chemical reduction plating and vapour phase deposition. A second class is that of formation from the substrate and examples are anodisation, gaseous anodisation and thermal growth.

It must be emphasised that there is often considerable overlap between the physical and chemical classifications, and also between the sub-classification of formation from the medium and from the substrate.

The methods summarised under the classifications given are often capable of producing films defined as thin films, i.e. 1 μm or less, and films defined as thick films, i.e. 1 μm or more. However, there are certain techniques which are only capable of producing thick films and these include screen printing (as used in hybrid thick film technology), glazing, electrophoretic deposition, flame spraying and painting.

However, these techniques will not be examined in the context of this chapter. (See Chapman and Anderson[3] for surveys of these techniques.)

2. Physical Deposition Techniques

2.1 Evaporation

2.1.1 Introduction

Evaporation or sublimation techniques are widely used for the preparation of thin layers. A very large number of materials can be evaporated and, if the evaporation is undertaken in a vacuum system, the evaporation temperature will be very considerably lowered, the tendency to form oxides will be considerably reduced, the amount of impurities included in the growing layer will be minimised and, finally, straight line propagation will occur from the source to the substrate which will allow for the reproduction of finely defined patterns by the introduction of a patterned screen between the source and the substrate. Substrates can be of a wide variety of materials and can be held at a temperature appropriate to the properties of the deposited films which are required.

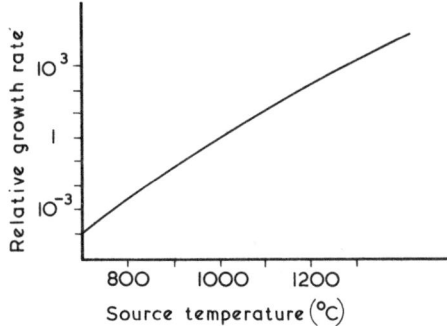

FIGURE 1. Relative growth rate of gold as a function of source temperature.

The pressures that are required in a vacuum system to obtain satisfactory deposition, in terms of the reduction of oxides, reduction of included impurities in the deposition, and the fabrication of a sharply defined pattern, is less than 10^{-4} torr with an ideal pressure for normal evaporation work being 10^{-5} torr.

Rates of evaporation and condensation can vary over very wide limits and are dependent upon the temperature of source and the material used: a typical relative growth-rate curve for gold as a function of source temperature is given in Fig. 1.

An average figure obtainable for growth rate can be reckoned as 10Å per second, but it is possible to obtain rates as high as 10^4 Å/second, using special boats[57]. Very low rates of growth are also possible (c.f. the work of Walton et al. on nucleation studies[4]).

Control of the deposition rate and hence film thickness, can be effected by several systems[5,6], basically either mechanical or electrical, but all of them depend eventually on a direct thickness measurement.

2.1.2 Basic principles

The thermodynamics of the evaporation processes have been considered in detail by various authors (c.f. Glang[7]). The rate of evaporation G from a surface at a temperature T K is given by the Langmuir expression:

$$G = p(M/2\pi RT)^{1/2} \qquad (1)$$

where p is the vapour pressure of the material at temperature T and M the molecular weight. R is the gas constant per mole. The relationship between

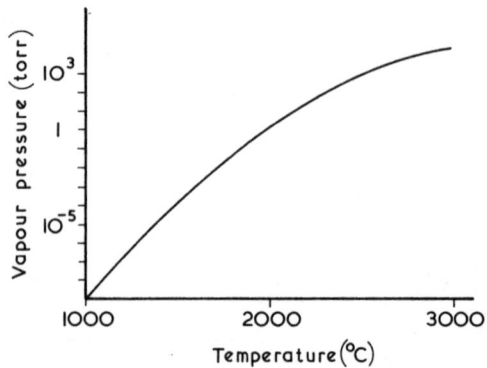

FIGURE 2. Vapour pressure of gold as a function of temperature.

p and T has been published by various authors for a wide variety of materials (Honig[8]) and Fig. 2 shows a typical curve for gold. The temperature which is normally considered suitable for evaporation is that at which the vapour pressure of the material is equal to 10^{-2} torr. As an example, for gold this corresponds to a source temperature of 1650°C at which value the rate of evaporation from the gold surface is 0.2×10^{-3} gm/cm² sec⁻¹, which is equivalent to 6×10^{17} atoms/cm² sec⁻¹. The average energy of the evaporant molecules ($3kT/2$) depends on the source temperature. For gold at 1650°C, this will be approximately 0.25 eV. The rate of arrival n, of molecules or atoms of molecular weight M of evaporant at the substrate is given by:

$$n = \frac{t\rho N}{M} \qquad (2)$$

where t is the thickness deposited and N is Avogadro's number. This value of n needs to be compared with the arrival rate of residual gas atoms at pressure p' in the vacuum, which is given by:

$$n_g = \frac{3p'}{4}(N^{2/3}\,RTM_g)^{1/2} \qquad (3)$$

where M_g is the molecular weight of the residual gas. For normal evaporation systems in which a psuedo point source may be assumed and with a source-substrate distance of around 15 cm, it is found that a ratio of the order of 1000:1 exists between the number leaving the source per unit area, and the number arriving. In these circumstances, an arrival rate of approximately $5 \times 10^{15}\,\Omega/\text{cm}^2\,\text{sec}^{-1}$ is obtained for a gold source at a temperature of 1750°C, and this corresponds to a growth rate of about 10 Å/sec. However, the value for the rate of arrival of residual gas atoms at room temperature and 10^{-5} torr, for nitrogen and/or oxygen is also 5×10^{15} torr, so that it can be seen that to be sure of obtaining a film with the least number of impurities, either the source temperature has to be considerably increased so as to raise the rate of evaporation, the source-substrate distance must be reduced (with the subsequent unavoidable heating of the substrate that will result), or the residual gas pressure must be reduced below 10^{-5} torr. It is this final solution which is normally used if very pure films are required, and the techniques for obtaining very low pressures in vacuum systems have been widely discussed by many authors in the literature[9]. It is now possible, at least in experimental systems, to work at pressures of 10^{-10} torr or less, but these are not used in commercial production.

2.1.3 Types of evaporation sources

In order to evaporate materials in a vacuum, a vapour source is required that will support the evaporant and supply the heat of vaporisation while allowing the charge of evaporant to reach a temperature sufficiently high to produce the desired vapour pressure, and hence rate of evaporation, without reacting chemically with the evaporant. To avoid contamination of the evaporant and hence of the growing film, the support material itself must have a negligible vapour pressure and dissociation pressure at the operating temperature. There are, therefore, two types of material that can be used for this, either refractory metals or certain non-metallic materials such as oxides, nitrides etc. The form in which these are used depends very much on the evaporant and tables are now available, summarising the best support materials used for the evaporation of elements, inorganic materials and alloys (Glang[7]), to which the reader is referred.

A. *Wire and metal foil structures.* Of the two basic types of material used for sources, wire and metal structures are very widely used for many evaporants.

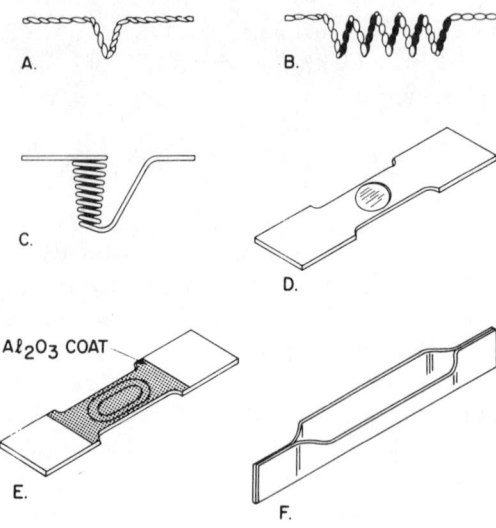

FIGURE 3. Wire and metal foil sources: (A) hairpin, (B) wire helix, (C) wire basket, (D) dimpled foil, (E) dimpled foil with aluminia coat, (F) trough type.

The simplest vapour sources are resistance-heated wires and metal foils of various types, examples of which are shown in Fig. 3. These are generally made from tungsten, molybdenum or tantalum. Wire sources are generally made from wire of diameter 0.02 to 0.06 in, and their use is limited to evaporants which wet the filament upon melting and are then held on by surface tension.

It is also possible to use wires of the material to be deposited, provided that the materials will sublime. This implies that a vapour pressure of 10^{-2} torr is reached before the melting point of the wire itself. Such a technique has been widely used for the deposition of nickel[10] and also for nickel-chromium[11], to mention but two, although in the latter case it is important to note that the composition of the deposited film will change with time because of the different sublimation rates of nickel and chromium from the wire.

Metal foil structures can take the form of open boats of various shapes, or, for materials that are apt to de-gas on evaporation, structures have been designed which prevent the ejection of solid particles directly from the source. A particularly well-known example of this type of source is that developed by Drumheller[12] for the evaporation of silicon monoxide (see Fig. 4).

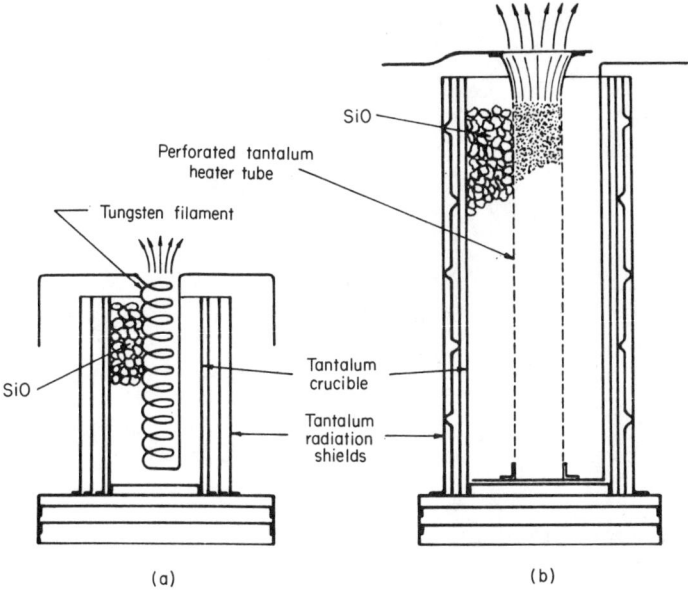

FIGURE 4. Chimney evaporation sources (after Drumheller [12]).

It is possible to use metal foil structures for the deposition of alloys as well as elements and oxides, providing enough information is available on the relative vapour pressures of the alloy constituents. In this context, two alloys have been examined in considerable detail; nickel-iron (Permalloy) and nickel-chromium. The nickel-iron alloy has been extensively used for magnetic memory elements whilst nickel-chromium is frequently used for resistors. Both are often evaporated from tungsten boats. However, the change of composition with time during the evaporation process, is a considerable disadvantage[13], particularly in the case of nickel-chromium, where the electrical properties are dependent on the nickel-chromium ratio of the final deposition. These problems have been overcome by using evaporation from two separate sources with the rates of evaporation separately controlled, or by the technique known as flash evaporation. Various arrangements for flash evaporation have developed[14,7] and one dispenser is shown in Fig. 5[13]. The principle is that a finely divided powder of the alloy is vibrated on to a very hot tungsten strip where it evaporates immediately on contact with the strip. In these circumstances the composition of the deposited film is the same as the original alloy powder. Care has to be taken, however, to shield the source in order to prevent

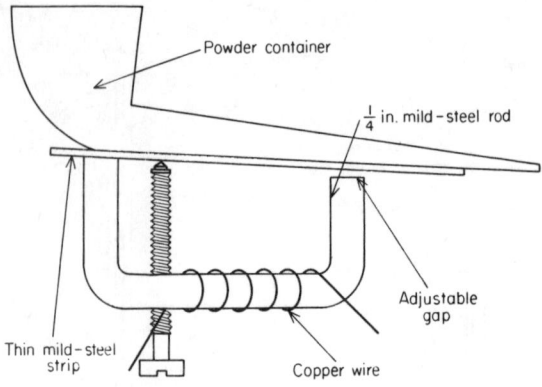

FIGURE 5. Dispenser for flash evaporation (after Campbell and Hendry[13]).

any loose powder from finding its way into the pumping system of the vacuum chamber.

B. *Non-metallic structures.* Crucibles of non-metallic materials are often used as evaporating sources[7]. These crucibles, which are electrically non-conducting, are supported in a suitable metal cradle. The cradle itself is heated by the direct passage of an electric current. The cradle can take the form of a wire coil directly wound around the crucible or of a foil structure.

Another way of heating crucibles is that of using an RF coil placed around the crucible but not actually touching it. Dielectric heating of the crucible material itself is obtained when the coil is energised.

Alumina is a widely used crucible material for temperatures up to 1900°C. It has a good thermal conductivity ($0.014 \, \text{cal sec}^{-1} \, \text{degree}^{-1} \, \text{cm}^{-2}$), enabling heat to be transferred easily from the heated metal coil or boat. An even better material than alumina, from the thermal conduction point of view, is beryllia, which has a thermal conduction over three times that of alumina. However, there are certain toxic disadvantages in the use of beryllia, which are important in any commercial use of such a source. Other oxide materials are available and all have been summarised by Glang[7].

Other materials which have been used for crucible sources. These include boron nitride and carbon, either as ordinary graphite, pyrolitic graphite or in the form of vitreous carbon.

C. *Electron-beam heating*[7]. The two main types of sources examined so far have been heated either by resistance heating or by induction heating. It is, however, possible to cause vaporisation of materials by using electron bombardment. A stream of electrons is accelerated up to 10 kV and

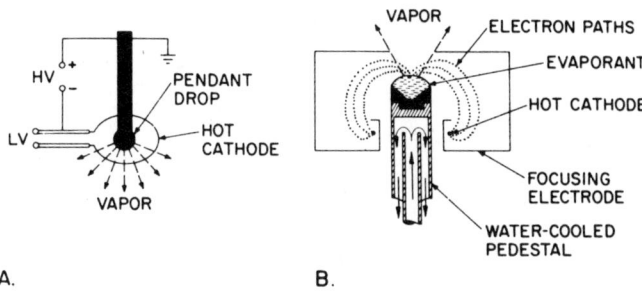

FIGURE 6. Electron gun sources: (A) pendant drop, (B) shielded filament.

focussed on to the evaporant surface. By this means, temperatures exceeding 3000°C may be obtained, enabling a variety of otherwise non-evaporatable materials to be deposited. Also, as only a very limited portion of the evaporant is heated, any interaction between the evaporant and the support materials is considerably reduced, and this technique is used in the preparation of very pure films.

A wide variety of electron gun structures has been used, and these have been classified by Glang[7], into work-accelerated guns, self-accelerated guns and bent-beam guns. Two typical electron beam sources are shown in Fig. 6. Fig. 6(A) shows a pendant drop configuration in which the metal to be evaporated is in the form of a rod or wire centred within a cathode loop. Evaporation takes place from the molten tip. The molten drop is held on the tip by surface tension and, as a result, careful control of the electrical energy supply is required to prevent the drop from falling off. The second source shown in Fig. 6(B) was developed by Unvala and Booker[15] and this uses a hearth, which can be water cooled, to contain the evaporant. Also, because of the electron beam configuration, the evaporating unit is self-contained.

D. *Other sources.* Over the last few years, modifications to sources have been made for specific purposes. One technique which has become of interest is that of the hot wall method[16]. In this technique, films are grown under conditions that are close to thermodynamic equilibrium. The wall of the small enclosure is held at a higher temperature than the substrate, so that deposition only occurs on the substrate itself. Figure 7 shows a diagram of an apparatus which has been used for the deposition of lead telluride films.

A further modification to evaporation sources that has become of interest involves the use of laser beams as the energy source. Laser beam evaporation

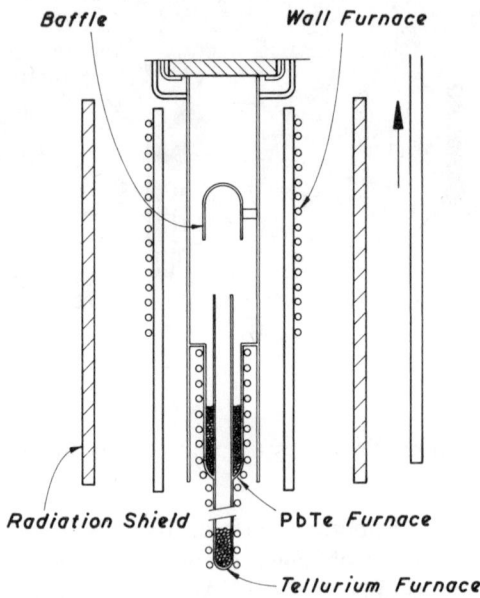

FIGURE 7. Hot wall evaporation system (after Lopez-Otero and Hass [16]).

has been found to be very useful for the preparation of polymer films[17].

E. *Reactive evaporation*[7]. Evaporation from metal wires or foils, from refractory oxide crucibles, or from electron beam sources, can be used to deposit oxides. However, it is also possible to evaporate metals under relatively high ($>10^{-4}$ torr) oxygen pressure and in these circumstances the oxide of the metal being evaporated will then be deposited on the substrate. The oxidation takes place in the main, at the surface of the film, and such techniques have been used successfully for the deposition of silicon oxide, tantalum oxide, aluminium oxide and even $BaTiO_3$. Nitrides can also be deposited if a nitrogen atmosphere is used in place of oxygen.

2.2. Sputtering

2.2.1 Introduction

If a surface is bombarded with energetic particles, it is possible to cause ejection of the surface atom: this is the process known as sputtering. The ejected atoms can be condensed on to a substrate to form a thin film. The process can be realised by forming positive ions of a heavy neutral gas such as argon, and causing these to bombard the surface of the target material by making the surface the cathode in an electrical circuit. This method of obtaining a film has various advantages over normal evaporation techniques, in as much as no container contamination will occur. It is also possible to deposit alloy films which retain the composition of the parent target material. High melting point materials can be deposited and finally, using R.F. techniques, dielectric films can be fabricated.

2.2.2 Basic principles[18-20]

When a charged particle strikes a surface, a variety of interactions are possible. The most important reactions are shown diagramatically in Fig. 8,

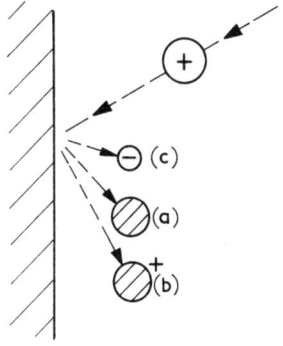

FIGURE 8.   Ejected species from an ion bombarded surface.

and are (a) the ejection of neutral atoms of the surface material, (b) the ejection of a small number of charged atoms of the surface material (usually only about 1% or less of the number of uncharged atoms) and finally, (c) the ejection of free electrons—the number of free electrons being greater than one for each arriving ion. The neutral ejected atoms (Process (a)) can be collected on suitably placed substrates to form a film. The electrons ejected can be accelerated away from the target cathode to an anode, and on their way to the anode they will cause further ionisation of the neutral gas in the surrounding space, and the positive ions so formed will then be accelerated towards the cathode target. A self-sustaining system can therefore be obtained and such a situation is known as a glow discharge condition. The pressure at which the system is self-sustaining depends on the cathode/anode spacing and on the residual pressure; typical values for spacing and pressure being 5 cm and 10^{-2} torr, respectively. Below 10^{-3} torr, a discharge cannot readily be sustained.

To obtain the reactions as shown in Fig. 8, the accelerating voltage must be limited. Figure 9 shows a typical graph of sputtering yield, i.e. the number of neutral atoms ejected by one incident ion as a function of voltage. The three types of interactions can be identified on this graph, namely:

(a) hard sphere ejection, below approximately 10 keV;
(b) a region in which the electron clouds of the incident ion and surface atoms begin to interact;
(c) a region in which the nuclei interact.

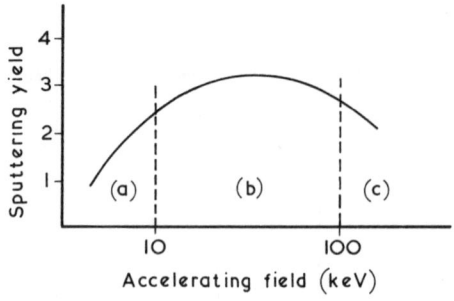

FIGURE 9. Idealised sputtering yield/bombardment field curve.

The majority of sputtering situations are concerned with a region of hard sphere ejection and studies on single crystal materials have shown that in this region, ejection from a target is a function of the crystallographic direction. On a polycrystalline target, however, the effect of crystallographic directions is averaged out.

An important feature of ejection under hard sphere conditions, from a polycrystalline target, is that the most probable energy obtained for ejected atoms is much higher than that obtained in the case of evaporation, e.g. 4 eV for Cu bombarded with Ar^+ at 1 keV. This average energy corresponds to an equivalent surface temperature of 4×10^4 K.

The rate of removal of atoms from the surface of a polycrystalline material is a function of the number of ions bombarding the surface, i.e. it is a function of current. A current of 1 mA cm^{-2} corresponds to an ion impingement rate of 5×10^{15} ions/cm^2 sec^{-1}. For a sputtering yield figure of approximately unity, this implies a sputtering rate of 5×10^{15} atoms/cm^2 sec. Comparison with evaporation situations previously discussed, implies that a current of around 100 mA cm^{-2} would be necessary to obtain rates of removal from the target, equivalent to that which can be obtained from a typical evaporation situation (e.g. gold at 1650 K).

The rate of arrival of atoms at the substrate surface will depend on the source-substrate distance, as with the case of evaporation, but it is not usually possible to consider a point-source approximation in sputtering. Furthermore, the source-substrate distance in a sputtering system is usually less than in an evaporation system (5 cm is a common distance) and, as a result of both these factors, the ratio of numbers of atoms leaving the target per cm^2/sec, to those arriving, is more generally of the order of 10:1 or less, rather than 1000:1 quoted for evaporation. Another major difference for glow discharge situations is that it is not possible to reduce the impingement rate of the residual gas atoms to anything like the level which is possible in evaporation, as the pressure of the residual gas must be maintained at a high value in order for the glow discharge to be self-sustaining.

2.2.3 Types of sputtering systems

A. *Glow discharge.* A normal glow discharge sputtering system operates under the conditions in which sufficient secondary electrons are generated to replace those lost to the anode or to the walls of the discharge chamber. Under these conditions, which are typified by a cathode/anode separation of approximately 5 cm and an accelerating field of 1.5 kV at a residual pressure of 10^{-2} torr, material will be sputtered from the metal cathode into the space between the cathode and the anode. Figure 10 shows an experimental arrangement. The best position for the substrate to collect this material would seem, at first glance, to be as close to the cathode as possible. However, such a position effectively prevents ions in the discharge from reaching the target cathode and, as a result, substrates placed close to the cathode receive only very thin coatings of sputtered materials. This screening effect can be avoided if the substrate is at some distance from the cathode and a very useful rule of thumb quoted by Maissel[21] is that the cathode-substrate distance should be about twice the length of the Crooke's dark space. This dark space is the region of low luminosity found adjacent to the cathode, and represents the acceleration distance required by the electrons, emitted from the cathode, in order to reach an energy at which ionisation of the residual gas can occur. It is also often found convenient to place the substrate directly on the anode.

Films grown in such an environment are often classified as "dirty" because of the high background pressure of gas, but this is not necessarily the case since the high pressure in a sputtering system is, in the main, due to an inert gas such as argon and the partial pressure of reactive gases could be as low as in evaporation systems.

Practical growth rates for substrates placed directly on the anodes in glow discharge systems are often limited by the need to avoid heating effects when

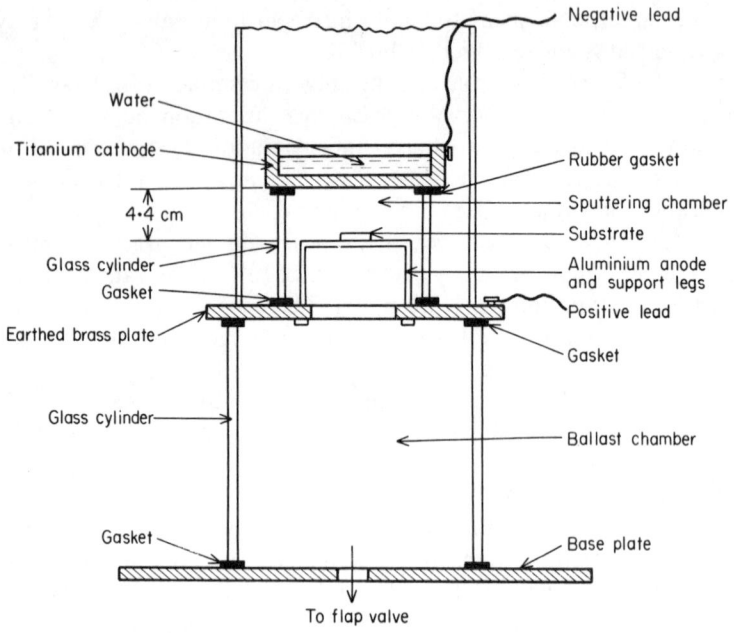

FIGURE 10. Glow discharge sputtering system.

high currents are involved. For gold sputtered by argon in the system shown in Fig. 10 at a cathode-substrate distance of 4.4 cm, with a voltage of 1.5 kV between anode and cathode and a total current of 2 mA over a cathode area of 48 cm^2, a growth rate of 0.5 Å/sec was obtained.

B. *Getter sputtering*[21]. Decreases in the partial pressure of reactive gases, beyond those usually associated with a normal glow discharge, can be effected by utilising the gettering action of the sputtering material to purify the argon before it reaches the part of the system where coating of the substrate occurs. This is accomplished by surrounding the cathode with an anode can (see Fig. 11). Under these conditions, the partial pressure of impurities can be reduced to levels as low as 10^{-10} torr.

C. *Bias sputtering*[22]. In the systems so far discussed the substrate has either been electrically floating, or kept at anode potential. It is, however, possible to give the film, assuming it is conductive, a small negative bias relative to the anode. The film is then subjected to ion bombardment throughout its growth and this process effectively cleans the film of absorbed gases which would otherwise become trapped in it as impurities. This technique is known as bias sputtering.

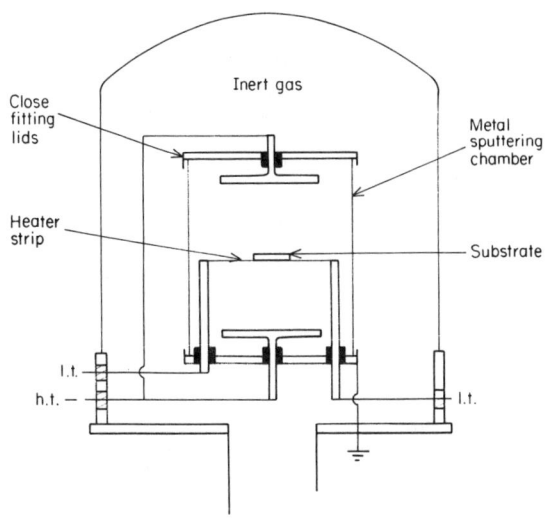

FIGURE 11. Getter sputtering apparatus.

D. *Triode sputtering.* Another method of reducing the effect of impurities in a glow discharge system is to work at lower pressures than 10^{-2} torr. This necessitates generating electrons by means other than the discharge itself and injecting them into the discharge. One such system is known as triode sputtering, and Fig. 12 shows a diagram of the apparatus. A heated filament at a voltage approximately 50 volts less than the anode acts as the electron source and a magnet, external to the vacuum chamber, is used to increase the path of the electrons prior to their collisions with the anode and hence to increase the probability of ionisation. In this situation pressures of 10^{-3} torr, or less, can be used in the discharge chamber.

E. *Ion beam sputtering.* If, instead of using a discharge system to generate the ions, either at 10^{-2} torr or at the lower pressures of triode sputtering, a separate ion source is used, then it is possible to grow films sputtered from targets in a residual pressure of less than 10^{-5} torr.

F. *R.F. supported sputtering.* Working pressures of less than 10^{-2} torr can also be achieved by applying an external R.F. field. This field, typically having a frequency of the order of 5 MHz, will cause both negative and positive ions to be formed, and the latter bombard the cathode in a normal manner. A typical apparatus is shown in Fig. 13.

G. *R.F. sputtering*[21,23]. The systems described so far have involved the use of a conducting cathode of the material to be sputtered. It is also possible to sputter directly from insulators by applying an R.F. potential between the

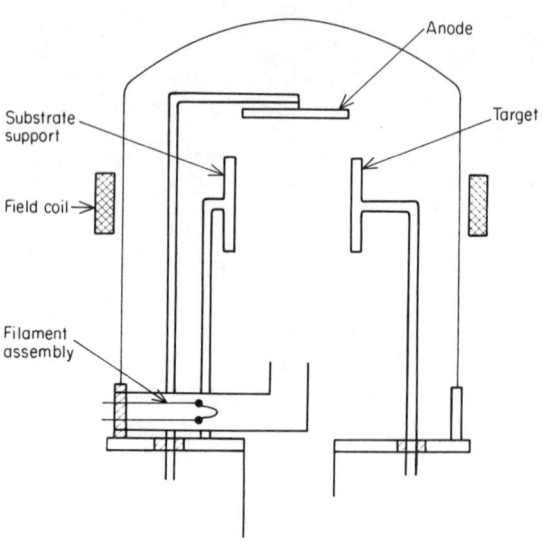

FIGURE 12. Triode sputtering apparatus.

cathode, on which is supported the insulating target, and the anode. This process is known as R.F. sputtering. A simple way of viewing the reactions which occur is to note that provided the frequency is large (greater than 50 kHz), the negative charge accumulated on the insulated target will not be sufficient, during the half cycle in which the target is positive, to prevent positive ions bombarding the target during the half cycle in which the target is negative. The actual behaviour of an R.F. sputtered system is, in fact, more complex when examined in detail[21].

H. *Reactive sputtering*[21]. If the residual gas in a glow discharge system reacts with the film being deposited, then deposits can be obtained of completely reacted materials such as oxides or nitrides. This system, known as reactive sputtering, has been used for the growth of silicon dioxide to prepare capacitors.

Reactive sputtering need not be confined to a glow discharge system, but can use any of the sputtering systems which have been discussed, e.g. triode sputtering, R.F. supported sputtering, etc.

I. *Plasma reactions.* Plasma reactions of various types have been used for the deposition of thin films. The reactions are difficult to classify but they are sometimes considered under sputtering. They are essentially vapour or thermal growth reactions, with the discharge supplying the energy necessary to effect the chemical changes.

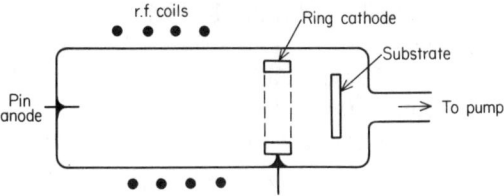

FIGURE 13. R.F. supported sputtering apparatus.

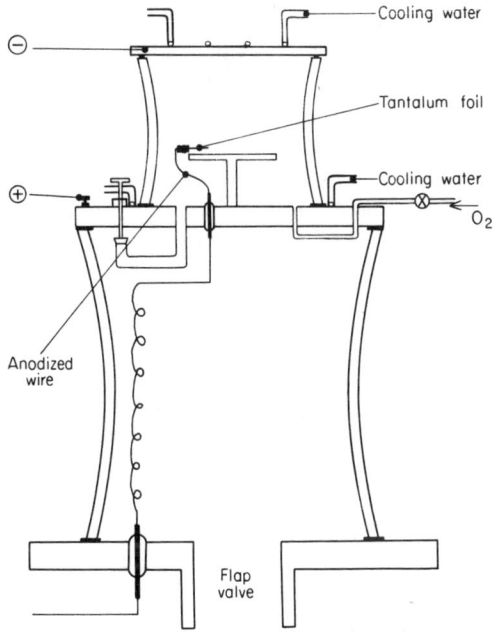

FIGURE 14. Gaseous anodisation apparatus (after N. F. Jackson[24]).

One such plasma reaction is that of gaseous anodisation[24]. Figure 14 shows a typical apparatus. Although a glow discharge is used, it effectively replaces the liquid electrolyte of a conventional wet anodisation process (see para. 3.2). It is important in such a system to use a non-reactive anode in the discharge circuit, otherwise all the voltage in the anodising circuit will be dropped across the oxide formed on the discharge anode. Various metals have been successfully oxidised in this way, including aluminium and tantalum, and more recently, titanium and molybdenum[25].

Plasma assisted thermal growth has been developed by Ligenza[26], who showed that using an R.F. excited discharge and an oxygen pressure of between 0.1 and 1 torr, silicon could be oxidised to a thickness of approximately 3500 Å at a temperature of about 300°C. If the silicon is the anode of a 50 volt system, then negatively charged oxygen ions will bombard the surface and thicknesses above 3500 Å can be obtained. However, in this latter case, material will be sputtered from the cathode as well, but this material can be prevented from landing on the anode by placing a suitable bend in the discharge tube, so that the cathode cannot "see" the anode.

If a normal glow discharge is used and the monomer of an organic or inorganic material is introduced into the discharge chamber, it has been found that the monomer will be polymerised in the glow discharge so that insulating films can be grown on suitably placed substrates[27]. Glow discharge conditions can also be used to effect vapour phase reactions such as the deposition of silicon nitride from a gaseous mixture of silicon hydride, ammonia and hydrogen[28]. R.F. excitation of the discharge has also been used and an apparatus has been described by Connell and Gregor[29] that can produce insulating films from styrene at high rates of deposition (20 Å/sec).

J. Ion plating[30]. The system of ion plating is essentially a combination of evaporation from a heated wire and the use of a discharge. The anode of the system is also the evaporation source, so that the evaporant is ionised in its passage towards the substrate. The latter is placed on the cathode and the evaporant is therefore accelerated before reaching the substrate. Using this technique, enhanced adhesion of the metal film to the substrate is obtained.

Mattox has recently reported improvements in the technique of ion plating[31], and has shown that it is possible to ion-plate alloys and ceramics, (in the latter case an R.F. plasma is used). The technique has also been extended to reactive ion plating, where carbide[32], nitride and boride films have been usefully deposited for wear resistance purposes. A system of vacuum ion plating has also been developed in which deposition again occurs at low vacuum conditions but in which ionisation of the evaporant is achieved using an ion gun.

K. Related techniques. Although not directly concerned with the deposition of films, two further techniques have become important which are based on ion beam sputtering. Firstly, ion beam techniques have become widely used for surface etching and indeed all sputtering processes result in an etching of the bombarded surface. Ion beam techniques can provide a controlled etching process[33]. Secondly, the use of ion implantation, in which ions are accelerated into a target, has become important not only with regard to doping of semiconductor films but for any system that requires the implantation of small amounts of impurities close to surfaces. Both of these processes are likely to play an increasingly important role in the future in basic studies of thin films and in the production of devices.

3. Chemical Deposition Techniques

3.1 Thermal Growth[34]

Films can be formed on a large variety of metal substrates by heating them in gases of the required type (oxygen for oxides, nitrogen for nitrides, CO for carbides). The films obtained, however, are limited in thickness because the reactions are generally self limiting and become very slow as the film thickness increases. If thickness is plotted against time, a characteristic relationship is obtained in which thickness approaches an asymptotic value (Fig. 15).

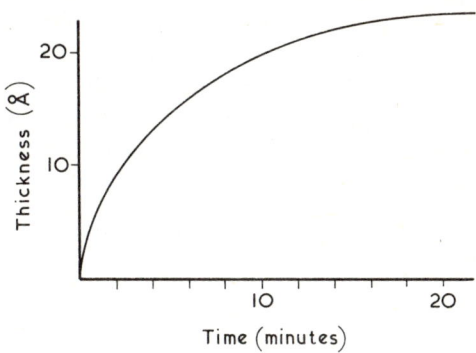

FIGURE 15. Thermal growth of Al_2O_3 on Al at 20°C as a function of time.

Since the mobility of both substrate and oxygen ions increases with temperature, higher temperatures will increase the thickness obtained in a given time. Only coherent layers are of use for practical applications and thus only a limited number of materials can be successfully oxidised. In other cases the film formed will continually flake off during growth, thus exposing a fresh surface to attack.

The process is widely used for the preparation of silicon dioxide films on silicon. Thermal growth, in these circumstances, is obtained using an atmosphere of either oxygen or water vapour at temperatures between 700 and 1400°C.

3.2 Anodisation[34, 35]

3.2.1 Introduction

As discussed above the thickness of oxide obtained in normal thermal growth is essentially limited by the ability of the ions to migrate through the film to the film/metal interface. The mobility, however, can be considerably enhanced if an electric field assists the ionic motion, i.e. if the films are grown from a metal substrate in an electrolytic bath. In such a system, the parent metal is made the anode of an electrolytic cell and a voltage is applied between the anode and the cathode.

3.2.2 Basic principles

The reactions that occur at the cathode and anode can be represented by the following equations:

$$M + H_2O \rightarrow MO + 2H^+ + 2\varepsilon \quad \text{(anode)}$$
$$2\varepsilon + 2H_2O \rightarrow H_2\uparrow + 2OH^- \quad \text{(cathode)}$$

These equations express the fact that oxide will grow on the metal/anode surface and hydrogen will be evolved at the cathode. The equations imply the presence of water and anodisation is usually effected in aqueous electrolytes, such as a solution of phosphoric acid. The acidity of the electrolyte must be carefully controlled if a coherent film is required. If the electrolyte is too acidic or too alkaline, the film can dissolve as it grows and porous oxide structures can result.

The final thickness of an oxide film will depend on the voltage applied across the electrolytic cell. A typical growth curve at constant applied voltage is shown in Fig. 16 and this is analgous to the growth curve previously shown (Fig. 15), with the thickness reaching an asymptotic value with time. The process can be characterised by the "anodisation constant", i.e. the thickness of the film which will be obtained after an infinite time for one volt applied. In the case of aluminium, this is 13.6 Å/volt, tantalum 16.0 Å/volt, silicon 3.5 Å/volt and niobium 36.3 Å/volt. In commercial practice, it is usually uneconomical to keep the metal substrate in the anodisation bath for sufficient time to obtain thicknesses corresponding to those expected from the anodisation constant. In these circumstances, the time in the bath is reduced so that the final thickness of oxide is between 70 and 80% of the maximum which can be obtained, and this thickness is realised in a relatively short period of time in the bath, which may be measured in minutes rather than hours.

The growth illustrated in Fig. 16 assumes that a constant voltage is applied to the cell. The disadvantage of this approach is that, in the initial stages of growth, very high current densities are required. This difficulty can be avoided by anodising under constant current rather than constant voltage conditions. Growth will then be as illustrated in Fig. 17, with a linear behaviour of thickness with time. The limiting condition in this case is that the required applied voltage rises as the thickness of oxide increases. This leads to breakdown in the film due to the inclusion of impurities in the

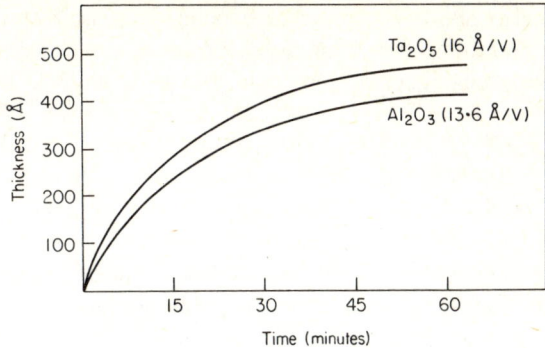

FIGURE 16. Constant voltage growth of Ta_2O_5 and Al_2O_3 by anodisation. Thickness/time for 30 V applied.

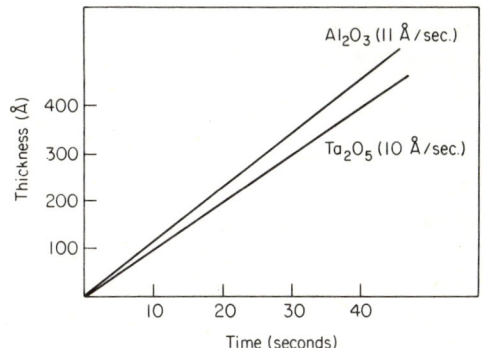

FIGURE 17. Constant current growth of Ta_2O_5 and Al_2O_3 by anodisation. Thickness/time for a current density of mA/cm².

structure which allow for avalanche effects to occur as the thickness of the dielectric increases. In the case of aluminium oxide, this limiting value occurs around 1100 V at a thickness of 1.5 µm, and in the case of tantalum oxide, at 700 V and a thickness of 1.1 µm. Since impurities in the film appear to be the initiators of the avalanche breakdown, the purity of the substrate and of the electrolyte is of great importance if the maximum values of breakdown voltage and hence thickness, are to be realised.

Growth rates obtained under constant current conditions are comparable to the previous figures for evaporation and sputtering. Aluminium can be anodised at a current density of 2 mA/cm² to give a growth rate of

11 Å/sec. This corresponds to the voltage being increased at the rate of 50 volts/min. In the case of tantalum at 2 mA/cm^2 current density, a growth rate of 10 Å/sec is obtained, corresponding to a voltage increase of 40 volts/min.

3.2.3 Practical aspects

Anodisation is widely used as a technique for obtaining amorphous, highly insulating films on the valve metals[35], semiconductors [35,38,39], and materials such as silicon carbide[37]. The mobility of ions in the anodising electrolyte is such that highly convoluted surfaces can be oxidised in a very even manner. This is illustrated by the fact that etched aluminium and tantalum, and porous tantalum structures can be easily anodised for use in electrolytic capacitors[36], and the oxide will be found to be formed successfully on all parts of the surface.

3.2.4 Gaseous anodisation

Gaseous anodisation has already been discussed in Section 2.2.3 under the heading of "Plasma reactions".

3.3 Vapour Phase Growth[27, 40]

3.3.1 Introduction

The deposition of a film on a surface, composed of the same or a different material, by means of a chemical reaction from the gas phase at the surface, is known as vapour phase growth or vapour phase plating. Usually the surface on which growth is required is at a higher temperature than the surroundings so that a heterogenous reaction occurs. However, other means of activating the chemical reaction may be used, such as glow discharge (see Sputtering—Plasma reactions) electron-beam excitation or ultraviolet radiation.

3.3.2 Types of vapour phase reactions

A. Disproportionation. The reaction is typified by the equation:

$$A + AB_2 \rightleftarrows 2AB$$

where A is the metal and B is usually a halide, and both AB and AB_2 are gaseous. A higher valency state is more stable at a low temperature so that if a hot gas of AB is passed into a colder region, AB_2 will be formed and a deposition of A will occur.

Figure 18 shows a typical closed tube system which has been employed for the deposition of germanium or silicon using disproportionation. For both of these materials, a halide is formed in the iodine vapour and the reaction is used

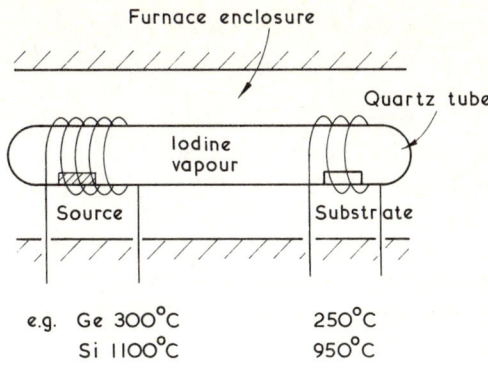

FIGURE 18. Disproportionation reaction vessel.

to transport silicon or germanium from the high temperature source zone, where the mono-iodine is the predominant gas, to the lower temperature source zone, where the di-iodine gas is stable.

Alternatively, a continuous flow open system can be used in which iodine vapour, usually diluted with hydrogen, is continuously passed over the source and then the substrate. Growth rates can be high; for example, germanium can be deposited at rates of up to 400 Å/sec in the above system and at the temperature shown.

B. *Polymerisation*[27]. Both organic and inorganic polymers may be prepared from a monomer vapour by the use of electron beam, ultraviolet or glow discharge irradiation. Insulating films prepared in this manner can have very desirable properties.

Ultraviolet polymerisation techniques are widely used in photo-resistor etching[41]. In the etching process a relatively thick layer of photo-sensitive material is spread evenly over the surface and then irradiated through a mask. The exposed areas polymerise to give a material that is insoluble in the solvent used for the un-polymerised film. A similar technique has been used to prepare insulating films. White [42], for example, exposed metal layers in butadiene vapour to irradiation and thus built up thin dielectric films. Various other vapours have also been used (e.g. acrolein) and growth rates of around 1 Å/sec are easily obtained.

Glow discharge techniques have already been referred to in the section Sputtering—Plasma reactions, using either a straightforward glow discharge or R.F. excited systems (para. 2.2.3, I).

C. *Oxidation*. This is usually undertaken using a halide (X) of the required metal oxide (A_2O), because of the high vapour pressure of the halide and the ease of removal of the by-products. The reaction used is:

$$2AX \xrightarrow{H_2O} A_2O + HX$$

Oxides may be deposited in this way using steam mixed with the halide and allowing the mixture to flow over a hot substrate.

To ensure thorough mixing and an even temperature in the reactor, a fluidised bed is sometimes used. Oxides which can easily be prepared in this way include those of aluminium, titanium, tantalum and tin, and high growth rates are possible (greater than 100 Å/sec).

D. Nitriding. If, instead of steam, an atmosphere of ammonia is used, then it is possible to grow layers of nitrides. This reaction has been used effectively for the growth of silicon nitride.

E. Reduction. Metal films can be prepared if hydrogen is substituted for steam in the oxidation reaction previously described. This type of reaction is widely used for the preparation of silicon and germanium. In the silicon case, $SiHCl_3$ or $SiCl_4$ is used, and high growth rates are obtained (200 Å/sec at 1100°C).

F. Decomposition. Decomposition of *AB* can be represented by the equation:

$$AB \rightarrow A + B$$

It can be effected both by heat (pyrolysis) and by glow discharge[43]. Pyrolitic reactions have been widely applied to the preparation of silicon (from SiH_4), nickel from nickel carbonyl and SiO_2 from the decomposition of silicon esters. Rates of growth for nickel can be very high at moderate substrate temperatures (1000 Å/sec at 200°C).

Glow discharges have also been used to prepare insulating films by decomposition of a vapour, either using a straightforward system, or with the discharge excited by radio frequency[43].

3.4 Electroplating[44, 45]

3.4.1 Introduction

Electroplating has been known for a considerable time, and many standard textbooks now exist on the subject. The apparatus involved is basically simple, consisting of a conducting anode and cathode immersed in a suitable electrolyte. Metal is deposited on the cathode, and the relationship between the weight of the material deposited and the various parameters can be expressed by the first and second laws of electrolysis. These state:

1. The mass of the deposit is proportional to the amount of electricity passed.
2. The mass material deposited by the same quantity of electricity is proportional to the electrochemical equivalent E.

Expressed as an equation, the mass deposited per unit area G/A is given by:

$$\frac{G}{A} = JtE\alpha \,(\text{g cm}^{-2}) \tag{4}$$

where J is the current density and t the time. This equation introduces another term, the current efficiency α, which is the ratio of the experimental to theoretical mass deposited; it can generally be expected to be between unity and 0.5.

The above equation can be written in a slightly different form to give the rate of deposition. If a thickness l is deposited in time t, then the rate of deposition l/t is given by:

$$\frac{l}{t} = \frac{JE\alpha}{\rho}(\text{cm sec}^{-1}) \tag{5}$$

where ρ is the film density.

The rate of deposition can be very high using high current densities. For example, silver will deposit at $10\,\text{Å sec}^{-1}$ at a current density of $1\,\text{mA cm}^{-2}$, and this will rise to $1\,\mu\text{m sec}^{-1}$ at $1\,\text{A cm}^{-2}$. The proportionality of the thickness to current density is valid if α remains unchanged. This can be expressed in another way by stating that no secondary reactions must occur.

3.4.2 Basic principles

Of the 70 metallic elements, it is found possible to plate only 33 successfully, and of these only 14 are used commercially. A large variety of baths can be used for the various elements to improve the adhesion, crystalline structure, current efficiency, etc. However, it is not possible to plate elements outside the group of 33, as other reactions (e.g. formation of hydrogen) tend to occur. This can be illustrated by considering the I-V characteristics of a plating solution. The characteristic of a simple system is shown in Fig. 19.

The equilibrium potential of the cathode in the solution is indicated by the intercept value on the voltage axis. A negative intercept implies that the cathode will dissolve in the electrolyte at zero voltage (i.e. it will corrode). Equilibrium potentials for the different metals vary from $+1.7$ to $-1.66\,\text{V}$. Saturation is seen in the curve because, at a high cathode voltage, ions cannot reach the cathode sufficiently rapidly, due to the repulsion effect of the charged ions for each other. If two reactions are possible two I-V curves can now be drawn for the two reactions, as shown in Fig. 20. Thus in the case of alloy plating this means that the composition of the alloy will depend on the voltage used, as illustrated by the dashed line, in Fig. 20. If one of the reactions involves formation of hydrogen, the curves will now be as shown in Fig. 21 and depending on the relative positions of the lines A and

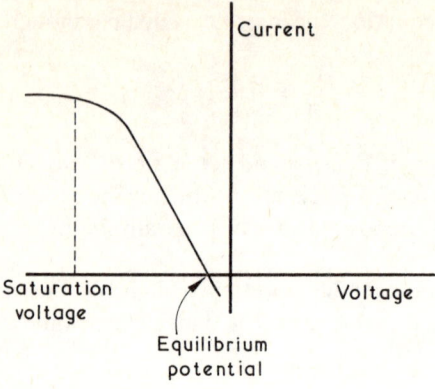

FIGURE 19. Ideal i/V curve for electroplating bath.

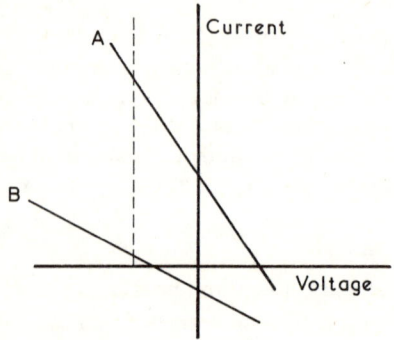

FIGURE 20. Ideal i/V curves showing effect of superposition of two plating reactions (A and B).

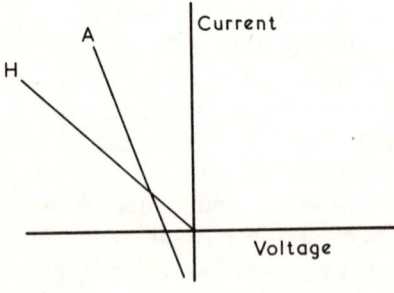

FIGURE 21. Ideal i/V curves showing effect of superposition of a plating reaction and hydrogen evolution (A and H).

H, it can be seen that hydrogen evolution can be the main reaction at a particular voltage.

The voltage drop in the bath must be reduced to as low a value as possible to reduce waste of power in heating, and this is usually achieved by the addition of conducting salts.

3.4.3 Practical aspects

In practice, care must be taken to control the current density. High current density will plate the films at such a rate as to include gas bubbles, etc. in the films and low current density will give crystalline, non-adherent deposits. The effect of solution temperature will not generally be important unless α changes with temperature and in general, if this is the case, α will increase with temperature.

As the deposition rate can be high, it is possible to use electrochemical deposition for forming thick layers—a process known as electroforming—and for refining. An example of electroforming is the preparation of master discs for gramophone records. In this case, the initial deposit used to form a suitable cathode is obtainable either by using a colloidal suspension of metal, or by chemical reduction plating (see para. 3.5).

The many alloys that have been successfully deposited (100 or so) are discussed in detail in Bremner's two volumes on electro-deposition[46]. It is not possible to deposit every alloy combination because of the characteristics of the separate elements, although it is often possible to slow one of the reactions down by suitable chemical complexing.

The effect of complexing is to lower the equilibrium potential to a more negative value and this results in a bunching together at low voltages of the I-V characteristics for the separate elements (Fig. 20). Cyanide ions provide a typical complexing ion for silver, cadmium, zinc and copper. However, it is not necessarily the case that the potentials of the separate metals can be brought to close enough values to permit co-deposition.

Ternary alloys can also be deposited by electroplating: Brenner lists 15 that can be easily formed.

It is possible to form oxides of elements successfully on the anode of an electroplating system and these are deposited from the solution (c.f. anodisation, where it is the oxide of the anode material that is formed). Films of Pb and Mn oxides have been grown in this way, but the method has little importance in thin film technology.

3.5 Chemical Reduction Plating[47, 48]

3.5.1 Introduction

Films of metal can be deposited directly without any electrode potentials being involved, by the chemical reduction of a suitable compound in solution. This technique is known as chemical reduction plating or electroless deposition and four different types of reaction may be distinguished which are summarised below.

A. Non-catalytic reactions. These take place at any surface submersed in the bath. Silver mirrors are usually formed in this way, by the use of a mild reducing agent such as formaldehyde in a solution of silver nitrate. Very thick layers may be grown.

B. Catalytic reactions. The ability of the metal to deposit on any surface can sometimes be a considerable nuisance, and controlled reactions are often more useful. In these, the metal will deposit only on certain surfaces of other metals and nowhere else. The deposition of nickel, for example, can be achieved by such techniques as the reduction of $NiCl_2$ by sodium hypophosphite, when the metal will grow on a surface of nickel itself, cobalt, iron and aluminium—the metal acts as the catalyst. (Note: the use of sodium hypophosphite as the reducing agent means that between 5 and 10% of phosphorous will become incorporated in the film.) This type of reaction has become so important that a complete book is available on the chemical reduction plating of nickel alone[49]. Other metals, particularly the Pt group, can also be deposited in this manner.

C. Catalytic reactions using activators. The number of metal surfaces that will catalyse deposition is limited. It is found, however, that it is possible to activate the surfaces of non-catalytic metals so that deposition will take place on these surfaces. The role of the activator is to lower the activation energy for the reduction reaction at points on the surface so that deposition will occur at these points. Islands of metal will thus grow and spread and eventually form a continuous film. The best activators for particular metals are listed in standard texts on the subject[47]; $PdCl_2$, for example, is often used for Cu and Ni. The activator is required only in very low concentrations and in the case of $PdCl_2$ a dip in a 0.01% solution, followed by a rinse in water, is adequate.

D. Catalytic reactions using activators and sensitisers. For non-metallic surfaces, a sensitisation before activation is required. For Ni a dip in a 0.1% solution of $SnCl_2$ is required and this is followed by a rinse in water. Activation is then carried out in the normal way. The advantage of such reactions is that it is possible to plate glass and other non-conducting surfaces. In addition, it is possible to plate surfaces to which access may be difficult, such as the inside of tubes.

Chemical reduction is often used for depositing metal films and is commercially important for resistor preparation[50].

3.6 Solution Deposition[51]

3.6.1 Oxide films

It is possible to deposit dielectric films on to non-metallic substrates, using organic solution techniques. The substrate to be coated is generally immersed in a solution of a suitable organic material so that a thin layer of material is formed on both sides of the substrate. Uniformity of the liquid film can often be improved by spinning the wetted surface, and after uniformity has been achieved, the substrate is baked to a temperature of between 200–500°C to convert the liquid layer to a solid, usable structure. For example, it has been found that a colloidal silicon dioxide hydrate can be deposited by immersing the substrate in a silicate solution to which acids have been added. The colloidal hydrate will then yield a film of SiO_2 on baking.

Film thicknesses are limited to 100 Å or less using this technique, but it is possible to form multiple layers of different materials. Two materials, SiO_2 and TiO_2 have been extensively prepared in this manner and the resulting films used for a variety of optical applications.

In certain cases oxides can be grown from metal salts and the most widely used technique in this system is that of the growth of manganese oxide on substrates which have been immersed in dilute solutions of manganese nitrate. Heating to temperatures of approximately 300°C results in an oxide layer[47].

3.6.2 Built up films (Langmuir films)[52]

It has been found possible to build up films of long-chain fatty acids such as barium stearate, by immersing substrates in water on top of which floats a layer of the acid. On withdrawing the substrate, a monolayer of the film is deposited. Consecutive immersions and withdrawals will build up the film in layers and it is found that films of up to 300 layers can be grown.

3.6.3 Liquid phase epitaxy[53]

Although not strictly a system for growing thin films, single crystal layers can be grown epitaxially from the liquid phase, provided that a suitable substrate is available. Normally, deposition from the liquid phase results in random nucleation and polycrystalline layers which are often of variable thickness in the initial stages of growth. However, it has been found possible to control the system sufficiently well so that the technique is now widely used for the growth of many semiconductor films.

One of the major advances of this type of technique is that it is possible to grow layers at temperatures which are several hundred degrees lower than the melting point of the compound thus reducing the number of defects in the crystalline structure of the film.

3.6.4 Growth of polymer films

Polymer films of materials such as polypropylene, polystyrene[54] and PVC[55], have been obtained by the technique of direct isothermal immersion of a substrate into a suitable solution of the polymer, or by allowing the evaporation of the solvent from the polymer solution placed on a substrate. The former technique appears to be very promising as a means of obtaining durable and useful films.

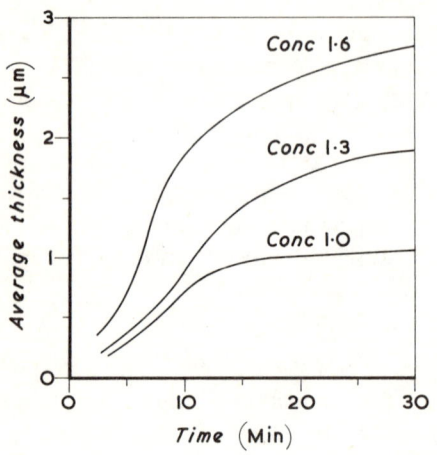

FIGURE 22. Thickness of PVC films as a function of time, solution grown from cyclohexanone (solution concentrations in g-weight PVC in 100 cm³ solvent). (After Rastogi and Chopra[55].)

Rastogi and Chopra[56] have recently shown that it is possible to obtain satisfactory insulating films of PVC from solutions of PVC in benzene/acetone mixtures or cyclohexanone. The growth of films on substrates of rock salt, copper or glass has been investigated and a typical growth curve is shown in Fig. 22 for PVC on glass using cyclohexanone solvent. It can be seen from this type of curve that an equilibrium thickness is achieved after periods of approximately ten minutes although this depends upon the solution concentration, in which the rate of growth is balanced by the rate of solution of the film. The dielectric behaviour of these films is now being actively investigated.

The growth model suggested for PVC allows initially for the PVC chain segments to be absorbed at suitable, unspecified sites on the surface. After the initial absorption, further growth occurs by absorption or attachment of new chain segments. However, the outer layer is also capable of being dissolved by the solvent so that a final thickness is reached when outer layer growth rate is exactly equal to its solution rate.

4. Summary

A large number of different deposition techniques has been surveyed in this chapter. Not all of them can be used for every device, but Table 1 shows the areas which are discussed in the following chapters and the techniques of deposition that are at present most appropriate to them.

Table 1
Deposition System used for Preparation of Films and Devices

Type of Film	Device	\multicolumn{8}{c}{SYSTEM}
		Evap	Sputt	Thermal	Anod	VP	EP	CRP	Sol
Metallic	Conductors	✓	✓			✓	✓		
	Resistors	✓	✓					✓	
	Magnetics	✓	✓					✓	✓
	Superconductors	✓							
Dielectric	Capacitors	✓			✓				
	Insulators	✓	✓	✓	✓	✓			✓
	Optics	✓	✓						
	Ferroelectrics	✓							
Metal/Dielectric Mixture	Resistors	✓	✓			✓			
	Optics	✓	✓						
Semi-conducting	Resistors					✓			
	Active	✓	✓			✓			✓
	Photoconducting	✓				✓			
	Photovoltaic	✓				✓			

References

1. D. S. Campbell, In "The Use of Thin Films in Physical Examinations", ed., J. C. Anderson, Academic Press, London and New York, (1966).
2. K. L. Chopra, "Thin Film Phenomena", McGraw Hill, New York, (1969).
3. B. N. Chapman and J. C. Anderson, "Science and Technology of Surface Coatings", Academic Press, London and New York, (1974).
4. D. Walton, T. N. Rhodin and R. Rollins, *J. Chem. Phys.*, **38**, 2695, (1963).
5. K. H. Behrndt, In "Physics of Thin Films", Vol. 3, eds., G. Hass and R. E. Thun, Academic Press, London and New York, (1966).
6. D. S. Campbell, In "Physics of Non-Metallic Thin Films", Plenum Press, New York, (1976).
7. R. Glang, In "Handbook of Thin Film Technology", eds., L. I. Maissel and R. Glang, McGraw Hill, New York, (1970).
8. R. E. Honig, *R. C. A. Review*, **23**, 567, (1962).
9. R. Glang, R. A. Holmwood and J. A. Kurtz, In "Handbook of Thin Film Technology", eds., L. I. Maissel and R. Glang, McGraw Hill, New York, (1970).
10. K. H. Behrndt, *J. Appl. Phys.*, **33**, 193, (1962).
11. P. Huijer, W. T. Langedam and J. H. Laby, *Philips Tech. Rev.*, **24**, (1963).
12. C. E. Drumheller, *Trans. 7th Nat. Symp. on Vacuum Technology*, (1960).
13. D. S. Campbell and B. Hendry, *Brit. J. Appl. Phys.*, **16**, 1719 (1965).
14. J. L. Richards, In "The Use of Thin Films in Physical Investigations", ed., J. C. Anderson, Academic Press, London and New York, (1966).
15. B. A. Uvala and G. R. Booker, *Phil. Mag.*, **9**, 691, (1964).
16. A. Lopez Otero and L. D. Hass, *Thin Solid Films*, **23**, (1974).
17. A. W. Stephens, A. W. Levine, J. Fech Jr., T. J. Zrebiec, A. V. Capiero and A. M. Garofalo, *Thin Solid Films*, **24**, 361, (1974).
18. G. K. Wehner and G. S. Anderson, In "Handbook of Thin Film Technology", eds., L. I. Maissel and R. Glang, McGraw Hill, New York, (1970).
19. M. Kaminsky, "Atomic and Ionic Impact Phenomena on Metal Surfaces", Springer-Verlag, Berlin, (1965).
20. L. I. Maissel, In "Physics of Thin Films", Vol. 3, eds., G. Hass and R. E. Thun, Academic Press, London and New York, (1966).
21. L. I. Maissel, In "Handbook of Thin Film Technology", eds., L. I. Maissel and R. Glang, McGraw Hill, New York, (1970).
22. L. I. Maissel and P. Schaible, *J. Appl. Phys.*, **36**, 237, (1965).
23. M. White, *Thin Solid Films*, **18**, 157, (1973).
24. N. F. Jackson, *J. Mat. Sci.*, **2**, 12, (1967).
25. K. Knorr and J. D. Leslie, *Thin Solid Films*, **23**, 101, (1974).
26. J. R. Ligenza, *J. Appl. Phys.*, **36**, 2703, (1965).
27. L. V. Gregor, In "Physics of Thin Films", Vol. 3, eds., G. Hass and R. E. Thun, Academic Press, London and New York, (1966).
28. H. F. Sterling and R. C. G. Swann, *Solid State Elec.*, **8**, 653, (1965).
29. R. A. Connell and L. V. Gregor, *J. Electrochem. Soc.*, **112**, 1198, (1965).
30. D. M. Mattox, *Electrochem. Tech.*, **2**, 295, (1964).
31. D. M. Mattox, *Proc. 6th Int. Vac. Conf.*, 443, (1974).

32. W. R. Stowell, *Thin Solid Films*, **22**, 111, (1974).
33. H. Dimigen and H. Luthje, *Thin Solid Films*, **27**, 155, (1975).
34. D. S. Campbell, In "Handbook of Thin Film Technology", eds., L. I. Maissel and R. Glang, McGraw Hill, New York, (1970).
35. C. J. Dell'Oca, D. L. Pulfrey and L. Young, In "Physics of Thin Films", Vol. 6, eds., M. H. Francombe and R. W. Hoffman, Academic Press, London and New York, (1971).
36. D. S. Campbell, *Rad. and Elec. Eng.*, **41**, (1971).
37. G. Restelli, *Thin Solid Films*, **23**, (1974).
38. R. A. Logan, B. Schwartz and W. J. Sunburg, *J. Electrochem. Soc.*, **120**, 1385, (1973).
39. H. Hasegawa, K. E. Forward and H. Hartnagel, *Thin Solid Films*, **32**, 65, (1976).
40. W. M. Feist, S. R. Steele and D. W. Ready. In "Physics of Thin Films", Vol. 5, eds., G. Hass and R. E. Thun, Academic Press, London and New York, (1969).
41. D. I. Gaffee, In "Thin Film Microelectronics", ed., L. Holland, Chapman and Hall, London, (1965).
42. P. White, *Elec. Reliab. and Micromin.*, **2**, 161, (1963).
43. L. L. Alt, S. W. Ing Jr., and K. W. Laendle, *J. Electrochem. Soc.*, **110**, 465, (1963).
44. F. A. Lowenheim, "Modern Electroplating", John Wiley, New York, (1965).
45. K. R. Lawless, In "Physics of Thin Films", Vol. 4, eds., G. Hass and R. E. Thun, Academic Press, London and New York, (1967).
46. A. Brenner, "Electrodeposition of Alloys", Vols. 1 and 2, Academic Press, New York, (1963).
47. W. Goldie, "Metallic Coating of Plastics", Electrochemical Publications Ltd., U.K., **1**, (1968).
48. M. Schlesginger, In "Science and Technology of Surface Coatings", eds., B. N. Chapman and J. C. Anderson, Academic Press, London and New York, (1974).
49. K. M. Gorbunova and A. A. Nikiforova, "Physichemical Principles of Nickel Plating", Israel Prog. for Scientific Translations, Jerusalem, (1963).
50. N. Feldstein, *Solid State Tech.*, **16**, 87, (1973).
51. H. Schroeder, In "Physics of Thin Films", Vol. 5, eds., G. Hass and R. E. Thun, Academic Press, London and New York, (1969).
52. V. K. Strivastava, In "Physics of Thin Films", Vol. 7, eds., G. Hass, M. H. Francombe and R. W. Hoffman, Academic Press, London and New York, (1973).
53. L. R. Dawson, In "Progress in Solid State Chemistry", eds., H. Reiss and J. O. McCaldin, Pergamon Press, Oxford, (1972).
54. K. A. Koutsky, A. G. Walton and E. Baer, *J. Polymer Sci.*, **B.5**, 177, (1967).
55. A. C. Rastogi and K. L. Chopra, *Thin Solid Films*, **18**, 187, (1973).
56. A. C. Rastogi and K. L. Chopra, *Thin Solid Films*, **27**, 311, (1975).
57. Y. Moriya, N. O. Kuma and K. Sugiura, *Trans. 8th Nat. Vac. Symp.* **2**, 1055, (1961).
58. D. S. Campbell, *Thin Solid Films*, **32**, 3, (1976).

Chapter 3†

Electrical Properties and Applications of Thin Metal and Alloy Films

T. J. Coutts

Industrial Science Research Unit,
Cranfield Institute of Technology,
Cranfield, Bedford, England

1. Introduction 58
2. Electron Transport in Bulk Metals and in Thin Metal Films 58
   2.1 The Boltzmann Transport Equation 58
   2.2 Electrical Conductivity of Bulk Metals 60
   2.3 The Origin of Resistivity in Bulk Metals. 61
      2.3.1 Thermal scattering 61
      2.3.2 Impurity scattering 63
      2.3.3 Resistivity of alloys 65
      2.3.4 Resistivity of transition metal alloys. 66
   2.4 Matthiessen's Rule 68
   2.5 Electrical Conductivity of Thin Metal Films 68
      2.5.1 Surface scattering 69
      2.5.2 Angular dependence of the specularity parameter 73
      2.5.3 Low angle scattering 74
      2.5.4 Grain boundary scattering 74
      2.5.5 The effect of heat treatment on resistivity 77
   2.6 Comparison of Theory and Experiment 78
      2.6.1 Electrical resistance 78
      2.6.2 Temperature coefficient of resistance 83
      2.6.3 The thermoelectric effect in thin films 84
      2.6.4 Conduction in a magnetic field 85
   2.7 Summary of the Properties of Thin Metal Films 86
3. Application of Thin Metal Films 88
   3.1 Thin Film Resistors 88
      3.1.1 Tantalum and associated materials 88
      3.1.2 Nickel/chromium alloys 93

† For a list of symbols used in this chapter and their definitions see p. ix.

		3.1.3	Other materials.	94
		3.1.4	Stability and ageing of thin film resistors	96
		3.1.5	Substrate considerations	98
	3.2		Thin Film Conductors and Contacts.	100
		3.2.1	Metal/semiconductor contacts	100
		3.2.2	Requirements of contacts	101
		3.2.3	Choice of materials	101
		3.2.4	Reliability of contacts and interconnections	103
	3.3		Thin Film Inductors and Capacitors.	105
	3.4		Thin Film Strain Gauges.	107
	3.5		Summary.	108
			References.	109

1. Introduction

In this chapter the electrical properties of thin metal and alloy films are related to the generally accepted framework of solid state physics which has of course been derived for simple crystalline bulk materials. In particular, both qualitative and quantitative approaches are used to illustrate the similarities and the differences in electron transport between thin films and bulk materials. Electronic conductivity will be the principal subject of interest rather than any of the other thermogalvanomagnetic effects as it has been more extensively investigated although mention will be made of one or two of these phenomena where appropriate. Thus, the aim is to indicate the nature of the intrinsic properties of a thin metal film although, as will be seen later, it is not necessarily the geometrical thinness which is exploited in resistors and other metal film devices.

In the following section the other basic properties of thin metal films which depend primarily on the manner in which the film has been produced, rather than on geometry, are discussed. Impurities and defects are generally incorporated into a thin metal film during deposition and in general these have an important effect on the film resistivity. In the case of at least one material, a completely new crystallographic phase is observed in film form which is not observed in bulk, and this is a function of the method of production rather than an inherent property of the film. The applications of thin metal and alloy films, particularly in the field of microelectronics, are of importance and are discussed together with design considerations which are highly relevant to the economics, performance and reliability of thin film devices.

2. Electron Transport in Bulk Metals and in Thin Metal Films

2.1 The Boltzmann Transport Equation

The Boltzmann transport equation is of fundamental importance to all physical phenomena whose origins lie in electron transport, whether under the influence of applied electric, magnetic or thermal fields. It is essentially an equation of conservation which describes the way in which the energy of electrons changes due to applied fields and collisions. Since large numbers of free (or nearly free) electrons in metals are dealt with, expressions are obtained which predict the way in which the electron energy distribution (i.e. the perturbed Fermi distribution in the case of metals) changes with time under the influence of fields and collisions. Let p_x, p_y, p_z, be the components of momentum of an electron in the x, y, z directions and let $f(p_x, p_y, p_z, x, y, z, t)\partial x\,\partial y\,\partial z\,\partial px\,\partial py\,\partial pz$ be the fraction of electrons in an elemental volume $\partial x\,\partial y\,\partial z$ which at time t have components of momentum in the range $px + \partial px$, $py + \partial py$, $pz + \partial pz$. If the system of particles is subjected to a set of forces X, Y, Z, the fraction of particles with the above components of momentum at a time ∂t later may have changed either because of changes in velocity or position. Thus the fraction of electrons within the appropriate volume element at a time ∂t later may be defined by:

$$f\left[px+X\,\partial t, py+Y\,\partial t, pz+Z\,\partial t, x+\left(\frac{px\,\partial t}{m}\right), y+\left(\frac{py\,\partial t}{m}\right), z+\left(\frac{pz\,\partial t}{m}\right), t+\partial t\right]$$
$$\partial px\,\partial py\,\partial pz\,\partial x\,\partial y\,\partial z$$

This expression may be expanded using a Taylor series and subtracting the result from the distribution at time t gives:

$$\left(\frac{df}{dt}\right) \text{ due to applied fields} = -\frac{X\,\partial f}{\partial px}-\frac{Y\,\partial f}{\partial py}-\frac{Z\,\partial f}{\partial pz}-\frac{V_x\,\partial f}{\partial x}-\frac{V_y\,\partial f}{\partial y}-\frac{V_z\,\partial f}{\partial z} \quad (1)$$

In the steady state the rate of change of the distribution due to applied fields must be exactly balanced by the rate of change in the opposite sense due to collisions of the electrons with the thermally agitated lattice atoms, impurities and lattice defects. Consequently:

$$\left(\frac{\partial f}{\partial t}\right)_{\text{total}} = \left(\frac{\partial f}{\partial t}\right)_{\text{fields}} + \left(\frac{\partial f}{\partial t}\right)_{\text{collisions}} = 0 \quad (2)$$

However, in the general case, without imposing steady state conditions, (1) may be written as:

$$\left(\frac{\partial f}{\partial t}\right)_{\text{total}} = -\left(\frac{X\,\partial f}{\partial px}+\frac{Y\,\partial f}{\partial py}+\frac{Z\,\partial f}{\partial pz}\right)-\left(\frac{V_x\,\partial f}{\partial x}+\frac{V_y\,\partial f}{\partial y}+\frac{V_z\,\partial f}{\partial z}\right)+\left(\frac{\partial f}{\partial t}\right)_{\text{collisions}}$$

which in vector notation becomes:

$$\left(\frac{\partial f}{\partial t}\right)_{\text{total}} = -\bar{F}\,\text{grad}_{Pr}f - \bar{V}\,\text{grad}_r f + \left(\frac{\partial f}{\partial t}\right)_{\text{collisions}} \quad (3)$$

This if the formal statement of the Boltzmann transport equation which under particular circumstances can be solved to obtain the distribution function. It is customary to assume a solution of the form:

$$\left(\frac{\partial f}{\partial t}\right)_{collisions} = -\frac{(f-f_0)}{\tau} \qquad (4)$$

i.e. the rate at which the perturbed distribution returns to its equilibrium state due to collisions with the lattice, once the perturbing field is removed, is proportional to the excess in the distribution $(f-f_0)$. The constant of proportionality τ is generally called the relaxation time. For a wide range of metals at about room temperature and above it is satisfactory to assume that τ is constant, it is in fact a function of direction and particle energy. However, an isotropic metal is assumed here.

2.2 Electrical Conductivity of Bulk Metals

The derivation of the expression for the electrical conductivity of a metal is based on the use of equation (3) from which, under certain simplifying assumptions, the perturbed probability distribution f under the influence of an applied electric field E can be calculated. Firstly, because the metal has been assumed to be isotropic, the distribution is not a function of direction. Consequently $\mathrm{grad}_r f = 0$. Secondly, steady state conditions are assumed so that $(\partial f/\partial t)_{total} = 0$. Thirdly, the argument will be restricted to the case of an applied electric field E_x in the x direction so that Y and $Z = 0$. Under these restrictions equations (3) and (4) may be combined to give:

$$X\frac{\partial f}{\partial p_x} = -\frac{(f-f_0)}{\tau}$$

$$\text{and since} \quad X = -eE_x \quad f = f_0 + eE_x\tau\frac{\partial f}{\partial p_x} \qquad (5)$$

The subsequent procedure for calculating the conductivity is based on the use of the expression:

$$\sigma = \frac{j}{E}$$

where j, the current density, is equal to the product of the total number of conduction electrons, their velocity in the x direction and their charge. Now the total number of conduction electrons is given by the product of the density of states (which is a function of all three wave numbers k_x, k_y and k_z) and the probability of a given energy state being occupied (given by equation (5)). In addition the velocity V_x is a function of the wave number k_x so that to evaluate the conductivity in the x direction it is necessary to carry out a triple integral of the simple expression for the current density given above over all k_x, k_y, k_z (the wave vectors). This procedure is clearly outlined in many books on solid state physics and it is not necessary to go through the detail of carrying out the mathematics here. Since the subject of cocern here is specifically the properties of thin films it is sufficient to indicate the reasons why their properties differ from those of a bulk metal. Having derived the perturbed distribution functions (equation (5) above) for a bulk material and a thin film (which will be carried out in the next section) the origin of the departures from bulk behaviour will then be indicated. The bulk conductivity is of course given by the well-known expression:

$$\sigma = \frac{N_{\text{eff}} e^2}{m} \tau(k_F) \tag{6}$$

where N_{eff} is the effective number of conduction electrons, e and m are the electronic charge and mass, respectively, and $\tau(k_F)$ is the relaxation time for electrons at the Fermi surface.

2.3 The Origin of Restivity in Bulk Metals

In the derivation of (6) the scattering of electrons by lattice atoms due to their vibration was mentioned. In a perfect metallic crystal the electrons, or, more correctly, the electron waves, could pass through the periodic potentials without being scattered, and therefore without suffering any opposition (or resistance) to their motion. However the thermal motion of the lattice atoms causes departures from perfect periodicity so that the electron waves are always, in practice, scattered. The mean distance travelled between successive collisions is called the mean free path. Of course, departure from periodicity in bulk metals is caused by other mechanisms and in this section these will be briefly discussed. The general conclusions will then be used later when dealing with the resistivity of thin films.

2.3.1 Thermal scattering

The formal theory of lattice vibrations was established by Debye[1] as part of his calculations on the specific heat of materials. Debye claimed that the vibrating lattice atoms could be represented by two standing waves travelling in the opposite direction to each other; the energy associated with the waves being quantised in units of hv where v is their frequency. This is directly analogous to the concept of quantised energy in the electromagnetic spectrum except that in this case the wave packets are called phonons and travel with the velocity of sound. It was suggested that the origin of thermal resistivity was due to the scattering of electron wave packets by interaction with phonon wave packets. Upon collision a phonon would either be annihilated or created and the total effect on the resistivity was calculated by integrating over the whole spectrum of phonon energies. The density function for the quanta of thermal energy was shown to be parabolic with frequency up to a minimum value v_{max} at which point it falls to zero. It was also shown that the maximum frequency of the distribution was a property solely of the material under consideration and involved the size of the unit cell of the lattice and the velocity of sound through the medium. The average energy of the quanta is related to their thermal energy through Boltzmann's constant by the expression $hv = kT$ and a quantity known as the "Debye temperature" can be defined as being that temperature corresponding to the energy of quanta at the cut-off point of the frequency spectrum divided by Boltzmann's constant, viz $hv_{max} = k\Theta_D$. Since Θ_D is proportional to v_{max} it must also depend on the size of the unit cell of the lattice and the velocity of sound through the medium. Since these quantities are temperature dependent it cannot be expected that Θ_D will be truly constant. Debye's theory leads to an integral in which specific heat is expressed as a function of T/Θ_D. From experimental data of specific heats at various temperatures, it is then possible to obtain empirically the values of Θ_D which give the best fit to the Debye theory. Such a procedure has been carried out for many materials and in general it has been established that Θ_D is in fact not a very sensitive function of temperature.

Using the same concept of scattering it is possible to obtain an expression for the mean free path of electrons where the latter is controlled by phonons. This is known as the Bloch-Gruneisen law which is obeyed quite well by many metals. Its predictions are that at high temperatures (i.e. $T \gg \Theta_D$) the resistivity should be proportional to T whereas at low temperatures (i.e. $T \ll \Theta_D$) the resistivity is proportional to T^5. This prediction can be examined by plotting $\log \rho$ against $\log T/\Theta_D$. For most metals, at low temperatures the slope of the resulting curve is five whilst at high temperatures it is unity. A temperature coefficient of resistance can also be defined as:

$$\alpha = \frac{1}{\rho}\frac{\partial t}{\partial T} \text{ for } T \gg \Theta_D$$

and it is possible to obtain the value of α from resistance measurements taken at temperatures, not necessarily very close together. Since resistivity is proportional to T, we have $\alpha = 1/T$ so for metals at about room temperature we can expect α to be of the order of $4000 \times 10^{-6} \text{ K}^{-1}$. Such a value is typical of observation even though for many pure metals Θ_D is about 300 K and therefore the condition $T \gg \Theta_D$ is not strictly obeyed. At lower temperatures $\alpha = 5/T$ and once again this is found to agree very well with observation.

Although the Debye theory of lattice vibrations can be criticised on several counts, in particular that longitudinal and transverse modes of vibration are not treated separately, the agreement of the theoretical predictions with experimental observation is excellent. In addition the theory of scattering gives us information about the use of a relaxation time τ (as used above). This cannot be treated in detail here but broadly the conclusions are:

(1) it is only by using the relaxation time approximation that a true solution for the conductivity can be obtained.

(2) the use of a constant relaxation time is only valid for elastic scattering.

It can be shown that in general this is true for $T \gg \Theta_D$ (this condition effectively says that the energy transferred to or from an electron upon collision must be small compared with kT and at temperatures less than Θ_D this is not always true.

Only the briefest outline of thermal scattering has been given but it should be clear that, at least in a bulk metal, the resistivity is due primarily to this mechanism. The reader is referred to the bibliography at the end of this chapter in which several books are listed where much more thorough treatments of the topic are presented.

2.3.2 Impurity scattering

It is important to understand the nature of impurity scattering since its effect on the resistivity of a thin film can be very significant due to the relatively high density of impurities which can be incorporated during deposition. Qualitatively, the mechanism underlying impurity scattering is easily understood. Impurity atoms are simply another source of disruption of the periodicity of the lattice, and in general they will carry a different electrical charge to that of the host metal. Even if the valency of host and impurity atoms are equal they will generally have different electrostatic fields around them so that additional scattering will occur. Consequently, even at absolute zero where lattice scattering does not occur, the metal will have a residual resistance due to impurities. For relatively low concentrations of impurities Ziman[2] has given an expression for the screened coulombic potential due to a charged impurity, viz:

$$V(r) = \frac{Ze^2}{r} \exp-(\lambda r) \tag{7}$$

where Z is the effective charge (or valence difference), r the distance from the impurity atom and λ is a simple function of the density of electrons within an incremental energy ∂E of the Fermi energy. From equation (7) it is possible to show that this potential gives rise to an effective scattering cross section:

$$\sigma(\theta) = \left(\frac{2mZe^2}{\hbar}\right)^2 \frac{1}{(k^2+\lambda^2)^2} \tag{8}$$

where $k = 2k_F \sin(\theta/2)$ (i.e. it is the magnitude of the vector causing scattering of a Fermi electron through an angle θ). Furthermore, Ziman gives the mean free path for impurity scattering as:

$$\frac{1}{\Lambda} = N_i 2\pi \int_0^\pi (1 - \cos\theta)\sigma(\theta) \sin\theta \, d\theta \tag{9}$$

so that from equations (7), (8) and (9) an expression can be obtained for the resistivity due to impurity scattering, viz:

$$\rho_{\text{imp}} = \frac{mV_F}{N_{\text{eff}} z e^2} N_i 2\pi \left(\frac{2mZe^2}{\hbar^2 \lambda^2}\right)^2 \int_0^1 \frac{8z^3}{\left[1 - \left(\frac{2k_F}{\lambda}\right)^2 z^2\right]^2} dz \tag{10}$$

where $z = \sin\frac{1}{2}\theta$. The important features of this equation are firstly that the charged impurity has an effective geometrical radius of:

$$\left(\frac{2mZe^2}{\hbar^2 \lambda^2}\right)$$

which depends linearly on the valence difference Z and, secondly, that the resistivity is linearly proportional to the concentration of impurity atoms, at least for relatively low concentrations.

Once again it can be shown that this expression is in general agreement with experimental observation. One further important point to note is that ρ_{imp} is not directly dependent on temperature, although because of the slight dependence of K_F on temperature it may be argued that there will be a weak dependence on temperature. However, as seen from Fig. 1 (the results of Linde[3]), for relatively low concentrations of impurities the variation of resistivity with temperature remains linear. The increase in resistivity per atomic percent of impurity is approximately $0.5 \,\mu\Omega\,\text{cm}$ for copper in gold whereas it is less than $0.1 \,\mu\Omega\,\text{cm}$ for copper in silver. However it is possible to obtain an increase of about $8 \,\mu\Omega\,\text{cm}$ by adding one atomic percent of iron to copper.

Of course, as stated above, scattering results whenever departures from perfect lattice periodicity occur so that contributions to the resistivity from lattice defects such as vacancies, interstitials, dislocations, etc. can also be expected. Overhauser and Gorman[4] have pointed out that not only do interstitials and vacancies cause scattering because of absent or excess charge but also because of the strain field that they impose on the lattice. In addition they suggest that it is also necessary to take into account the effect of interference between vacancy or interstitial scattering and the strain field. Their calculations suggest that strain scattering due to one atomic percent of interstitials in copper can cause an additional resistivity of about $10 \,\mu\Omega\,\text{cm}$, which illustrates the importance of this effect. The total extra resistivity per

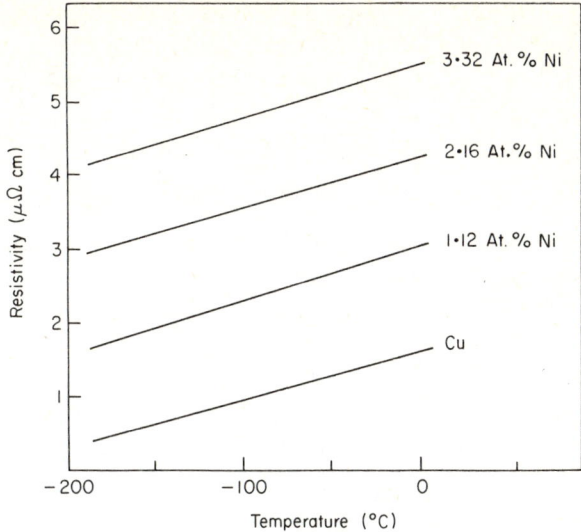

FIGURE 1. The figure shows that for low concentrations of impurities the variation of resistivity with temperature has a constant slope (from Linde[3]).

atomic percent of interstitials due to the three sources of scattering was estimated to be 10.5 $\mu\Omega$ cm whilst that for one atomic percent of vacancies was only 1.5 $\mu\Omega$ cm. In addition, electrons will also be scattered by grain boundaries although in bulk metals the mean free path is generally very much smaller than the grain size so that this contribution is small. However, it will be seen later that this is not necessarily the case for thin films.

2.3.3 Resistivity of alloys

The above discussion refers specifically to the case of a relatively low concentration of impurity atoms or defects but, since many of the materials used as thin film conductors or resistors are in fact alloys, it is worthwhile presenting the theory underlying their resistivity. Mott and Jones[5] derive d_n, the resistivity of an alloy, in the following way. An alloy consisting of two types of atom, A and B, present in the ratio $x:(1-x)$ has deviations in periodicity due to both types of atom leading to an increased probability of scattering. In the region of each A atom there is a potential $V_A(r)$ and in the neighbourhood of each B atom a potential $V_B(r)$. Thus the periodic potential within the alloy was taken to be:

$$V = xV_A(r) + (1-x)V_B(r) \qquad (11)$$

so that in the vicinity of an atom A, the magnitude of the difference in potential is given by:

$$V - V_A(r) = (1-x)[V_B(r) - V_A(r)] \tag{12}$$

and in the vicinity of a B atom

$$V - V_B(r) = x[V_A(r) - V_B(r)] \tag{13}$$

Now the probability of scattering is proportional to the square of the difference in potential so that this will be:

$$P_A \propto (1-x)^2[V_B(r) - V_A(r)]^2$$
$$P_B \propto x^2[V_B(r) - V_A(r)]^2$$

for A and B atoms respectively. The total probability of scattering by all the atoms of each specimen must therefore be given as:

$$P_A \propto x(1-x)^2[V_B(r) - V_A(r)]^2$$
$$P_B \propto (1-x)x^2[V_B(r) - V_A(r)]^2$$

so that the overall probability of scattering is:

$$P \propto x(1-x)^2 + x^2(1-x) = x(1-x) \tag{14}$$

Consequently the resistivity of a totally disordered alloy was taken as:

$$\rho = K x(1-x) \tag{15}$$

where K is a constant of proportionality. This result was found to be generally in good agreement with resistance measurements on a variety of alloys. Equation (15) predicts that a maximum value of resistivity will occur for a disordered system containing 50% of each type of atom and as seen in Fig. 2 (from Barrett[6]) this is the case. The results shown in this figure are much quoted and in fact wholly support the contention that scattering results from departures in periodicity. The dome shaped curve represents the totally disordered material (resulting from quenching from 650°C) whilst the solid curve represents the resistivity of the same range of composition after annealing at 200°C thereby forming ordered alloys at specific compositions.

2.3.4 Resistivity of transition metal alloys

To complete the discussion of the origins of resistivity in metals electron scattering in the transition metal alloys will be considered. The reason for this is not that the scattering mechanism in these metals is distinct from the processes described so far, but rather that these materials find frequent use as thin film resistor elements because of their unusually high resistivity. As a general rule it can be said that the density of states in metals follows, to

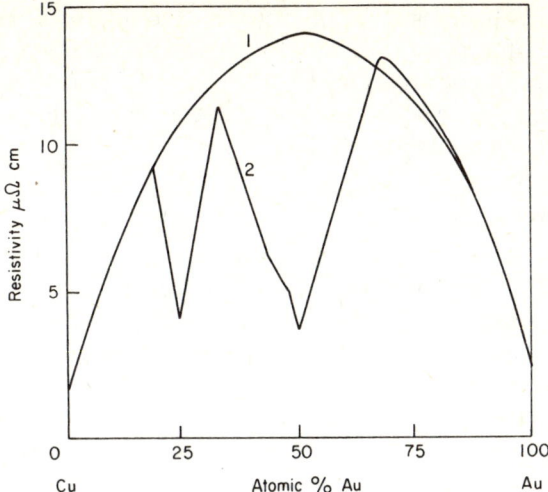

FIGURE 2. Variation of resistivity of Cu/Au alloys with composition. Curve 1 refers to alloys which have been quenched from 650°C, whilst curve 2 refers to alloys which have been annealed and therefore possess a periodic variation in resistivity. The minima correspond to positions of maximum crystallographic order in the ordered alloy lattice (from Barrett[6]).

quite a good approximation, the free electron distribution. In fact there are significant departures from the free electron model but these only become evident when measuring, for example, the thermoelectric effect which depends sensitively on the precise shape of the Fermi surface. However, the transition metals; chromium, manganese, iron, cobalt and nickel, etc. cannot be represented by a free electron model because of the manner in which the electrons of these elements are distributed in their atomic orbits. The transition metal atoms are characterised by the filling of the 4s shell before completely filling the 3d shell. Consequently, in the solid state where the discrete levels have split into bands there is an energy overlap between the 3d and 4s bands. It is in fact the scattering from the 4s to the 3d band which is responsible for many of the unusual physical properties of the transition element alloys. Because the 3d band is nearly full its electrons contribute very little to conduction so that the conductivity of these materials is thought to be due to the 4s electrons. As the temperature is increased the probability of scattering into the 3d band increases but because not many states are available such an event still tends to be comparatively unusual. As a consequence the temperature coefficient of resistivity is lower for the transition metals than for other simpler metals. In fact over limited ranges of temperature it is possible for certain transition metal alloys to have a temperature coefficient very nearly equal to zero. This, of course, together with the associated high resistivity of the transition metal alloys makes these materials of practical importance.

2.4 Matthiessen's Rule

It has now been indicated that there are several mechanisms which can cause loss of lattice periodicity and consequent electron scattering. Matthiessen's rule states that the individual resistivities due to each source can be simply added together to give the overall resistivity of the material. The reason for this lies in the interpretation of the term "relaxation time". Having decided that a relaxation time can be defined for each scattering process it is readily shown that one interpretation of this quantity is that it is the mean time between collisions, or, that its reciprocal gives the probability of a collision taking place within unit time. If the various scattering mechanisms take place independently then the net probability of scattering taking place in unit time is simply the sum of the individual probabilities, viz

$$\frac{1}{\tau} = \frac{1}{\tau_{phonons}} + \frac{1}{\tau_{imp}} + \frac{1}{\tau_{defects}} \qquad (16)$$

Since resistivity is proportional to the reciprocal of the relaxation time, the individual resistivities can thus be added viz:

$$\rho = \rho_{phonons} + \rho_{imp} + \rho_{defects}$$

Many deviations from this rule have in fact been observed and these will be discussed later. However, most of these are relatively small and certainly for the present purposes it is safe to continue to use the rule. It must, however, be remembered that its validity depends on the assumption of a constant relaxation time and that this is not always realistic.

2.5 Electrical Conductivity of Thin Metal Films

It will now be appreciated that "real" metal resistivity is influenced by collisions with not only lattice atoms but also with impurity atoms and lattice defects. Additional scattering processes only become significant when the mean free path of the conduction electrons is of a comparable magnitude to, or greater than, the separation between the centres responsible for scattering. The mean free path of conduction electrons in metals at room temperature is of the order of several hundreds of Angstroms and since continuous metal films of this thickness or less can be prepared it must be expected that their surfaces will cause additional scattering of the electrons

It is only when such scattering is diffuse that the resistivity will be affected. If specular scattering occurs then the component of electron velocity in the direction of the field is unaffected and the scattering event does not affect the magnitude of the resistivity.

2.5.1 Surface scattering

It was Thomson[7] who first postulated the concept of surface scattering when he was studying the variation of the conductivity of thin metal films with a view to verifying the Drude-Lorentz theory (almost identical with equation (6) above). Thomson formulated the thickness dependence of the conductivity and although this was inadequate in several respects, his formulation was a good approximation over certain ranges of thickness as well as being conceptually acceptable.

An accurate analysis of the size effect was first carried out by Fuchs[8] who used as his starting point the Boltzmann transport equation (equation (3)). The fundamental difference between the derivation of the distribution function for a bulk material and for a thin film lies in the description of surface scattering given above. Whereas in arriving at equation (5), $\text{grad}_r f$ was equated to zero, this is no longer valid in the case of a film whose thickness is of the order of, or less than, the mean free path since the distribution at right angles to the plane of the film is affected by the process of surface scattering. Consequently, although $\partial f/\partial x$ and $\partial f/\partial y$ can be equated to zero, it must be assumed that the distribution f is a function of position in the thickness of the film and therefore that $\partial f/\partial z$ is finite. Once again we take $\partial f/\partial t = 0$ in the steady state and consider the case where an electric field is applied only in the x direction. Under these conditions equation (3) reduces to:

$$X\frac{\partial f}{\partial p_x} + V_z\frac{\partial f}{\partial z} = \left(\frac{\partial f}{\partial t}\right)_{\text{coll}} = \frac{-(f-f_0)}{\tau} \tag{17}$$

Equation (17) is identical to equation (5) except for the term

$$V_z\frac{\partial f}{\partial z}$$

which, mathematically, expresses Thomson's description of surface scattering. A solution for the distribution function will now be obtained by putting $X = -eE$ and assuming a solution for f of the form $f = f_0 + f_1(V,z)$: substituting into equation (17) yields:

$$V_z\frac{\partial f_1}{\partial z} + \frac{f_1}{\tau} = eE\frac{\partial f_0}{\partial p_x} \tag{18}$$

This equation can readily be solved for f_1 to give:

$$f_1 = eE\tau \frac{\partial f_0}{\partial p_x}[1 + \phi \exp(-z/\tau V_z)] \quad (19)$$

where ϕ is an arbitrary function of V which can be determined from the boundary conditions as follows. For totally diffuse scattering of electrons at the surface, $z = 0$, $f_1 = 0$ so that $\phi = -1$ whilst for electrons scattered diffusely at the surface $z = d$, $f_1 = 0$ and $\phi = \exp(d/\tau V_z)$. Thus two equations for f_1 must be used depending on whether motion of electrons towards the surface $z = 0$ or $z = d$ is being considered, (i.e. whether $V_z < 0$ or $V_z > 0$) viz:

$$f_1 = eE\tau \frac{\partial f_0}{\partial p_x}[1 - \exp(-z/\tau V_z)] \quad V_z > 0 \quad (20)$$

and

$$f_1 = eE\tau \frac{\partial f_0}{\partial p_x}\left[1 - \exp\left(\frac{d-z}{\tau V_z}\right)\right] \quad V_z < 0 \quad (21)$$

Now $f = f_0 + f_1$ so that the two distribution functions are identical with that for the isotropic bulk material (equation (5)) except for the exponential terms in both cases, which must reduce the value of f. Consequently it can now be seen quite clearly that the effect of diffuse surface scattering is to reduce the probability of occupancy of any given quantum state (i.e. N_{eff}) and it is perfectly reasonable to anticipate that this reduction will have the effect of reducing the conductivity of the films as its thickness decreases. Once again the procedure for calculating the conductivity is to carry out a triple integral of the product of the density of states, the distribution function and the electron velocity over all k_x, k_y and k_z. However, by deriving the various distribution functions the validity of Thomson's original hypothesis has been illustrated and it is sufficient to quote Fuchs' final result. This was expressed as the ratio of the film conductivity σ to that of the bulk material σ_0, in terms of a dimensionless quantity $k = d/\lambda_0$, where d is the film thickness and λ_0 the mean free path of electrons in the bulk material, viz:

$$\frac{\sigma}{\sigma_0} = 1 - \frac{3}{8k} + \frac{3k}{4}\left(1 - \frac{k^2}{12}\right)E_1(k) + \left(\frac{3}{8k} - \frac{5}{8} - \frac{k}{16} + \frac{k^2}{16}\right)e^{-k} \quad (22)$$

If k is somewhat greater than unity then equation (22) can be approximated by:

$$k > 1 \quad \frac{\sigma}{\sigma_0} = 1 - \frac{3}{8k} \quad (23a)$$

and

$$k < 1 \quad \frac{\sigma}{\sigma_0} = \frac{3k}{4}\ln\left(\frac{1}{k}\right) \quad (23b)$$

from which it can be seen that for values of k much greater than unity the conductivity of the film approaches asymptotically the conductivity of the bulk material. The above model can be extended to allow for a fraction p of electrons being specularly reflected from the film surfaces. In this case the conductivity can be expressed by:

$$\frac{\sigma}{\sigma_0} = 1 - \frac{3}{8k}(1-p) + \frac{3}{4k}(1-p)^2 \times$$

$$\sum_{v=1}^{\infty} p^{v-1} \left[E_1(kv)\left(k^2v^2 - \frac{k^4v^4}{12}\right) + e^{kv}\left(\frac{1}{2} - \frac{5kv}{6} - \frac{k^2v^2}{12} + \frac{k^3v^3}{12}\right) \right] \quad (24)$$

which reduces to:

$$\frac{\sigma}{\sigma_0} = 1 - \frac{3}{8k}(1-p) \quad \text{for} \quad k > 1 \quad (25)$$

Consequently it can be seen that for electrons which are completely specularly reflected (i.e. $p = 1$) the conductivity is not thickness dependent. In practice, because the surface of films are "rough" compared to the wavelength of conduction electrons (given by the de Broglie relationship), the scattering tends to be almost completely diffuse (i.e. $p = 0$) although evidence will be presented in a later section which in certain cases suggests the reflection coefficient can have finite values. The magnitude of the surface scattering effect can be appreciated by referring to Fig. 3 in which the computed curves of Campbell[9] are shown. Suppose a continuous film 200 Å thick and an electron mean free path of 500 Å are assumed (i.e. $k = 0.4$) then the film conductivity can be expected to be a factor of about half that of the bulk material σ_0.

The discussion in this section has centred on what was earlier termed the "inherent" property of thin metal films. That is that the conductivity is thickness dependent. In fact it is not primarily the conductivity which is thickness dependent but rather the distribution function in the z direction. Having recognised this it must be anticipated that other phenomena depending on electron transport will also be sensitive to film thickness. This has been shown to be the case for temperature coefficient of resistance, Hall effect, thermoelectric power, magnetoresistance, optical properties and many other physical phenomena. Since thin film resistors are considered in detail later it is only the first of these quantities which is relevant here. Those wishing to pursue a detailed discussion of the other phenomena may wish to consult the book by Chopra[10].

Savornin[11] showed that it is possible to derive an expression for the size dependence of temperature coefficient of resistance (T.C.R.).

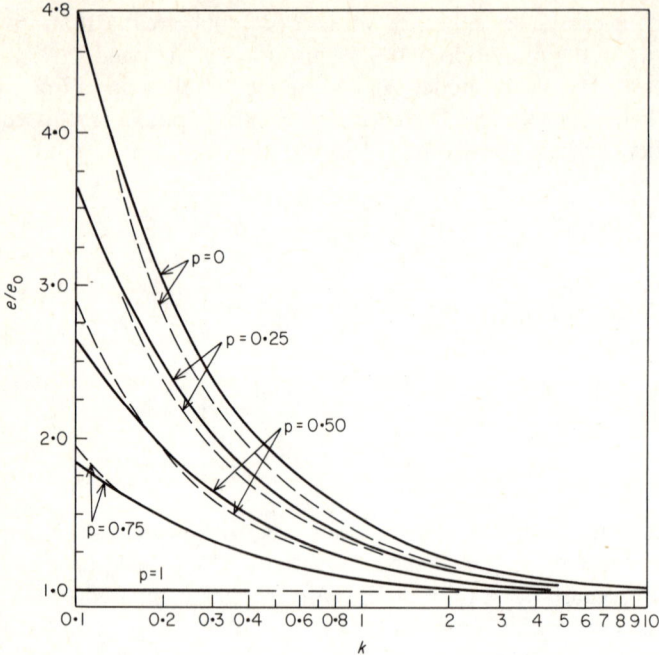

FIGURE 3. Theoretical size dependence of film conductivity. The solid curves are exact as calculated from equation 24 whilst the dotted curves are calculated from the approximate form (equation 25) (from Campbell[9]).

Equation (22) may be written as:

$$\frac{\rho_0}{\rho} = f(k) \tag{26}$$

and the T.C.R. of the film and bulk material are generally defined as:

$$\alpha = \frac{1}{\rho}\frac{\partial \rho}{\partial T} \tag{27}$$

and

$$\alpha_0 = \frac{1}{\rho_0}\frac{\partial \rho_0}{\partial T} \tag{28}$$

Now
$$\rho = \rho_0/f(k)$$

$$\therefore \quad \frac{\partial \rho}{\partial T} = \frac{f(k)\dfrac{\partial \rho_0}{\partial T} - \rho_0 \dfrac{\partial f(k)}{\partial T}}{[f(k)]^2} \tag{29}$$

This procedure can readily be shown to give:

$$\frac{\alpha}{\alpha_0} = \frac{f(k) - k\,\partial f(k)/\partial k}{f(k)} \tag{30}$$

In equation (30), $f(k)$ may be represented by the form in equation (22) for totally diffuse scattering or by (24) for partially specular scattering. In either case a plot of α/α_0 against k demonstrates that whereas surface scattering leads to an increase in resistivity, it leads to a decrease in the T.C.R. Experimental data concerning the size dependence of resistivity and T.C.R. will be given in a subsequent section.

In formulating the effect of surface scattering, Fuchs tacitly assumed that the metal under consideration had a spherical Fermi surface or that this was effectively the case due to randomly oriented grains creating isotropic scattering. Certainly much of the vast quantity of experimental data published over the years relate to metals which do not have a spherical Fermi surface and occasionally, because of epitaxial growth, is highly single crystal. Consequently, some caution should be exercised in interpreting such data on the basis of the Fuchs model of surface scattering. Other workers have criticised the use of a simple specularity parameter and alternative theories will now be discussed.

2.5.2 Angular dependence of the specularity parameter

Cottey[12] argued that electrons travelling normally to the surface of a film would be far more likely to reach the surface than would electrons travelling at an oblique angle. Furthermore he considered it unlikely that the surface of a film would be smooth compared to the electron wavelength so that electrons actually reaching the surface would in all probability be diffusely scattered. However the layers just beneath the surface may well exhibit a high degree of crystalline perfection so that it would be reasonable to expect the obliquely incident electrons which do not reach the surface to be specularly reflected from these lower layers. This qualitative argument suggests that the specular reflection coefficient may indeed be a function of the angle of approach to the surface of the conduction electrons.

Parrott[13] also concluded that the specularity parameter should be dependent on angle of incidence and drew an analogy with the reflection of light. The angular dependence of p was obtained in terms of the change of wave-vector of the electron upon scattering. It was shown that electrons in metals would almost certainly be diffusely scattered unless they were incident at grazing angles.

Ziman[2] pointed out that because of the lack of detailed knowledge of surfaces, reflection could only be characterised by statistical parameters and derived a collision operator in terms of wavelength and surface roughness. His derivation was based on certain simplifying assumptions but demonstrated once again that reflection from the surface is a function of angle of incidence.

Several other models have been proposed which support the concept of angular dependence and it is encouraging that all are in agreement for perfectly rough or smooth surfaces.

2.5.3 Low angle scattering

At low temperatures, electrons are scattered through relatively small angles by phonons and as a consequence such collisions do not contribute greatly to the resistivity. In fact the large angle Umklapp processes (i.e. the creation or annihilation of a phonon simultaneously with a Bragg reflection of an electron from a zone boundary) should be accounted for even in the low temperature regime but are frequently neglected. In a thin film however, low angle events will contribute to the resistivity because they will cause the electrons to strike the surfaces of the film thereby bringing about large angle scattering. This is known as the Olsen effect[14] and it results in the film resistivity being greater than predicted by the Fuchs theory. Calculations of the mean free path incorporating small angle scattering have been carried out by Azbel and Gurzhi[15] who suggest that the film resistivity at low temperature should be proportional to T^3 rather than to T^5 as implied by the Bloch-Grineissen law. Although there now exists experimental evidence which appears to support the Azbel-Gurzhi theory, Bass[16] points out that a definitive test of the model has not yet been carried out. Nevertheless, resistivity measurements on films at low temperature certainly do indicate size-dependent deviations from Matthiessen's rule and this phenomenon is therefore another example of an inherent thin film property.

2.5.4 Grain boundary scattering

In a bulk material, the grains are generally considerably greater in size than the electron mean free path so that scattering at the boundaries does not contribute significantly to the resistivity. However the structure of a thin metal film does not approximate to a plane parallel slab, but rather consists of an array of randomly oriented polycrystallites. Although preferred orientation is occasionally observed it tends to be the exception rather than the rule. The size and orientation of the microcrystallites of a film are governed by the deposition conditions such as residual gas pressure, rate of deposition, the nature of the substrate, etc. Consequently the grain size and associated effects are a function of the method of preparation of the film rather than being an inherent thin film property. Because the typical grain size is often comparable to the electron mean free path it is to be expected that scattering at the boundaries will make a significant contribution to the resistivity.

Mayadas and Shatzkes[17] formulated the effect of this mechanism in thin films and pointed out that it was only necessary to consider those boundaries actually at an angle to the applied field. Although their analysis is valid for a bulk metal it cannot be tested for the reason given above. It was assumed that the grain boundaries could be represented by δ-functions and the additional resistivity was calculated by solving the transport equation with the additional scattering mechanism included. This required that an estimate be made of the relaxation time for grain boundary scattering so that a new total relaxation time for all scattering mechanisms could be calculated using Matthiessen's rule. Mayadas and Shatzkes obtained the new relaxation time τ_{grains} in terms of a parameter α which described the geometry of a grain and the scattering power of its boundary, viz:

$$\alpha = \frac{\lambda_0}{r^1} \frac{R}{(1-R)} \tag{31}$$

where r^1 is the average grain diameter, λ_0 is the mean free path in the bulk material and R is the grain boundary reflection coefficient. The ratio of the resistivity, inclusive of grain boundary scattering, to that in a bulk material was found to be:

$$\frac{\rho_0}{\rho_g} = 3\left[\frac{1}{3} - \frac{1}{2}\alpha + \alpha^2 - \alpha^3 \ln\left(1 + \frac{1}{\alpha}\right)\right] \tag{32}$$

This function is shown in Fig. 4 where it is seen, as expected, that for small values of α i.e. large r, small λ_0 or small R, $\rho_g \to \rho_0$. The resistivity of a thin film in which both grain boundary and surface scattering occur could not be obtained analytically but the various integrals involved were evaluated numerically and the results are compared with the Fuchs model in Fig. 5. It can be seen that, for a given value of p, the ratio of the film to bulk resistivity (ρ_g if grain boundary scattering is included, ρ_0 if not) is less when grain boundary scattering is included in the latter. For films of a practicably realisable thickness which are continuous (i.e. $d/\lambda_0 \simeq 1$) the effect is relatively small for the values of the parameters chosen. On the other hand it is clearly possible to obtain very large values of α in which case the effect would of course become much more pronounced. Mola and Heras[18] have undertaken calculations of resistivity for a wide range of values of α and these do indeed demonstrate that for small

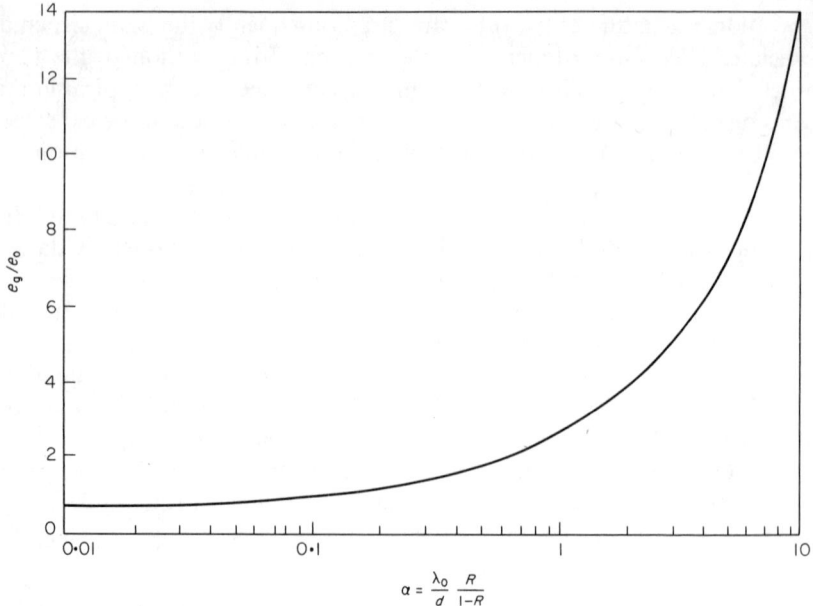

FIGURE 4. The effect of grain boundary scattering on a bulk material (from Mayadas and Shatzkes[17]).

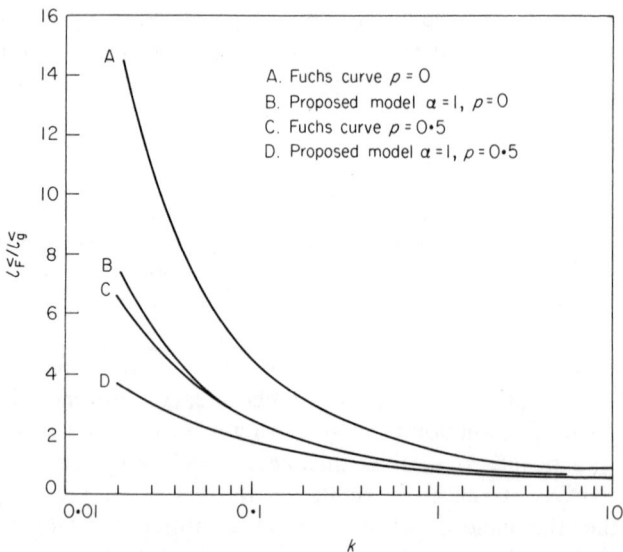

A. Fuchs curve $p = 0$
B. Proposed model $\alpha = 1$, $p = 0$
C. Fuchs curve $p = 0.5$
D. Proposed model $\alpha = 1$, $p = 0.5$

FIGURE 5. The effect of grain boundary scattering on the resistivity of a thin film already subject to surface scattering (from Mayadas and Shatzkes[17]).

grain sizes the dominant contribution to the additional resistivity arises from boundary rather than surface scattering.

2.5.5 The effect of heat treatment on resistivity

In its deposited state, a thin film very often contains a great deal of structural imperfection, impurities, strain, etc. In practice, it is quite often necessary to anneal films at elevated temperature to remove long term drift in resistivity due to self annealing of defects. By annealing it should be appreciated that the removal of defects is implied rather than growth of grains. The density of defects in a film, as in a bulk specimen, is a function of the preparation parameters and of the thermal history. To characterise a particular deposition process it would be desirable to calculate the density of defects immediately after deposition, or the effect of a particular type of defect on the film resistance. In this section it will be seen that a considerable amount of progress has been made in this area.

Early work was prompted by the observation that the resistance of a thin film decreased irreversibly when heated to a temperature greater than that at which it had been deposited. It was also observed to decrease spontaneously even at constant temperature immediately after deposition. Vand[19] considered the origins of this effect and concluded that lattice vacancies, interstial atoms or combined vacancies and interstials could be responsible. However, vacancies and interstitials can only be removed by long range diffusion to the surface or by chance annihilation upon one type meeting the other. If the vacancies and interstitials happen to be more closely linked (a combined distortion) then only relatively short diffusion is necessary and Vand concluded that it was this form of defect which was responsible for the observed changes in resistance. It was assumed that each distortion had a characteristic, or critical, decay energy and that when any atom involved in the distortion achieved the critical energy, decay would be initiated. Furthermore, Vand stated that the rate of decay was proportional to the product of the density of distortions and a Maxwell-Boltzmann exponential factor containing the characteristic decay energy. After a certain amount of mathematical approximation Vand obtained a characteristic function for the decay process, viz:

$$F_0[E_0(t)] = -\frac{t}{kT}\frac{\partial R_i}{\partial t} \qquad (33)$$

where $F_0(E)$ is the product of the density of distortions of decay energy E at time equals zero and the additional resistivity $r(E)$ per distortion; $E_0(t)$ is a characteristic energy for the system as a whole; R_i is the additional resistance due to combined distortions, T is absolute temperature and k is

Boltzmann's constant. In general it was found that $F_0[E_0(t)]$ showed a maximum in the region of 1 eV for films prepared by evaporation.

The procedure outlined above applies to the so-called method of ageing in which the resistance was measured as a function of time at constant temperature. Vand pointed out that this method had certain limitations and developed a second method whereby the resistance was measured as a function of time under conditions of a linearly increasing temperature. This enabled Vand to extend his measurements over a much wider range of energies from which he found that a maximum in $F_0[E_0(t)]$ always occurred. The corresponding value of energy E_m was found to depend on the state of the substrate and the thickness of the film. This was one of the earliest attempts to relate the annealing behaviour of a film to its deposition conditions and although the work can be criticised on a number of grounds it must be regarded as an important contribution to the subject.

2.6 Comparison of Theory and Experiment

In this section a very small fraction of the literature published on electrical measurements of thin metal films will be reviewed. In particular an indication will be made of how well or badly the data can be represented by Fuchs' theory of surface scattering, and a description of methods by which estimates can be made of the various parameters in the size effect will be given. It will also be shown how use can be made of thin films to obtain information of fundamental importance to solid state physics. Because of the theory presented previously on conduction in bulk metals, and on thin films, it will be evident that studies of the latter can yield information such as the mean free path, the effect of the surface on the conduction process and the effect of impurities and structure. In order that the thin films used in practice can be fabricated in a controllable and reproducible manner, it is important to have some knowledge not only of these topics, but also of the manner in which they may be altered by changes in, for example, the deposition procedure. It was the lack of this understanding which was responsible for the characteristic lack of reproducibility of the results of many early workers in the field.

2.6.1 Electrical resistance

Appleyard and Lovell[20] studied the thickness dependence of resistivity of very thin alkali films which should follow the free electron theory and thereby be well represented by Fuchs' theory. By careful preparation of the substrates and the use of very clean vacuums, continuous films only a few tens of Angstroms thick were produced. It was found, however, that the data could not be represented by a single value of the specular reflection coefficient. The reason for this may be due to a change of structure (perhaps larger grains) with thickness and it will be seen later that other authors have considered this possibility. Further studies on alkali films, Mayer[21], Nossek[22] and Worden and Danielson[23] also indicate that the resistivity of films whose thickness was of the order of the electron mean free path approached that of the pure bulk material. In particular the latter authors assume that potassium has one free electron per atom and from this calculate that their data can best be represented by $p = 0$. At lower thicknesses there is again some divergence from the theoretical curves, and this may again be due to lack of continuity. For alkali films one may conclude that predominantly diffuse scattering occurs from the film surface whilst the resistivity and mean free path approach the bulk values. Reynolds and Stilwell[24] also demonstrated the possibility of obtaining bulk parameters from thin film measurements. In the thick film limit for diffuse scattering, the dependence of resistivity on thickness is given by:

$$\frac{\rho}{\rho_0} = 1 + \frac{3}{8k} \text{ or } \log\left(\frac{\rho}{\rho_0} - 1\right) = \log\left(\frac{3\lambda_0}{8}\right) - \log d$$

By plotting $\log[(\rho/\rho_0) - 1]$ against $\log d$ these authors obtained λ_0 for the bulk material. Thus the mean free paths for copper and silver were found to be 450 Å and 250 Å respectively. These values are surprisingly high and, in the light of later evidence, suggest that the bulk resistivities were assumed rather than the corresponding values for very thick films exhibiting no size effect. This assumption has often been made even though it is hardly ever likely to be justifiable because of the additional contributions to the resistivity from grain boundary, impurity and defect scattering. Coutts and Matthews[25] have avoided making this assumption but for a given set of deposition conditions have assumed that, for continuous films, ρ_0 is independent of thickness. By depositing two films of known but different thickness simultaneously it was thus found possible to evaluate the mean free path and the true value of ρ_0 from the film resistances. Thus, for copper films deposited onto baked Pyrex substrates at a pressure of approximately 10^{-8} torr, Coutts[26] found an electron mean free path of 270 Å and $\rho_0 \simeq 2\rho_b$ where ρ_b is the pure bulk resistivity. For copper films deposited in argon at 10^{-8} torr Coutts estimated the mean free path as 200 Å although the grain size appeared to be greater than for films deposited at 10^{-8} torr. This result suggests that argon was incorporated in the films and reduced the mean free path due to one of the scattering mechanisms discussed earlier. The above technique relied on the iterative solution of the Fuchs expression for a pair of films using a digital computer. Borrago and Heras[27] have criticised this technique on the grounds that it is rather cumbersome and have demonstrated that the range of validity of equations (23a) and (23b) is much greater than might be expected. Subsequently, they show that these approximate forms lead to results of acceptable accuracy without resorting to the time consuming iterative solution described above.

The assumption that ρ_0 is independent of thickness was shown to be doubtful from the measurements of Mayadas[28] who demonstrated that grain size increased with thickness thereby reducing ρ_g and ρ_0. However it would be difficult to use the more general theory of Mayadas and Shatzkes discussed earlier to obtain the correct values of λ_0 and ρ_0 since this involves the introduction of even more parameters. In principle it is possible to use the method of Coutts and Matthews to obtain an estimate of the specular reflection coefficient, as well as λ_0 and ρ_0, by depositing three films of different thickness simultaneously. However, calculations show that small errors in the measurement of thickness can lead to values of p greater than unity or even less than zero. Thus, it would appear that the only way in which the relevant parameters can be obtained is to measure the size dependence of three physical properties of one film.

Gilham, Preston and Williams[29] deposited gold films onto nucleating layers of bismuth oxide to obtain small grained, but relatively thin, continuous films. Upon annealing the films a large decrease in resistivity was observed and it was concluded that this was due, not only to the removal of structural defects, but also to an increase of the specularity parameter from $p = 0$ to $p = 0.9$. Alternatively, if grain growth took place without loss of continuity then the resistivity would decrease due to a reduction in grain boundary scattering. It is not of course impossible that several changes could take place simultaneously and it is difficult to interpret these results unambiguously.

Lucas[30] provided what appeared to be conclusive evidence of specular reflection in gold films deposited onto bismuth oxide. These polycrystalline films were annealed, and since the resistivity was close to the bulk value, it was concluded that specular reflection was occurring at both surfaces. However, when an overlayer of gold of several tens of Angstroms thickness was deposited over the upper surface an increase in resistance to a maximum value was found (Fig. 6). This was then followed by the usual thickness dependent decrease. Lucas interpreted these changes as being due to artificial roughening of the upper surface so that its specularity parameter decreased from $p = 1$ to $p = 0$. A new theory of surface scattering allowing for different specular reflection coefficients at the upper and lower surface was developed and this was in reasonable agreement with the observations. However, Chopra and Randlett[31] have carried out similar experiments and

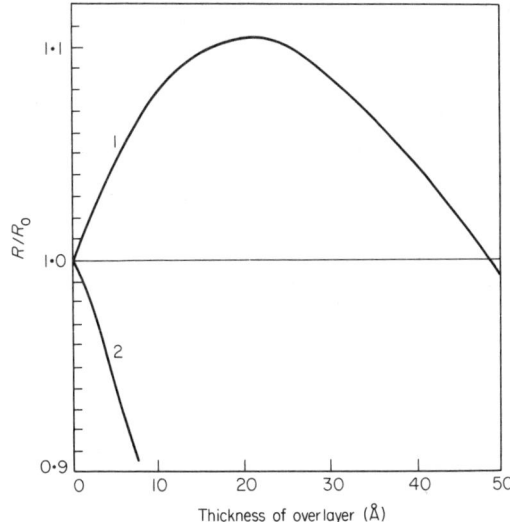

FIGURE 6. The figure shows how the resistance of a thin film exhibiting specular reflection (curve 1) increases initially when an overlayer is deposited whereas that of a film exhibiting diffuse reflection (curve 2) decreases immediately upon deposition of the overlayer (from Lucas[30]).

these cast a certain amount of doubt on the conclusions of Lucas. The latter authors found that an increase in resistance was only obtained for certain film/overlayer combinations. For example, silicon monoxide or Permalloy overlayers increased the resistance of gold films but had almost negligible effect on silver. The resistance of the latter however was increased by an overlayer of germanium which itself had little effect on aluminium. The conclusion of Chopra and Randlett was that overlayers can behave rather similarly to absorbed gases and modify the density and mobility of conduction electrons near the surface. Thus, it must be concluded that the evidence for specular reflection in polycrystalline films is still rather uncertain even though the experimental work discussed above cannot be lightly dismissed. The evidence for partially specular reflection in single crystal films is considerably more conclusive and in particular the work of Chopra, Bobb and Francombe[32] is an important contribution. These authors demonstrated that it was possible to deposit structurally perfect, single crystal gold films by sputtering onto a hot mica substrate. For films greater than 250 Å thick, the resistivity was found to be independent of thickness and only 10% greater than that of pure gold. Consequently, it is reasonable to conclude that these films exhibited almost perfectly specular reflection. Their results are shown in Fig. 7.

It is interesting to note that many estimates of the specularity parameter have been made at room temperature and these generally indicate $p = 0$. On the other hand there is a certain amount of evidence suggestive of specular reflection for measurements made at low temperature. Larson and

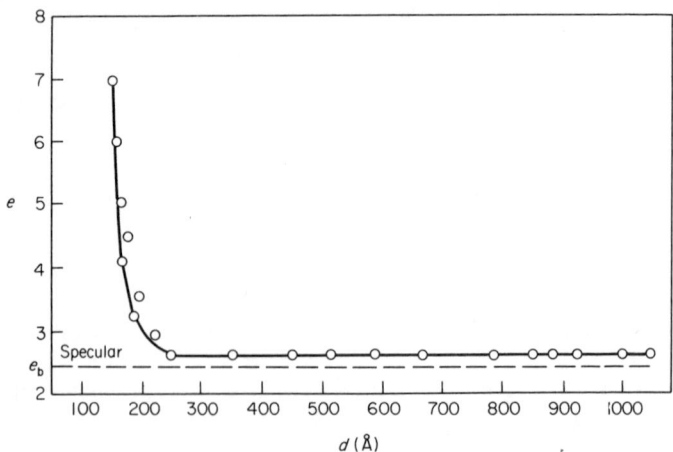

FIGURE 7. The figure shows that electrons in gold films, deposited by sputtering onto hot mica substrates, are specularly reflected for thicknesses greater than 250 Å (from Chopra, Bobb and Francombe[32]).

Boiko[33] deposited silver films on mica substrates at room temperature and measured the resistivity ratio $\rho_{300}/\rho_{4.2}$. Their data were best represented using Fuchs' theory with a specularity parameter $p = 0.5$ and a mean free path of 31 200 Å (at low temperature).

Isaeva[34] also measured the resistivity ratio for copper whiskers formed by reduction from copper iodide and these data suggested a value of $p = 0.6$. However, it was recognised that because the measurements had not been made in vacuum the possibility of errors due to oxidation could not be neglected. Niebuhr[35] demonstrated that partially specular reflection occurred for films deposited at very low temperatures but totally diffuse reflection was evident when deposition was carried out at 200 K or higher. This evidence strongly suggests that the surface of a film is dependent on the manner in which the film nucleates.

2.6.2 Temperature coefficient of resistance

In Section 2.5.1 the effect of surface scattering on the T.C.R. of thin metal films was discussed and it was seen that this quantity is in general reduced below the pure bulk value. In addition, it was indicated above that the bulk resistivity of the film material was invariably greater than that of the pure starting material. Young and Lewis[36] demonstrated that the T.C.R. of the bulk film material would be less than that of the pure starting material, in the following way. Matthiessen's rule states:

$$\rho_0 = \rho_b + \rho_{\text{impurities}}$$

where ρ_0 is the resistivity of the bulk impure film material and ρ_b is that of the pure starting material

$$\therefore \quad \frac{\partial \rho_0}{\partial T} = \frac{\partial \rho_b}{\partial T}$$

(since $\rho_{\text{impurities}}$ can be regarded as temperature independent, although as seen earlier this is not strictly correct)

$$\therefore \quad \frac{\rho_0}{\rho_0}\frac{\partial \rho_0}{\partial T} = \frac{\rho_b}{\rho_b}\cdot\frac{\partial \rho_b}{\partial T}$$

$$\therefore \quad \alpha_0 \rho_0 = \alpha_b \rho_b \tag{34}$$

where α_0 and α_b are the T.C.R. of the impure and pure materials respectively. Now from equations (26) and (30) we have:

$$\frac{\alpha \rho}{\alpha_0 \rho_0} = \frac{f(k) - k f(k)/dk}{f(k)^2}$$

Substitution for the product $\alpha_0 \rho_0$ from (34) gives:

$$\frac{\alpha \rho}{\alpha_b \rho_b} = \frac{f(k) - k f(k)/dk}{f(k)^2} \tag{35}$$

Since α and ρ can be measured for a thin film and α_b and ρ_b refer to the pure bulk material, the left hand side of this equation is known for a given film so that an iterative process can be used to find k and hence λ. Consequently, assuming diffuse scattering, λ_0 can be obtained from measurements of resistivity and T.C.R. of a single film. If the assumption can be made that the value of ρ_0 does not vary rapidly with thickness, then in principle it should be possible to combine the method of Coutts and Matthews with that of Young and Lewis, and thereby obtain values of ρ_0, λ_0 and p from measurements of resistivity and T.C.R. of only two films. This however has not as yet been attempted. Young and Lewis calculated λ_0 from equations (22) and (30) alone, for gold films deposited by Holland and Siddall[37] and found inconsistent results. However using equation (35) resulted in a value of $\lambda_b = 360\,\text{Å}$ (in good agreement with the pure bulk value at room temperature) and $\lambda_0 = 200\,\text{Å}$. Thus $\rho_0/\rho_b \simeq 1.8$ so that $\alpha_0/\alpha_b \simeq 0.6$. Consequently it must always be expected that the T.C.R. of a metal film will be considerably less than the pure bulk value since both surface scattering and impurity effects tend to reduce its value. From the practical point of view, this of course is a very favourable situation since two of the most important design considerations of a film resistor are a high resistivity and a low T.C.R. Singh[38] has calculated the effect of grain boundary scattering on the T.C.R. of metal films using as a starting point the model of Mayadas and Shatzkes. His conclusion is that the T.C.R. of films in which $d < \lambda_0$ is largely dependent on the mean grain size rather than the thickness. For films of a very small grain diameter the T.C.R. becomes very small, as indeed it does for very thin films.

2.6.3 The thermoelectric effect in thin films

The thermoelectric effect in metals arises when a temperature gradient exists along the conductor. In order to formulate the effect it is necessary to solve the Boltzmann transport equation in the presence of a perturbing electrical field and a thermal gradient. The effect manifests itself by the presence of a voltage between the ends of a conductor even when there is no applied field but only a thermal gradient. The magnitude and sense of the thermoelectric voltage depends on the shape of the Fermi surface and it has been shown that studies of the effect in thin films can yield useful information about the electronic structure of metals.

For a bulk metal the thermoelectric power is given by:

$$S = \frac{\pi^2 k^2 T}{3e}\left[\frac{\partial \log \sigma}{\partial E}\right]_{E=E_F} \quad (36)$$

The physical size dependence of the thermoelectric power can be readily understood from that of the conductivity σ. It is a relatively straightforward matter to show that for a thin film the thermoelectric power can be expressed by:

$$S = \frac{\pi^2 k^2 T}{3e}\left[\frac{\partial \log A}{\partial \log E} + \frac{\alpha}{\alpha_0}\frac{\partial \log \lambda_0}{\partial \log E}\right]_{E=E_F} \quad (37)$$

where A is the area of the Fermi surface. Thus the thermoelectric power is sensitive to the way in which the area of the Fermi surface varies with energy, and any curvature in the Fermi surface can be expected to have a pronounced effect on the thermopower. Mayer[21] has arranged equation (37) in the form:

$$Sd = S_0 d - \frac{\pi^2 k^2 T \lambda_0}{8eE_F} \left[\frac{\partial \log \lambda_0}{\partial \log E} \right]_{E=E_F} \qquad (38)$$

where S_0 is the thermoelectric power of the bulk metal (this should refer to the bulk impure metal). Mayer took the results of Nossek[22] on alkali metal films and plotted Sd against d. This procedure gave a straight line whose intercept was very simply related to the energy dependence of the mean free path. The derivative was almost exactly equal to two (i.e. $\lambda_0 \propto E^2$) which is as expected for the free electron theory. Once again therefore it is confirmed that the alkali metals are quite well represented by the free electron model.

On the other hand Huebener[39] made thermocouples from thin films and wires of gold and because these have a different size dependence he was able to obtain a thermoelectric output. Analysis of the results showed that the free electron theory cannot explain the positive sign of the thermoelectric power of gold and it was concluded that the Fermi surface must be distorted and actually touch the zone boundaries. Consequently, although the free electron theory can account for the resistivity of gold it is quite unable to account for the thermoelectric power which is clearly more sensitive to the shape of the Fermi surface. Thakoor et al.[40] have recently measured both the thermoelectric power and resistivity of thin copper films and have invoked the relationship:

$$S_F \rho_F = S_0 \rho_0 + S_i \rho_i + S_S \rho_S$$

(where the suffices i and S refer to the impurities and structural defect contributions to the resistivity and thermoelectric power) to evaluate S_i and S_S. They demonstrate that S_i is thickness dependent and that this dependence is able to account for the apparent enhancement of S_S at low thickness.

2.6.4 Conduction in a magnetic field

When a metallic conductor carrying a current is placed in a magnetic field at right angles to the current, charged particles are deflected sideways and a charge builds up between the sides of the conductor. Equilibrium is reached when the field due to the deflected charges (the Hall voltage) exactly balances the force deflecting further charge. It may not therefore be imagined that the conductivity would be influenced by the magnetic field. However, even for a good free electron metal, like sodium, a transverse magneto-resistance is always observed and this phenomenon is a useful tool with which to study the band structure of metals and semi-conductors. In the case of thin films exhibiting diffuse surface scattering Sondheimer[41], Chambers[42] and MacDonald and Sarginson[43] all predict a finite magneto-resistance even for perfect free electron materials and Sondheimer has analysed the problem for the case of a magnetic field perpendicular to the plane of the film. His solution of the appropriate form of the transport equation predicted that the conductivity of the film should oscillate as a function of applied field. Chambers solved the equations of motion of an electron in perpendicular electric and magnetic fields and showed that the velocity was also an oscillatory function of β where $\beta = d/r^*$ (d = film thickness and r^* = radius of electron spiral). When substituting the velocity into an expression for the electron density a further oscillatory solution results and therein lies the origin of the conductivity oscillations.

To observe the Sondheimer oscillations it is necessary that the film thickness, the electron mean free path and the radius of the electron spiral are all approximately equal. This dictates the use of relatively thick films (or foils) and low temperatures as it is not possible to obtain a large enough magnetic field to reduce the radius of the helix to a value comparable to the mean free path at room temperature. Försvoll and Holwech[44] studied the transverse magneto-resistance in very pure annealed aluminium and, after correcting their results for the bulk magneto-resistance, obtained the predicted oscillations. From the positions of the maxima they calculated the momentum of the Fermi electrons and obtained a value in good agreement with previous estimates.

2.7 Summary of the Properties of Thin Metal Films

In the preceding sections the various sources of resistivity in metals have been discussed and it has been shown that thin metal films behave in a very similar manner to the parent metals. The primary source of resistivity in bulk metals is due to the interaction of conduction electrons with the quanta of thermal energy (phonons) associated with lattice vibrations. This mechanism gives rise to a linear dependence of resistivity on temperature for $T \gg \Theta_D$. It is only in this case that the relaxation time approximation can validly be used because it is only in this situation that elastic scattering of electrons occurs. In a thin film, the resistivity is found to have a T^3 dependence at low temperature rather than T^5 and this is one of the two inherent thin film effects to have been considered. The other of course, is the additional contribution to the resistivity from surface scattering of electrons. It is these two effects which give rise to the departures from Matthiessen's rule which have provided the bulk of the discussion.

The solution of the transport equation leads to the concept of a size dependence for all thermo-galvano-magnetic effects and it was demonstrated that it is possible to obtain fundamental information from thin film measurements. Fuchs' theory of surface scattering allows for totally diffuse or partially specular reflection of electrons, and it was indicated that a theory utilising an angular dependent specularity parameter was probably more realistic than one using a constant value for a given film. On the other hand it is more difficult to apply the alternative theories to actual measurements because they appear to introduce further adjustable parameters. Consequently despite its limitations Fuchs' theory is still widely used to explain, qualitatively and quantitatively, size dependent phenomena. The majority of films appear to exhibit diffuse scattering although some evidence now exists for specular scattering in polycrystalline films whilst certain single crystal films have been shown not to exhibit a size effect (i.e. totally specular reflection). For films exhibiting diffuse scattering the resistivity can be increased by about two or three times for thicknesses approximately equal to the mean free path. Increases greater than this have been observed for much thinner films but it is necessary to deposit these under specially controlled conditions to ensure that the films are continuous. It was also shown that an increase in resistivity brings about a similar decrease in the temperature coefficient of resistance.

In a bulk metal the additional resistivity introduced by impurity and defect scattering is relatively small but in a thin film these effects can be quite important. Impurity scattering depends on the density of impurities, their effective scattering area and their effective charge. In a thin film, impurities can be introduced from the background gas so that from a practical point of view it is important to control the constituents and relative partial pressures of the residual gases. The effect of grain boundary scattering has been shown to be significant in thin films but not in bulk metals; therefore it is again necessary to control parameters such as rate and temperature of deposition in order to achieve reproducibility. Estimates of impure bulk resistivity are approximately twice as large as the resistivity of the pure material. Thus for a typical single metal film we could expect, at most, a resistivity approximately ten times greater than that of the pure bulk metal. On the other hand it is possible to achieve considerably greater resisitivities using alloy films containing one or more transition metals. These will be discussed further in a subsequent section. Finally, "as-deposited" films always show an irreversible decrease in resistance upon ageing (when measured in vacuum) and this is apparently due to the annihilation of closely linked vacancies and interstitials. It is normal in practice to subject films to heat treatment before use so that they will become structurally and chemically stable, since this tends to form a protective oxide on metal films which prevents further chemical attack.

3. Application of Thin Metal Films

In this, the second half of the chapter, some of the many applications of thin metal and alloy films will be discussed and where appropriate, the advantages of using devices in thin film form, will be indicated. Although some of the design requirements such as choice of substrate, reliability, stability, etc. will be mentioned these topics are very extensive themselves so the amount of detail will, of necessity, be kept to a minimum.

3.1 Thin Film Resistors

Before discussing the various materials suitable for use as thin film resistors one of the fundamental requirements will be considered. Clearly, the resistance of a thin film depends on its thickness, length and breadth. If the latter two dimensions are equal, the resistor is a square, and the sheet resistance is measured in ohms per square. The sheet resistance is given by:

$$R_\square = \rho/d$$

Take an example: a film 1000 Å thick of resistivity 20 $\mu\Omega$ cm, would have a sheet resistance of 2 ohms per square. Other considerations apart for the moment, if a 10 kΩ resistor is required then 5000 squares in series of this particular material must be used. If the resistor track is 10^{-3} in wide (25 μm) then it must be approximately 12 cm long. In many circuits this would require using an undesirably large proportion of the available substrate area so the primary requirement is to find a material with a large sheet resistance due to either a large resistivity or to the use of very thin films.

In practice the minimum usable film thickness consistent with reliability requirements is approximately 200 Å and although this represents a five fold increase in sheet resistance for the above film, a further increase via the resistivity would be desirable. The need for a high degree of control over the film properties would appear to necessitate using single component metal films. On the other hand the resistivity of metals tends to be too low for the values mentioned earlier. Consequently a great deal of development work has been carried out on alloys, particularly of transition metals. However single component resistors based on tantalum have been successfully developed and these will now be discussed.

3.1.1 Tantalum and associated materials

In its bulk form, tantalum would not appear to be greatly attractive as a potential candidate for use as a thin film resistor material since it has a resistivity of only 13 $\mu\Omega$ cm together with a typical and unacceptably large, metallic T.C.R. Indeed the earlier studies on tantalum at the Bell Telephone Laboratories were concerned with the prospect of using its oxide as a dielectric for thin film capacitors (McClean[45]). Since this early work, subsequent studies have shown that tantalum also has potential advantages as a thin film resistor material and, of course, the prospect of a universal material for capacitors and resistors is economically very attractive. In practice, it has been demonstrated that tantalum can exhibit three characteristic structures in thin film form: (i) the body centred cubic (bcc) or α structure, (ii) the tetragonal or β structure and (iii) a low density form which can be either bcc or tetragonal. The α phase is rather similar to the bulk except that its resistivity is approximately $25-50\,\mu\Omega\,\text{cm}$ and its T.C.R. $\simeq +500 \to 2000\,\text{ppm}°\text{C}^{-1}$. Clearly this phase is more useful than the bulk material but is nevertheless not wholly acceptable as a thin film resistor material.

The β phase is exclusive to the thin film form and is not exhibited in bulk tantalum. It is a much more useful material since it has a resistivity and T.C.R. of approximately $250\,\mu\Omega\,\text{cm}$ and $-100 \to +100\,\text{ppm}°\text{C}^{-1}$ respectively. The low density form has an apparent resistivity of perhaps $5000\,\mu\Omega\,\text{cm}$ and a similar T.C.R. to the β phase. Although this might be expected to be the most useful form of tantalum, in practice, because of the rather porous nature of the low density films, the resistors tend to drift in value at an unacceptable rate. In practice β tantalum is the most desirable material for resistors by virtue of its large resistivity, near zero T.C.R. and acceptable stability. However the precise reasons for the formation of β rather than α tantalum are, even today, not completely understood and some of the rather confusing experimental work on this topic will now be discussed. It is, of course, vitally important that a controllable structure can be produced consistently so that it is desirable to understand the reasons for the formation of β tantalum (or at least, to determine the deposition conditions required). Because of its refractory nature, tantalum is generally deposited using either D.C. or R.F. sputtering although it can also be evaporated using electron beam heating.

Baker[46] R.F.-sputtered films in oxygen-doped atmospheres and observed mixed phases of α and β tantalum. For 1–10% oxygen in the sputtering atmosphere the resistivity was found to rise rapidly to a typical value for β tantalum alone whereas at lower partial pressures of oxygen the value corresponded to a mixture of both α and β phases. Baker suggested that β tantalum is an impurity phase which grows when tantalum is unable to accommodate an excessively large concentration of impurities (possibly water vapour) which is in general agreement with the model suggested by Westwood et al.[47]. Schauer and Roschy[48] R.F.-sputtered tantalum onto glass and onto Ta_2O_5-coated glass in pure argon, in argon/oxygen and in argon/nitrogen mixtures. For deposition in pure argon onto substrates which had been heated to about 300°C, a very pure α phase tended to be formed whereas for substrates which had not been heat treated (or had only been heated to about 150°C) only the β phase was formed. Thus they concluded that the α phase was formed as a consequence of the cleaner deposition conditions. In support of this conclusion it was found that β tantalum was always observed when sputtering in oxygen (unless very high partial pressures were used in which case Ta_2O_5 tended to be formed). Very low partial pressures of nitrogen promoted the growth of β tantalum whilst at higher pressures Ta_2N and TaN were found. Das[49] has discussed the role of substrate bias on the structure of tantalum films and argued that this parameter, too, is important in determining the impurity content of the films. Both oxygen and nitrogen are typical impurities in a vacuum system and in a plasma they are ionised to form N_2^+, N^+ and NO^+ ions which are accelerated towards a negatively biased substrate. Of these three species, it can be anticipated that NO^+ will bond well to tantalum and will tend to be sputtered off the film leading to an effective increase in the N/O ratio. For improved vacuum conditions the N/O ratio again tends to be greater so that, Das argued, negative bias is equivalent to higher vacua. This argument was supported by ion probe analysis which showed that negatively biased films contained up to five times more nitrogen than unbiased films. Since the biased films showed the tetragonal structure Das concluded that it was nitrogen rather than oxygen which is responsible for the formation of β tantalum. Here then is an indication of the confusion which exists regarding the growth of the β phase: these workers all believed that impurities were responsible but there is no agreement as to whether nitrogen or oxygen was the primary agent. On the other hand, there is also experimental evidence which suggests that the nature of the substrate is the dominant factor.

Feinstein and Hutteman[50] studied the formation of α and β tantalum on several substrates and discovered that:

(a) oxide substrates always nucleate β tantalum;

(b) non-oxide substrates, such as gold, always nucleate α tantalum;

(c) substrates which are non-oxidisable at low temperature but which oxidise at high temperature nucleate α and β tantalum respectively.

They claim that oxygen or OH ions are necessary for the formation of β tantalum but once nucleation has started the deposition conditions are unimportant and the initially nucleated phase will continue to grow. If α tantalum is nucleated on gold it will continue to grow even if a substantial quantity of oxygen is present in the sputtering atmosphere. Unbaked sapphire substrates nucleate β tantalum but the α phase grows when the substrates have been heat treated. The above evidence all suggests that β tantalum grows as a consequence of impure rather than very clean conditions and in particular the nature of the substrate is of vital importance. However there has been a considerable amount of evidence to the contrary and the work of

Feinstein and Gerstenberg[51], Cook[52] and several other authors all suggests that β tantalum forms under very clean conditions. In general, it is difficult to draw a positive conclusion from the literature and so the conditions necessary for the formulation of β tantalum still remain something of a mystery. It is fortunate that the majority of workers appear to be able to grow the material they wish even though there is a lack of consistency in the conditions specified. In any event, β tantalum has largely been superseded by associated compounds, some of which stem from the above investigations, and which have superior long-term ageing characteristics.

As seen above, the structure and even the chemical composition of tantalum films is influenced by the partial pressure of impurities in the sputtering gas. There has been a considerable volume of work on the effect of nitrogen-doped tantalum films and Gerstenberg and Calbick[53] have shown that as the partial pressure is increased so the resistivity rises to approximately 250 μΩ cm and thereafter remains constant. This effect is illustrated in Fig. 8 by the results of Berry et al.[54]. In principle, it would appear that the partial pressure of nitrogen should be held at a value corresponding to any position along the plateau region, and, in practice, it has been found that the most

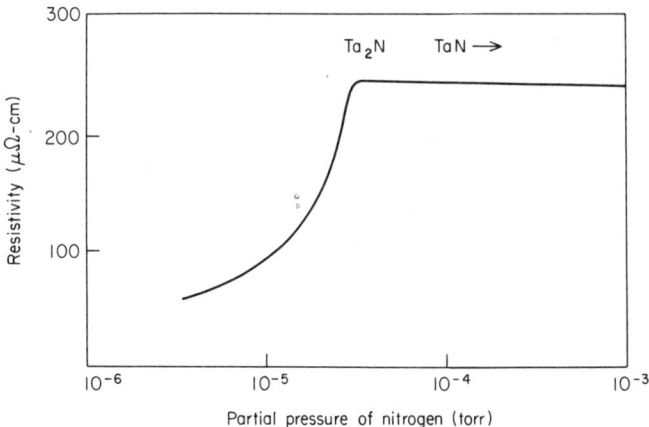

FIGURE 8. The figure shows the variation in resistivity of reactively sputtered Ta films. For partial pressures of nitrogen of approximately 5×10^{-4} torr the composition of the films corresponds to that of Ta_2N which exhibits excellent long term stability. A similar curve to that above is found for the variation of T.C.R. with the partial pressure of nitrogen. In this case the plateau occurs for a T.C.R. of about -75 p.p.m. $°C^{-1}$ (from Berry et al.[54]).

stable films are those consisting of Ta_2N. Nakamura et al.[55] for example, consider that the ratio of the sputtering current I_s to the partial pressure of nitrogen P_{N_2} is an important parameter in determining the film composition. They argue that since the number of sputtered tantalum atoms is closely related to I_s and the number of nitrogen atoms available for chemical combination is simply related to P_{N_2}, the ratio I_s/P_{N_2} must be indicative of a particular chemical composition. In practice, this was found to be the case and it was possible to normalise several curves of resistivity against partial pressure, for a series of sputtering currents, by plotting resistivity against I_s/P_{N_2}. This is a good example of the dependence of the electrical properties of films on a combination of deposition parameters. It does not seem unreasonable that an explanation along similar lines might account for the apparent inconsistencies in the growth of β tantalum. Nakamura et al. anodised their films in phosphoric acid, for protection, and then stabilised them in air at 270°C for 5 hours. Under stability tests, films corresponding to Ta_2N composition only drifted by about 0.1% after 3000 hours in air at 150°C. Since ageing takes place much more rapidly at elevated temperature, it is clear that the Ta_2N system is excellent as regards long term stability. This factor, taken together with a resistivity and T.C.R. approximately equal to the corresponding values for β tantalum suggests that Ta_2N meets most of the material requirements. In addition, Nakamura et al. found that it was readily possible to achieve zero T.C.R. or any other target value of T.C.R. they desired. For ladder networks, where precise values of resistance are required, they were therefore able to use Ta_2N resistors most advantageously. As the temperature coefficient of tantalum capacitors is approximately -225 ppm°C^{-1}, by using Ta_2N resistors with a T.C.R. opposite in sign to this value they were able to achieve R.C. networks with a temperature independent time constant. This system was therefore used very effectively in connection with oscillators for telephone systems.

It has been suggested that the long term drift of tantalum resistors is due to oxidation at the grain boundaries. Scaible and Maissel[56] have demonstrated that this effect can largely be overcome by depositing a thin layer of gold (onto the tantalum) which can then be diffused into the tantalum using heat treatment. The gold is thought to agglomerate in the grain boundaries and its presence prevents subsequent oxidation. These films were found to be extremely stable.

Further work has been carried out on other tantalum compounds and the results of Duckworth[57] at Ultra Electronics Ltd., on tantalum/aluminium alloy films is particularly important. Duckworth has shown that tantalum/aluminium alloy films formed by co-sputtering from a composite target are much more stable than tantalum alone. The behaviour of the resistivity and T.C.R. are very similar to nitrogen-doped tantalum films and maximum stability was found for an aluminium concentration in the range 7.5–14 at.%. At the optimum doping level the T.C.R. is approximately $-150\,\text{ppm}\,°\text{C}^{-1}$ for a sheet resistance of 110 ohms per square or less. These values refer to a fully aged film. In the initial phases of this work Corning (7059) aluminoborosilicate glass substrates were used but because of the somewhat irreproducible quality of this material (in particular the variable alkali content) high quality alumina is now preferred. The alloy film is deposited and aged in air for at least sixteen hours at 270°C. This procedure leads to the formation of a protective oxide which results in films of exceptional long-term stability having a nominal sheet resistance of 100 ohms per square. The resistor pattern is then defined using photoresist and unwanted film is removed using R.F. sputter etching. Contact is made to the resistor network via nichrome/gold films which are later thickened by dipping briefly in a solder bath. These films are superior to nitrogen-doped tantalum in that they can be trimmed more readily using anodisation. Because of their high stability and well-defined behaviour tantalum/aluminium resistors are now used in demanding applications where their qualities are a prerequisite.

3.1.2 Nickel/chromium alloys

It was shown earlier that one of the characteristics of the transition metal alloys is their unusually high resistivity due to their peculiar band structure. In bulk form the alloy nichrome (80% nickel 20% chromium) has a resistivity of $110\,\mu\Omega\,\text{cm}$ and a T.C.R. of approximately $85\,\text{ppm}\,°\text{C}^{-1}$ and it has been successfully used as a bulk resistor material. Consequently, it was an obvious choice for a thin film resistor particularly when one considers the additional sources of resistivity in a thin film.

The first nichrome resistors were deposited onto cylindrical ceramic substrates and these were cut in a spiral to trim to the required resistance. Control over the electrical properties depends on the method of preparation and since the vapour pressures of chromium and nickel are widely different the films generally tend to be chromium-rich. Individual manufacturers appear to have developed their own deposition procedures and in some ways the degree of control and reproducibility is surprisingly good. The bulk material may be either evaporated (or more accurately sublimed) from a wire, flash evaporated or evaporated using electron beam heating. Additionally, it may be R.F. or D.C. sputtered. In the latter cases the variation in the film composition is not so problematic although a great deal of care must be taken to obtain consistent deposition conditions. In general, oxidation of the chromium and the extent to which gas is trapped in the body of the film are believed to be of importance. The resistivity and T.C.R. are also often found to be thickness dependent and these variations have been correlated with a continually varying composition. The precise values of thickness, at which the best compromise between sheet resistance and T.C.R. occurs appears to vary slightly between various authors depending on the deposition conditions used. However a plateau region is often found corresponding to a resistivity of about 2–300 $\mu\Omega$ cm and a T.C.R. of approximately -50 ppm°C^{-1}; the films being upwards of 200 Å in thickness. The films tend to show a larger drift in air if unencapsulated, than do tantalum films, presumably because of the natural protective oxide formed on the latter material. In humid conditions under D.C. load the drift of nichrome resistors is usually associated with electrolytic action and to avoid this the films are generally stabilised in air at elevated temperature and then passivated with an inert material such as silicon monoxide. Zinsmeister[58], for example, reports that ageing after 1000 hours at temperatures up to 175°C amounts to a few tenths of a percent for unencapsulated resistors whilst for passivated films the drift is of the order of only a few hundredths of a percent. For most applications these values would be quite acceptable. Mayer[59] has also published related data on nichrome films which here obtained for a variety of film thicknesses, substrate temperatures and background gas compositions and these again appear to be in agreement with the other results reported above. Of course the important point in reporting these data is that nichrome film resistors are now in widespread commercial usage, as are tantalum films. Neither type of resistor appears to have an outstanding advantage over the other except, perhaps, that tantalum forms its own protective oxide quite naturally rather than requiring deposition of an additional layer of silicon monoxide. Ultimately it seems likely that the choice between tantalum and nichrome will be governed by detailed user requirements and the cost per item.

3.1.3 Other materials

Many other alloys have been tested in thin film form with a view to their use as resistors and a brief résumé of these will be given although their commercial exploitation is, to date, relatively limited in the face of the opposition from tantalum and from nichrome, both of which now appear to meet most user requirements.

Titanium, for example, has been evaporated from a tantalum boat (Hueber[60]) and the resultant films were found to have a similar sheet resistance to Ta$_2$N resistors. The T.C.R. was found to be thickness dependent and lay in the range ± 1000 ppm°C^{-1} for a sheet resistance range of 5–2000 ohms per square. Resistors with near zero T.C.R.'s were found for sheet resistances of about 50 ohms per square. However the stability under load was not good and after an initial decrease of 1%, further changes were of the order of 0.5% in 100 hours; the applied power being 0.5 W/cm^2. Reactive sputtering of titanium in an argon-oxygen mixture was carried out by Lackshmanan et al.[61] who also found a thickness dependent T.C.R. In this case they interpreted their results in terms of a parallel conduction process by proposing that the lower layers were highly oxidised and had a negative T.C.R. whilst the upper layers had a positive T.C.R. characteristic of metallic titanium. Once again it was possible to obtain films with near zero T.C.R.'s. Osadnik and Das[62] have prepared films of titanium nitride by evaporation of titanium in a partial pressure of nitrogen. These can be passivated and trimmed using anodisation and it is claimed that the resistors are of very high stability. However this is not entirely justifiable since stability tests appear to have only been carried out for approximately 100 hours at 25°C. Furthermore the upper limit of sheet resistance which was consistent with satisfactory performance was only 100 ohms per square. Chromium is another example of a potentially suitable resistor material and it has been studied by a large number of workers. As expected, the electrical properties are very sensitive to the deposition conditions and in particular the substrate temperature, rate of deposition and background pressure are found to have a radical effect. Because chromium tends to form an oxide bond with glass substrates, its adhesion is particularly good and it has often been used as a seeding layer upon which other materials, which would not themselves adhere well to glass, may be deposited, although as will be seen later this has disadvantages which have been overcome using alternative combinations of materials.

A very wide range of materials has been examined for use at very high temperature (5–600°C). Hemmer et al.[63] have deposited rhenium films onto silica substrates at a temperature of 500°C and have found a zero T.C.R. for a sheet resistance of about 250–300 ohms per square. However, to achieve good stability it was necessary to passivate the films with a layer of silicon monoxide and to hermetically seal the resistor. Clearly, this would not always be feasible. Neubert[64] has examined a wide range of alloys with a view to manufacturing thin foil strain gauges for use at 600°C in connection with long term creep measurements. He concluded that a number of platinum alloys may be suitable and an alloy based on 92% platinum, 8% tungsten was regarded as particularly promising. However, Preston[65] has D.C.-sputtered thin films from a composite target of platinum and tungsten and reports that the films were not stable at 600°C as required. In connection with this application Procter[66] claims that high stability films of nichrome can be produced on glass substrates and these, when passivated with several microns of R.F.-sputtered 7059 glass, remain almost totally free of drift even at 600°C. The sheet resistance and T.C.R. of the films, which were deposited by R.F. sputtering, are approximately 10 ohms per square and several hundred ppm°C^{-1} so that clearly the films would not be suitable as simple resistors. However, this work was undertaken with the aim of developing very stable high temperature strain sensors and as such the requirements are not the same as for a straightforward resistor. At the time of writing, development in this area is still continuing, and although the drift at 600°C appears to be acceptable, as might be expected, contact problems are rather severe and continue to present difficulties.

Wasa et al.[67] have sputtered composite targets of titanium/zirconium/ aluminium in a nitriding atmosphere and have obtained sheet resistances in the range 600–7800 $\mu\Omega$ cm for T.C.R.'s of ± 200 ppm°C^{-1}. These workers consider that the structure of the films is similar to that of cermets, which will be discussed in a later chapter, in which metallic conduction occurring in a ternary metallic alloy of titanium/zirconium/nitrogen takes place in parallel with semiconduction in an aluminium/nitrogen compound. Because the T.C.R. of the two processes is opposite in sign, the overall T.C.R. of the films is relatively small. Information on the stability of these films is rather limited but is claimed to be $<0.1\%$ after 1000 hours on test at 80°C without an applied load. Consequently, it would appear that in situations where a very high sheet resistance and low T.C.R. are required, films of the mixed nitride composition may be useful provided that extreme stability is not required. Calow et al.[68] have studied the production of thin film circuits for microwave applications which require that bonding of silicon semiconductors to gold conductors be carried out at temperatures of the order of 400°C at which a gold/silicon eutectic is formed. Although the reliability of the bonds is increased as a consequence of this technique, it has been found that conventional NiCr thin film resistors behave erratically at the high temperatures involved. However, they have found that sputtered films of zirconium diboride have excellent stability even at 400°C and consider that the NiCr could be replaced by this compound. The resistivity of ZrB_2 is approximately 300 $\mu\Omega$ cm compared with 100–150 $\mu\Omega$ cm for NiCr. Consequently, thicker films are used to achieve a particular sheet resistance and this alone leads to improved stability. The T.C.R. of ZrB_2 is negative and lies between 20–100 ppm°C^{-1} which is again something of an improvement on NiCr films. The data on stability indicate that not only do the ZrB_2 films drift less than do NiCr films but also such drift as there is occurs in a much more predictable manner. No information is available at the time of writing on the long term stability under load but since there is an equivalence between thermal and load stability, it is unlikely that ZrB_2 films will behave worse under load than NiCr films.

3.1.4 Stability and ageing of thin film resistors

One of the primary advantages of thin film resistors is that they can be made with very great precision. It is usual to deposit a film of a resistance somewhat less than that actually required and to trim the resistance up to the target value using one of the several available techniques (anodisation, thermal and mechanical trimming). In the case of anodisation of, for example, Ta_2N resistors the extraordinarily high degree of precision of 0.01% can be achieved. However, there is little point in manufacturing close tolerance resistors if the latter is not complemented by very high stability. This topic has been reviewed by Berry, Hall and Harris[54] particularly in relation to tantalum based resistors although many of the remarks apply equally well to nichrome.

Tantalum films react very readily with oxygen to form an oxide film, of great resilience to chemical attack, which is formed by anodisation or by heat treatment in air. For thicker oxide films the degree of protection will be increased. Consequently, heating the films will bring about an ageing process after which the resistors will have an improved stability. Experimentally it has been shown that the fractional change in resistance at a given temperature is proportional to the square root of time and this is in accordance with the well-known parabolic law of oxidation. The fractional change increases with temperature and a linear relationship has been obtained between log $\Delta R/R$ and the reciprocal of absolute temperature $(1/T)$. These results were all obtained on Ta_2N films which had been pre-aged at 200°C for five hours. Combining the time and temperature relationship gives:

$$\frac{\Delta R}{R} = \sqrt{\frac{t}{t_0}} \left(\exp -\frac{T_0}{T} \right) \qquad (39)$$

where t_0 and T_0 are constants. T_0 can of course be thought of as an activation energy divided by Boltzmann's constant and for Ta_2N the activation energy has the value 0.6 eV corresponding approximately to the oxidation energy for tantalum.

It has also been demonstrated experimentally that a D.C. load applied to a thin film resistor gives rise to a temperature gradient along its length. Thus the centre of the resistor ages faster under a D.C. load than any other part but a plot can still be made of log $\Delta R/R$ against log t and a family of parallel straight lines is obtained; each line corresponding to a particular applied electrical power. Once again these results show the parabolic dependence and suggest that a given electrical power is equivalent to an average temperature of the resistor. This was verified by depositing three resistor tracks very close to each other but with electrical power applied only to the two outside resistors, which, because of their proximity, produced a temperature gradient along the centre resistor almost identical to that along their own length. All three resistors exhibited almost identical ageing characteristics which demonstrated that Joule heating due to a D.C. load and external thermal heating have a similar effect on the films. Because the effect of the latter is much greater than the former it is possible to carry out an accelerated drift test and in fact Berry et al. show data which demonstrate the equivalence of 27 hours at 290°C and 1 year at 1 watt (i.e. there is an acceleration of 320). In practice it is most useful to have this technique available since it provides an excellent quality control check. It is also possible to use more severe pre-ageing procedures by increasing the temperature at which ageing is carried out. Although this results in greater proportional changes during pre-ageing, the change thereafter under less severe conditions is greatly reduced. For example, a film aged 250°C for five hours will change in resistance by approximately 2–3% whilst in a standard stability test thereafter (290°C for 24 hours) its resistance changes by approximately 4–5%. However if the pre-ageing schedule is 500°C for 0.5 hour then the subsequent change during the standard test is approximately 0.1%. After the latter schedule the change in resistance is virtually zero after about 1500 hours under 1 watt D.C. load. Although high temperature stabilisation therefore appears most advantageous there are drawbacks. Firstly, most of the advantage is lost if the resistors are anodise trimmed after stabilisation and secondly, oxidisation of the contact areas can occur which can prevent good soldering. Consequently, it is common practice to use a less severe pre-ageing schedule.

3.1.5 Substrate considerations

Although the above points concerning stabilisation are all very relevant to resistor systems used in practice, it is also evident that as well as oxidation at the upper film surface interactions which may take place at the film/substrate interface should also be taken into account. Duckworth[69] has pointed out that it is more appropriate to refer to the stability of a particular "system" rather than a particular type of film. Drift tests carried out on tantalum/aluminium films indicated that alumina and silica were significantly better than various glasses although, in some cases, the results on 7059 alumino-borosilicate glass were excellent. The main criticism of the latter is its apparently inconsistent alkali ion content which appears to concentrate at the film/substrate interface. In the case of silica substrates there is some evidence to suggest that free silicon is produced at the interface during ageing since, with inadequate pre-ageing, the resistor films showed a negative, rather than the much more common positive, drift. It was suggested that this was due to additional low resistance conduction paths. For these reasons Duckworth chose to use high quality alumina substrates. There are many factors to be taken into account in choosing a substrate and it is generally considered that no single material meets all the requirements. The reader is again referred to the appropriate chapters in the book by Berry et al.[54] and that by Maissel and Glang[70] for a detailed discussion of substrate properties.

These are summarised in the table below, reproduced from the former reference.

Desired property	Reason
Atomically smooth surface	Provide film uniformity
Perfect flatness	Provide mask definition
No porosity	Prevent excessive outgassing
Mechanical strength	Prevent breakage
Thermal coefficient of expansion equal to that of film	Prevent film stress
High thermal conductivity	Prevent heating of circuit components
Resistance to thermal shock	Prevent damage during processing
Thermal stability	Permit heating during processing
Chemical stability	Permit the unlimited use of process reagents
High electrical resistance	Provide insulation of circuit components
Low cost	Permit commercial application

In order to obtain good adhesion of the film to the substrate it is also necessary that the surface of the latter be rigorously cleaned. The cleaning procedure actually selected appears to vary considerably from one laboratory to the next but so long as standardised process is used it appears to be possible to obtain reproducible results.

In concluding this section on thin film resistors, it can be seen that, principally, tantalum based materials or nichrome are as good or better than the other materials discussed so that the latter are unlikely to be used in normal practice. However, there may occasionally be the situation where one wishes to exploit a particular material's property. The platinum/tungsten alloy discussed earlier has, for example, approximately twice the strain sensitivity of most metals and there may be circumstances where this is a useful advantage. The thickness of the films used as resistors is generally of the order of, or greater than, typical electron mean free paths so that in practice little use is made of the additional resistivity due to surface scattering. Even with nichrome where films as thin as 200 Å are used, it is believed that the mean free path is approximately 50 Å so that surface scattering is comparatively small.

The principal aim is to obtain films with a high sheet resistance and low T.C.R. and to meet these requirements the properties of β tantalum (which exists only in thin film form), Ta_2N and Ta/Al alloys, which coincidentally appear to have very similar electrical properties to β tantalum, and the intermetallic alloy nichrome whose properties are governed by its electronic structure are exploited. It is not therefore the inherent thin film property which is of advantage but rather the fact that resistors can be made in film form. As such they have the advantage of being small and therefore permit an increase in circuit complexity, can lead to improvement in reliability, have a lower assembly cost and have a higher tolerance than conventional resistors. The advantages of reliability and assembly cost would not be so evident were it not possible to reduce the number of interconnections by using thin film conductors, some aspects of which will be discussed in the next section. The stability of the resistors is related to the chemical processes occurring at both film surfaces and consequently both the nature of the ageing process (oxidation) and substrate requirements have been discussed.

3.2 Thin Film Conductors and Contacts

Since one of the major applications of thin film conductors is in connection with semiconductor integrated circuits, it is clearly a prerequisite that the contacts between conductor and semiconductor are ohmic. Integrated circuits consist of single crystal silicon substrates into which active areas have been formed by diffusion doping. It is between these areas that thin film resistors and capacitors and any externally added components with high reliability contacts must be made. No only must the contacts to the individual devices be reliable but also they must be capable of having external leads attached to their terminations, which for convenience generally consist of larger areas of the conductor material itself. In this section the nature of the metal/semiconductor contact, conductor materials used in practice and the factors affecting the reliability of conductor tracks and contacts will be given.

3.2.1 Metal/semiconductor contacts

The topic of metal/semiconductor contacts is very extensive and even now not completely understood. However it is possible to lay down a few guidelines for making low resistance ohmic contact to silicon devices even though the best materials in practice do not necessarily conform to the theoretical predictions. When contact is established between any two materials it is a requirement of band theory that the Fermi level is continuous across the interface. This is true of metal/semiconductor contacts. The mechanism of alignment of the Fermi levels is the flow of electrons into or out of the semiconductor, the direction of flow being governed by the relative magnitude of the two work functions. At the interface the bands of the semiconductor bend to accommodate the difference in the work functions and an accumulation or depletion of electrons is formed within the semiconductor. Although an accumulation layer is not problematic for electron transport it is quite possible that a depletion layer will cause rectification. However, this is not a general rule and ohmic contacts can still occasionally result even when a depletion layer is present, due to tunnelling through the barrier, provided that the latter is very thin. The problem is further complicated by the fact that the work functions of materials are very sensitive to the state of their surface, on the crystal orientation (Herring and Nichols[71]), by the presence of an oxide film on the semiconductor (particularly in the case of silicon) and by the existence of surface states whose effect is unpredictable. For a complete review of metal/semiconductor contacts the reader is referred to Tauc[72] or to Chapter 4 of this book for a fuller discussion than that given here.

3.2.2 Requirements of contacts

For obvious reasons interconnections between active and passive devices on a silicon surface are generally formed from the same material as the contacts. Blech et al.[73] have summarised the properties required:

(i) The materials should be able to be deposited in thicknesses up to 1 μm and must be capable of being etched to line widths of about 10 μm using photolithography.
(ii) The resistivity of the film must be 4 $\mu\Omega$ cm or less so that very low sheet resistances can be achieved as required for some integrated circuit applications.
(iii) The metal must give low resistance ohmic contact to both n- and p-type silicon.
(iv) The metal must be capable of withstanding current densities of the order of 3×10^5 A cm^{-2}.
(v) The metal film must adhere well to silicon and silicon dioxide and be compatible with externally bonded leads.
(vi) The metal must be chemically stable.
(vii) The film must be scratch resistant.
(viii) The metal must remain stable during the high temperatures used during processing the integrated circuit.
(ix) Assuming that the material chosen can meet the above requirements it must still be economically acceptable on the grounds of deposition and fabrication.

In practice the choice of a suitable contact and interconnection material is something of a compromise since no single material can meet all the above requirements.

3.2.3 Choice of materials

Aluminium is the most widely used contact material since it appears to meet all but two of the above requirements. It can be evaporated and condensed onto the integrated circuits and then photolithographically processed to give the required interconnection pattern. The silicon slice and aluminium film are then heat treated to bring about inter-diffusion and alloying which results in good adhesion and low contact resistance to the silicon. Because aluminium tends to be rather prone to scratching and corrosion, alternative systems have been proposed. Molybdenum can be sputter-deposited onto the silicon wafer to which it both adheres well and forms an ohmic contact. A gold film of perhaps 1 μm thickness can then be deposited onto the molybdenum to provide the low sheet resistance required. This system can withstand relatively high temperatures and only fails when gold diffuses through the molybdenum to form a eutectic compound with silicon. Occasionally a thin layer of platinum or aluminium is deposited prior to deposition of the molybdenum to ensure ohmic contact. It is claimed that this system has a higher scratch resistance and is less prone to degradation due to high current effects than aluminium alone. Another alternative is the titanium/platinum/gold system. Initial contact is made by depositing a thin layer of platinum onto the silicon and this is then heat treated to form platinum silicides. A thin layer of titanium provides good adhesion to the silicides whilst the platinum provides a diffusion barrier between the gold and titanium layers. This system has a high corrosion resistance and is very suitable up to 350°C. Both the above systems, however, tend to be rather more costly than aluminium alone.

In addition to the application of thin films to silicon integrated circuits a large number of thin film circuits are available on ceramic or glass substrates to which discrete devices are attached. In this case the resistors are generally formed from one of the systems discussed earlier (whereas with silicon integrated circuits the resistors are formed by suitable doping of appropriate areas of silicon). The advantage of the thin film resistor of course is the precision with which it can be fabricated. Interconnections on thin film circuits present similar demands to those made by silicon integrated circuits and in some cases the same materials can be used. Mattox[74] has reviewed the subject of providing metallisation to oxide substrates and has considered a number of factors which may influence performance and reliability. He suggests that improved adhesion is likely to result from the use of sputtering because impurities tend to be sputtered off the substrate surface and because local defects which act as nucleating centres can be created. On the other hand gases can be incorporated into the film during sputtering and the film properties can be affected if small bubbles agglomerate to form large voids. Similarly when using interconnections based on two- or three-layer systems care must be taken to ensure that weaknesses do not develop due to the formation of undesirable intermetallic compounds. Mattox states that of the various metallisation systems available, the titanium/platinum/gold or titanium/palladium/gold combinations are most stable for the same reasons as given earlier. If it is required to solder to the contacts then the combination nichrome/copper/palladium is very suitable. This system is not prone to the effect known as scavenging in which preferrential dissolution of one of the components of the metallisation into the solder occurs. Mattox also concludes that chemical etching may ultimately be replaced by sputter etching in which it is possible to achieve far greater line resolutions. Indeed, one example of the practical application of sputter etching as used for the delineation of resistor and conductor patterns has already been quoted.

3.2.4 Reliability of contacts and interconnections

In order for contact and interconnection materials to be acceptable it is necessary that they remain able to meet most of the requirements outlined earlier for a very long period of time. In other words the advantages inherent to integrated circuit technology cannot be exploited unless the conductors and bonds are completely stable. There are a number of causes of unreliability of conductors and these will be discussed now.

Firstly, it is possible for the resistance of the metal/semiconductor contact to increase due to silicon dissolving in aluminium (for example). Secondly, the semiconductor junction can be shorted out by diffusion of the metal through the emitter. Thirdly, the resistance of the conductor itself may increase with time for any one of a number of reasons:

(a) Aluminium can react with insulating layers of silicon dioxide and as a consequence its sheet resistance increases. However rather high temperatures are required to promote this reaction and in general it is not problematic.

(b) In interconnection systems using two or more components interdiffusion of metals can again lead to an increase in sheet resistance. It is this mechanism which gives rise to the well known "purple plague". Fortunately it can to a large extent be prevented using diffusion barrier layers such as platinum or palladium.

(c) Unless the circuits are hermetically sealed, chemical action can degrade conductors although the primary mechanisms of oxidation and hydration are relatively small.

(d) Electro-transport is a mass transfer mechanism by which not only can a significant increase in resistance take place but also a total open circuit can occur. In circuits where high current densities are used ($> 10^5$ A cm^{-2}) this failure mechanism can be very significant. The effect itself is due to the transfer of momentum from electrons to the atoms of the conductor which as a consequence tend to migrate towards the anode. Whenever an atom moves from its location in the conductor a vacancy is created and the tendency is for many vacancies to agglomerate and eventually this causes conductor failure. The topic has been reviewed by d'Heurle and Rosenberg[75].

Attardo et al.[76] point out that the effect depends on the difference between the rate of flow of vacancies into and out of a given point. If the flux difference is positive then accumulation of vacancies occurs. Since mass transport tends to occur along grain boundaries Attardo argued that the rate of accumulation and the mean time to failure (M.T.F.) (a more widely used term) must depend on a number of macroscopic structural parameters. In particular it is now well established that the conductor length and width, the mean grain size and the standard deviation about the mean, the crystallographic orientation and the nature of the grain boundaries are of special significance. It was considered that three types of structural inhomogeneities could cause flux divergence. These were due firstly to variations in grain boundary mobility. The mobility of ions in grain boundaries depends on the local atomic structure of the grain which in turn depends on the misorientation of the boundary. Hence in a polycrystalline film having grains of varying misorientation, flux divergence can occur. Secondly, they were due to variations in mobility caused by the variation in the angle between the grain boundaries and the applied field. That is, it is possible for two grain boundaries of differing misorientations to have the same mobility but because they are at different angles to the applied field, ionic transport along them will not occur at the same rate. Thirdly, they were due to grain size divergence; conduction electrons passing from large to small grains can sweep ions from the interface and cause the accumulation of vacancies. Attardo claims that the latter mechanism has been identified as the primary cause of failure in large-grained well-oriented films. A model was developed which combined the three mechanisms and from a knowledge of the statistical distribution associated with each, it was possible to simulate the behaviour of conductors in a range of conditions. The main conclusions of this work were that:

(i) M.T.F. decreases with increasing conductor length.
(ii) Standard deviation of M.T.F. distribution decreases with increasing conductor length.
(iii) M.T.F. increases linearly with conductor width (at least in the range 0.1–1 mm).
(iv) M.T.F. increases with mean grain size if standard deviation about the mean is constant.
(v) M.T.F. decreases with standard deviation of grain size for a constant value of the mean.
(vi) M.T.F. distribution closely approximates to log normal.

Although one or two of the above statements may appear to be self evident, the work of Attardo *et al.* is important because it enables predictions about M.T.F. to be made purely on the basis of a measurement of grain size distribution in a conducting film. Anderson[77] has arrived at a similar conclusion to Attardo *et al.* and claims that for high current densities the titanium/platinum/gold system should be used in conjunction with annealing to create larger mean grain size.

Failure of circuits can also occur due to contact resistance between conductors and external leads. Once again this is due to inter-diffusion and the effect occurs when the contact areas are of different material to the leads. It is also important that the conductors are deposited under conditions which do not lead to their being under excessive stress due, for example, to the effects of thermal mismatch. If stress in the films is too large they can actually flake from the silicon or silicon dioxide and cause open circuits. In addition to electrotransport, diffusion of atoms in a conductor also occurs in the presence of a temperature gradient. This topic has been reviewed by Hehenkamp[78]. Johns and Blackburn[79] have recently studied the effect of grain size on thermomigration, and have established that mass transport always occurred in the same direction as that of the flow of heat. By comparing the relative magnitudes of the effect in single crystal and polycrystalline lead, they were able to establish that the mobility of migrating atoms was much greater along grain boundaries than in the volume of discrete crystallites. Since this conclusion is very similar to high current electrotransport it is clear that every effort should be made to grow films of large crystallites.

Whilst the above causes of failure can all be significant it is clear that they can be made small by careful choice of materials, use of low current densities and ensuring a large grain size of the conductor by suitable heat treatment. Failure due to inter-diffusion and the subsequent formation of high resistance or mechanically weak alloys occurs more rapidly at elevated temperature but is generally not to be anticipated in normal operational conditions. In general the use of thin film conductors in both integrated and thin film circuits can be expected to give increases in reliability for the reasons given earlier but clearly it is necessary to be aware of the circumstances under which failure can occur. In practice failure of circuits can occur not only due to conductor failure but also because of shortcomings in passivation or encapsulation.

3.3 Thin Film Inductors and Capacitors

Because of obvious economic advantages a great deal of effort has been devoted to the development of entire circuits consisting only of thin film components. It has often been said that the reason for the dominance of the silicon integrated circuit over the thin film circuit is the lack of a reliable thin film active device. A great deal of progress in this area has been made however, and this subject will be discussed in a later chapter. Inductors and capacitors can readily be fabricated in thin film form, although the extent to which they are used in practice appears to be rather limited. Thin film capacitors essentially consist of a metal/dielectric/metal sandwich. These are discussed in greater detail in Chapter 4. The metal layers are generally deposited by conventional means and are compatible with circuit interconnections. The choice of the dielectric and its thickness depend of course on a number of factors including the values of capacitance required, but the bulk of the work on thin film dielectrics has concerned silicon, aluminium and tantalum oxides. The metal layers are thin film conductors and it is desirable that they have as low a sheet resistance as possible in order to achieve a reasonably high Q. In thin film and integrated circuits, crossovers of conductors constitute capacitances and thin film circuit designers normally attempt to minimise the number of such junctions. Crossovers deserve special consideration in their own right and Siddall[80] considers that they should be regarded as vital components. It is not, however, appropriate to discuss the properties of crossovers here.

Thin film inductors can be made by depositing a low sheet resistance conductor in a spiral track; the centre connection being made using an insulated crossover. The rather restricted use of thin film inductors arises partly from the fact that values of inductance are limited to a few microhenries because of geometrical considerations. Inductances up to a few tens of microhenries can be obtained by depositing the conductor spiral onto a ferrite substrate (Gleason[81]). However it is unlikely that values much greater than this will be possible because of size restrictions. Because miniature inductors wound on ferrite cores are available up to a few millihenries it is more customary to attach these to thin film circuits separately, rather than to use thin film inductors. In the case of silicon integrated circuits it is possible to simulate inductance using active devices in a region of their characteristics where they exhibit negative resistance.

Thin film inductors and capacitors are examples of the combined use of thin film conductors and insulators. One further example which has had limited application in research topics rather than a major impact on the electronics industry, is the thin film thermocouple. The measurement of the surface temperature of a substrate is very difficult to carry out accurately and since this quantity plays a major role in determining the structure and electrical properties of a thin film, many workers have examined the possibilities of using thin film thermocouples. This device is very simply formed from two metal (or semiconducting) films chosen to give a suitable thermoelectric output. Because the thermal mass of the films is extremely low, their presence on a substrate should not distort the thermal distribution and, furthermore, they should react very rapidly to transients. On the other hand, if the substrate is at a high temperature then the thermal distribution could be affected by virtue of the fact that the emissivity of the substrate and film will, in general, not be equal. For a complete review of the literature on this topic the reader is referred to the appropriate chapter in the book by Chopra[10]. Bastius and Hentschke[82] in their studies of the behaviour of noble metal alloy thin film resistors have concluded that, because of their thermal behaviour, the resistors can provide protection against overload of miniature circuitry. Indeed, from examination of failed circuits, it would appear that film resistors have been providing this service for some time albeit without this necessarily being the intention of the designers. These workers consider that these resistors are particularly well suited for use in the output stages of telecommunications equipment, and for protection of relays, low voltage equipment, etc. during overloads of 15–100 W. However the turn-off time is measured in seconds or even minutes and the failure mechanism appears to be more like that of thermal drift discussed earlier rather than that of fast thermal rupture (taking perhaps tens of milliseconds) which would normally be associated with fusing. It is, however, an interesting application and one which does not previously appear to have been exploited.

3.4 Thin Film Strain Gauges

A conventional metal foil strain gauge operates on the principle that a change in the dimensions of a conductor or resistor causes a change in its resistance. In addition, there is a change in the resistance due to the effect of strain on the electron mean free path. It is a relatively straightforward matter to calculate the strain in the device from its change in resistance. Strain gauges are generally bonded to the surface of the test piece using a suitable adhesive. Unfortunately the use of a glue layer limits the accuracy of measurement and also limits the range of temperature over which the gauge can be used. The thin film gauge was originally proposed by Ball[83] with a view to overcoming these difficulties. Witt[84] has recently reviewed the subject of the electromechanical properties of thin films, particularly as related to the resistance strain gauge. Thin film strain gauges may be formed from discontinuous, semi-continuous or totally continuous films although for the present only the third type will be considered.

The resistance of a plane parallel foil can be written in the usual form as:

$$R = \frac{l}{\sigma b d} \quad \therefore \quad \frac{\partial R}{R} = \frac{\partial l}{l} - \frac{\partial \sigma}{\sigma} - \frac{\partial b}{b} - \frac{\partial d}{d}$$

where l, b and d are the length, breadth and thickness respectively of the conductor, so that for a longitudinal strain $\partial l/l$ a gauge factor or strain sensitivity can be defined by:

$$\gamma = \frac{\partial R/R}{\partial l/l} = -\frac{\partial \sigma/\sigma}{\partial l/l} + 2v$$

where v is Poisson's ratio for the film material. The term representing the fractional change in conductivity with strain is, for a bulk material, due to the change in the electron mean free path due to phonon scattering.

In the case of a thin film it is due not only to this effect but also to a change in the degree of surface scattering. The latter effect may, or may not, be significant depending on whether the film thickness is greater or less than the electron mean free path. Meiksin[85] has analysed the effect of surface scattering on the strain sensitivity and has shown that in general the gauge factor of a thin film will be less than that of a foil of the same material. The factors are however not problematic. It is usual to fabricate gauges from high sheet resistance alloys, rather than from a single metal, as these have several advantages. Firstly, the resistance of the gauge is compatible with conventional measuring circuitry. Secondly, alloys of the transitional metals have relatively small T.C.R.'s so that it is somewhat easier to discriminate between changes in resistance due to temperature and those due to strain. Thirdly, these alloys have a gauge factor of $\gamma \simeq 3$–4 whereas the simple metals exhibit values of about 1.5–2 (this is probably due to the complex band structure of the transition metals and their alloys). Fourthly, resistors of nichrome, for example, have excellent stability so that in very long term measurements changes in strain do not become confused with resistor drift. As with foil gauges, it is usual to fabricate film gauges in the form of a relatively long meandering path so that the gauge can exhibit good directionality with a poor transverse strain response. Witt lists ten potential applications for thin film resistance gauges and considers that thin film technology is likely to make a very large impact on the strain transducer industry. It has been independently confirmed by the author that several companies are actively developing devices based on thin film strain gauges and it would indeed appear that several new products will appear on the market within the next few years.

3.5 Summary

In this chapter it has been seen that thin metal and alloy films exhibit size and structure effects which change their bulk physical properties. The size effect, or the inherent property, does not yet appear to have been exploited in a practical application although, as pointed out, it can be most useful in determining fundamental physical quantities. The improvements in vacuum technology and deposition procedures have led to a much greater degree of control than was possible only a few years ago and there is no doubt that fundamental research on metal and alloy films is still continuing with a view to gaining a better understanding of the solid state behaviour of thin metals.

The practical advantages of thin metal/alloy films have been discussed by a number of authors and it is clear that considerations such as size and mass, which in space and military applications are vital, are no longer of paramount importance. Rather, the commercial success of thin film technology is due to the applicability of large scale production techniques which makes the devices economically viable. Of course, there are other situations in which the vital factors are not the cost but the quality and reliability brought about by thin films. Biological implant devices represent one such example and this is a field in which a considerable expansion can be expected. There are other examples; the close tolerance with which thin film networks can be made represents a further major advantage over other techniques. For example, A–D convertors require very accurately known ratios of resistances and thin film networks can be ideal for this application.

It has also been seen that in the fields of temperature and strain measurement the use of thin film devices can be an outstanding advantage. In the former case the sensor can be made very small and of vanishingly small heat capacity. Whole arrays can be deposited on a given surface so that if required a whole temperature distribution can be determined. In the case of strain measurement it is possible to deposit a passivating layer directly onto the workpiece and then deposit the strain-sensitive element. Hysteresis and other inaccuracies can thus be avoided. At least one major strain gauge manufacturer already produces thin film devices and several others are committed to development programmes.

References

1. P. Debye, *Ann. Physik*, **39**, 789, (1912).
2. J. M. Ziman, "Electrons and Phonons", Clarendon Press, Oxford, (1960).
3. J. O. Linde, *Ann. Physik*, **15**, 219, (1932).
4. A. W. Overhauser and R. L. Gorman, *Phys. Rev.*, **102**, 676, (1956).
5. N. F. Mott and H. Jones, "The Theory of the Properties of Metals and Alloys", Oxford Univ. Press, London, (1936).
6. C. S. Barrett, "Structure of Metals", McGraw Hill, New York, (1952).
7. J. J. Thomson, *Proc. Camb. Phil. Soc.*, **11**, 120, (1901).
8. K. Fuchs, *Proc. Camb. Phil. Soc.*, **34**, 100, (1938).
9. D. S. Campbell, In "The Use of Thin Films in Physical Investigations", ed. J. C. Anderson, Academic Press, New York, (1966).
10. K. L. Chopra, "Thin Film Phenomena", McGraw Hill, New York, London, (1969).
11. F. Savornin, *C. R. Acad. Sci.*, **248**, 2458, (1959).
12. A. A. Cottey, *Thin Solid Films*, **1**, 297, (1968).
13. J. E. Parrott, *Proc. Phys. Soc.*, **85**, 1143, (1965).

14. J. L. Olsen, *Helv. Phys. Acta.*, **31**, 713, (1958).
15. M. Ya. Azbel and R. N. Gurzhi, *Sov. Phys. JETP*, **15**, 1133, (1962).
16. J. Bass, *Advances in Physics*, **21**, 431, (1972).
17. A. F. Mayadas and M. Shatzkes, *Phys. Rev. B.*, **1**, 1382, (1970).
18. E. E. Mola and J. M. Heras, *Thin Solid Films*, **18**, 137, (1973).
19. V. Vand, *Proc. Phys. Soc.*, **55**, 222, (1942).
20. E. T. S. Appleyard and A. C. B. Lovell, *Proc. Roy. Soc.*, (*London*), **A158**, 718, (1937).
21. H. Mayer, In "Structure and Properties of Thin Films", eds., C. A. Neugebauer, J. B. Newkirk and D. A. Vermilyea, Wiley, New York, (1959).
22. R. Nossek, *Z. Phys.*, **142**, 321, (1955).
23. D. G. Worden and G. C. Danielson, *J. Phys. Chem. Solids*, **6**, 89, (1958).
24. F. W. Reynolds and G. R. Stilwell, *Phys. Rev.*, **88**, 418, (1952).
25. T. J. Coutts and G. G. Matthews, *Proc. Phys. Soc.*, **90**, 1175, (1967).
26. T. J. Coutts, *J. Phys. D.*, **1**, 1071, (1968).
27. J. Borrajo and J. M. Heras, *Thin Solid Films*, **18**, 267, (1973).
28. A. F. Mayadas, *J. Appl. Phys.*, **39**, 4241, (1968).
29. E. J. Gilham, J. S. Preston and B. E. Williams, *Phil Mag.*, **46**, 1051, (1955).
30. M. S. P. Lucas, *Appl. Phys. Letts.*, **4**, 73, (1964).
31. K. L. Chopra and M. R. Randlett, *J. Appl. Phys.*, **38**, 3144, (1967).
32. K. L. Chopra, L. C. Bobb and M. H. Francombe, *J. Appl. Phys.*, **34**, 1699, (1963).
33. D. C. Larson and B. T. Boiko, *Appl. Phys. Letts.*, **5**, 155, (1964).
34. R. V. Isaeva, *J. E. T. P. Letts.*, **4**, 209, (1966).
35. J. Niebuhr, *Z. Phys.*, **132**, 468, (1952).
36. I. G. Young and C. W. Lewis, *Trans. Tenth Natl. Vac. Symp.*, Macmillan, New York, (1963).
37. L. Holland and G. Siddall, *Vacuum*, **3**, 375, (1952).
38. A. Singh, *J. Appl. Phys.*, **45**, 1908, (1974).
39. R. P. Huebener, *Phys. Rev.*, **136**, A1740, (1964).
40. A. P. Thakoor, R. Suri, S. K. Suri and K. L. Chopra, *App. Phys. Letts.*, **26**, 160, (1975).
41. E. H. Sondheimer, *Phys. Rev.*, **80**, 401, (1950).
42. R. G. Chambers, *Proc. Roy. Soc.* (*London*), **A202**, 378, (1950).
43. D. K. C. MacDonald and K. Sarginson, *Proc. Roy. Soc.* (*London*), **A203**, 223, (1950).
44. K. Försvoll and I. Holwech, *Phil. Mag.*, **9**, 435, (1964).
45. D. A. McLean, N. Schwartz and E. D. Tidd, *Proc. IEEE*, **52**, 1450, (1964).
46. P. N. Baker, *Thin Solid Films*, **6**, R57, (1970).
47. W. D. Westwood and F. C. Livermore, *Thin Solid Films*, **5**, 407, (1970).
48. A. Schauer and M. Roschy, *Thin Solid Films*, **12**, 313, (1972).
49. G. Das, *Thin Solid Films*, **12**, 305, (1972).
50. L. G. Feinstein and R. D. Hutteman, *Thin Solid Films*, **16**, 129, (1973).
51. L. G. Feinstein and D. Gerstenberg, *Thin Solid Films*, **10**, 79, (1972).
52. H. C. Cook, *J. Vac. Sci. Technol.*, **4**, 80, (1967).
53. D. Gerstenberg and C. J. Calibick, *J. Appl. Phys.*, **35**, 402, (1964).
54. R. W. Berry, P. M. Hall and M. T. Harris, "Thin Film Technology", D. van Nostrand, Princeton, (1968).

55. M. Nakamura, M. Fujimori and Y. Nishimura, *Jap. J. Appl. Phys.*, **12**, 30, (1973).
56. P. M. Scaible and L. I. Maissel, *Trans. Ninth Natl. Vac. Symp.*, Macmillan, New York, (1962).
57. R. G. Duckworth, *Thin Solid Films*, **26**, 77, (1975).
58. G. Zinsmeister, Personal Communication, (1975).
59. G. Mayer, paper presented at *Semi-conductor International 1975*, Wiesbaden, March (1975).
60. F. Huber, *IEEE Trans. Components Parts*, **CP-11**, 38, (1964).
61. T. K. Lackshmanan, C. A. Wysocki and W. J. Slegesky, *IEEE Trans. Component Parts*, **CP-11**, 14, (1964).
62. S. J. Osadnick and M. B. Das, *Microelec. Reliability*, **11**, 71, (1972).
63. F. J. Hemmer, C. Fieldman and W. T. Layton, *Proc. Natl. Elec. Conf.*, **20**, 201, (1964).
64. H. K. P. Neubert, "Strain Gauges: Kinds and Uses", Macmillan, London, (1967).
65. D. W. Preston, paper presented at *Congrès International sur les Couches Minces*, Cannes, (1970).
66. E. Procter and T. Strong, paper presented at *Annual Conf. B. S. S. M.*, Warwick, (1974).
67. K. Wasa and S. Hayakawa, *Thin Solid Films*, **10**, 367, (1972).
68. J. T. Calow, K. G. Knauff and G. Luttke, *IEEE Conference on Hybrid Microelectronics*, Canterbury, (1974).
69. R. G. Duckworth, *Thin Solid Films*, **10**, 337, (1972).
70. R. Brown, In "Handbook of Thin Film Technology", eds., L. I. Maissel and R. Glang, McGraw Hill, New York, (1970).
71. C. Herring and M. H. Nichols, *Rev. Mod. Phys.*, **21**, 185, (1949).
72. J. Tauc, "Photo and Thermoelectric Effects in Semi-conductors", Pergamon Press, Oxford, (1962).
73. I. Blech, H. Sello and L. V. Gregor, In "Handbook of Thin Film Technology", eds., L. I. Maissel and R. Glang, McGraw Hill, New York, (1970).
74. D. M. Mattox, *Thin Solid Films*, **18**, 173, (1973).
75. F. M. d'Heurle and R. Rosenberg, *Phys. Thin Films*, **7**, 257, (1973).
76. M. J. Attardo, R. Rutledge and R. C. Jack, *J. Appl. Phys.*, **42**, 4343, (1971).
77. J. C. Anderson, *Thin Solid Films*, **12**, 1, (1972).
78. Th. Hehenkamp, "Vacancies and Interstitials in Metals", North-Holland, Amsterdam, (1969).
79. R. A. Johns and D. A. Blackburn, *Thin Solid Films*, **25**, 291, (1975).
80. G. Siddall, In "Thin Film Microelectronics", ed., L. Holland, Chapman and Hall, London, (1965).
81. F. R. Gleason, *Proc. Natl. Elect. Conf.*, **20**, 197, (1964).
82. S. Bastius and P. Hentschke, *Components Rep.*, **9**, 39, (1974).
83. L. M. Ball, U.S. Patent 2556132, (1961).
84. G. R. Witt, *Thin Solid Films*, **22**, 133, (1974).
85. Z. H. Meiksin, *Thin Solid Films*, **1**, 61, (1964).

Chapter 4

Thin Film Dielectrics

D. S. Campbell

Department of Electronic and Electrical Engineering,
Loughborough University of Technology,
Loughborough, England

1. Introduction 113
2. Deposition Methods 114
   2.1 Introduction 114
   2.2 Evaporation 115
   2.3 Sputtering 116
   2.4 Thermal Growth 118
   2.5 Anodisation 118
   2.6 Vapour Phase Deposition 119
   2.7 Solution Deposition 120
3. Electrical Properties 122
   3.1 Introduction 122
   3.2 Basic Properties 124
4. Applications of Dielectric Films 143
   4.1 Capacitors and Related Applications 144
   4.2 Insulators 150
   4.3 Conductors 152
5. Summary 154
   References 157

1. Introduction

The aim of this chapter is to give a summary of the position regarding the preparation, properties and electronic applications of dielectric films. The preparation of films has already been extensively covered in an earlier chapter of this book and therefore all that needs to be done in this area is to discuss those techniques which are used for deposition of dielectrics. The properties and uses of dielectric films have both been widely reviewed and it is unnecessary to include this material. Therefore, it is felt best to highlight the main areas which have been studied, and to reference material which the reader can use in order to obtain detailed information. Major reviews have been given by Campbell and Morley[1], Harrop and Campbell[2], Simmons[3,4], Jonscher and Hill[5], and Hill[6].

Dielectric films are also widely used for optical purposes[7,8], and these are discussed later in this book.

2. Deposition Methods

2.1 Introduction

Deposition methods for dielectric films are many and varied and a summary of the techniques which can be used is given in Table 1. All these techniques are capable of producing films 1 µm thick or less. Other techniques, such as glazing, electro-phoretic deposition, flame spraying, painting and screen

Table 1
Summary of Preparation Methods
Applicable to Dielectric Films

Technique			Major references
Evaporation (using all source types)			(2) (9) (10)
	Reactive		(11)
Sputtering	Reactive		(2) (9) (12) (13) (11)
	R.F.		(2) (9) (12) (13)
	Ion implantation		(14)
	Plasma reactions		
		Gaseous anodisation	(11) (13) (15) (16) (17)
		Assisted thermal growth	(11) (13) (15)
		Polymerisation	(11) (13) (15)
		Assisted vapour phase	(11) (13) (15)
Thermal growth			(2) (11) (16)
Anodisation			(11) (16) (17)
Vapour phase—Polymerisation (E.B. and U.V.)			(2) (11) (16) (18)
	Oxidation		(2) (11) (16) (18)
	Nitriding		(16)
Solution Deposition—Oxides			(19)
—Langmuir			(20) (21)
—Polymer			(22)
—Casting			(23)
—Liquid chemical reaction			(24)

printing can be used to produce thick films. These techniques do not come within the scope of this chapter and readers interested in this area will find general summaries in "Science and Technology of Surface Coating"[25], and sections dealing with the screen printing of dielectric layers in the "Handbook of Thick Film Hybrid Microelectronics"[26].

2.2 Evaporation[9]

Evaporation techniques are extensively used in the preparation of thin dielectric layers and a very large number of materials can be evaporated, either directly or in an oxygen atmosphere (reactive evaporation), under extremely clean conditions in the vacuum chambers. Rates of evaporation can vary over very wide limits and are dependent on the type of source used. Figure 4 in Chapter 2 shows a particular source developed by Drumheller[27] for the evaporation of silicon monoxide. The main point of the system is that no macroscopic particles are capable of being ejected from the source on to the substrate and therefore the film should not suffer from defects of this type.

A further technique which has found considerable application in the deposition of dielectric films, is that of flash evaporation which is discussed in detail in Chapter 2 (see Fig. 5 for a diagram of a typical source originally developed for deposition of nickel-chromium alloys[28], but which has also been used to prepare dielectric films, e.g. complex oxides such as barium titanate[29]). The advantage of this type of source is that the composition of the final film can be the same as that of the original starting material, and fractionation of the different constituents can be eliminated. The major disadvantage is that the material can be evaporated so quickly from the hot filament, that "spitting" can occur and particles, although not necessarily landing on the substrate, can be ingested into the vacuum pumping apparatus unless great care is taken to prevent this from happening, by the use of baffles, etc.

Recently, laser beam evaporation has been found to be useful for the preparation of polymer dielectric films[23]. A wide variety of polymers has been prepared using this technique, including such complicated materials as polyvinylidene difluoride and polystyrene.

Reactive evaporation[11,30], as its name implies, involves the deposition of films in a reactive atmosphere such as oxygen or nitrogen. Materials such as silicon monoxide and tantalum oxide have been successfully deposited using this technique. One such system, used for the preparation of titanium dioxide films, involves the evaporation of titanium monoxide in an oxygen atmosphere of between 10^{-3} and 10^{-4} torr, using a source to substrate distance of 40 cm and a substrate temperature of 300°C. Kerner[31] has recently published his work on the kinetics of active evaporation with particular regard to alumina, silicon monoxide and chromium oxide. He has shown that in the case of silicon monoxide, one hundred times the amount of oxygen must impinge on the surface, compared with that which would be required if every oxygen atom actually reacted. Similar figures were found for aluminium, but in the case of chromium oxide, the necessary impingement ratio was only 2 : 1.

2.3 Sputtering[13]

Glow discharge sputtering has been used for a long time and Fig. 10 of Chapter 2 shows a typical apparatus which can be used. A discharge is maintained at a pressure of approximately 10^{-2} torr between the anode and the cathode, and material is ejected from the bombarded cathode and deposited on a suitably placed substrate. The apparatus mentioned above was designed for the deposition of titanium, but if the argon atmosphere is replaced by an argon-oxygen mixture, then a reactive sputtering system is obtained which allows the deposition of an oxide on the substrate. The site of the oxidation process has been discussed by various workers. In the case of sputtering of tantala[32], it has been shown that there is a critical oxygen pressure (6×10^{-5} torr for a 1 cm^2 target), below which tantalum metal was sputtered and the oxidation reaction occurred at the substrate and above which tantalum oxide was sputtered from a tantalum surface as a result of oxidation at the target.

R.F. sputtering[13] is an important process for preparing dielectric films and two typical electrode systems for R.F. sputtering are shown in Fig. 1. Not only oxides but also polymers are now being deposited using this technique[33].

Ion implantation with O_2^+ ions can be used as a method of preparing dielectric films[14]. The rate of oxidation of the target material must be greater than the sputtering rate, but it is generally found that this can be achieved, and high resistivity (up to $10^{12}\,\Omega$cm) dielectric layers in the case of Si[34], Al[35], Ta[36] and Ti[35] prepared. If the dosage is insufficient for total oxidation to occur then cermet structures can be obtained[14,37]. Another technique is to ion-implant during growth of a dielectric film, and encouraging results have been obtained by ion implantation of SiO during growth from an evaporation source[38]. A fully oxidised film of SiO_2 was obtained with a resistivity of $10^{14}\,\Omega$cm.

Gaseous anodisation has been used for the preparation of dielectric films[17]. Figure 14 in Chapter 2 shows a typical apparatus in which the material to be anodised is placed in the discharge at a slightly higher bias voltage than the discharge anode. The bias voltage can be D.C. or a combination of D.C. and R.F.[39]. The system has the advantage of enabling films to be grown at anodising temperatures well above the approximate limit of 100°C set by the use of conventional aqueous electrolytes. However, the

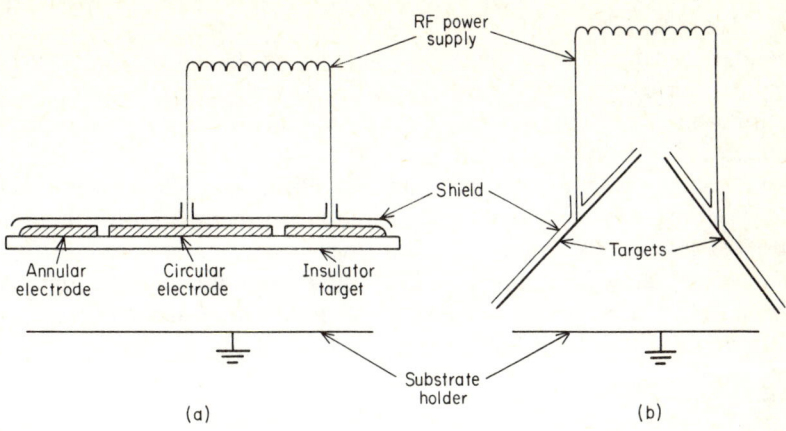

FIGURE 1. R.F. sputtering systems, (a) using concentric electrodes; (b) using adjacent electrodes at right angles (Ref. 13).

penetration of the plasma into any interstices in the substrate is extremely small, due to the shielding effect of the parent metal and so only flat surfaces can be successfully gaseously anodised, in contrast to the anodisation of highly convoluted surfaces, which is possible with liquid systems. Successful anodic films have been obtained on Al[40], Ta[39], Ti and Mo[41].

Another plasma reaction that has been used for growing oxides, is that of plasma assisted thermal growth[11, 13, 15, 42]. Ligenza, in his original work[42], used an R.F. excited discharge and an oxygen pressure of between 0.1 and 1.0 torr, and found that silicon could be oxidised to a thickness of around 3500 Å at around 300°C.

Polymerisation techniques have become widely used with a glow discharge acting as the polymerisation medium. If a suitable monomer is introduced into the discharge chamber, it has been found that the monomer will be polymerised so that insulating films can be grown on suitably placed substrates[15]. A wide variety of monomers have been polymerised and deposited in this way.

A glow discharge can also be used to effect vapour phase reactions such as the deposition of silicon nitride from a gaseous mixture of silicon hydride, ammonia and hydrogen[43]. R.F. excitation of the discharge has also been used and an apparatus has been described by Connel and Gregor[44] which can produce insulating films from styrene at the rate of 20 Å per second.

2.4 Thermal Growth[16, 45]

Thermal growth has been discussed briefly in the chapter on preparational methods. It can be enhanced by heating the substrate and, in the case of silicon, an atmosphere of either oxygen or water vapour is used with temperatures between 700 and 1400°C.

In the case of aluminium, tantalum and silicon and other valve metals, a coherent oxide film is formed, thus preventing further oxide growth on the substrate. However, with a great many other metals, a non-coherent film is obtained which will continually flake off during preparation, thus re-exposing the substrate. Such behaviour is of little use in preparing dielectric films, and explains the limited applicability of thermal growth techniques.

2.5 Anodisation[16, 17]

In order to avoid the necessity for heating substrates to high temperatures, it is possible to enhance oxidation by immersing parent metals in an electrode medium with the material to be oxidised being made the anode of an electrolytic cell. The medium may be aqueous, non-aqueous or a fused salt. As in the case of thermal growth, only a limited number of materials form coherent oxides. These include the valve metals (Al, Ta, Nb, Zr, Hf), either alone or as alloys within the group, Group IV and Group III-V semiconductors, and materials such as $Bi^{[46,47]}$, $W^{[48]}$, $TaN^{[49]}$ and $SiC^{[50]}$. The rate of formation of the oxide is a function of the voltage applied to the electrolytic cell. Growth curves for tantalum oxide and aluminium oxide have been shown in Fig. 16, Chapter 2. The asymptotic thickness values are again obtained, as was the case for thermal growth, and for the case of aluminium an anodisation constant of approximately 13 Å per volt and for tantalum 16 Å per volt, is found. The disadvantage of this approach is that the initial stages of anodisation require very high current densities, and one way around this difficulty is to anodise initially using a constant current rather than a constant voltage.

The limiting condition of the thickness of the film is the breakdown under high voltages, and in the cases of aluminium and tantalum this occurs at oxide thicknesses of 1.5 μm and 1.1 μm respectively. Other factors can, however, limit the thickness to values less than those quoted, and in particular the purity of the substrate and composition of the electrolyte are important.

Anodisation is a widely used technique for obtaining amorphous, highly insulating films and there is a very large commercial usage of these in electrolytic capacitors, both those based on tantalum and on aluminium. Anodisation systems for capacitors are often constructed on a massive scale and in the case of aluminium, continuous rolls of etched foil, 50 cm or so in width, are anodised to voltages ranging from 20 to 600 volts. One of the features of large-scale anodisation is that in order to avoid too lengthy a stay of the foil in the electrolytic bath, immersion may only be for a relatively short time. Thus, the limiting thickness will not be obtained and although the foil may have been anodised with a known applied voltage, the final thickness will not be that determined from the anodisation constant, but is likely to be only 60% or 70% of the expected value. Figure 2 shows a

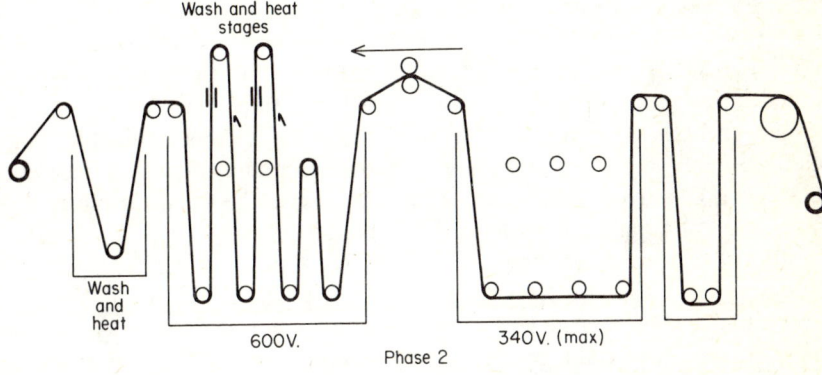

FIGURE 2. Schematic diagram of large scale anodisation apparatus for high voltage film.

diagram of a typical anodisation plant for aluminium foil, arranged for anodising to the largest voltage normally used; washing tanks and a two-stage anodisation process are shown.

2.6 Vapour Phase Deposition[2, 11, 16, 18]

Vapour phase growth is widely used in the preparation of single crystal silicon for semiconductor purposes. In the case of dielectric films, however, the techniques of disproportionation or decomposition which are normally used for metallic or semiconducting films, are not applicable. However, dielectrics may be prepared by oxidation, usually using the halide of the required metal oxide (e.g. alumina can be deposited from a vapour of aluminium trichloride mixed with steam at substrate temperatures of 600°C). Difficulties are encountered in ensuring thorough mixing of the halide and the water vapour and in maintaining a uniform temperature in the reaction vessel. This can be effected using a fluidised bed[16], although in these circumstances the particles of the bed will themselves become coated with oxide.

Polymerisation systems can also be used, as has been indicated, in the case of glow discharge vapour phase growth. Other polymerising agents are electron beam irradiation[15] and ultraviolet irradiation[15,33].

2.7 Solution Deposition

Over recent years, it has been found possible to deposit satisfactory dielectric films by solution systems.

A. Deposition of oxide films from colloidal suspensions[19]. In this technique a colloidal suspension of the material is prepared and the substrate to be coated is immersed in the suspension so that a thin layer of material is formed on both sides of the substrate. Uniformity of this liquid film can often be improved by spinning the wetted surface and after spinning, the substrate is baked, often to a temperature of between 200–500°C, to convert the liquid colloidal layer to a solid, usable structure. Satisfactory silicon dioxide films have been prepared in this way using a colloidal silicon dioxide hydrate. Film thicknesses can be as low as 100 Å and multiple layers can be built up by repeated immersion and baking, a technique which has been applied for making multi-layer silicon dioxide/titanium dioxide structures.

B. Deposition from non-colloidal solutions. Non-colloidal systems can also be used, in which the baking process converts the solution into an oxide film. A typical example of this type of film growth is the preparation of manganese oxide layers from dilute solutions of manganese nitrate, the liquid film being pyrolised at a temperature of approximately 400°C[51]. This technique, which is widely used in tantalum electrolytic capacitors, can be used for the preparation of counter-electrodes on the anodised tantalum surface.

C. Langmuir films[20,21]. Multi-monomer layers of long-chain, fatty acids (Ba and Ca salts of stearic and arachic acid), can be built up on a substrate by repeated immersion of the substrate in liquid, on the surface of which are floating long-chain, fatty acids. Barium stearate has been widely used in this connection and the layers obtained at each immersion are around 20 Å thick. Figure 3 shows a schematic diagram of the process. Successive layers can be obtained by repeated immersion and withdrawal of the substrate and useful films, up to 2 μm thick, have been prepared. The continuity of even monomolecular layers prepared by this technique has been demonstrated by the preparation of successful tunnelling devices[52], but recent studies have shown that a molecularly flat substrate must be used, as the films do not contour the surface[53].

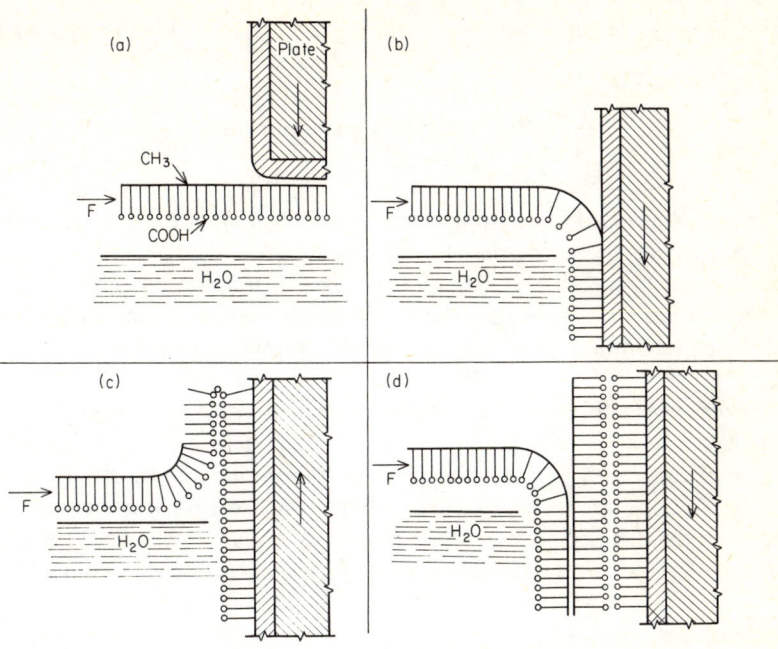

FIGURE 3. Diagrams showing the build up of monomolecular layers by the Langmuir process (Ref. 20). (a) to (d) is the sequence of growth.

D. *Polymer deposition from solution*[22]. Polymer films from materials such as polypropylene, polystyrene and PVC, have been obtained by the simple technique of direct isothermal immersion of a substrate into a suitable solution of the polymer. Material will deposit on the immersed substrate until an equilibrium is reached between deposition rate and re-solution rate. Typical growth curves for solutions of PVC in cyclohexanone at 40°C are shown in Chapter 2, Fig. 22.

E. *Solution casting*[23]. It has been found possible to prepare satisfactory film by allowing the evaporation of a polymer-containing solution placed on a substrate. Very satisfactory films of a fairly wide variety of materials have been obtained by this technique, e.g. polystyrene films from polystyrene dissolved in chloroform, but the layers obtained are usually relatively thick (greater than 1 µm).

F. *Liquid chemical reactions*[24]. It has been found possible to grow very thin (10–20 Å thick) continuous oxide films on silicon, by the action of concentrated nitric acid. Oxidation was effected at temperatures from 20 to

110°C in concentrations from 0 to 100% (100% ≡ 68% HNO_3 by weight in water).

3. Electrical Properties

3.1 Introduction

Having reviewed the techniques which are available for the preparation of dielectric films, the properties of the layers obtained will now be discussed. To study both basic properties and applications, it must be recognised that various criteria of stability must be fulfilled. These criteria can be identified in that the preparation process must produce films with the following characteristics:

A. Continuity. Preparation techniques which produce discontinuous or non-uniform films will lead to structures with unsatisfactory performances. Lack of continuity is often alleviated by the use of thin (less than 1000 Å) counter-electrodes, usually of aluminium, which will melt above the weak point in the dielectric when a field is applied so that excessive current flows through the film. This melting process isolates the weak, or discontinuous area and thereby allows the rest of the film to operate satisfactorily and this technique is widely used in the preparation of thin film, non-polar capacitors. A second useful technique involves the immersion of the dielectric film in a re-forming medium. In the case of solid tantalum capacitors[51], a counter-electrode of manganese oxide can provide this re-forming characteristic, since any weak point (and hence region of high current flow in the tantalum oxide dielectric) which has been prepared on the tantalum metal, will cause reduction on the manganese oxide counter-electrode at that point. The tantalum oxide film will then become fully oxidised and non-conducting, whilst at the same time the manganese oxide which was initially a semiconductor, will be reduced and will become an insulator, thereby further isolating the initially weak area in the film. In the case of wet electrolytic capacitors[51], the presence of an anodising solution in contact with the film, will cause re-anodisation of any weak area in the film during use, provided the polarity of the device is maintained such that the tantalum or aluminium base material is always an anode (i.e. positive), relative to the other electrode.

B. Mechanical stability. It has been shown over the past few years that all films in their as-deposited state, are either in compressive or tensile stress and extensive reviews on the type of stress obtained and the mechanism involved, have been given by various authors, both with regard to thin films in general[54,55] and to dielectrics in particular[56]. Films in tensile stress are basically unstable and if the tensile stress is of sufficient magnitude, this can cause the film to crack and become an electrical short circuit. A wide variety of dielectric films are therefore comparatively useless because of this problem. On the other hand, films in compressive stress are much more stable, provided the film stress is not sufficiently high to cause buckling of the film on the substrate. The stability of a wide variety of used dielectrics arises from this property, particularly materials such as alumina, silicon monoxide and silicon dioxide.

C. *Homogeneity.* Even with continuous films in a state of slight compressive mechanical stress, satisfactory behaviour will not be obtained, particularly in basic terms, unless the films are homogeneous.

The crystal structure of materials can be divided into three classes, namely, single crystal, polycrystalline and amorphous. Single crystal materials are characterised by a well-defined lattice spacing between the constituent atoms over a large volume, and dielectric films are not easily obtained in this form. However, materials have been prepared in this manner and some of these are summarised in Table 2.

Table 2
Dielectric Thin Films obtainable in Single Crystal Form

Material	Single crystal substrate	Deposition method	Notes	Ref.
MgO	Ag	Evaporation		(57)
βGa_2O_3	Sapphire	Vapour phase		(58)
PVC	Rock-salt	Soln. deposition		(59)
$Bi_4Ti_3O_{12}$	MgO & $MgAl_2O_4$	R.F. sputtering	Ferroelectric properties	(60)

Polycrystalline materials, as the name implies, consist of a collection of individual crystallites of varying sizes and shapes with grain boundaries between them. Dielectric films of this type can sometimes be satisfactory, although in the majority of cases the presence of grain boundaries, both in terms of the accumulation of impurities in the grain boundaries or of the easy access of water vapour etc. to the grain boundary positions, are unsatisfactory, with electrical conduction occurring easily along the grain boundaries themselves.

Amorphous films have crystallite sizes which are only atomic in dimensions and it is not possible to describe these films in terms of the normal lattice spacings associated with single crystal materials, or the crystallites of polycrystalline materials. The most satisfactory physical description of an amorphous material can be given in terms of a radial distribution analysis in which the probability of finding an atom at a given distance from a fixed point is described. The majority of satisfactory dielectric films consist of layers of this type. They have the advantage of lack of grain boundaries and also because of this, impurities up to certain limits can be scattered randomly through the amorphous matrix without affecting the dielectric properties to any considerable degree. To obtain reproducible basic properties, however, it should be noted that the amorphicity must itself be homogeneous and any impurities present must be distributed through the matrix in a homogenous sense, and it is this lack of homogeneity in the amorphous matrix which has bedevilled the detailed physical examination of films in the past. Uniformity of the amorphicity is very much a function of the preparation conditions used.

As an example of the importance of impurity and defect effects, it has recently been shown that it is inadvisable to sputter-clean substrates on to which dielectric layers have already been deposited, as the sputtering will change the electrical characteristics of the layers by the introduction of impurities and ion migration effects[61].

3.2 Basic Properties

To examine a preparation technique and to determine whether reproducible, satisfactory films have been obtained, all the above criteria outlined must

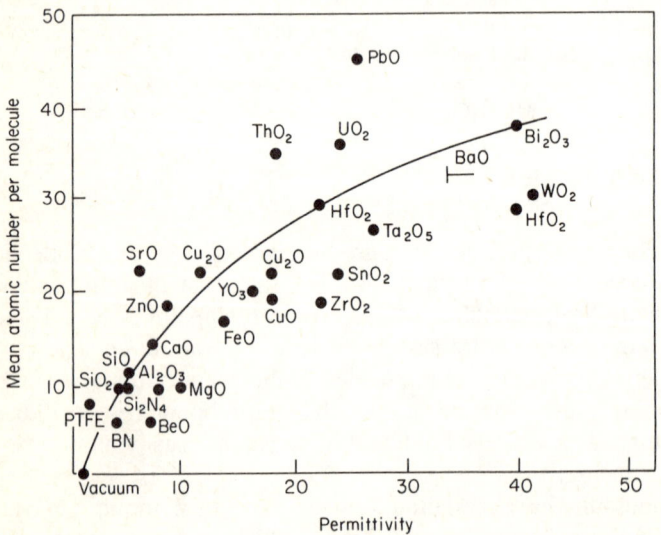

FIGURE 4. Mean atom number per molecule vs. permittivity for thin film dielectrics (Ref. 2).

be fulfilled. On top of this, however, various characteristics will need to be examined and some of these can be identified. Detailed discussions of electrical characteristics of thin films have been given by various authors and the reader is particularly advised to consult the work of Anderson[62], Harrop[63], Harrop and Campbell[2] and Fonash[64] on permittivity and polarisation behaviour, Hill[6], Jonscher and Hill[5] and Simmons[3,4] on D.C. conduction processes, Simmons[3,4] and Jonscher[65] on A.C. behaviour, and Hill[66] on electrode effects.

3.2.1 Permittivity and related phenomena

A. Permittivity. Permittivity of thin film dielectrics varies for non-polarisable materials from values around 2 for plastics to 45 or more for oxides. Figure 4 shows the range for oxides.

Table 3
Some Medium and High Values of Permittivity obtained in Thin Films (See also Ref. 2 p. 16.31)

Material	ε	Prep. method	Notes	Ref.
ZnS	8	Evaporation		(67)
AlN	8.5	Reactive evap.		(68)
NiO	14	Thermal oxy.	Annealed at 600°C to diffuse in the Li	(69)
NiO/Li	22	Thermal oxy. and soln. dep. or evap. of Li		
Ta_2O_5	27	Anod. on sputt. Ta or Ta/Al	Increased Al content will lower ε value	(70) (71)
	27	Anod.		(49)
	27	Anod. on sputt. Ta or Ta/N	Increasing N_2 content will lower ε value	(72)
Er_2O_3	30	Evap.		(73)
Nb_2O_5	45	Anod.		(74)
TiO_2	100	Evap.		(75)
$BaTiO_3$	820	Evap.		(76)

Polarisable materials can have high values of permittivity and some ferroelectric films of BaTiO$_3$ have been obtained with $\varepsilon = 820$. Table 3 references some of the medium and higher values obtained.

The permittivity of a non-polarisable dielectric is almost independent of structure or electrical conduction. It is almost entirely an intrinsic property of the constituent ions, as has been summarised by Harrop and Campbell[2]. Permittivity has an extrinsic and an intrinsic element. The extrinsic element is that part of the permittivity which does not arise from the constituent ions and electrons and is usually very small. The intrinsic permittivity can depend on the electronic contribution, particularly where the constituent ions are small and relatively non-deformable. It also dominates if there is a surfeit of conduction electrons. An intrinsic contribution to permittivity will also arise due to ionic contributions, particularly in the case of non-deformable, fully ionic compounds. A final intrinsic contribution is due to the deforming of ions and this effect is particularly important in oxides, since the oxide ion is large and very deformable.

B. *Temperature coefficient of permittivity.* A further permittivity effect is that of the behaviour of permittivity with temperature. The temperature coefficient of permittivity,

$$\gamma_p, \left(\frac{1}{\varepsilon}\frac{\partial \varepsilon}{\partial T}\right)$$

is related to the temperature coefficient of capacitance

$$\gamma_c, \left(\frac{1}{C}\frac{\partial C}{\partial t}\right)$$

for a thermally isotropic material, by:

$$\gamma_c = \gamma_\varepsilon + \alpha_L \tag{1}$$

where α_L is the linear expansion coefficient.

As in the case of permittivity itself, γ_C can be examined in terms of extrinsic and intrinsic components. Both elements can be analysed in detail and Equation (2) summarises the situation:

$$\gamma_c = f(\varepsilon, \alpha_m, T, \alpha_L) + A \tan \delta \tag{2}$$

where the first term represents a contribution from intrinsic behaviour, due to permittivity ε, polarisability α_m, temperature T and linear expansion coefficient α_L, and the second term that due to extrinsic behaviour, with A being a constant. For most solids ($\varepsilon > 2$), this equation can be simplified to:

$$\gamma_c = A \tan \delta - \alpha_L \varepsilon \tag{3}$$

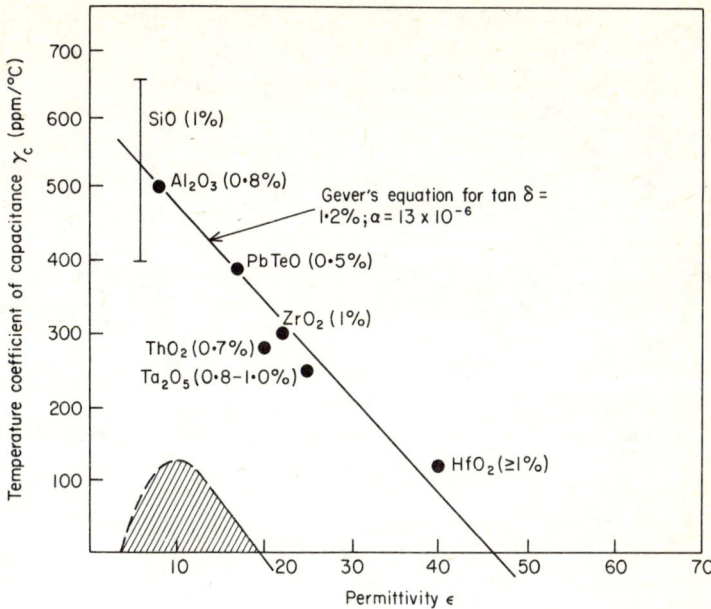

FIGURE 5. Temperature coefficient of capacitance, γ_s, vs. permittivity for high loss films (tan $\delta > 0.1\%$) (Ref. 2).

Figure 5 shows how this relationship [Equation (2)] operates for various thin film dielectrics; the temperature coefficient of capacitance is plotted against permittivity for materials with losses greater than 0.1%. For low loss where the intrinsic term is all that matters, the relationship shown in Fig. 6 is obtained, and this curve defines the minimum value of temperature coefficient of capacitance that can be obtained in a dielectric film.

C. *Dielectric loss (tan δ) and frequency dependent behaviour.* The dielectric loss is basically a measure of the series or parallel resistance associated with the dielectric film. As the values are usually small in practical devices, it is usually approximated as the tangent of the loss angle and in the case of parallel components of resistance capacitance, will be given by:

$$\tan \delta = 1/\omega R_p C_p \qquad (4)$$

where C_p and R_p are the parallel values of capacitance and resistance and ω is the frequency. Tan δ is usually measured as a percentage.

The dielectric loss is an important quantity when considering A.C. conduction mechanisms in dielectrics. However, as has been pointed out by

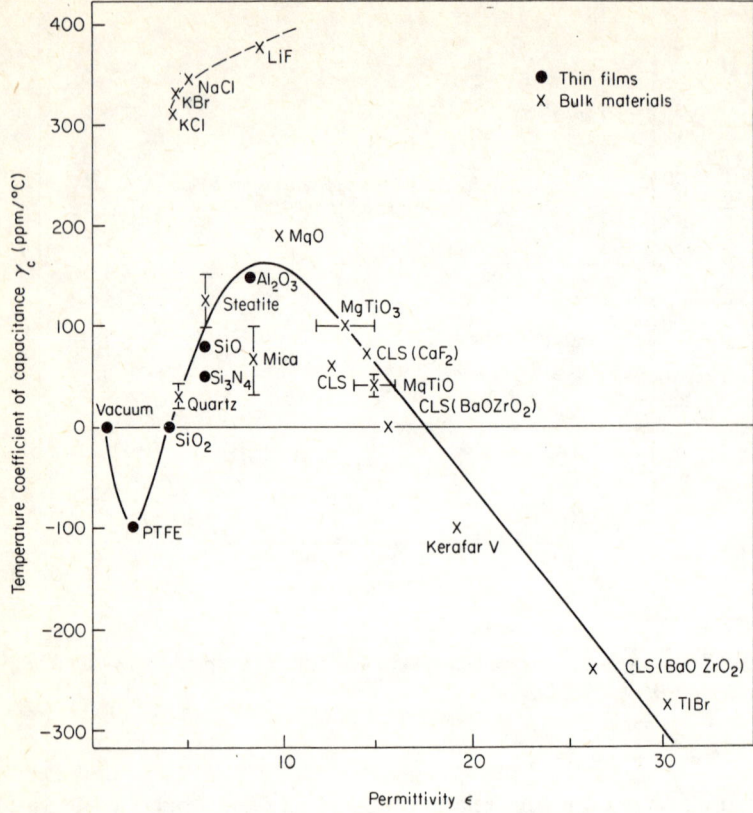

FIGURE 6. Temperature coefficient of capacitance, γ_c, vs. permittivity for low loss thin film materials (Ref. 2).

Jonscher and Hill[5], this area of study, in which frequency and also time dependence of electrical conductivity is examined, has not yet been related in a quantifiable manner to the specific structural characteristics of materials, although it has received a considerable amount of attention from both theorists and experimentalists.

When one considers frequency dependent phenomena, it must be realised that D.C. contributions will always be present. These D.C. contributions will be seen as a parallel resistance, R_p, which is constant with frequency and a value of tan δ which decreases with frequency. However, A.C. effects will be superimposed on this, so that the resultant general loss vs. frequency plot can often take the form as shown in the generalised graph of Fig. 7. Peaks of the type shown schematically in Fig. 7 are often found in the study of real dielectric films. These relaxation peaks can be due to a variety of causes; for example, interfacial polarisation effects for which there is a relaxation of charge build-up at the metal-dielectric interface or due to structural vibrations of the lattice itself. Peaks in loss are rarely found between 100 Hz and 1 MHz and in all cases, the peaks move to higher

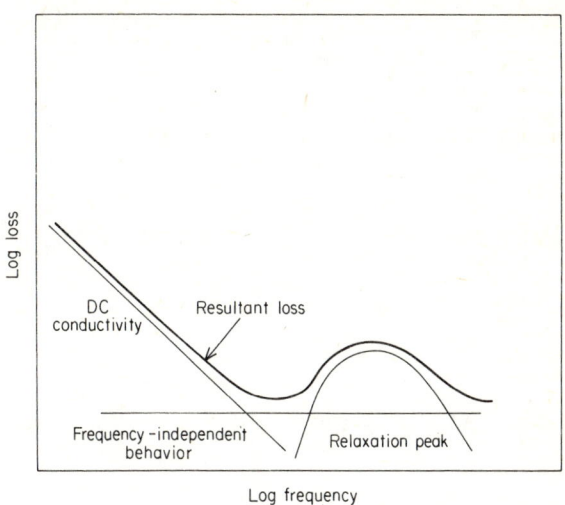

FIGURE 7. Schematic diagram of loss (tan δ) vs. frequency showing the effect of the various A.C. conduction processes (Ref. 2).

frequencies as the temperature is raised. Their behaviour is often best described by the Debye relaxation equations[62].

Interfacial polarisation effects themselves can be analysed in terms of a two-layer model using Maxwell-Wagner theory[1]. Such a model predicts a loss peak which is often at very low frequencies (0.001 Hz).

Most dielectrics outside the region in which relaxation peaks occur, have a considerable region in which loss is nearly invariant with frequency (at room temperatures). This frequency independent loss has been proved in a great many materials, including silicon monoxide. Present understanding of the phenomena is that the effect occurs at low fields under conditions in

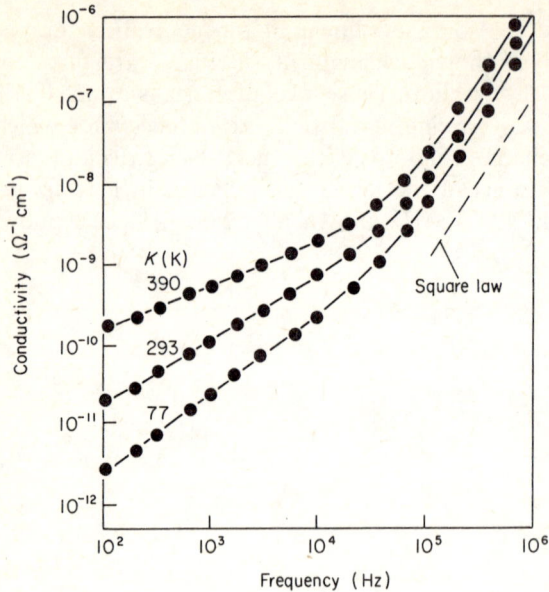

FIGURE 8. A.C. conductivity vs. frequency for SiO films (Ref. 5).

which Ohm's law is obeyed. It is present when flaws in films carry the major part of the current and therefore, for typical dielectric films with losses of 1% or more, the effect is probably due to a distribution of relaxation processes in the flaws. The effect also occurs with bulk single crystals and with flaw-free amorphous films. Here it is suggested[2] that the most likely mechanisms at low fields are impurity and ionic conduction. Impurity conduction has been shown to give invariant loss at audio frequencies because of the energy distribution of the impurity sites. Ionic conduction can also give near invariant losses at low temperatures, although theoretical analyses in the literature are still incomplete.

D. *Resistivity.* The resistivity associated with a useful dielectric will be 10^6 Ωcm or greater, and values of 10^{15} Ωcm have been obtained. Furthermore, since an activation process of some type is associated with charge transfer through the dielectric, the value of resistivity will decrease with increase in temperature—i.e. the temperature coefficient of resistivity, α_p, will be negative[1]. This contrasts with the positive value of α_p associated with metallic conduction, so that a measure of α_p or the temperature coefficient

Thin Film Dielectrics 131

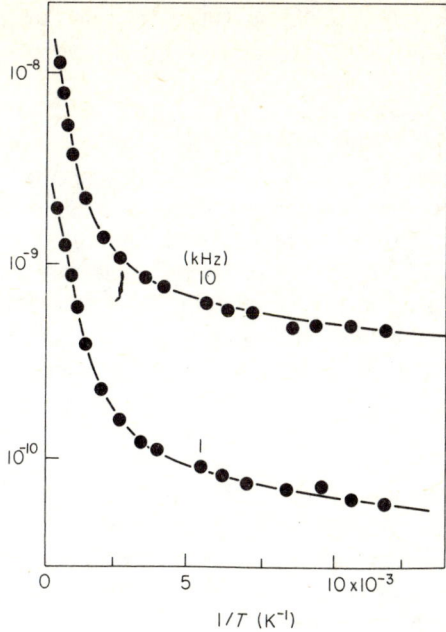

FIGURE 9. A.C. conductivity vs. reciprocal temperature for SiO films (Ref. 5).

of resistance, α_R, can be used in composite metal-dielectric structures (cermets), to determine the importance of the dielectric contribution to the conduction system. This aspect is discussed in greater detail in Chapter 5.

As will be shown in the section on D.C. conduction processes, a non-linear relationship is often found between i and V. Thus, a dependence of resistivity on voltage is often obtained. Apart from direct ρ vs. V measurements, non-linearity can also be detected in terms of the third harmonic index, a measurement which is normally discussed in terms of resistors[77,78], but has recently been measured for thin film capacitors[79]. Third harmonic measurements essentially involve the measurement of the A.C. resistivity. A.C. resistivity or conductivity can depend on the same type of conduction mechanisms as does the D.C. value. Calculations on A.C. conductivity, $\sigma_{A.C.}$, have been given by Simmons[4] on the basis of a hopping conduction model, [see para. 3.2.2.1 (a)], and Figs 8 and 9 show the results obtained for $\sigma_{A.C.}$ as a function of frequency and of temperature respectively, for silicon monoxide layers.

Surface resistivity is sometimes measured on a dielectric film and is very sensitive to moisture. It is of importance in transistor type applications, and measurements on SiO_2 at 20% humidity have given values of $> 10^{18}\,\Omega/\text{sq}^{(80)}$.

3.2.2 D.C. conduction processes

The basic experimental fact which is observed in thin film dielectrics is that much larger currents flow through the films than would be expected from the bulk properties. It is found necessary in examining this phenomenon, to discuss in detail the energy level structure of amorphous solids and when this is done, it is found that the conduction behaviour can be explained in terms of a whole variety of different mechanisms. The two possible charge carriers are ions and electrons; the mechanisms which are mainly discussed are those which refer to transport of electrons under the influence of an applied field. Ion movement is also possible and has been recently reviewed by Fonasch[81].

There are two methods of distinguishing experimentally between ionic and electronic charge carriers. If the carrier is ionic the application of a D.C. voltage for a sufficiently long length of time would mean that material would be deposited at the cathode or anode. Also, the large transition time involved in the transport of ions can be observed by studying the effect of applying rectangular voltage pulses.

Having determined the time dependence of the phenomenon and therefore ascertaining the rapidity of the response to field changes, various conduction phenomena can be postulated.

A. Hopping conduction. At low fields ($< 10^4$ V/cm), studies over the last few years have demonstrated the importance of hopping conduction to explain conduction behaviour. A large number of localised states are postulated in the forbidden band gap of the dielectric and electrons can pass through the material by hopping from one localised trap to the next, given that the energy difference between adjacent traps is sufficiently low for activation to be possible and that the traps are spatially close together. It is recognised that the distribution of these traps is peculiar to the individual materials, but in general a conduction mechanism can be detailed using this type of model. Such studies have been very adequately reviewed by Jonscher and Hill[5]. One of the most important relationships which is established by considerations of hopping conduction is that the conductivity, σ, is given by a $T^{-1/n}$ law. With some materials $n = 4$, so that one can write:

$$\sigma = \sigma_0 \exp(-A/T^{\frac{1}{4}}) \qquad (5)$$

where σ_0 and A are generally treated as constants.

This type of characteristic is most easily seen in amorphous elemental semiconductors and Fig. 10 shows a typical plot for amorphous carbon, as

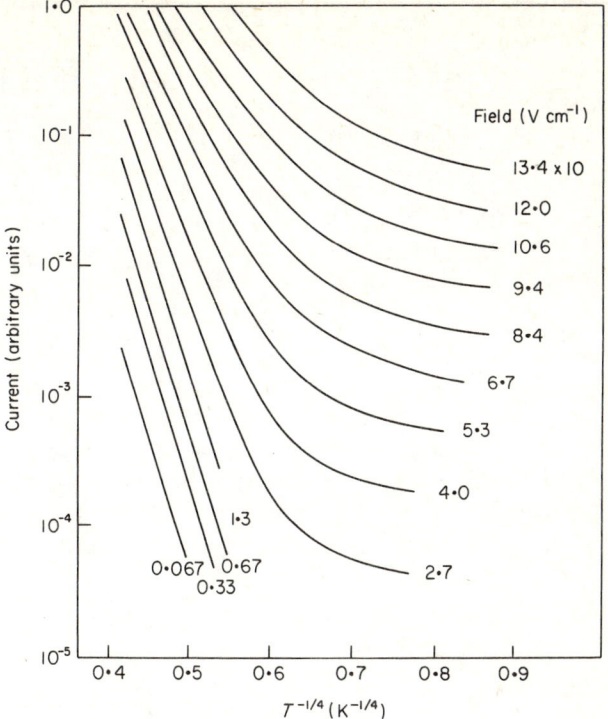

FIGURE 10. Current vs. $1/T^{1/4}$ for amorphous Ge (Ref. 5).

analysed by Hill. Curves of this type, often with $\frac{1}{4}$ replaced by $1/n$, allow for analysis to obtain a postulated trap distribution in the forbidden band gap.

B. *Conduction at high fields* (*greater than* $10^4 V/cm$). At high fields conduction may be considerably modified from that of simple hopping conduction. Electrons may be activated out of trapping sites in the dielectric into the conduction band, giving rise to the Poole-Frenkel and Poole effects, or carriers may be introduced into the dielectric by injection from the contacts (Schottky effect). An analysis of these effects gives the relationship that the logarithm of the current is proportional to $\beta E^{\frac{1}{2}}/kT$, where E is the applied field, β is a constant, the value of which differs between Poole-Frenkel and Schottky phenomena, and T is the absolute temperature. However, it has been pointed out by Jonscher and Hill[5] that this

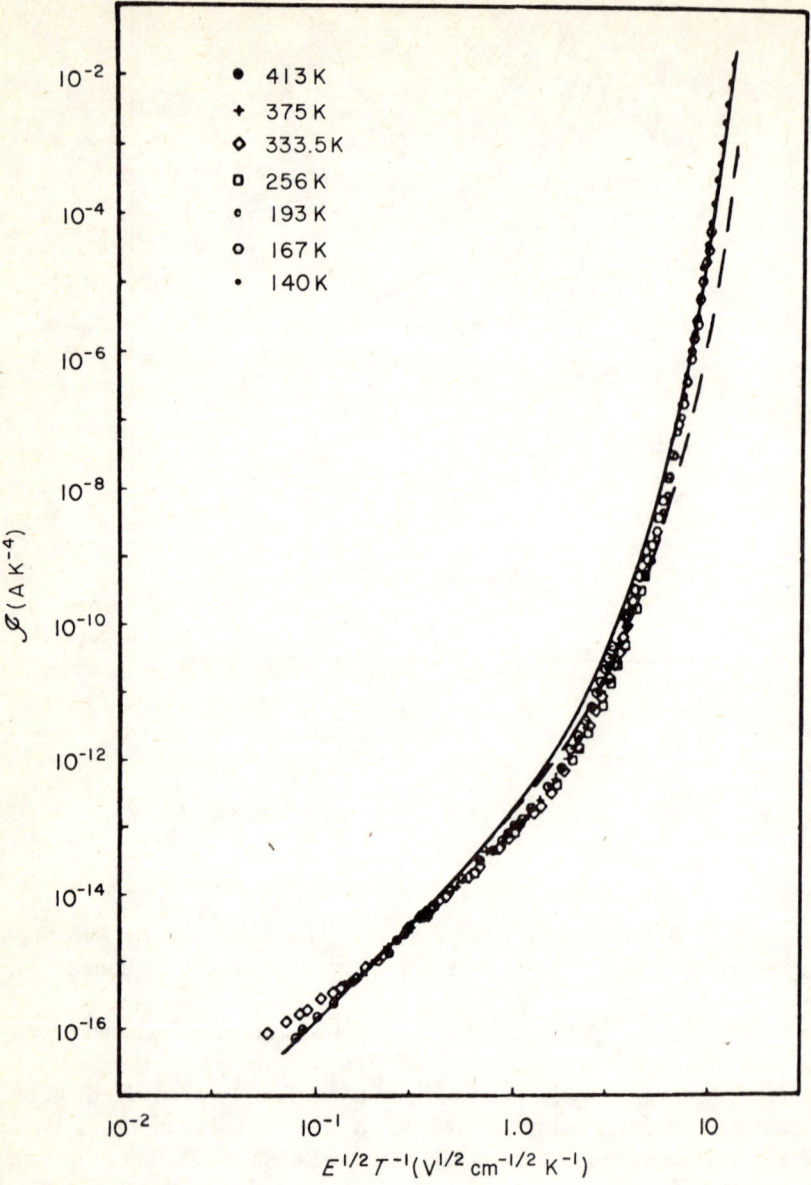

FIGURE 11. Normalized plot of current vs. $E^{1/2}/T$ for SiO. This plot shows the Poole-Frenkel effect applies as the trap density is relatively low (Ref. 5).

relationship is not entirely satisfactory, and a more complicated relationship should be involved, such that the normalised current I is given by a relationship of the form:

$$I = \gamma \cosh \gamma - \sinh \gamma \qquad (6)$$

where γ is given by:

$$\gamma = e\beta E^{\frac{1}{2}}/kT \qquad (7)$$

with e being the charge on an electron. Figure 11 shows a typical plot for results on silicon monoxide over a range of temperature from 140 K to 413 K[5]. This type of behaviour has been found to be followed by silicon monoxide, boron nitride, zirconium dioxide and chromium oxide in particular. Other modified forms of Equation (6) have also been found to give good fits with experimental results, i.g.

$$I = \gamma^2 \sinh \gamma \qquad (8)$$

$$I = \cosh \gamma - \gamma^{-1} \sinh \gamma \qquad (9)$$

If the impurity density and hence trapping sites density in the material is particularly high, then the Poole-Frenkel effect behaviour will be modified, and in these circumstances log current is proportional to E/kT (the Poole effect). For the most satisfactory results, however, a similar plot to Equation (6) can again be obtained, with γ being the modified function $e\beta E/kT$. Figure 12 shows such a plot for the case of silicon monoxide and Jonscher and Hill's calculations show that this behaviour corresponds to a trapping site spacing of 80 Å, a trapping site density of 2×10^{-18} cm^3, and an activation energy of 0.35 eV.

C. *Tunnelling behaviour.* For films less than 50 Å, it is possible for electrons to tunnel through the dielectric from one electrode to another, thereby giving a current which can be plotted as a function of film thickness. An analysis by Simmons[4] has shown that characteristics of the type given in Fig. 13 should be obtained where the conductivity is plotted as a function of applied voltage for various film thicknesses.

3.2.3 Contact phenomena

Dielectric films are usually used sandwiched between two metal or semiconducting layers and these contact layers are important in defining the behaviour of the dielectric film. The interfacial states which result from the presence of these contact layers can spread some distance into the film and therefore, if the film is thin enough, these interfacial states will affect the behaviour of the total structure.

Contact phenomena have been reviewed by Campbell and Morley[1],

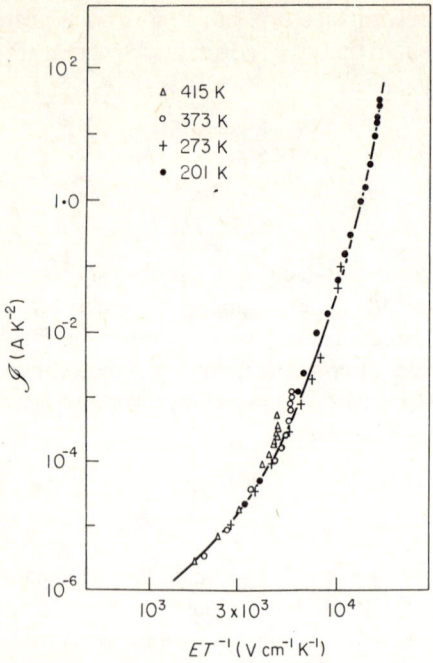

FIGURE 12. Normalised plot of current g vs. E/T for SiO. This plot shows Poole's law applies as the trap density is high (Ref. 5).

Simmons[3,4], Jonscher and Hill[5], and in theoretical terms by Mott[82]. Unfortunately, the nomenclature associated with the type of contact obtained tends to vary and the reader should therefore be on guard as to the meaning of the actual terms which he may find in different references. Contact behaviour is determined by two factors, namely that the vacuum and Fermi levels of electrode and insulator must be continuous across the interface, and that the energy difference between the top of the conduction band of the insulator and the Fermi level of the metal at a distance far removed from the interface, will be equal to the effective work function of the insulator. Given these two criteria, three separate contacts can be distinguished.

(i) *Ohmic contact*.

This is defined as one which is capable of supplying and removing charge carriers from the insulator at exactly the required rate, without any accumulation of excess carriers or

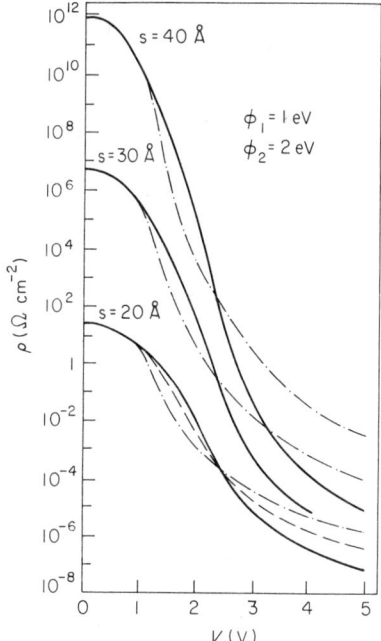

FIGURE 13. Resistivity vs. voltage for thin dielectric layers of different thicknesses, s. Work functions of electrodes, ψ and ψ_2 of 1 and 2 eV respectively. Full line is the reverse and the chain line the calculated forward characteristics (Ref. 4).

inhibition of their passage. As Jonscher and Hill[5] have noted, this means that in the case of crystalline insulators in which conduction occurs effectively only in a free band, the requirement for an ohmic contact is the flat band situation shown diagramatically in Fig. 14 (c) and (d).

(ii) *Blocking contact.*

If the situation is such that there are far fewer free carriers at the metal interface than in the bulk material, the supply of carriers from the metal is restricted and a blocking contact, or Schottky barrier, exists. [Fig. 14 (e) and (f)].

(iii) *Injecting contact.*

When the accumulation region of charge at the metal interface serves as a ready source for excess electrons, then an injecting contact is obtained. [Fig. 14 (a) and (b)].

From the diagrams shown as Fig. 14, it can be seen that an ohmic

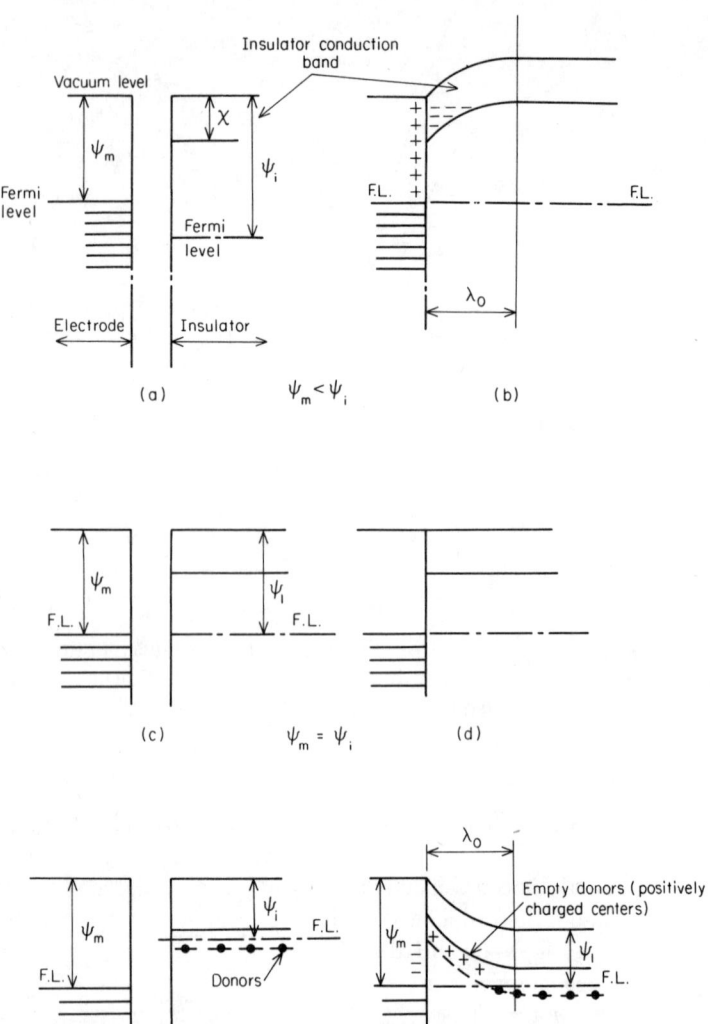

FIGURE 14. Energy level diagrams before and after joining a dielectric to a metal electrode: (a) and (b) injection contact; (c) and (d) neutral contact; (e) and (f) blocking contact (Ref. 4).

contact arises when the work function of the metal, ψ_m, and that of the insulator, ψ_i, are the same. An injecting contact will arise when ψ_m is less than ψ_i and a blocking contact when ψ_m is greater than ψ_i. It can also be seen that the effect of the contact extends for some distance into the insulator and this distance is determined by the density of ionisable, fixed impurities in the case of a depletion barrier (blocking contact for electrons), or by the free electron density in the case of an accumulation barrier (injecting contact).

For a blocking contact the width of the depletion region, λ_0, has been given by Simmons[4] as

$$\lambda_0 = \left(\frac{2(\psi_m - \psi_i)\varepsilon\varepsilon_0}{e^2 N_d}\right)^{1/2} \qquad (10)$$

where N_d is the donor density within the insulator and ε_0 is the permittivity of free space. In practical terms, for a value of $\psi_m - \psi_i = 2\,\text{eV}$ and a permittivity, ε, of 5, the relationship between the depletion depth and the impurity density, N_d, is as shown in Fig. 15. It can thus be seen that the depletion depth can be of the same order of magnitude as the dielectric film thickness which is normally considered in the context of thin films.

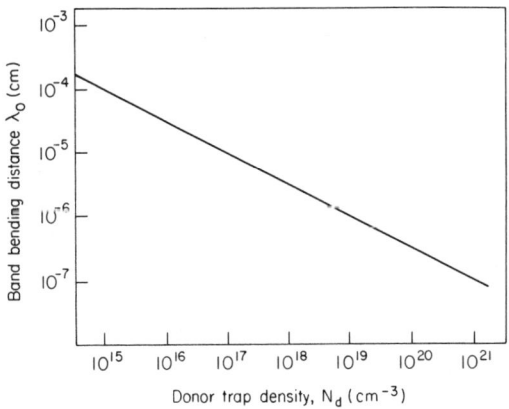

FIGURE 15. Band bending distance λ_0 as a function of donor trap density N_d, for a blocking contact ($\psi_m - \psi_i = 2\,\text{eV}$; $\varepsilon = 5$). (After Simmons[4].)

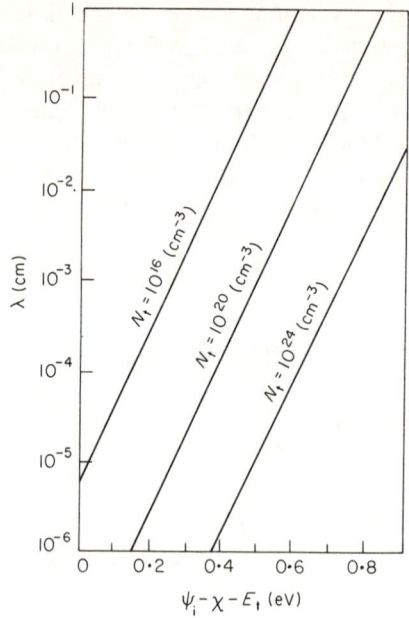

FIGURE 16. Band bending distance, λ, as a function of the energy term, $\psi_i - X - E_t$ for various values of the N_t, the electron trap density. (After Simmons[4].)

The width of the accumulation region, λ, in the injecting contact case is given by:

$$\lambda = \frac{\pi}{2}\left(\frac{2kT\varepsilon\varepsilon_0}{e2N_t}\right)^{1/2} \exp\left[\frac{\psi_i - X - E_t}{2kT}\right] \quad (11)$$

for a density of shallow electron traps N_t, positioned at an energy E_t, below the bottom of the conduction band where X is the electron affinity of the dielectric. As with the case of the blocking contact, this width can be calculated. Typical values have again been obtained by Simmons[5]. (Fig. 16).

When the metal-insulator-metal system is considered, then the effect of both contacts must be examined. In these circumstances, band diagrams of the type shown in Fig. 17 (a) and (b) for two blocking contacts or Fig. 17 (c) and (d), for two injecting contacts, will result, and the effect of

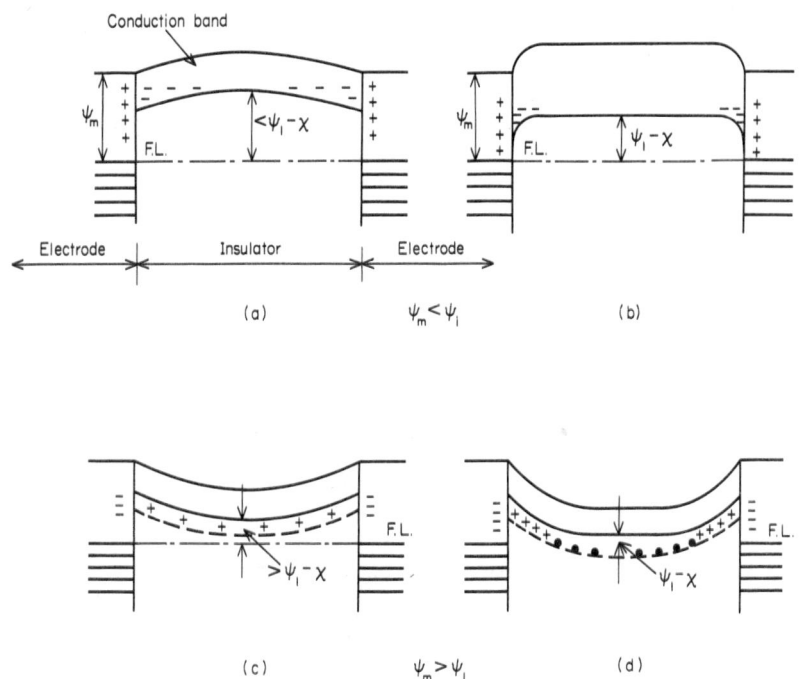

FIGURE 17. Energy level diagrams for two metal contacts and an insulator: (a) and (b) show the effect of different band bending distances with injection contacts and (c) and (d) similar situations with blocking contacts (Ref. 4).

the two regions can well overlap, enhancing the importance of the actual contact on the conduction process in a thin layer. In fact, it can be seen from these diagrams (Fig. 17) that dielectric conduction processes can be electrode-limited rather than being dependent on the bulk properties of the dielectric itself.

The band bending situation at the electrode interface can also have an effect on A.C. properties so that the equivalent circuit diagram for a capacitor can well be represented as shown in Fig. 18. The barrier (Schottky capacitance) is shown as C_s and the interior capacitance as C_b. E_d is the activation energy of the trapping centres of the dielectric and R_0 is a resistive parameter whose magnitude depends on mobility, film thickness, donor density and temperature. Results of apparent change in dielectric constant with thickness at a particular frequency[83], can often be explained in terms of this type of equivalent circuit.

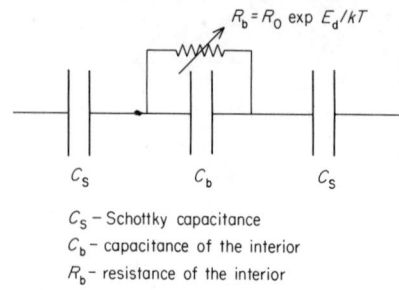

C_S – Schottky capacitance
C_b – capacitance of the interior
R_b – resistance of the interior

FIGURE 18. Equivalent circuit showing effect of interfacial barriers (Schottky barriers) (Ref. 4).

3.2.4 Space charge effects

Space charge limited conduction may occur in certain cases. A necessary condition of its predominance over other effects in any material is the availability of excess free carriers from some form of injecting contact. It is found that the classic Mott-Guerney characteristic in which current is proportional to V^2, is not apparent in many amorphous materials such as Si, Ge, SiO (Jonscher and Hill[5]). However, Jonscher and Hill note that clear evidence exists for such behaviour in plastic materials such as polythene[84]. Other workers have shown that non-linear effects in thin film polymer capacitors can also be explained in terms of electromechanical behaviour[85].

3.2.5 Breakdown[2]

It is found that breakdown in thin films is a more simple process than in bulk materials because the geometry and structure of the materials precludes a number of mechanisms. This is reflected by the fact that the best insulating films have high breakdown fields near 10 MV/cm, whilst the best bulk materials, with the exception of mica and a few plastics, break down at 1 MV/cm at room temperature.

Various types of breakdown have been identified in practice and these are dependent on the impedance of the voltage supplies. For low impedance sources, it is found that the breakdown points will travel randomly across the electrodes and these self-propagating breakdowns give a characteristic "spider" pattern. Such behaviour is extrinsic in that it occurs at relatively low fields and depends on the electrodes. For high impedance sources, the dielectric can short at weak points due to dust, for example, on the surface of the substrate, or other defects incorporated during deposition. At breakdown, the electrodes can often melt and a characteristic "splashing" pattern is obtained. This type of process can be avoided if the electrodes are thin enough so that the weak points may be isolated by local melting of the electrodes. In these circumstances, higher fields can be achieved before overall extrinsic breakdown occurs.

Methods of self-healing films are not necessarily confined to the use of thin electrodes. In practical electrolytic capacitors, re-formation of the oxide dielectric is possible because of the medium surrounding the film. In the case of so-called "wet" structures, an electrolyte is always present (aluminium and tantalum) and in dry structures a layer of manganese oxide forms the contact and supplies oxygen to the field to repair any defects (solid tantalum and aluminium capacitors).

The cause of intrinsic breakdown, which occurs under the conditions of a high impedance source and with electrodes or some similar technique which eliminates local defects, has been studied by various workers, notably Klein and his school[86,87,88]. The two most important mechanisms identified are those of electron avalanching by impact ionisation of the lattice, and joule heating breakaway. Avalanche effects in amorphous solids are still a matter for speculation. It has been pointed out by Jonscher and Hill[5] that there is no well-established experimental evidence for avalanche multiplication and further work is necessary in this field.

Joule heating breakaway implies that although the conduction mechanism prior to breakdown is not specified, the attendent joule heating begins directly the current starts to flow, causing the current to increase exponentially. The heat lost to the medium will balance out that gained from the current flow at breakdown, and an analysis of this type of behaviour therefore involves a knowledge of the thermal conductivity of the medium. Such a study is necessarily complex because of the anisotropy of the thin film.

An extensive bibliographical survey on breakdown and related phenomena in thin dielectric films has recently been published by Agarwal[89] for the period from 1960 to 1974, and the field has also been recently surveyed for anodic films by Dell'Oca et al.[17].

4. Applications of Dielectric Films

These are not necessarily related to the basic properties discussed in the previous section, although reproducibility of characteristics is obviously important. Basic properties which are of relevance in device applications are such characteristics as permittivity and temperature coefficient of capacitance for applications involving insulation, and D.C. conduction at low fields, together with tunnelling, for applications concerned with conduction processes such as cermet resistors, switches and tunnelling devices. Three areas of applications can be identified.

4.1 Capacitors and Related Applications

4.1.1 Capacitors

From a production point of view, only a limited number of materials are used as capacitor dielectrics although a wide variety has been examined experimentally. A recent review by Gerstenberg[90] contains a useful survey of the different materials which have been used and more recently, Walter and Johnson[91] have established selection lists of thin film dielectrics using a computer program to analyse 74 materials with 11 properties, and drawing up lists using 13 different figures of merit.

Table 4 summarises the main materials which have been used for the production of capacitors.

By far the most important, if only in volume terms, is that of alumina prepared by wet anodisation on tantalum foil[92]. The foil is usually heavily etched in the form of long, cylindrical tunnels, often of a length of approxi-

Table 4
Typical Properties of Thin Film Dielectrics
of Major Importance in Capacitor Structures

Material	Preparation method	Typical capacitance value ($\mu F/cm^2$)	Loss (% at 1 kHz)	Permittivity	Ref.
Al_2O_3	Wet anodisation	0.02–0.4	0.5	10	(2) (17) (90) (51) (93) (92)
Si_3N_4	R.F. sputtering	0.02	0.1	9	(2) (90) (94)
SiO	Evaporated	0.01	0.1	6	(2) (90)
SiO_2	R.F. sputtered thermally grown	0.1	0.2	4	(2) (90) (94)
Ta_2O_5	Wet anodisation	0.1	1.0	27	(90) (95)

mately 30 μm with a diameter of 1 or 2 μm. Tunnel densities are of the order of $10^8/cm^2$ and various electron microscopy studies which have been undertaken have shown the convoluted nature of the etched surface[51,92] [Fig. 19 (a) and (b)]. The films which are prepared by the process of wet anodisation on these convoluted surfaces are usually relatively lossy ($\pm 5\%$), often due to the inclusion of impurities from the anodising electrolyte into the film. However, this loss is usually overshadowed by the series resistance of the wet electrolyte used as the re-forming medium and the conducting medium between the dielectric and the cathode of the structure.

Capacitors based on tantalum are also an important business[95] and here again, the oxide is prepared by wet anodisation. The surface can either be a plain foil, an etched foil (in this case the etched surface is much less convoluted than in the case of etched aluminium), or sintered tantalum powder. Sintered tantalum powder usually has a particle size of around 10 μm, and Fig. 19(c) shows an example of the porous structures which are obtained. Capacitance values obtained in such structures can be as high as

FIGURE 19. Scanning electron micrographs of aluminium and tantalum surfaces used for capacitors (Jackson and Campbell[92]).
(a) Etched Al surface (underneath of original surface after removal of Al after etching, visible between etched surfaces).

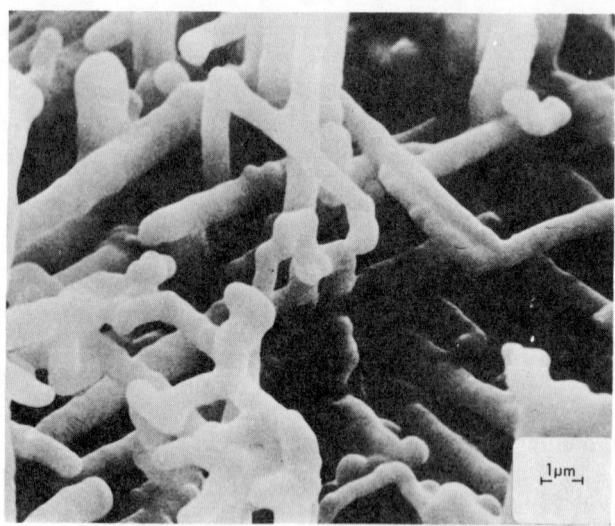

(b) Internal structure of etched Al (Al subsequently removed). (By permission of the Plessey Co. Ltd.) (Ref. 49).

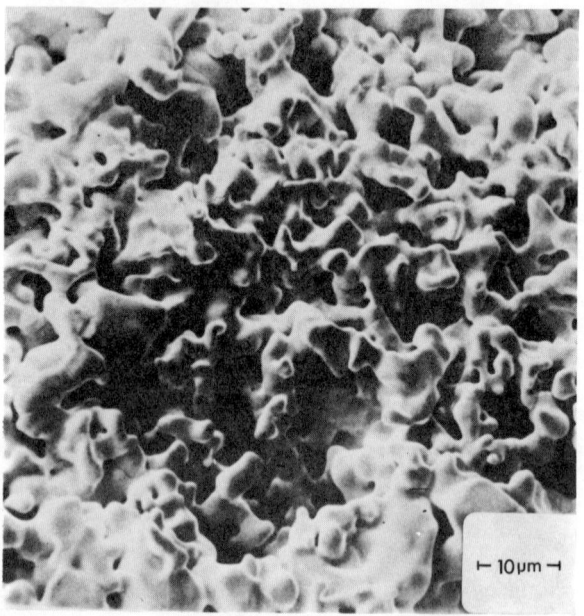

(c) Sintered Ta powder for use in Ta electrolytic capacitors. (By permission of the Plessey Co. Ltd.) (Ref. 49).

10000 μFV/g. Many electron microscopy studies have been undertaken on capacitors produced on foil or sintered tantalum blocks, and Fig. 20 shows a typical electron micrograph of fractured tantalum capacitor structure. This particular micrograph is of interest because not only does it show a fractured section of the tantalum metal together with the oxide film grown on by

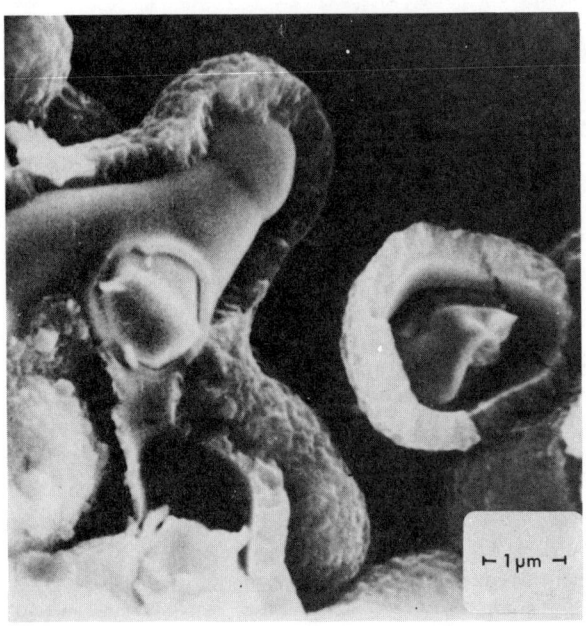

FIGURE 20. Internal structure of Ta electrolytic capacitor with MnO counter-electrode.
(By permission of the Plessey Co. Ltd.) (Ref. 51).

anodisation, but it also shows a manganese oxide layer deposited by solution pyrolysis and acting as the second electrode. This manganese oxide layer gives the self-healing properties discussed previously.

Aluminium and tantalum foil capacitors are usually wound in spirals, together with an absorbent paper spacer in between the foils to contain the electrolyte. In the case of sintered tantalum structures, however, the electrolyte is either contained in the pores, the whole body being inside a metal case, (Fig. 21), or the sinter is totally impregnated with manganese oxide to form a dry structure.

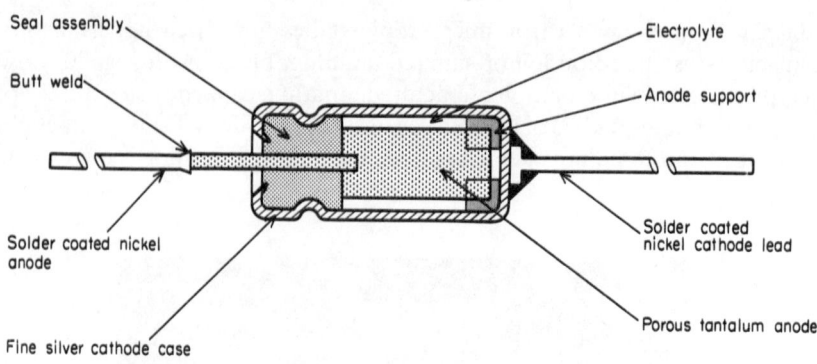

FIGURE 21. Diagram of Ta capacitor using sintered anodised Ta powder and a wet electrolyte.

Evaporated silicon monoxide has been used as capacitor material for some considerable time[90], and the compressive stresses of these films, as has already been pointed out, are of importance in determining the stability of the capacitor structure. Film thicknesses are usually greater than 3000 Å and less than 10000 Å for the best stability. Recently it has been shown that the properties of SiO can be improved if 5% B_2O_3 is incorporated into the structure by co-evaporation[96,97]. A loss of 0.1% at 1 kHz is obtained, the behaviour being explained by the increase in amorphicity that results from the addition of the B_2O_3[98].

Silicon dioxide capacitors have applicability in integrated circuit technology[94] and discrete silicon dioxide capacitors have been prepared on glass substrates for commercial applications by the technique of R.F. sputtering. Provided sufficiently low loss capacitors are prepared, these structures have the advantage of a very low temperature coefficient of capacitance associated with their permittivity value of around 4 (see Fig. 6). High permittivity ferroelectric structures have been used in capacitors[76] although the problem is to obtain a sufficiently high permittivity associated with a low enough loss for device purposes.

Electrodes in thin film capacitors in circuits are often of aluminium, in order to give the self-healing effect already referred to in the comments on continuity. The shape of the electrodes is usually rectangular, although subsidiary electrode structures are sometimes incorporated into the top electrode shape in order to give a trimming facility. Interdigitated structures have also been used[99] to improve the high frequency response.

4.1.2 Temperature sensitive devices

Thin film capacitor structures which are sensitive to temperature changes have been investigated as devices. These can take the form of bolometers, in which impedance is measured as a function of temperature, and pyroelectric detectors, in which the polarisation change with temperature is observed (see also Chapter 12). One type of bolometer[100] consists of a temperature sensitive Sb_2O_3 layer sandwiched between barrier layers of SiO. These high resistance, temperature-insensitive barrier layers reduce the behavioural dependence of the Sb_2O_3 on evaporation conditions. Electrodes of NiCr are used and the whole structure is prepared by evaporation.

Pyroelectric detectors using a variety of thin film materials have been examined[76] including $LiTaO_3$, TGS, PVF_2 and a wide variety of other plastics[23] prepared by hot pressing, solvent casting and laser evaporation but these will not be discussed here since they are fully covered in Chapter 12.

4.1.3 Pressure sensitive devices

The change of capacitance with pressure has been utilised to produce transducers. Since ferroelectric materials are also piezoelectric, thin ferroelectric films have been investigated in this context, especially $LiNbO_3$ and PZT[76] prepared by triode sputtering. Other materials utilised have included CdS, ZnO and AlN deposited in fibre-orientation, polycrystalline form by evaporation (again, see Chapter 12).

4.1.4 Thin film circuits

Thin film circuits consist of layers of metallic conductors and connector pads, resistive tracks, capacitor electrodes, capacitor dielectrics and crossover layers. They are deposited by sequential processes on to glass or ceramic and discrete active devices are attached afterwards. In certain circuits, higher values of R or C are needed than can be conveniently obtained in thin layer form and these are also added as discrete devices.

Two thin film systems can be identified; those based on Au conductors, NiCr resistors and SiO capacitors[101], and those based on tantalum technology[102] with Ta conductors, TaN resistors and Ta_2O_5 capacitors. Hafnium-based circuits have also been recently investigated using the same technology as with tantalum[104]. In the case of Au/NiCr/SiO circuits in particular, these can have associated with them thin film transistors leading to "all thin film" circuits[105,106]. Recent authors have discussed the importance of thin film circuits in commercial electronics[107], professional electronics[108] and in comparison with thick film technology[109]. In this last paper it is commented that thin film technology has reached maturity, whereas thick film technology is still realising its potential, so that judgements will have to be made for many years to come, in the light of developments, on the best system for any particular application.

4.2 Insulators

4.2.1 Crossovers

When dielectric films are used as insulators, the loss requirements are often not as stringent as in the case of capacitors. Silicon monoxide is often used in this context for isolating conducting lines in thin film circuitry and silicon dioxide in integrated circuit technology. However, to reduce interconductor capacitance to very low levels (< 0.01 pF) and hence allow very high frequency operation, techniques have been devised for producing conductors which are sufficiently robust mechanically to be self-supporting so that the insulation is an air gap[110,111].

4.2.2 Transistor and diode applications (see also Chapter 6)

An insulating dielectric is an integral part of the structure of a thin film field effect transistor[105,112] (TFT) separating the conductivity channel from the gate (Fig. 22). The same situation applies in metal-oxide-semiconductor field effect transistors (MOSFET) [Fig. 22(a)] and metal-nitride-oxide-semiconductor field effect transistors (MNOS devices) [Fig. 23(b)][111,113]. For the TFT the oxide is normally deposited by evaporation, whereas for MOS devices the oxide is thermally grown at high temperatures ($>1000°C$) or deposited using low temperature vapour phase systems[113,114,115]. The silicon nitride in the MNOS devices is often deposited by vapour phase deposition, increasing the permittivity of the insulation layer (ε for $Si_3N_4 = 7.5$) and also acting as a passivation layer.

The quality of the dielectric in MOS devices is such that these configurations can also be used as capacitor elements in integrated circuits[94]. Furthermore, if the oxide is sufficiently thin (< 50 Å) then tunnelling can occur directly through the oxide, and the system can be used as a MOS tunnel diode[113].

MOS structures can be used as memory systems either as read-only memories (ROM) or in random access memory (RAM) configurations[94]. For non-volatile memory applications the storage of charge in the insulator is sensed by the degree of depletion or inversion of the silicon surface underneath this stored charge. Figure 24 shows diagrams of such devices. The mechanism of charge transfer across the insulating layer is by tunnelling. Such systems are often made in silicon on sapphire complementary MOS structures (SOSCMOS)[113]. Other memory systems using oxides utilise

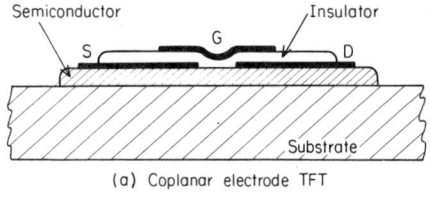
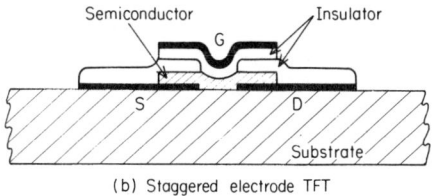

FIGURE 22. Cross-sectional views of two thin film transistor structures.
(a) Coplanar electrode.
(b) Staggered electrode.

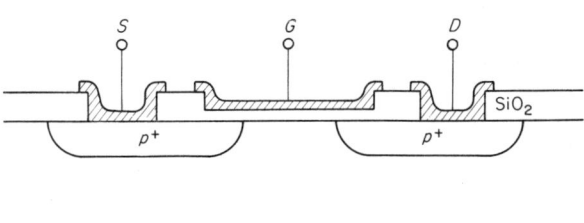
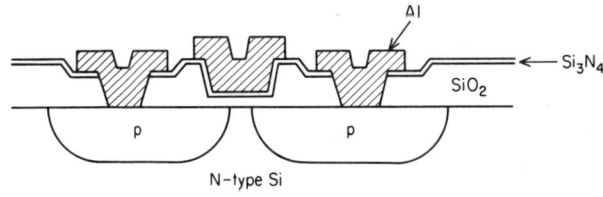

FIGURE 23. Transistor structures using SiO_2/Si_3N_4 insulation.
(a) Integrated channel MOSFET.
(b) MNOS structure.

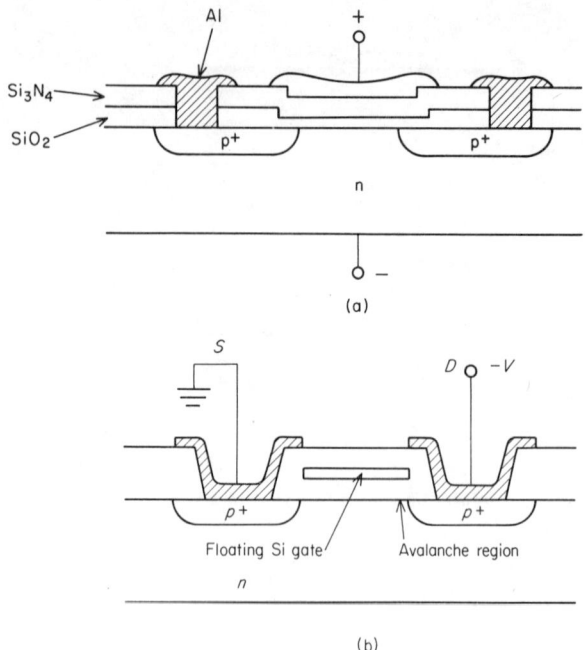

FIGURE 24. Memory structures based on MOS devices.
(a) MNOS ROM structure.
(b) FAMOS ROM structure.

avalanche injection from the silicon into the insulator with a floating silicon gate (FAMOS)[116,94,113] [Fig. 24(b)].

4.3 Conductors

4.3.1 Tunnelling devices

The conducting aspects of dielectric films have already been noted with regard to MOS tunnel diodes, and Crowell and Sze[117] have surveyed the situation with regard to hot electron transport and electron tunnelling in thin film structures. More recent work by Dittmer[118] has been concerned with metal-insulator-metal, (MIM), tunnelling structures of the form Al–Al$_2$O$_3$–Au. He has shown that the stability can be greatly improved if the Al$_2$O$_3$ layer is made up of a thermally grown layer followed by an anodically grown layer to a total thickness of up to 100 Å. The Au layer must always be biased negatively to prevent diffusion of the Au into the oxide. Thin film (20 Å) tunnel junctions have also been investigated in superconducting tunnelling devices[119].

4.3.2 Cermets

Cermet resistors (metal-dielectric mixtures)[1,120,121], have been discussed in detail in another chapter of this book. These resistors may be of a relatively simple type, prepared by the co-evaporation of a metal and a dielectric (Au–SiO_2) or more complicated as in the case of SiO–Cr, where an annealing

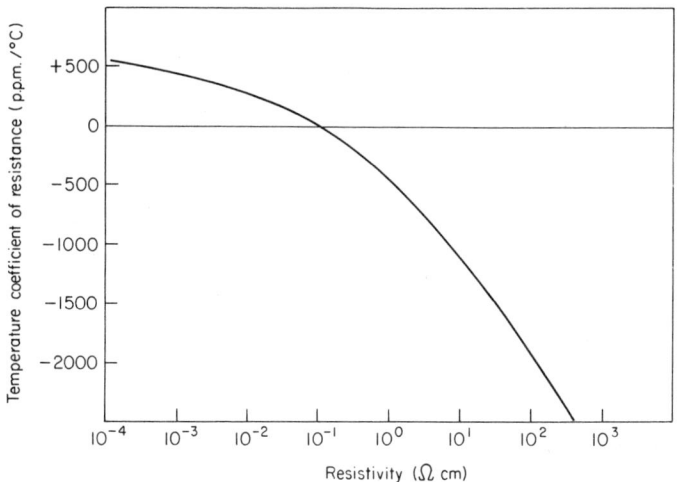

FIGURE 25. Temperature coefficient of resistance vs. resistivity for an Au–SiO_2 cermet (Ref. 121).

treatment is involved and the conduction phase is then a silicon–chromium compound. Figure 25 shows a typical temperature coefficient of resistance, α_R, vs. resistivity plot for an Au–SiO_2 cermet, and it can be seen that the value of α_R, becomes increasingly negative as the resistivity of the mixture increases, showing the increasing importance of the activated, negative α_R, dielectric conduction process. Similar behaviour is found when cermets are prepared by the implantation of metal films with oxygen (e.g. O_2^+ into Al) when the α_R value changes from positive to negative at the same time as the resistivity changes by four orders of magnitude when the dose time is increased[14].

4.3.3 Thermistors

Thin film thermistors, in which the variation in resistivity of the dielectric with temperature is the important characteristic, have been prepared using $BaTiO_3$, $SrTiO_3$, doped $BaTiO_3$ and $(BaSr)TiO_3$[76]. For 10 μm thick lanthanum doped $(BaSr)TiO_3$ prepared by sputtering and then firing to 1400°C in argon and cooling to room temperature in oxygen, a resistivity of 10^6 Ωcm was obtained with a positive temperature coefficient of resistivity between +20 and +70°C of 6%/°C.

4.3.4 Switches

A wide variety of dielectrics as well as the amorphous materials discussed in Chapter 7 have been shown to exhibit switching and memory effects[4,122,123]. The essential feature of behaviour is that the materials need to be "formed" — often referred to as "electroforming". This consists of raising the metal-insulator-metal (MIM) sandwich to a certain threshold voltage. On removal of the voltage a voltage-controlled negative resistance (VCNR) state is found to be induced in the material [Fig. 26(a)]. Subsequent low frequency behaviour is as shown in Fig. 26(b). The forming process may involve a repeated voltage cycling to a value above the threshold value[4]. At low voltages a formed device will then be in a low impedance state. However, if the voltage is raised above the threshold level and then rapidly removed, a high impedance state will be re-introduced, which will only be removed by voltage cycling [Fig. 26(c)]. Thus, a switch or memory device is available.

Explanations have been offered in terms of the creation of defect or impurity bands in the dielectric[123] or the formation of conducting filaments[124,125,126,127]. Whether the defects or impurities be distributed on an atomic or semi-macroscopic scale, it has been shown by numerous workers[118,123] that impurities must be present.

5. Summary

This chapter has surveyed the preparation of dielectric films, has noted the characteristics which are of importance in dielectric conduction processes and has briefly discussed some of the applications. It is impossible to characterise dielectric films so as to identify the best method of preparation which is applicable for all situations. Stability and reproducibility are the major characteristics required for studies of conduction processes and in the case of the applications of films, these requirements are coupled to those of the economics of large-scale production. Table 5 summarises some of the major thin film dielectrics, identifying the preparation technique used and the application.

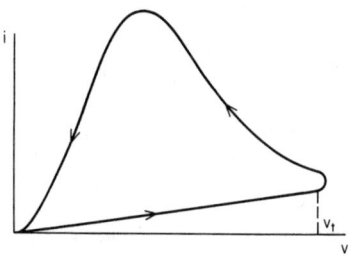

(a) Establishment of a VCNR region in a dielectric film

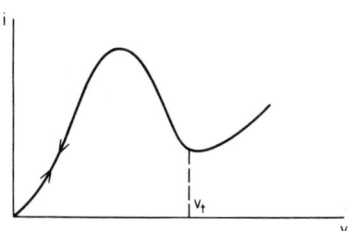

(b) i/v characteristic of a formed dielectric

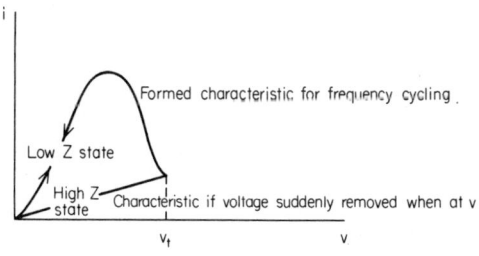

(c) Low and high impedance states in a dielectric switch

FIGURE 26. *i/V* characteristics for dielectric switches.
(a) Establishment of a VCNR region in a dielectric film.
(b) *i/V* characteristic of a "formed" dielectric.
(c) Low and high impedance states in a dielectric switch.

Table 5
Uses of Thin Dielectric Films

Material	Preparation	Application	Reference
Al_2O_3	Anodisation	Capacitors Thin film transistors	(17)(112)
	Gaseous anodisation Thermal growth	Tunnelling devices Switches	(3)(117)(4)
$BaTiO_3$; $PbTiO_3$	Flash evaporation	Thermistors, Transducers, High ε capacitors	(76)(90)
Glasses (complex oxides)	Sputtering	Capacitors, passivation, threshold and memory switches	(10)(4)
Halides (Alkali)	Evaporation	Capacitors	(90)
Polymers	Glow discharge, U.V. or E.B. Polymerisation	Capacitors (research)	(90)(15)(33)
	Solution deposition Solvent casting	Capacitors (research) Capacitors (research)	(22) (23)
Stearates	Langmuir	Capacitors (research)	(20)
SiO	Evaporation	Switches, capacitors, insulators, thin film transistors	(4)(90)(112)
SiO–Cr	Flash evap.	Resistors (cermets)	(120)
SiO–Au	Evaporation	Resistors (cermets)	(120)
SiO_2	Reactive sputt. R.F. sputt.	Low T.C.C. capacitors Insulators	(90)(2)
	Thermal growth	Passivation, capacitors MOS etc. Tunnelling devices	(90) (94) (3)
	From solution	Capacitors	(19)

Table 5—*continued*.

Material	Preparation	Application	Reference
Si_3N_4	Vapour phase	Capacitor layers, passivation, MNOS structures	(94)
Ta_2O_5	Anodisation of Ta	Capacitors; insulators	(95)
	Anodisation of Ta_2N	Capacitors; insulators	(95)

References

1. D. S. Campbell and A. R. Morley, *Reports on Progress in Physics*, **34**, 308, (1971).
2. P. J. Harrop and D. S. Campbell, in "Handbook of Thin Film Technology", eds., L. I. Maissel and R. Glang, McGraw Hill, New York, (1970).
3. J. G. Simmons, in "Handbook of Thin Film Technology", eds., L. I. Maissel and R. Glang, McGraw Hill, New York, (1970).
4. J. G. Simmons, *J. Phy. D.*, **4**, 613, (1971).
5. A. K. Jonscher and R. M. Hill, in "Physics of Thin Films", Vol. 8, eds., G. Hass, M. H. Francombe and R. W. Hoffman, Academic Press, London and New York, (1975)
6. R. M. Hill, in "Physics of Non-Metallic Thin Films", eds., C. H. S. Dupuy and A. Cachard, Plenum Press, New York, (1976).
7. F. Abeles, in "Physics of Thin Films", Vol. 6, eds., M. H. Francombe and R. W. Hoffman, Academic Press, London and New York, (1971).
8. E. Ritter, in "Physics of Thin Films", Vol. 8, eds., G. Hass, M. H. Francombe and R. W. Hoffman, Academic Press, London and New York, (1975).
9. R. Glang, in "Handbook of Thin Film Technology", eds., L. I. Maissel and R. Glang, McGraw Hill, New York, (1970).
10. A. Pliskin, D. R. Kerr and J. A. Perri, in "Physics of Thin Films", Vol. 4, eds., G. Hass and R. E. Thun, Academic Press, London and New York, (1967).
11. C. Weissmantel, *Thin Solid Films*, **32**, 11, (1976).
12. L. I. Maissel, in "Physics of Thin Films", Vol. 3, eds., G. Hass and R. E. Thun, Academic Press, London and New York, (1966).
13. L. I. Maissel, in "Handbook of Thin Film Technology", eds., L. I. Maissel and R. Glang, McGraw Hill, New York, (1970).
14. P. T. Stroud, *Thin Solid Films*, **11**, 1, (1972).
15. L. V. Gregor, in "Physics of Thin Films", Vol. 3, eds., G. Hass and R. E. Thun, Academic Press, London and New York, (1966).
16. D. S. Campbell, in "Handbook of Thin Film Technology", eds., L. I. Maissel and R. Glang, McGraw Hill, New York, (1970).
17. C. J. Dell'Oca, D. L. Pulfrey, and L. Young, in "Physics of Thin Films", Vol. 6, eds., M. H. Francombe and R. W. Hoffman, Academic Press, London and New York, (1971).

18. W. M. Feist, S. R. Steele and D. W. Ready, in "Physics of Thin Films", Vol. 5, Academic Press, London and New York, (1969).
19. H. Schroeder, in "Physics of Thin Films", Vol. 5, eds., G. Hass and R. E. Thun, Academic Press, London and New York, (1969).
20. V. K. Srivastava, in "Physics of Thin Films", Vol. 7, eds., G. Hass, M. H. Francombe and R. W. Hoffman, Academic Press, London and New York, (1973).
21. V. K. Agarwal, *Elec. Comp. Sci. and Tech.*, **2**, 1, (1975).
22. A. C. Rastogi and K. L. Chopra, *Thin Solid Films*, **18**, 187, (1973).
23. A. W. Stephens, A. W. Levine, J. Fech. Jnr., T. J. Zrebiec, A. V. Cafiero and A. M. Garofalo, *Thin Solid Films*, **24**, 361, (1974).
24. R. E. Oakley and G. A. Godber, *Thin Solid Films*, **9**, 287, (1972).
25. "Science and Technology of Surface Coating", eds., B. N. Chapman and J. C. Anderson, Academic Press, London and New York, (1974).
26. "Handbook of Thick Film Hybrid Microelectronics", ed., C. A. Harper, McGraw Hill, New York, (1975).
27. C. E. Drumheller, *Trans. 7th Nat. Vac. Symp.*, 306, (1960).
28. D. S. Campbell and B. Hendry, *Brit. J. Appl. Phys.*, **16**, 1719, (1965).
29. A. Moll, *Z. Angew Phys.*, **10**, 410, (1956).
30. R. F. Bunshah, in "Science and Technology of Surface Coatings", eds., B. N. Chapman and J. C. Anderson, Academic Press, London and New York, (1974).
31. K. O. T. Kerner, *Proc. 6th Int. Vac. Congress, Jap. J. Appl. Phys., Suppl. No. 2*, Part 1, 463, (1974).
32. E. Hollands and D. S. Campbell, *J. Mat. Sci.*, **3**, 544, (1968).
33. M. White, *Thin Solid Films*, **18**, 157, (1973).
34. J. Dylewski and M. C. Joshi, *Thin Solid Films*, **35**, 327, (1976).
35. J. G. Perkins, *Thin Solid Films*, **9**, 257, (1972).
36. I. H. Wilson, K. H. Goh and K. G. Stephens, *Thin Solid Films*, **33**, 205, (1976).
37. P. T. Stroud, *Thin Solid Films*, **10**, 205, (1972).
38. J. Dudonis and L. Pranevicuius, *Thin Solid Films*, **36**, 117, (1976).
39. G. Olive, D. L. Pulfrey and L. Young, *Thin Solid Films*, **12**, 427, (1972).
40. N. F. Jackson, *J. Mat. Sci.*, **2**, 12, (1967).
41. K. Knorr and J. D. Leslie, *Thin Solid Films*, **23**, 101, (1974).
42. J. R. Ligenza, *J. Appl. Phys.*, **36**, 2703, (1965).
43. H. F. Sterling and R. C. G. Swann, *Solid State Elec.*, **8**, 653, (1965).
44. R. A. Connel and L. V. Gregor, *J. Electrochem. Soc.*, **112**, 1198, (1965).
45. I. M. Ritchie, *Thin Solid Films*, **34**, 83, (1976).
46. S. Ikonopisov, E. Klein, A. Stanchev and T. S. Nikolov, *Thin Solid Films*, **26**, 99, (1975).
47. T. S. Nikolov, M. Aroyo, E. Klein and S. Ikonopisov, *Thin Solid Films*, **30**, 37, (1975).
48. J. Sarakinos and J. Spyridelis, *Thin Solid Films*, **27**, 239, (1975).
49. W. D. Westwood, N. Waterhouse and P. S. Wilcox, "Tantalum Thin Films", Academic Press, London and New York, (1975).
50. G. Restelli, A. Ostidich and A. Manara, *Thin Solid Films*, **23**, 23, (1974).
51. D. S. Campbell, *Rad. and Elec. Eng.*, **41**, 5, (1971).

52. A. Leger, J. Klein, M. Belin and D. Defourneau, *Thin Solid Films*, **8**, R51, (1971).
53. W. J. Plieth and W. Hopfner, *Thin Solid Films*, **28**, 351, (1975).
54. D. S. Campbell, in "Handbook of Thin Film Technology", eds., L. J. Maissel and R. Glang, McGraw Hill, New York, (1970).
55. K. Kinosita, *Thin Solid Films*, **12**, 17, (1972).
56. R. W. Hoffman, in "Physics of Non-Metallic Thin Films", eds., C. H. S. Dupuy and A. Cachard, Plenum Press, New York, (1976).
57. M. O. Aboelfotoh, *Thin Solid Films*, **33**, 373, (1976).
58. G. V. Chaplygin and S. A. Semiletov, *Thin Solid Films*, **32**, 321, (1976).
59. K. L. Chopra, G. L. Malhotra and A. C. Rastogi, *Thin Solid Films*, **24**, 125, (1974).
60. S. Y. Wu, M. H. Francombe and W. J. Takei, *Thin Solid Films*, **36**, 509, (1976).
61. D. V. McCaugham and R. A. Kushner, *Thin Solid Films*, **22**, 359, (1974).
62. J. C. Anderson, "Dielectrics", Chapman Hall, London, (1966).
63. P. J. Harrop, "Dielectrics", Butterworths, London, (1972).
64. S. J. Fonash, in "Physics of Non-Metallic Thin Films", eds., C. H. S. Dupuy and A. Cachard, Plenum Press, New York, (1976).
65. A. K. Jonscher, *Thin Solid Films*, **36**, 1, (1976).
66. R. M. Hill, *Thin Solid Films*, **15**, 369, (1973).
67. A. Goswami and A. P. Goswami, *Thin Solid Films*, **16**, 175, (1973).
68. S. Winsztal, B. Wnuk, H. Majewska-Minor and T. Niemyski, *Thin Solid Films*, **32**, 251, (1976).
69. N. Fuschillo, B. Lalevic and B. Leung, *Thin Solid Films*, **24**, 181, (1974).
70. A. Schauer, M. Roschy and W. Juergens, *Thin Solid Films*, **27**, 111, (1975).
71. S. Luby, *Thin Solid Films*, **32**, 61, (1976).
72. R. T. Simmons, P. T. Morzenti, D. M. Smyth and D. Gerstenberg, *Thin Solid Films*, **23**, 75, (1974).
73. U. Saxena and D. N. Srivastava, *Thin Solid Films*, **33**, 185, (1976).
74. N. Fuschillo, B. Lalevic and N. K. Annamalai, *Thin Solid Films*, **30**, 145, (1975).
75. Y. Katsuta, A. E. Hill, A. M. Phahle and J. H. Calderwood, *Thin Solid Films*, **18**, 53, (1973).
76. M. H. Francombe, *Thin Solid Films*, **13**, 413, (1972).
77. P. L. Kirby, *Electronic Eng.*, **37**, 722, (1965).
78. J. C. Anderson and V. Rysanek, *Rad. and Elec.*, **39**, 321, (1970).
79. S. Hellström and H. Wesemeyer, Paper No. E2, *Int. Symp. on Vacuum and Thin Film Technology*, Uppsala, Sweden, Aug. 31–Sept. 3, (1976).
80. R. Castagne, P. Hesto and A. Vapaille, *Thin Solid Films*, **17**, 253, (1973).
81. S. J. Fonash, in "Physics of Non-Metallic Thin Films", eds., C. H. S. Dupuy and A. Cachard, Plenum Press, New York, (1976).
82. N. Mott, in "Metal-Insulator Transitions", Taylor and Francis, London, (1974).
83. G. S. Nadkarni and J. G. Simmons, *J. Appl. Phys.*, **41**, 545, (1970).
84. F. Argall and A. K. Jonscher, *Thin Solid Films*, **2**, 185, (1968).
85. J. A. Valles-Abarca and J. C. Anderson, *Thin Solid Films*, **11**, 113, (1972).
86. N. Klein and H. Gafni, *I.E.E.E. Trans. Electronic Devices*, **ED13,** 281, (1966).
87. N. Klein and N. Levanon, *J. Electrochem. Soc.*, **116**, 963, (1969).

88. P. Solomon, N. Klein and M. Albert, *Thin Solid Films*, **35**, 321, (1976).
89. V. K. Agarwal, *Thin Solid Films*, **24**, 55, (1974).
90. D. Gerstenberg, in "Handbook of Thin Film Technology", eds., L. J. Maissel and R. Glang, McGraw Hill, New York, (1970).
91. D. J. Walter and S. Johnson, *Thin Solid Films*, **16**, 325, (1973).
92. N. F. Jackson and D. S. Campbell, *Thin Solid Films*, **36**, 331, (1976).
93. D. S. Campbell and A. R. Morley, *Rad. and Elec. Eng.*, **43**, 421, (1973).
94. D. J. Hamilton and W. G. Howard, in "Basic Integrated Circuit Engineering", McGraw Hill, New York, (1975).
95. W. D. Westwood, N. Waterhouse and P. S. Wilcox, "Tantalum Thin Films", Academic Press, London and New York, (1975).
96. H. Vardhan, G. C. Dubey and R. A. Singh, *Thin Solid Films*, **8**, 55, (1971).
97. P. A. Timson and C. A. Hogarth, *Thin Solid Films*, **10**, 321, (1972).
98. G. C. Bubey, R. P. Mall and K. L. Chaudhary, *Thin Solid Films*, **24**, 261, (1974).
99. L. Biriotto and G. F. Piacentini, *Thin Solid Films*, **12**, 325, (1972).
100. M. C. Lancaster and R. J. Myton, *Thin Solid Films*, **13**, 243, (1972).
101. N. Schwartz and R. W. Berry, in "Physics of Thin Films", Vol. 2., eds., G. Hass and R. E. Thun, Academic Press, London and New York, (1964).
102. R. W. Berry, P. M. Hall and M. T. Harris, in "Thin Film Technology", Van Nostrand, New York, (1968).
103. H. Basseches and D. Gerstenberg, *Thin Solid Films*, **13**, 295, (1972).
104. S. Leppavuori, I. Suni and T. Stubb, *Thin Solid Films*, **36**, 365, (1976).
105. J. C. Anderson, *Thin Solid Films*, **12**, 1, (1972).
106. A. E. Hill and P. A. Rigby, *Thin Solid Films*, **13**, 21, (1972).
107. G. Krüger, *Thin Solid Films*, **12**, 335, (1972).
108. J. T. Law, *Thin Solid Films*, **36**, 232, (1976).
109. F. Forlani, *Thin Solid Films*, **36**, 313, (1976).
110. A. Pfahnl and R. W. Berry, *Thin Solid Films*, **13**, 51, (1972).
111. H. C. Narthanson and J. Guldberg, in "Physics of Thin Films", Vol. 8, eds., G. Hass, M. H. Francombe and R. W. Hoffman, Academic Press, London and New York, (1975).
112. P. K. Weimer, in "Handbook of Thin Film Technology", eds., L. I. Maissel and R. Glang, McGraw Hill, New York, (1970).
113. V. LeGoascoz, in "Physics of Non-Metallic Thin Films", eds., C. H. S. Dupuy and A. Cachard, Plenum Press, London and New York, (1976).
114. W. B. Glendinning and D. W. Yarborough, *Thin Solid Films*, **18**, 321, (1973).
115. F. Leuenberger, *Thin Solid Films*, **22**, 245, (1974).
116. C. A. Neugebauer, J. F. Burgess, R. E. Joynson and J. L. Mundy, *Thin Solid Films*, **13**, 5, (1972).
117. C. R. Crowell and S. M. Sze, in "Physics of Thin Films", Vol. 4, eds., G. Hass, M. H. Francombe and R. W. Hoffman, Academic Press, London and New York, (1967).
118. G. Dittmer, *Thin Solid Films*, **9**, 141, (1972).
119. W. T. Band, *Thin Solid Films*, **34**, 225, (1976).
120. Z. H. Meiksin, in "Physics of Thin Films", Vol. 8, eds., G. Hass, M. H. Francombe and R. W. Hoffman, Academic Press, London and New York, (1975).

121. L. I. Maissel, in "Handbook of Thin Film Technology", eds., L. I. Maissel and R. Glang, McGraw Hill, New York, (1970).
122. H. K. Kenisch and C. Popescu, in "Physics of Non-Metallic Thin Films", eds., C. H. S. Dupuy and A. Cachard, Plenum Press, London and New York, (1976).
123. T. W. Hickmott, *Thin Solid Films*, **9**, 431, (1972).
124. G. Dearnley, A. M. Stoneham and D. V. Morgan, *Rep. Prog. Phys.*, **33**, 1129, (1970).
125. D. P. Oxley, R. E. Thurstans and P. C. Wild, *Thin Solid Films*, **20**, 23, (1974).
126. R. E. Thurstans, P. C. Wild and D. P. Oxley, *Thin Solid Films*, **23**, 253, (1974).
127. D. P. Oxley and P. C. Wild, *Thin Solid Films*, **23**, 253, (1974).

Chapter 5†

Electrical Properties and Applications of Metal/Dielectric Films

T. J. Coutts

Industrial Science Research Centre,
Cranfield Institute of Technology,
Cranfield, Bedford, England

1. Introduction 163
2. Conduction in Discontinuous Metal Films 165
   2.1 The Influence of the Dielectric 179
   2.2 Discussion of Conduction in Discontinuous Films 180
3. Potential Applications of Totally Discontinuous Metal Films . . . 186
3.1 Gas Detectors 186
   3.2 Thermistors 187
   3.3 Strain Gauges 187
   3.4 Light Emitters 188
4. Percolation and Conduction in Semi-continuous Films 189
5. Properties and Applications of Films of Mixed Structure and Composition 195
   5.1 Cermet Resistors. 196
   5.2 Strain Gauges 201
   5.3 Other Applications 202
6. Conclusion 203
   References. 205

1. Introduction

In the previous two chapters the properties and applications of continuous metal and dielectric films have been discussed and to complete the review of these topics a third class of materials, mixed metal and dielectrics structures will now be considered. The physical properties of mixed metal/dielectric

† For a list of symbols used in this chapter and their definitions see p. x.

films are governed partly by the metal and partly by the dielectric component. The resistivity, for example, of films in this category is an extremely sensitive function of the relative proportions of the conducting and non-conducting phases and of the way in which these are distributed throughout the volume of the film. Many examples of mixed phase systems have found useful applications and these have been extensively studied over the course of the last century. These include antistatic coatings, resistors, screens for power cables, and high conductivity surface coatings. Attempts to develop a theory of conduction in thin films of mixed metal/dielectric composition have only relatively recently been made and as yet the theory is incomplete. The extent of the problem in reconciling theory and experiment can be appreciated by recalling that the ratio of the resistivities of the conducting and insulating phase is of the order of 10^{22} so that the electrical properties are very sensitive to both small changes in composition and structure. During the growth of a thin metal film on an inert substrate there are three readily identifiable structural regimes. In the initial stages of growth discrete nuclei are formed and these are generally stable once they consist of several atoms. The nuclei grow by capturing migrating surface adatoms or atoms direct from the vapour phase and when the island separation is reduced to a few tens of Angstroms it is found that direct electrical current can pass through the film with much less resistance than would be expected. With further island growth the stage is reached where coalescence occurs. This is generally accompanied by a more rapid decrease in electrical resistance. Eventually, island coalescence leads to the formation of an interconnected network of capillaries which conduct like a normal metal. The important feature of this class of structure is that the overall film resistance is dominated not by the resistivity of the capillaries but by the manner in which they are connected. It is quite usual to observe very wide differences in resistance between films having nominally identical masses of metal per unit area and this is due entirely to the distribution of material on the substrate. The third structural class, the continuous film, is formed when the holes between the capillaries are filled in and the film approximates to a plane parallel slab, generally of polycrystalline metal. The resistivity in this regime is governed by surface and grain boundary scattering as was discussed in Chapter 3. In this chapter the subject of discussion will be those films having an island and/or capillary structure in two and in three dimensions. The dielectric may be the substrate itself or an oxide round the boundaries of metal grains such as is sometimes found in thin film cermets.

The problem is particularly complicated because conduction can occur simultaneously between the discrete nuclei or through the interconnected capillaries so it is necessary to consider two distinct mechanisms of charge transport. That between individual nuclei is generally believed to occur by activated quantum mechanical tunnelling and in the following section the relevant theories of this phenomenon will be discussed. On the other hand, conduction in capillaries is rather more straightforward and follows the usual laws of metals with the modifications imposed by surface scattering discussed in Chapter 3. In both cases it is the statistical distribution of material on the substrate which dominates the conductivity and this is particularly relevant to these classes of film if they are to be considered for practical applications. In the case of charge transfer between discrete nuclei there is a close analogy with impurity conduction in a semiconductor. When an electron is transferred to an island, the latter remains charged until such time as the electron is thermally activated to a higher energy state from which it can tunnel to an adjacent island. At any time the number of electrons with the required activation energy follows the usual Boltzmann distribution (at least, for energies greater than a few times kT) in the same way as does the density of charged impurity atoms in a semiconductor. Thus metal islands on an inert substrate can be regarded as a macroscopic version of donors and acceptor atoms in a semiconductor. Transfer between the islands is not merely influenced by obvious geometrical quantities such as their size and separation but also, and probably of equal importance, by the dielectric within or upon which they are located. It has been suggested that traps in the dielectric can influence the conduction process and, in general, promote somewhat higher current densities than for a dielectric without traps. Even so the existence of traps is not able to account for the several orders of magnitude which are usually observed between the theoretical and measured film conductivities.

2. Conduction in Discontinuous Metal Films

In the totally discontinuous state individual islands or nuclei exist in a relatively stable form provided that they are not subjected to changes in environmental conditions such as temperature, etc. Because metallic links between the islands do not exist it might be expected that this class of film would not conduct electricity. However, early structural examinations established that even discontinuous films could permit the passage of current In addition it was found that the T.C.R. of these films was generally negative, as is also the case for semiconductors. The observations of significant conductivity and negative T.C.R. led to a great deal of experimental and theoretical study and these phenomena can be regarded as the inherent properties of discontinuous films. Although thermionic emission of electrons between islands has been proposed (van Steensel[1]) as a possible mechanism of charge transfer, this theory will not be discussed here since it is found to be applicable only to those structures having large island separations and low potential barriers. The predicted current densities are much lower than those observed in practice. Hartman[2] proposed a model which invoked the observation that in islands consisting of relatively few atoms, the energy levels do not spread into continuous bands but rather into many separate bands each of a very small energy width. It was suggested that an activation energy was required to raise the energy of an electron in an island from the ground state to the first excited level, since the probability of overlap between ground states is very small, but as the excited state is much wider the corresponding probability would be much larger. Therefore tunnelling was supposed to occur between the first excited states of adjacent particles. However, with increasing applied voltage, the stage must be reached where the first excited levels of the two particles uncross and at this stage it would be expected that the conductivity would fall. This has never been observed in practice. In addition there is a further more fundamental reason why the model cannot be accepted. The activation energy is calculated on the basis of the separation of the lowest and next highest states of the potential well formed by the island. In fact it should be based on the energy difference between a state at the Fermi level and the next highest unoccupied state. This correction gives activation energies much smaller than Hartmann's value or of those observed in practice.

Quantum mechanical tunnelling has been studied for many years in connection with current flow between plane parallel electrical contacts when their separation is of the order of a few tens of Angstroms and because the separation of islands, when significant current flows in a discontinuous film, is of the same approximate magnitude it was natural to attempt to apply the theory to thin films. Objections to doing so have been made on the grounds that tunnelling is not very sensitive to temperature whereas the T.C.R. of discontinuous thin films can be up to 1% per °C negative. Darmois[3] however, realised that when an electron is transferred from one neutral island to another a certain amount of work must be done against electrostatic forces, and as a consequence only electrons with an energy of the order of e/r above the Fermi level of the initial island could move to a nearby island. The quantity e/r is the work done in transferring an electron from the surface of an island of radius r to infinity. This energy can be supplied thermally and, if Boltzmann statistics are obeyed, then the number of charged islands can be written as:

$$n = N \exp\left(-\frac{\delta E}{kT}\right) \quad (1)$$

where N is the total number of islands: δE is the activation energy which must be supplied to overcome the electrostatic forces and it can be expressed approximately by:

$$\delta E = \frac{e}{4\pi\varepsilon_0\varepsilon_r r} \quad (2)$$

where δE is measured in electron volts if ε_0, e and r are expressed in MKS units. Neugebauer and Webb[4] extended these ideas to develop the first theory of conduction in a discontinuous film. The probability that an electron will tunnel from one island to another depends firstly, on the transmission coefficient D, which will be discussed in greater detail later, but at this time it will be regarded as a constant, and secondly, on the product of the fraction of full states in the first island and the fraction of empty states in the second island. This product will be calculated using the potential diagram in Fig. 1 and will then be used to formulate an expression for the conductivity of

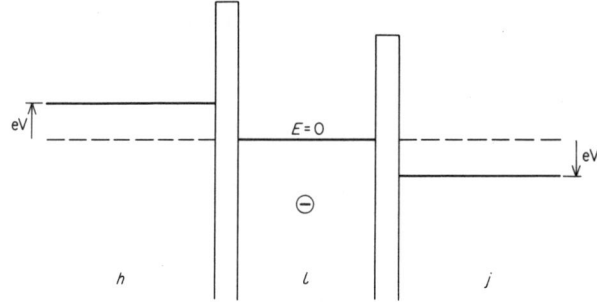

FIGURE 1. Simple potential diagram for three particle tunnelling model of Neugebauer and Webb[4].

the island array. The figure shows the potential diagram for three metal particles between which there is an applied field which leads to a relative displacement of the respective Fermi levels. Without an applied field the rates of electron transfer by tunnelling to and from the central island are equal and there is no net flow of current. The presence of the field reduces the rate of tunnelling from i to h and increases it from i to j so that there is a net flow in the field direction. The net probability of transition is simply the difference of the two tunnelling terms viz:

$$P_{net} = P_{ij} - P_{ih} \quad (3)$$

where

$$P_{ij} \propto D \int_{-\infty}^{\infty} f_i(1-f_j)\,dE \quad \text{and} \quad P_{ih} \propto D \int_{-\infty}^{\infty} f_i(1-f_h)\,dE$$

Since the particles are metallic the functions f_{ijh} are given by the appropriate Fermi distributions;

$$f_i = (1+\exp[E/kT])^{-1} \tag{4a}$$

$$f_j = (1+\exp[E+eV]/kT)^{-1} \tag{4b}$$

$$f_h = (1+\exp[E-eV]/kT)^{-1} \tag{4c}$$

In arriving at these expressions it has been assumed that the potential of the central island is zero and that islands j and h are raised to potentials $+eV$ and $-eV$, respectively, above that of island i. It is reasonably straightforward to substitute the Fermi functions given by equations (4) into equation (3), to carry out the integration and to show for the practical case of $eV \ll kT$ that:

$$P_{net} \propto DeV \tag{5}$$

An appropriate constant of proportionality can be inserted into equation (5) and an island of area of the order of r^2 can be assumed so that the probable number of electron transitions per second is:

$$P = Kr^2 DeV \tag{6}$$

Since the average time τ per transition is equal to $1/P$, the average velocity of electrons during transition must be given by:

$$s/\tau = Kr^2 s\, DeV \tag{7}$$

where s is the separation of the islands. The charge mobility is given by the average velocity divided by the field, or

$$\frac{s/\tau}{V/s} = Ks^2 r^2 De \tag{8}$$

from which the conductivity can be obtained by combining equation (1) and (8) to give:

$$\sigma = n_r e\mu = Ks^2 e^2 D \exp\left(-\frac{\delta E}{kT}\right) \tag{9}$$

where $n_r = n/r^2$.

At first sight this equation appears incorrect since it predicts that conductivity increases as the square of island separation. However, the quantity D is a far more sensitive function of s and decreases very rapidly with increasing s so that the conductivity decreases as well. Since transfer of electrons is between islands separated by a distance s, the activation energy is less than that required to move an electron to infinity, by a factor

$$\left(\frac{r+s}{2r+s}\right)$$

and with an electric field applied it is further reduced so that the expression for the activation energy becomes:

$$\delta E = \frac{e}{4\pi\varepsilon_0\varepsilon_r r}\left(\frac{r+s}{2r+s}\right) - Fs \tag{10}$$

By differentiating this expression with respect to s it can be shown that the maximum in the activation energy is given by:

$$\delta E_{max} = \frac{e}{4\pi\varepsilon_0\varepsilon_r r} - \frac{2(eF)^{\frac{1}{2}}}{(4\pi\varepsilon_0\varepsilon_r)^{\frac{1}{2}}} + 2rF \tag{11}$$

which is very similar to the expression for Schottky lowering of a potential barrier to thermionic emission.

Although the theory of Neugebauer and Webb was able to account for several of the experimental observations its major shortcomings were that the activation energy is part of the potential barrier through which the electrons must tunnel and so should be included in the Fermi functions, and that the transmission coefficient was not regarded as a function of energy.

Hill[5] overcame these objections by combining the model of Neugebauer and Webb with a detailed theory of tunnelling developed by Simmons[6]. Based on the model shown in Fig. 2 the appropriate Fermi functions were given as:

$$f_i = \{1 + \exp[(E_F - E_x - E_r + \delta E)/kT]\}^{-1} \tag{12a}$$

$$f_j = \{1 + \exp[(E_F - E_x - E_r + \delta E + eV)/kT]\}^{-1} \tag{12b}$$

$$f_h = \{1 + \exp[(E_F - E_x - E_r + \delta E - eV)/kT]\}^{-1} \tag{12c}$$

It is convenient to separate the electron energy into two components, that in the x direction E_x and the sum of the components in the y and z directions E_r, because it is the transmission coefficient as a function of E_x which is of interest since it is in this direction that current flow occurs. The current density can then be written as:

$$J(V, T) = \frac{4\pi m e}{h^3} \int_0^\phi D(E_x)\,dE_x \int_0^\infty f_i(f_h - f_j)\,dE_r \tag{13}$$

where $D(E_x)$ is given by:

$$D(E_x) = \exp-\left(\frac{4\pi}{h}\int_{s_1}^{s_2}\{2m[E_F + \phi(x) - E_x + \delta E]\}^{\frac{1}{2}}\,dx\right) \tag{14}$$

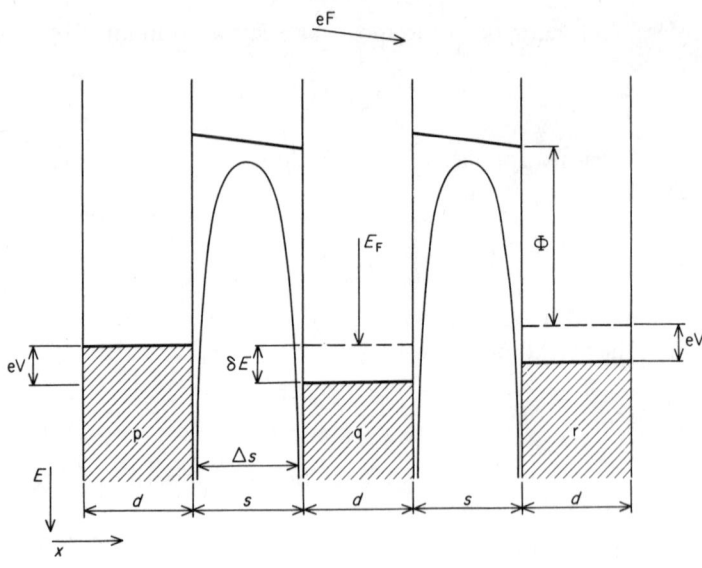

FIGURE 2. Potential diagram for three particle model as modified by image force and activation energy effects (Hill[5]).

Equation (14) is known as the WKB approximation to the transmission coefficient which is derived from the Schroedinger wave equation. The integral over E_r effectively gives the number of electrons which can participate in tunnelling whilst the integral over E_x gives us the probability that they will tunnel. Equation (13) can be solved using further approximations and for $\delta E < kT$ gives:

$$J(V, T) = \frac{8\pi me}{h^3 B^3} \sinh \frac{eV}{kT} \exp(-A\bar{\phi}^{\frac{1}{2}}) \frac{\pi BkT}{\sin \pi BkT} \exp\left(-\frac{\delta E}{kT}\right)[1 - \exp(-B\,\delta E)] \quad (15)$$

where $\bar{\phi}$ is the average barrier height

$$\left[\text{i.e. } \bar{\phi} = \frac{1}{s_1 - s_2} \int_{s_1}^{s_2} \phi(x)\,dx, \quad A = \frac{4\pi(s_1 - s_2)}{h}(2mm^*)^{\frac{1}{2}} \text{ and } B = A/2\bar{\phi}^{\frac{1}{2}}\right]$$

The main feature of equation (15) is that the principle temperature dependence arises from the activation energy term and as a consequence a linear Arrhenius plot may be expected. This almost invariably turns out to be the case but Hill points out that at high temperatures some curvature is found which is due to the term $\pi BkT/\sin \pi BkT$. It is possible to extract the term B from the curvature and hence to calculate the effective mass of tunnelling electrons and the average barrier height $\bar{\phi}$. Kiernan and Stops[7] further generalised the mathematics of this model and expressed the tunnel current density as:

$$J(V, T) = \frac{4\pi me}{h^3 B^2} \exp(-A\bar{\phi}^{\frac{1}{2}}) \frac{\pi BkT}{\sin \pi BkT}$$

$$\times \left[\frac{1 - \exp(-B(\delta E + eV))}{1 - \exp(\delta E + eV)/kT} \pm \frac{1 - \exp(\pm B(\delta E - eV))}{1 - \exp(\delta E - eV)/kT} \right] \quad (16)$$

By choosing as their energy zero the position of the Fermi level in the metallic island, rather than the bottom level of the conduction band, Barr and Finney[8] were able to remove the rather unsatisfactory positive and minus signs used in equation (16). This procedure led to a more generalised form of the equation derived by Hill and showed that as $\delta E \to 0$ so the tunnel current density approached the value predicted by Simmon's theory for plane parallel metal-insulator-metal sandwiches. Other tunnelling models have been developed and invoke such possibilities as doubly charged islands (Swanson et al.[9]) but unfortunately these do not really fit the facts better than the less complicated theories. All the theories discussed so far enable a calculation of the current density through an array of three particles to be made but they do not permit a direct calculation of the resistance of a thin film. The test of any theory is the extent to which it agrees not merely qualitatively but also quantitively with experimental observation and unfortunately the above theories are quite unable to account for the relatively high currents found in practice. One reason for this may be that in a typical thin film there is a wide range of island sizes and separations rather than a uniform array and this would lead to relatively few preferential paths carrying most of the current.

Recently, Abeles et al.[10] have made some progress in this area. Their work was concerned with the temperature and field dependence of the conductivity of several metal/dielectric cermet combinations at right angles to the plane of the film as a function of the volume fraction of metal.

Now although within a given film, r, s, and δE are not constant it was suggested that the quantity $s \, \delta E$ is constant and depends only on the volume fraction of metal. If the film can be represented by a simple cubic lattice with a constant $(s + 2r)$, then the volume fraction of metal is given by:

$$x = \frac{4\pi r^3 / 3}{(s + 2r)^3} = \frac{4\pi}{3} \div \left(\frac{s}{r} + 2\right)^3$$

$$\therefore \quad \frac{s}{r} = 2\left[\left(\frac{\pi}{6x}\right)^{\frac{1}{3}} - 1\right] \quad (17)$$

now
$$\delta E = \frac{e}{4\pi\varepsilon_0\varepsilon_r r}\left[\frac{r+s}{2r+s}\right]$$

$$\therefore \quad s\,\delta E = \frac{e}{4\pi\varepsilon_0\varepsilon_r}\left(\frac{s}{r}\right)\frac{\left(1+\dfrac{s}{r}\right)}{\left(2+\dfrac{s}{r}\right)} \tag{18}$$

Now because of the processes of diffusion and capture by which an island structure is formed, large islands are to be associated with large gaps and it can therefore reasonably be assumed that in any macroscopic volume of the film the ratio of s/r will be equal to the average value given by equation (17). Thus, equations (17) and (18) show that the quantity $s\,\delta E$ can indeed be expected to be a constant for all regions of the film even though the individual structural parameters may vary from one region to the next. This result is of vital importance to the following argument and will presumably be of significance to any reader who has worried about structural non-uniformities in films. The low field, finite temperature conductivity relies upon thermal activation and on tunnelling and will now be derived according to the theory of Abeles et al.[10].

Firstly, consider the density of charges with a charging energy $\delta E'$; this is assumed to be proportional to the Boltzmann factor

$$\exp\left(-\frac{\delta E'}{kT}\right)$$

One of the problems which has been an obstacle to progress in this field is the apparently vast number of conduction paths available to a given charge. However Abeles et al. attempted to simplify the situation using the following argument. A charge with energy $\delta E'$ cannot tunnel to a nearby smaller island with $\delta E'' > \delta E'$ since it has insufficient energy. Furthermore, it is unlikely to tunnel to an island with $\delta E'' < \delta E'$ since the rule $s\,\delta E = $ const. would imply a much wider gap and therefore a very much lower transmission probability. Consequently the optimum tunnelling path is likely to be between islands of similar sizes and separations and Abeles et al., therefore, neglected interaction between dissimilar sized islands. In the case of hopping conduction in amorphous materials, transitions between sites of different energies are permitted and despite the above arguments the possibility of transitions between islands of different sizes should not be neglected. This will be discussed again later.

Now the tunnelling probability is given by $p = \exp(-2\chi_0 s)$ which is a further simplification of the WKB approximation mentioned earlier and in which χ_0 is given by $\chi_0 = (2m\phi/\hbar^2)^{\frac{1}{2}}$. (Note that in this treatment the energy dependence of χ is disregarded.) χ_0 was taken as proportional to the mobility of the charges. Since there will be a distribution of values of s there will also be a large number of possible conduction paths, the number density which is described by a function $\beta(s)$. Thus the conductivity of the film is proportional to be product of number of paths, charge, number of charges and their mobility, integrated over all s, viz:

$$\sigma_L \propto \int_0^\infty \beta(s) \exp\left(-\frac{\delta E}{kT}\right) \exp(-2\chi_0 s) \, ds \qquad (19)$$

After considerable simplification and a number of approximations equation (19) gives:

$$\sigma_L \propto \frac{\sqrt{\pi}}{2\chi_0} \left(\frac{C}{kT}\right)^{\frac{1}{4}} \beta(s_m) \exp\left[-2\left(\frac{C}{kT}\right)^{\frac{1}{2}}\right] \qquad (20)$$

In this equation, $C = 2\chi_0 s \, \delta E$. It is only because $s \, \delta E = $ constant that this substitution can be made and all the foregoing results hinge on this point. The quantity C is of vital importance to the whole theory. For a given value of x it is independent of grain size and the nature of the particular metal/dielectric combination affects its value only through the barrier height and dielectric constant. It is thus a most useful parameter by which to characterise a given combination. s_m is the position of the maximum in the function $\exp[f(s)]$ where $f(s) = -2\chi_0 s - C/2\chi_0 s kT$, so that $\beta(s_m)$ is the number of conduction paths corresponding to this island separation. Finally, by neglecting any temperature dependence in the pre-exponential factors (20) reduces to:

$$\sigma_L = \sigma_0 \exp\left[-2\left(\frac{C}{kT}\right)^{\frac{1}{2}}\right] \qquad (21)$$

This important result is clearly quite distinct from those of other theories because of the predicted $\exp(-T^{-\frac{1}{2}})$ dependence which was observed in several metal/dielectric combinations. It has always been difficult to understand why films with wide ranges of island sizes and separations should exhibit a unique activation energy and the above theory suggests that in fact this should not be the case. Abeles et al. state "Some previous workers have plotted $\log \sigma_L \sim 1/T$ in order to extract a single activation energy δE. However, closer inspection shows that either a better fit to the data can be obtained with $\log \sigma_L \sim 1/T^{\frac{1}{2}}$ or the temperature range of measurement was too small to distinguish from the $1/T^{\frac{1}{2}}$ behaviour." The authors also claim that the results are readily distinguishable from the $1/T^{\frac{1}{4}}$ law for hopping conductivity. There are however several assumptions in the above theory and these should be appreciated.

(i) It is assumed that $\beta(s)$ can be replaced by $\beta(s_m)$ which is then taken outside the integral. This would only be valid for a very sharply peaked distribution.

(ii) It is tacitly assumed that all tunnelling takes place at the Fermi level since the probability term involves the value of χ at that particular position, although the product $f(1-f)$, where f is the Fermi function, has a maximum at $E = E_F$. Neglecting contributions from nearby energy states may not be wholly justifiable.

(iii) χ is not regarded as a function of s but since it involves ϕ which, because of image force effects, depends on s this is not strictly correct.

(iv) The $1/T^{\frac{1}{2}}$ dependance arises partly from the fact that the function $f(s)$ is expanded around the maximum s_m and then truncated at the second term. Presumably a quite different dependance would result if extra terms were taken into account so that the validity of this assumption is governed by the standard deviation of the appropriate distributions. Once again this approximation assumes a sharply peaked distribution.

Despite these criticisms the agreement with experiment was reasonable as shown in Fig. 3 for $Ni-SiO_2$ cermet films over a wide range of metallic volume fraction.

Following the above theory of low field, thermally activated conduction Abeles *et al.* then go on to derive the high field, low temperature conductivity and then to extend this to finite temperature. The relative importance of the two processes can be judged by the ratio $kT/e\Delta V$. When this quantity decreases to less than unity the conductivity becomes non-ohmic.

At $T = 0$ a simple Boltzmann factor cannot be used to represent the number of available charges since they are not thermally activated. This must be done using a density of states function $\rho(E)$ which is multiplied by G, the volume of the grain and v, the frequency with which the electrons strike the potential barrier, and then integrated over all E to obtain the number of carriers per unit time striking the barrier between islands. Two additional functions, $\theta(E)$ and $\theta(E + e\Delta V - \delta E)$, are used to describe whether the initial and final states are full or empty, and must also be included within the integral sign. These are defined such that $\theta(\mu) = 0$ for $\mu > 0$ and $\theta(\mu) = 1$ for $\mu < 0$; which appears to be identical to the Fermi-Dirac condition.

The conductivity measurements made by Abeles *et al.* were across the plane of the films and their structure was assumed to consist of layers of islands, each layer being at a fixed potential.

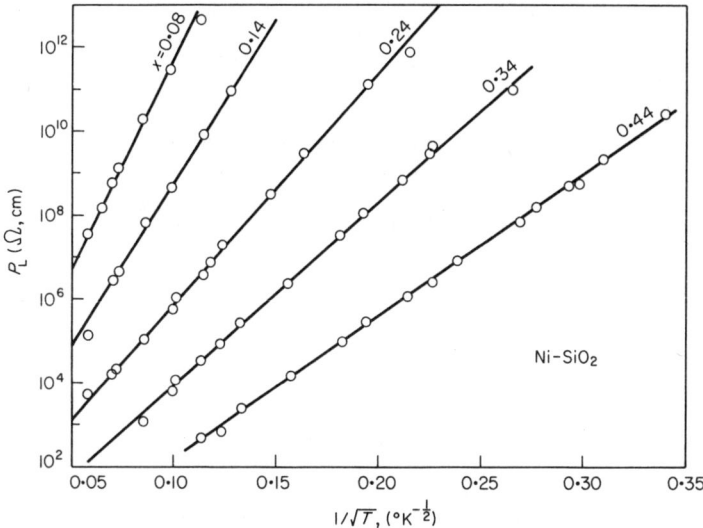

FIGURE 3. Log resistivity $\sim 1/T^{\frac{1}{2}}$ for nickel/silicon dioxide cermets. (Abeles et al.[10].)

The rate of tunnelling from one island to another in the next layer is then given by:

$$r(s) = \gamma G \int_{-\infty}^{\infty} v \exp(-2\chi s) \theta(E) \theta(E + e\Delta V - \delta E) \rho(E) \, dE \quad (22)$$

γ is a factor which takes into account the irregular shape of the grains, and the factor $\exp(-2\chi s)$ gives, as before, the transmission coefficient between two equipotential layers. Once again because of the relationship $s \, \delta E =$ constant we can make the substitution $2\chi s \, \delta E = C$ and this leads to a result of fundamental importance. Thus (22) and the subsequent comments give:

$$r(s) = 0 \text{ for } \frac{C}{e\Delta V} > 2\chi s$$

and

$$r(s) = \frac{\gamma}{h} \exp(-2\chi s) \left[e\Delta V - \frac{C}{2\chi s} \right] \text{ for } \frac{C}{e\Delta V} < 2\chi s \quad (23)$$

Equation (23) implies that tunnelling will not occur unless the voltage drop between islands is greater than a defined critical value. The product $vG\rho(E)$ was shown to be equal to $1/h$ although it is not necessary to derive this

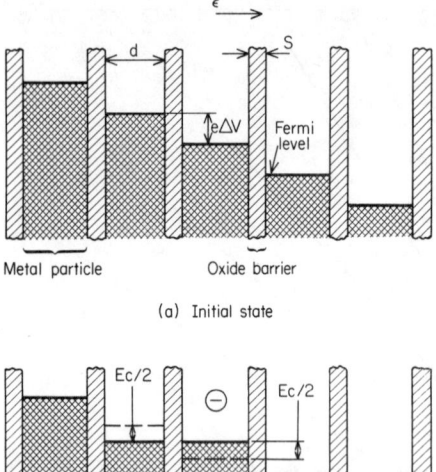

FIGURE 4. Potential diagram used by Abeles et al.[10].

relationship here. The important feature of (23) is that for a given structure there is a critical voltage below which tunnelling cannot occur. This is consistent with the band diagram representation of the structure in Fig. 4. When equation (23) is integrated over all possible values of s, the result is an expression for the high field conductivity at $T = 0$ as:

$$\sigma_H = \sigma_\infty \exp\left(-\frac{F_0}{F}\right) \tag{24}$$

where σ_∞ and F_0 are constants expressable in terms of the various parameters of the film.

To extend the above arguments to the case of finite temperature the functions $\theta(E)$ and $\theta(E + e\Delta V - \delta E)$ are replaced by the appropriate Fermi functions. After the required simplifications Abeles et al. show that:

$$\sigma_H = \sigma_\infty \exp\left(-\frac{F_0}{F}\right) \int_{-F_0/F}^{\infty} z \exp(-z) \frac{g(s)}{1 - \exp\left[-\frac{z}{(z + F_0/F)} \cdot \frac{C_0}{F_0 kT}\right]} dz \tag{25}$$

where z is a function of s and the film parameters and $C_0 \simeq C$ above.

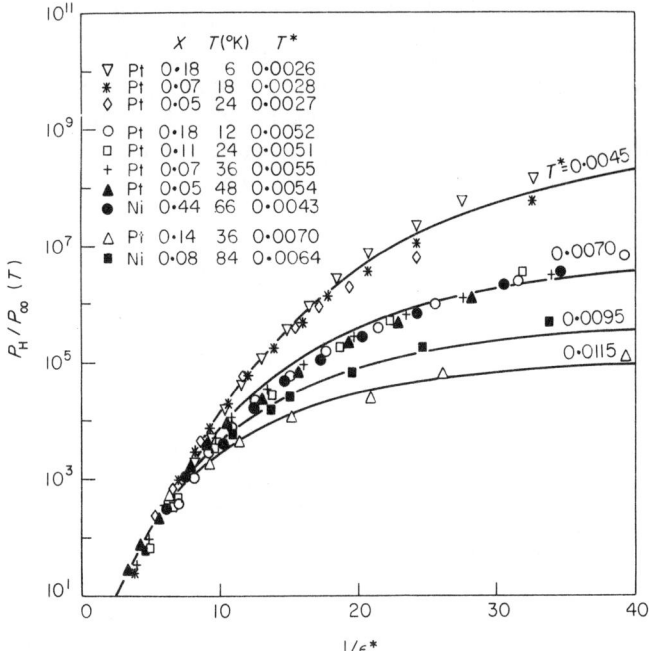

FIGURE 5. Log of the ratio of high to low field resistivities for various cermet films as a function of normalised field — normalised temperature as a parameter (Abeles et al.[10]).

$g(s)$ is the probability distribution of s which for the purpose of comparing theory with experiment was equated to unity. Equation (25) can be written as:

$$\cong \sigma^* = \frac{\sigma_H}{\sigma_\infty} = \exp\left(-\frac{1}{F^*}\right) \int_{-1/F^*}^{\infty} z \exp(-z) \frac{dz}{1 - \left[\exp\left(-\frac{z}{z+1/F^*}\right) \cdot \frac{F^*}{T^*}\right]} \quad (26)$$

where F^* and T^* are normalised field and temperature given by:

$$F^* = \frac{F}{F_0} \quad \text{and} \quad T^* = \frac{kT}{C_0} = \frac{kT}{2\chi s \, \delta F_0} \quad (27)$$

The most important feature of equation (26) is that it predicts that the dimensionless high field conductivity is a function only of the normalised field and temperature. Consequently, a wide range of structures of various metal/dielectric combinations will have a unique set of curves of $\sigma^* \sim F^*$ for the range of T^*. Alternatively, the variables which determine T^* could all change from one film to the next but provided that the value of T^* did not change significantly, the normalised high field conductivity would follow a single curve. The validity of this prediction is admirably illustrated in Fig. 5. Inset in this diagram are the compositions of the films and the values of T^* calculated from the low field conductivity. The values of T^* beside each curve are those which give the best fit to the particular set of data and it can be seen that there is reasonable agreement.

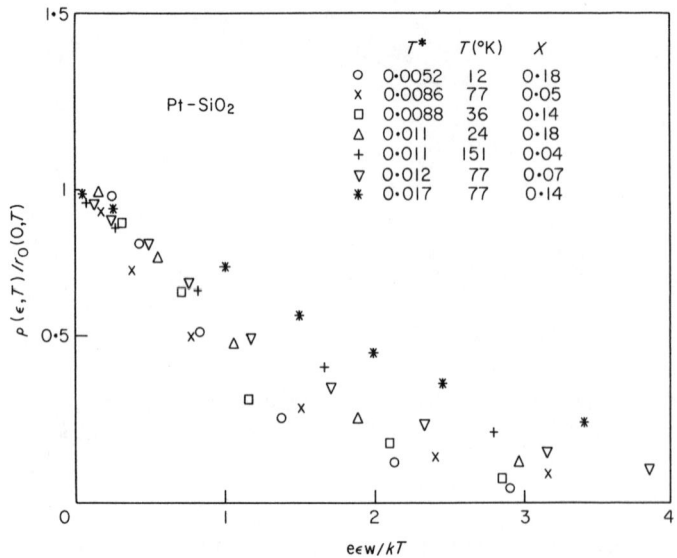

FIGURE 6. Log of the ratio of high to low field resistivities of platinum/silicon dioxide films as a function of the ratio of normalised field to normalised temperature (Abeles et al.[10]).

Further justification of the theory was provided by determining the position at which the transition from the ohmic low field to the non-ohmic high field conductivity occurred. It was stated earlier that this could be expected to occur when $(e\Delta V/kT) \simeq 1$. However, the transition can be expressed in terms of the normalised variables as $F^*/T^* \simeq 1$. In Fig. 6 σ_H/σ_L is plotted against F^*/T^* for the combination platinum/silicon dioxide and it can be seen that for a given value of T^* the deviation from linearity does indeed occur approximately at $F^*/T^* = 1$.

2.1 The Influence of the Dielectric

In developing his theory of charge transfer between islands in a discontinuous film Hill[5] pointed out that electrons were very much more likely to tunnel via the substrate than through the vacuum separating the particles. The reason for this is that the potential barrier to electrons tunnelling through the substrate is lower than the metal/vacuum interface; the difference in the height of the two barriers being the electron affinity of the insulating substrate. Therefore it is apparent that any variations in the composition of the substrate between islands will distort the potential barrier presented to tunnelling electrons. It seems reasonable to assume that free ions near the substrate surface could lead to an increase or a decrease in the tunnel current but in fact the various workers who have invoked this model have only done so with a view to explaining the fact that theoretical current densities are always less than those observed in practice. Schmidlin[11], for example, showed that under certain conditions ions could lead to a substantial increase in the tunnel current but some doubt was cast on this work by Gadzuk[12] whose theory of resonance tunnelling via ions suggested that randomly positioned ions would be unlikely to affect the conductance radically.

On the other hand Hill[13] demonstrated conclusively that the conductance of discontinuous films could be modulated by an electric field applied to the back of the substrate and interpreted his results in terms of sodium ion traps in the substrate between which charge transfer took place.

Fehlner and Irving[14] repeated these experiments and also concluded that sodium ions were responsible although these authors invoked a rather different physical model to that used by Hill to explain his results. The details are not important here since the purpose was merely to indicate the active role of the substrate. Further models based on the active role of the substrate have been developed by Herman and Rhodin[15] and by Milgram and Lu[16]. Once again the general form of the predicted temperature dependences was qualitatively correct. It is interesting to note in the former case, the parallel drawn between conduction between islands and impurity band conduction in certain crystalline materials. It is quite clear that the substrate does play an important part in the charge transfer process although its precise role is by no means clear.

Van Steensel[1] deposited films onto barium-titanate substrates which are ferroelectric and change their dielectric constant radically at a critical temperature. Van Steensel argued that the electrostatic activation energy is inversely proportional to the permittivity so that if there is a sudden change in the latter at a particular temperature then the conductivity should also change dramatically. This however was not observed. This experiment cannot be taken as ruling out the active role of the substrate since it is only the low frequency dielectric constant which changes its value suddenly and it can be argued that, because tunnelling of electrons takes place in a very short time, it is the high frequency value which should be employed. There is clearly some uncertainty in this area and possibly measurements on discontinuous films made over a wide frequency and temperature range could clarify the situation. It is an interesting point that nowhere in the literature can any reference to the temperature coefficient of the dielectric constant and its likely effect on conductivity be found.

This is another complication somewhat similar to that introduced by thermal expansion and mechanical strain effects which cannot easily be incorporated into the rather idealised physical models of discontinuous films. It is particularly interesting to note despite the obvious importance of the geometry of the islands that the charge transfer process is, as in the case of metal-insulator-metal sandwiches, almost certainly limited by the dielectric itself.

2.2 Discussion of Conduction in Discontinuous Films

At this stage it will be appreciated that there are essentially two theories of conduction in discontinuous films. The first, due to Hill[5], predicts a T^{-1} characteristic, typical of a singly activated process and a uniform distribution of island sizes and separations, whilst the second due to Abeles et al.[10], predicts a $T^{-\frac{1}{2}}$ behaviour which is characteristic of a non-uniform structure. Although the two laws appear to be fundamentally different they are in fact simply separate facets of the same problem and it will be the object in this section to show that this is indeed so.

The theory of Abeles et al. contains many assumptions and there is no obvious fundamental reason why it should be preferred to Hill's theory if the latter could be modified in some way to incorporate the spread in island sizes and separations to calculate the conductivity of the entire array of islands rather than just that of a pair of supposedly typical particles. The rule $s\,\delta E$ can be applied to Hill's theory with some interesting consequences. The principle structural dependance of the conductivity in equation (15) is contained in the terms $\exp-(A\bar{\phi}^{\frac{1}{2}})$ and $\exp-(\delta E/kT)$. Although the quantity B appears several times outside these exponential terms its effect is less pronounced. Now A is directly proportional to Δs or to a good approximation to s therefore the first term can be written as $\exp-(K's\bar{\phi}^{\frac{1}{2}})$. Similarly because s/r is assumed constant throughout the film, leading to $s\,\delta E$ = constant, the second term may be written as $\exp-(K/s)$. The two structure sensitive terms can now be linked as:

$$f(s) = \exp-\left(K'\bar{\phi}^{\frac{1}{2}}s + \frac{K''}{skT}\right) \qquad (28)$$

This is clearly very similar to the procedure used by Abeles et al. Now the exponent of equation (28) has a maximum in the value of s, given by:

$$s_m = \left(\frac{K''}{K'\bar{\phi}^{\frac{1}{2}}kT}\right)^{\frac{1}{2}} \qquad (29)$$

so it is reasonable to expect the tunnel current to flow mainly through these gaps. In fact the probability distribution of island separations should also be taken into account because even though s_m is the preferred gap on the basis of the opposing effects of activation energy and tunnelling probability, if there are very few gaps of this value then the corresponding current may still be relatively small. However, without the precise distribution it is not possible to proceed with this approach. If the bulk of the current flows through the gaps corresponding to s_m, then s_m may be substituted for s (equation 28 in 29) and this yields a $T^{-\frac{1}{2}}$ law for the temperature dependence of conductivity.

To investigate the $T^{-\frac{1}{2}}$ law in greater detail, Hill and Coutts[17] have carried out an analysis of the data published by Abeles et al.[10] which was taken as verification of the $T^{-\frac{1}{2}}$ law. Abeles et al. plotted $\ln \sigma \sim T^{-\frac{1}{2}}$ and because the data lay on a straight line this was taken as verification of the $T^{-\frac{1}{2}}$ law. However, Hill[18] has previously shown that this procedure is not ideal and can lead to misleading results. By plotting $\ln \sigma \sim T^{-\frac{1}{2}}$ it has tacitly been assumed that the relationship is obeyed. The alternative method proposed by Hill[18] was to define an activation energy:

$$\delta E = -\frac{\partial \ln \sigma}{\partial (1/T)} \qquad (30)$$

which for the generalised relationship:

$$\sigma = \sigma_0 \exp - \left(BkT\right)^n \qquad (31)$$

will take the form:

$$\delta E = nB(kT)^{1-n} \qquad (32)$$

Therefore a plot of $\log \delta E \sim \log T$ gives a straight line whose gradient is equal to $(1-n)$ and from which the power of the temperature dependance can be determined without any prior assumption. This was carried out for all the data published by Abeles et al. and the results of this analysis are shown in Table 1. If a gradient of 0.5 ± 0.05 and a regression coefficient of greater than approximately 0.8 is accepted as significant then it can be seen that six of the twenty films exhibit a temperature dependance supporting the $T^{-\frac{1}{2}}$ law. This evidence suggests that the law is not applicable to all cermet films (and presumably, two dimensional, discontinuous films) but is limited to specific cases.

More recently, Hill and Coutts[19] have published a generalised theory of conduction in discontinuous two and three dimensional structures which takes into account the statistical distribution of island sizes and separations. This shows that the power of the temperature dependance is a consequence of a variable range transport process and its magnitude is a function of the distributive properties of the system. Furthermore its magnitude is also governed by the temperature range over which the conductivity is measured and, as will be seen, it is this feature which reconciles the original theory of Hill[5] and that of Abeles et al.[10].

In their treatment of the problem Hill and Coutts[19] considered three particle size distributions, these being, firstly independant of particle size, secondly proportional to r^{-1} and thirdly proportional to r^{-2}. Furthermore, the range of particle sizes was taken to be finite; the maximum and minimum values being used to normalise the total probability of the distribution to unity. The evidence presented by Abeles et al.[10] tends to support the first hypothesis above that the size distribution is independent of particle size.

The distribution of the centre to centre spacing, R, of the particles was assumed to be random and therefore to follow the distribution function:

$$f(R) = \rho_2 R^p \tag{33}$$

where ρ_2 is equal to 4π times the volume density of particles and $p = 2$ for a three dimensional system, or where ρ_2 is equal to 2π times the area density of particles and $p = 1$ for a two dimensional system. Using the distribution functions for r and R, a similar function $g(E_s)$ for the distribution of energies required to transfer an electron from one island to another was then generated and was expressed in a simplified form in terms of the maximum and minimum energies of the distribution as calculated from the corresponding values of r and R.

Similarly, a distribution function for the total number of inter-island spacings $g(s)$ consistent with the restrictions of finite energy ranges of the system, was also derived. The distribution functions $g(E_s)$ and $g(s)$ were derived for both two and three dimensional systems having any one of the three particle size distributions discussed earlier.

It was then assumed that transfer of charge between islands took place by a hopping process and that the probability of hopping was proportional to:

$$\exp-\left(\frac{E_s}{kT} + 2\alpha s\right) \tag{34}$$

where E_s is the energy range over which the hop takes place, between two islands separated by a distance $s \cdot \alpha$ is the decay distance of the electronic wave functions. The above term is identical to that first used by Miller and

Abrahams[20] to describe impurity band conduction and later used by Mott and Davis[21] to describe hopping conduction between localised sites in amorphous semiconductors.

The array of particles was considered to be divided into two classes; firstly, a sub-matrix with a probability of connectedness greater than a critical minimum value and secondly, a disconnected portion which takes no part in the D.C. conduction process.

If the critical hopping probability is $\exp(-h_c)$ then we can write:

$$\exp-\left(\frac{E_s}{kT}+2\alpha s\right) \geqslant \exp(-h_c) \qquad (35)$$

for hopping to occur between two sites. For overall connectivity it is necessary that the average number of neighbours to which hopping can take place is greater than a critical value given by:

$$N_c = \frac{D}{D-1} \qquad (36)$$

where D is the dimensionality of the system. Now the average number of sites accessible from a datum particle can be expressed in terms of the distribution functions discussed earlier as:

$$N = \int g(s) \int g(E_s)\, ds\, dL_s \qquad (37)$$

and for connectivity to be established this must be equal to or greater than N_c. The range of energy and separation over which equation (37) was integrated were obtained from the straight line on an s against E_s diagram defined by equation (35) and shown in Fig. 7. E_u and E_v are taken as the lower and upper limits of E_s. From this diagram it can be seen that there are two areas of consideration. Firstly, the rectangle bounded by the abscissae

$$s = 0 \quad \text{and} \quad s = \frac{1}{2\alpha}\left(h_c - \frac{E_v}{kT}\right)$$

and the ordinates $E_s = E_u$ and $E_s = E_v$: and secondly the triangular region bounded by the abscissa:

$$s = \frac{1}{2\alpha}\left(h_c - \frac{E_v}{kT}\right)$$

the ordinate $E_s = E_u$ and the line

$$s = \frac{1}{2\alpha}\left(h_c - \frac{E_s}{kT}\right)$$

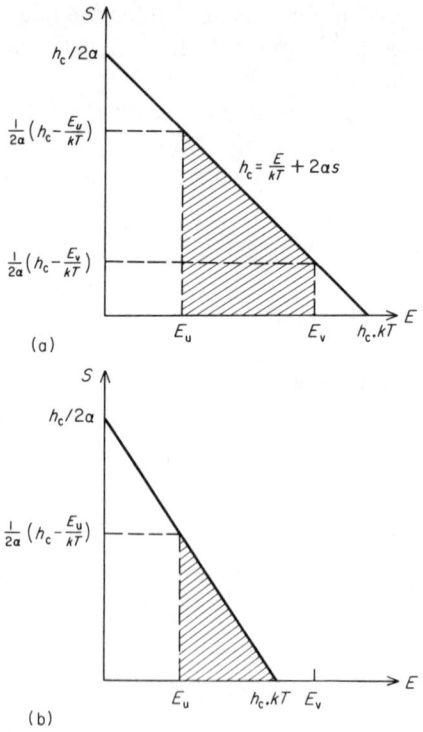

FIGURE 7. $s \sim E$ diagrams for percolation conduction in island structures (Hill et al.[19]).

Now the area of the rectangular region was shown to correspond to simply activated conduction and from the diagram it can be seen that this is dominant at high temperatures or if E_u is only slightly less than E_v. Thus simply activated conduction will be expected at sufficiently high temperatures or for those films where there is a high degree of structural uniformity corresponding to a very small range of values of E_s.

The area of the triangle in Fig. 4b leads to an expression for the critical co-ordination number, N_c, of a radially independent size distribution:

$$N_c = C[0.45kTh_c^2 a\alpha^{-1} + 0.075kTh_c^3 b\alpha^{-2} - 0.0208(kT)^2 h_c^4 b\alpha^{-2} E_c^{-1} \\ - 0.1667(kT)^2 h_c^3 a\alpha^{-1} E_c^{-1}] \qquad (38)$$

where a and b are constants dependent on the particular size distribution used and E_c is the energy corresponding to the minimum particle size of the distribution. Since the conductivity is proportional to $\exp(-h_c)$ it is necessary to solve equation (36) for h_c. However it is more appropriate to look for dominant terms and to determine the ranges over which they remain dominant. When the first term is dominant we have:

$$h_c = \left[\frac{N_c \alpha}{0.45aC} \cdot \frac{1}{kT}\right]^{\frac{1}{2}} \quad (39)$$

when the conditions:

$$\frac{a\alpha}{b} > \frac{h_c}{6} \quad \text{and} \quad \frac{E_c}{kT} \gtrsim \frac{h_c}{3}$$

are obeyed but as the temperature decreases h_c increases until the second term is dominant leading to:

$$h_c = \left[\frac{40 N_c \alpha^2}{3bC} \cdot \frac{1}{kT}\right]^{\frac{1}{3}} \quad (40)$$

over the range:

$$\frac{4E_c}{kT} > h_c > \frac{6a\alpha}{b}$$

Thus, we now see that the $T^{-\frac{1}{2}}$ condition in equation (39) is equivalent to that proposed by Abeles et al.[10] but it is now clear that this is simply a transition region between the predicted low temperature $T^{-\frac{1}{3}}$ behaviour of equation (40) and the simply activated high temperature process. At very high temperatures the negative terms in (38) dominate and the analysis breaks down. Although the above analysis applies to the radially independent size distribution, Hill and Coutts[19] point out that the same features are evident for the $1/r$ and $1/r^2$ distributions mentioned earlier. However, the temperature ranges over which the predicted behaviour can be expected will depend on the form of the distribution.

In view of the basis used for the above theory it is worthwhile pointing to the evident similarities between conduction in discontinuous films and in amorphous semiconductors. The islands may be regarded as macroscopic versions of sites in amorphous materials and as they are present in roughly similar densities it is not surprising that the two systems exhibit similar behaviour. Monte Carlo methods have been used by Maschke et al.[22] to simulate percolation in amorphous materials and it seems clear that these techniques could be useful in the present context.

Determination of film characteristics over a wide range of temperature, the transition temperatures and the high temperature curvature would undoubtedly be of great value in studies of discontinuous film conduction. It has also been suggested that during deposition an amorphous compound is formed between the metal and the substrate. Sumner[23] has referred to this as a semiconducting carpet and with the benefit of the wide range of surface analytical techniques now available it may be possible to identify such a phase. If this can be shown to exist then the theories of conduction based on tunnelling between islands may need modification. In any event the subject of conduction in discontinuous films is anything but a closed book and is likely to be a fruitful field of research for physicists for some time to come.

3. Potential Applications of Totally Discontinuous Metal Films

In this section the discussion is restricted to two dimensional arrays of metal islands on inert substrates whose properties were discussed in detail earlier in this chapter. In practice there are no current applications of discontinuous films and even the word "potential" in the title of this section must be used with considerable caution. The reason for this is that electron transport in these structures is so sensitively dependant on the interface properties of the islands that any slight changes in these tend to have an adverse and large effect on their stability. Therefore, although it is a simple matter to produce films of a very large sheet resistance, as soon as these are exposed to normal atmospheric, or indeed almost any other gaseous environment, their electrical resistance changes by up to several orders of magnitude. Even passivation of the films with an inert material such as glass cannot wholly prevent the drift so they are generally regarded as totally unsuitable for use as resistors. This is reinforced by the large negative T.C.R. almost always exhibited by these structures. In addition to these limitations, discontinuous films are unsuitable as resistors because of their unusually high strain sensitivity so that if used in a dynamic environment they would exhibit fluctuations in their resistance. However, the very existence of these sources of drift suggest that discontinuous structures could form the basis of the devices discussed below provided that associated problems of linearity, reversibility, hysteresis, etc. could be overcome.

3.1 Gas Detectors

When a gas is adsorbed upon the surface of the islands of a discontinuous film, its resistance changes. The usual reason given to explain this effect is that the absorbed gas modifies the work function of the metal islands and this is reflected in the resistance because of a change in the height of the inter-island potential barrier presented to tunnelling electrons. However it was concluded earlier that the electrons actually tunnel through the substrate and if this is the case it is more difficult to understand why the change occurs at all. Morris[24] however has suggested an absorption mechanism whereby gas atoms diffuse to the island/substrate interface thereby causing the observed changes. In either event it is possible to envisage using such a film to detect changes in the total pressure of a known gas mixture or of a single gas. If however varying gas mixtures are involved it is difficult to envisage any application, except perhaps by utilising in some way the time dependance of the observed changes.

3.2 Thermistors

The principal temperature dependance of the resistance of discontinuous films is due to the term $\exp(-\delta E/kT)$, and this gives rise to a T.C.R.

$$\alpha = -\frac{\delta E}{kT^2}$$

Assuming a value of $\delta E = 0.1\,\text{eV}$ then at 300 K the T.C.R. would be approximately -1.3% per degree change. Such a large change in resistance implies that these films could be used as temperature sensors and indeed this would seem to be a valid application provided that the device is used under conditions of vacuum in which the quantity and nature of ad- (or ab-) sorbed gases would not change, and provided that the temperature variations do not promote a permanent change in structure. Thus, such a device could find limited application in some situations where it may be necessary to monitor small variations in surface temperatures but even here thin metal film thermocouples or even semiconducting thin films would appear to hold considerably greater promise.

3.3 Strain Gauges

The principal dependance of the resistance of discontinuous thin films on the inter-island separation is contained in the term $\exp(-A\bar{\phi}^{\frac{1}{2}}\Delta s)$. Thus, when the film is strained it is the change in this term due to the change in Δs which affects the resistance. By defining strain sensitivity as:

$$\gamma = \frac{\partial R/R}{\partial \varepsilon/\varepsilon}$$

where ε = strain, and following Knight[25], equation (15) can be differentiated to obtain:

$$\gamma \simeq A\bar{\phi}^{\frac{1}{2}} \Delta s$$

Now the constant A is approximately equal to unity so that for an average barrier height of 1 eV, the gauge factor is numerically equal to the inter-island separation Δs. Since this quantity would be expected to be of the order of several tens of Angstroms it is clear that gauge factors up to 50 or more can be achieved. Several years ago this was considered to be an outstanding potential advantage although with the availability of modern temperature compensated silicon strain gauges this must now be dubious in view of the sensitivity of the device to gas adsorption and temperature variations.

3.4 Light Emitters

The phenomenon of light emission from discontinuous metal films has been known for about ten years and was first observed by Borzjak[26]. Bischoff and Pagnia[27] have recently offered an explanation of the physical mechanisms involved and have correlated the current-voltage characteristics with the spectral emission from the films. Gold films were deposited onto a variety of substrates in a planar configuration, a low constant voltage was applied and the current was observed to increase for several seconds. During this process it was noted that small spot-like areas of the film emitted light. This process was referred to as electroforming. At higher voltages however a number of the "spots" were "switched off", the current decreased and the brightness of the remaining spots increased. It was found possible to correlate the intensity of the emitted radiation with the applied voltage but only if the number of spots could be kept constant. In this case the intensity was found exponentially dependent on the reciprocal of the voltage (or, on other occasions, the reciprocal of the square root of the current-voltage product).

It was postulated that the initial electroforming process was caused by Joule heating which led to electrotransport of some of the gold islands. This was assumed to lead to low resistance chains with only a few high resistance links across which all the applied volts would appear. It was thus concluded that hot electrons would be produced in the high field regions, and that these would come to equilibrium with the film by losing part of their energy by radiation; this being the observed emitted light. Figielski et al.[28] have shown that the energy radiated by a Maxwellian electron gas at a particular frequency is given by:

$$W_v \simeq \exp\left(-\frac{h_v}{kT_{el}}\right) \qquad (41)$$

where T_{el} is the electron temperature. Furthermore Tomchuk et al.[29] have shown that the electron temperature can be related to that of the islands in a high field region and the applied power by the expression:

$$kT_{el} = (k^2 T_{Au}^2 + \alpha VI)^{\frac{1}{2}} \qquad (42)$$

where T_{Au} is the temperature of the gold islands and α is a constant depending on the power exchange with the lattice. Combining these two equations yields:

$$W_v \simeq \exp-\left(\frac{h_v}{(k^2 T_{Au}^2 + \alpha VI)^{\frac{1}{2}}}\right) \qquad (43)$$

This reduces to:

$$W_v \simeq \exp-\left(\frac{h_v}{(\alpha VI)^{\frac{1}{2}}}\right) \tag{44}$$

for values of $k^2 T_{Au}^2 \ll \alpha VI$ which was the inverse exponential dependence of intensity on voltage, for those regions where current was proportional to voltage, observed in practice.

There are clearly many practical difficulties likely to present themselves before any use could be made of the light emitting properties of discontinuous films. Indeed, this particular topic appears to have received very little attention and as yet there does not appear in the literature any suggestion for specific application. If efficient light emitting devices could be fabricated there is the possibility that they could compete on economic grounds (because of ease of fabrication) with the now widely used light emitting diodes. This stage is clearly somewhat distant and, as with the other devices mentioned in this section, it is quite likely that it will not be achieved.

4. Percolation and Conduction in Semi-continuous Film

In this section electrical conduction in thin metal films at the stage where they consist of a random interconnected array of capillaries will be discussed. This system is of particular interest because of its relevance to the mathematical topic of percolation theory. The semi-continuous films provide but one of many pieces of information relevant to the topic and a brief description will be given of several sets of results, not in themselves necessarily related to thin films, but certainly related to percolation in mixed metal/insulator systems, amorphous films and impurity conduction in crystalline semiconductors. Firstly, then, what is percolation? To explain, it is sufficient to quote from a recent paper by Pike and Seager[30] "...percolation models are composed of sites and of bonds between sites...". In general their relation to physical problems is made by identifying the site with sources of interaction and the bonds with interactions of some minimum strength or greater.... The essence of percolation theory is to determine how a given set of sites, regularly or randomly positioned in some space, is interconnected. Consider for example a regular crystallographic lattice of a transitional metal alloy. Atoms located at the nodes or sites of the lattice may or may not interact with adjacent atoms, via for example electron spins, depending on their concentration and the lattice temperature. When the concentration of transition atoms which interact, or form bonds, exceeds a minimum, or critical value, then an infinite cluster is formed and the lattice becomes ferromagnetic. It is possible using the mathematics of percolation theory to predict the value of the Curie temperature in such lattices.

It is also possible to provide information relevant to other critical, co-operative, phenomena such as phase changes and order/disorder transitions.

The problem of particular interest here can be described as follows. Consider a substrate as an orthogonal lattice upon which metal nuclei can grow, but only at the nodes (sites). As growth proceeds so the stage will be reached where adjacent nuclei actually come into contact and coalesce. Although in the high temperature limit a single nearly spherical nucleus may be expected to be formed upon coalescence of two smaller nuclei, in general (as known from electron microscopy) elongated capillaries tend to be formed. Thus, the nuclei constitute the sites, their size governs the strength of interaction and the capillaries are equivalent to the bonds. There are basically two problems in percolation theory. The "site" problem is concerned with the variation of a physical property (such as ferromagnetism) in terms of the proportion of occupied or unoccupied sites which is why it may be applicable to conduction in discontinuous films. The "bond" problem is concerned with the variation in terms of the proportion of bonds present; and this is the more applicable to the present problem. From the point of view of electrical conduction it is of interest to determine the stage at which the transition from discontinuous tunnelling conduction occurs. In fact, for a thin metal film this will not be a sharp transition since there will be a range of structures in which conduction between islands will occur simultaneously with metallic capillary conduction, and initially the resistance contributions may be of similar values. Eventually the tunnel contribution will be completely shorted by the capillaries. However, it is necessary at this stage to restrict the discussion to the situation in which only capillaries contribute so that the film resistance would switch at a critical bond occupancy from infinity to some finite value governed by the configuration of the capillary network. It is of primary interest to predict the critical proportion of bonds needed for capillary conduction to occur and to determine the variation in resistance of the film with a gradually increasing bond occupancy. By considering the various configurations (clusters) of capillaries which can be formed, Sykes and Essam[31] have derived series expansions in terms of the fractional bond occupancy and used these to predict the critical percolation probability for all the common two and three dimensional regular crystallographic lattices. In addition, Pike and Seager[32] and Kirkpatrick[33] have carried out Monte Carlo computer studies of percolation on simulated regular and random lattices. In general, the values of the critical percolation probability derived by these two methods are in reasonable agreement and one of the interesting features of their work is the "verification" of an empirical rule first stated by Frisch et al.[34]. This states that the average number of bonds per site at the critical probability is a constant governed only by the dimensionality and not the crystallography, of the lattice. For a three dimensional lattice the constant is approximately equal to $\frac{3}{2}$ whilst for a two dimensional lattice it is approximately two. Since the average number of bonds per site at percolation is simply equal to the co-ordination number of the lattice multiplied by the critical percolation probability, this is a reasonably accurate technique for calculating the latter quantity for a particular lattice. Coutts and Hopewell[35] have also carried out computer simulation of random networks of various sizes and showed that the empirical rule was quite accurately obeyed as a means of calculating the critical probability p_c. Having established at least the approximate validity of the empirical law it will now be applied to real systems. For a mixture of conducting and non-conducting spheres conduction across the array may be expected when the volume fraction of the conducting species is equal to the critical probability, p_c, multiplied by the packing fraction of equal sized conducting spheres then the maximum volume which can be occupied by the spheres is $f = \pi/6$.

A random close packed system has a co-ordination number $Z = 6$ and therefore $p_c = 0.25$. The critical volume V_c should be a fixed quantity given by $V_c = fp_c$ which for random packing gives $V_c = (\pi/6) \times 0.25 \simeq 0.13$. Scher and Zallen[36] have derived values of $V_c(A_c)$ for all the common three and two dimensional lattices and conclude that these quantities should be almost independent of the lattice; A_c having a value of approximately 0.44, the corresponding value of V_c being approximately 0.15. In real systems the situation is complicated by non-uniform size distributions and irregularly shaped particles and Pike and Seager[30] have advised caution in using the critical volume or area concept in practice. However results from several systems enabled a determination of the critical fraction from experimental measurements rather than by calculation. It will be seen that several apparently quite distinct physical systems appear to follow a single law which describes their electrical behaviour.

Essam et al.[37] have calculated the conductivity of disordered branching networks and have shown that the conductivity is given by:

$$\sigma(p)_{p \to p_c^+} \simeq \frac{2(1+p_c)}{(1-p_c)p_c^2}(p-p_c)^\beta \tag{45}$$

where p is the fraction of occupied bonds in the lattice, p_c is the critical percolation probability and β is a constant whose value for a Cayleigh tree (i.e. a network in which closed loops are not permitted) is approximately two. For larger values of p Maxwell[38] introduced the so-called "effective medium" theory and again predicted that the conductivity of a mixture should be proportional to $(p-p_c)^\beta$ but β should be approximately unity.

It is more common to express (45) as:

$$\sigma_{p \simeq p_c^+} = \sigma_0 (p - p_c)^\beta \tag{46}$$

for simulated lattices or models in which the volume or area of the conducting species is negligible. If, however, real physical systems are considered then the quantities p and p_c would be replaced by A and A_c, or V and V_c. The object now will be to apply (46) to several systems and to demonstrate its apparent universal applicability. Consider firstly the case of simulated lattices; the data of Coutts and Hopewell[35] and those of Kirkpatrick[33] yield values of $\beta = 1.35 - 1.40$. In the former case a random network was generated in which the resistance of the capillaries between the nodes of the lattice was either infinite or one ohm, the resistance being calculated using standard network analytical techniques and being expressed as a function of the number of capillaries with finite resistance. In the latter case the resistances were selected randomly from a logarithmic distribution. In both cases the models can be criticised as being unrealistically small and therefore subject to very significant edge effects.

Watson and Leath[39] carried out an elegant table-top experiment to study the percolation problem. This involved taking a wire-mesh of 137×137 sites, removing sites at random and measuring the resistance of the network as a function of the fraction of sites removed. By making the usual plot of $\log \sigma \sim \log (p - p_c)$ they obtained a value for the critical exponent β of 1.38 in good agreement with the results of computer simulation.

There are literally dozens of examples of mixed conductor/insulator systems to be found in the literature but reference to two or three of these should be sufficient to make the point. The most important feature is that the electrical properties of these mixed systems, which are essentially equivalent to the random networks discussed above, can be accounted for by percolation theory and in particular by a modified form of equation (46).

Scarisbrick[40] and van Loan[41] have both studied the sheet resistance of silver loaded, screen printed conductors, as a function of the volume fraction of silver. Their two sets of data clearly illustrate the weakness of the "critical volume fraction" rule of Zallen and Scher[36]. For two nominally identical systems the volume fraction at which the resistance dropped very rapidly was found to be 0.25 and 0.10. If however this discrepancy, which is presumably due to differences in particle size distribution, method of preparation or other unknown causes, is accepted then it is found that the results are encouraging. Re-writing equation (46) in terms of volume fraction gives:

$$\sigma = \sigma_0 (V - V_c)^\beta \tag{47}$$

In Fig. 8 both sets of data have been plotted and as seen these are admirably represented by a single line the slope of which (β) is equal to 1.45. This

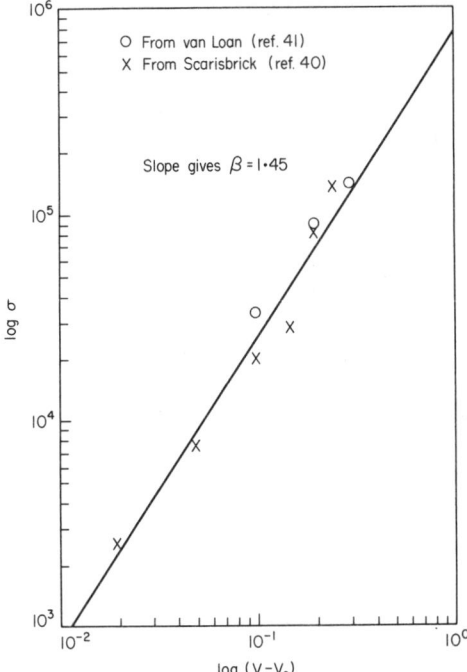

FIGURE 8. Log conductivity against log $(V - V_c)$ for silver loaded paints (Van Loan[41] and Scarisbrick[40]).

example of the applicability of (46) is further reinforced by Scarisbrick's data on conducting carbon black/insulator mixtures. The data are shown in Fig. 9 and once again the scatter about the straight line (of slope $\beta = 1.38$!) is remarkably low. Here then, are two macroscopic examples illustrating agreement with theory and it will now be shown that certain microscopic systems, also follow the same laws.

Lightsey[42] studied conduction in sodium tungstate (Na_xWO_3) and found that the conductivity followed the law $\log \sigma \sim \log(x - x_c)$ where x is the atomic fraction of sodium. His results gave $x_c = 0.16$ and $\beta = 1.8 \pm 0.2$. It is interesting to note that not only is the value of β in agreement with the theoretical values stated earlier but also that x_c corresponds almost precisely with the critical volume fraction for three dimensional lattices calculated by Zallen Scher[36].

Thin film cermets, which are of particular interest in this chapter, can also be regarded as a three dimensional, microscopic mixed conducting system. At this stage no attempt will be made to specify the nature of the conducting and

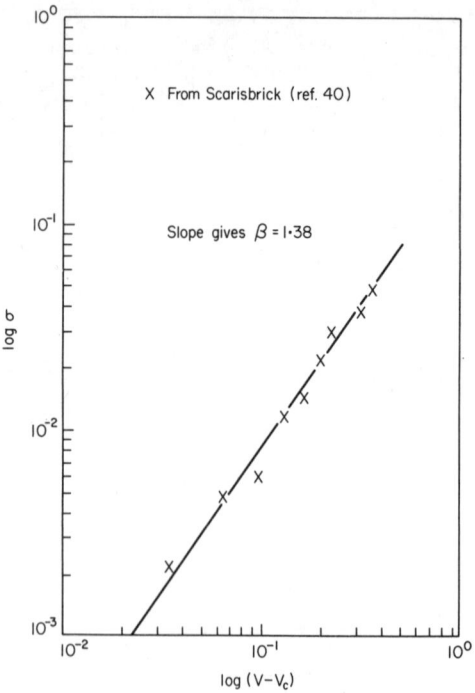

FIGURE 9. Log conductivity against $\log(V - V_c)$ for carbon black/insulator mixtures (Scarisbrick[40]).

insulating phases since to do so would be an unnecessary complication. Neugebauer[43] analysed a large number of results on cermet thin films and showed that the electrical properties of many metal/insulator combinations can best be represented by plotting resistivity against the volume fraction of metal; a universal curve being obtained for each insulator. Taking for example, the results for chromium/silicon dioxide cermets a critical volume fraction of metal of 50% can be inferred (once again in contradiction with the hypothesis of Zallen and Scher) and, at least for volume fractions not much greater than this, a plot can be made of $\log \sigma$ against $\log(V - V_c)$. The quantity of data is very limited and therefore the value of β obtained is rather inaccurate ($\beta \simeq 1.2 - 1.8$) but does correspond to the values derived earlier. Abeles[44] has fabricated tungsten/alumina cermets and has extracted a value of $\beta = 1.9 \pm 0.2$ for this system. In this instance a value of $V_c = 0.47$ was used and the data can be regarded as much more accurate than those pertaining to chromium/silicon dioxide because of the care taken in determining the metallic volume fraction.

It is really because of the applicability of equation (46) to thin film cermets that Neugebauer found a universal law for each insulator used. In fact it may even be expected that most metal/insulator combinations would to a first approximation, follow a single curve. A summary of the values of β for the various systems discussed is given in Table 2.

5. Properties and Applications of Films of Mixed Structure and Composition

There is in practice not a sharp transition in the structure from discontinuous to semi-continuous films even though the electrical resistance decreases very dramatically in this region. The reason is that over a limited compositional range both capillary and island structures can be observed. This behaviour is modified by the deposition conditions and by the mutual reactivity of the conducting and insulating phases. From the practical point of view it is the region of mixed structure which is likely to be of greatest interest since it is here that it may be possible to obtain a relatively high resistivity film together with a nearly zero T.C.R. These qualities are two of the principal requirements of a thin film resistor although there are, of course, several others.

Conduction in the island structure is as stated earlier, an activated process which gives rise to a negative T.C.R. whilst conduction in the capillaries is of a normal metallic type subject to the limitations of surface scattering discussed in Chapter 3. The effect of this will be explained later in this chapter.

Certain polycrystalline metal films have been shown to form oxide layers around the crystal edges and although these structures do not conform to the island/capillary model they can also be expected to exhibit a dual conduction mechanism, i.e. metallic within individual grains and semiconduction or dielectric tunnelling in the oxide layers between grains. In all cases the combination of positive and negative T.C.R.'s gives rise to a smaller absolute T.C.R. than would be obtained were either mechanism present alone.

It is possible to postulate several models to explain the mixed conduction properties. Firstly, if capillaries and islands do not interact then the two processes can be considered to act in parallel and Coutts[45] has formulated this situation. Second, if there is an island/capillary interaction then the processes must be taken as acting in series as proposed by Hill[46]. It is possible to make certain predictions using one or other of these models although their value is doubtful since the models themselves are too idealised. A much more realistic, but intractable model would involve both series and parallel interactions. Thus although it is possible to predict qualitatively the behaviour of the films in the totally discontinuous regime or even in the critical capillary connectivity regime it is not clear how to take account quantitatively of the presence of the islands immediately before and after the transition to a random network of capillaries. This statement applies equally to other films in which the relative magnitudes of the separate contributions to the conduction process cannot be specified.

Materials consisting of a physical or chemical mixture of metals and dielectrics are generally classified under the general heading of cermets and thin films of these have received a considerable amount of attention, mainly over the last decade and primarily with a view to developing resistors of improved characteristics. It was noted in Chapter 3 that to minimise the surface area required by high value resistors it is necessary to maximise the sheet resistance of the materials of which they are formed. It is this requirement which has led to the use of Ta_2N and NiCr resistors.

However, any further increase in sheet resistance beyond that offered by these materials would be welcome to the circuit designer and it is to meet this need that much of the research on cermets has been undertaken. When a metal is deposited simultaneously with an insulator, or ceramic, electrical conduction in the resulting film is generally believed to take place simultaneously through at least two structural or compositional, categories. As stated already these categories may be metallic capillaries and discrete islands, or metallic grains and tunnel junctions formed by the oxide at grain boundaries, etc. The overall effect of the co-existence of two species whose T.C.R.'s are opposite in sign is to reduce the T.C.R. of the film as a whole whilst, because of the disperse nature of the conducting components, possessing a much greater than normal sheet resistance. Indeed, as will be seen in the forthcoming sections, it has not been unusual to read reports in the literature of thin film cermets with near zero values of T.C.R. and with sheet resistances of several kilohms per square. These properties alone have stimulated extensive investigation into thin film cermets. Before discussing the properties of several particular cermets it should be pointed out that in this context the name cermet is taken to imply a three dimensional array of discrete islands and/or capillaries, or any other chemical or physical mixture of metallic and insulating phases. The discussion will be limited to those films whose thickness is less than about 0.1 μm and which have been prepared by the means of deposition discussed in Chapter 2. This restriction, of course, precludes the vast class of thick film resistors although the composition and models of electrical conduction in many of these materials are similar to those observed and postulated to occur in thin film materials.

5.1 Cermet Resistors

The applications of cermet films are essentially the same as discontinuous films but in the former case the possibility of an improvement in stability due to the encapsulating effect of the dielectric is a considerable potential advantage. In practice, however, their stability has rarely been shown to be a good as that of the alloy films discussed in Chapter 3 so that they have not as yet achieved anything approaching even a modest usage. At the time of writing there appears to be very little attempt to overcome any of the outstanding problems and, in fact, if anything, the term cermet appears to be positively unfashionable. However, their state of development is such that a fairly rapid rate of progress could be anticipated should an obvious industrial requirement be evident. No attempt will be made here to present a comprehensive review of research and development on cermet films but, rather, a very limited discussion will be given of several types of cermet which exhibit two or more conduction mechanisms simultaneously. A comprehensive list of some forty papers on cermets is presented by Campbell and Morley[47] which covers references published up to 1970.

(a) Chromium/silicon oxide
This combination is possibly the most extensively studied cermet and a wealth of literature is available. Glang et al.[48] have made a careful investigation of the structure and composition of films prepared by flash evaporation and were able to correlate, at least qualitatively, these properties with their electrical resistivity. These authors published a phase diagram of the Cr/Si system which showed that the components of the cermet would produce reaction species depending on their relative proportions. The resulting films were found to consist of Cr islands and at least two chromium silicide phases (also in island form) separated by SiO_2. It was shown that the oxide phase was responsible for the negative part of the conduction mechanisms because it resulted in activated quantum mechanical tunnelling as discussed earlier, whilst the silicides lead to a metallic like linkage between the chromium particles thus producing a positive contribution to the T.C.R. These two processes were assumed to take place in parallel.

The magnitude of the positive contribution depended on the relative abundances of the various silicides which themselves each possessed a different T.C.R. Thus in this rather complex system it is possible to produce equal values of T.C.R. in films of quite different structures and compositions, whilst also having widely differing values of sheet resistance. This is precisely the sort of situation which is found to occur in certain thick film systems based on molybdenum and its several oxides[49]. Although the Cr/SiO system might be expected to be very stable because of the reaction of the individual components, nothing was reported of the long term stability of the films.

One interesting application suggested for the cermets was that of maximum thermometers. It was found that the films stabilised at a particular annealing temperature, thus to determine the previous maximum temperature experienced by a film it was necessary only to heat it at an increasing temperature until its resistance showed a decrease due to further annealing. Although this is a potentially useful application for cermet films it must in principle be possible to use other materials based on simple metals or alloys since these are also known to change in resistance upon annealing.

(b) Tantalum/alumina cermets

As was the case with the cermet combination discussed above, the Ta/Al_2O_3 system has been examined by many researchers. The object of using a tantalum based cermet was its supposedly good thermal stability and the fact that it can be trimmed and passivated by anodic oxidation as discussed in Chapter 3. Henrickson et al.[50] prepared Ta/Al_2O_3 films by sputtering from a composite target and studied the variation of their sheet resistance and T.C.R. and also carried out compositional and structural analyses. The electrical properties were, as would be anticipated, found to be extremely sensitive to the deposition conditions and to the proportion of tantalum as well as to the thickness of the films. The maximum sheet resistance obtained was for a relatively thick film (2400 Å), and was 1000 Ω/square with a T.C.R. of $-420\,\text{ppm}°C^{-1}$. The minimum value of sheet resistance was approximately 10 Ω/square with a T.C.R. of between $+30 - +80\,\text{ppm}°C^{-1}$; the film thickness being 2100 Å. The T.C.R. in the latter case is surprisingly low and suggests considerable suppression of the bulk value due to grain boundary scattering as discussed in Chapter 3. The films were examined using transmission electron microscopy (which is perhaps surprising in view of their relatively large thickness) and were found to consist of metallic tantalum grains surrounded by alumina. It was proposed that the metallic grains extended through the entire film thickness and that the conduction mechanism was due to tunnelling between the grains through the alumina boundary layer. The conductivity/temperature Arrhenius plots indicated an activation energy for charge transfer between the metal grains of about 0.002 eV which was not found to be consistent with the size of the grains measured from micrographs. Thus Henrickson et al. analysed their results in terms of the temperature dependence of the tunnel current of a metal-insulator-metal sandwich; it is upon this theory, developed by Stratton[51] and Simmons[6], that Hill's theory of conduction in discontinuous structures is based.

In the absence of an activation energy term, the principal temperature dependence of the tunnel current is due to the term $\pi BkT/\sin\pi BkT$ in equations (15) and (16). To a first approximation therefore the temperature coefficient of the tunnel resistance is:

$$\alpha_T = \pi Bk \cot \pi BkT - \frac{1}{T}$$

Henrickson et al. obtained a value of the quantity B from conductivity measurements and, using this value, the T.C.R. at 300 K is found to be approximately $-400\,\text{ppm}°C^{-1}$; in excellent agreement with their results

Although it is tempting to conclude that this evidence is decisive, it must be remembered that the metallic grains are certain to make a positive contribution to the T.C.R. and equally, it could be incorrect to calculate an activation energy or a value of B, from a current/temperature characteristic which was in fact representative of at least two conduction mechanisms. This criticism is typical of those which can be made quite validly of theories of conduction in cermets. This point aside, the films did appear to be potentially useful although once again no information was given concerning long term drift or noise measurements.

(c) Gold/insulator cermets

Cermets based on a mixture of gold and an insulator such as silicon oxide, glass or alumina have been studied by a number of workers with the objective of gaining information regarding the conduction mechanisms rather than to produce useful thin film resistors. The reason for the choice of gold was based on its virtual non-reactivity so that the electrical properties would be dependent solely on observable and measurable structural characteristics rather than on compositional variations which would be expected to be much more difficult to quantify. The presence of the various silicide phases in Cr/SiO cermets is typical of the complexities which it was thought might be avoided by using a physical rather than a chemical mixture of metallic and insulating components.

Pollard et al.[52] prepared Au/SiO cermets by evaporation from separate sources and used electron microscopy, electron and x-ray diffraction to determine the dispersion of the gold in the silicon oxide, and attempted to correlate the electrical properties with the compositional and structural analyses. The usual Arrhenius plots of log resistivity against reciprocal absolute temperature were made and, in general, these exhibited some curvature despite the fact that Pollard *et al.* derived from them an activation energy which they considered was characteristic of a singly activated process. The importance of this paper is not so much that the films were particularly useful but rather that the compositional and structural analyses led to a new conception of the dispersion of the gold. Pollard *et al.* cite the particular case of a film of 550 Å thickness deposited at a substrate temperature of 240°C. Transmission electron microscopy indicated that the film consisted of discrete gold islands of 20 Å mean diameter, dispersed in an insulating matrix of silicon oxide. The volume fraction of gold was estimated from the structural measurements as 2%, whilst x-ray fluorescence spectroscopy implied a value of 21 v/o. Since the resolution of the transmission microscope used was 15 Å it was concluded that most of the gold was dispersed throughout the silicon oxide in the form of islands of less than this size. It was considered that it was these "invisible" particles which dominated the electrical properties and the authors rightly pointed out that it is necessary to undertake further work to determine the effect of the deposition parameters on the nucleation and growth of two phase systems. On the other hand, the values of activation energy obtained from the Arrhenius plots were typically less than 0.02 eV which implies that much larger islands are responsible. This is clearly a most interesting area for further research although, because of the anticipated instability of the films (due to structural changes), this particular combination is unlikely to be of value as a practical resistor material.

Miller et al.[53] prepared Au/SiO_2 cermets by co-sputtering and again found that the films consisted of gold islands dispersed in an insulating matrix. Interestingly, however, these workers claimed to find excellent agreement between their results and theoretical values of resistivity predicted by the relatively simple activated tunnelling model of Neugebauer and Webb[4]. However, as has almost inevitably been the case with two dimensional discontinuous films, there was found to be a factor of two between the activation energy based on measurements of structural parameters and that obtained from an Arrhenius plot. Thus, to claim agreement between theory and experiment is perhaps rather optimistic.

Neither Pollard et al. nor Miller et al.[53] attempted to analyse their results on the basis of anything other than a singly activated mechanism and certainly did not consider the possibility of the co-existence of two or more processes of charge transfer. Coutts[45] however prepared gold/glass cermets by co-sputtering and found extremely wide variations in the structure of even a single film. Electron microscopy showed that with a high gold content the island structure was absent having formed, presumably by coalescence, meandering interconnected capillaries. Conduction in the capillaries can be admirably described by the percolation model discussed previously and for a fully branched network it is possible to approximate the resistance of the entire film by that of a single link of the network. The T.C.R. of these films was found to be approximately $+400$ ppm$°C^{-1}$ and Coutts interpreted this as being due to suppression of the bulk T.C.R. by impurities and by the surface scattering mechanism discussed in Chapter 3. The sheet resistance of fully interconnected films lay between 50–100 Ω/square and so could not be considered useful.

High sheet resistance films with a low gold content were also found to be more uniform and consisted, predictably, of discrete gold islands. These gave rise to an activated process and negative T.C.R.'s which were too large to enable useful resistors to be made. Between the two extremes the region of non-uniform structures was found and it was only occasionally possible to produce a film with a near zero T.C.R. and a sheet resistance of approximately 10^4 Ω/square. These electrical properties would be ideal were it not for a number of disadvantages. The films generally drifted unpredictably in resistance either positively or negatively, immediately after deposition and

Coutts suggested that this may be due to the break-up of capillaries and the coalescence of islands. If these processes occurred simultaneously then their relative magnitudes would determine the overall direction of drift. In addition, even films passivated with about 1 μm of glass were found to drift too rapidly for commercial application when exposed to the atmosphere. Thus, although the three cermets based on gold described immediately above have provided some useful basic information, they have no apparent value as resistor materials. The precise reasons for the instability of most of the cermet systems examined to date is not entirely clear although both Morris[24] and Williams and Stone[54] have shed some light on the situation with their work on the effect of absorption and/or adsorption of various gases on the resistance of discontinuous films. Tick and Fehlner[55] have made an interesting suggestion regarding the possible improvement of the long-term stability and short-term temperature sensitivity of discontinuous films. It has always been found extremely difficult to produce films consisting of large islands and small gaps because of structural changes caused by surface mobility and island coalescence. These structures would be desirable in a cermet resistor because the activation energy of the islands would be low and therefore they would be relatively temperature insensitive. Tick and Fehlner suggested that it may be possible to achieve stability by depositing a nucleating layer of palladium, or any other metal which exhibits a high density of very small nuclei, and then depositing on this a second layer of a relatively mobile metal, such as gold. It was considered that the gold would form large nuclei and would grow preferrentially upon the palladium nuclei which would in turn prevent their surface mobility. If such a film could be adequately passivated with an impervious layer of glass or other suitable insulator then reasonably stable films may result. Even so these films are unlikely to be accepted in view of the generally superior performance of other cermet systems.

General reviews of the performance of cermet resistors have been given by Head[56] and Pitt[57] and although these are the most recent reviews they are nevertheless about ten years old which indicates that the subject has not developed significantly during the last few years. This is indicative of the current lack of industrial interest in, or possibly even requirement for, high sheet resistance components, rather than the excessive difficulty in producing good quality resistors. If a positive need for these arose then undoubtedly their development would proceed in much the same way as did that of the metal and alloy resistors discussed in Chapter 3.

5.2 Strain Gauges

The use of thin cermet films as strain sensitive elements may not appear to warrant a separate section because they are, in essence, identical to a resistor.

However, it is of interest to note that their strain sensitivity as well as T.C.R. is modified to a value intermediate between that of the totally discontinuous and totally continuous states. Witt[58] has pointed out that a gold/glass cermet structure with a sheet resistance of approximately $5\,\mathrm{k\Omega}$/square has a near zero T.C.R. and a gauge factor of approximately twenty. In addition, their drift rate was found to be relatively low. However, the lack of data in the critical transition region does not confirm this finding.

Stolinski et al.[59] examined the application of Cr–SiO cermet films as strain gauges and as expected, found gauge factors of about ten. It was noted that the gauge factor remained stable over a period of four weeks although the resistances did drift by amounts which depended on their initial value.

5.3 Other Applications

Other applications for cermet films have been quoted but in the main these are identical to those for two dimensional discontinuous films. However, at least two novel uses have recently been proposed. The first concerns the application of cermets of a structure graded in depth from all metal (adjacent to the substrate) to all dielectric at the upper boundary. Films of this sort can be tailored to absorb or emit incident electromagnetic radiation over specific wavelength bands and it has been suggested therefore that they could be used as selective filters for solar thermal collectors. This topic is of course currently of great interest and the award of a large development contract to a U.S.A. glass manufacturer has recently been announced in the press[60]. The properties of these films could doubtless be obtained using the multi-layer stacks discussed in Chapter 8, but the virtue of the graded cermet structure is that it can be prepared in relatively large areas by chemical vapour deposition, rather than by more costly vacuum deposition techniques. C.V.D. of aluminium/alumina cermet resistors has been carried out using as a starting material triethyl-aluminium[61]. Although the performance of the resistors was not particularly good, it is interesting in the present context to note that the structure and composition of the films could be readily controlled.

Secondly, Wronski et al.[62] have developed a vidicon in which the blocking contact to a thin photoconductive layer of cadmium selenide is provided by a thin cermet film of gold/silicon dioxide. Each of the metal grains making contact with the semiconductor forms a Schottky barrier and becomes charged as it is scanned by the electron beam. If light is incident on the opposite surface of the semiconductor electron/hole pairs are generated and the latter drift to the interface with the cermet where they recombine with the stored electrons. A stored image is thus produced and this can be converted to an electrical signal by the electron beam which re-charges the image on its next scan. This system has the potential of a very high resolution since the lateral resistance of the cermet is high thus avoiding the necessity of a patterned array which increases costs. On the other hand, no comment regarding the stability of the device was made and in view of the known tendency of thin film cermets to drift, some change of the metal/semiconductor interface may be expected.

6. Conclusion

It will be evident from the discussion of both the properties and applications of mixed metal/dielectric films that there is a great deal yet to be said about this class of film before their electrical behaviour is understood theoretically and before they can be usefully applied as reliable materials. On the theoretical side, it is unlikely that their variation of resistance with temperature will be fully understood until the statistical distributions of island sizes and separations can be taken into account. Quite clearly additional and detailed measurements of these are required so that some attempt to incorporate them into the formulation of the charge transfer effect can be made. Recent information draws the interesting parallel between conduction in cermet films and in amorphous films: it seems likely that the critical percolation approach used so successfully to predict the behaviour of the latter class, will prove useful in the present context. However, at the present time, the problem of charge transfer in cermet films is not at all as well understood as a great deal of the published literature indicates. In addition to understanding the fundamentals of the problem, there are many other effects such as the variation of the dielectric constant and activation energy with temperature and the effect of the contacts which may yet prove to be important but which have not in fact been quantified. Thus, the theoretical aspects of the topic are likely to be of interest for some time yet and possible new lines of investigation have been suggested in a recent review by Morris and Coutts[63].

Although the unusual structure of cermet films presents the possibility of developing a number of potentially useful devices, the limitation in stability has not yet been adequately overcome and unless extensive research is devoted to this problem it appears that reliable devices will not be produced. The stimulus to such research often results from a real need to exploit one or other of the properties of a novel material or device, or because it is possible even if only in principle, to duplicate the function of existing devices, but at a lower cost. At present, neither appears to apply to cermet films and it seems likely therefore that if this class of thin film does ultimately form the basis of a widely used device it will be because of some aspect of their behaviour yet to be researched or because of an almost fortuitous improvement in their stability resulting from the relatively low priority research presently being undertaken.

Table 1

System	Reference	β
Simulated	Coutts and Hopewell[35]	1.38
Simulated	Kirkpatrick[33]	1.33
Wire mesh	Watson and Leath[39]	1.33
Thick silver films	Scarisbrick[40]	1.45
Thick silver films	Van Loan[41]	1.45
Carbon black	Scarisbrick[40]	1.38
Thin film cermets	Neugebauer[43]	1.2–1.8
Thin film cermets	Milgram and Lu[16]	$\doteq 1.5$
Sodium tungstate	Lightsey[42]	1.8 ± 0.2

Table 2

Cermet system	Metallic concentration (x)	Gradient (m)	Coefft. of regression (r^2)	$M+1$
Ni/SiO$_2$	0.44	−0.45	0.85	0.55
	0.34	−0.70	0.68	0.30
	0.24	−0.48	0.87	0.52
	0.14	−0.68	0.86	0.32
	0.08	−0.95	0.95	0.05
Au/Al$_2$O$_3$	0.42	−0.64	0.76	0.36
	0.32	−0.49	0.79	0.51
	0.28	−0.17	0.36	0.83
W/Al$_2$O$_3$	0.40	−1.06	0.90	−0.06
	0.34	−0.58	0.79	0.42
	0.26	−0.52	0.83	0.48
	0.22	−0.53	0.93	0.47
	0.18	−0.49	0.81	0.51
	0.15	−0.76	0.89	0.24
	0.10	−1.48	0.78	−0.48
Pt/SiO$_2$	0.18	−0.41	0.80	0.59
	0.14	−0.45	0.17	0.55
	0.11	−0.83	0.51	0.17
	0.07	−0.065	0.002	0.93
	0.05	−1.32	0.09	−0.32

References

1. K. van Steensel, *Philips Res. Repts.*, **22**, 246, (1967).
2. T. E. Hartman, *J. Appl. Phys.*, **34**, 943, (1963).
3. G. Darmois, *J. Appl. Phys. Radium*, **17**, 211, (1956).
4. C. A. Neugebauer and M. B. Webb, *J. Appl. Phys.*, **33**, 74, (1963).
5. R. M. Hill, *Proc. Roy. Soc. (London)*, **A309**, 377, (1969).
6. J. G. Simmons, *J. Appl. Phys.*, **35**, 2655, (1964).
7. R. Kiernan and D. W. Stops, *Nature*, **224**, 907, (1969).
8. A. Barr and R. D. Finney, *Thin Solid Films*, **24**, 511, (1974).
9. J. G. Swanson, D. S. Campbell and J. C. Anderson, *Thin Solid Films*, **1**, 325, (1967).
10. B. Abeles, Ping Sheng, M. D. Coutts and Y. Arie, *Advan. Phys.*, **24**, 407, (1975).
11. F. W. Schmidlin, *J. Appl. Phys.*, **37**, 2823, (1966).
12. J. W. Gadzuk, *J. Appl. Phys.*, **41**, 286, (1970).
13. R. M. Hill, *Nature*, **204**, 35, (1964).
14. F. P. Fehlner and S. M. Irving, *J. Appl. Phys.*, **37**, 3313, (1966).
15. D. S. Herman and T. N. Rhodin, *J. Appl. Phys.*, **37**, 1594, (1966).
16. A. A. Milgram and C. Lu, *J. Appl. Phys.*, **37**, 4774, (1966).
17. R. M. Hill and T. J. Coutts, *Thin Solid Films*, **35**, L17, (1976).
18. R. M. Hill, *Phys. Stat. Sol.* (a), **34**, 601, (1976).
19. R. M. Hill and T. J. Coutts, (to be published).
20. A. Miller and E. Abrahams, *Phys. Rev.*, **120**, 475, (1960).
21. N. F. Mott and E. A. Davis, "Electronic Processes in Non-Crystalline Materials", Clarendon Press, Oxford, (1971).
22. K. Maschke, H. Overhof and P. Thomas, *Phys. Stat. Sol.*, **62**, 113, (1974).
23. G. G. Sumner, *Surface Sci.*, **4**, 313, (1966).
24. J. E. Morris, *Thin Solid Films*, **5**, 339, (1970).
25. M. J. Knight, *Proc. Joint IERE-IEE Conf. on Applications of Thin Films in Electronic Engineering*, (1966).
26. P. G. Borzjak, O. G. Sarbei and R. D. Fedorovich, *Phys. Stat. Sol.* (a), **8**, 55, (1965).
27. M. Bischoff and H. Pagnia, *Thin Solid Films*, **29**, 303, (1975).
28. T. Figielski and A. Torum, *Proc. Int. Conf. on Semi-conductor Physics*, Exeter, (1962).
29. P. M. Tomchuk and R. D. Fedorovich, *Soviet Physics–Solid State*, **8**, 226, (1966).
30. G. E. Pike and C. H. Seager, *Phys. Rev. B.*, **10**, 1421, (1974).
31. M. F. Sykes and J. W. Essam, *J. Math. Phys.*, **5**, 1117, (1964).
32. G. E. Pike and C. H. Seager, *Phys. Rev. B.*, **10**, 1434, (1974).
33. J. Kirkpatrick, *Phys. Rev. Letts.*, **27**, 1722, (1971).
34. H. L. Frisch, V. A. Vyssotsky, S. B. Gordon and J. M. Hammersley, *Phys. Rev.*, **123**, 1566, (1961).
35. T. J. Coutts and B. Hopewell, *Thin Solid Films*, **9**, 37, (1971).
36. R. Zallen and H. Scher, *J. Chem. Phys.*, **53**, 3759, (1970).
37. J. W. Essam, C. M. Place and E. H. Sondheimer, *J. Phys. C.*, **7**, 1258, (1974).
38. J. C. Maxwell, "A Treatise on Electricity and Magnetism", Vol. 1, Art. 314, (1881).
39. B. P. Watson and P. L. Leath, *Phys. Rev. B.*, **9**, 4893, (1974).

40. R. M. Scarisbrick, *J. Phys. D.*, **6**, 2098, (1973).
41. P. R. van Loan, *Insulation/Circuits*, **18**, 35, (1972).
42. P. A. Lightsey, *Phys. Rev. B.*, **8**, 3586, (1973).
43. C. A. Neugebauer, *Thin Solid Films*, **6**, 443, (1970).
44. B. Abeles, H. L. Pinch and G. I. Gittleman, *Phys. Rev. Letts.*, **35**, 247, (1975).
45. T. J. Coutts, *Thin Solid Films*, **4**, 429, (1969).
46. R. M. Hill, *Proc. of Conference on Hybrid Microelectronics*, Loughborough, (1975).
47. D. S. Campbell and A. R. Morley, *Rep. Prog. Phys.*, **34**, 283, (1971).
48. R. Glang, R. A. Holmwood and S. R. Herd, *J. Vac. Sci. Technol.* **4**, 163, (1967).
49. A. S. Laurie, *Proc. Electronics Components Conference*, Washington, (1973).
50. J. F. Henrickson, G. Krauss, R. N. Tauber and D. J. Sharp, *J. Appl. Phys.*, **40**, 5005, (1969).
51. R. Stratton, *J. Phys. Chem. Solids*, **23**, 1177, (1962).
52. J. K. Pollard, R. L. Bell and G. G. Bloodworth, *J. Vac. Sci. Technol.*, **6**, 702, (1969).
53. N. C. Miller, B. Hardiman and G. A. Shirn, *J. Appl. Phys.*, **41**, 1850, (1970).
54. J. L. Williams and I. L. Stone, *Thin Solid Films*, **11**, 329, (1972).
55. P. A. Tick and F. P. Fehlner, *J. Appl. Phys.*, **43**, 362, (1972).
56. K. Head, *Thin Solid Films*, **4**, 153, (1969).
57. K. E. G. Pitt, *Thin Solid Films*, **1**, 173, (1967).
58. G. R. Witt, *Thin Solid Films*, **22**, 133, (1974).
59. E. J. Stolinski and Z. H. Meiksin, *IEEE Trans. Electron Devices*, **ED-22**, 102, (1975).
60. Press release, Optical Spectra, p15, Sept., (1975).
61. A. Gurev, *Thin Solid Films*, **18**, 275, (1973).
62. C. R. Wronski, B. Abeles, and A. Rose, *Phys. Rev. Letts.*, **27**, 91, (1975).
63. J. E. Morris and T. J. Coutts, *Thin Solid Films*, **47**, 63, (1977).

Chapter 6†

Thin Film Transistors

J. C. Anderson

Department of Electrical Engineering,
Imperial College,
London, England

1. Introduction 207
2. Theory of the Device 208
   2.1 Enhancement Theory 208
   2.2 Depletion Theory 210
   2.3 Barrier Theory 211
3. Barrier Enhancement Theory 216
4. Slow States 219
5. Choice of Semiconductor/Insulator Combination 225
   5.1 Fast States. 225
6. CdSe Transistors 227
   6.1 Fabrication 228
   6.2 Characteristics 232
   6.3 Transconductance 235
   6.4 Contacts 237
   6.5 Noise Characteristics 238
   6.6 TFT Design Parameters 239
7. Conclusions 241
   Acknowledgements 242
   References. 242

1. Introduction

The field-effect transistor exploits the variation of the conductivity of a semiconductor layer by the application of an external electric field. Such a device was described by J. E. Lilienfeld in American patents granted in 1930 and 1935[1,2]. Significant conductivity modulation was not achieved, however, and this was ascribed by W. Shockley[3] to the trapping of induced charge at surface states, which he attributed to "dangling" bonds at the

† For a list of symbols used in this chapter and their definitions see p. xii.

semiconductor surface. A long period of development led ultimately to the discovery by M. M. Atalla et al.[4] that surface state densities were considerably reduced at the interface between thermally grown oxide and a silicon single crystal. Successful silicon metal–oxide–semiconductor field effect transistors (MOSFET) were first reported by S. R. Hofstein and F. P. Heiman[5] and these are now well-established in semiconductor technology.

The evaporated thin-film FET remained attractive, if only for the reason that a wide range of semiconductors can be prepared in thin-film form without undue difficulty. This, in principle, should allow the use of high mobility materials and consequent achievement of better transistor performance. However, the first successful thin-film FET used CdS, which has a low mobility, and was reported by P. K. Weimer[6]. Since then, there has been a large number of attempts, with varying degrees of success, reported in the literature, but all the most successful have involved wider band-gap, low mobility semiconductors.

In this paper an explanation is offered as to why this comes about and an account of the problems that have been met, and overcome, together with their theoretical background, in the development of a successful thin-film F.E.T. is given. Full data are given on the present generation of CdSe transistors being produced in the author's laboratories for use in large-area applications, such as the addressing of matrix display panels (as an alternative to the possibility of using the amorphous thin-film latches discussed in Chapter 7).

2. Theory of the Device

In Fig. 1(a) a schematic diagram of the device is given, showing the symbols used for its dimensions. In Fig. 1(b) a sectional view is shown which gives the voltages for an *n*-channel enhancement device, whilst in Fig. 1(c) the shape of a typical I–V characteristic is given.

2.1 Enhancement Theory

In the simple theory developed by Borkan and Weimer[7] the assumption is made that the energy bands do not bend at the semiconductor/dielectric interface at zero gate voltage. Surface scattering and trapping effects are ignored and the theory applies only over the region OA of the curve of Fig. 1(c), for which $V_g > V_d$.

In this approximation $V(x)$ is a linear function of distance, x, from the source and the drain current density J_D is given by

$$\int_0^L J_D \, \partial x = \int_0^{V_D} \frac{G_s(x)}{h} \, \partial V_x \tag{1}$$

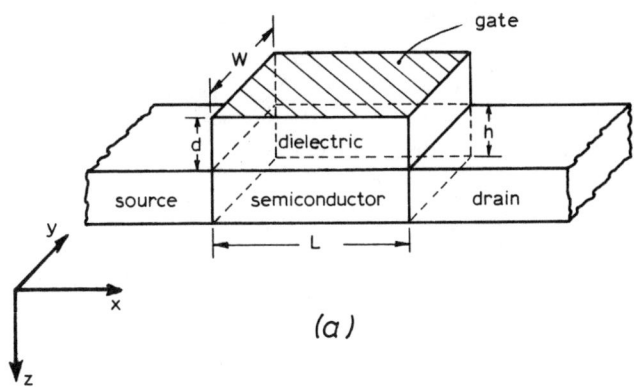

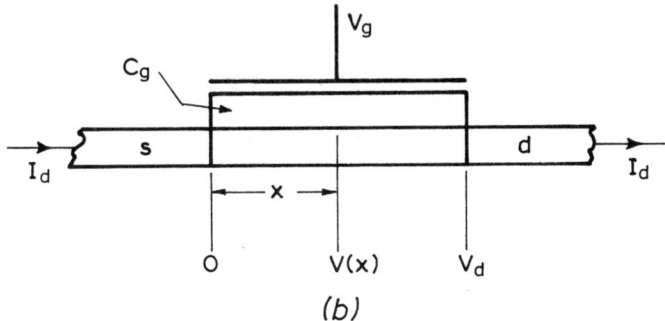

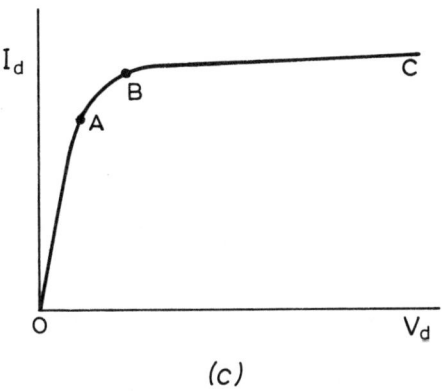

FIGURE 1. (a) Geometry of the TFT.
(b) Cross-section of TFT showing voltages.
(c) Typical $I_d - V_d$ curve.

where $G_S(x)$ is the sheet conductivity of the semiconductor which is given by

$$G_S(x) = q\mu \left[n_c + \left(\frac{V_g - V_x}{q}\right) C_g \right] \quad (2)$$

Integration gives the source-drain current as

$$I_d = \frac{W}{L} C_g \mu_c \left[(V_g + V_0) V_d - \frac{V_d^2}{2} \right] \quad (3)$$

where C_g is the gate capacitance per unit area, μ_c is the conductivity mobility of the carriers in the semiconductor and $V_0 = n_c q / C_g$ where n_c is the charge per unit area in the semiconductor film for zero gate voltage, defined by the bulk carrier density multiplied by film thickness, and q is the charge on the electron.

As V_d is increased, for fixed V_g, the stage will be reached when $V_d \geqslant V_g$. The gate electrode is then effectively negative with respect to the drain and the film begins to deplete in the vicinity of the drain, giving the portion AB of the characteristic in Fig. 1(c). Saturation is reached [BC in Fig. 1(c)] when the drain current is pinched off. This condition corresponds to depletion of all the carriers in the vicinity of the drain which occurs when

$$V_d = V_g + V_0 \quad (4)$$

V_0 is called the pinch-off voltage and is determined by the original carrier density, n_c, in the semiconductor. The drain current at saturation is given by substituting equation (4) in equation (3) giving

$$I_{ds} = -\frac{W}{L} C_g \mu_c \frac{V_d^2}{2} \quad (5)$$

Beyond the point A on the characteristic equation (1) ceases to apply.

2.2 Depletion Theory

Neumark[8] and Johnson[9] have treated the depletion case, which occurs when $V_g < V_d$. The modification to the physical picture is that the gate field penetrates into the semiconductor where a depletion region forms. This is illustrated in the band diagram of Fig. 2 in which the semiconductor depletion layer is represented by upward bending of the bands (n-type semiconductor) to a depth λ. All donors in the semiconductor are assumed to be ionized so that the charge density in the depletion region (the space charge) is given by qN_d, where N_d is the donor density, and $N_d = n_c/h$ in the absence of bulk trapping.

By solving Poisson's equation in the dielectric and space-charge regions an expression for λ, and hence for the sheet conductance of the film, can be

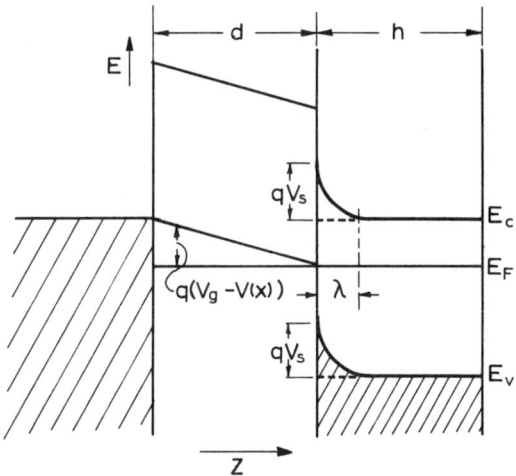

FIGURE 2. Band structure for the depletion case.

found. Integration of this conductance over the voltage range 0 to V_d gives an expression for I_d. Using the parameters for a CdSe/SiO$_2$ combination in which $N_d = 10^{22}$ m^{-3}, $\mu_c = 0.01$ m^2 V^{-1} sec^{-1}, $V_g = 5$ V, $d = 1400$ Å, $h = 1000$ Å, the variation of I_d with V_d has been plotted in Fig. 3, for (a) the simple theory of equations (1) and (3) and (b) for the partial depletion model, in which the source-drain channel is considered to consist of enhancement and depletion regions in series. It will be seen that the fuller theory, taking account of partial depletion, makes little difference to the $I_d - V_d$ curve with this set of parameters.

2.3 Barrier Theory

The foregoing theory makes the assumption of a homogeneous conducting channel, between source and drain, such as would exist in a single crystal layer. Practical thin-film transistors generally employ a polycrystalline semiconductor layer with many grain boundaries intersecting the conducting channel. These boundaries represent a physical interruption of the mean free path of electrons and also may provide traps—usually acceptor-like— in the band gap. Trapping at the grain boundaries depletes the bulk carrier density and causes a trapped space charge which can be represented as a potential barrier to an electron endeavouring to cross the boundary. Conduction under these conditions has been analysed by Petritz[10] and has been reviewed by Anderson[11,12].

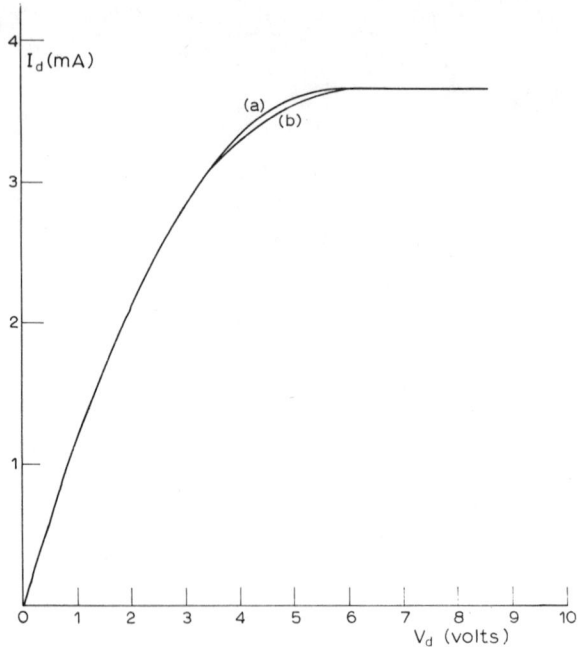

FIGURE 3. Theoretical $I_d - V_d$ characteristics.
(a) Simple theory.
(b) Partial depletion model.

In Fig. 4(a) such a barrier in n-type material is shown, whilst Fig. 4(b) represents the result of applying a bias voltage V across it. Following Miles[13] and Green and Miles[14] it is assumed that the Fermi level is virtually pinned at the crystallite interface by the high density of trapping states at the intercrystalline boundary. The height of the barrier, ϕ_0, is assumed to be constant at constant temperature and the barrier can be regarded as two back-to-back diodes. In the absence of applied voltage the net current in either direction across the barrier will be thermionic and given by

$$J_{10} = J_{20} = J_0 = \frac{1}{4} n \bar{c} q \exp\left(-\frac{q\phi_0}{kT}\right) \qquad (6)$$

where $\bar{c}$ is the thermal velocity of carriers, given approximately by $(2kT/\pi m_i^*)^{1/2}$, where m_i^* is the inertial effective mass, and n the carrier density in the bulk crystallite.

On applying the bias V [Fig. 4(b)] a portion ΔV will be dropped across

Thin Film Transistors

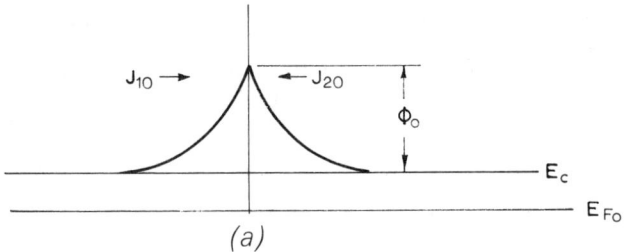

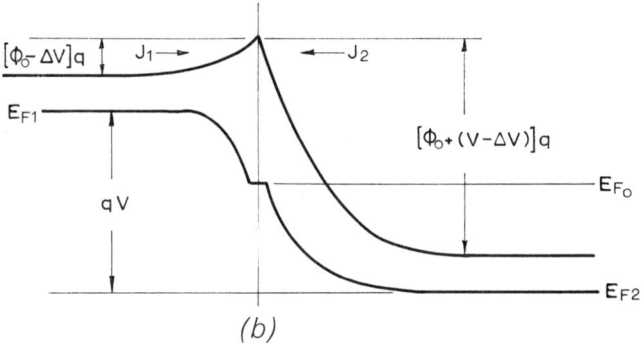

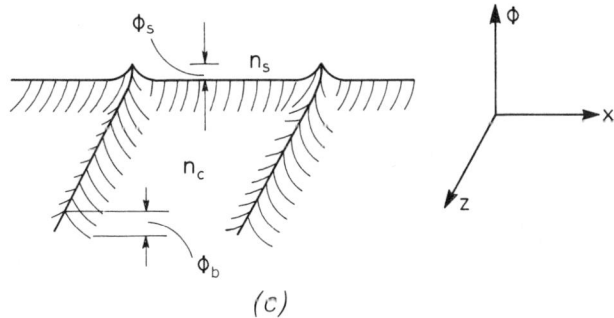

FIGURE 4. The barrier model.
(a) Potential barrier in an *n*-type material in the absence of a field.
(b) The barrier modified by an applied potential *V*.
(c) Two-dimensional barrier model.

the "forward" diode and $(V-\Delta V)$ across the "backward" diode so that we have

$$J_1 = J_0 \exp\left(-\frac{q\Delta V}{kT}\right) \tag{7a}$$

and

$$J_2 = J_0 \exp\left(-\frac{q(V-\Delta V)}{kT}\right) \tag{7b}$$

The total current is then given by

$$J = J_1 + J_2 = J_0 \exp\left(\frac{q\Delta V}{kT}\right)\left[1 - \exp\left(-\frac{qV}{kT}\right)\right] \tag{7c}$$

This leads to a current which saturates, at high values of V, at the value of J corresponding to the maximum possible value of ΔV. This clearly occurs when $\Delta V = \phi_0$ when the current becomes

$$J = \frac{1}{4} nq\bar{c}$$

and corresponds to elimination of the barrier. The J–V characteristic will then simply follow Ohm's law determined by the resistivity of the film.

However, in thin film transistors with small-grained polycrystalline films, the p.d. across each barrier for practical values of source-drain voltage, V_d, will be such that $qV \ll kT$. It can then be taken that $\Delta V \simeq 0$ and the term $\exp[-(qV)/(kT)]$ can be replaced by the first term of the expansion. Thus,

$$J = \frac{1}{4} n \frac{q^2 \bar{c}}{kT} V \exp\left(-\frac{q\phi_0}{kT}\right) \tag{8}$$

where $V = V_d/N_1 L$ and N_1 is the number of barriers per unit length of film and L is the source-drain distance. The carrier density in the bulk crystallite is given by $n = (n_c)/(h)$, where h is the film thickness and n_c the carrier density/unit area in the crystallite. The carrier mobility, μ_g, due to boundary scattering alone is given by $q\bar{c}/4N_1 kT$.

The effects of trapping at the semiconductor/insulator interface will now be taken into account. This will produce a depletion layer in the semiconductor crystallites which has the effect of reducing the carrier density at the surface of the crystal below its bulk value. Now the intercrystalline potential barrier height in the bulk is given by

$$\phi_{0b} = \frac{kT}{q} \ln\left(\frac{n_c}{n_b}\right) \tag{9}$$

where n_b is the sheet carrier density at the top of the barrier and n_c is the bulk sheet carrier density. At the surface, the carrier density $n_s < n_c$. The two-dimensional potential energy diagram up to the interface is shown in Fig. 4(c) in which the co-ordinate z is the distance below the interface and x is the direction of current flow; the third co-ordinate is potential energy.

The current density given in equation (8) is now written as

$$J = qn(z)\mu_g \frac{V_d}{L} \exp[-\phi_0(z)] \qquad (10)$$

where $\phi_0(z)$ is in units of kT.

For simplicity we take

$$n(z) = n_{0s} + \frac{(n_{0c} - n_{0s})}{L_c} \qquad (11)$$

$$\phi_0(z) = \phi_{0s} + \frac{(\phi_{0b} - \phi_{0s})z}{L_c} \qquad (12)$$

where L_c is the characteristic penetration depth of the surface space charge (Many, Goldstein and Grover[15]) and $n_{0s} = n_s/h$, $n_{0c} = n_c/h$.

Using equations (11) and (12) in equation (10), the current dI through an elementary slab of thickness dz and width W is given by

$$dI = W_q \mu_g \frac{V_d}{L} n(z) \exp[-\phi(z)] dz \qquad (13)$$

The total current I_{Lc}, through the slab is found by integrating equation (13) between the limits $z = L_c$ to $g = h - L_c$, where h is semiconductor thickness, is

$$I_b = qn_{0c}\mu_g e^{-\phi_{0b}}(h - 2L_c) W (V_d/L) \qquad (14)$$

Combining these equations Anderson[16] shows that the source-drain current is given by

$$I_D = I_b + 2I_{Lc} = V_d W q \mu_g \frac{L_c}{L} \Biggl\{ n_{0c} e^{-\phi_{0b}} \left(\frac{h}{L_c} - 2 \right)$$

$$+ 2 \left[\frac{n_{0s}}{(\phi_{0b} - \phi_{0s})} + \frac{(n_{0c} - n_{0s})}{(\phi_{0b} - \phi_{0s})^2} \right] (e^{-\phi_{0s}} - e^{-\phi_{0b}})$$

$$- 2 \frac{(n_{0c} - n_{0s})}{(\phi_{0b} - \phi_{0s})} e^{-\phi_{0b}} \Biggr\} \qquad (15)$$

Anderson[16] shows that, by assuming $L_c \geqslant (h/2)$ and that the carrier density, n_{0b}, at the top of the barrier is constant throughout the film, equation (15) reduces to

$$I_d \frac{Wh}{L} = V dq\mu_g n_{0c} e^{-\phi_{0b}} \frac{2(\cosh \Delta\phi - 1)}{(\Delta\phi)^2} \quad (16)$$

where $\Delta\phi = (\phi_{0b} - \phi_{0s})$. The conductivity of the film is then given by

$$\sigma = \frac{l}{2l_D} n_{0c} q \mu_g e^{-\phi_{0b}} \frac{(\cosh \Delta\phi - 1)}{(\Delta\phi)^2} \quad (17)$$

where l is the mean crystallite size and l_D is the effective Debye length of the grain boundary barrier given by

$$l_D = \left(\frac{\varepsilon_r \varepsilon_0 kT}{q^2 n_{0c}} \right)^{1/2} \quad (18)$$

In a typical experiment on a CdSe TFT ϕ_{0b} and ϕ_{0s} are found to be $5.2\,kT$ and $1.4\,kT$ respectively using equation (17).

The application of barrier theory to the thin film transistor can now be considered.

3. Barrier Enhancement Theory

When a gate voltage is applied the carrier density per unit area at the top of the barrier will be given by

$$n_b = n_{b0} + \Delta n - \Delta n_{bT} \quad (19)$$

where n_{b0} is the density per unit area for $V_g = 0$, Δn is the sheet charge density induced by the gate and is equal to $C_g V_g / q$, and Δn_{bT} is the amount of induced charge which is trapped at the intergrain boundaries. A barrier trapping parameter, θ_b is now introduced and is defined as the ratio of the charge remaining free to the charge induced, i.e.

$$\theta_b = \frac{\Delta n - \Delta n_{bT}}{n} \quad (20)$$

Graeffe[17] assumes that θ_b is a constant, for varying V_g, which implies from equation (19) that Δn_{bT} varies proportionately to V_g, since Δn is proportional to V_g through $\Delta n = C_g V_g / q$. Since Δn_{bT} will depend on the spatial and energy distribution of the trapping states in the barrier region, a constant θ_b would imply an energy distribution in the band gap such that the variation of trap density with energy was the same as the field effect curve which relates surface potential to the applied gate voltage. There is absolutely no evidence for the nature of the trapping states' energy distribution in the grain boundaries but, experimentally, a constant value of θ_b would give a conductivity which is a linear function of V_g. Van Heek[18] found such linearity in CdSe transistors for $V_g \simeq 5$ to 15 V, and, as is seen in Fig. 5, the transistors produced in the author's laboratory show the same linearity for $V_g = 0$ to 5V. As an approximation θ_b is therefore treated as constant. Using the assumption that n_b, the sheet charge density at the barrier, is a constant for $V_g = 0$, then

$$n_b = n_c \exp(-\phi_b) = n_s \exp(-\phi_s) \tag{21}$$

where n_c and n_s are sheet charge densities.

The experimental values of ϕ_s are always $< 2kT$ and to a good approximation therefore $\phi_s \doteq 0$. Using this together with equation (21) in equation (17), the sheet conductivity is given by

$$G_s = \frac{l}{2l_D} q\mu_g \frac{(n_c - n_b)^2}{n_c [\ln(n_c/n_b)]^2} \tag{22}$$

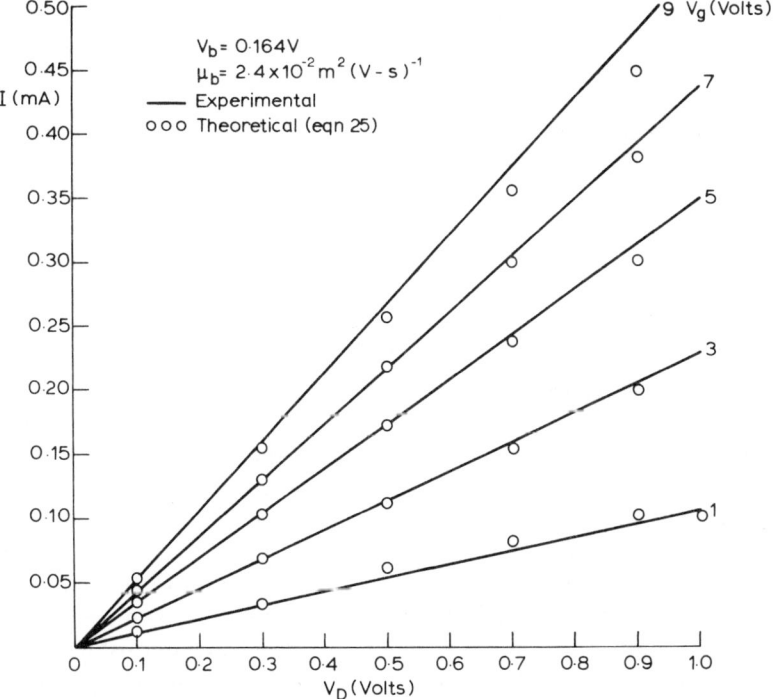

FIGURE 5. $I_d - V_d$ characteristics for a CdSe/SiO$_2$ transistor.

The use of $\phi_s = 0$ (which means $\Delta\phi = \phi_b$) implies that the density of trapping states at the interface is equal to the density of trapping states in the grain boundary regions and that the surface band-bending is the same as the intergrain potential barrier.

When a gate voltage is applied, an amount $\Delta n\theta_b$ of charge will be induced in both the interface and grain boundary regions. The conductivity will then be given, from equation (22), by

$$G_s(x) = \frac{l}{2l_D} \frac{q\mu_g(n_c - n_b)^2}{(n_c + \Delta n\theta_b)\left[\ln\frac{(n_c + \Delta n\theta_b)}{(n_b + \Delta n\theta_b)}\right]^2} \quad (23)$$

Integration of equation (23) gives the source-drain current as

$$I_d = \frac{W}{L}\frac{l}{2l_D} q\mu_g (n_c - n_b)^2$$

$$\times \int_0^{V_d} \frac{dV_x}{\left[n_0 + \frac{C_g\theta_b}{q}(V_g - V_x)\right]\left\{\ln\left[\frac{n_c + (C_g\theta_b/q)(V_g - V_x)}{n_b + (C_g\theta_b/q)(V_g - V_x)}\right]\right\}^2} \quad (24)$$

This integral cannot be solved analytically. It can however be written in a form suitable for numerical integration as follows:

$$I(V_d, \theta_b) = \frac{1}{\theta_b}\int_{\theta_b(V_g - V_d)}^{\theta_b V_g} \frac{du}{(a + bu)\left[\ln\left(1 - \frac{c}{a + bu}\right)\right]^2} \quad (25)$$

where $u = \theta_b(V_g - V_x)$, $a = n_c$, $b = C_g/q$, $c = (n_c - n_b)$.

The results of numerical integration for the transistor whose experimental characteristics are given in Figs 5 and 6 are shown as circles. It is seen that a relatively good fit is obtained, thereby justifying the assumptions involved in the development of equation (25). Figure 7 shows the variation of θ_b with gate voltage for these results and it can be seen that θ_b varies by only a small amount over the range $V_g = 0$ to $V_g = 10$ V.

The important result of the above is that the approximation $\phi_s \simeq 0$ is justified. This means that the device characteristics are dominated by the CdSe/SiO$_2$ interface, as in the single-crystal field-effect transistor. Thus in modelling the TFT it is reasonable to use the usual FET equivalent circuit but with a field-effect mobility which is reduced compared with the single-crystal value and is also a slowly-varying function of gate voltage.

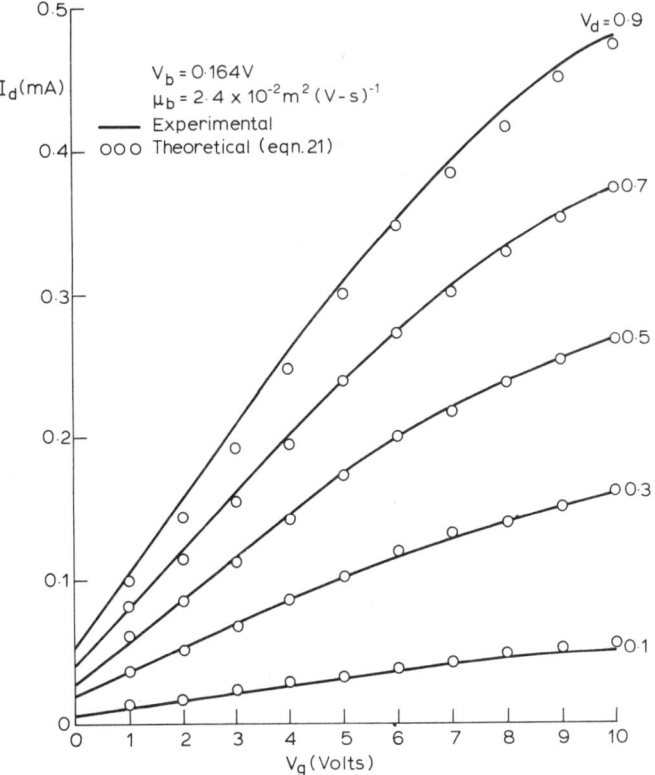

FIGURE 6. $I_d - V_g$ characteristics for a CdSe/SiO$_2$ transistor.

4. Slow States

In the previous section the effects of trapping states at the semiconductor/insulator interface and at the inter-crystalline boundary, the so-called "fast" states, were introduced via the trapping parameter θ_b. Thin film transistors are notoriously subject to a second type of trapping of relatively long time constant due to so-called "slow" states. The practical effect of these is that a sudden change in gate voltage producing, say, a sudden increase in source-drain current, I_d, is followed by a slow decay of I_d towards the value obtaining before the initial step.

A number of decay types are observed and these are illustrated in Fig. 8(a) as graphs of V_{SD} for constant I_d against time for an n-type semiconductor, the gate voltage step occurring at $t = 0$. These curves can be understood by reference to the field effect curve in Fig. 8(b), showing the surface conductivity

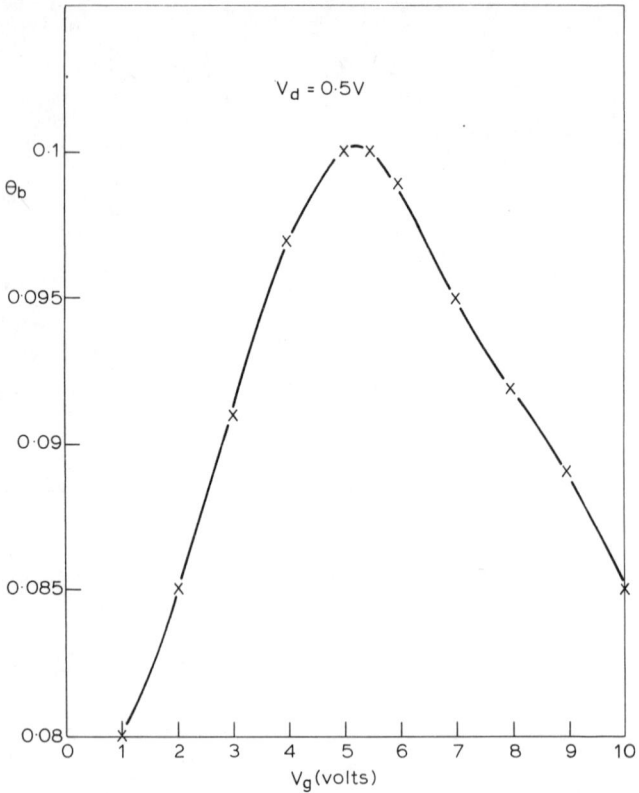

FIGURE 7. Trapping parameter θ_b as a function of V_g.

against gate voltage for an n-type semiconductor. Type (a) decay corresponds to enhancement produced by a positive gate voltage step whilst type (b) represents depletion due to a small negative gate voltage step. In type (c) a larger negative step takes the semiconductor into inversion from where it decays back to depletion whilst in type (d) a large negative step takes the semiconductor into heavy inversion from where it decays but remains inverted. It will be noted that type (b) decay is of the same form as observed in commercial MOSFETs due to positive ion movement in the dielectric. Hofstein[19], from a study of MOS devices, concluded that there were three decay mechanisms due to (i) ion drift in the insulator, (ii) temperature dependent deep trapping at the insulator-silicon interface and (iii) generation of donor-type interface states. H. Ishii and K. Yamada[20] showed the slow

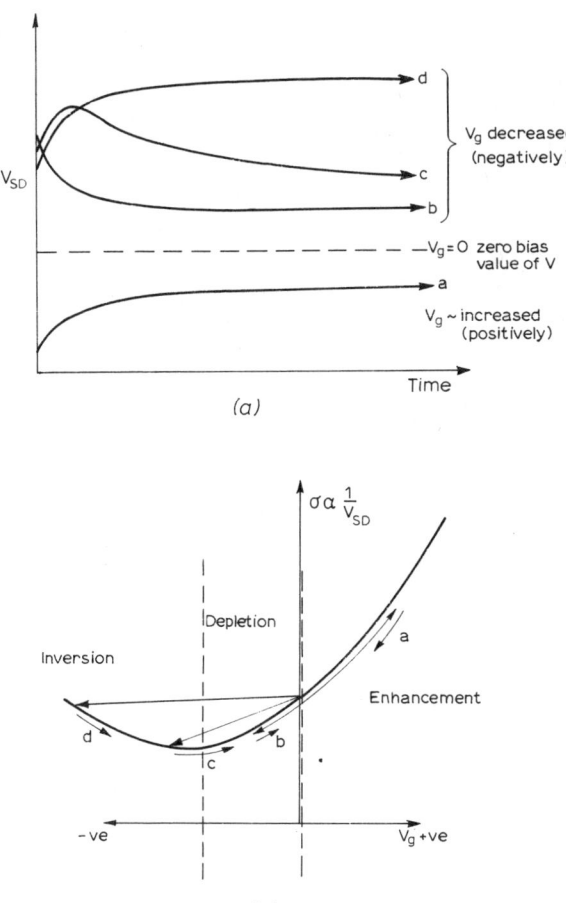

FIGURE 8. (a) Types of decay of V_{SD} for constant current as a function of time. (b) Field effect curve of conductivity σ against gate voltage V_g.

states in CdS/SiO and CdSe/SiO structures to be due to electron trapping in the dielectric. Sewell and Anderson[21] showed that in the InSb/SiO$_x$ combination the slow states are due to electron trapping centres in the insulator adjacent to the interface. Since in all practical thin-film transistors the insulator is amorphous, it is highly likely that this mechanism is the dominant cause of the slow decay generally observed in these devices, due to the large density of localised states in the band gap of an amorphous insulator.

Sewell[22] has shown that, for this model, the equivalent density of slow states per unit area at the interface in the insulator is given by

$$N_{ss} = \left\{\frac{V_g x_m^2}{2d\,\Delta v_s\,kT} + L_t\left[1-\exp\left(-\frac{x_m}{L_t}\right)\right]\right\}N_t \tag{26}$$

In this, V_g is the gate voltage and d the gate insulator thickness; Δv_s is the change in surface band bending in the semiconductor from the flat band condition, where $v_s = V_s/kT$ (Fig. 2), due to the application of V_g. L_t is the Debye length in the insulator given by

$$L_t = \left(\frac{\varepsilon_r \varepsilon_0 kT}{q^2 N_t}\right)^{1/2} \tag{27}$$

where N_t is the trap density per unit volume at an energy E_t in the insulator band gap. x_m is the maximum depth to which charge penetrates into the insulator in the measurement time T_m. It is given by (Sewell[22]) as

$$x_m = \frac{1}{2K_0}\ln\left[T_m \Sigma_0 \bar{v}\,(n_s+p_s+n_1+p_1)\right] \tag{28}$$

where K_0 is the wave function decay constant (Heiman and Warfield[23]) and is given by $K_0^2 = 2m^*/h^2\,(\psi_0-\varepsilon_x)$ where ψ_0 is the energy barrier at the interface and ε_x is the x-directed kinetic energy of the electron. n_s and p_s are the surface densities of electrons and holes respectively; n_1 and p_1 are the numbers of electrons and holes in the conduction and valence bands of the semiconductor respectively, for the case in which the Fermi energy coincides with the trap energy E_t, and are the carriers emitted from the trap level to the conduction and valence bands; Σ_0 is the capture cross-section of an insulator trap at the interface; $\bar{v}$ is the mean electron velocity in the x direction.

Returning to equation (26), the first term in the brackets represents the traps that are filled due to field lowering of the trap energy levels and the second term represents the traps that are filled due to the change in surface band bending of the insulator. This is illustrated in Fig. 9 in which a uniform density of traps is assumed; those below the Fermi level being full in the absence of gate field. The shaded area represents the slow state traps which are activated by the application of the gate field. Sewell and Anderson[23] found for the case of n-InSb/SiO$_x$, $x_m = 20$ Å with $N_t = 10^{20}$ cm^{-3} and obtained the slow state density as a function of gate field as shown in Fig. 10.

Anderson[24] has examined slow states in CdSe/quartz and CdSe/7059 glass transistors, in which the quartz was deposited by sputtering and the glass by electron beam evaporation. The effective energy gap for quartz, from the ultraviolet absorption edge, is ~ 6.2 eV. The CdSe had a carrier density of

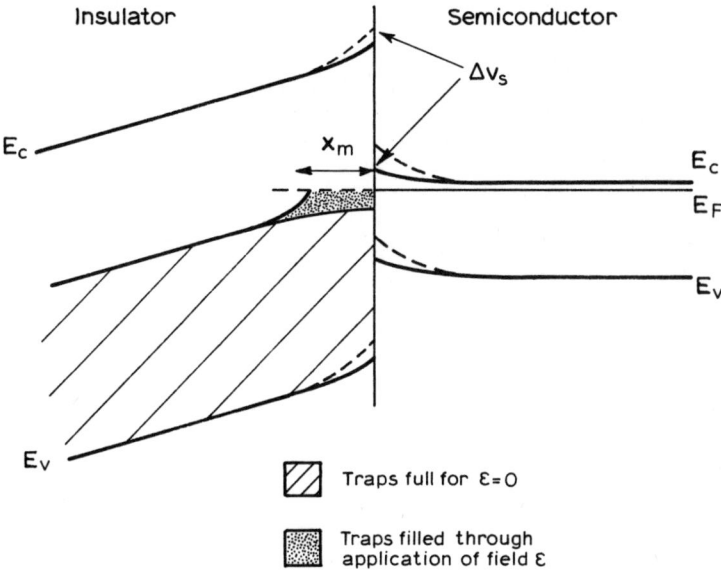

FIGURE 9. Band diagram for *n*-type TFT with positive gate field ε illustrating slow state trapping.

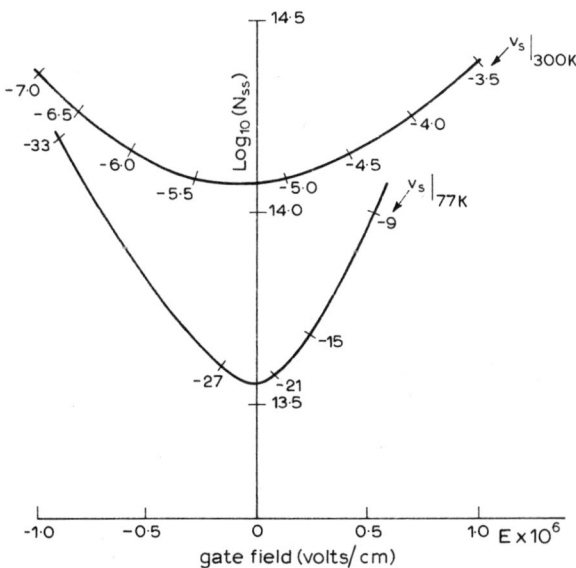

FIGURE 10. Slow state density as a function of gate field for an InSb/SiO$_x$ TFT.

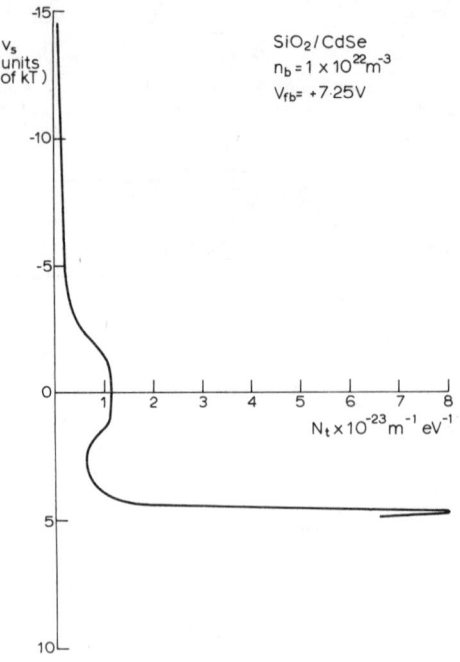

FIGURE 11. Experimental slow state trap density as a function of reduced surface potential V_s.

4.5×10^{16} cm^{-3} with the Fermi level at 0.17 eV below the conduction band edge, giving an interface energy barrier of ~ 2.93 eV. Under these conditions x_m, from equation (28), is proportional to ln n_s. The experimental method adopted was, with a source-drain voltage of 0.5 V, to apply a given gate voltage, allow the slow states to come to equilibrium, and then to apply a fixed step in gate voltage of ~ 1.5 volts. The conductivity of the semiconductor immediately after the step was noted and the gate voltage steadily increased to maintain the conductivity constant at this value. Conductivity was monitored by measuring the voltage across a 1 kΩ resistor in series with the device. The increase ΔV_g, after a fixed measurement time was divided by the value of the voltage step and this is a measure of the slow-state density at that particular gate voltage. Using equation (26) the trap density variation over a region of the insulator band-gap has been deduced. The results for a typical transistor with quartz dielectric is shown in Fig. 11, in which the trap density is plotted against interface potential v_s in units of kT at 300 K. It is of interest to note that the results are in agreement with the predictions of the Davis and Mott[23] model for the distribution in energy of states across the band gap of an amorphous semiconductor, in which the presence of a peak in the density of states at midband effectively pins the Fermi level at this position.

From the practical point of view the slow state density with the quartz insulator is relatively low for band bending between the flat band and just short of the Fermi energy. This is a gate voltage range of ~ 15 volts and represents a device with extremely good long-term stability.

These results are confirmed by measurement of the differential capacitance of the device. In this technique the source and drain are shorted together and treated as one terminal of a parallel-plate capacitor of which the other terminal is the gate electrode. The capacitance is measured by a standard bridge operating at $\omega = 10^4$ rad/sec. Following the same technique, equilibrium is obtained at a given gate voltage and then a gate voltage step is applied and the decay in capacitance in a fixed measurement time is noted. The ratio of the decay ΔC to the step in capacitance C_{step} corresponding to the step in gate voltage, is a measure of the slow state density over the range of surface potential represented by the gate voltage step. This technique permits measurement in the heavily depleted region where the conductivity of the semiconductor is too low to observe, but yields no more information than is obtainable from the conductivity measurements (Anderson[24]).

5. Choice of Semiconductor/Insulator Combination

5.1 Fast States

In the case of the single-crystal field effect transistor it is necessary, in order to achieve conductivity modulation, to reduce the fast state density at the semiconductor/insulator interface to a low value. This value is related to the maximum amount of charge that can be induced in the semiconductor from the gate electrode which is, in turn, determined by the properties and thickness of the gate insulator. For thermally grown quartz the maximum breakdown field is $\geqslant 10^6$ V/cm so that a 1000 Å layer can safely support 10 volts on the gate. The gate capacitance for such a film would be $\simeq 5 \times 10^{-8}$ F cm^{-2} so that the sheet charge density that could be induced would be $\simeq 3 \times 10^{12}$ electrons/cm^2. Thus the trapping state density should preferably not be greater than 10^{11}–10^{12} cm^{-2}. In the case of the polycrystalline film there is the additional complication of fast trapping states at the intercrystalline boundary. Barrier modulation occurs by virtue of penetration of the gate field into the space charge region at the boundary and modifies the barrier height by alteration of the sheet charge density in the boundary region [see equation (9)]. The extent to which trapped carriers on the surface of the semiconductor screen the inter-crystalline space charge region from the gate field is uncertain. However, from the results of Egerton[26], Lile[27], Ling et al.[28] and Sewell[22], all of whom obtained fast state densities of $10^{12}-10^{13}$ cm^{-2} and yet observed conductivity modulation in TFT structures, it appears that barrier modulation persists beyond the level at which the trapped carriers at the surface will have largely screened the gate field from the space-charge region in the bulk of the crystallite. The fast state density at the intercrystalline boundaries themselves can be expected to be determined mainly by the dislocation density at the boundary since edge dislocations give rise to dangling bonds. This is determined by the angle of the boundary, for low-angle boundaries, and has been discussed by Matare[29]. Thus, in choosing a semiconductor/insulator combination from the point of view of fast states, the requirements for TFTs operating by barrier modulation appear to be somewhat less stringent than in the case of single crystal devices. So far as *surface* fast states are concerned, most practical semiconductors, evaporated in normal vacuum, will form a layer of surface oxide which takes up the majority of dangling bonds. This oxide may become an integral part of the gate insulator film and a smooth transition from semiconductor to insulator is achieved with a surface state density determined mainly by defects at the interface. For thermally grown SiO_2 on single-crystal Si density may be as low as 10^9 cm^{-2}. However, in evaporated thin films the density of defects is generally higher than in equivalent bulk material. For this reason the interface state densities tend to be rather high unless they can be reduced by a suitable subsequent heat treatment. Thus the choice of semiconductor should be limited to materials in which grain growth and annealing out of defects can be promoted by heat treatment at reasonable temperatures after deposition, or else by deposition on a heated substrate. The author has found InSb, PbTe, PbSe, CdS and CdSe all meet this requirement, but this list is by no means exhaustive.

Slow states have been a much greater problem in thin film transistors, frequently occurring at such great density ($\sim 10^{13}$ cm^{-2}) as to virtually eliminate conductivity modulation of other than a fast transient type. These states are invariably associated with the insulator and may take the form of electron traps, hole traps and mobile ions. It is therefore necessary to choose an insulator in which these are reduced to a low level. The density of localized states in the mobility gap of an amorphous insulator is, according to the model proposed by Mott and Davis[38], a function of the co-ordination number and the specific volume of the material. In general, the higher the co-ordination number and the lower the specific volume (volume per atom) the higher the state density. The most useful guide to a good insulator, however, is to choose one with low optical obsorption below the absorption edge, i.e. one which is transparent and colourless when deposited. This narrows the range considerably and certainly rules out silicon monoxide, i agreement with the findings of Sewell[22]. In addition it is necessary to ensur that the insulator is not contaminated with ions which can move througl it under the influence of a field. In general sodium ions are the most likel) contaminant of this sort in oxides whilst alkali halides have a naturally occurring density of mobile ions. It is also important to ensure that the insulator is stoichiometric so that there are few incomplete bonds to act as trapping centres.

In accordance with the theory of Davis and Mott[25] and with the results shown in Fig. 11 there will always be a sharp rise in slow state density in the vicinity of the Fermi level. Thus the insulator/semiconductor combination must be such that there is a useful range of band bending without reaching the Fermi level. This rules out the use of low band-gap semiconductors such as InSb (Lile[27]) and at the same time explains the success of the wide-gap materials such as CdS and CdSe.

To summarize, whilst at first sight it seems reasonable to choose a high-mobility (and therefore low gap) semiconductor, for good transistor performance, the overriding considerations will necessarily be reduction of ·fast and slow state densities. This tends to favour higher-gap semiconductor materials combined with high-quality insulator films.

6. CdSe Transistors

There exist many reports in the literature on thin film transistors employing various semiconductors, particularly CdS, CdSe, InAs, InSb, PbTe, PbSe and Te. In this section a comprehensive report is given on the measurable properties of CdSe thin film transistors. The results of these are relevant to thin-film transistors in general and, so far as the author is aware, represent the most complete investigation of the device to be reported. All the measurements, unless otherwise stated, have been made with the structure shown in Fig. 12.

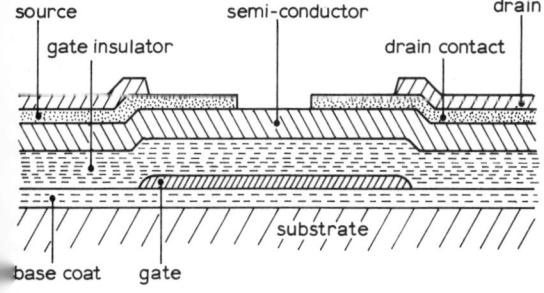

FIGURE 12. TFT structure.

6.1 Fabrication

Corning 7059 glass substrates are cleaned ultrasonically in detergent and then vapour cleaned in IPA. A thin layer of quartz is sputter deposited first to give a smooth, reproducible substrate surface and to act as a barrier to diffusion of ions out of the glass substrate. This is followed by:

(1) evaporation of aluminium gate electrode strips,
(2) sputtering of quartz to a thickness of 1400 Å,
(3) evaporation of CdSe to a thickness of 1300 Å,
(4) evaporation of chromium source and drain contacts to a thickness of 1000 Å,
(5) thickening up of source-drain contacts with evaporated aluminium.

The flow diagram of the process is given in Fig. 13, a photomicrograph of the actual device is given in Fig. 14 whilst Fig. 15 shows a general view of the carousel in the production plant. The sputtering is carried out with

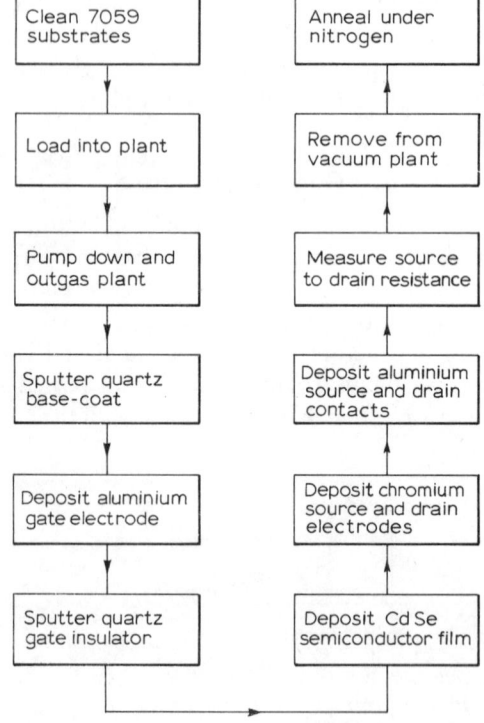

FIGURE 13. Flow diagram for TFT production.

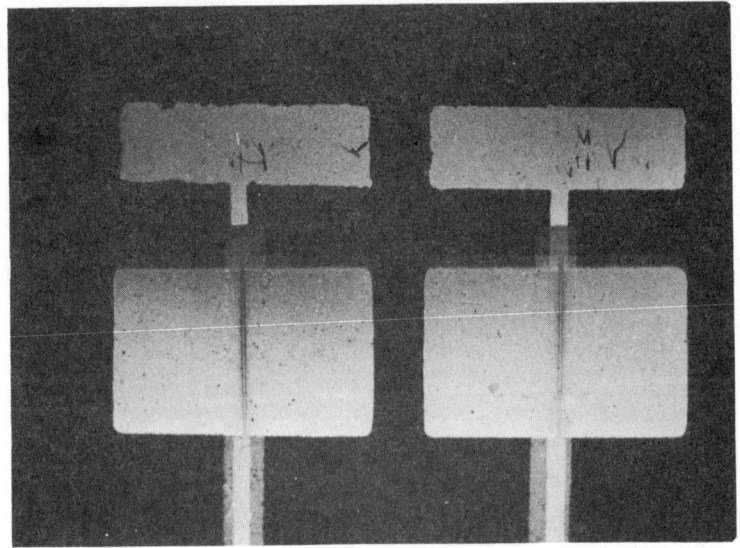

FIGURE 14. Photomicrograph of CdSe/SiO$_2$ TFT with the structure of Fig. 12.

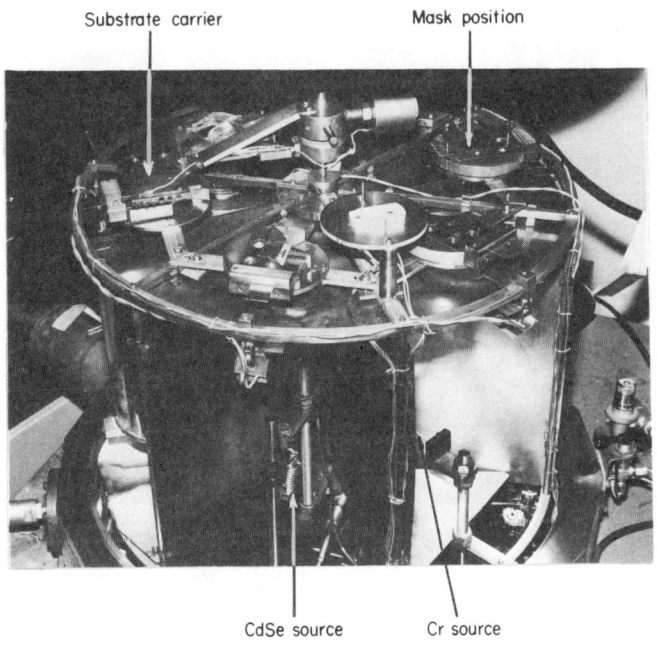

FIGURE 15. The carousel of the TFT production plant.

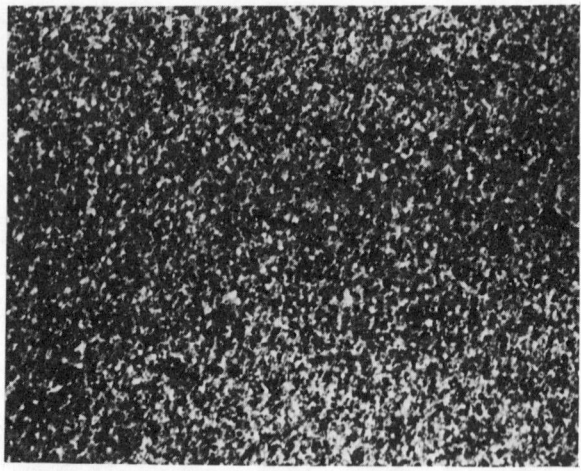

(a) CdSe film as deposited

→| |← 2500 Å

(b) CdSe film after anneal

FIGURE 16. CdSe films; (a) before and (b) after annealing in dry nitrogen (courtesy J. H. Fisher[34]).

300 W R.F. power in 3 μm pressure of argon with a 5% partial pressure of oxygen.

All patterns are defined by means of contact masks at the various stations and the substrate is moved from one to the next on the carousel. The masks are located using pins on the substrate carrier which locate in holes in the mask. The source-drain gap is defined by a 1.6×10^{-3} in wire which gives a gap width of 40 μm.

As they emerge from the plant, the semiconductor films are in a low resistance state with resistivities in the range 0.1–1 Ω cm. They are then annealed in dry nitrogen at 350°C for $1\tfrac{1}{2}$ hrs, which promotes crystallite growth and the resistivity rises to $10^5 \rightarrow 10^6$ Ω cm. Figure 16(a) shows a micrograph of the film before annealing and Fig. 16(b) shows the crystallite growth which results from the anneal. A histogram of crystallite size after annealing is shown in Fig. 17.

Devices were produced in arrays of ten on a standard microscope slide. The yield figure of workable transistors was 85.3% over a total of 290 made.

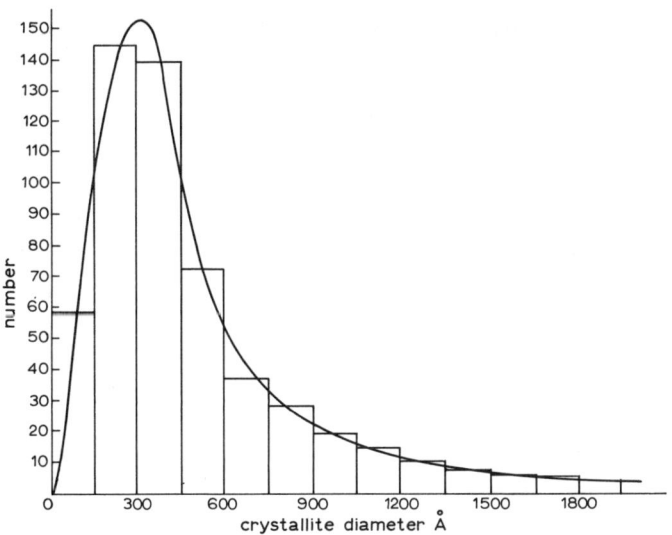

FIGURE 17. Histogram of crystallite size distribution for the film of Fig. 16(b) (courtesy J. H. Fisher[34]).

6.2 Characteristics

Typical $I_d - V_d$ characteristics for the transistors are shown in Fig. 18. The transfer characteristics, I_{DS} against V_g with $V_d = 10$ V are shown in Fig. 19 for two different temperatures.

CdSe is well known to be photo-sensitive and the effect of illumination of the saturated drain current of an unencapsulated transistor is shown in Fig. 20.

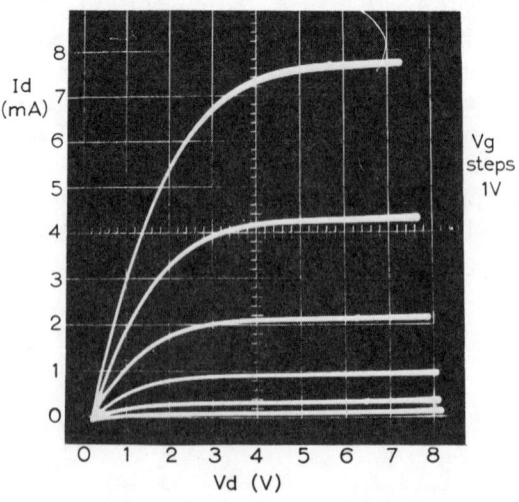

FIGURE 18. Characteristics of a CdSe/SiO$_2$ TFT.

The lack of an analytical solution to equation (25) means that there is not a general equation to describe the characteristics. However, taking the case of flat bands at the interface [$\Delta\phi = 0$ in equation (16)] the sheet conductivity is given by

$$G_s = \frac{l}{2l_D} q \mu_g n_c e^{-\phi_b} = q \mu_b n_b \tag{29}$$

where

$$\mu_b = \mu_g \frac{l}{2l_D}$$

Thus, when a gate voltage is applied

$$G_s(x) = q \mu_b \left[n_b + \frac{C_g \phi_b}{q}(V_g - V_x) \right] \tag{30}$$

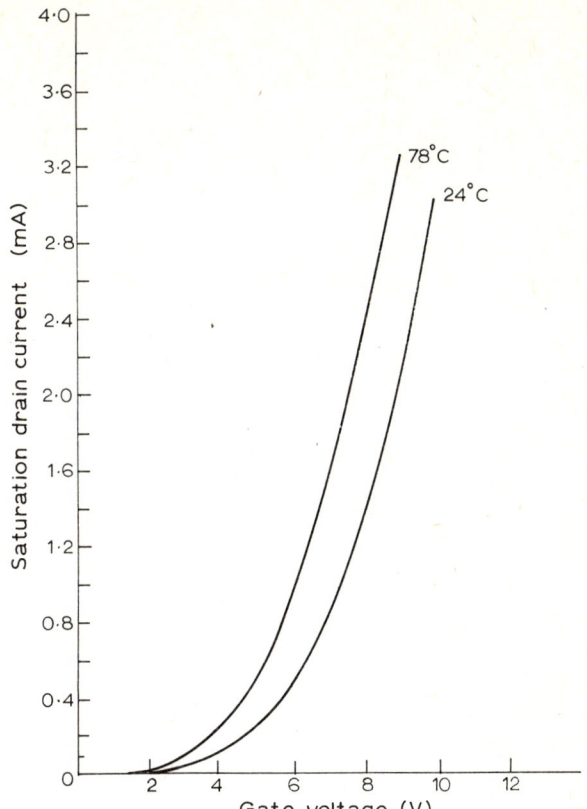

FIGURE 19. I_d as a function of gate voltage for 24°C and 78°C (courtesy J. H. Fisher[34]).

Integration then yields

$$I_D = \frac{W}{L} q\mu_b \theta_b \left[(V_g + \frac{V_b}{\theta_b}) V_D - \frac{V_D^2}{2} \right] \quad (31)$$

where

$$V_b = \frac{n_b q}{C_g}$$

Equation (31) is, of course, of the same form as equation (3) for the single crystal FET but with different effective mobility and pinch-off voltage. The

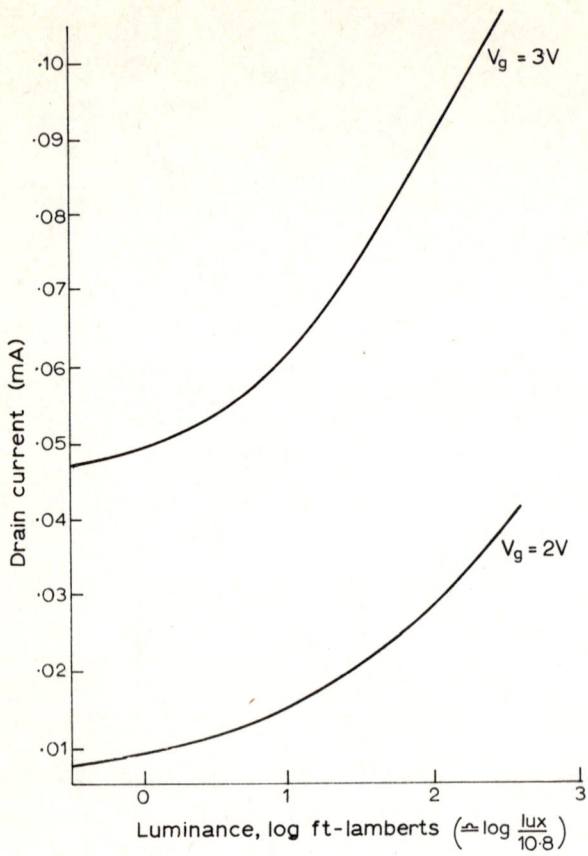

FIGURE 20. I_d as a function of illumination for different gate voltages (courtesy J. H. Fisher[34]).

same form will be adopted for the general case where μ_b, θ_b and V_b are now parameters to be determined from the experimental characteristics.

Using the results presented for θ in Fig. 7, together with the experimental curves of Fig. 5, it is found that $\mu_b = 2.4 \times 10^{-2}$ m² (V-sec)⁻¹ and $V_b = 0.164$ V. The results calculated from equation (31) fit Figs 5 and 6, from which it is concluded that an expression of the form of equation (31) can be applied generally over the whole range of FET's, both polycrystalline and single crystal, without too much error. Defining $\mu_b \theta_b$ as a field-effect mobility μ_{eff} it will be seen that this is dependent on carrier density (through I_D), grain size and gate voltage (through θ_b).

The temperature dependence of I_D, assuming all donors ionized, will be a complex function, depending on μ_{eff} and on V_b [which contains terms of the kind $\exp(-qV/kT)$].

6.3 Transconductance

A useful parameter in assessing the operation of a field-effect transistor is its transconductance g_m, given by

$$g_m = \left(\frac{\delta I_D}{\delta V_g}\right)_{V_d \text{const}} \tag{32}$$

For the straight-forward enchancement case, with barrier modulation, this is given, from equation (31) viz;

$$g_m = \frac{W}{L} C_g \mu_b \theta_b V_d \tag{33}$$

if θ_b is assumed to be independent of gate voltage. This is independent of V_g over the linear part of the characteristics. However, the transistor is normally operated in the saturation region. At pinch-off,

$$V_d = \left(V_g + \frac{V_b}{\theta_b}\right)$$

and, substituting this in equation (31) and differentiating, yields

$$g_m^{\text{max}} = \left(\frac{\partial I_{DS}}{\partial V_g}\right)_{V_d} = \frac{W}{L} C_g \mu_b \theta_b \left(V_g + \frac{V_b}{\theta_b}\right) \tag{34}$$

assuming constant θ_b. Thus, in this approximation it is expected that $g_m^{(\text{max})}$ will be linearly proportional to V_g.

In the present devices a plot of $g_m^{(\text{max})}$ against V_g showed a quasi-linear region starting at a value of V_g in the general region of a typical flat band voltage of ~3 V, as shown in Fig. 21, but the linearity is poor which demonstrates that the assumption of θ_b being independent of gate voltage is incorrect.

Snejdar et al.[31] have suggested that, due to barrier modulation, a polycrystalline TFT should be capable of giving g_m values equal to or greater than those attainable with a single-crystal, field-effect transistor of the same material. They define an effective mobility, μ_{eff}, (the field-effect mobility) and show that, if this is a function of gate voltage, the transconductance can be expressed as

$$g_m = \frac{Wd}{L} V_D \left(\mu_{\text{eff}} \frac{\delta n}{\delta V_g} + n \frac{\delta \mu_{\text{eff}}}{\delta V_g}\right) \tag{35}$$

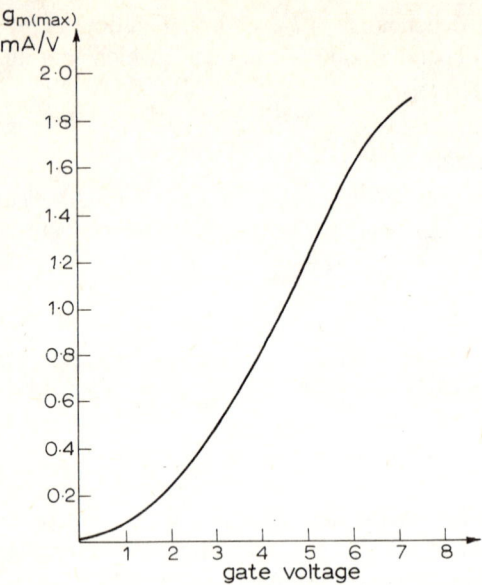

FIGURE 21. $g_{m(\text{max})}$ as a function of gate voltage for a typical TFT of the type shown in Fig. 14.

where d is the semiconductor film thickness and n the carrier density. They show that, if scattering in the space-charge region is negligible,

$$\frac{\partial \mu_{\text{eff}}}{\partial V_g} = -\frac{q}{kT} \frac{\mu_b}{\left(1+\frac{\mu_b}{\mu_c}\right)^2} \frac{\partial \phi_b}{\partial V_g} \tag{36}$$

where θ_b is the intercrystalline barrier height. In the absence of barrier modulation ($\partial \mu_{\text{eff}}/\partial V_g$) is zero and using this fact, combining equations (35) and (36) they show that if

$$\frac{\partial \phi_b}{\partial V_g} < -\frac{kT}{q} \frac{1}{n} \frac{\partial n}{\partial V_g} \left(1+\frac{\mu_c}{\mu_b}\right) \tag{37}$$

g_m will be higher for a polycrystalline than for a single-crystal device.

Now $\partial n_c/\partial V_g$ at a given V_g is simply given by C_g/q, (where $n_c = nd$), for $\theta_b = 1$, so an estimate can be made of the magnitude of the right-hand side of equation (28) using the typical values of a CdSe transistor with $kT/q = 0.025$ eV, $n = 10^{22}$ m^{-3}, $C_g = 3 \times 10^{-4}$ F m^{-2} and $V_g = 1$ V; thus the right-hand side is approximately $0.1 [1+(\mu_c/\mu_b)]$.

$\partial \phi_b / \partial V_g$ is obtained as follows:

$$\phi_b = \frac{kT}{q} \ln\left(\frac{n_c + \Delta n \theta_c}{n_b + \Delta n \theta_b}\right) \tag{38}$$

where $\Delta n = C_g V_g / q$, θ_c and θ_b are the bulk trapping and barrier trapping parameters.

Assuming $\Delta n \theta_c / n_c < 1$ and $\Delta n \theta_b / n_b > 1$, it can thus be shown that

$$\frac{\partial \phi_b}{\partial V_g} = -\frac{q n_c}{\theta_b C_g V_g^2} \exp \frac{q n_c}{\theta_b C_g V_g} \tag{39}$$

Using the values given above with $\theta_b = 1.0$ gives $\partial \theta_b / \partial V_g = 0.3$. This implies that $\mu_c \doteqdot 2\mu_b$ for the above to be fulfilled.

For CdSe $\mu_c = 500 \text{ cm}^2/\text{V} \cdot \text{sec}$ and μ_b is given by the equation $\mu_b = (q\bar{c}/4N_1 kT)$, where no correction is now made for bulk conductivity. For CdSe, $m_i^* = 0.13 m_0$, and with $kT = 0.025$ eV, $\bar{c}_\mu = 7.3 \times 10^4$ m/sec and, using a mean crystallite diameter of 500 Å (Figs 16 and 17) and a source-drain gap of 40 μ, this gives $\mu_b = 1460 \text{ cm}^2/\text{V} \cdot \text{sec}$. Thus the required condition is far from being fulfilled in the extreme case of no barrier trapping considered here. If the crystallite size were considerably smaller μ_b would decrease, but so also would θ_b; experimentally θ_b is found to be generally in the region of 0.1 in the present films. Also the decreased crystallite size leads to much increased source-drain resistance so that transistor current is reduced.

In general, in the opinion of the present author, it is unlikely that a practical polycrystalline TFT can be produced with a g_m value as high as that attainable in a good, single-crystal FET. However, from the device point of view, perfectly adequate g_m values can be obtained from a polycrystalline transistor.

6.4 Contacts

It is well known that the effects of resistive contacts at source and drain is to degrade the FET characteristics (see, for instance, Wallmark and Johnson[32]). In general a parasitic source resistance has a greater effect on the characteristics than has a similar drain resistance. Physically, this is because a resistance in series with the source affects not only the source drain potential drop but also reduces the effective value of the gate voltage with respect to the film, i.e. it gives rise to negative feedback. The drain resistance, on the other hand, simply reduces the effective source drain voltage and leads to saturation at lower gate voltages.

When the contact resistances are non-linear with voltage, such as for the case of Schottky barriers, similar effects are observed but with the non-ohmic contact characteristic superimposed on the transistor characteristics

6.5 Noise Characteristics

In the case of the silicon/SiO_2 IGFET a noise voltage is generated which has an amplitude inversely proportional to frequency. This $1/f$ noise has been shown by Fu and Sha[33] to originate from trapping of charge in the oxide, i.e. slow states. In Fig. 22 a plot is shown of spot noise voltage e_n against frequency for a commercial MOSFET (MT 101) and a CdSe/SiO_2 TFT. It will be seen that $1/f$ noise is also generated in the TFT.

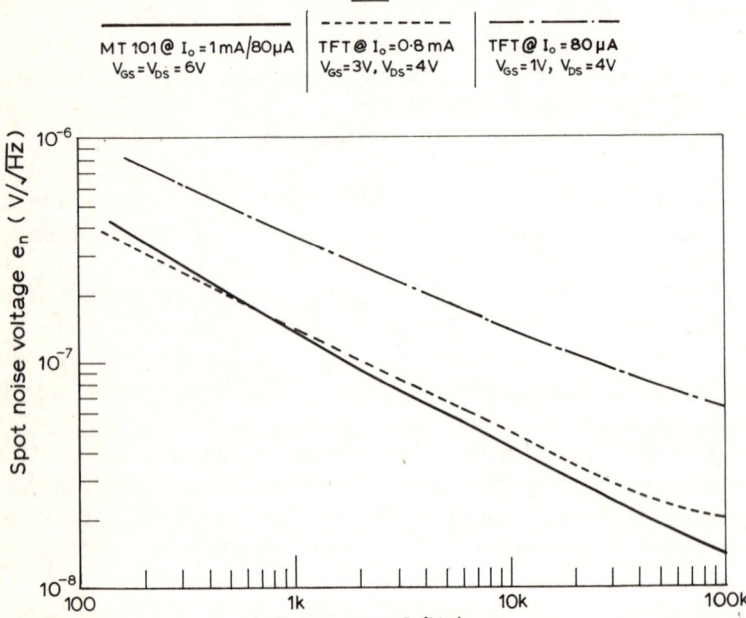

FIGURE 22. Spot noise voltage as a function of frequency for a commercial MOSFET and a TFT.

In Fig. 23 the spot noise voltage for different frequencies is plotted as a function of the gate voltage. The marked rise in low frequency noise for gate voltages in the region of 18 V is consistent with the slow state density shown in Fig. 11 for these transistors. The presence of this peak provides further confirmation of the Davis and Mott model of the density of states in the amorphous insulator.

It is found that at frequencies above 1 kHz the peak is absent, in agreement with the definition of slow states as those having a trapping time $\leqslant 1$ m/sec.

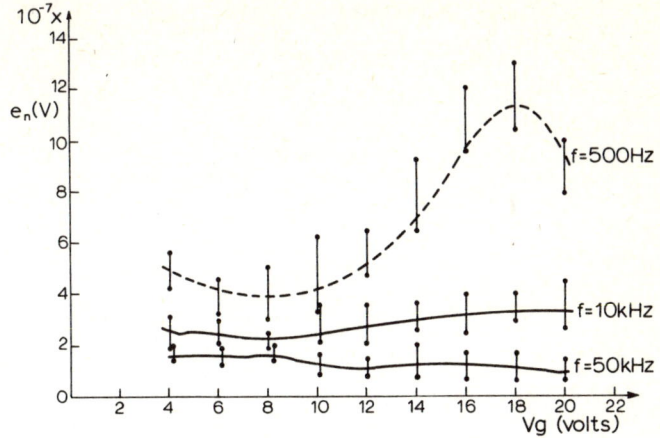

FIGURE 23. Spot noise voltage as a function of gate voltage.

6.6 TFT Design Parameters

For a given semiconductor-insulator pair the electrical characteristics of the TFT will be determined by its geometry. The factors that can be varied are [see Fig. 1(a)], source-drain gap L, width W, dielectric thickness d, semiconductor thickness h and gate electrode width g. The latter is usually taken to be equal to L but in the most simple technology it is convenient to make $g \gg L$ to overcome mask registration problems. This however introduces larger gate-source and gate-drain capacitances which affect the dynamic behaviour of the device.

The electrical parameters of most interest are: saturation drain current I_{DS}, transconductance g_m, maximum allowable gate voltage $V_{g\max}$, "on"-resistance corresponding to I_{DS}, R_{on} and "off"-resistance for $V_g = 0$, R_{off}. The relationships between these parameters and the geometrical factors are obtained from equations (19) and (20), and for fixed V_D, are given by

$$I_{DS}, g_m \propto w/Ld$$
$$R_{on} \propto Ld/w$$
$$R_{off} \propto L/w$$
$$V_{g\max} \propto d$$

It will be seen that decrease of R_{on} by decrease of the ratio L/w will also increase I_{DS} and g_m, but will decrease R_{off} which, in switching applications, may be undesirable. To avoid this, the gate dielectric thickness may be reduced which will allow decrease of R_{on} without affecting R_{off}. However, the maximum allowable switching voltage, $V_{g\max}$ will also be reduced whilst the input capacitance is increased. In an enhancement device using a single-carrier

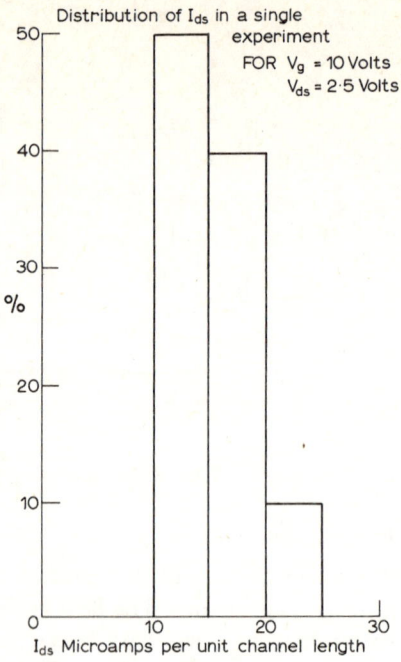

FIGURE 24. Histogram of I_{ds} for 10 devices.

material; like CdSe, which cannot be inverted, it is always possible to increase R_{off} by applying a negative gate voltage instead of simply reducing V_g to zero.

Typical results from 10 identical CdSe transistors in an 8-cell shift register deposited in one pump-down are presented in Figs 24 and 25. In Fig. 24 a histogram is shown of I_{DS}, expressed as μA per square. To obtain the value of I_{DS} for any given transistor, the value given in the figure is multiplied by W and divided by L. It will be seen that the spread in characteristics is such that 90% of the transistors had a value of I_{DS} within the range 10–20 $\mu A/\square$, whilst all fell in the range 10–25 $\mu A/\square$. This corresponds to a range for R_{on} of 830 Ω to 2 kΩ for a transistor of 3 mm width and 25 μm source drain gap. Figure 25 shows a histogram of R_{off} in ohms per square for the same value of V_D. A transistor of the same dimensions would have R_{off} in the range 20 kΩ to 1 MΩ with 50% in the range 100–500 kΩ.

These results were obtained with relatively crude technology and are quoted merely to indicate the orders of magnitude found in practice with CdSe/SiO$_2$ transistors having $h \doteq 1300$ Å and $d \doteq 1200$ Å.

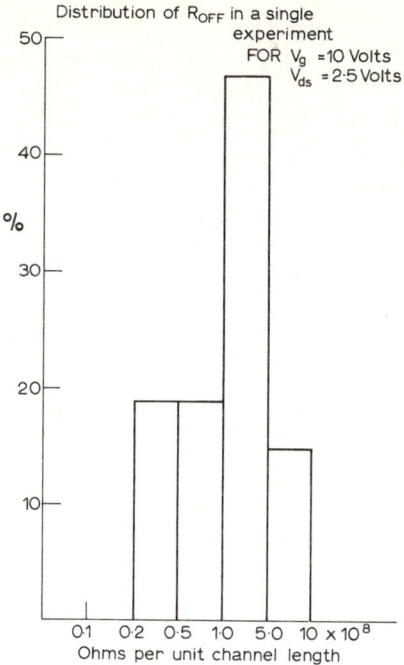

FIGURE 25. Histogram of R_{off} for 10 devices.

7. Conclusions

The viability of the thin film transistor as an active device is now well established and it is appropriate to consider its potential applications.

At the present stage of development its most useful characteristic is that it provides a switch with a very high on/off ratio. It is therefore an obvious contender as a switching element at each picture point in a matrix-addressed display panel. TFTs of $25\,\mu\text{m} \times 25\,\mu\text{m}$ area with an on/off ratio of 10^5 to 1, suitable for a liquid-crystal matrix, have already been produced in the author's laboratories. The potential advantages over the amorphous switch in this application is that the TFT is a three-terminal device, which enables control of the on and off conditions and also that it can be expected to have a longer life than the amorphous switch.

Matrix-addressed display panels being investigated and developed include electroluminescent (EL); plasma panel (PP) and light emitting diode (LED) elements. The first two share the common problem that the voltage required to excite the element is much greater than the logic level of 5 volts available from TTL integrated circuits. Thus amplification is required between the logic element and the display panel. In the case of EL and PP displays the amplifier/driver circuit can form a major part of the cost of the system since high-voltage, discrete transistors must be used in the present technology.

All matrix-addressed displays present considerable interconnection difficulties, which arise from the fact that the logic/driving circuits are necessarily separate from the display panel and require a large number of discrete wiring connections to it. Even if an integrated circuit with the necessary high drive-voltage capability is developed to replace the present discrete drive transistors, the interconnection problem will remain.

The thin film transistor offers a possible solution to both these difficulties. It can readily be designed to handle the high voltage/high current capability necessary for driving an EL display and can be deposited directly on the margin of the matrix display panel. Logic circuits using TFTs can be realised in integrated form, also on the margins of the panel, with direct connection to the matrix by evaporated conductors. In a fully integrated, TFT-addressed and refreshed display system, the only external circuitry would be the information and memory integrated circuits, giving an enormous reduction in bonded connections to the display panel.

It is considered that the TFT will ultimately be widely used in the new generations of display panels which will be developed in the future.

Acknowledgements

It is a pleasure to acknowledge the work of my staff and research students over the period of ten years during which many aspects of TFTs have been studied. In particular, the major contributions of J. H. Fisher and M. J. Lee are gratefully recognised.

References

1. J. E. Lilienfeld, U.S. Patent 1745175, (1930).
2. J. E. Lilienfeld, U.S. Patent 1900018, (1935).
3. W. Shockley, *Phys. Rev.*, **56**, 317, (1939).
4. M. M. Atalla, E. Tannenbaum and E. J. Scheibner, *Bell Systems Tech. J.*, **38**, 749, (1959).
5. S. R. Hofstein and F. P. Heiman, *Proc. IEEE*, **51**, 1190, (1963).
6. P. K. Weimer, *Proc. I.R.E.*, **52**, 1462, (1962).
7. H. Borkan and P. K. Weimer, *R.C.A. Review*, **24**, 153, (1963).
8. G. F. Neumark, *Solid State Electron.*, **7**, 725, (1964).
9. J. E. Johnson, *Solid State Electron.*, **10**, 657, (1967).

10. R. L. Petritz, *Phys. Rev.*, **104**, 1508, (1956).
11. J. C. Anderson, *Advan. Phys.*, **19**, 311, (1970).
12. J. C. Anderson, *Thin Solid Films*, **18**, 239, (1973).
13. R. E. Miles, PhD Thesis, London University, (1973).
14. M. Green and R. E. Miles, *J. Phys. D.*, **6**, 145, (1973).
15. A. Many, Y. Goldstein and N. B. Grover, "Semi-conductor Surfaces", North-Holland, Amsterdam, (1965).
16. J. C. Anderson, *Thin Solid Films*, (1976).
17. R. Graeffe, PhD Thesis, Helsinki Technical University, (1969).
18. H. F. Van Heek, *Solid State Electron.*, **11**, 459, (1968).
19. S. R. Hofstein, *Solid State Electron.*, **10**, 657, (1967).
20. H. Ishii and K. Yamada, *Solid State Electron.*, **10**, 1201, (1967).
21. H. Sewell and J. C. Anderson, *Solid State Electron.*, **18**, 641, (1975).
22. H. Sewell, PhD Thesis, London University, (1973).
23. F. P. Heiman and G. Warfield, *IEEE Trans. Electron. Devices*, **ED19**, 273, (1972).
24. J. C. Anderson, *Phil. Mag.*, **30**, 839, (1974).
25. E. A. Davis and N. F. Mott, *Phil. Mag.*, **22**, 903, (1970).
26. R. F. Egerton, PhD Thesis, London University, (1968).
27. D. L. Lile, PhD Thesis, London University, (1968).
28. C. H. Ling, J. H. Fisher and J. C. Anderson, *Thin Solid Films*, **14**, 267, (1972).
29. H. F. Mataré, "Defect Electronics in Semi-conductors", John Wiley, New York, (1971).
30. N. F. Mott and E. A. Davis, "Electronic Processes in Non-crystalline Materials", Clarendon Press, Oxford, (1971).
31. V. Snejdar, J. Jerhot and D. Berlsova, *Thin Solid Films*, **13**, 47, (1972).
32. J. T. Wallmark and H. Johnson, "Field Effect Transistors", Prentice-Hall, U.S.A., (1966).
33. H. S. Fu and C. T. Sha, *IEEE Trans. Electron. Devices*, **ED19**, 273, (1972).
34. J. H. Fisher, M. Phil. Thesis, London University, (1972).

Chapter 7

Amorphous Semiconducting Films

J. R. Bosnell

*R.S.R.E. Malvern,
Worcestershire, England*

1. Introduction 246
2. Preparation of Amorphous Semiconductors 252
3. Material Characterisation 258
4. Contact Effects 261
5. Switching in Amorphous Semiconductors 265
   5.1 Switching Phenomena 265
   5.2 Threshold Switching in Chalcogenide Glasses 270
   5.3 The Forming Process 273
   5.4 The ON State of the Threshold Switch 275
   5.5 Switching Theories 278
   5.6 Memory Devices 282
   5.7 Switching in Transition Metal Oxides 285
   5.8 Device Fabrication 286
6. Applications of Switching Devices 288
   6.1 Memory Switch Applications 288
      6.1.1. Radiation hardness 289
      6.1.2 Telephone applications 291
   6.2 Threshold Switching Applications 292
      6.2.1 Display switching 292
      6.2.2 Logic circuits 294
      6.2.3 Amplification 295
      6.2.4 Oscillators 296
      6.2.5 Thin film inductive elements 297
      6.2.6 Temperature sensors 297
      6.2.7 Pressure sensors and strain gauges 297
      6.2.8 Others 298
      6.2.9 Three terminal devices 298
   6.3 Other Switching Devices 299

7. Mass Memory Proposals using Amorphous Materials 299
8. Acousto-optical Properties 303
9. IR Detectors 304
10. Imaging Systems 304
11. Photographic Effects 307
12. Xerography 307
13. Conclusions 309
    Acknowledgements 309
    References. 310

1. Introduction

There is little doubt that the present interest in the amorphous and liquid states stems principally from two factors. Firstly, the success of quantum theory in explaining and predicting phenomena in the crystalline state[1] led physicists to wonder what would happen when crystal order was absent. Typical of the pioneering theoretical work is that of Mott and Twose[2], who, using a random Kronig-Penney model, suggested that all states in a one-dimensional lattice were localised. The question of the existence of some equivalent of the energy gap of the crystalline state has been reviewed by Halperin[3]. It seems that a gap can exist if limits are placed on the magnitude of the fluctuations from the mean interatomic distance in one dimension at least. The reader is referred to the excellent text of Mott and Davis[4] for further discussion on this matter. Secondly, the commercial success of xerography[5] using amorphous selenium films and the use of amorphous oxides in the semiconductor integrated circuit (SIC) industry, particularly MOS (metal–oxide–silicon) devices, gave fresh impetus to the search for exploitable amorphous meterials. In general, as with xerography, the proposed device has come before its theoretical understanding and has not grown on the theoretical base outlined above. This is particularly true of the electronic switches manufactured from amorphous chalcogenide alloys following Ovshinsky's announcement of the two effects, memory and threshold action, in 1966[6]. Indeed, as Mott has often stressed[7,8], for three dimensions it has not been possible to translate the earlier quantum mechanical theories and show under what conditions an energy gap exists in a glassy material, yet we know from common experience, glass transparency for example, that such a pseudogap exists.

Perhaps some may say that it is a sad reflection on the scientific community that the debate which followed Ovshinsky's revelations was at times so acrimonious but the situation is now more sober even if a completely satisfactory theoretical description of the phenomena involved is not to hand. However, it is the view of the author that progress in science is not as devoid of personal bias as the much vaunted image of the unemotional scientist would lead us to believe. A study of the book by Watson[9] on the discovery of DNA would soon show the reader how involved with and committed to a theory a scientist can be. Indeed this commitment appears necessary if science is to progress. A subject thrives on controversy and atrophies on consensus. This deeply subjective element in research seems to surround all thriving lines (as Mitroff's recent study on the "moon scientists" emphasises[10]), and has marked many discussions regarding amorphous switching devices. Furthermore, scientists distrust the media and unfortunately many first read about switching phenomena in amorphous systems from the press conference following Ovshinsky's paper in *Physical Review Letters* in 1968[11]. Furthermore, because of patent secrecy in announcements at that time, device constructional details and material compositions were sparse.

The comparison, often made in the early literature, with the development of the transistor aided this euphoria but it was an unfortunate analogy. There was a genuine size reduction, improvement in reliability and power saving to be gained by replacing vacuum tubes. Amorphous switches do not offer such clear advantages but could have enormous potential in areas other than direct replacements for crystalline silicon based devices. The recent scientific emphasis on amorphous switching devices should not be allowed to cloud the fact that semiconducting glasses, for example As_2Se_3, As_2S_3, have been used in such commercial devices as infrared windows and vidicon tube elements for many years. It is the purpose here to review these possibilities. However, space will not allow the inclusion of comment on the use of amorphous or glassy oxides in SICs and electrolytic capacitor manufacture or the production of amorphous silicon films as a stage in the silicon gate process for SIC production.

What is the interest of equipment manufacturers in amorphous materials? There are two main areas. Firstly, single crystal technology constrains the device area to approximately 10 cm diameter in the case of silicon, for example, whereas amorphous materials can be produced over large areas and hence lower costs by a variety of techniques to be outlined later. Secondly, the constraints of phase rules lead to immiscibility gaps in crystalline alloys which may be overcome in the amorphous solid by rapid quenching methods from the liquid. It should be emphasised that these ideas will in general only be employed when conventional technology has hit fundamental barriers. There is always a sailing ship effect. The new material must either offer a combination of improved properties more cheaply before the capital already invested in older product lines will be amortised and the new technology adopted, or, as with the Xerox process, make possible a completely new approach to fulfil a well recognised market.

There is in the literature a widespread interchange of terms; liquid-like, non-crystalline, glass, vitreous and amorphous are often taken as synonymous although it is usual to define a glass as an amorphous solid formed by continuous solidification of a liquid. Whether or not there is any significant difference between these terms is beyond the scope of this review but this matter is treated by Stevels[12]. It is proposed to take as a definition that a non-crystalline solid is one in which there is no long range periodicity in the lattice and this implies one or two diffuse rings in electron and x-ray diffraction patterns. Adopting this definition, the various terms can indeed be taken as synonymous. Conductivities in the semiconducting glasses, particularly the chalcogenides (a term usually taken to refer to the sulphur group S, Se, Te together with Si, Ge, As and Sb), cover a wide range from 10^{-13} to $10^{-2} \Omega^{-1} cm^{-1}$, whilst the Hall mobility is low ($10^{-2} cm^2 V^{-1} sec^{-1}$) and generally n-type, whereas the sign of the Seebeck coefficient implies that the same material is p-type. This controversy is slowly being reconciled (Queisser[13], de Wit[14]), but it is relevant here since one of the major models for switching effects in these materials is electronic in nature relying upon carrier injection. In some cases the conductivity of the glassy state is greater than that of the crystalline state of the same material, for example As_2S_3, and this emphasises the complexity of the observed phenomena.

Since the amorphous state is essentially metastable, this property can be utilised to produce materials which are easily crystalliseable so that a memory transition is produced. This phase transition can be induced by laser irradiation, heating from an external source or switching within the material. Glasses containing several constituents can phase-separate so that the experimenter may not be dealing with a homogeneous material. The phase difference may amount to minute composition changes or more gross structural changes, even crystalline, leading to glass ceramics. Differential thermal analysis (DTA) (Mackenzie[15]) is the tool commonly used to obtain information about the phase changes occurring in the glass as a function of temperature. Figure 1 shows the DTA trace of an easily crystalliseable glass, $Ge_{15}Te_{85}$, compared with a glass from near the centre of a glass forming region, $Ge_{10}Si_{12}T_{48}As_{30}$, together with their respective electron diffraction patterns. It can be seen that although both show a glass transition temperature, T_g, only the crystalliseable material exhibits any further characteristics and in this case a crystallisation exotherm results, which is followed by a melting point endotherm. Materials conventionally designated memory compositions all undergo such a phase change.

Because the disordered state generally does not allow doping in the accepted crystalline silicon technology sense, variations in physical parameters, e.g. conductivity, are not so dependent upon composition. In single crystal semiconductor systems, the energy gaps tend to be a monotonic function of the mean atomic number of the system. This large spread in energy gap up to 3eV is further multiplied in the amorphous state since, if

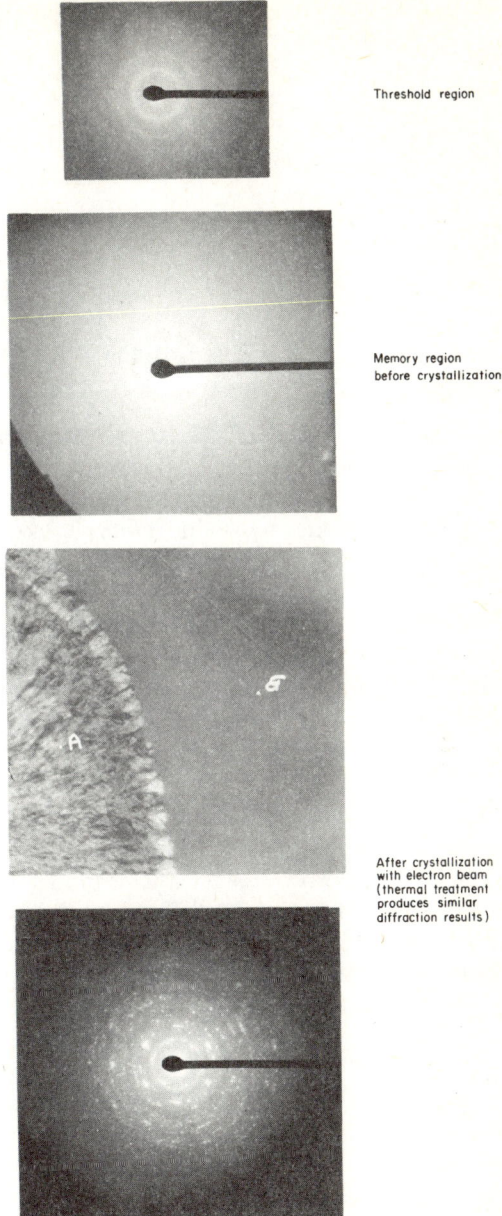

FIGURE 1(a). Transmission electron diffraction.

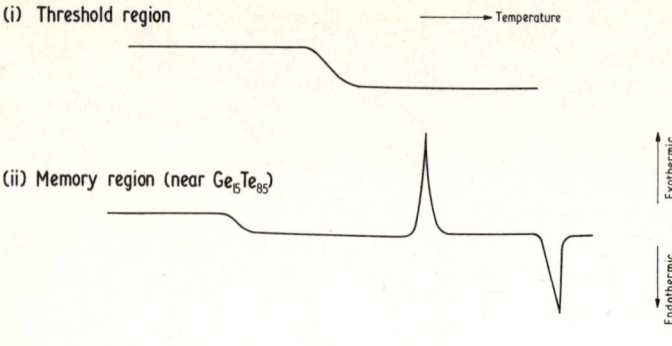

FIGURE 1(b).

the deposition rate is high enough and the substrate temperature cool enough, virtually any combination of the elements can be made to produce amorphous films. This point will be treated later, but the recent use of sputtering by IBM workers to produce amorphous magnetic films for bubble domain stores and higher critical temperature superconductors by Bell Laboratories, illustrates the versatility of the technique. Applications such as electrolytic electrodes, magnetic cores, low temperature heaters and even cutlery are forecast for the Allied Chemical Corporation's "Metglas" ferro-metallic glasses which can be cast to provide continuous ribbon. It is possible that eventually it will be in these more unconventional areas, rather than in fields where they are in direct competition with silicon products, that amorphous films will find their major future utilisation.

Semiconducting amorphous material is generally split into two major classes; the chalcogenides and the oxides, mainly of the transition metals. This division, apart from the obvious one, stems from a difference in conductivity mechanism. Table 1 lists some of the possibilities. Within the chalcogenides there is a clear demarcation between the elemental films and the multi-element alloys, as well as in the nature of the bonds involved.

The third area noted in Table 1, that of dielectric films, was discussed in Chapter 4 and need only concern us here in so far as certain phenomena observed during breakdown of such films are sometimes referred to as switching (see Klein[16]), and certain aspects of "forming" in these metal–insulator–metal (MIM) sandwiches are akin to forming in some chalcogenide alloy films (see p. 273).

It is important to stress three important differences between amorphous materials and their crystalline counterparts:

(1) In the crystalline state it is possible to quantify the deviations from perfect crystalline order of the material due to impurities, defects,

Table 1

Classification and Examples of Non-crystalline Semiconductors

(1) Covalent non-crystalline solids
(often referred to collectively as CHALCOGENIDES)

  (a) Tetrahedral non-crystalline solids
    Si, Ge, SiC, InSb, GaAs, GaSb, ...

  (b) Two-fold co-ordinated and two-dimensionally bonded non-crystalline solids
    Se, S, Te, As_2Se_3, As_2S_3, ...

  (c) Cross-linked network non-crystalline solids
    Ge–Sb–Se, Ge–As–Se,
    Si–Ge–As–Te, As–Se–Te,
    As_2Se_3–As_2Te_3, Tl_2Se–As_2Te_3, ...

(2) Semiconducting oxide glasses

V_2O_5–P_2O_5	MnO–Al_2O_3–SiO_2
V_2O_5–P_2O_5–BaO	CoO–Al_2O_3–SiO_2
V_2O_5–GeO_2–BaO	FeO–Al_2O_3–SiO_2
V_2O_5–PbO–Fe_2O_3	TiO_2–B_2O_3–BaO
...	...

(3) Dielectric films

    SiO_x, Al_2O_3, ZrO_2, Ta_2O_3, Si_3N_4, BN, ...

---

dislocations and so on. No such measure is possible for amorphous materials even on a run-to-run reproducibility basis. As pointed out above the term "amorphous" is used very loosely.

(2) It is dangerous to generalise the results from a particular amorphous material to the whole field because the range of possible materials combinations is enormous.

(3) Material homogeneity is a much more intangible quantity than it is in the crystalline state. Phase separation is a well known phenomenon in glasses and it is exploited in many applications, e.g. glass ceramics. One phase can be crystalline embedded in a glassy matrix. Hence, there may be anisotropy of the physical properties of the material, like easy current paths, for example, which were also postulated in Chapter 5 to explain abnormally high current densities in discontinuous metal films.

These points should be considered carefully before directly adopting crystalline state theories since these can be misleading when applied universally to the "amorphous" state. There is a lack of complete data on a wide range of phenomena against which to test the theoretical models. The most complete set exists for electrical conduction as a function of various parameters, viz., electric field, temperature and measurement frequency. In some cases these data are augmented by drift mobility and photoconductivity information. But measurements on the same sample, preferably *in situ* in the preparation chamber, e.g. in the case of thin films without breaking the vacuum, of these and other parameters like Hall effect, thermopower, optical absorption etc. are not extant. Therefore a complete picture of the energy level structure in the material, for example, is not available and hence several conflicting theoretical models have been proposed. Understanding is beginning to emerge but the position is still similar to that of crystalline semiconductors before the role of dopants was appreciated. The trend recently[17] has been to stress the need for careful sample preparation and characterisation. For further details of the earlier experimental and theoretical studies the reader is referred to Adler[18] for a review of amorphous semiconductors in general, and to Owen[19] for semiconducting glasses in particular. A detailed account of electrical conduction in disordered nonmetallic films has been given recently by Jonscher and Hill[20]. Theoretical models are discussed in the recent paper of Anderson[21], whilst Bagley[22] has reviewed the structural aspects of the amorphous state.

2. Preparation of Amorphous Semiconductors

As implied by the definition above, true glasses can be prepared by supercooling the liquid. Generally upon cooling a liquid below the material's melting point, T_m, it will either crystallise or form a glass.

During crystallisation, physical properties like viscosity, volume and entropy change discontinuously, a first order transformation. On the other hand, as can be seen from the temperature dependence of the volume of a material, (shown schematically in Fig. 2 for a material which can either crystallise or form a glass) these physical properties change continuously although there may be a more rapid change in the vicinity of the glass transition temperature, T_g. Glass formation is a widespread phenomenon involving a wide variety of bonding types: covalent (e.g. As_2S_3), ionic [$KNO_3 - Ca(NO_3)_2$], metallic (Pd_4Si), van der Waals (*o*-terphenyl) and hydrogen ($KHSO_4$) which cover a wide conductivity range from dielectric to metallic. The ease with which glasses are formed varies from As_2S_3, which is such a good glass former that crystals are very hard to grow from the melt, to some monatomic metals which have not been prepared even using

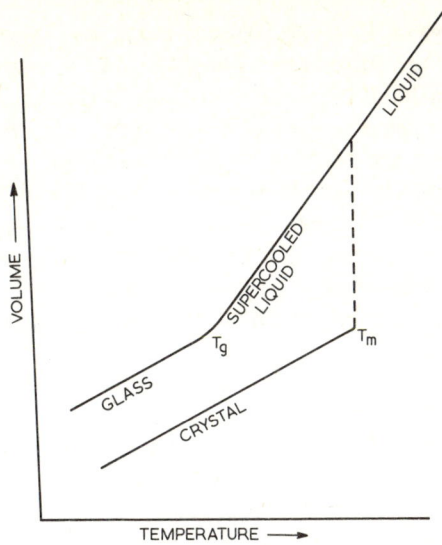

FIGURE 2. Schematic of volume-temperature dependence of a material in various states.

rapid quench rates onto cold substrates. It is not proposed to discuss the reasons for these differences here; the reader is referred to Turnbull[23] for further information, and for general information on glass to the book by Rawson[24].

Generally, liquids with small crystallisation kinetic constants form glasses directly from the melt. This can be done with many chalcogenide materials, but there is a wide range of possibilities. For example, the crystallisation kinetics of As_2S_3 are so slow that a large mass of glass can be obtained with very slow cooling. As the crystallisation rate increases, as for the less stable As_2Se_3, the liquid must be cooled more quickly through the region of maximum crystallisation rate to prevent the formation of microcrystals. As this rate increases, a point is reached where even a few grams of material cannot be cooled directly from the melt fast enough to form a homogeneous glass; As_2Te_3 is an example of such a material. Very fast quenching is often required for some chalcogenide alloys using iced brine, or even liquid nitrogen. However, care must be taken because of the high vapour pressure of these materials, the immiscibility of certain of them, particularly Si, and the ease with which oxidation can take place. Hence, preparation is often done in silica tubes which have been well evacuated, and sealed under vacuum. The tubes are then heated in a rocking furnace at about 1000°C for many hours to achieve a homogeneous glass (for example, Savage[25]). They are then subjected to rapid quenching.

If fast quenching of small samples is insufficient to prevent crystallisation, more drastic methods must be employed with cooling rates in excess of 1000°C per second. This technique is known as "splat cooling". In general, this involves propelling a molten drop of the material at high speed onto a cooled thermally conducting substrate. For example, Chen and Miller[26] have reported a splat cooling machine which is capable of quench rates of greater than 10^5 °C sec^{-1} and produces films of amorphous material down to 10 μm thickness by dropping the molten fillet of material between two high-speed rollers. A comprehensive review of splat cooling, which is especially useful for extending the range of solid solution and the production from metastable phases has been given by Jones[27].

In the case of pure elements, the susceptibility to formation of amorphous solids is greater than for metals, greater for chalcogens than for metalloids amongst the non-metals, and greater for the transition elements than others amongst the metals. The latter usually require evaporation onto substrates held at liquid helium temperatures[28] and possibly some stabilisation by impurities[29,30] to maintain their amorphous state. The chalcogens form amorphous solid phases most readily for S and Se by fairly rapid cooling from their melts but not Te, which requires vapour deposition at liquid air temperature[24]. In thin film form, where a much greater apparent quench rate is achieved than even splat cooling, amorphous Se films resist crystallisation up to at least $0.6T_m$ compared with Te the films which crystallise above $0.4T_m$[30]. For the metalloids, splat cooling has had to be used to produce amorphous solids from the melt, e.g. boron [see Ref. (27)] Ge and Te[31].

It is clear from this latter work that the metallic-like liquid co-ordination is not retained in these materials. For the remaining metalloids Be, Ga, C, Si, Ge, As, Sb, Bi vapour deposition is required (see Chopra[32], Behrndt[30]), although the substrate temperature can be as high as 400 K for Si and Ge. (e.g. Walley[33]). Almost all the usual thin film deposition techniques (see the texts of Maissel and Glang[34] or Chopra[32]) have been employed in the production of amorphous thin films. The metalloids form amorphous films less readily with increasing displacement from S and Se in the periodic table, but B, Si, Ge, As and Sb resist rapid crystallisation up to at least $0.3T_m$ (Rawson[24], Behrndt[30], Bosnell and Voisey[35]).

However, there are significant differences in the properties of films of the same element made by different techniques. This is highlighted in the work of Chittick et al.[36] on the properties of amorphous Si films produced by glow discharge techniques compared with the evaporated films of Walley[33]. The resistivity of films prepared by glow discharge is much greater; being typically $10^9 \Omega$ cm compared with $10^2 \Omega$ cm. Based on the absence of the strong silicon–oxygen infrared absorption spectrum of the glow discharge prepared films, this technique is judged to minimise oxygen contamination. Amorphous Ge has also been made by electro-deposition, usually from a solution of $GeCl_4$ in propyline glycol (see Chen and Turnbull[37], and Tauc et al.[38]). These variations clearly require that more careful film characterisation be performed if theoretical models are to be adequately tested.

For multicomponent films even greater care is required, because of preferential vaporisation for example. It is often difficult therefore to prepare films of chalcogenide glasses with any degree of control. On the other hand, co-evaporation of two or perhaps three elements to grow the film has been carried out as, for example, the work of Saks et al.[39] on GaAs and Ga–Mg–As amorphous films has shown. Additionally, some compounds can be evaporated intact with reasonable retention of stoichiometry (see Newbury and Kirk[40], for example, on crystalline ZnSe films). For certain experiments, particularly with very thin films or high electric fields the role of the substrate is extremely important since many substrate glasses contain Na^+ ions in particular, leading to ionic currents which can mask the physical effects under investigation in the film. For more complex materials R.F. sputtering has proved to give better compositional control for stable glasses, e.g. the threshold switching types. The extent to which this can be achieved is illustrated by Bosnell and Thomas[41]. In general one-to-one correspondence between sputtering target and film composition is difficult to achieve because of the low sputtering yield of Si but this can be compensated for in the target. Sputtering yield data (Carter and Colligan[42]) is not available for argon bombardment of As and Te to the author's knowledge. However, the inference from the above work[41] is that in the chalcogenide alloys Ge, Si, As, Te—As and Te have a sputtering yield comparable to Ge. Some alloys, particularly close to the crystalline-amorphous phase boundary, like the memory composition $Ge_{15}Te_{85}$ eutectic are difficult to produce in the glassy form but can be made into crystalline targets sufficiently conducting to allow D.C. triode sputtering to be used. Care must be taken to avoid incorporating large amounts of sputtering gas in the film, and a low gas pressure and sputtering rate is one means of minimising the problem. If high pressures and rates are used, bubbles can appear in the film due to argon incorporation. Figure 3 illustrates this point for a $Si_{12}Ge_{10}Te_{48}As_{30}$ starting composition, sputtered at a rate of 10 Å sec^{-1}; a reduction in the gas pressure to produce a sputtering rate of 2 Å sec^{-1} alleviated this problem. However, it is clear that some gas will still be incorporated, see for example the work of Winters and Kay[43] and on chalcogenide films the work of Fagen[44]. Gas incorporation is particularly

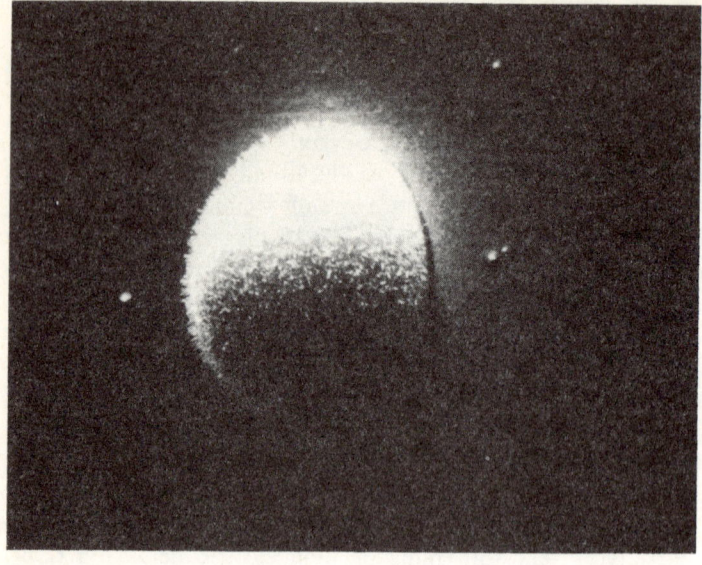

FIGURE 3. $Si_{12}Ge_{10}Te_{48}As_{30}$ film sputtered at a rate of 10 Å sec^{-1}. SEM micrograph at 2000 times magnification showing argon bubble formation

important because of the relatively high electric fields involved in switching (of the order of 10^5 V cm^{-1}) and this cannot be stressed too strongly. Premature breakdown rather than switching can occur with the rupture of the electrode system.

Screen printing has become a major manufacturing method in the electronics field generally but very little work has been reported on the printing of amorphous semiconducting films. The chalcogenides have proved to be difficult to handle in the thick film printing process because of the ease with which they oxidise and volatise. However, there is on the market a thick film ink "TYoX" (Trademark of DuPont)[45] which has several variants used to produce a thermistor, a thick film switch, and a varistor. The properties of this thick film switch indicate that it is based on the VO_2 insulator–metal transition. More recently Higgins et al.[46] have described a thick film screen printing method again for Vanadate glasses, based on $V_2O_5 - P_2O_5$ with modifying oxides Li_2O, Na_2O, K_2O MgO, CaO, SrO, or BaO. As a manufacturing process, thick film printing technology is much more cost-effective than vacuum deposition processes and compatibility with conductor inks is reported by Higgins et al.[46]. In general films of the vanadate, titanate and niobate glasses have been prepared by both R.F. and reactive sputtering. Various additions to VO_2 have been proposed with a view to moving the insulator–metal transition in VO_2 to higher temperatures than the usual 67°C. This work has been reviewed by Balberg et al.[47]. Using two platinum wires embedded in a bead of the glass, Regan and Drake[48] have shown that even large percentages (up to 16% in one case) of the oxides of Co, Ni, Fe, Mn, Li, Nb and Cr added to a V_2O_5/P_2O_5 glass (75/25) did not alter the trend that the switching threshold voltage of the material tends to zero at about 67°C. This was despite the fact that the x-ray diffraction studies indicate that ordered regions, if present at all, were smaller than 100Å. It seems that in amorphous films of the vanadates as prepared to date many of properties are still dominated by the transition at 67°C. Pyrolysis has also been extensively employed for producing oxide films (e.g. Ryabova and Savitskaya[49]) but not in the manufacture of the multicomponent chalcogenide films.

More recently, attention has been turned to ion-implantation which at sufficiently high implant doses can cause crystalline to amorphous transitions. For example, Kräutle[50] has implanted Ge, Sn, Pb, Pt, Pd, In, Sb and Tl into Si and Ge single crystals with energies in the range 35 to 60 keV. For room temperature implants, the resultant layer is amorphous for doses greater than 10^{15} ions cm^{-2}, in agreement with theory[51]. Re-ordering was studied by Rutherford backscattering[52] and conductivity measurements. The damage in Si began to anneal at 600°C and in Ge at 400°C, nearly independently of the ion species causing the damage. The variety of phenomena induced in non-metallic solids by energetic heavy ions has been reviewed by Naguib and Kelly[53]. Switching effects have been reported in Si layers amorphised in this way[54]. They used p-type, 1.5 Ω cm Si single crystals and implanted them with Ar^+ ions at energies up to 200 keV and doses between $2 \times 10^{14} - 1.5 \times 10^{17}$ per cm^2. Specimens having the highest dose showed threshold switching similar to the chalcogenides. However, the dose required for the layers to exhibit switching is about two orders of magnitude greater than the amorphisation dose above. These authors' conjecture is that broken bonds in the amorphous layer arise from additional irradiation giving rise to further localised states in the energy gap. Ion implantation into amorphous films produced in the normal way has been performed on Ge by Anderson et al.[55] and into both amorphous Ge and Si by Beyer et al.[56]. In the latter experiments, bombardment with 4×10^{16} the He^+ ions cm^{-2} at 20 keV led to an increase in the hopping conductivity by an order of magnitude, attributed by the authors to the enhancement in the number of defects. The work of Beyer et al.[56] also emphasises the effect of substrate temperature and gas content on the film properties. Indeed even outgassing the substrate before deposition, which is good thin film practice, can change the film properties significantly.

The data of Anderson et al.[55] using ^{11}B implants into evaporated amorphous Ge films indicated that some of the B was electrically active since the conductivity increased by a factor of 40 but also changed from weakly n-type, as deposited, to p-type, as determined from the thermoelectric power. The peak concentration level was about 10^{21} per cm^3. Hence, changes in the electrical properties, although not doping in the crystalline semiconductor sense, are possible. The gas discharge process allows ready doping of the material by the use of phosphine[36] in the gas mixture as well as the production of the oxide and nitride[57]. For the chalcogenide glasses, the quantities involved tend to look more like alloying than doping in the accepted sense[7,19].

Sintering of bulk materials, particularly metallic oxides has been used in varistor manufacture, particularly ZnO/Bi_2O_3 with small additions of other metal oxides. Recent work[58] indicates that at least the material between the ZnO grains is amorphous and this helps to explain some of the earlier electrical data[59].

3. Material Characterisation

One of the most common methods for analysing glasses, differential thermal analysis (DTA) or differential scanning calorimetry (DSC) was briefly mentioned in Section 1. The technique can also be applied to thin films[60] if they can be floated off the substrate. This is standard practice in thin film investigations. Several methods are used, including deposition of the film onto rock salt which is then dissolved in water, or onto a glass substrate coated with boric oxide which can also later be dissolved, whilst collodion substrates have been favoured by electron microscopists. The combination of DSC and a mass spectrometer to analyse the thermal decomposition products was used by Fagen[44] (see above). Bosnell and Savage[60] indicate that DTA can be a valuable adjunct to electron diffraction especially in crystallisation phenomena when electron beam crystallisation in the electron microscope can be a difficulty. Thermogravimetric analysis (TGA) has been used to investigate the vaporisation of As_2Te_3 by Northrop[61]. These studies are important because of the fact that switching between two electrodes on the surface of a sample can cause vaporisation[62].

Apart from this problem, great care must be taken if experiments are performed in air. For example, Phillips et al.[63] have reported the formation of a crystalline layer on the surface of a chalcogenide threshold glass when heated at about 300°C in air, making measurements of resistivity versus temperature difficult unless done in vacuo. These problems of surface nucleation or crystallisation are emphasised in the DTA work of Takamori et al.[64]. The presence of water on the surface severely influences the conductivity[65] and plays an important role on the persistence of photoconduction in As_2Se_3[66]. The film composition has been ascertained by a variety of techniques but because of the small amount of material present, mass spectrometry and x-ray fluorescence have been found to be most satisfactory. Structural determinations have generally been performed in the classical manner: low angle x-ray diffraction, electron diffraction and light scattering which in glasses is a valuable aid to highlight long range inhomogeneities. The reader is referred to the reviews of Dove[67], Grigorovici[68] and Apling[69] for further details on the analysis of the data from these experiments. The likely heterogeneities in amorphous elemental films have been emphasised by Grigorovici [Ref. (68) page 206]. However, gross heterogeneities like multiphase separation occur similarly in glasses as in multicomponent alloys[70]. Etching and replication techniques are then required as illustrated by the observation of di-phasic structure in the switching glasses by Roy and Caslavska[71]. Such structures have been invoked to explain conductivity data[72,73] in both semiconducting liquids and glasses. The use of the 1 MeV electron microscope has enabled thicker pieces of material to be examined[74], which has been extremely valuable providing care is taken to ensure that electron-beam-induced crystallisation is not taking place (see p. 249). Apart from amorphous supporting materials, significant enhancement of the image contrast for amorphous specimens in the electron microscope has been reported if the material is supported on thin crystals and viewed in the vicinity of bend contours in the crystal support[75]. Many earlier results, particularly in the elemental materials have now been attributed to voids. However virtually void free material has been reported by Donovan and Heineman[76] by evaporation in UHV at low rate onto a substrate not far below the crystallisation temperature. There is a need to be constantly on guard for such problems and the need for a variety of techniques, e.g. structure, resistivity, thermopower and composition, to be employed on the same films is illustrated in the work of Nath et al.[77] on GeTe and Barna et al.[78]. The latter authors had an in situ experimental arrangement, allowing galvanomagnetic measurements to be made on films deposited in UHV in an electron microscope. They were able to elucidate three types of crystallisation: (i) surface crystallisation (both the free surface and the film substrate interface) if films are prepared and annealed in UHV ($<10^{-9}$ torr); (ii) volume crystallisation if the films are contaminated before beginning crystallisation, e.g. increasing the pressure up to 10^{-7} torr 5–10 minutes before heat treatment; (iii) highly contaminated volume crystallisation if the films are exposed to air between preparation and crystallisation.

This work is largely confirmed by the use of Auger Electron Spectroscopy (AES) recently reported by Knotek[79]. The ease with which amorphous Ge films are contaminated was amply demonstrated and the effect of substrate temperature on film density seems to confirm the conclusions of Donovan and Heineman[76]. The problem of contamination is not confined to elemental films, and Apling et al.[80] for example, have reported the production of As_2O_3 on the surface of As_2S_3 films stimulated by x-ray irradiation. The authors postulate that the photo-induced oxidation mechanism may involve secondary (band gap) radiation following the absorption of the primary x-ray photon. AES has also been used to analyse the contact–film–interface in an Al–C system by Morisaki et al.[81]. Depth profiles were obtained by ion milling. It was noted that a 100Å oxide film was present together with some contamination from the tungsten heater used for evaporating the Al. Electron spin resonance (ESR) has been used to show that in amorphous elemental films prepared in the normal manner, the signal strength, and hence the number of free spins, was proportional to the sample thickness[82] whilst no detectable signal was obtained from the chalcogenide glasses. Amorphous Ge and Si films, prepared in the author's laboratory[35], having a density inferred by x-ray fluorescence counts equal to that of crystalline films, or not much lower, showed only a small ESR signal on the limits of the resolution of the system (10^{15} spins cm^{-3} compared with 10^{20} spins cm^{-3} observed above). Chalcogenide films of $Ge_{10}Si_{12}Te_{48}As_{30}$ or As_2S_3 gave no signal (unpublished RRE data).

A wide range of film densities has been reported for amorphous Ge films compared with the bulk density of 5.325 g cm^{-3}. For example, Clarke[83] reported a density of 3.9 ± 0.4 g cm^{-3} for films prepared by evaporation, compared with 4.6 ± 0.2 g cm^{-3} for those prepared by glow discharge, (Chittick[84]), who also reports a value of 1.9 ± 0.15 g cm^{-3} for amorphous Si compared with the crystalline value of 2.23 g cm^{-3}. Wu and Luo[85], from measurements of pressure induced crystallisation, have estimated that the upper limit of the amorphous state density in germanium is 97.4% of the crystalline value, the corresponding values for As and GeTe being 92.6% and 92.4%, although they actually achieved 89.7%, 85.1% and 79.4% of the crystalline density in the as-deposited states, respectively. On the other hand, heavy dose neutron irradiation of GaAs, whilst introducing a large number of defects, produced a vary small decrease in density (0.447%)[86], so that defects *per se* are not the problem, consistent with the above data.

Valuable information relating to the atomic bonding and oxidation state of a sample can also be obtained from electron spectroscopy, sometimes called ESCA (electron spectroscopy for chemical analysis) or x-ray photoelectron spectroscopy (XPS)[87,88]. More recently, the technique has been extended to give density of states information[89]. Such data as well as bonding information can also be obtained from nuclear magnetic resonance (NMR)[90].

Apart from the use of spectrographic techniques to elucidate pseudo-band gaps in amorphous materials (for reviews of the earlier data see Edmond[91] and Fritzche[92] and, more recently, by Theye[93] and Davis[94]), they have also been employed to investigate the oxygen content of films[84] and the water content of glasses[95] in the usual manner. In the transition metal glasses the ratio of ions in different bonding states may be of importance in elucidating the conduction mechanism. Since the electron motion may be of those for a $V^{4+}-O-V^{5+} \rightarrow V^{5+}-O-V^{4+}$, XPS can be used for this, but more commonly titration techniques have been employed[96].

Because of the problems associated with amorphous material characterisation outlined above, only a general discussion of some physical properties will be given. The reader is referred to the reviews already quoted for further information. Furthermore, it is not the purpose of this review to incorporate a critique of the status of the theory on amorphous materials. In general, the conductivity of the material can be expressed in the form $\sigma = C \exp(-E/kT)$, where C is a constant, E is the energy gap in eV, k = Boltzmann's constant, T = temperature in K. The values of the coefficients are given Figure 4. Figure 5 indicates the relationship between D.C. conductivity and density of states, both figures after Davis and Mott[97]. The optical absorption of these materials is of great interest both from an academic and practical point of view. The absorption spectra for the two most studied materials, As_2S_3 and As_2Se_3, are given in Figure 6. A review of the IR transmission of the chalcogenide glasses has been given by Savage and Neilsen[98] and, more recently, by Hilton[99]. Table 2 gives some indication of the wide range of properties available in these materials.

Following this brief description of the physical properties of some amorphous semiconductors, the problem of contacts will be emphasised before the potential applications of these materials are reviewed.

4. Contact Effects

The critical nature of electrode materials on the properties of amorphous materials, particularly during annealing or temperature cycling has been emphasised by Bosnell and Voisey[35] who found that crystallisation of elemental amorphous semiconductors was induced, particularly with noble metal electrodes, at temperatures much lower than that found with material on its own. Sigurd et al.[100] infer, from 2 MeV He$^+$ ion backscattering techniques, that this crystallisation comes about by the dissolution of the semiconductor solute into the metal solvent followed by transport of that solute and crystal growth perhaps because of the higher free energy of the

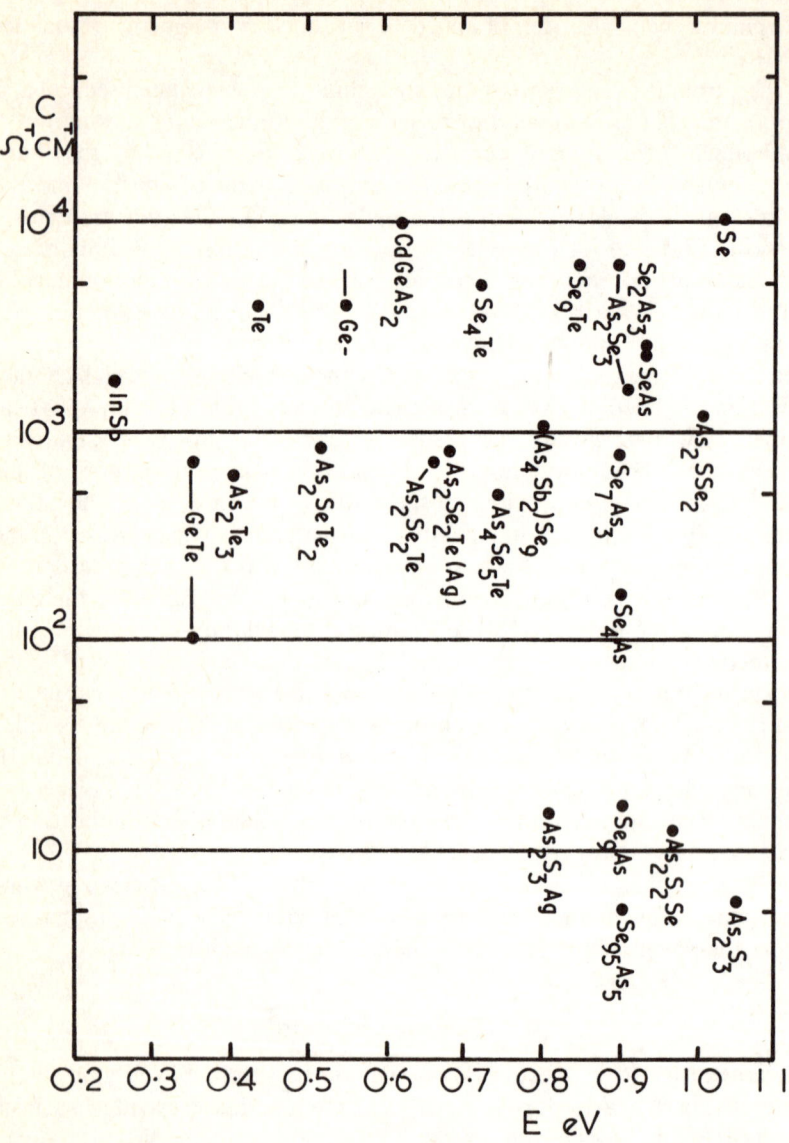

FIGURE 4. Experimental values of C and E for amorphous semiconductors, the conductivities of which are given by $\sigma = C\exp(-E/kT)$ (from Davis and Mott[97]).

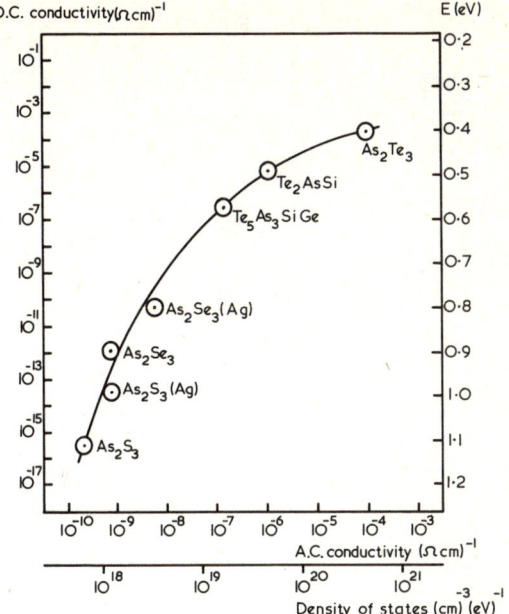

FIGURE 5. A.C. conductivity (at $\omega = 10^6 \sec^{-1}$ and room temperature) versus D.C. conductivity in various chalcogenide semiconductors (from Davis and Mott[97]).

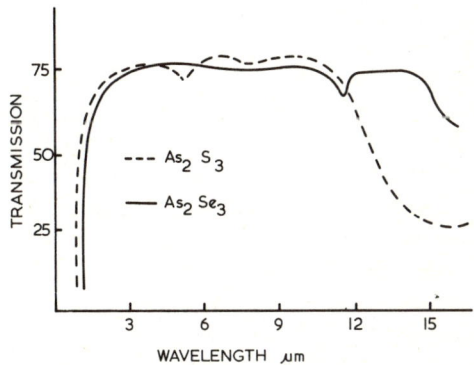

FIGURE 6. Official absorption spectra of As_2S_3 and As_2Se_3.

Table 2

Some Properties of Glasses in The System As_2Se_3–As_2Te_3

Composition	Resistivity (ohm cm)	Dielectric constant (relative)	Refractive index
Se	10^{12}	7.20	2.69
As_2Se_3	10^{13}	12.25	3.50
$4As_2Se_3$–As_2Te_3	3.6×10^9	14.10	3.76
$3As_2Se_3$–As_2Te_3	1.1×10^9	14.90	3.87
As_2Se_3–As_2Te_3	1.8×10^7	17.7	4.22
$2As_2Se_3$–$3As_2Te_3$	3.3×10^6	18.7	4.34
As_2Se_3–$2As_2Te_3$	1.0×10^6	20	4.48

amorphous state compared with the crystalline one. Hence, apart from the care required with the film deposition, attention must also be given to the electrode material. Molybdenum has been highlighted as a useful material[35] since Si showed no interaction with it up to 800°C[101]. Furthermore, specimens with Al electrodes deposited under a high background pressure can show marked dispersion in the A.C. characteristics with frequency and also a high drift with time[102]; possibly associated with oxygen migration. It will be recalled that Morisaki et al.[81] reported an interface oxide film of about 100 Å thickness for Al electrodes on amorphous C, consistent with a proposal by Morgan and Jonscher[102]. This interface, if present, does however prevent device degradation due to electrode ingress.

On the multicomponent chalcogenide materials diffusion of metals into As_2Se_3 was reported by Freeman et al.[103] and into As_2S_3 by Maruno et al.[104]. Again, in general, the noble metals present most of the difficulties. Ag contacts can convert a material which normally exhibits threshold switching into a memory device by metal ingress (see the early work of Pearson[105] particularly with the low melting point materials, and more recently, Petrillo and Kao[106]). Some metals are glass formers with the chalcogenides in small quantities, e.g. Al and In[107] and can also change a threshold to a memory behaviour in switching terms, or from a relatively stable glass to one close to the glass-crystalline boundary. Bosnell and Thomas[41] indicate that Cu, Ag and Au contacts are an anathema to device lifetime and note that there is a strong correlation between the electrodes and switching of amorphous elemental materials[108]. These problems of obtaining low resistance ohmic contacts therefore appear as serious in relation to amorphous devices as in the case of contacting the II–VI compounds (see for example Ref.[109]). The role of contact effects on the physical properties of a variety of materials has been reviewed by Jonscher[110] and it may well be that contact-less techniques like that of Strom and Taylor[111] are desirable in some cases. If good quality contacts are necessary, one recommended practice[112] has been to use a sandwich structure comprising a low resistance material Al, Au and a thin layer of a refractory metal, e.g. Ti, so that blocking contacts with materials like Al are avoided and a diffusion barrier is established for materials like Au and Cu. The thin layer prevents diffusion and reaction of the chalcogenides, and good electrical contact ensues with few associated strain problems.

Hence it can be concluded that as much care is required with the preparation of amorphous semiconducting samples and devices as with their crystalline counterparts. There is no place for anything other than the best semiconductor and thin film laboratory clean room practice.

5. Switching in Amorphous Semiconductors

5.1 Switching Phenomena

There are numerous early reports of switching from a low to a high conductivity state in amorphous semiconductor films, but the recent wave of interest stems from the paper by Ovshinsky[11] in which two forms of switching were reported. In the threshold switch (Fig. 7a), on application of a voltage greater than a specific threshold voltage V_T, the material is transformed from a high resistance to a low resistance state provided a holding current, i_H and a commensurate holding voltage V_H are applied. The memory switch (Fig. 7b) is similar to the threshold device except that no holding current is required, in other words the device is SET in a permanent ON state. This can be RESET by application of high current pulse, typically 100 mA for a few microseconds. In both cases the threshold switching field is of the order of 10^5 V cm^{-1} at which point the device switches from, for example, $10^6 \Omega$ cm to $2 \times 10^2 \Omega$ cm. The memory switch can be operated as a threshold device provided short duration pulses are used. Lock-on only occurs after about 100 µsec (Bosnell and Thomas[41], Thomas et al.[113]). In both cases there is a delay to switching τ_D which is sensitive to pulse amplitude (Fig. 8) and can be expressed in the form:

$$\tau_D \propto \exp(-V)$$

In general, memory switches are manufactured from alloys near the boundary of the glass forming regions whilst threshold devices consist of the more stable

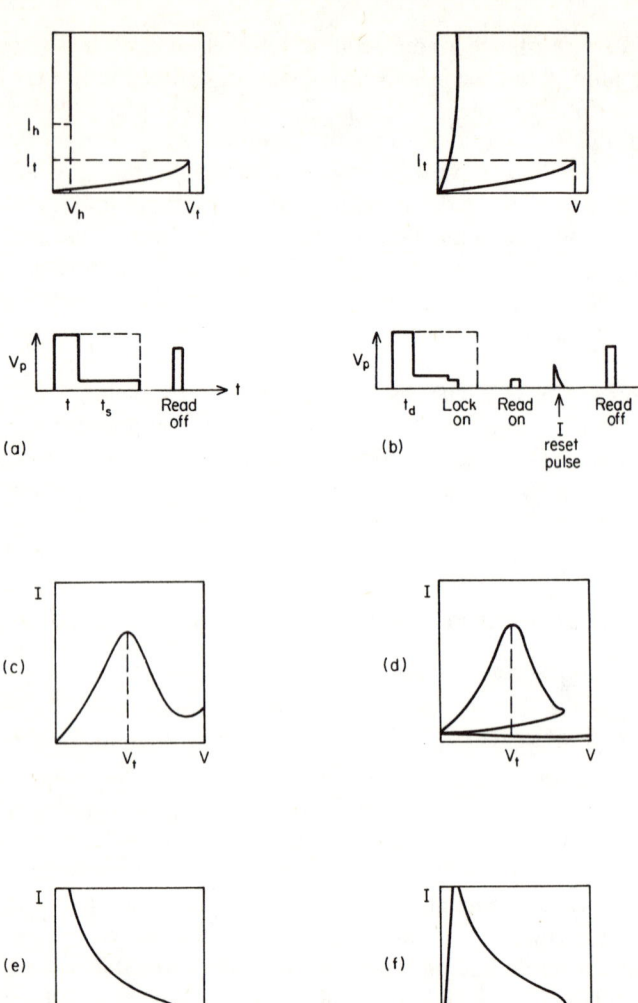

FIGURE 7.
(a) Typical chalcogenide threshold switch characteristics (including pulse response).
(b) Chalcogenide memory characteristics with write–read–erase–read pulse response.
(c) Typical schematic I–V for an oxide switch (N shaped or VCNR).
(d) VCNR with memory.
(e) CCNR.
(f) CCNR with memory.

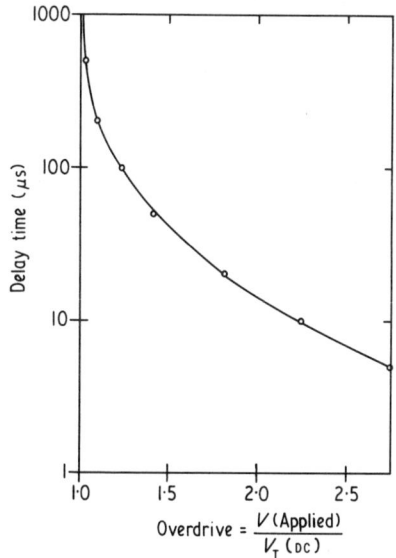

FIGURE 8. Switching delay time plotted against voltage overdrive for a 10 μm thick OMS.

glassy compositions. Figure 9, after Savage[114] shows schematically the Ge Te As system. The basic differences in these glasses under thermal treatment has been illustrated in Figure 1. Similar results have been obtained from certain transition metal oxide glass systems, e.g. Regan and Drake[48], see Section 5.7 for further details.

However, many reports of switching in simple oxides, e.g. Al_2O_3, SiO, Ta_2O_5, and ZrO_2, show that the results are not as reliable with a strong tendency to dielectric breakdown or (as the thin film capacitor manufacturers call it) self-healing breakdown, where the electrode system can be progressively destroyed from one operation to the next. Dearnaley et al.[115] have proposed a model for switching in this type of material involving the formation of high conductivity filaments which has been expanded by Ralph and Woodcock[116]. These devices, although usually shown schematically having clean $I-V$ characteristics (Fig. 7c) generally show a great deal of spikey noise on an actual trace. They belong to a class of characteristics usually termed voltage-controlled negative resistance (VCNR) or N-shaped instability, the type of characteristic utilised in mocrowave oscillators based on GaAs. Memory can also exist in this type of material (Fig. 7d). A further possibility, again with memory, is illustrated in Figs. 7e and 7f. This is

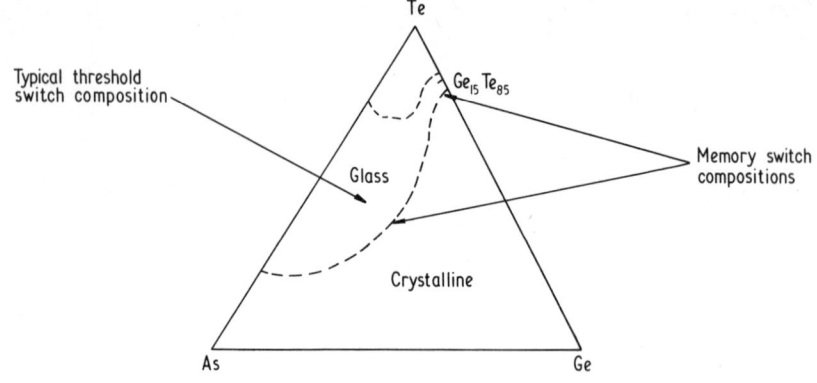

FIGURE 9. As, Ge, Te phase diagram.

S-shaped instability or current-controlled negative resistance (CCNR) mechanism. Unlike the switching characteristic illustrated in Fig. 7a, the CCNR device can be switched along the load line from the high to the low resistance state and hence it should be possible to stabilise a CCNR switch in its negative resistance region provided a sufficiently large load resistance is used, whereas the switching device would only oscillate. However, recent work of Shaw and co-workers[117,118,119] has shown that device capacitance and inductance are sufficient to lead to instabilities in the CCNR case so that it would be difficult experimentally to distinguish between the two phenomena.

On an oscilloscope, dielectric breakdown or even air breakdown under pulse conditions can look similar to switching if the noise in the on-state is suppressed. Just how many reports of switching in various materials are due to either open or short-circuit breakdown it is impossible to say. Klein[120,121,122] has reviewed in detail the wealth of data on this problem and this review will be confined to a few illustrative references. Switching has been reported in a diverse range of materials. These include organic films (Carchano et al.[123]) but circular holes about 5 µm in diameter were formed in the gold electrodes, an observation also made on oxide devices (e.g. Collins et al.[124]). Furthermore, as in the work of Sliva et al.[125] on oxide switches, sparks were seen at each switching operation. These characteristics seem more consistent with a breakdown phenomenon than switching. On the other hand, Pinto[126] has observed switching and memory action in anodic Ta_2O_5 and Nb_2O_5 films, which he attributes to a predominantly metallic filament caused by the orientation of metallic ions under the action of high electric fields. Some aspects of these results are akin to "forming" in chalcogenide devices (Section 5.3). Agarwal[127] has provided a bibliography of dielectric breakdown recently and many features of the thermal breakdown models for insulators are common with thermal models for switching[128,129] (see Section 5.5). Furthermore, the switching characteristics obtained from chalcogenide or oxide threshold switches bear a strong resemblance to the phenomenon of second breakdown in p–n junctions in single crystal technology, particularly power transistors. For a survey of the early data in this field see, for example, IEEE Electron Devices Special Issue[130] (1966) and Sze's book[131]. More recent work, particularly that of Budenstein and his coworkers[132,133] has illustrated the filamentation occurring in the breakdown of the p–n junction. A thermal model to describe the phenomena based on the work on amorphous switching materials has been given by Popescu[134] which shows good agreement with experimental data.

Several of the I–V characteristics illustrated in Fig. 7 can be produced by simple circuits involving only two transistors[135] as well as pn–pn structures such as thyristors or triacs, (see Reference 131). Hence some type of electronic mechanism for the switching cannot be excluded. These arguments will be developed in Section 5.5. Similar results have been obtained in silicon multi-layer structures[136] comprising 14 silicon layers of alternate doping on single crystal silicon each about 1000 Å thick. At low temperatures, strongly doped and compensated Ge shows similar switching behaviour[137] as would be expected because of the possibility of localised and extended states as in amorphous materials (see Mott and Davis[4]). It is possible that the reported switching in some liquid semiconductors, particularly Se up to $1000°C$[138], and the chalcogenide alloys $Se_{0.77} Te_{0.13} S_{0.1}$[139], may be electronic in origin. However, in a wider range of materials, viz. Se, As_2Se_3, Sb_2S_3 and $GeSe_2$, Andrev et al.[140] conclude that there may be an electronic component which is suppressed by thermal effects. At higher temperatures, thermal effects were reported to be paramount.

Similar switching effects have also been observed in single crystal layered chalcogenides by Lee et al.[141] which are of the CCNR type. Thompson[142] concludes that this switching is associated with the decomposition of the layered material into a filament, probably pure tin. Similar filamentary behaviour has been invoked by Newbury and Kirk[143] to explain the memory effect in p-type Ge–n-type ZnSe heterojunctions. Switching effects were also observed in organic polymer films by Henisch and Smith[144] who considered that the memory effect may be due to electrode filaments, although these workers had not completed their investigations. The majority of workers have concentrated upon the chalcogenide glass or transition metal oxide glass devices to date. For this reason these materials will be reviewed in more detail.

5.2 Threshold Switching in Chalcogenide Glasses

Switching has been reported in a wide range of chalcogenide glasses (see review by Bosnell[145] for a compilation). The threshold switching field for both mono- and bi-stable switches is less for bulk samples than for thin film devices, typically less than 5×10^3 Vcm^{-1} for the former compared with 10^5 Vcm^{-1} for the latter. However, many early experiments were reported which had involved the use of coplanar electrodes on bulk samples in air, and since air breakdown (3×10^4 Vcm^{-1}) can yield similar characteristics on pulse excitation to threshold switching, great care must be taken. Kolomiets et al.[146] have reported the variation of this threshold field with thickness for an unspecified Si Ge Te As glass (Figure 10). Furthermore, Thomas et

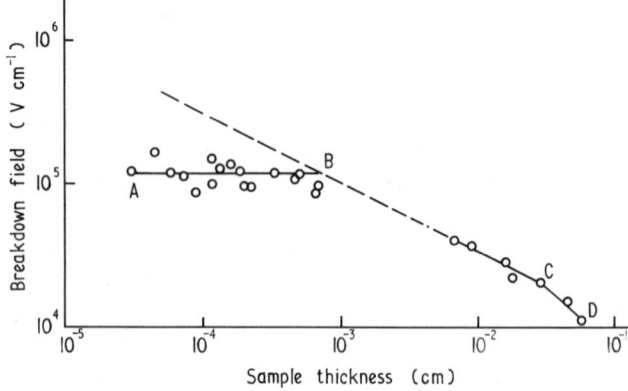

FIGURE 10. Dependence of threshold switching field on sample thickness (after Kolomiets et al.[146]).

al.[113] show that thick samples give a qualitatively different temperature dependence to V_T than do thin samples (Figure 11). It appears that the data on thick bulk samples (> 10 μm) can be fitted to standard thermal runaway theories (see later). For thin films the holding voltage is typically 1 to 3 volts, that is 10^4 V cm^{-1}, with a holding current in the range 300 μA to 1 mA. However, as shown in Table 3, very low holding currents are sometimes observed using semiconducting electrodes, although in this case the holding voltage is anomalously high[147]. In this early work there seems to be some dependence of V_H on electrode material but this is by no means categoric since no systematic survey for one glass composition in one laboratory has been reported (see Section 5.4). It is clear that the power at holding is roughly constant at a few mW (Table 3), even when semiconducting electrodes are

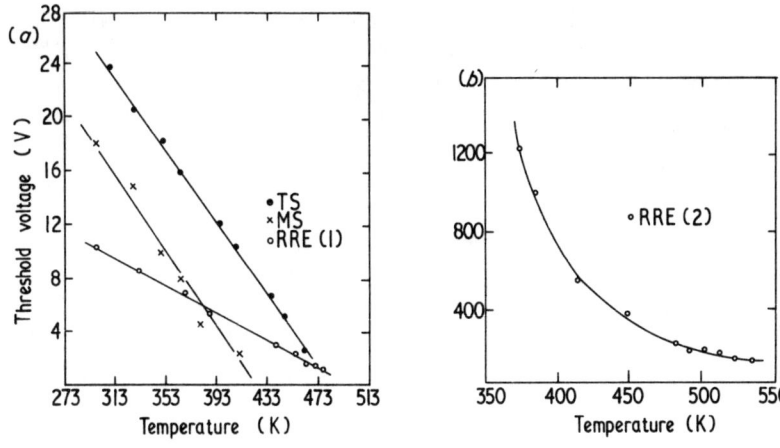

FIGURE 11.
(a) Threshold voltage of thin film chalcogenide devices versus temperature.
(b) Threshold voltage of a bulk chalcogenide glass switch versus temperature.

Table 3

**Holding Current and Voltage
(After Owen and Robertson[147])**

V_H (V)	I_H (mA)	$V_H I_H$ (mW)	Electrodes
1.3	0.6	0.8	C
3	0.3	0.9	Mo
12^a	0.1	1.2	Mo
2.5	0.4	1	Au
~7	~0.3	~2	Sn, Pt
2.8	0.4	1.1	Cu
3.0	0.4	1.2	Brass
1.8	0.3	0.6	V
1.3	1.9	2.5	C
20	0.06	1.2	Si
1–3	1	1–2	Au, Mo
10^b	0.1	1^b	Ag

^{a} gap cell 15 μm coplanar electrodes
^{b} 100 μm thick device

employed. However, if one electrode is semiconducting and the other ohmic (say carbon) then the device becomes highly asymmetric (Henisch et al.[148]). The sensitivity of all these properties to composition is not very clear from most studies so far reported even though their resistivities and inferred band gaps are spread over a wide range. The data of Thomas et al.[113] and Bosnell and Savage[60] on the other hand show that there is a very high degree of correlation between Tg and composition.

The mechanisms involved when a relatively high electric field is applied to the sample are complex. The major switching parameters, threshold voltage V_T, holding voltage V_H and holding current I_H, delay time to switching τ_D, switching time τ_S, and recovering time τ_R are all subject to changes with variation in physical parameters like temperature, pressure and composition as would be expected. One of the most widely used compositions, $Si_{12}Te_{48}As_{30}Ge_{10}$, the so-called STAG material, can be rewritten as basically As_2Te_3 with 7% excess Te in a solid solution with Ge and Si thus: [0.14 $(As_2Te_{3.2})$ plus 0.59 (Ge+Si).] As Thornburg[149] has pointed out in review of the properties of the $As_2(Se,Te)_3$ system, there is a decrease in glass stability for the Te-rich samples. Furthermore, both memory and threshold switching can be observed in As_2Te_3 (see Section IX of Thornburg's review). Hence, although under normal conditions as assessed by differential scanning calorimetry, the STAG glass looks a stable threshold material, it may be that even though great care is taken in the preparation of the glass, phase separation, which is so common in glasses, can occur. Recent evidence from both resonance absorption measurements[150] and switching in bulk STAG samples[151] points this way aside from any "forming" process (see Section 5.3). It would seem that if forming is not essential to the switching process then any tendency towards phase separation must be suppressed and any other switches, as Adler[152] states, are inferior. Indeed, Adler et al.[153] appear to have achieved a device fabrication route in which the first switching V_T is the same as subsequent ones. It is not clear, apart from minor additions to the starting composition, how this was achieved. However, it may well be that many of the controversies and inconsistencies stem from this problem.

There is broad agreement about the nature of the memory state, that is, that a crystallisation process is taking place (see Section 5.6). If the applied switching pulse time is short enough, then even for a memory material only threshold switching takes place, i.e. memory action is a second order effect in this sense. Similarly, the "off" state of the threshold device is relatively well understood. In a previously unswitched or virgin device, several possible conduction mechanisms have been invoked, the $I–V$ characteristic changing from ohmic at low voltages to a more sensitive function of voltage at values up to a maximum of just less than V_T; varying from $I \propto V^2$[154] to $I \propto \exp(V)$[155]. The reader is referred to the review by Jonscher and

Hill[20] for a detailed discussion of the mechanisms involved, whilst Owen and Robertson[147] have reviewed their relevance to switching. Generally, it has been necessary to include the electric field dependence of the conductivity in thermal breakdown models for switching in order to achieve the observed switching speed of less than 150 picoseconds.

Since the monostable device is a precursor to the bistable device it appears that the ON state of the threshold device is the key to the understanding of switching phenomena. However, before looking at the threshold ON state in detail it is pertinent to consider "forming" processes since these may influence the characteristics of the device.

5.3 The Forming Process

The "first fire" threshold voltage of a virgin device is usually higher than subsequent values and this process has been referred to as "forming". A plateau in the threshold voltage may be reached in anything from 1 to 10^3 switching operations depending upon the material and its previous thermal history[41, 151, 156, 157]. The early work of Pearson and Miller[158] indicated the formation of a filamentary liquid phase but the tests were generally destructive, more like the "self-healing" breakdown in thin film capacitors. Forming has been a notable feature of the $I-V$ characteristics of metal–insulator–metal (MIM) structures, and has been associated with the physical formation of filamentary paths or at least a region capable of conversion into a high conductivity state if destructive breakdown his not to occur[115, 116, 159]. Very often, insufficient attention has been paid to the devices, particularly oxide films, after forming. Pivot et al.[160] in Al–SiO–Al structures show, by electron microscopy techniques, that electrode material diffuses into the SiO under high electric fields, and can cause irreversible modifications of the physical and electrical properties of the insulator, including the formation of aluminium compounds. Recent work by Bidadi and Hogarth[161] on Au–SiO/TiO–Al structures supports the view that forming in insulators follows the model of Dearnaley et al.[115] involving the development of metallic filaments. The reproducibility from device to device in MIM structures is poor with up to 50% of devices failing to form[162]. This is not true of the transition metal oxide devices or the chalcogenide glasses. In the latter case, however, as the resistivity of the material increases, there is a strong tendency to breakdown with consequent electrode destruction. As pointed out above, Adler et al.[152,153] doubt that forming is an essential feature of threshold switching but note that it always occurs in the memory switch (see for example References (163) and (164) together with Section 5.6). Furthermore, in STAG films, Bosnell and Thomas[165] have isolated a filament in "formed" devices. This work has been confirmed and extended by Ormondroyd et al.[166] using conductivity probe measurements. Electrode area influences the results since the formed filament can[46] be as small as 5 μm in diameter. Some changes in conductivity were observed outside the electrode pattern by the careful profile measurements used by Ormondroyd et al.[166].

The filament as elucidated by Bosnell and Thomas[165] consisted of Te crystallites and some other polycrystalline compound in an amorphous matrix. The Te crystallites appeared to have the C-axis parallel to the applied electric field. Around this central core is a graded structure. The low conductivity activation energy observed by Ormondroyd et al.[166] of 0.29 eV compared with 0.59 eV for unformed material results from this complex mixture. These latter authors indicate that low activation energies can be obtained in STAG films on annealing and clearly more work requires to be done in this area of possible phase separation under thermal cycling of the chalcogenide glasses, particularly under high electric field strength. Ormondroyd et al.[167] have recently extended their work to include compositional changes in the virgin films as well as the effect of pulse ON time during forming. Although the Si content has a large influence on the conductivity of the virgin film, once formed Si no longer controls this parameter. They find that relaxation of the glass structure back to a more stable glass is possible during the ON time if the ON-state temperature is high enough.

Because of this filament size, temperature measurements on the device during switching have been difficult to perform. Bunton[168], using a liquid crystal technique has shown that in the OFF state, the surface temperature of the switch rises less than 1 K covering the whole electrode area. On switching, only a filamentary area was seen being relatively warmer than the surrounding electrode area which does not rise significantly.

Further confirmatory evidence for a devitrified filament can be found in the work of McMillan and Nesvadba[169] who invoke microstructural changes in the glass during switching to explain fluctuations in switching parameters occurring from one operation to another. Saji and Kao[151] also explain their data on the basis of two phases in the glass, one crystalline and the other glassy; the formation of this filament being initiated by a current-aided thermal process. Furthermore, they show that there is a critical maximum current in the ON state of the device, stability and consistency only being achieved if the current is kept below this critical value.

The relationship between dark current and photocurrent yields the same universal curve for both formed and virgin devices[152], indicating that formation affects both in the same way. Hence, the conductivity increase on formation must be an increase in average mobility, not in carrier concentration. This would be expected from a partial crystallisation model, since crystalline material, in general, has a higher mobility than the glass.

In the vanadate glasses it appears that forming leads to the production of a filamentary path dominated by VO_2 so that the properties are influenced by the metal–insulator transition at 67°C (see Section 5.7).

5.4 The ON State of the Threshold Switch

The fact that the threshold ON state is a precursor to the memory state and that a memory material can be switched in a purely monostable manner, providing the pulse length is short enough, has been pointed out above. The reversability of both memory and threshold effects in the device, because of the existence of the ON state, is the one property which distinguishes switches from dielectric breakdown or metallic bridging between the electrodes. The nature of the ON state is therefore critical to a theoretical description of switching phenomena. Male, in his excellent descriptive article[170] makes this point strongly, going on to draw analogy with switching in air where the ON state is a channel of hot gas, where conductivity has been greatly enhanced by thermal ionisation. The stability of the arc depends upon the electrical power input balancing the heat losses. Male points out that the electrical conductivity-temperature curves for chalcogenide glasses and air have a similar shape, so that thermal effects would be expected in both cases. Calculations show (e.g. Reference 170) (see Section 5.5) that the development of a thermal runaway channel is as follows:

The channel starts by being hottest at the centre (only 7°C above the remainder in the OFF state), but splits into two areas, each continuing to grow whilst at the same time moving outwards to the electrodes, reaching over 400°C each. In two dimensions the calculations indicate that to carry a few tens of mA, the channel centre temperature needs to be between 450 and 500 degrees above ambient, in a radius of about 1 μm. At about 1100°C, the electrical conductivity of the chalcogenide begins to saturate and a minimum voltage across the whole device exists which is predicted to be about 1V, i.e., typical of the experimental values observed (see Table 3 above). The pressure involved in the hot channel is not unreasonable, although there may be some difficulty in making devices from high vapour pressure materials. Thus on a purely thermal model a hot channel is sufficient to sustain the ON state.

If the cross-sectional area of the device is small, then I_H is greater than I_T, and a current gap exists therefore in the $I-V$ characteristics. For large area electrodes I_H is less than I_T, i.e. the current is non-uniform in the ON state, whilst V_H is independent of the electrode area. The sausage-shaped filament seen in the memory device[171] and the shape of the formed filament in threshold devices, particularly from the electrical conductivity data of Ormondroyd et al.[166] would be consistent with this observation. V_H is independent of temperature whilst I_H decreases linearly with increasing temperature[172].

If I_H is not maintained, the device relaxes back into the OFF state. If a new voltage pulse is applied sufficiently rapidly only the holding voltage is necessary to recapture the ON state. As the time after the removal of V_H increases, the voltage necessary to regain the ON state increases until ultimately V_T is required[173]. Various double pulse experiments have been devised to measure the characteristics of this relaxation, starting with the Henisch and Prior[173] experiments outlined above. Thomas et al.[171] have measured the manner in which the resistance of the device relaxes from the ON state to the OFF state. These data show three major regions, (a) an initial time independent portion observed by Henisch and Prior[172] as well, which lasts for less than or of the order of 1 μsec after the removal of the holding voltage. This is a difficult area to address experimentally and is interpreted by Adler[152] as possibly being an artefact of the experiment. However, Lee and Henisch[174] appear to have overcome some of the experimental difficulties and present data in this region showing that no interruption in the ON state occurs for a given ON state current up to a time t_s(max). The free carrier concentration corresponding to I_{ON} decays exponentially with time thus:

$$N = N_{ON} \exp(-t/\tau_f)$$

giving a free carrier lifetime at room temperature of 0.13 μsec; (b) a relatively short temperature independent portion (of the order of 5 μsec), associated with the thermal time constants of the device. It is dependent upon the ON time of the device; (c) a long time constant (seconds) which was temperature dependent and had an activation energy of 1.6 eV which is consistent with crystallisation energies (Sugi et al.[175]). It was associated with structural rearrangement by Thomas et al.[172]. However, a model, dependent upon the electronic properties of the contacts, particularly for large electric fields existing near point contacts, has been proposed by Thompson and Allison[176] on the basis of similar data. Hence, even these experiments are open to several interpretations. As Vezzoli and Doremus[177] point out, providing the time interval is short enough, the holding current can be essentially zero, confirming the above data. These latter authors also invoke an electronic process with carrier lifetimes in the ON state of the order of 10^{-7} sec, which they associate with decay of trapped carriers. Lee[178] analyses his experimental data using a model involving thin free-carrier space change regions at the electrodes. Carriers are assumed to tunnel through the corresponding barriers which are maintained by the electron and hole flow, i.e. the fields at the electrodes are within barrier regions, which Prior[179] terms a "blocked ON state". He reports that: (a) the ON state current is independent of the superficial electrode areas and (b) the ON voltage is only slightly dependent on the electrode spacing (i.e. film thickness). To some extent the first observation appears to conflict with the result of Thompson and Allison[176].

Bosnell and Thomas[165] note that there is a variation in the "lock-on" time in a memory device from one switching process to the next, which they associate with the possibility of different percolation paths in the formed filament. Furthermore, the ON state of a threshold device is often noisy, involving multiple branching[180]. This effect is also observed, however, during switching in liquid semiconductors (see Fig. 3 of Vezzoli et al.[139]). These latter authors conclude that thermal effects must have an influence on the electronic properties of the ON state. Ling[181] recently has also associated multiple branching with differing percolation paths between Te crystallites in the formed device.

Radiation has also been observed from devices during the ON state[182] at approximately 1.2–1.4 µm, but only if the ON-state current was greater than 50mA. A second band between 0.75 µm and 3 µm was detected less frequently. The analysis of these authors indicates that for the radiation tube of thermal origin a temperature in excess of 1000 K would be required. They favour the origin being from band-gap electron-hole radiative recombination with a gap compatible with OFF-state measurements.

The critical importance of the external circuit to the ON-state characteristic has been emphasised by Hughes et al.[183] and Ling[184]. The latter author shows that in order to obtain low holding currents, a large series resistance and minimum parallel capacitance are required. Hughes et al.[183] achieved an I_H of 10 µA by taking such care, but with a commensurate high V_H. Ling[184] emphasises this point by showing how the external resistor can affect the slope of the I–V characteristic. If allowance is made for the influence of the load resistor no minimum is observed in the I–V characteristic. Although different workers have found little correlation between V_H and the electrode material, recent work[180] indicates a monotonic increase of V_H with the increase in thermal conductivity of different electrode materials. The fit to $V_H \propto K_e^{\frac{1}{2}}$ predicted from thermal models is seen to be good. Mott[185] has reviewed the ON state data recently, however, and favours a double-injection model; a conclusion supported by the review of Vezzoli et al.[186]. However, much rests upon the forming process which it is claimed is absent in "good" devices. No manufacturing method to achieve this step has yet been outlined. It would seem that even now the controversy regarding the switching mechanism remains. The major contenders are reviewed in the next section. As Ling[184] emphasises, earlier data not taking account of the circuit aspects outlined above may be suspect. Bosnell and Thomas[165] using the magneto-resistance effect in the ON state of a formed device have shown that material resembles the "set" state of a memory switch with a mobility of about 15 cm^2 V^{-1} sec^{-1}, and independent of ON state current. Hence, several mechanisms may occur, depending upon the glass and its susceptibility to forming, upon the external circuit and the electrode material.

5.5 Switching Theories

As Adler[152] has indicated, a great deal of effort has been expended on the controversy of the origin of the threshold switching mechanism, much of it fruitless, since it has been based on questionable experimental procedures. Theoretical models for the observed switching action fall into three categories: a totally electronic process, thermal or electrically initiated thermal breakdown, and finally, microstructural changes due to heating. The application of high electric fields to dielectrics gives rise to an unstable situation, which at a high enough electric field results in catastrophic breakdown of the material, either in the short circuit (conducting) mode or open circuit. It has been pointed out above that this general property of low conductivity solids and arc plasma can resemble the switching action in the glassy chalcogenide alloys. In some way, however, a feedback mechanism exists in these latter materials which allows this breakdown to be stabilised, so that relatively large currents at low voltages are possible in the device and repetitive switching results. To date, the most successful quantitative models for this action have been those invoking a thermal mechanism solving the heat transport equation:

$$\frac{C\partial T}{\partial t} = \nabla(K\nabla T) + \sigma E^2$$

where C is the specific heat of the material of thermal conductivity, K, and σ is the conductivity, $\sigma = \sigma(T,E)$ where E is the electric field. These thermal runaway and subsequent stabilisation modes are reviewed in the earlier work of Kaplan and Adler[187], Popescu and Croitoru[188], Kroll and Cohen[189], Warren[190], Male and Thomas[191] and Burton and Brander[192]. In general, to take account of the non-linear I–V characteristic in the OFF state $\sigma \propto \exp(E/E_0)$ is used, (Warren[190], but also see reference (129)). Typical solutions for three forms of $\sigma(T)$ are shown in Fig. 12 (after Fritzche[193]).

It is sufficient to add that the feedback loop implied by these treatments and hence the device stability requires the internal temperature of the device during operations to rise substantially. This leads to an increase in conductivity, hence greater current and so on. Stability is only attained in the earlier calculations because it is a property of the chalcogenides that the conductivity reaches a plateau at temperatures in excess of 1000 K (Male[194]). More recently, however, by taking into account non-uniformities introduced

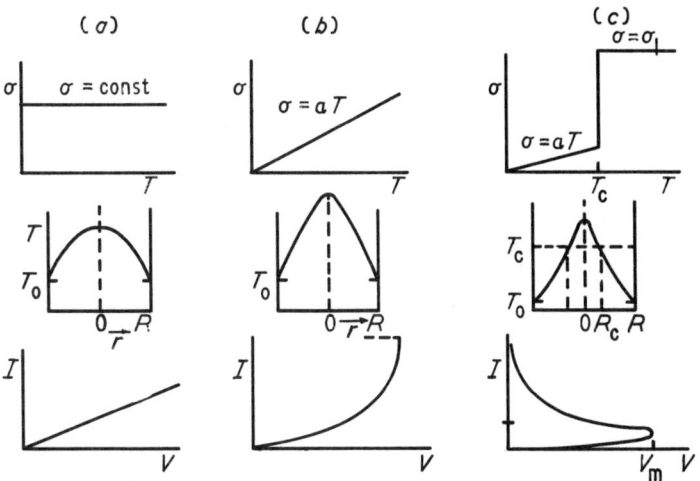

FIGURE 12. Temperature profiles and static I–V curves resulting from model calculations for a material of conductivity σ sandwiched between two electrodes assuming radial heat flow only. The cross at the end of the I–V curve represents material destruction. The three cases depicted differ in the functional form of $\sigma(T)$ (after Fritzsche[190]).

by forming, Popescu[195] has indicated that the temperature may not be so high. Furthermore, Thomas et al.[196] obtained an improved fit of these data to the thermal equation if the properties of the formed filament were used in the analysis, with a very low temperature rise required to initiate switching. However, Kroll[197] has recently used an electro-thermal model, applying it to essentially virgin devices, i.e., no gross heterogeneities or formed filaments. He concludes that electronic processes are central to the switching mechanism. If no filamentation has occurred on forming then, under uniform flow conditions current controlled negative differential resistance should be observed[198]. Whether CCNR or threshold switching is seen, it is governed by the geometry of the device. Thornburg's data[198,199] confirm this.

The data are consistent on the point that V_T decreases with increasing temperature (see above). There is some evidence that V_T may be related to Tg, see, for example, Thomas et al.[172] and Iizima et al.[200]. The latter authors find:

$$V_T \propto \exp(C\,\mathrm{Tg})$$

More recently, Thornburg[149] on different glasses has reported that the conduction activation energy related to Eg is proportional to Tg (see Fig. 9 of his review). Hence the relationship predicted by Sheng and Westgate[201] on the thermal model, viz:

$$V_T \exp(V_T/2V_0) = \frac{16Kk}{CEg} \text{Ta} \exp(Eg/4k\text{Ta})$$

where V_0 is the over-voltage and Ta the ambient temperature may account for the observed behaviour of V_T with temperature. It should be noted additionally, however, that there is a change in carrier lifetime in the vicinity of Tg in Se at least[202], the data being explained in terms of structural changes near Tg.

Electronic processes form the second major set of theories. The early attempts were based upon the band model of amorphous semiconducting alloys suggested by Cohen, Fritzche and Ovshinsky[203] (CFO), for example Henisch and Prior[173]. In broad outline, these models suggest that a double injection space charge process is taking place in the material. The neutral plasma thus produced creates an instability which is propagated through the material, resulting in a voltage drop across the device and the observed S-shaped negative resistance in the monostable switch. Thus the potential breakdown mechanism is stabilised. More recently Mott[185] has extended these ideas, particularly to describe the ON state. Change injection phenomena explain the OFF-state characteristic well[20,204], and Sergent[205] has extended the work of Lampert, and this leads to a well defined V_T with a temperature dependence close to the experimental data, particularly when applied to a well defined system, in this case neutron irradiation silicon. Van Roosbroeck[206], on the other hand, invokes his relaxation semi-conductor approach since in amorphous semiconductors, the dielectric relaxation time is greater than the recombination time of diffusion-length lifetime. A large space charge region exists near an injecting contact which grows towards the other electrode under the applied voltage. This gives rise to a rapidly increasing electric field across the remainder of the material, which breaks down into the ON state. Once again though, a uniform film is implied. Klein[16] has attempted to differentiate between electronic and thermal events (Table 4). It should be noted that again uniform planar specimens are implied.

If forming has occurred, or heterogeneities are present, then the above considerations need modification. Additional high field points will be observed. For this reason a third approach was proposed[165,172] namely, thermally induced structural changes. The orientated Te islands in the formed device (above) can give rise to preferred current paths with local heating on high field regions. Thus the ON state is established. If the holding current is not maintained, the phase separation becomes unstable, and some re-

Table 4

D.C. Switching Properties of Uniform, Thin, Planar Specimens (After Klein[16])

	Thermal events	Electronic events
Initiated by:	Temperature rise due to Joule heat and/or dielectric loss	Electronic processes
Temperature rise on instability:	Tens of °C	May be insignificant
D.C. V–I characteristic:	Continuous (with negative differential resistance range)	Discontinuous
Current flow:	Uniform	Mostly filamentary
Switching voltage:	Precise	May be subject to statistical scatter

vitrification occurs, so that the material reverts to the OFF state. Potential breakdown is thus avoided. Many properties, for example the temperature dependence of V_T, are then explicable. Gross heating in the filament is not required, and some of the minor switching events can be explained[207]. The observed switching speed of 150 picoseconds can be explained in terms of a phase change since it will propagate at roughly the speed of sound. Since the distance is small, the time taken is of the correct order of magnitude[105]. Furthermore, McMillan and Nesvadba[208] have analysed the statistical variation in V_T and T_0 for a large number of events. On the basis of these data, they suggest that structural changes were taking place in the nominally stable threshold glass; a similar suggestion to that of Bosnell and Thomas[165]. With the complexity of the materials and the high electric fields involved, it is doubtful if, at this stage, any one model alone can explain the data. Some combination of effects could well occur. (See also the references in the review by Vezzoli et al.[186] and that of Mott[185].)

5.6 Memory Devices

As indicated above, there is general agreement on the mode of operation of memory devices. The material for these devices is generally located near the glass-crystalline boundary of the phase diagram (see Fig. 9) and upon heating undergoes crystallisation (see Fig. 1). The switching procedure outlined in Fig. 7b is typical, a relatively long (msec) SET pulse being required to achieve "lock-ON" into the memory ON state of the device. This can be read or sensed many times without loss of information. To RESET the device, a short (μsec) high current spike is utilised. Figure 13, after Bosnell and Thomas[41] and

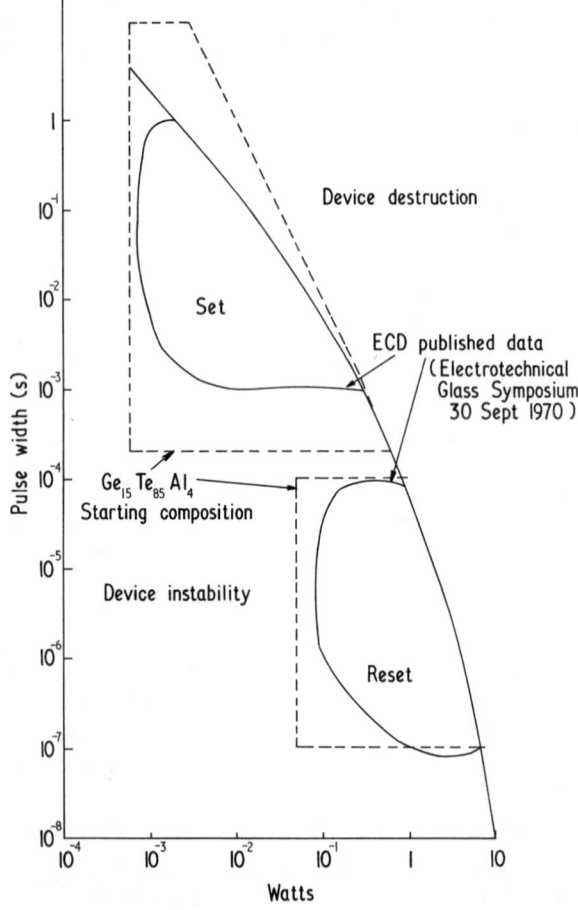

FIGURE 13. Pulse width–power input criteria during SET and RESET of a memory switch.

Neale et al.[209], illustrates the energy balance in these SET–RESET conditions to avoid device instability. If the SET pulse is short enough then the device switches in the threshold mode, i.e. although a well formed filament is observed following RESET normal threshold operation if the conditions are carefully established. This SET-state filament has been isolated, and is barrel-shaped and polycrystalline[210]. Hence, the RESET pulse simply destroys this filament. So in a very real sense, the memory effect is a second order one. The work of Dargan et al.[211] extends the information on the SET–RESET conditions and supports the thesis of Te-rich islands remaining after the RESET pulse which subsequently rejoin during the SET pulse[165]. In the main, the majority of workers have studied compositions around the $Ge_{15}Te_{85}$ eutectic. Dargan et al.[211] have compared the lifetime of devices produced with S, Si and Sb additions up to 14% of this basic composition. The further from $Ge_{15}Te_{85}$, the more unstable the device became. Since V_T is generally less than 20 V, and many device applications require at least twice that value, Dargan et al.[212] attempted to build up the thickness of a single device, but at around 4 μm the curve of V_T versus thickness began to saturate. Hence these authors resorted to connecting three devices in series when in excess of 45 V for V_T was achieved or when trying new compositions. In some cases, because it was felt that films could not support a highly conducting phase, breakdown occured. Hence it was required to find a material which would separate into a highly conducting phase embedded in a good insulating matrix. A wide variety of materials were reviewed[213]. A V_T of 30 V was achieved using BiAsSe films, but the lock-ON and RESET characteristics were unsatisfactory.

As illustrated above, there has been a widespread search for suitable memory materials, usually involving DTA. Typical of this work is the detailed study of Savage[107]. DTA and x-ray work allied have been utilised by Shimakawa et al.[214] on the TeGeSe system. Metallic additions to the basic AsSe glass have also been reported (Asahara and Izumitani[215]). These latter authors claim that heating during switching causes aluminium crystals to precipitate from the glasses; a precursor to lock-ON. Memory effects have also been reported in SiO films[216] and Ge–ZnSe[217] heterojunctions, and in both these cases filamentation is the proposed SET state which is disrupted upon RESET. In no case in these materials is an electronic memory effect observed. One difficulty with the reported chalcogenide systems is their sensitivity of V_T to temperature. For example, in the S–Se–Te system, Hirose and Kunioka[218] report the disappearance of the threshold voltage at 85°C in their memory switches: this temperature is equal to Tg within the limits of experimental error. Similarly in the GeTe system, V_T approaches zero at Tg, which is of the order of 125°C.

Apart from the indirect evidence for the switched filament, several direct observations of filamentary growth have been made, e.g. Chaudhari and Laibowitz[219] by transmission electron microscopy on a bistable material. They found that, using gap cells, filaments started on the electrodes at those points where a local field concentration was present. The starting alloy was an amorphous GeTe chalcogenide and the crystalline filaments formed were Te-rich. It appears that it may be possible to look at the movement of Te directly in a switch using the Mössbamer effect (see Hafemeister and de Waard[220]) although much depends upon the volume of material which actually participates in the switching of formed devices. Experimentally, the maximum velocity of the crystallisation form has been reported[221] to be 10 cm sec^{-1} for GeTe. This rate is clearly dependent upon the current and the material composition. For example, time before lock-ON in bulk samples of $Ge_{10}Te_{90-x}As_x$[222] was found to be as long as 10 sec for $x = 50$ at %. In As_2Se_3, a material with well characterised parameters, Thornburg and White[223] propose a model involving field assisted spinodal decomposition to produce dispersed As clusters which then grow into an As filament, in preference to an earlier model involving Joule heating effects[224]. However, more recently, Thornburg[225] has proposed that the chalcogenide glasses be divided into three classes: (I) stoichiometric compounds which do not undergo significant changes in local order on crystallisation (e.g. As_2Se_3), (II) stoichiometric compounds which undergo significant changes in local order on crystallisation (e.g. GeTe) and (III) non-stoichiometric alloys which phase separate on crystallisation. It is Thornburg's[225] view that the spinodal decomposition model will only apply to type I materials, not as suggested by Bosnell and Thomas [165], to all types including threshold material, and Dargan et al.[212] on the $Ge_{15}Te_{85}$ based memory materials.

Uniaxial pressure can induce the memory effect[165] even from the ON state of threshold devices. In general, the density of amorphous materials increases monotonically with pressure, crystallisation occurring at a certain pressure, around 20 kbar for amorphous germanium[85]. Shock-crystallisation has also been observed in sputtered films with a propagation velocity of about 100 cm sec^{-1} (Mineo et al.[226]). Hence care must be taken with certain probe measuring systems.

The reliability of device operation is critically dependent upon the mode of use (see Fig. 13). Recently, Steventon[227] has shown that the statistical distribution of the lock-ON time can predict the reliability of the SET operation. This in turn depends upon the RESET parameters. Two modes of degradation of V_T were observed. The first could be eliminated by the use of multiple RESET pulses and associated with remnant crystalline debris following imperfect RESET. The second is permanent and associated with electromigration. It is accelerated by multiple RESETs and so a compromise of 6 RESET pulses was recommended. 10^5–10^6 SET–READ–

RESET–READ cycles are in general possible. Presumably, electromigration is the root cause of reports of devices displaying memory switching with one polarity and threshold switching with the other[228].

5.7 Switching in Transition Metal Oxides

There is a close analogy between switching of a thermal nature in VO_2 crystalline films through the insulator–metal transition at 67°C and switching in the chalcogenide glasses, as the papers of Berglund[229] and Cope and Penn[230] illustrate. Glasses based on V_2O_5–P_2O_5 have been made with a variety of additives and the former show a decrease in V_T with temperature which extrapolates to zero at 67°C[231]. The delay in switching can vary from operation to operation in these devices[232] but in general the properties are similar to the chalcogenides once the device is formed. Both threshold and memory action have been reported in transition metal oxide films[233,126], generally in a formed filament which Pinto[126] considers is predominantly metallic in nature and is produced under high electric fields. The conduction mechanisms in the OFF state of the switch are likely to be different from those found in the chalcogenides. Naively, hopping processes of the type: V^{4+}–O–V^{5+} → V^{5+}–O–V^{4+} can be involved. It is out of place to review such theories here. The reader is referred to the paper by Austin and Sayer[234]. Preparation techniques may alter the ratio of V^{4+}/V^{5+} and hence the conductivity of the virgin state. However, in switches produced from the DuPont "Tyox" material[45], by thick film techniques, the temperature dependence relates to the insulator–metal transition in VO_2 at 68°C. The earlier work of Regan and Drake[232,233] has recently been extended by Higgins et al.[46] using thick film techniques. These authors also review glass praparation techniques in some detail. In general, they obtain resistance ratios (R_{ON}/R_{OFF}) which are less than those usually observed in chalcogenide systems but with higher values of V_T. In order to overcome the temperature sensitivity of V_T, attempts have been made to use materials in which the insulator–metal transition is at a higher temperature, e.g. Nb_2O_5, although Park et al.[235] feel that it is unlikely that crystalline Nb_2O_5 is involved in the memory switching device. Lalevic et al.[236], for threshold devices made anodically, indicate that the long delay time to switching they observe supports a double-injection switching mechanism. However, there appears little doubt that in most, if not all, transition metal oxide devices made to date, some forming has taken place. Co-sputtering TiO_2 and VO_2 in crystalline films does change the transition temperature[237] but so does the deposition of SiO_2 onto crystalline VO_2[238]. TiO_2 films themselves appear to operate by the growth of Ti_2O_3 grown by reduction of TiO_2 film[239], although no temperature performance of V_T is given. The addition of 30 mole % TiO_2 to VO_2 moves the transition temperature to over 90°C[237]. Hence, whilst the properties of transition metal oxide devices are similar to those of the chalcogenides, the most investigated systems based on VO_2 are more temperature sensitive than the chalcogenides, with the indication that TiO_2 and Nb_2O_3 based systems will have improved temperature performance.

5.8 Device Fabrication

The silicon industry has become used to large scale integration, and now produces thousands of devices on a single chip of only one square millimetre. Clearly, the aim of the memory device, if it were to compete with silicon based products, was to achieve this integration density, whereas for large area applications this is not so necessary. Early devices, particularly from Ovshinsky's laboratory, consisted of the active material flash-evaporated onto graphite electrodes which were spring loaded together in the so-called DO7 package (Fig. 14a). This device generally had a self capacitance of 1–2 pF

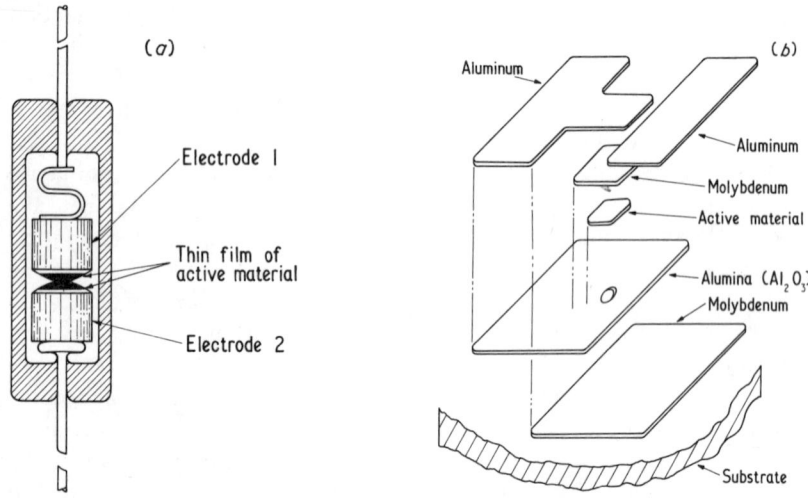

FIGURE 14.
   (a) DO7 package device.
   (b) Thin film device (exploded view after Neale[241]).

but a 50–100pF capacitor across such a device can cause failure. The role of capacitative energy in switching has been discussed by Thornburg[240], who uses a thermal model, and experimental data are given by Bosnell and Thomas[41]. Some other sources of possible device instability are listed in Table 5 (after Neale[241]). These considerations have led to a design called the pore device (Fig. 14b), in which the active area is confined to a 20 μm diameter pore etched in a suitable oxide (Neale[241]). Successful 256-bit memory arrays using this structure have been constructed. Details of the fabrication procedure are given by Bosnell[145] with a typical device processing schedule.

A process schedule for a silicon diode isolated 16 × 16 bit memory array is given by Dargan et al.[211]. As indicated in Table 5, high quality dielectric and metallisation steps are required. A dielectric system for use with amorphous semiconductors has been reported by Haden et al.[242] and the diode array problems are dealt with by Buckley and Moss[243] together with some contact difficulties. The importance of suitable contacts was dealt with earlier (Section 4). The problems of the Al_2O_3 interface using Al contacts

Table 5

Sources of Device Instability (After Neale[241])

Electrode-originated

    Crystallisation of active material by electrode material
    Active material dissolving the electrode material
    Structural mismatch (amorphous electrodes are most desirable)

Process-originated

    Fractional distillation of active material during evaporation
    Partial re-evaporation or devitrification of active material during process steps occurring after active material is deposited
    Introduction of electrode-active material interface layers

Substrate-originated

    Free ions (more serious in coplanar gap devices)

Structure-originated

    Lack of thermo-mechanical stability
    Self capacitance (avoid large discharge current $i = C dV/dt$)
    Electrode hot spots (thin-film edge effects)

Extraneous

    Circuit capacitance
    Formation of switching glass in the device structure (e.g. GaAs with Te electrode)

for memory switches is well illustrated in the paper by Vezzoli and Pratt[244]. Many workers have made use of "gap" devices in which a suitable gap between two planar electrodes is filled with the active material. In general, these devices have only been successful when the active area has been overcoated with a dielectric, e.g. SiO_2. This device structure has been successfully exploited recently by Hughes et al.[183] who gave fabrication details.

6. Applications of Switching Devices

Glassy switching devices first came to the notice of the author under the name "Quantrol" in the U.K., and "Ovonics" in the U.S.A. in January 1963, originating from S. R. Ovshinsky of Energy Conversion Devices (ECD) Detroit, and Electronic Machine Control Ltd. in the U.K. The Quantrol application notes of that time relate to monostable or threshold devices with applications ranging from transient surge suppression to staircase and square wave generation. Very few data on device performance or processing were given, but the devices appear to have consisted mainly of GeTe pellets with As of Ga sandwiched between two graphite pressure contacts. One property of these devices was a tendency to lock-ON in the high conductivity state. This property has now been developed into a 256-bit array referred to above. Because potential applications for the memory devices are well defined, these will be reviewed first.

6.1 Memory Switch Applications

A conventional read-only-memory (ROM), as the name suggests, is a data store that can be interrogated an indefinite number of times, but the stored information cannot be changed whilst in normal operation. ROM's are used typically for microprogram stores, look-up tables, code conversion, and in emulation. The features which make ROM's attractive in these applications are high speed and security of the stored information. In general, in current systems, ROM's have been made from fuseable metallisation or diode links in integrated circuit arrays (often called programmable ROM's or PROM's) or custom-masked chips, wired in ferrite cores or resistor or capacitor cards. A limitation is the permanent nature of the data, i.e. it is impossible, or very expensive, to make changes. Hence a genuine need exists for an electrically alterable ROM or EAROM or read-mainly-memory (RMM). Such a non-volatile store could find applications in, for example, airborne control systems (e.g. fuel injection control), machine tool control, communications (especially secure coding and processing character generators) and automatic testing systems. If the store were radiation hard then certain military areas of application arise which may be inaccessible to conventional semiconductor stores, although the full military temperature specification of $-55°C$ to $+125°C$ may be a problem at present. The sensitivity of V_T to temperature, outlined above, needs careful consideration. Two factors may help here. Firstly, Nicolaides and Doremus[245] report data on switches based on As_2Se_2Te with 0.3–0.4% Al or 0.3–0.5% Ag whose V_T was relatively temperature insensitive: furthermore, the memory diodes were stable in their ON state even to neutron doses of 2×10^{17} n cm^{-2}. Secondly, Quilliam et al.[246] have reported a writing technique which instead of using a wide set pulse, as Figure 7b, employs a short switching spike ($\sim$ a few μsec) at sufficient voltage to switch the device, followed immediately by a longer pulse to set the device which is at the minimum set current level. It is claimed that writing and reading can then be performed at least over the range $-40°C$ to $+85°C$ without alteration of the external circuitry. The ECD product of an integrated memory array of 256 or more bits has been reviewed by Neale and Aseltine[247]. During the development of this device, the silicon industry has been galvanized into the production of a major competitor, the metal–nitride–oxide–silicon (MNOS) memory (see for example Dill and Toombs[248]) as a direct alternative, which, being based on Si technology, may well be cheaper than the amorphous RMM in volume production because of the amortisation of processing costs over a wider product range. Furthermore, the problem of drivers and decoders on the same chip is much more complex than in the case of the amorphous device. MNOS, for example, are reported to be used in TV tuners as well as in point-of-sale equipment[249]. The FAMOS (floating gate avalanche injection MOS[250]) has achieved a density of 8192 bits per chip with a write time of 50msec. The device suffers from having to be erased using UV or x-rays and is thus more like a PROM than a true EAROM. However, Verwey and Kramer[251] have reported what they call an adjustable threshold MOS (ATMOS) transistor which can act as a true EAROM, i.e. the charge on the floating gate is electrically erasable, unlike the FAMOS. The so-called BEAMOS-beam addressed MOS[252] is another contender for this market. In some applications, CMOS (complementary MOS) with a battery has been used, where an EAROM would have been favoured and, it is said, gave reasonable system performance. In some areas, charge-coupled devices (CCD)[253] and magnetic bubble memories[254] are competitors of the amorphous memory technology. It is ironic that one of the cornerstones of amorphous state theory, the Anderson transition, has recently been applied to explain certain aspects of the MNOS device[255,256].

6.1.1 Radiation hardness

The attraction of amorphous materials in a radiation environment was alluded to above. A comprehensive review on the radiation sensitivity of amorphous materials has been given by Holmes-Siedle[257]. In this context, there is a clear advantage in using the glassy RMM. Early data (Kolomiets[258]) indicate that high γ-ray doses up to 10^9 rad can cause changes in the electrical conductivity of vitreous semiconductors. However, these dose rates are very high. Later work by Ovshinsky et al.[259] on threshold switching devices under x-ray levels of 1.8×10^{11} rad sec^{-1} and neutron fluences of 1.2×10^{17} n cm^{-2} showed that the devices function well with no changes in their electrical characteristics. These levels are probably far in excess of those required in space and military environments. It should be recalled that one of the first applications for vitreous semiconductors was as a thermometer in a heavy radiation environment (Edmond et al.[260]). The effect of 30 MeV electrons on threshold devices is reviewed by Flanagan and Wyatt[261]. Some small changes were observed, but again these were relatively heavy doses, and no switching was induced in devices biased near their threshold voltage. The effect of neutron fluences of 5×10^{17} n cm^{-2} on V_2O_5–P_2O_5 glasses is given by Hench[262]. Again, except for tan δ where the characteristic loss peak of the pre-irradiated sample was destroyed, little effect on conductivity was observed. More recent data, presented at the Ann Arbor Amorphous and Liquid Semiconductors Conference, confirm these results unless heterogeneities are present (Hench et al.[263]). Nicolaides and Doremus[245] show that both their doped and undoped memory devices, when subjected to neutron irradiation up to 10^{15} n cm^{-2}, are, in most cases, unaffected. Their procedure was to irradiate the devices both in their OFF and ON states, and they observed no detectable changes. The more conventional memory material from ECD, $Te_{81}Ge_{15}X_4$ (where X is quoted as a multi-element additive), was studied by Smith et al.[264] of the Naval Ordnance Laboratory, who showed that this all thin film memory will tolerate 10^{16} n cm^{-2} and 3×10^7 rad of γ-irradiation without damage to its key properties unless radiation-induced thermal effects occur. Heavy ion bombardment, on the other hand, caused marked changes (Olley and Yoffe[265]) but this is less important from an operational point of view. None of these reports has anything to say about the electromagnetic pulse effects associated with nuclear explosions, which may be more significant in that fast transients could be injected into the system (see Hays[266]). Like most silicon devices the MNOS is sensitive to radiation effects well below the above dose levels. It is possible that the bipolar silicon diode array in the commercial amorphous RMM could be affected by such radiation, and schemes for an all-glassy system have been put forward in which the silicon diode is replaced by a glassy threshold switch (Henisch[267], Nelson[268], Adams[269]). Such an array is shown in Figure 15. Dargan et al.[211] report that their array withstood 10^{18} n m^{-2} with no change in V_T or R_{ON} or R_{OFF}. The author's own data confirm this. Similarly, a rate of change of γ-dose of 10^{10} rad sec^{-1} did not change the state of individual memory devices.

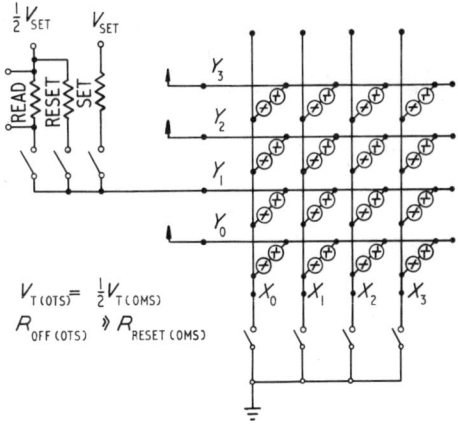

FIGURE 15. Threshold switch isolated memory array (after Nelson[268]).

6.1.2 Telephone applications

The British P.O. system, "Callmaker", and Bell system, "Touch-a-matic", are both examples of the use of look-up tables. In the latter system, a 15 kbit semiconductor store enables pre-recorded numbers to be dialled by touching a single number, whilst the former relies on a punched card system. Both amorphous semiconductor and bubble memories have previously been mooted for these applications, but it appears that conventional technology has now been adopted. The move to automatic exchanges, in particular those with stored program control (SPC), could employ large non-volatile stores of the EAROM type. To perform the speech switching function in telephone systems, the following device specification seems to be required: at least 10^7 operations, though preferably many more, with a relatively high threshold voltage around 40 volts, which should not vary by more than $\pm 10\%$, about a resistance which at maximum should be only $5\,\Omega$ with a minimum OFF-resistance in the region of 50 MΩ at no greater than 5 pF and preferably 1 pF of stray capacitance.

Devices which compare closely with the glassy diode in characteristics are the family referred to as PNPN diodes, silicon controlled rectifiers or thyristors. These have been used in telephone circuits but generally compare unfavourably on the grounds of cost and reliability, with mechanical devices. Furthermore, for bell-ringing applications, relatively high currents need to be passed (say, about 0.25 A) at possibly mains voltages (240 V), so the above tentative specification precludes this application. The fact that the glassy switch is two-terminal is an anathema to many engineers who are used to mechanical devices, such as relays, which have independent control circuits completely isolated from the switching circuit.

6.2 Threshold Switching Applications

Well characterised memory arrays made by amorphous semiconductor technology have been on the market for some time competing, as can be seen in the above section, with more conventional devices. A comparison can therefore be made in terms of the cost and systems trade-off if these devices are used. The same cannot be said for threshold devices. Applications for these devices are much more speculative.

6.2.1 Display switching

Several years ago, there was a great deal of publicity in the fringe journals regarding flat screen TV. There are several motives behind this move: the annoying flicker on a conventional CRT can be tiring, and the general lack of storage is a shortcoming, especially from the military standpoint. Furthermore, the CRT is not well suited to displaying alphanumeric information. Many competing technologies have emerged: liquid crystals, electroluminescent panels, plasma panels, light emitting diodes, electrophoretic and electrochromatic devices. (See Proceedings of the SID, Vol. 16/2, Second Quarter 1975 for reviews of these.) It is likely that if a memory could be incorporated with the display transducer, the electronics costs would fall, and with a control device associated with each display element, some of the problems of brightness and contrast may be overcome. Some display technologies achieve bistable performance: LED's gas discharge panels, electrochromics and some liquid crystals for example, but electroluminescent panels generally do not, and the problems of assembling a large array of discrete LED's are numerous. The use of thin film transistors to control display panels has been reported by Westinghouse, achieving 120×120 display points on a 6-inch-square panel, and an up-to-date account of developments in this area is given in Chapter 6 of this volume. The use of amorphous switching devices in X–Y display addressing applications is an area where the large area deposition capability is an obvious advantage. In principle, a threshold switch and an electroluminescent panel element in series could be used, but this would require a current of greater than 0.5mA to be passed continuously through the glass switch for long periods. The reliability of most devices in this mode of operation is suspect and requires much research effort. Hence, even here it is likely that a memory switch would be required. In the case of D.C. electroluminescent panels for a $2 \text{ mm} \times 2 \text{ mm}$ area element, the OFF-resistance may be as high as 1 MΩ so that the OFF state resistance of the threshold device before switching must approach this value. At present high voltages (100 V) are required. Emphasis has been placed upon tailoring the chalcogenide material to meet this either by compositional changes[213] or by using a number of devices in series[212]. However, it may be necessary to move towards higher resistivity glasses, e.g. the vanadates. A typical arrangement with D.C. EL panels would be to hold the panel on with a memory switch, which can be erased through a threshold device (Fig. 16a) as

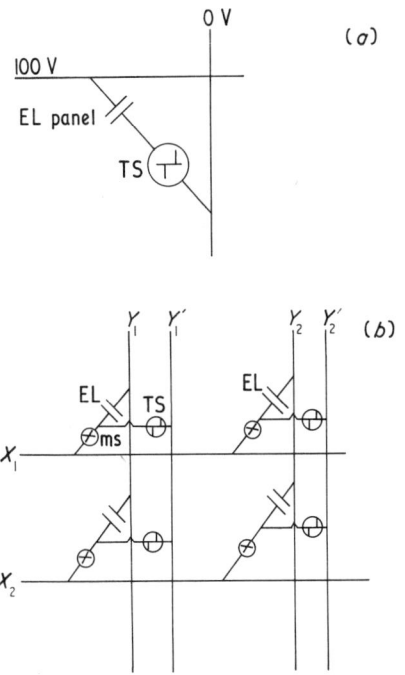

FIGURE 16. Latching an EL panel.
(a) Threshold switch only.
(b) Threshold-memory switch combination.

opposed to the straight EL panel threshold switch combination (Fig. 16b). Clearly, efforts are also being made to reduce the required drive voltage to the panel for other reasons, such as compatability with integrated circuit logic families. However, if this were achieved, the requirement for a switching device at the individual element would be removed. Furthermore, a memory effect has been observed in the EL material itself[270]. The A.C. electroluminescent element can be regarded as a capacitor which is on when given bias charge and off when acted on by a reverse pulse. This is the basis of the OVEL device described by Fleming[271], see Figure 17. The device specifications are much the same as for the D.C. EL panel.

Apart from the switching requirements, display devices may need their contrast ratio element to element enhanced, because in an X–Y addressed mode they suffer from a brightness-voltage characteristic which allows for insufficient discrimination when half-voltage selection techniques are em-

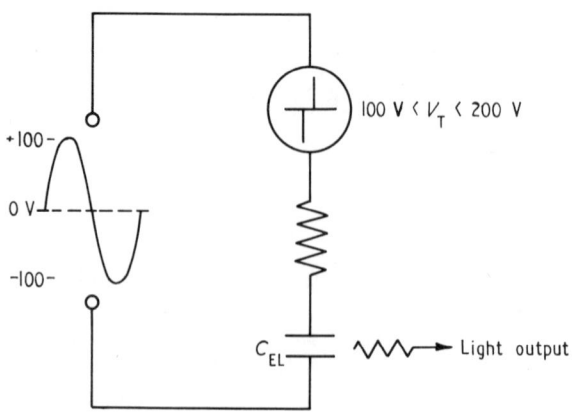

FIGURE 17. Basic series OVEL display element.

ployed: this is the so-called cross-talk problem. A variety of techniques have been employed (O'Connell and Narken[272]) and essentially a switch performs this task admirably. However, in some requirements these may not be necessary, and a simple non-linear element may be used to sharpen up the brightness-voltage curve, or the transmission-voltage curve in the case of liquid crystals. Modern electroluminescent panels have much sharper brightness-voltage characteristics in this respect. Amorphous materials can provide these non-linear elements, in the OFF-state without switching.

6.2.2 Logic circuits

Because of the radiation hardness outlined above, there may be some interest in logic circuits based on glassy switches. Several such circuits are described by Nelson[198], and Van Laningham[273]. Figure 18 shows a flip-flop based on threshold switches, and Nelson reports successful operation of a logic gate, a shift register and oscillators. It is projected that switching speeds up to 5 MHz could be achieved with an all-thin-film design, but for individually packaged devices a speed of 0.5 MHz had been achieved experimentally. These withstood a flash x-ray irradiation of 3×10^{11} rad sec^{-1}. The efficiency of this technology, which is in direct competition with silicon integrated circuits, will depend upon the detailed cost-performance of any system in which they are to be employed. With SIC's price dropping dramatically for complex functional

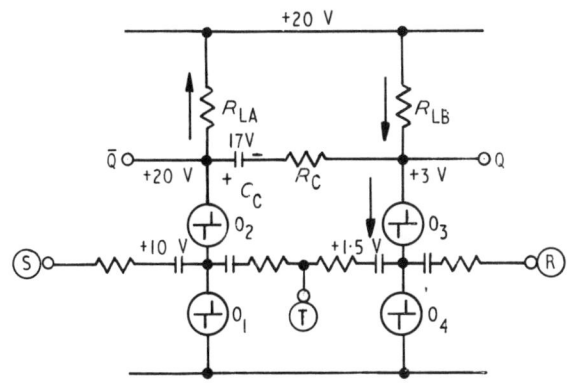

FIGURE 18. Threshold switch flip-flop (after Nelson[268]).

modules this is going to be a very difficult area in which to compete. Kobylarz[274] has proposed the possibility of ternary logic with glassy threshold switches because that device inherently possesses two thresholds. The algebra associated with such an element is outlined. A difficulty would seem to be persuading engineers of the advantages over their well-established binary system.

6.2.3 Amplification

Kobylarz[274] has reported linear amplification with a simple dual threshold switch circuit based on their ON-OFF switching action, whilst Rockstad[275] described the observation of direct amplification in a Te_2AsSi device. These devices had a pronounced differential negative resistance in their ON state which is utilised to provide A.C. amplification. Modest A.C. gains around 3 were achieved depending upon the load resistor used provided short pulses, typically 1 msec, were used. Longer pulse lengths were unsuccessful since the device holding voltage changed with time.

6.2.4 Oscillators

Since we have a negative resistance device, the immediate question that arises is, does it oscillate? We must not forget the time delay to switching. Assuming a simple equivalent circuit for the device, Fig. 19, then simple network analysis shows that oscillations depending upon the load resistance, R_L, will set in if $R_L < |R|$ and $L/R_L < |R|C$. These are usually relaxation oscillations, but can be sinusoidal depending upon the magnitude of the damping term,

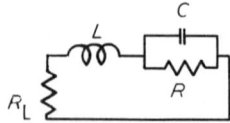

FIGURE 19. Small signal equivalent circuit of a threshold device (simplified). C is device plus stray capacitance; L is the load resistance; R_L is the load resistance; R is the device resistance.

and generally preclude investigation of the negative resistance portion of the curve. Typical values for such oscillations are between 1 kHz and 1 MHz, (see in oxide glasses Hed and Freud[276], Lipiscus et al.[277]). A comprehensive equivalent circuit analysis is given by Shaw et al.[119] which indicates that the frequency of the relaxation oscillations is voltage-tunable. Recently a different type of oscillation has been reported by Haden et al.[278], which is of high frequency, and associated not with the load resistor but with the so-called minor switching event which is sometimes observed in threshold switches. Minor switching is essentially a precursor switching event before the major switching takes place. Two main frequencies were reported, 13.7 MHz and 11.4 MHz, that is, generally higher than the relaxation oscillations frequency, although some earlier Russian work by Zaliva and Zakharov[279]) appears to report similar oscillations with a broader frequency range, from 5 $\times 10^5$ to 5×10^7 Hz. Like the amplifier case, the present range of possibilities appears restrictive compared with SICs but such results may well give an insight into the switching mechanism.

6.2.5 Thin film inductive elements

Berglund and Walden[280] have shown that it is possible to construct thin film inductive elements based on filamentary thermal switching in VO_2 crystalline films. These authors quote inductance values in the range 0.1 to 10 H per cm length of filament with high Q at low frequencies (< 1 MHz). Good fits to the experimental data for negative capacitance in the chalcogenide devices have been obtained by the use of thermal models by Sheng and Westgate[201] and Allison and Dawe[281]. These latter authors show that the effect is frequency sensitive, being almost masked at 1 MHz. A rival model for the negative capacitance effect has been proposed by Rockstad[282] invoking impact ionisation phenomena. So once again there appear to be two separate explanations of the same phenomenon. Whether or not such devices will be used in real circuits is doubtful since, increasingly, SIC designers have overcome the problems of circuit design without the use of inductances, as was indicated in Chapter 3. The fact that a device exists which at a specific voltage changes the sign of its effective capacitance from positive to negative or from capacitative to inductive behaviour can lead to novel circuits, but the practical value seems to be limited.

6.2.6 Temperature sensors

The form of the V_T versus temperature plot (Fig. 11 for the chalcogenides and Ref. (231) for the vanadates), indicates that one of the disadvantages of the present range of threshold devices could be turned to advantage as a fast-acting fire alarm switch by using a suitable bias to the device so that it switches at a particular temperature. One of the original uses of the chalcogenide glass was not to use the switching effect, however, but to use the rapid changes in conductivity with temperature (Edmond et al.[260]) as a thermometer, particularly in the liquid state, in the harsh environment of a nuclear reactor. Even in neutron doses of 1.8×10^{20} cm^{-2} for 100 days the devices performed well as resistance thermometers with very little hysteresis due to irradiation.

6.2.7 Pressure sensors and strain gauges

The effect of pressure on the various switching parameters has been reported by Calella[283]. Thomas and Warren[284] show reasonable correlation between a thermal model and experimental results of hydrostatic pressure, particularly in the pre-switching I–V characteristics. However, the effects of uniaxial pressure are more difficult to explain. Bosnell and Thomas[165] have shown that it is possible to induce a threshold device to go into the memory state by uniaxial pressure alone. Whilst this pressure is being applied, a RESET current pulse can return the device to its OFF state. Relaxation of the stress also returned the device to the OFF state. The procedure was cyclable. Mathur and Arntz[285] observe a steady decrease in the delay to switching with applied hydrostatic pressure and observe a rate of decrease of V_T of 1.2 mV per atmosphere pressure. Hence the possibility exists of developing active pressure gauges biasing the device which switch when a particular applied pressure exists.

6.2.8 Others

The two major possible applications still appear to be memory elements and latches in displays. The other possibilities outlined above are much more speculative. Many further applications are outside present reported device handling capabilities, for example, the control and routing of power loads where high voltages, about 500 V, and high currents, about 5 A, would need to be handled. Devices capable of suppressing the large voltages which can occur when switching on inductances in D.C. circuits, the so-called diode-quenching elements for circuit breaker coils, are of interest but relatively high voltages (250 V, say) at high currents (3 A, say) prohibit the use of glassy semiconductors currently. Applications currently occupied by devices such as selenium and silicon carbide voltage-transient suppressors usually require a capability to handle relatively high currents. Such a device is the GE–MOV metal–oxide varistor[286]. The general area of circuit breakers, usually gas discharge or vacuum arc usually requires high power-handling capabilities so that much development will be required for the glassy switches to compete in that market.

The general area of applications currently met by thermistors[287]; temperature control, power control, surge suppression, liquid level and so on, could be serviced by some amorphous device properties, but again the manufacturing processes for the conventional devices are well established.

6.2.9 Three terminal devices

Between Lilienfeld's FET patent (1930)[288] and Weimer's thin film transistor[289] (1962), a whole series of three terminal solid state amplifying devices were announced. In the past fifteen years it appears that most new devices have been diodes of which the monostable and bistable switches are two. As indicated above, engineers prefer to control high currents with a low current, as in the electro-mechanical relay. The temperature sensitivity of V_T has been utilized by ECD[273] to produce a four terminal device where a thin film nickel-chromium resistor element can be used to heat up the switching element, thus obtaining control on V_T. A device similar to the three-terminal concept has been reported by Madan et al.[290] using $Si_{18}Te_{45}As_{28}Ge_9$ glass. An extra electrode is included between the usual two electrodes in a planar geometry. The device can be turned on or be prevented from switching by application of an appropriate pulse to this third electrode. Thus, unlike the ECD device or that of Schuöcker[291], the devices do not rely upon heating, but the formed filament geometry across the gaps does influence the performance. Egerton[292] has shown a field effect exists in thin chalcogenide films but the effect is small. Measurements and theory by Tick et al.[293] confirm this, as do data on amorphous Si films[294]. Hence, the possibility of FET's in the chalcogenides looks remote unless new materials are found. The threshold switching glass ($Te_{29}As_{36}Si_{17}Ge_7P_1$) has been used as the emitter of a heterojunction transistor[295] which, apart from indicating that in the ON state the conduction species is electrons rather than holes (it is the opposite way round in the OFF state in general), may have device application as a ternary logic system. Three states of the device exist; the OFF state as well as the normal transistor states. However, with binary systems favoured, there may be some engineer resistance to this proposal.

6.3 Other Switching Devices

Evans et al.[296] have reported a device they name the Ovonic Adaptive Memory, such that a continuous range of resistance values between the ON and OFF states is achieved. The amount of energy at RESET determines this resistance value. This appears therefore related to the model of switching action proposed by Bosnell and Thomas[165] and memory action in particular proposed by Cohen et al.[164], the resistance reached being an intermediate stage of the reset process. Shaw[297] has reported the fabrication of a thin film tunnel diode using p-type $As_{34}Te_{28}Ge_{16}S_{21}S_1$. It is inferred from the data that switching speeds of 10^{-8} sec should be achievable. Here again there is no immediately obvious particular advantage over the silicon counterpart.

The chalcogenides may have applications in R.F. switching. Their conductivity increases with increasing frequency[298]. Stone et al.[299] have reported the use of bulk pieces of $Si_3Ge_4As_{38}Te_{55}$ as a resistance load in $50\,\Omega$ stripline as an R.F. switch. An ON/OFF ratio of 10 is reported at an R.F. level of 500 mV r.m.s. at 300 MHz. At X band (9.4 GHz) Staiger et al.[300] have shown that the microwave conductivity remained constant at field strengths below a critical value near $10^4 V\,cm^{-1}$. Above this value, which depends upon the microwave pulse length, switching to a low resistance state was observed, with thermal breakdown as the switching initiation mechanism.

7. Mass Memory Proposals using Amorphous Materials

Table 6 (after Ovshinsky and Klose[301]) indicates possible mechanisms whereby the amorphous to crystalline phase transition can be affected, and

Table 6

Mechanism for Information Retrieval and Display by Structure Transformation (After Ovshinsky and Klose[301])

Energy barrier can be reduced by any of the following—applied singly or in combination:	Amorphous/crystalline transformation produces changes in:
Light 　Heat 　Electric field 　Chemical catalyst 　Stress-tension pressure	Resistance Capacitance Dielectric constant Charge retention Index of refraction Surface reflection Light absorption, transmission and scattering Differential wetting and sorption Others

the resultant changes in physical properties which can be utilised for information storage and display. In this section concentration will be on the electron beam and light, particularly laser light, interaction with amorphous semiconductors. Selenium in its vitreous state can have a polymeric structure, consisting of disordered chains with an average chain length of 10^4 atoms. Heating alone will crystallise Se at above about 70°C but there is an increase in growth rate of the crystallites under light irradiation. Dresner and Stringfellow[302] demonstrate that this enhancement arises because the light produces electron-hole pairs in the amorphous state so that the growth of the crystallites is controlled by the free hole flux at the crystal boundary. They postulate that the effect should be apparent in other polymeric and amorphous semiconductors. However, it should be noted that the addition of certain elements to Se, for instance sulphur, can lead to a diminution in the crystallisation rate (Paribok-Aleksandrovich[303]).

Later work by Feinleib et al.[304] shows that sputtered amorphous thin films of $Te_{81}Ge_{15}Sb_2S_2$ undergo rapid crystallisation and revitrification when exposed to short laser pulses (1–16 μsec) with peak intensities at 5145 Å of 10 to 100 mW. They state that the crystallisation of the bulk material by purely thermal means is of the order of minutes, whereas the light-induced crystallisation is less than or of the order of microseconds, compared with normal electrically induced crystallisation in the memory switch which takes milliseconds. Following Dresner and Stringfellow[302], they postulate that the light has a twofold role in enhancing crystallisation. Firstly, the recombination of light-produced carriers raises the temperature of the glass well above T_g where atomic mobility is high. Secondly, the very creation of such carriers gives an efficient way of weakening the bonding. There is a substantial shift in the absorption edge to lower photon energy when the material undergoes an amorphous to crystalline phase change with a resulting change in reflectivity and transmission (Feinleib and Ovshinsky[305]). This effect therefore forms the basis of an optical memory system which is potentially of high density, a value of 2×10^7 bits cm^{-2} being reported by Ovshinsky and Fritzche[306] using a mechanical system akin to magnetic disc storage techniques. This is clearly not the maximum theoretical bit density achievable. Neale and Aseltine[247] indicate that a mass store in the region of 10^{11} bits with an access time of 1 sec at a cost of 0.0004 cents per bit should be achievable. Erasure of the crystalline regions, or revitrification, has also been demonstrated by Hamada et al.[307]. The energy pulse width data for writing and erasure using a He–Cd laser is similar to the memory switch energy balance problem. Similar results have been reported on $As_{25}Se_{45}Ge_{30}$ (Igo and Toyoshima[308]) using an He–Ne laser (6328 Å). It is interesting to note that Smith[309] has proposed a similar storage system using VO_2 films. The addition of As to Se films, reduced the photosensitivity whilst Te doping caused only minor changes[310]. Modifications to the crystallisation rate have been noted by Clement et al.[311], particularly caused by the nature and temperature of the substrate during deposition. For example, on gold, crystallisation is inhibited near the gold interface. Asahara and Izumitani[312] have shown that additions of Cu to As–Se glasses increases the concentration of the photo-darkening effect, related to the decrease in transmission in the visible region upon irradiation. Further work[313] has shown that the darkening of As–Se glasses containing additions of the elements Zn, Sn, Pb, and Cd was lower than in systems containing Cu, Ag and Tl. A study of the structure of such laser-induced crystalline areas has been reported by Chaudhari et al.[314]). Whilst agreeing that reversible amorphous to crystalline phase transformations are achieved by laser irradiation, they conclude that the difference in optical properties, reflection and transmission, is not entirely due to such a phase change but is partly associated with the film morphology. Density fluctuations in the film lead to light scattering in transmission giving rise to a difference in transmissivity between the two states. Indeed, voids, cavities, and Te crystallites were observed in a $Ge_{15}Te_{81}As_4$ alloy. None of the above papers is clear on the number of write-erase cycles which are possible in such a system. Data reported by Feinleib et al.[315] on chalcogenide films under laser irradiation show that bubbles or voids can be formed in the interior of the film which lead to a change in the transmissivity, in agreement with Chaudhari et al.[314]. Von Gutfeld and Chaudhari[316] have shown that, starting with crystalline $Te_{80}Ge_{15}As_5$ films, they have been able to switch to the amorphous state with 5 μsec laser pulses, a system of operation they have termed "reverse mode". When operated in this fashion the chalcogenides appear to have promise as an optical fast-write slow-erase memory, with bit rates of 2×10^8/sec being possible, given that the laser beam can be deflected sufficiently fast (10^5 cm/sec). After 250 full cycles the medium remained essentially unaltered. On the other hand, in a variety of materials, Korsakov et al.[317] report a marked decrease in diffractive efficiency following the second and subsequent erasures of stored information. It has been pointed by de Neufville et al.[318] that the reversible photo-darkening effect is accompanied by negligible refractive index change compared with the irreversible polymerisation changes in As_2Se_3 and As_2S_3 films and their review is a particularly valuable one. The stores proposed, based on the effects described above, generally involve the use of a secondary detector to observe changes in reflectivity or transmissivity. A further optical storage technique is the hologram whereby an interference grating of the radiation field is produced. A review of holography in information storage is given by Ramberg[319]. Such a grating can be produced if a phase transformation involves a change in absorption coefficient and refractive index in a local area under, say, laser irradiation. Generally these have not been recyclable media, but a cyclable system was proposed by Roach[320] using the phase change in VO_2 thin films, which, as pointed out above, is an analogous system to the chalcogenides. The latter have been utilised (Keneman[321]) to give relatively high holographic efficiencies (80% for 10 μm thickness read at 6328 Å). The exact phase transformation in As_2S_3 favoured by Keneman[321] was the following reaction:

$$2As_2S_3 \leftrightarrow 2S + As_4S_4$$

as opposed to an amorphous-crystalline transition. Work by Pearson and Bagley[322] on the mechanism of hologram formation in As 7 mole%, S 93 mole% glass shows that after hologram formation, the material consists of a glassy matrix with microcrystals of almost pure α-S. The photo-induced change is again markedly faster than the equivalent thermal one. Laser-induced refractive index changes have been reported (Ohmachi and Igo[323]) for the glasses As_2S_3, $As_{30}S_{70}$, $As_{15}Se_5$, $As_{30}S_{60}Ge_{10}$. The maximum change in refractive index (0.056) was obtained for As_2S_3. Adem and Bismith[324] report holographic recording in a thin amorphous film of an unspecified Se–Te alloy written at 4880 Å and reconstructed at 6328 Å. Similar work, but moving into the infrared, using a ruby laser at 0.69 μm and a neodymium laser at 1.06 μm is reported by Belokrinitskii et al.[325] on GeTe and InSb amorphous films. Similar work has been reported by Gurevich et al.[326] on

As–Se and As–Se–I films. Their data show that the diffraction efficiency is constant after the first erasure cycle, i.e. even here some forming is taking place.

Dakss and Sadagopan[327] have outlined devices similar to those above, based upon the amorphous-to-crystalline phase changes in $As_2Se_3 As_2S_3$ and $As_{1.4}Sb_{0.6}Se_3$. They also indicate that writing with an electron beam is equally possible. This idea is developed in more detail by Chen et al.[328] using thin films of $Ge_{11.5}Te_{57.5}As_{31}$ and $Ge_{12}Te_{50}As_{38}$. The latter authors propose using the difference in conductivity of the two states of the material by measuring, at a lower beam current than the crystallisation current, the difference in secondary electron yield from the two phases, in much the same way as with one of the modes of the scanning electron microscope, can be used for fault-finding in integrated circuits. Experimentally, this technique is demonstrated, but no details of the maximum number of write-erase cycles is given. Experiments by Mèe[329] on $Se_{80}As_{20}$ and $Se_{64}As_{16}Au_{20}$ thin films show that in the former system localised electron-beam heating failed to induce crystallisation prior to melting and/or vaporisation. However, the gold-doped films were crystallised at current densities of $1.8\ A\ cm^{-2}$ into α-AuSe and As platelets. Above $2.4\ A\ cm^{-2}$ rapid vaporisation of the Se and As occurred, followed by flash evaporation at $3.8\ A\ cm^{-2}$.

8. Acousto-optical Properties

Dakss and Sadagopan[327] briefly take the pattern of amorphous and crystalline phases produced by light irradiation a stage further. They propose that the difference in velocity of surface sound waves in these two regions can provide a waveguide effect in desired directions. It is of interest to note in this context that acoustic loss in some chalcogenide glasses, particularly the Ge–As–S system, for instance $Ge_{30}As_5S_{65}$, can be as low as in fused silica (Krause et al.[330]). These authors point out that the materials also have a high acoustic loss and high figure of merit, approximately 150 times that of fused silica. This combination of low-loss acoustic loss and high figure of merit, especially an isotropic medium would, they believe, make very attractive acousto-optic devices such as infrared modulators, deflectors and optical correlators. This is because of the high transmission of these glasses in the IR (Savage and Nielsen[98], Hilton[99,331] and Ohsaka[332]) and their extensive use as IR windows in a variety of applications which are not reviewed in this article. Furthermore, the low values of the sound velocity (approximately $2.5 \times 10^5\ cm\ sec^{-1}$ for longitudinal waves) lead to a shorter length delay line for a given delay time than the fused silica. These proposals have been given some experimental verification by Ohmachi and Uchida[333] who used As_2Se_3

as the modulator of 1.153 µm laser radiation at an acoustic carrier frequency of 140 MHz. It is interesting to note that some chalcogenide crystals have very low acoustic velocities and high acousto-optic figures of merit[334] so that ultimately the range of useful amorphous chalcogenide glass materials in this field may be large. Furthermore, the use of chalcogenide glass films on lithium niobate has been reported by Zigel et al.[335] as yielding ultrasonic dispersive waveguides with a centre frequency of 56 MHz and an insertion loss of less than 60 dB across a 32 MHz band. Also, the properties of sputtered films of As_2S_3, $Ge_{28}Sb_{12}Se_{60}$, and $Ge_{33}As_{12}Se_{55}$ as waveguides at 1.064 µm have been reported by Watts et al.[336]. Propagation losses of less than 1 dB/cm have been achieved for As_2S_3 films. With the increased activity in integrated optics these phenomena appear to have great potential.

9. IR Detectors

The strong interest in the chalcogenides as materials for use in the IR, outlined above, is further strengthened by the potential of some chalcogenide materials as IR detectors. Bishop and Moore[337] have reported a thin film thermistor bolometer based on $Tl_2SeAs_2Te_3$. Although mica, glass and sapphire substrates were used, the best results were achieved using 10 µm thick chalcogenide on mica, achieving a detectivity D^*, of about 10^7 W^{-1} $Hz^{\frac{1}{2}}$ cm, which is an order of magnitude less than current production ceramic pyroelectric detectors. The major problem with the chalcogenide bolometer is the slow response time of about 1 second at room temperature. However, using amorphous Bi–Ge alloys, Vass and Anderson[338] have reported response times of the order of 1–2 msec. When cooled to 162 K they report similar D^* values to those of the thermistor device. Hence, the composition does not yet appear to have been optimised. The reader is referred to Chapters 9 and 12 for further discussions of IR detectors.

10. Imaging Systems

The amorphous chalcogenides have traditionally been strong in the area of image-to-video signal converters or transducers in television-like systems, for example, the vidicon (see also Chapter 12 for a discussion of more traditional photoconductors in this connection). These may be sensitive to visible or IR or, in the medical field especially, x-rays. Sensitivity to electrons is required in electron microscopy for example. The general mode of operation is that the image becomes a stored charge pattern on a suitable target, and the resulting required video signal is achieved by scanning this charge pattern with a low energy electron beam. Assuming a homogeneous target, homogeneously charged, the rate of charge leakage through the film at a specific place will depend upon the brightness at that point. The charge pattern or electrical image so-formed is raster scanned with the electron beam which thus replenishes leaked charge. The recharging current generates an equivalent voltage across the load resistor R_L, hence scanning gives a pulse train across the resistor, depending upon the stored charge pattern which constitutes the video output signal. Each image is effectively a small capacitor C whose leakage current is light modulated. The $R_D C$ (where R_D is the dark resistance of the target) time constant of this arrangement must be matched to the scan interval (0.04 sec). The leakage current, i_L, is related to the dark resistivity ρ thus:

$$i_L = \frac{AE}{\rho}$$

where A = target area, and E = bias field. For typical materials, film thicknesses and areas, the capacitance is of the order of 5 nF so that R_D must be 10 MΩ for compatibility, that is, a dark resistivity of greater than 5×10^{11} Ωcm giving a leakage current at reasonable bias field (say 20 V cm^{-1}) of less than 20 nA. Hence the material for the target must meet this specification and, furthermore, good target/film homogeneity is required if high spots are to be avoided. Several possible modes of operation exist involving the use of decelerating grids, this system became known as the "orthicon" in the U.S.A., or the "emitron" in the U.K. Sometimes excess conductivity is induced in the target by energetic electrons penetrating through the back plate, exciting long-lived carriers, the so-called EBIC mode, or electron bombardment induced conductivity, giving rise to the "Ebicon" tube. Another development has been the use of the high secondary emission yield of some materials which can have long storage times, the SEC (or secondary electron conduction) tube. In some cases, ohmic contacts have not been used but blocking electrodes used instead, as in the "Plumbicon" based on a PbO target. These developments are reviewed in detail by Gordon[339] and Koelmans[340]. Table 7 (after Gordon[339]) illustrates the range of materials which have been employed. The high value of dark-resistivity required has necessitated the use of the amorphous semiconducting glasses Sb_2S_3, As_2Se_3 and As_2S_3. The PbO films used in this application have grain sizes less than 100 Å. "Burn-in" relates to exposure of the target brightness, for example full sunlight, particularly under zoom conditions and highly focused shots. Not surprisingly, the relatively low T_g of some of the glass, particularly Se (42°C), leave them very susceptible to this problem. The material in the plumbicon appears to be very difficult to work with because of its structural sensitivity and the number of possible oxidation states, both of which are awkward to control. Any new target material must have a good photo-

Table 7

Summary of Vidicon Photoconductors (After Gordon [339])

Response	Material (amorphous film)	Comments
Visible	Sb_2S_3 Se PbO	Most common Poor burn-in p–n junction, poor red response
IR	Sb_2Se_3 PbS PbO–PbS	 Straight photoconductor, poor sensitivity in IR
UV	Se Sb_2S_3 As_2Se_3	Poor burn-in Insensitive Laggy, good burn-in
X-ray	Se PbO	Can have large area targets
EBIC	Se As_2S_3 PbO ZnS	Poor burn-in p–n junction
SEC	KCl	Extremely long storage

response, a high dark-resistivity and good long-term charge storage. These requirements appear to be met by glow discharge deposited amorphous silicon (Chittick et al.[36]) but not by evaporated or sputtered silicon, since the resistivity is some orders of magnitude lower (Walley[33], Brodsky et al.[341]). However, all new developments must be viewed in the light of the new range of targets based on silicon technology which have sprung up under the impetus of the picture-phone requirement. Large-scale integration of over a quarter of a million back-biased p–n diodes on a silicon slice has been achieved such that the light modulates the leakage current of the diodes, thus producing an acceptable vidicon (Crowell et al.[342]). Even these latter devices require electron beam scanning. However, newer techniques using developments of MOS technology, like charge coupled devices, offer wide scope (see, for example, Proceedings CCD, 74, Edinburgh University) with on-chip readout capabilities. One prognosis (Allan[343]) is that the CCD dark currents of 5–10 nA cm^{-2} will eventually be reduced to less than 1 nA cm^{-2}, which is already exhibited by the Plumbicon tubes, with the potential of 500 × 500 elements by 1980. Hence, although amorphous semiconductors are well entrenched in this area, new developments in silicon processing are making the field increasingly competitive.

11. Photographic Effects

As indicated in Table 6, upon exposure to light, changes take place in the transmission of amorphous films. This gives rise to the possibility of producing a photographic film from amorphous material deposited upon glass or flexible substrates like paper of mylar. A wet process aimed at microfilm and the graphic arts field has been announced by ECD[334] which requires wet-processing but has the advantage of image contrast enhancement. The resolution is determined by the number and size of crystallites in the exposed area, which in turn depends upon the film thickness and composition. Typically, 10^7 per cm^2 crystallite grains of 1 to 10 μm have been achieved, giving a resolution of 500 lines per millimetre. The possibility of dry processing, based on exposure to heat gives the potential of an instant-process camera. The production of a multilayer photo sensor, thermally developable, has been reported by Shimizu et al.[345]. This consists of an acceptor layer of $As_{10}Se_{80}Te_{10}$ and a barrier layer of As_2S_6 100 Å thick with Ag deposited on top. Exposure to light causes Ag to migrate into the barrier layer, and subsequent thermal treatment causes the Ag to diffuse into the acceptor. Changes in the optical properties of the material result, which form the basis of a sensor.

12. Xerography

No review of amorphous semiconductor applications would be complete without some reference to the largest, in terms of turnover (> £600M), application: that of Xerography or "dry writing" process. This topic has been reviewed in great depth by Dessauer and Clark[346] and by Vincent in Chapter 12 of this book. It provides an example of a technology which was in advance of a complete understanding of the material used, that is, vitreous Se, and also used electrostatic discharge techniques when they had been neglected for some time. For a review of the properties of Se, see Zingaro and Cooper[347]. Briefly, the process is as follows. In the dark, a positive electrostatic charge is applied to the Se surface. An optical image of the matter to be printed is projected onto the Se layer, causing the charge to leak away from the illuminated areas. Thus an image composed of charged and uncharged areas is produced as with the vidicon above. Aspects of photoconduction in amorphous materials have been reviewed by Bube[348] and Mott et al.[349]. Data on the mechanisms of discharge of such photoreceptors are reported by Mort et al.[350] and Chen[351]. This image is now "developed" or made visible by sprinkling over it positively charged particles of "toner" from which the final printed image is formed. Obviously the toner is only attracted to the uncharged areas which have been illuminated. Negatively charged plain paper then attracts the toner as it passes over it and the image is made permanent by fusing the toner to the paper thermally. The Se surface is then prepared for re-use by illuminating the surface evenly all over and brushing away any unused toner particles. The advantages of these processes are primarily that the photoconductive surface is only a temporary carrier of the image and hence can be re-used many times. Also, the toner is transferred to unprepared paper thus reducing operating costs. Alternative systems, apart from ones based on the chalcogenides, will not be reviewed here but are the subject of a recent review of electrophotographic systems by Comizzoli et al.[352]. The potential of organics in this area should not be ignored[353]. Charge retention characteristics again require high dark resistivities greater than $10^{12}\,\Omega$ cm, and a move to oxide glasses with high photoconductivities (e.g. Strickler and Roy[354]) may be required.

Three major problems are extant with the xerox system in current use. These are a poor panchromatic photoresponse, relatively poor thermal stability of the image layer, and a relatively poor reproduction of half-tones. Some steps have been taken to tackle the first two problems by the use of Bi additions to the Se, which broadens wave-length response compared with pure Se, and increases the glass transition temperature a little to 55°C (Schottmiller et al.[355]). During the charging process, breakdown of the Se layer can occur in a filamentary manner[356] akin to the forming process outlined above. Hence care must be taken to use uniform films.

Some data have been reported (Donovan[357]) on the x-ray discharge of Se which indicate that, in the typical medical x-ray spectrum, the overall efficiency of the xerox process is relatively high. This obviously leads to on-line rapid x-ray images which would be a useful diagnostic tool. Xeroradiography has been comprehensively reviewed by Boag[358].

ECD have proposed an alternative process which they call ovnographic printing (Henisch[267]), using a memory material so that the primary image is written, say, with a laser beam as in the store case above. Erasure can similarly be effected locally, or total erasure using general heating could be employed. Printing would then proceed as with the electrostatic system outlined above. Direct control of the laser to take data from a computer is proposed. Such a so-called line printer was also outlined by the Rank Xerox

Company but employed standard xerox printing, using a CRT as the converter from computer output. They proposed a linear paper speed of 40 feet per minute, equivalent to about 5000 characters per second. A similar result is achieved by de Rosa[359] using chalcogenide glasses. The potential applications in banking, mail-order business and general filing indicate a large new market for such a copying system. The potential for facsimile reproduction (FAX) has been reviewed by Kaplan[360].

13. Conclusions

If this review is weighted towards switching it reflects the author's bias and, to some extent, the area in which most applications work has been concentrated recently. However, many potential applications areas, for example, amorphous read mainly memories and amorphous vidicon targets, have forced the silicon technologist to develop new products. It is ironic that an explanation for some of the properties of a rival to the amorphous RMM, the MNOST, should be found in the physics of amorphous materials (Ref. (255) and (256)). But this increased understanding of solid state physics, brought about by the study of the amorphous state, is one of its greatest contributions. The field of glass technology has benefited immensely by the spotlight thrown on that area. It is clearly going to be a matter of commercial judgement, having reviewed the systems trade-offs, to decide which technology will be used for a particular application. But where large areas of material are required, as in the printing technologies, or where amorphous materials offer a physical property like high infrared transmittance, that they have their major advantage. It may well be that these, rather than switching, will be the major applications areas. However, new effects are continually being noted, for example, Spear and Le Comber[361] have reported substitutional doping of amorphous Si and the use of a chalcogenide glass in sensing ferric ion concentration[362]. These, and the wide range of available properties, make this a very exciting field. From work on amorphous Si deposited on crystalline Si, Brodsky et al.[363] have shown that there may be device potential in the silicon technology field, as Schottky barriers, for example, so that rather than replace silicon devices, amorphous films may be an adjunct to them. Already such films are being used as photosensitive elements in solar cells and, as discussed in Chapter 10, this approach may provide an opportunity for the large-scale exploitation of solar energy for terrestrial uses.

Acknowledgements

This chapter is published with permission Director R.S.R.E., copyright H.M.S.O.

References

1. See for example C. Kittel, "Quantum Theory of Solids", Wiley, New York, (1963).
2. N. F. Mott and W. D. Twose, *Adv. in Phys.*, **10**, 107, (1961).
3. B. I. Halperin, *Adv. Chem. Phys.*, **13**, 123, (1967).
4. N. F. Mott and E. A. Davis, "Electronic Process in Non-crystalline Materials", Oxford University Press, (1971).
5. J. H. Dessauer and H. E. Clark, "Xerography and Related Processes", Focal Press, London, (1965).
6. S. R. Ovshinsky, U.S. patent 3271591, (1966).
7. N. F. Mott, *Electronics and Power*, **19**, 321, (1973).
8. N. F. Mott, *Inst. of Physics Bulletin*, **7**, 451, (1974).
9. J. D. Watson, "The Double Helix", Weidenfeld and Nicolson, London, (1968).
10. I. Mitroff, "The Subjective Side of Science", Elsevier, North Holland, (1974).
11. S. R. Ovshinsky, *Phys. Rev. Letts.*, **21**, 1450, (1968).
12. J. M. Stevels, *J. Non-Cryst. Sol.*, **6**, 307, (1971).
13. H. J. Queisser, *J. Appl. Phys.*, **42**, 5567, (1971).
14. M. J. de Wit, *J. Appl. Phys.*, **43**, 908, (1972).
15. R. C. MacKenzie, "Differential Thermal Analysis", Academic Press, London, (1970).
16. N. Klein, Proc. Low Mobility Conf., Taylor and Francis, London, 229, (1971).
17. W. Beyer and J. Stuke, *Phys. Stat. Sol.*, **30**, 511, (1975).
18. D. Adler, *Crit. Rev. Solid State Sci.*, **2**, 17, (1971).
19. A. E. Owen, *Contemp. Phys.*, **11**, 227, (1970).
20. A. K. Jonscher and R. M. Hill, in "Physics of Thin Films", Vol. 8, eds., G. Hass, M. H. Francombe and R. W. Hoffman, Academic Press, London, (1975).
21. P. W. Anderson, *Phys. Rev. Letts.*, **34**, 953, (1975).
22. B. G. Bagley, in "The Nature of the Amphorous State", ed., J. Jane, Plenum Press, New York, (1974).
23. D. Turnbull, *Contemp. Phys.*, **10**, 473, (1969).
24. H. Rawson, "Inorganic Glass-forming Systems", Academic Press, London, (1967).
25. J. A. Savage, *J. Mater. Sci.*, **6**, 964, (1971).
26. H. S. Chen and C. E. Miller, *Rev. Sci. Inst.*, **41**, 1237, (1970).
27. H. Jones, *Rep. Prog. Phys.*, **36**, 1425, (1973).
28. J. R. Bosnell, *Thin Solid Films*, **3**, 233, (1969).
29. J. C. Suits, *Phys. Rev.*, **131**, 588, (1963).
30. K. H. Behrndt, *J. Vac..Sci. Tech.*, **7**, 385, (1970).
31. H. A. Davies and J. B. Hull, *Scripta Metallurgica*, **7**, 637, (1973).
32. K. L. Chopra, "Thin Film Phenomena", McGraw Hill, New York, (1969).
33. P. A. Walley, *Thin Solid Films*, **2**, 327, (1968).
34. R. Glang, in "Handbook of Thin Film Technology", eds., L. I. Maissel and R. Glang, McGraw Hill, New York, (1970).
35. J. R. Bosnell and U. C. Voisey, *Thin Solid Films*, **6**, 161, (1970).
36. R. C. Chittick, J. H. Alexander and H. F. Stirling, *J. Electrochem. Soc.*, **116**, 77, (1969).
37. H. S. Chen and D. Turnbull, *J. Appl. Phys.*, **40**, 4214, (1969).

38. J. Tauc, A. Abraham, R. Zallen and M. Slade, *J. Non-Cryst. Sol.*, **4**, 279, (1970).
39. N. S. Saks, D. F. Barber and G. W. Anderson, IEEE Trans. Parts Hybrids and Packaging *PHP*-10, 244, (1974).
40. D. M. Newbury and D. L. Kirk, *Microelectronics*, **6**, 17, (1974).
41. J. R. Bosnell and C. B. Thomas, *Solid State Electron.*, **15**, 1261, (1972).
42. G. Carter and J. S. Colligan, "Ion Bombardment of Solids", Spon, London, (1968).
43. H. F. Winters and E. Kay, *J. Appl. Phys.*, **38**, 3928, (1967).
44. E. A. Fagen, *Mat. Res. Bull.*, **7**, 279, (1972).
45. See Du Pont Manufacturers Literature, (1973).
46. J. K. Higgins, J. E. Lewis, M. Lower and F. V. Roue, *J. Non-Cryst. Sol.*, **18**, 77, (1975).
47. I. Balberg, B. Abeles and Y. Aric, *Thin Solid Films*, **24**, 307, (1974).
48. M. Regan and C. F. Drake, *Mat. Res. Bull.*, **7**, 1559, (1972).
49. L. A. Ryabova and Ya. S. Savitskaya, *J. Vac. Sci. Tech.*, **6**, 934, (1969).
50. H. Krautle, *Rad. Effects*, **24**, 255, (1975).
51. F. F. Morehead and B. L. Crowder, *Rad. Effects*, **6**, 27, (1970).
52. M. A. Nicolet, J. W. Mayer and I. V. Mitchell, *Spectrometry Science*, **177**, 841, (1972).
53. H. M. Naguib and R. Kelly, *Rad. Effects*, **25**, 1, (1975).
54. G. Gotz and R. Eudter, *J. Non-Cryst. Sol.*, **13**, 286, (1973/4).
55. G. W. Anderson, J. E. Davey, J. Comas, N. S. Saks and W. H. Lucke, *J. Appl. Phys.*, **45**, 4528, (1974).
56. W. Beyer, J. Stuke and H. Wagner, *Phys. Stat. Sol. (a)*, **30**, 231, (1975).
57. R. J. Joyce, H. F. Sterling and J. H. Alexander, *Thin Solid Films*, **1**, 481, (1967/8).
58. J. Wong, P. Rao, and E. F. Koch, *J. Appl. Phys.*, **46**, 1827, (1975).
59. L. M. Levinson and H. R. Philipp, *Appl. Phys. Lett.*, **24**, 75, (1974).
60. J. R. Bosnell and J. A. Savage, *J. Mater. Sci.*, **7**, 1235, (1972).
61. D. A. Northrop, *Solid State Comms.*, **7**, 147, (1972).
62. R. T. Johnson, D. A. Northrop, and R. K. Quinn, *Solid State Comms.*, **9**, 1397, (1971).
63. S. V. Phillips, R. E. Booth, and P. M. McMillan, *J. Non-Cryst. Sol.*, **4**, 510, (1970).
64. T. Takamori, R. Roy, and G. J. McCarthy, *J. Appl. Phys.*, **42**, 2577, (1971).
65. S. Iizima, N. Sugi, M. Kikuchi and K. Tanaka, *Solid State Comms.*, **9**, 795, (1971).
66. N. Utsumi and M. Wada, *Jap. J. Appl. Phys.*, **10**, 79, (1971).
67. D. B. Dove, in "Physics of Thin Films," Vol. 7, eds., G. Hass, M. H. Francombe and R. W. Hoffman, Academic Press, London, (1973).
68. R. Grigorovici, in "Electronic and Structural Properties of Amorphous Semiconductors," eds., P. G. LeConber and J. Mort, Academic Press, London, (1973).
69. A. J. Apling, in "Electronic and Structural Properties of Amorphous Semiconductors," eds., P. G. LeComber and J. Mort, Academic Press, London, (1973).
70. M. C. Flemings, *Scientific American*, **231**, 88, (1974).
71. R. Roy and V. Caslavska, *Solid State Comms.*, **7**, 1467, (1969).
72. P. C. Taylor, S. G. Bishop and D. L. Mitchell, *Phys. Rev. Letts.*, **27**, 41, (1971).
73. R. J. Hodgkinson, *Phil. Mag.*, **22**, 1187, (1970).
74. H. A. Davies, J. Ancote and J. B. Hull, *Nature*, **246**, 13, (1973).
75. R. L. Hines, and A. Howie, *Phil. Mag.*, **32**, 257, (1975).

76. T. M. Donovan and K. Heineman, *Phys. Rev. Letts.*, **27**, 1794, (1974).
77. P. Nath, S. K. Suri and K. L. Chopra, *Phys. Stat. Sol.* (a), **30**, 771, (1975).
78. A. Barna, P. B. Barna, Z. Bodo, J. F. Pocza, I. Pozsgai and G. Radnoczi, *Thin Solid Films*, **23**, 49, (1974).
79. M. L. Knotek, *J. Vac. Sci. Tech.*, **12**, 117, (1975).
80. A. J. Apling, M. F. Daniel and A. J. Leadbetter, *Thin Solid Films*, **27**, L11, (1975).
81. H. Morisaki, H. Iwasaki and K. Yazawa, *Appl. Phys. Letts.*, **26**, 294, (1975).
82. S. C. Agarwal, *Phys. Rev. B*, **7**, 685, (1973).
83. A. H. Clark, *Phys. Rev.*, **154**, 75, (1967).
84. R. C. Chittick, *J. Non-Cryst. Sol.*, **3**, 255, (1970).
85. C. T. Wu and H. L. Luo, *J. Non-Cryst. Sol.*, **18**, 21, (1975).
86. R. Coates and E. W. J. Mitchell, *Adv. Phys.*, **24**, 593, (1975).
87. R. M. Waghorne and J. R. Bosnell, *J. Non-Cryst. Sol.*, **15**, 107, (1974).
88. B. A. Hatt, J. A. Savage, J. R. Bosnell and R. M. Waghorne, Proc. 5th Intl. Conf. on Amorphous and Liquid Semi-conductors. Sept. 1973, Vol. 1, p. 413, eds., J. Stuke and W. Bremig, Taylor and Francis, London, (1974).
89. S. G. Bishop and N. J. Shevdick, *Phys. Rev. B*, **12**, 1567, (1975).
90. D. Brown, D. S. Moore and E. F. W. Seymour, *J. Non-Cryst. Sol.*, **8-10**, 256, (1972).
91. J. T. Edmond, *J. Non-Cryst. Sol.*, **1**, 39, (1968).
92. H. Fritzche, *J. Non-Cryst. Sol.*, **6**, 49, (1971).
93. M. L. Theye, Proc. 5th Intl. Conf. on Amorphous and Liquid Semi-conductors. Sept. 1073, Vol. 1, pp. 479–498, eds., J. Stuke and W: Bremig, Taylor and Francis, London, (1974).
94. E. A. Davis, Proc. 5th Intl. Conf. on Amorphous and Liquid Semi-conductors, Sept. 1973, Vol. 1, pp. 519–532, eds., J. Stuke and W. Bremig, Taylor and Francis, London, (1974).
95. A. I. Kolyadin and A. N. Kashintseva, *Optical Technology*, **40**, 759, (1973).
96. A. W. Dozier, L. K. Wilson, E. J. Friebele and D. L. Kinser, *J. Am. Ceram. Soc.*, **55**, 373, (1972).
97. E. A. Davis and N. F. Mott, *Phil. Mag.*, **22**, 903, (1970).
98. J. A. Savage and S. Nielsen, *Infrared Physics*, **5**, 195, (1965).
99. A. R. Hilton, *J. Electronic Materials*, **2**, 211, (1973).
100. D. Sigurd, G. Ottaviani, V. Marrello, J. W. Mayer and J. O. McCaldin, *J. Non-Cryst. Sol.*, **12**, 135, (1973).
101. S. R. Herd, P. Chaudhari and M. H. Brodsky, *J. Non-Cryst, Sol.*, **7**, 309, (1972).
102. M. Morgan and A. K. Jonscher, *Thin Solid Films*, **9**, 67, (1971).
103. L. A. Freeman, R. F. Shaw and A. D. Yoffe, *Thin Solid Films*, **3**, 367, (1969).
104. S. Maruno, T. Yamada, M. Noda and Y. Kondo, *Jap. J. Appl. Phys.*, **10**, 653, (1971).
105. A. D. Pearson, *IBM J. Res. & Dev.*, **13**, 510, (1969).
106. G. A. Petrillo and K. C. Kao, *J. Non-Cryst. Sol.*, **16**, 247, (1974).
107. J. A. Savage, *J. Non-Cryst. Sol.*, **11**, 121, (1972).
108. J. R. Bosnell and C. B. Thomas, *J. Phys. D: Appl. Phys.*, **5**, L29, (1972).
109. C. de May, *Electrocomponent Science & Technology*, **1**, 39, (1974).
110. A. K. Jonscher, *J. Phys. C: Solid State Phys.*, **6**, L235, (1973).

111. U. Strom and P. C. Taylor, *Phys. Rev. Letts.*, **30**, 13, (1973).
112. IBM Technical Disclosure **15**, 577, (1972).
113. C. B. Thomas, A. F. Fray and J. R. Bosnell, *Phil. Mag.*, **26**, 617, (1972).
114. J. A. Savage, *J. Mater. Sci.*, **6**, 964, (1971).
115. G. Dearnaley, D. V. Morgan and A. M. Stoneham, *J. Non-Cryst. Sol.*, **4**, 593, (1970).
116. J. E. Ralph and J. M. Woodcock, *J. Non-Cryst. Sol.*, **7**, 236, (1972).
117. M. P. Shaw and I. J. Gastman, *Appl. Phys. Lett.*, **19**, 24, (1971).
118. H. K. Rockstad and M. P. Shaw, *IEEE Trans. Electron Devices*, **ED 20**, 593, (1973).
119. H. L. Grubin, M. P. Shaw and P. R. Solomon, *IEEE Trans. Electron Devices*, **ED 20**, 63, (1973).
120. N. Klein, *Thin Solid Films*, **7**, 149, (1971).
121. N. Klein, Proc. Low Mobility Materials Conference, Taylor and Francis, London, 299, (1971).
122. N. Klein, *Adv. Phys.*, **21**, 605, (1972).
123. M. Carchano, R. Lacoste and Y. Segui, *App. Phys. Letts.*, **19**, 414, (1971).
124. R. A. Collins, I. A. Edge and K. O. Legge, *Phys. Stat. Sol.*, **9**, 309, (1972).
125. P. O. Sliva, G. Div and G. Griffiths, *J. Non-Cryst. Sol.*, **2**, 316, (1970).
126. R. Pinto, *Phys. Letts.*, **35A**, 155, (1971).
127. V. K. Agarwal, *Thin Solid Films*, **24**, 55, (1974).
128. L. Altcheh, N. Klein and I. N. Katz, *J. Appl. Phys.*, **43**, 3258, (1972).
129. L. Altcheh and N. Klein, *IEEE Trans. Electron Devices*, **ED 20**, 801, (1973).
130. H. A. Schafft and J. C. French, *IEEE Trans. Electron Devices*, **ED-13**, 613, (1966).
131. C. Sze, "The Physics of Semi-conductor Devices", Wiley, New York, (1969).
132. D. N. Pontius, W. B. Smith and P. P. Budenstein, *J. Appl. Phys.*, **44**, 331, (1973).
133. W. B. Smith, D. N. Pontius and P. P. Budenstein, *IEEE Trans. Electron Devices*, **ED-20**, 731, (1973).
134. C. Popescu, *IEEE Trans. Electron Devices*, **ED-21**, 428, (1974).
135. C. K. Sharma and S. C. Dutta Roy, *Microelectronics and Reliability*, **11**, 499, (1972).
136. E. Lange and N. Mader, Paper B7.2 ESSDRC 1974, IOP Conference, Nottingham University.
137. A. G. Zabrodskii, S. M. Ryvkın and I. S. Shilimak, *JETP Letts.*, **19**, 290, (1973).
138. H. Gobrecht and F. Mahdjuri, *J. Phys. C: Solid State Phys.*, **5**, 366, (1972).
139. G. C. Vezzoli, A. Napier and L. W. Doremus, *J. Non-Cryst. Sol.*, **13**, 80, (1973/4).
140. A. A. Andrev, V. A. Alexev, E. A. Lebedev, M. Mamadaliev, B. T. Melekh, A. R. Regel and Yu. F. Ryzhkov, *Sov. Phys. Semi-conductors*, **6**, 570, (1972).
141. P. A. Lee, G. Said and R. Davis, *Solid State Comms.*, **7**, 1359, (1969).
142. A. H. Thomson, *Solid State Comms.*, **10**, 581, (1972).
143. D. M. Newbury and D. L. Kirk, *Microelectronics*, **6**, 17, (1974).
144. H. K. Hehisch and W. R. Smith, *App. Phys. Letts.*, **24**, 589, (1974).
145. J. R. Bosnell, *Physics in Technology*, **4**, 113, (1973).
146. B. T. Kolomiets, E. A. Lebedev and I. Takasami, *Sov. Phys. Semicond.*, **3**, 267, (1969).

147. A. E. Owen and J. Robertson, *IEEE Trans. Electron Devices*, **ED-20**, 105, (1973).
148. H. K. Henisch, R. W. Prior and G. T. Vendura, *J. Non-Cryst. Sol.*, **8-10**, 415, (1972).
149. D. D. Thornburg, *J.Electronic Materials*, **2**, 495, (1973).
150. M. A. Barakat, K. C. Kao and E. Bridges, *J. Non-Cryst. Sol.*, **18**, 209, (1975).
151. M. Saji and K. C. Kao, *J. Non-Cryst. Sol.*, **18**, 275, (1975).
152. D. Adler, *J. Vac. Sci. Tech.*, **10**, 728, (1973).
153. D. Adler, F. O. Arntz, L. P. Flora, B. P. Mathur and D. K. Reinhard, Proc. 5th Intl. Conf. on Amorphous and Liquid Semi-conductors, p. 859, (1973), Taylor and Francis, London, (1974).
154. N. Croitou, L. Vescan, C. Popescu and M. L. Lazarescu, *J. Non-Cryst. Sol.*, **4**, 493, (1970).
155. P. J. Walsh, R. Vogel and E. J. Evans, *Phys. Rev.*, **178**, 1274, (1969).
156. L. A. Coward, *J. Non-Cryst. Sol.*, **6**, 107, (1971).
157. G. V. Bunton, S. C. M. Day, R. M. Quilliam and P. H. Wisbey, *J. Non-Cryst. Sol.*, **6**, 251, (1971).
158. A. D. Pearson and C. E. Miller, *App. Phys. Letts.*, **14**, 280, (1969).
159. D. V. Morgan, M. J. Howes, R. D. Pollard and D. G. P. Waters, *Thin Solid Films*, **15**, 123, (1973).
160. J. Pivot, M. Bondeulle, A. Cachard and C. M. S. Dupuy, *Phys. Stat. Sol. (a)*, **2**, 319, (1970).
161. N. Bidadi and C. A. Hogarth, *Thin Solid Films*, **27**, 319, (1975).
162. J. P. A. Williamson and R. A. Collins, *Int. J. Electronics*, **38**, 413, (1975).
163. C. L. Dargan, P. Burton and R. M. Redstall, *Int. J. Electronics*, **38**, 711, (1975).
164. M. H. Cohen, R. G. Neale and A. Paskin, *J. Non-Cryst. Sol.*, **8-10**, 885, (1972).
165. J. R. Bosnell and C. B. Thomas, *Phil. Mag.*, **27**, 665, (1973).
166. R. F. Ormondroyd, J. Allison and M. J. Thomson, *J. Non-Cryst. Sol.*, **15**, 310, (1974).
167. R. F. Ormondroyd, M. J. Thomson and J. Allison, *J. Non-Cryst. Sol.*, **18**, 375, (1975).
168. G. V. Bunton, *Thin Solid Films*, **14**, 249, (1972).
169. P. W. McMillan and P. Nesvadba, *J. Non-Cryst. Sol.*, **17**, 189, (1975).
170. J. Male, *New Scientist*, **62**, 18, (1974).
171. C. H. Sie, M. P. Dugan and S. C. Moss, *J. Non-Cryst. Sol.*, **8-10**, 877, (1972).
172. C. B. Thomas, A. F. Fray and J. Bosnell, *Phil. Mag.*, **26**, 617, (1972).
173. H. K. Henish and R. W. Prior, *Solid State Electron*, **14**, 765, (1971).
174. S. H. Lee and H. K. Henisch, *Solid State Electron*, **16**, 155, (1973).
175. M. Sugi, S. Iizima, N. Kikuchi and K. Tanaka, *J. Non-Cryst. Sol.*, **5**, 358, (1971).
176. M. J. Thomson and J. Allison, *J. Phys. D: Appl. Phys.*, **7**, L53, (1974).
177. G. C. Vezzoli and L. W. Doremus, *J. Appl. Phys.*, **44**, 3245, (1973).
178. S. H. Lee, *App. Phys. Letts.*, **21**, (1972).
179. R. W. Prior, *Solid State Electron.*, **16**, 425, (1973).
180. J. L. Williams and A. Y. Irfan, *J. Phys. D: Appl. Phys.*, **7**, 1287, (1974).
181. C. H. Ling, *Thin Solid Films*, **29**, L39, (1975).
182. G. C. Vezzoli, P. J. Walsh, P. J. Kisatsky and L. W. Doremus, *J. Appl. Phys.*, **45**, 4534, (1974).

183. A. J. Hughes, P. A. Holland and A. M. Lettington, *J. Non-Cryst. Sol.*, **17**, 89, (1975).
184. C. H. Ling, *J. Phys. D: Appl. Phys.*, **8**, L189, (1975).
185. N. F. Mott, *Phil. Mag.*, **32**, 159, (1975).
186. G. C. Vezzoli, P. J. Walsh and L. W. Doremus, *J. Non-Cryst. Sol.*, **18**, 333, (1975).
187. T. Kaplan and D. Adler, *J. Non-Cryst. Sol.*, **8-10**, 538, (1972).
188. C. Popescu and D. Croitoru, *J. Non-Cryst. Sol.*, **8-10**, 531, (1972).
189. D. M. Kroll and M. H. Cohen, *J. Non-Cryst. Sol.*, **8-10**, 544, (1972).
190. A. C. Warren, *IEEE Trans. Electron Devices*, **ED-20**, 123, (1973).
191. J. C. Male and D. J. Thomas, *J. Non-Cryst. Sol.*, **13**, 409, (1973/4).
192. P. Burton and R. W. Brander, *Int. J. Electronics*, **27**, 517, (1969).
193. H. Fritzche, Proc. Low Mobility Materials Conf., p. 279, Taylor and Francis, London, (1971).
194. J. C. Male, *Electronics Letters*, **6**, 91, (1970).
195. C. Popescu, *Solid State Electronics*, **18**, 671, (1975).
196. C. B. Thomas, R. Carew-Jones and J. Bosnell, *Electronics Letts.*, **8**, 447, (1972).
197. D. M. Kroll, *Phys. Rev.*, **B11**, 3814, (1975).
198. T. M. Hayes and D. Thornburg, *J. Vac. Sci. Tech.*, **10**, 744, (1973).
199. D. D. Thornburg, *J. Non-Cryst. Sol.*, **17**, 9, (1975).
200. S. Iizima, M. Sugi and M. Kikuchi, *Solid State Comms.*, **8**, 153, (1970).
201. W. W. Sheng and C. R. Westgate, *Solid State Comms.*, **9**, 387, (1971).
202. W. Herms, H. Karsten and U. Zerrenthin, *Phys. Stat. Sol. (a).*, **23**, 479, (1974).
203. M. H. Cohen, H. Fritzche and S. R. Ovshinsky, *Phys. Rev. Letts.*, **22**, 1065, (1969).
204. P. Mark and M. Allen, "Electrical Properties: Charge Injection Phenomena" pp. 111–146, Ann. Rev. of Materials Sci. Vol. 3, (1973), eds., R. A. Huggins, R. M. Bube and R. W. Roberts. Annual Reviews, Inc., Palo Alto, California, (1974).
205. J. E. Sergent and A. R. Carlson, Charge Injection in Neutron Irradiated Silicon pp. 32–35. *Proc. South Eastcon.*, **74**, IEEE, (1974).
206. W. Van Roosbroeck, *J. Non-Cryst. Sol.*, **12**, 232, (1974).
207. M. P. Shaw, S. H. Holmberg, and S. A. Kostylev, *Phys. Rev. Letts.*, **31**, 542, (1973).
208. P. W. McMillan and P. Nesvabda, *J. Non-Cryst. Sol.*, **17**, 189, (1975).
209. R. G. Neale, D. Nelson and G. E. Moore, *Electronics*, **43**, 56, (1970).
210. C. H. Sie, M. P. Dugan and S. C. Moss, *J. Non-Cryst Sol.*, **8-10**, 877, (1972).
211. C. L. Dargan, P. Burton and R. M. Redstall, *Int. J. Electronics*, **38**, 711, (1975).
212. C. L. Dargan, P. Burton and C. P. Bloomer, *Thin Solid Films*, **23**, 343, (1974).
213. C. L. Dargan, P. Burton, S. V. Phillips, A. S. Bloor and P. Nesvadba, *J. Mat. Sci.*, **9**, 1595, (1974).
214. K. Shimakawa, Y. Inagaki and T. Arizumi, *Jap. J. Appl. Phys.*, **11**, 1319, (1972).
215. Y. Asahara and T. Izumitani, *J. Non-Cryst. Sol.*, **11**, 407, (1973).
216. S. Manhart, *J. Phys. D: Appl. Phys.*, **6**, 82, (1973).
217. D. L. Kirk and D. M. Newbury, *Microelectronics*, **5**, 2, (1973).
218. Y. Hirose and A. Kunioka, *J. Appl. Phys.*, **44**, 1706, (1973).
219. P. Chaudhari and R. B. Laibowitz, *Thin Solid Films*, **12**, 239, (1972).

220. D. Hafemeister and H. de Waard, *J. Appl. Phys.*, **43**, 5205, (1972).
221. V. I. Zaliva and V. P. Zakharov, *Sov. Phys. Semi-conductors*, **6**, 1095, (1973).
222. K. Tanaka, Y. Okada, M. Sugi, S. Iizima and M. Kikuchi, *J. Non-Cryst. Sol.*, **12**, 100, (1973).
223. D. D. Thornburg and R. M. White, *J. Appl. Phys.*, **43**, 4609, (1972).
224. D. D. Thornburg, *J. Non-Cryst. Sol.*, **11**, 113, (1972).
225. D. D. Thornburg, *J. Electronic Mat.*, **2**, 3, (1973).
226. A. Mineo, A. Matsuda, T. Kurosu and M. Kikuchi, *Solid State Comms.*, **13**, 1307, (1973).
227. A. G. Steventon, *J. Phys. D: Appl. Phys.*, **8**, 1869, (1975).
228. R. Nicolaides, *App. Phys. Letts.*, **24**, 331, (1974).
229. C. N. Berglund, *IEEE Trans. Electron Devices*, **ED-16**, 432, (1969).
230. R. G. Cope and A. W. Penn, *J. Phys, D: Appl. Phys.*, **1**, 161, (1968).
231. M. Regan and C. F. Drake, *Mat. Res Bull.*, **7**, 1559, (1972).
232. M. Regan and C. F. Drake, *IEEE Trans. Electron Devices*, **ED-20**, 144, (1973).
233. S. Basavaiah and K. C. Park, *IEEE Trans Electron Devices*, **ED-20**, 149, (1973).
234. I. G. Austin and M. Sayer, *J. Phys. C: Solid State Phys.*, **7**, 905, (1974).
235. K. C. Park, M. Berkenblit, D. J. Herrell, T. Blight and A. Reisman, *J. Electronic Materials*, **2**, 201, (1973).
236. B. Lalevic, N. Fuschillo and W. Slusark, *IEEE Trans. Electron Devices*, **ED-22**, 965, (1975).
237. I. Balberg, B. Abeles and Y. Arie, *Thin Solid Films*, **24**, 307, (1974).
238. J. Duchene, M. Terraillon and M. Pailly, *Thin Solid Films*, **12**, 231, (1972).
239. S. Tanifufi, K. Matsunaga and K. Yahagi, *Jap. J. Appl. Phys.*, **12**, 150, (1973).
240. D. D. Thornburg, *Phys. Rev. Letts.*, **27**, 1208, (1971).
241. R. G. Neale, *J. Non-Cryst. Sol.*, **2**, 558, (1970).
242. C. R. Haden, J. L. Barrett and J. L. Stone, *IEEE Trans. Solid State Circuits*, **SC-9**, 118, (1974).
243. E. D. Buckley and S. C. Moss, *Solid State Electronics*, **15**, 1331, (1972).
244. G. C. Vezzoli and I. M. Pratt, *Thin Solid Films*, **14**, 161, (1972).
245. R. G. Nicolaides and L. W. Doremus, *J. Non-Cryst. Sol.*, **8-10**, 857, (1972).
246. R. M. Quilliam, S. C. M. Day, C. M. Morter and G. V. Bunton, GEC Marconi unpublished, (1972).
247. R. G. Neale and J. A. Aseltine, *IEEE Trans. Electron Devices*, **ED-20**, 195, (1973).
248. H. G. Dill and T. N. Toombs, *Solid State Electron.*, **12**, 981, (1969).
249. TV uses MNOS Memory, Electronics, May 15, 1975, 40-41.
250. D. Frohman-Beutchkowsky, *IEEE Trans. Solid State Circuits*, **SC-6**, 301, (1971).
251. J. F. Verwey and R. P. Kramer, *IEEE Trans. Electron Devices*, **ED-21**, 631, (1974).
252. W. C. Hughes, C. Q. Lemmond, H. G. Parks, G. W. Ellis, G. E. Possin and R. H. Wilson, *Proc. IEEE*, **63**, 1230, (1975).
253. W. F. Kosonocby and J. E. Carnes, *RCA Review*, **36**, 566, (1975).
254. Magnetic Bubble Memories, Ovum Ltd., (1975).
255. N. Mott, M. Pepper, S. Pollitt, R. H. Wallis and C. J. Adkins, *Proc. Roy. Soc.*, **A345**, 169, (1975).
256. M. Pepper, S. Pollitt, C. J. Adkins and R. A. Stradling, *Critical Reviews in Solid State Sciences*, **5**, 375, (1975).

257. A. G. Holmes-Siedle, *Rep. Prog. Phys.*, **37**, 699, (1974).
258. B. T. Kolomiets, *Phys. State Sol.*, **7**, 713, (1964).
259. S. R. Ovshinsky, E. J. Evans, D. L. Nelson and H. Fritzsche, *IEEE Trans. Nuclear Science*, **NS-15**, 311, (1968).
260. J. T. Edmond, J. C. Male and P. F. Chester, *J. Phys. E. Sci. Instrum.*, **1**, 373, (1968).
261. T. M. Flanigan and M. E. Wyatt, *J. Non-Cryst. Sol.*, **2**, 229, (1970).
262. L. L. Hench, *J. Non-Cryst. Sol.*, **2**, 250, (1970).
263. L. L. Hench, A. E. Clark and H. F. Shaake, *J. Non-Cryst. Sol.*, **8-10**, 837, (1972).
264. R. A. Smith, R. Sanford and F. E. Warnock, *J. Non-Cryst. Sol.*, **8-10**, 862, (1972).
265. J. A. Olley and A. D. Yoffe, *J. Non-Cryst. Sol.*, **8-10**, 850, (1972).
266. J. B. Hays, *IEEE, Spectrum*, **1**, 115, (1964).
267. H. K. Henisch, *Scientific American*, **221**, 30, (1969).
268. D. L. Nelson, *J. Non-Cryst. Sol.*, **2**, 528, (1970).
269. R. T. Adams, U.S. Patent, **3573**, 757, (1971).
270. H. Hasegawa, S. Nakagawa and K. Takagi, *Jap. J. Appl. Phys.*, **12**, 153, (1973).
271. G. R. Fleming, *J. Non-Cryst. Sol.*, **2**, 540, (1970).
272. J. A. O'Connell and B. Narken, *IBM Journal*, **4**, 426, (1960).
273. K. E. Van Laningham, *IEEE Trans. Electron Devices*, **ED-20**, 178, (1973).
274. T. J. Kobylarz, *J. Non-Cryst. Sol.*, **2**, 515, (1970).
275. H. K. Rockstad, *J. Appl. Phys.*, **53**, 238, (1972).
276. H. Z. Hed and B. J. Freud, *J. Non-Cryst. Sol.*, **2**, 484, (1970).
277. J. Lipsicus, D. C. Mattis, E. Lord and A. Kornblit, *J. Non-Cryst. Sol.*, **2**, 550, (1970).
278. C. R. Haden, J. T. Stone, J. L. Barnett and R. W. Gill, *J. Non-Cryst. Sol.*, **8-10**, 432, (1972).
279. V. I. Zaliva and V. P. Zakharov, *JETP Letts.*, **13**, 133, (1971).
280. C. N. Berglund and R. K. Walden, *IEEE Trans. Electron Devices*, **ED-17**, 137, (1970).
281. J. Allison and V. R. Dawe, *Electron Letts.*, **7**, 707, (1971).
282. H. K. Rockstad, *J. Appl. Phys.*, **42**, 1159, (1971).
283. P. Callela, Picatinny Arsenal Technical Report, 3935, (1970).
284. D. L. Thomas and A. C. Warren, *Electron Letts.*, **6**, 62, (1970).
285. B. F. Mathur and F. O. Arntz, *J. Non-Cryst. Sol.*, **8-10**, 445, (1972).
286. L. M. Levinson and H. R. Philipp, *App. Phys. Letts.*, **24**, 75, (1974).
287. C. G. Smith and M. E. Closland, *Electronics and Power*, **18**, 404, (1972).
288. J. E. Lilienfeld, U.S. Patent, **1745**, 175, (1930).
289. P. K. Weimer, *Proc. IRE*, **50**, 1462, (1962).
290. A. Madan, M. J. Thomson and J. Allison, *Electronics Letts.*, **11**, 496, (1975).
291. D. Schuöcker, *J. Appl. Phys.*, **44**, 310, (1973).
292. R. F. Egerton, *App. Phys. Letts.*, **19**, 203, (1971).
293. P. A. Tick, J. H. P. Watson and N. K. Hindley, *J. Non-Cryst. Sol.*, **13**, 229, (1973/4).
294. G. W. Neudeck and A. K. Malhotra, *J. Appl. Phys.*, **46**, 239, (1975).
295. K. E. Petersen, D. Adler and M. P. Shaw, *App. Phys. Letts.*, **25**, 585, (1974).
296. E. J. Evans, J. H. Helbers and S. R. Ovshinsky, *J. Non-Cryst. Sol.*, **2**, 334, (1970).
297. R. F. Shaw, *App. Phys. Letts.*, **20**, 241, (1972).

298. P. C. Taylor, S. G. Bishop and D. L. Mitchell, *Solid State Comms.*, **8**, 1783, (1970).
299. J. L. Stone, W. A. Porter, J. S. Linder and C. R. Halden, *Proc. IEEE*, **59**, 323, (1971).
300. E. H. Staider, R. J. Gutmann and J. M. Borrego, *J. Non-Cryst. Sol.*, **17**, 273, (1975).
301. S. R. Ovshinsky and P. H. Klose, *J. Non-Cryst. Sol.*, **8-10**, 892, (1972).
302. J. Dresner and G. R. Stringfellow, *J. Phys. Chem. Solids*, **29**, 303, (1968).
303. I. A. Paribok-Aleksandrovich, *Sov. Phys. Solid State*, **11**, 1629, (1970).
304. J. Feinleib, J. P. de Neufville, S. C. Moss and S. R. Ovishinsky, *App. Phys. Letts.*, **18**, 254, (1971).
305. J. Feinleib and S. R. Ovshinsky, *J. Non-Cryst. Sol.*, **4**, 564, (1970).
306. S. R. Ovshinsky and H. Fritzche, *Metallurg. Trans.*, **2**, 641, (1971).
307. A. Hamada, T. Kurosu, M. Saito and M. Kikuchi, *App. Phys. Letts.*, **20**, 9, (1972).
308. T. Igo and Y. Toyoshima, *Jap. J. Appl. Phys.*, **11**, 117, (1972).
309. A. W. Smith, *App. Phys. Letts.*, **23**, 437, (1973).
310 K. S. Kim and D. Turnbull, *J. Appl. Phys.*, **45**, 3447, (1974).
311. R. Clement, J. C. Carballes and B. de Cremoux, *J. Non-Cryst. Sol.*, **15**, 505, (1974).
312. Y. Asahara and T. Izumitani, *J. Non-Cryst. Sol.*, **16**, 407, (1974).
313. Y. Asahara, Y. Ishibashi and T. Izumitani, *Jap. J. Appl. Phys.*, **14**, 289. (1975).
314. P. Chaudhari, S. R. Herd, D. Ast, M. H. Brodsky and R. J. von Gutfeld, *J. Non-Cryst. Sol.*, **8-10**, 900, (1972).
315. J. Feinleib, S. Iwasa, S. C. Moss, J. P. de Neufville and S. R. Ovshinsky, *J. Non-Cryst. Sol.*, **8-10**, 909, (1972).
316. R. J. von Gutfeld and P. Chaudhari, *J. Appl. Phys.*, **43**, 4688, (1972).
317. V. V. Korsakov, V. I. Nalivaiko, V. G. Remesnik and V. G. Tsukerman, *Sov. Phys. Tech. Phys.*, **19**, 565, (1974).
318. J. P. de Neufville, S. C. Moss and S. R. Ovshinsky, *J. Non-Cryst. Sol.*, **13**, 191, (1973/4).
319. E. G. Ramberg, *RCA Rev.*, **33**, 5, (1972).
320. W. R. Roach, *App. Phys. Letts.*, **19**, 453, (1971).
321. S. A. Keneman, *App. Phys. Letts.*, **19**, 205, (1971).
322. A. D. Pearson and B. G. Bagely, *Mater. Res. Bull.*, **6**, 1041, (1971).
323. Y. Ohmachi and T. Igo, *App. Phys. Letts.*, **20**, 506, (1972).
324. M. Adem and G. Bismith, *Opt. Comms.*, **3**, 2334, (1971).
325. N. S. Belokrinitskii, A. V. Gnatouskii, M. V. Danileiko, V. P. Zakharov, A. V. Kozlov and M. T. Shpak, *Sov. Phys. JETP Letts.*, **15**, 137, (1972).
326. S. B. Gurevich, N. N. Il'yashenko, B. T. Kolomiets, V. M. Lyubin and M. V. Sukharev, *Sov. Phys. Tech. Phys.*, **19**, 152, (1974).
327. M. Dakss and V. Sadagopan, *IBM Tech. Disclosure Bulletin*, **13**, 96, (1970).
328. A. C. M. Chen, J. F. Norton and J. M. Wang, *J. Non-Cryst. Sol.*, **8-10**, 917, (1972).
329. P. B. Mee, *Thin Solid Films*, **26**, 227, (1975).
330. J. T. Krause, C. R. Kurkjian, D. A. Pinnow and E. A. Sigety, *App. Phys. Letts.*, **17**, 367, (1970).
331. A. R. Hilton, *Appl. Optics.*, **5**, 1877, (1966).
332. T. Ohasaka, *J. Non-Cryst. Sol.*, **17**, 121, (1975).

333. Y. Ohmachi and N. Uchida, *J. Appl. Phys.*, **43**, 1709, (1972).
334. M. Gottlieb, T. J. Isaacs, J. D. Feichtner and G. W. Roland, *J. Appl. Phys.*, **45**, 5145, (1974).
335. V. V. Ziegel, A. A. Litvimenko, G. K. Ul'yanov and G. A. Chalabyan, *Sov. Phys. Accoust.*, **21**, 77, (1975).
336. R. K. Watts, M. de Wit and W. C. Holton, *Appl. Optics.*, **13**, 2329, (1974).
337. S. G. Bishop and W. J. Moore, *Appl. Optics.*, **12**, 80, (1973).
338. R. W. Vass and R. M. Andersen, *J. Appl. Phys.*, **45**, 3463, (1974).
339. E. I. Gordon, *Trans. Met. Soc. (AIME)*, **245**, 517, (1969).
340. H. Koelmans, *Thin Solid Films*, **8**, 19, (1971).
341. M. N. Brodsky, R. S. Title, K. Weiser and G. D. Petit, *Phys. Rev.*, **B1**, 2632, (1970).
342. M. H. Crowell, T. M. Buck, E. F. Labuda, J. V. Dalton and E. J. Walsh, *Bell Syst. Techn. J.*, **46**, 491, (1967).
343. R. Allan, *IEEE Spectrum*, **12**, 56, (1975).
344. G. Lapidus, *IEEE Spectrum*, **10**, 44, (1973).
345. I. Shimizu, H. Kokado and E. Inoue, *Photographic Science and Engineering*, **19**, 136, (1975).
346. J. H. Dessauer and H. E. Clark, "Xerography and Related Processes", Focal Press, London, (1965).
347. R. A. Zingaro and W. C. Cooper, "Selenium", Van Nostrand-Reinhold, New York, (1975).
348. R. H. Bube, *RCA Rev.*, **36**, 467, (1975).
349. N. F. Mott, E. A. Davis and R. A. Street, *Phil. Mag.*, **32**, 961, (1975).
350. J. Mort, I. Chen, R. L. Emerald and J. H. Sharp, *J. Appl. Phys.*, **43**, 2285, (1972).
351. I. Chen, *J. Appl. Phys.*, **43**, 1137, (1972).
352. R. B. Comizzoli, G. S. Lozier and D. A. Ross, *Proc. IEEE*, **60**, 348, (1972).
353. E. P. Goodings, *Endeavour*, **34**, 123, (1975).
354. D. W. Strickler and R. Roy, *J. Mater. Sci.*, **6**, 200, (1971).
355. J. C. Schotmiller, T. W. Taylor and F. W. Ryan, *Appl. Optics.*, **Suppl. 3**, 55, (1969).
356. B. Petretis, K. Baceviciute, E. Montrimas and S. Tamosiunas, *J. Non-Cryst. Sol.*, **16**, 418, (1974).
357. J. L. Donovan, *J. Appl. Phys.*, **41**, 2109, (1970).
358. J. W. Boag, *Phys. Med. Biol.*, **18**, 3, (1973).
359. L. A. de Rosa, U.S. Patent 3550155, (1970).
360. G. Kaplan, *IEEE Spectrum*, **11**, 77, (1974).
361. W. E. Spear and P. G. Le Comber, *Solid State Comms.*, **17**, 1193, (1975).
362. R. Jasinski and I. Trachtenberg, *J. Electrochem. Soc.*, **120**, 1169, (1973).
363. M. H. Broadsky and G. H. Dohler, *Critical Reviews in Solid State Sciences*, **5**, 591, (1975).

Chapter 8†

Thin Film Optical Devices

H. A. M. Macleod

*Department of Physics and Physical Electronics,
Newcastle upon Tyne Polytechnic,
Newcastle upon Tyne, England*

1. Introduction 322
2. Optical Coatings 323
   2.1 Theoretical Treatment 326
   2.2 Antireflection Coatings 334
   2.3 Reflection Increasing Coatings 340
   2.4 Edge Filters 344
   2.5 Band Pass Filters 349
      2.5.1 Fabry Perot filters 350
      2.5.2 Angular behaviour of the Fabry Perot filter . . 356
      2.5.3 Multiple cavity filters 358
      2.5.4 Induced transmission filters 360
   2.6 Polarisers 361
   2.7 Dichroic Beam Splitters 363
   2.8 Factors which Limit Performance 364
3. Thin Film Light Guides 367
   3.1 The Simple Thin-film Light Guide 369
   3.2 Power Flow in a Thin-film Light Guide 376
   3.3 Losses in Thin-film Light Guides 380
   3.4 Metal-clad Guides 381
   3.5 Strip Guides 381
   3.6 Coupling 382
      3.6.1 Prism output coupler 383
      3.6.2 Prism input coupler 384
      3.6.3 Grating couplers 391
      3.6.4 Thin phase-grating 393
      3.6.5 Bragg grating couplers 394
      3.6.6 Tapered film coupler 399
      3.6.7 Interconnection in integrated optical circuits . 399
      3.6.8 Fibre-waveguide coupling 400
   3.7 Two-dimensional Optics 402

† For a list of symbols used in this chapter and their definitions see p. xiv.

3.8 Waveguide Filters and Beam Splitters 404
  3.8.1 Narrow-band grating filters 404
  3.8.2 Beam splitters 406
3.9 Second Harmonic Generation 408
3.10 Materials 411
3.11 Active Devices 415
  3.11.1 Electro-optic devices 415
  3.11.2 Magneto-optical devices 421
  3.11.3 Acousto-optical devices 421
  3.11.4 Thin film lasers 421
References. 422
Note Added in Proof 427

1. Introduction

Thin film optics is an essential division of modern optics. Without thin film optical coatings most modern optical systems could not function. They play a vital part in enhancing or reducing reflectance, absorptance or transmittance, in splitting or combining beams, in polarising or analysing beams, in colour separation, in rejecting or accepting spectral bands, in making adjustments to phase and, in short, in almost all the operations of optical devices. Recently, increased interest has been shown in optical waveguides for communication, and in this field, too, thin films are likely to be indispensable. The most promising techniques for processing the optical signals involve thin films. These act as light guides, containing beams so that propagation is along the films, parallel to the boundaries, rather than across them. The guides can be modified to enable switching, modulation, mode selection and the like. Combinations of numbers of these units into optical signal processors, known as integrated optical circuits, are foreseen.

Thus, thin film optics can be divided into two broad divisions. The first is optical coatings and the second light guides and associated devices. Optical coatings are characterised by propagation across the films. Interference effects are immensely important and multilayers of fifty or more films on a single surface are not uncommon. The technology is well established. Optical light guides are characterised by propagation along the films parallel to the boundaries and the number of layers involved is usually small; often only a single layer being used. Light guides are still primarily research items. The technology is only now being developed and is changing rapidly. Although light guides can be fabricated using the thermal evaporation techniques already established for optical coatings other techniques have been found to show more promise. Differences between the two fields are accentuated because the approach to the thin film light guide is often similar to that in microwaves and papers are frequently written in the terminology of communications rather than optics.

This account of thin films in optics is therefore split into two. That part dealing with optical coatings is the more nearly complete since the field is more mature and major developments less frequent. The section on light guides is brief and only an introduction to the subject. The field is developing too quickly at present for any review to remain definitive for long and in this section the emphasis is on the introductory theory and application of the devices.

2. Optical Coatings

Almost all thin film optical coatings depend on interference for their operation. The colours of soap bubbles or of oil films on water are good examples of interference effects in single films. Light reflected at the outer and inner surfaces of the films interferes, constructively if the path difference is an integral number of wavelengths, destructively if it is an odd number of half wavelengths. Since the path difference between the beams depends on the layer thickness and the angle of incidence, the colours are seen to vary with layer thickness and with the angle at which they are viewed. In a thin film optical coating, interference effects of this type, but usually more complex because more films are involved, are used to produce a wide range of useful characteristics.

A thin film optical coating, whilst behaving similarly in principle to the soap bubble or oil film, is very different in construction. The films, or layers, together with their support or, as it is normally called, their substrate, are solid. The substrate can be an optical element having an important role of its own, such as a lens, or it can simply be the support for the films consisting perhaps of a straightforward window. If the coating is to be exposed in use, as is the case with an antireflection coating, then the thin films must be chosen to have great mechanical robustness and resistance to atmospheric attack. In such cases it is often necessary to choose materials more for their strength than for their optical performance. Sometimes it is possible to protect the thin films by cementing a cover over them and in such cases the optical performance of the films can be the prime consideration.

It is possible to construct assemblies of thin films which will reduce the reflectance of a surface and hence increase the transmittance of a component, or increase the reflectance of a surface, or which will give high reflectance and low transmittance over part of a region and low reflectance and high transmittance over the remainder, or which will have different properties for different planes of polarisation, and so on. Thin film coatings are often known by names which describe their function, such as antireflection coatings, beam splitters, polarisers, long wave pass filters, short wave pass filters, band pass filters, band stop or minus filters, or which describe their construction, such as quarter wave stack, quarter-half-quarter coating and so on. Some of the different types of filter are shown in Fig. 1 together with some of the terms used to describe important features.

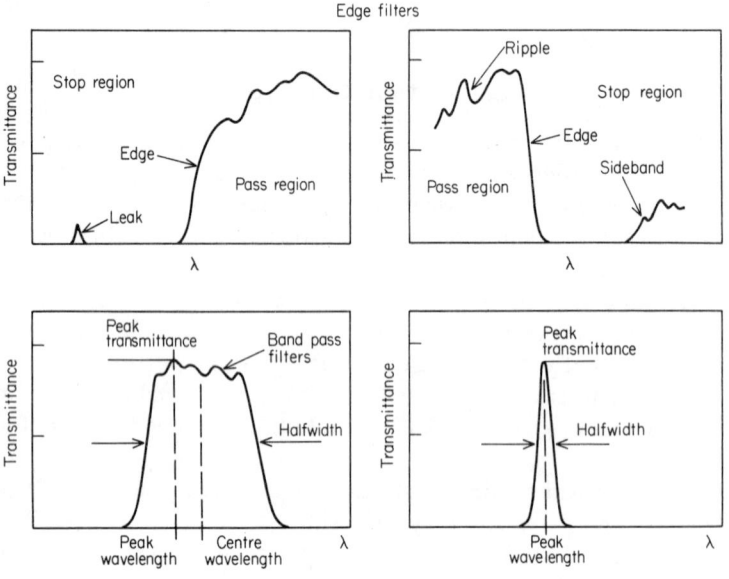

FIGURE 1. Typical performance curves for thin-film filters. Unwanted transmission in a stop region is usually referred to as a leak if small, or as a sideband if more extensive.

In a thin film assembly, or multilayer, the intensity of the light reflected at each interface depends on the refractive indices of the materials on either side and thus the intensities of the various beams involved in the interference can be adjusted by choosing the refractive indices of the films. The phases of the beams on the other hand can be altered by changing the layer thicknesses. There are thus two parameters associated with each layer, thickness and refractive index, which can be chosen to give the required performance, the number in a multilayer being twice the number of layers. Complete freedom of choice is not possible since the number of suitable coating materials is limited. There are also reasons related to ease of production which make it desirable to use layer thicknesses which are related to each other in a simple manner. Although multilayers of greater than 100 layers can be produced, it is better if possible to keep the total number of layers in a coating rather less than this value, because the probability of an unacceptable error during construction is very high when the number of layers becomes very large. Also the greater the number of layers the more expensive and less robust the coatings tend to be.

Thus the number and range of independent variables associated with a coating are effectively limited, and there are therefore limits to optimum theoretical performance apart from any drop in actual performance caused by the inevitable errors within the production tolerances when the coatings are manufactured. These limits can take different forms, such as the spectral width over which a particular performance can be attained, the minimum bandwidth of a narrow band filter, the sharpness of the edge of a long wave or short wave pass filter, the ripple in the pass band, the range of usable angles of incidence and so on. There is always a limit to the width of the spectral region over which acceptable performance is achieved. The regions outside are often referred to as sidebands and it is frequently necessary to add subsidiary filters to correct the performance in the sidebands. An important example of this is the suppression of transmission in the sidebands of band pass filters: this will be discussed later.

A film in an optical coating is said to be thin when interference effects can be detected in the light which it transmits or reflects, and thick when they can not. Of course, whether or not interference effects are detected depends just as much on the nature of the incident light and of the detector as it does on the film itself. Even without changing the wavelength, the same film can appear thick or thin depending entirely on the illumination and detection arrangements. However, there is seldom any confusion. In the coatings we shall be considering, the maximum thickness of the films will not usually exceed several wavelengths while the substrates which carry the films will usually be at least one millimetre in thickness. Under these conditions the films will be almost invariably thin while the substrates can be considered thick. Thick implies that the path difference between the various beams exceeds the coherence length of the light so that the resultant intensity is simply the sum of the intensities of the individual beams and no interference effects can be observed and the thick, or incoherent, case can be shown to be consistent with the thin, or coherent, case integrated over a sufficiently large wavelength interval or range of angles of incidence or range of thicknesses.

In a stack of layers of different refractive index, light will be reflected at every boundary and the interference will involve an enormous number of beams. If absorption is present this complicates matters still further. Even with just a few layers the calculation of performance on the basis of straightforward multiple beam interference becomes exceedingly involved and so multiple beam calculations of this type are rarely used to determine the performance of a multilayer. Either the assumption is made that the performance is adequately represented by just one beam reflected from each interface, so that the total number of beams is one greater than the number of layers, or, for accurate calculation, a different solution to Maxwell's equations is used which involves an elegant matrix product, each layer being represented by a 2 × 2 matrix. In the next section the most important features of this theory are listed.

2.1 Theoretical Treatment

The following is not an exhaustive treatment of the theory but simply a collection of some of the most important results for optical thin film calculation. More details will be found in Born and Wolf[1], Knittl[2], Heavens[3] and Macleod[4].

It is usual to assume that optical thin films consist of parallel sided slabs of homogeneous isotropic material and that the electromagnetic wave is described by Maxwell's equations and the appropriate material equations. Any real wave is considered to be a combination of simple harmonic waves and so the fundamental optical properties of an assembly of thin films are described in terms of a single plane harmonic wave. For ease of manipulation this is normally assumed to be a complex wave, the physical significance being attached to the real part. Complex waves give results identical with those for real waves, provided the operations carried out are linear, but expressions have to be adapted to suit complex waves when the operations are non-linear.

The phase factor for a complex harmonic wave is

$$\exp[i(\omega t - \mathbf{k} \cdot \mathbf{r})]$$

where $\mathbf{k}$ is the wave vector, and $\mathbf{r}$ is the position vector for the point in question. An electromagnetic wave is characterised by a magnetic vector $\mathbf{H}$ and an electric vector $\mathbf{E}$ with $\mathbf{E}$, $\mathbf{H}$ and $\mathbf{k}$ in almost all cases mutually perpendicular and forming a right-handed set. In most of what follows this will be assumed to be true. The simplest form of wave is a plane polarised harmonic wave with

$$\mathbf{E} = \mathscr{E} \exp[i(\omega t - \mathbf{k} \cdot \mathbf{r})]$$

and

$$\mathbf{H} = \mathscr{H} \exp[i(\omega t - \mathbf{k} \cdot \mathbf{r})]$$

where $\mathscr{E}$ and $\mathscr{H}$ are complex constant vector amplitudes. Simple logical considerations show that the invariant characteristic of such a wave, as it passes through different media, is the angular frequency, ω, which is therefore the obvious choice of fundamental parameter to characterise the wave. The parameter which similarly directly describes the properties of the medium is the velocity of propagation, v. This is found to be complex and a function of ω, except in free space where the velocity is invariant and is the fundamental constant, c. Unfortunately ω is not a particularly convenient quantity for characterising a wave since it is of the order of $10^{15}\,\mathrm{sec}^{-1}$ and cannot be measured directly. The free space wavelength

$$\lambda = \frac{2\pi c}{\omega}$$

or the wavenumber

$$v = \frac{\omega}{2\pi c}$$

are frequently used instead. Similarly the velocity of propagation is inconvenient and the quantity, N, given by c/v is normally used. Since v is complex, N is complex and is written

$$N = n - i\kappa$$

N is known as the complex refractive index, n is the real part of the refractive index or, as it is sometimes known, simply the refractive index, and κ is the imaginary part of the refractive index, or the extinction coefficient. n and κ which are functions of λ are known as the optical constants of the medium. The wave vector $\mathbf{k}$ can then be shown to be given by

$$\mathbf{k} = \frac{2\pi N}{\lambda}\hat{\mathbf{s}}$$

where $\hat{\mathbf{s}}$ is the unit vector in the direction of propagation. This permits a further relationship to be derived from Maxwell's equations,

$$\mathbf{H} = Y\hat{\mathbf{s}} \times \mathbf{E} \tag{1}$$

where Y is a further property of the medium and is known as the optical admittance. At optical frequencies μ_r is always unity and if this simplification is used it can be shown that

$$Y = NY_0 \tag{2}$$

where Y_0 is the admittance of free space (1/377 siemens). Therefore, provided we know the optical constants of the medium we can always find $\mathbf{H}$ given $\mathbf{E}$ and the direction of propagation $\hat{\mathbf{s}}$. As far as most physical effects are concerned $\mathbf{E}$ is the more important quantity and thus light waves are normally specified in terms of the electric vector and the direction of propagation. The term polarisation is used to describe the orientation of **E**. A wave in which **E** and $\hat{s}$ are confined to a plane is known as plane polarised.

The Poynting vector, given by

$$\mathbf{P} = \mathbf{E} \times \mathbf{H} \tag{3}$$

is a measure of the instantaneous rate of flow of energy per unit area. This expression involves a non-linear operation and hence should strictly be used only with waves in the real representation. For a harmonic wave the rate of flow fluctuates at twice the frequency of the wave and it is the mean rate which interests us. The mean rate is known as the intensity, **I**, of the wave. To obtain the correct result for the intensity of a harmonic wave in the complex formulation we must alter the expression (3) to read

$$\mathbf{I} = \langle \mathbf{P} \rangle = \tfrac{1}{2} \text{ Real Part } (\mathbf{E} \times \mathbf{H}^*) \tag{4}$$

From (1) and (4),

$$\mathbf{I} = \tfrac{1}{2} \operatorname{Re}[\mathbf{E} \times (\psi^* \hat{s} \times \mathbf{E}^*)]$$

Now provided **E** is perpendicular to $\hat{s}$

$$I = \frac{nY_0}{2} \mathbf{E} \cdot \mathbf{E}^* = \frac{nY_0}{2} \mathscr{E} \cdot \mathscr{E}^* \tag{5}$$

where $\mathscr{E}$ is the amplitude of the wave, and I is the magnitude of **I**.

We can now consider what happens when an optical wave is incident on a simple interface between two media. The wave splits into a reflected wave and a transmitted wave and we can define four quantities to describe the situation. These are the amplitude reflection coefficient ρ, the amplitude transmission coefficient, τ, both referring to the electric vector, the intensity reflection coefficient, or reflectance, R, and the intensity transmission coefficient, or transmittance, T. ρ and τ are usually complex. Then from (5)

$$R = \rho \rho^*$$

$$T = \frac{n_1}{n_0} \tau \tau^* \tag{6}$$

where the suffixes 0 and 1 refer to the two media, 0 being on the incident side of the boundary. Maxwell's equations can be solved using the boundary conditions that the components of **E** and **H** which are parallel to the interface are continuous across it. Then we find that for normal incidence, writing N_0 and N_1 for the complex refractive indices of the incident and exit media respectively

$$\rho = \frac{N_0 - N_1}{N_0 + N_1}$$
$$\tau = \frac{2N_0}{N_0 + N_1} \quad (7)$$

The sign convention implied in these expressions is that the positive direction for **E** in the reflected and transmitted waves is the same as in the incident wave. Reflectance and transmittance are then given by

$$R = \rho\rho^* = \left(\frac{N_0 - N_1}{N_0 + N_1}\right)\left(\frac{N_0 - N_1}{N_0 + N_1}\right)^*$$

$$T = \frac{n_1}{n_0} = \frac{4n_1 n_0}{(N_0 + N_1)(N_0 + N_1)^*} \quad (8)$$

When the incident medium is absorbing there are some apparent anomalies when equations (8) are applied. These are due to a cross-coupling between the incident wave and the reflected wave which adds an extra term to the normal expression for energy flow at a simple boundary ($R + T = 1$). Provided the incident medium is non-absorbing then these equations hold without difficulty. Equations (7) hold whether or not the incident medium is absorbing.

When the wave is incident at an angle other than normal to the boundary, there is considerable complication. The state of polarisation of the waves must also be considered and it is found that the polarisation is not altered if the wave is either plane polarised with the electric vector in the plane of incidence (*p*-polarised) or plane polarised with the electric vector perpendicular to the plane of incidence (*s*-polarised). Any other form of polarisation must be reduced first to a combination of these two forms. The sign convention is shown in Fig. 2 and is chosen so that it corresponds to the normal incidence convention already stated when the angle of incidence, θ_0, becomes zero. We introduce a new parameter, η, which is related to the components of **E** and **H** parallel to the boundary in the same way that N is related to the values of **E** and **H** normal to $\hat{s}$:

$$\eta = \frac{1}{Y_0} \frac{H_{\|\text{boundary}}}{E_{\|\text{boundary}}}$$

The various relations now become

$$\rho = \frac{\eta_0 - \eta_1}{\eta_0 + \eta_1}$$
$$\tau = \frac{2\eta_0}{\eta_0 + \eta_1} \quad (9)$$

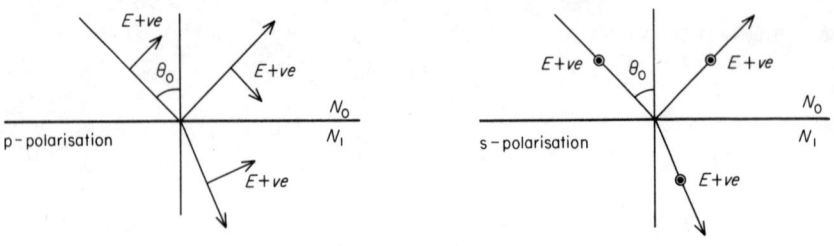

FIGURE 2. Sign convention for the direction of the electric vector.

where the relationships have to be evaluated for each plane of polarisation separately using

$$\eta_s = N/\cos\theta$$
$$\eta_p = N/\cos\theta$$

where

$$N_0 \sin\theta_0 = N_1 \sin\theta,$$

i.e. $(\eta_0)_s = N_0 \cos\theta_0, (\eta_1)_s = N_1 \cos\theta$, etc.

There are again difficulties in interpreting R and T if the incident medium is absorbing. These expressions have been used in deriving the reflectance curves in, for example, Fig. 16.

A thin film on a substrate is equivalent to the addition of a second interface to the single surface (Fig. 3). The most direct way of estimating the properties is simply to introduce a value of ρ and τ for each of the interfaces and to sum the multiply reflected beams taking account of the phase shifts due to each traversal of the film. This technique, used in deriving the expressions well known in multiple beam interferometry, leads to very difficult and cumbersome expressions when further films are added to increase the number of interfaces and so a different approach, leading to expressions which can be easily extended to assemblies of many layers, is normally adopted. The tangential values of **E** and **H** at interfaces are related to the reflectance and transmittance as follows, where I_i is the incident intensity:

$$\tfrac{1}{2} Re(\mathbf{E}_b \times \mathbf{H}_b^*) = \text{transmitted intensity} = TI_i$$
$$\tfrac{1}{2} Re(\mathbf{E}_a \times \mathbf{H}_a^*) = \text{intensity entering assembly} = (1-R)I_i$$

Let us denote the scalar values of $\mathbf{E}_a$ by E_a, of $\mathbf{E}_b$ by E_b and so on. We can

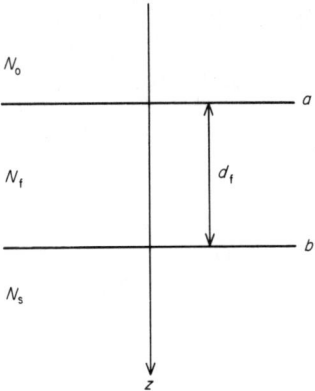

FIGURE 3. Two interfaces making up a single film.

show that $E_a \, H_a$ and $E_b \, H_b$ are related by

$$\begin{bmatrix} E_a \\ H_a \end{bmatrix} = \begin{bmatrix} \cos \delta_f & (i \sin \delta_f)/Y_f \\ iY_f \sin \delta_f & \cos \delta_f \end{bmatrix} \begin{bmatrix} E_b \\ H_b \end{bmatrix} \quad (10)$$

where

$$Y_f = \eta_f Y_0$$

and

$$\delta_f = \frac{2\pi N_f d_f \cos \theta_f}{\lambda}$$

By analogy with equation (1) we can define an equivalent admittance of the assembly as H_a/E_a and then the reflection and transmission coefficients become:

$$\rho = \frac{\eta_0 - \dfrac{1}{Y_0} \cdot \dfrac{H_a}{E_a}}{\eta_0 + \dfrac{1}{Y_0} \cdot \dfrac{H_a}{E_a}}$$

$$\tau = \frac{2\eta_0}{\eta_0 + \dfrac{1}{Y_0} \cdot \dfrac{H_a}{E_a}}$$

Now we can choose any one of E_a, H_a, E_b, H_b as we wish without altering the expressions for the transmission and reflection coefficients. It is most convenient if we choose E_b as unity. We can also choose to use

$$\frac{H_a}{Y_0} \text{ and } \frac{H_b}{Y_0}$$

instead of H_a and H_b. But,

$$\frac{H_b}{Y_0} = \eta_s E_b = \eta_s$$

and so we can write, instead of equation (10)

$$\begin{bmatrix} B \\ C \end{bmatrix} = \begin{bmatrix} \cos\delta_f & (i\sin\delta_f)/\eta_f \\ i\eta_f \sin\delta_f & \cos\delta_f \end{bmatrix} \begin{bmatrix} 1 \\ \eta_s \end{bmatrix} \quad (11)$$

B and C representing E_a and H_a/Y_0 respectively in this special arrangement.

Then

$$\rho = \frac{\eta_0 B - C}{\eta_0 B + C} \qquad \tau = \frac{2\eta_0}{\eta_0 B + C}$$

and if η_0 is real

$$R = \rho\rho^* = \frac{(\eta_0 B - C)(\eta_0 B - C)^*}{(\eta_0 B + C)(\eta_0 B + C)^*}$$

and

$$T = \frac{4\eta_0 \eta_s}{(\eta_0 B + C)(\eta_0 B + C)^*} \quad (12)$$

This can simply be extended to any number of layers, r, as

$$\begin{bmatrix} B \\ C \end{bmatrix} = \left\{ \prod_{j=1}^{r} \begin{bmatrix} \cos\delta_j & (i\sin\delta_j)/\eta_j \\ i\eta_j \sin\delta_j & \cos\delta_j \end{bmatrix} \right\} \begin{bmatrix} 1 \\ \eta_s \end{bmatrix} \quad (13)$$

where layer $j = r$ is next to the substrate, s.

Equations (12) and (13) are the most commonly used expressions for calculating the performance of thin film coatings.

Equation (13) is valid for any thin film whether or not it is absorbing, i.e. whether or not the extinction coefficient κ is non-zero. In most coatings use is made of either metal films with $\kappa \gg n$, for instance silver, or of dielectric films where $\kappa \ll n$. Films intermediate between these two extremes are seldom used, except in some special applications. Many of the coatings and filters we shall be considering use dielectric films only. The theory of these all dielectric designs is simplified by the assumption that $\kappa = 0$. In what follows dielectric films are assumed to have $\kappa = 0$ unless it is explicitly stated otherwise.

From the matrix expression we can immediately derive some useful expressions for thin dielectric films having optical thicknesses which are an integral number of quarter wavelengths. If p is the number of quarter wavelengths then δ is given by $p\pi/2$ and the matrix for the film becomes particularly simple. It can readily be shown that a single quarter wave optical thickness of index n_f on a substrate of index n_s acts as a single surface of a medium of effective index given by n_f^2/n_s. A single half wavelength on the other hand acts as if it did not exist. The index of the combination of film and substrate acts like a single surface of a medium of index n_s, and the reflectance is simply that of the uncoated substrate. A half wavelength film is sometimes known as an absentee layer because of this. For a single thin film on a substrate these special cases mark the extrema of reflectance and transmittance. In between these values the reflectance or transmittance behaves in a manner not unlike a sine or cosine function. This behaviour is illustrated in Fig. 4 where the reflectance of various thin films of various indices on a glass substrate of index 1.52 corresponding to that of crown glass is shown.

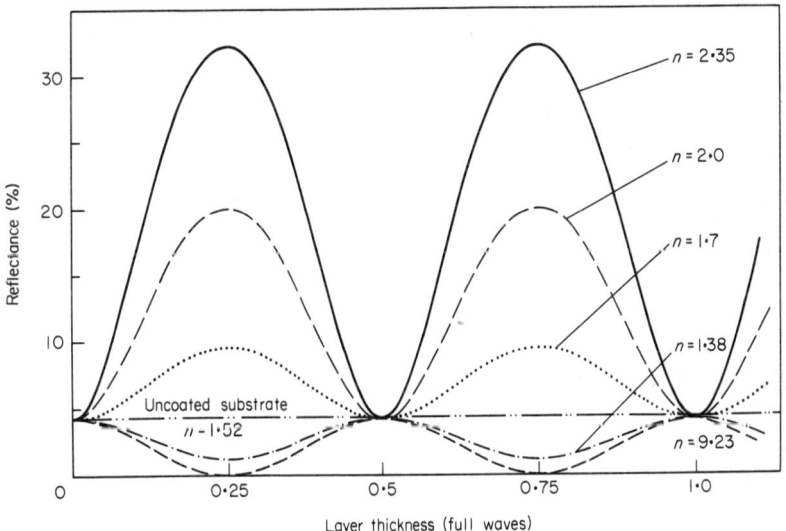

FIGURE 4. Calculated reflectance of single films on a single surface of glass ($n = 1.52$).

Because layers which are an integral number of quarter waves thick, are simple to handle theoretically and are also more straightforward for deposition control, they are used in preference to layers of non-integral thickness wherever possible. A useful shorthand notation represents a quarter wave of high index material by H, of low index by L, and of intermediate index by M. Half waves then become HH, LL, MM and so on. A filter design can then be written as, for example,

$$\text{Glass/HLHHLHLHLHHLH/Air}$$

representing an 11-layer double cavity filter (see 2.5.3) on a glass substrate. This could also be written as

$$\text{Glass/(HLH)}^2\text{L(HLH)}^2\text{/Air}$$

or

$$\text{Glass/HLH(HL)}^3\text{HHLH/Air}$$

2.2 Antireflection Coatings

An important application of thin film optics, and probably the largest commercially, is the reduction of the reflection losses which transmitting optical components suffer at each surface due to the difference in refractive index between the component and the surrounding media. Thin film optics really began with the development, in the late nineteenth century, of processes to reduce the reflection loss from glass surfaces by producing thin films of lower index upon them. In spite of the early start, antireflection coating remains one of the least straightforward areas of optical thin films largely due to the lack of suitable low index thin film materials which forces the adoption of rather unsatisfactory compromises in design.

The simplest form of antireflection coating is a quarter wavelength film of index n_f between the index of the substrate, n_s, and that of the surrounding medium, n_0. The coated surface will have a reflectance of

$$\left[\frac{n_0 - n_f^2/n_s}{n_0 + n_f^2/n_s}\right]^2$$

at the reference wavelength, which will be zero when

$$n_f^2 = n_0 n_s$$

but will always be less than the reflectance

$$\left[\frac{n_0 - n_s}{n_0 + n_s}\right]^2$$

of the uncoated substrate if $n_0 < n_f < n_s$ or $n_s < n_f < n_0$. In the visible region, crown glass of index around 1.52 bounded by air of index 1.0 is probably the commonest optical material and a single layer antireflection coating would ideally have an index of $(1.52)^{\frac{1}{2}} = 1.233$. An antireflection coating must be reasonably robust especially if it is on the outside of an optical component where it will have to withstand abrasion. No suitable material is known which combines such a low index with a sufficient degree of robustness. Over

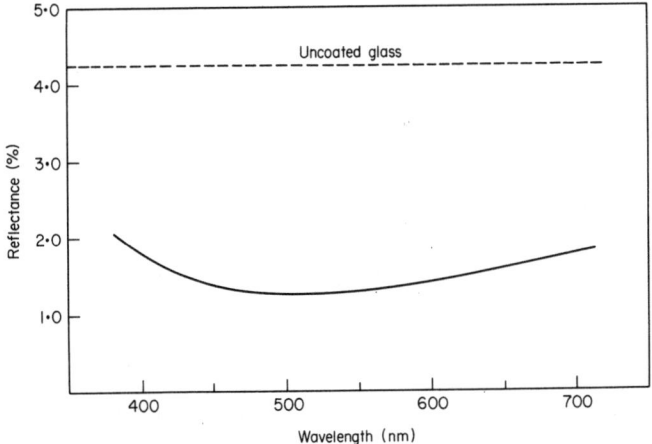

FIGURE 5. Calculated reflectance of single layer of MgF_2 ($n = 1.38$) on glass ($n = 1.52$).

many years the material which has been found to be most satisfactory is magnesium fluoride which has an index of 1.38 and if deposited on a substrate at a temperature of around 300°C, it is extremely durable. The minimum reflectance of a quarter wave on crown glass is 1.25% per surface. Figure 5 shows a curve of reflectance over the visible region for such a coating. The reflectance is slightly higher in the red and the blue than in the green giving the coating its well-known magenta appearance in reflection, which has led to the use of the term "bloomed" when describing a coated component. Magnesium fluoride is also used for coating higher index substrates. For these the residual reflectance is rather lower and is zero for a substrate of index 1.90.

For very high index substrates, such as silicon with index around 3.5, or germanium with index around 4.0, higher index coating materials must be used (Cox and Hass[5], and Cox et al.[6]). Zinc sulphide ($n \simeq 2.3$) and zinc selenide ($n \simeq 2.5$) are popular materials. These combinations are used if transmittance is required beyond 10 μm. Otherwise, oxide materials such as zirconium dioxide can be used with the advantage of increased robustness. Absorbing substrates such as metals can also be antireflection coated, although in this case, unless the metal is in the form of a very thin layer, the light not reflected will be absorbed. We can readily show that a metal of index $N = n - ik$ will have zero reflectance in air at a particular wavelength if coated with a layer whose index n_f and phase thickness δ, are given by

$$n_f = \sqrt{\frac{n(n-1) + \kappa^2}{(n-1)}} \quad \text{and} \quad \tan\delta = \frac{n_f(n-1)}{\kappa} \qquad (14)$$

provided that n_f is real. This requires either $n > 1$ or $\kappa^2 < n(1-n)$ otherwise two or more layers are necessary. This accounts for the colours obtained when metals such as steel are heated in air. The oxide layers which form on the metal act as reflection reducing coatings.

The single layer coating is clearly not an ideal antireflection coating although for most applications involving higher index glasses it is perfectly satisfactory. Improved performance requires greater complexity. The most effective coating, from the theoretical point of view, is a film which has an index varying smoothly from that of the substrate to that of the medium. This gives an antireflection coating characteristic stretching from wavelength $\lambda = 2D$ (where D is the total optical thickness of the layer) right through to $\lambda = 0$. Such a thin film is impossible to produce in perfect form when the medium is air because of the lack of very low index materials. The index must always terminate at the lowest available value. Jacobsson and Martensson[7] have constructed graded index films in the infrared which are mixtures of germanium and magnesium fluoride the proportions varying through the thickness of the layer. The outermost index is that of magnesium fluoride ($n = 1.38$) and in air the coatings have a residual reflectance of 2.5% so that they are really useful only for high index substrates. Inhomogeneous films are reviewed by Jacobsson[8].

Recently Clapham and Hutley[9] have proposed a technique for overcoming this problem involving a completely new type of thin film known as the "moth's eye coating". Consider a regular array of cones of diameter a made up of a material of refractive index n_c. Light incident on the cones will be scattered but for wavelengths not much smaller than a the scattering will be into the various orders of a two dimensional diffraction grating of grating constant a. For λ greater than a, therefore, the scattering will disappear except for the zero order, i.e. the specularly reflected or the transmitted beam, and for such wavelengths the array of cones will appear to be a simple thin film of index varying smoothly from that of the medium at the tips of the cones, to n_c at the bases. If the cones have physical height h_c and the surrounding medium is air of index 1.0, then the total optical thickness is approximately

$$D = h_c \frac{(n_c + 1)}{2}$$

Now the cones cannot be too long compared with their bases or they will be very weak mechanically. If h_c is approximately a, say, and n_c is 1.5 then $D = 1.25a$ and the film acts as an antireflection coating for crown glass of very good performance from $\lambda = 2.5a$ to $\lambda = a$, comfortably spanning the

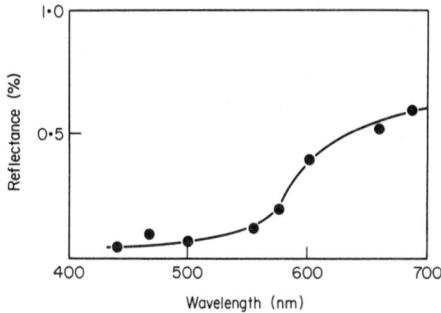

FIGURE 6. Measured specular reflectance of experimental "moth's eye coating". (After Clapham and Hutley[9].)

visible region if a is between 300 and 400 nm. Clapham and Hutley's cones were produced in hardened photoresist by exposing a layer of it to an array of interference fringes of the correct spacing. Two exposures were made, the second after rotating the fringes through 90° with respect to the first. The index of the hardened photoresist was around 1.6 and the glass substrate was chosen to have this refractive index also. Such a coating is less robust than a solid homogeneous thin film and it would be difficult to clean, but it does have the advantage that high performance could possibly be achieved cheaply with a good yield. It could be used for surfaces inside sealed units where abrasion resistance is not essential and where it would not be exposed to dirt. Figure 6 shows the achieved performance.

The principal technique used commercially to achieve these results is the replacement of the single graded index film with a descending staircase of layers all of the same optical thickness. If these are arranged to follow certain regular rules then extremely low reflectance over wide regions can be obtained. For a given number of layers the width of the characteristic can be traded against the height of the residual ripple in the pass band. The greater the number of layers, the higher is the potential performance. Tables which enable rapid design of such coatings are given by Young[10], and these are based on results derived originally for microwave filters. A method giving similar results but with the entire analysis in optical terms is included by Musset and Thelen[11]. For high index materials the technique can be applied directly and the reflectance curve for such a coating is shown in Fig. 7. For lower indices, especially crown glass, again there is the lack of really suitable low index materials, and the final coating must be magnesium fluoride giving, in the limit, 2.5% reflectance. The normal practice is to employ two staircases, the first, nearest the substrate, rising to an even higher index value so that the second, descending, can have several regularly spaced steps even although the bottom one is magnesium fluoride. The width of the antireflection zone of such a coating is much smaller than that of a single graded layer but it is adequate to cover the visible region with reasonable performance. Such a coating is illustrated in Fig. 8. A number of design techniques exist. A quantitative version of the staircase approach described here is given by Musset and Thelen[11], while a completely different approach giving identical results is described by Cox et al.[12]. The number of layers can be increased with consequent improvement in performance (Young[10], Musset and Thelen[11]).

An obvious difficulty in such coating designs is the large number of different materials required. It is possible to limit the number of materials by making use of a most useful property possessed by any symmetrical assembly of layers (Berning[13]). The structure of the thin film matrix ensures that the resultant matrix which expresses the combination of any number of individual layers will always have a form similar to that of a single layer, provided only that the arrangement of layers is symmetrical (i.e. ABA ABCBA etc.). Any symmetrical combination acts therefore as a single layer having an equivalent index and thickness which can be derived from the product matrix. Of course the analogy must not be pushed too far and in particular it cannot be used for estimating the angular behaviour of the combination. For arrangements of the form ABA at long wavelengths, i.e. those for which the total optical thickness is less than $\lambda/2$, the equivalent index is intermediate between that of A and B and the equivalent optical thickness is approximately

$$D_A + D_B + D_A$$

where D_A, D_B denote the optical thicknesses of A and B respectively. For $D_A = 0.5 D_B$ the equivalent index tends to $(n_A n_B)^{\frac{1}{2}}$ as the wavelength tends to

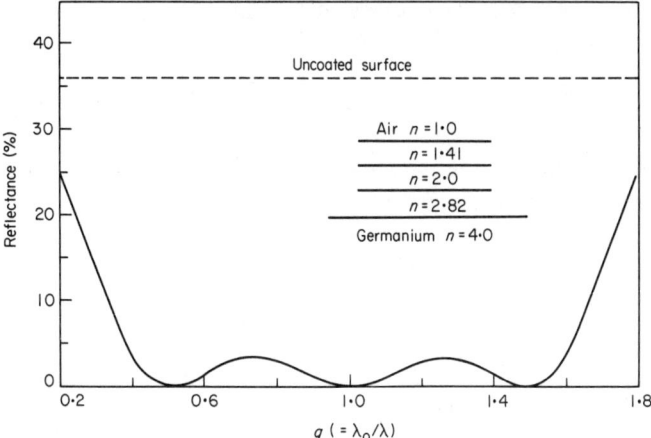

FIGURE 7. Calculated performance of a 3-layer antireflection coating for germanium. All layers are quarter waves.

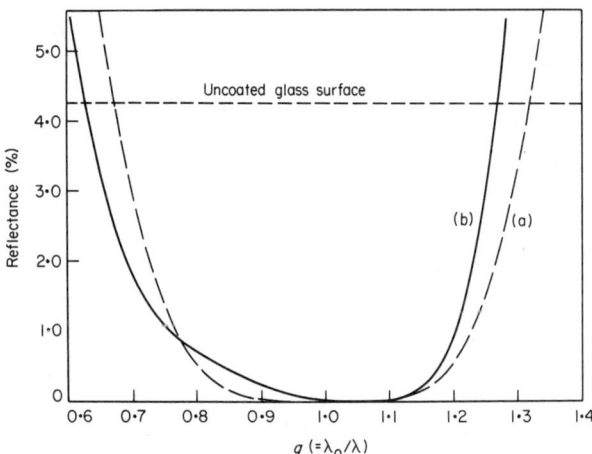

FIGURE 8. (a) $\frac{1}{4}$–$\frac{1}{2}$–$\frac{1}{4}$ antireflection coating. Glass $n = 1.52/(0.25\lambda_0; n = 1.70)$ $(0.5\lambda_0; n = 2.35)$ $(0.25\lambda_0; n = 1.38)$/Air $n = 1.0$.
(b) Coating as (a) but with the $\frac{1}{4}$ wave of index 1.70 replaced by a Herpin equivalent, the new design being Glass/$(0.072\lambda_0; n = 2.35)$ $(0.096\lambda_0; n = 1.38)$ $(0.572\lambda_0; n = 2.35)$ $(0.25\lambda_0; n = 1.38)$/Air $n = 1.0$.

infinity. For greater relative amounts of A or of B the index tends to a value nearer that of A or B respectively. If

$$\frac{2D_A}{D_B} = \zeta$$

then the equivalent index is given by

$$n_A \left[\frac{\zeta + \frac{n_B}{n_A}}{\zeta + \frac{n_A}{n_B}} \right]^{\frac{1}{2}} \quad (15)$$

This property can be used to produce equivalents for the intermediate layers in antireflection coating designs. Figure 8 is an example of such a coating. An alternative method of using two materials to produce an equivalent single layer in antireflection coatings is given by Vermeulen[14]. A particularly useful guide to practical coatings on glass is given by Ward[15].

More information on antireflection coatings will be found in Cox and Hass[16], Musset and Thelen[11] or Macleod[14]. Materials for antireflection coatings and for all other types of coatings are reviewed by Ritter[17].

2.3 Reflection Increasing Coatings

The results quoted for the single quarter wave show that a series of quarter waves can be replaced by a single surface of index

$$\frac{n_1^2 \, n_3^2 \, n_5^2 \, n_7^2}{n_2^2 \, n_4^2 \, n_6^2 \, n_5}$$

where 1 indicates the layer remote from the substrate. If, in the above expression, n_1, n_3, n_5, n_7 are all high index and n_2, n_4, n_6 are low, then the index of the surface will be high, and can be made higher still simply by increasing the number of high and low index films. A high value gives high reflectance and, theoretically at any rate, we can make it as high as we wish simply by increasing the number of layers. For example, 5 quarter waves of ZnS ($n = 2.35$) and cryolite (Na_3AlF_6 $n = 1.35$) of the form HLHLH on glass ($n = 1.52$), gives a reflectance in air of 89%. An alternative explanation for this behaviour is that the phase shift suffered by any ray, which penetrates the multilayer and then re-emerges at the front interface, is the same regardless of the particular route taken, so that an enormously efficient constructive interference takes place amongst all the beams. This type of assembly is known simply as a quarter wave stack.

Our calculation of this high reflectance of course is only valid for those wavelengths where the layers are either exact quarter waves or odd multiples of quarter waves. For wavelengths where the layers are integral numbers of half waves the reflectance of the combination will be that of the uncoated substrate. Calculation of the complete performance of the coating shows that the reflectance remains high over a specific region and then outside that region it falls rapidly to a low oscillatory value. A typical characteristic is shown in Fig. 9. The width of the high reflectance zone is a function of the ratio of the two refractive indices making up the stack. If we express the

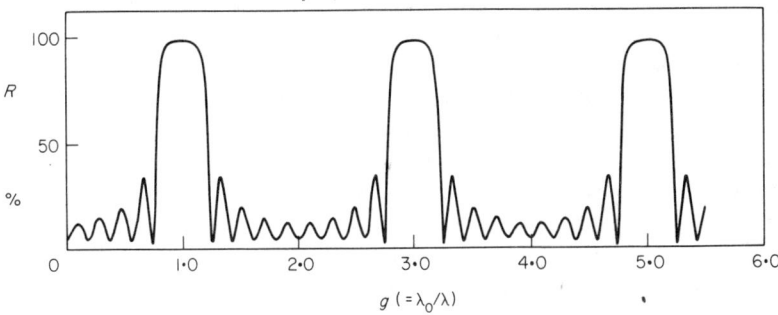

FIGURE 9. Quarter wave stack of ZnS ($n = 2.35$) and cryolite ($n = 1.35$) on a glass ($n = 1.52$) substrate. Design: Substrate/HLHLHLHLH/Air.

width in terms of g, where $g = (\lambda_0/\lambda)$, λ_0 being the wavelength for which the layers are quarter waves then the edges of the high reflection zones are $1 \pm \Delta g$, $3 \pm \Delta g$, $5 \pm \Delta g$, and so on with

$$\Delta g = \frac{2}{\pi} \sin^{-1}\left(\frac{n_H + n_L}{n_H + n_L}\right) \quad (16)$$

The limiting reflectance which can be achieved from any quarter wave stack is a function of the residual losses in the layers. These losses may be scattering or absorption. Scattering is usually caused by dust including particles of the coating material embedded in the layers or by defects of the substrate. Less often it can be due to the structure of the layers if the evaporation conditions are not ideal or if the layers are particularly thick. The incidence of dust particles can be reduced by taking very careful precautions, (Behrndt and Doughty[18], Heitmann[19] and Perry[20]) so that the losses due to this effect can be made very small indeed. Substrates can be better finished by using special polishing techniques (Bennett and Bennett[21]). Absorption in the layers is another problem. A number of workers have derived expressions for the performance of weakly absorbing films in multilayers, see for example Hemingway and Lissberger[22], and the residual absorption in a quarter wave stack of $2P+1$ layers with high index layers outermost is given by

$$A = 2\pi \frac{n_L}{(n_H - n_L)} \cdot \frac{(\kappa_H + \kappa_L)}{(n_H + n_L)} \qquad (17)$$

At the red end of the spectrum and in the near infrared measurements on combinations of zinc selenide and cryolite, and zinc selenide and thorium fluoride, show that residual losses can be less than 0.001% (Heitmann[19]). Results calculated from measurements of narrow band filters show that the absorption loss for zinc sulphide and cryolite must be of similar order or less. The very hard and robust materials which are usually refractory oxides are inclined to be rather poorer in this respect.

Apart from its use as a sub-unit of many optical coatings one of the commonest uses of the quarter wave stack as a component on its own is as a reflector in gas lasers. The simplest high reflectance coating is a single metal layer but even silver, which has the highest reflectance of all metals in the visible region, has a loss of some 4% and a maximum reflectance of 96%. The HeNe system has an extremely low gain at the 6328 Å line and such losses would be much too great to permit oscillation. Quarter wave stacks are therefore invariably used as the cavity reflectors. Quarter wave stacks are also used as coatings for Fabry Perot plates. The Fabry Perot interferometer has a peak transmittance of

$$T_{peak} = \frac{1}{[1 + (A/T)]^2}$$

where A and T are absorptance and transmittance respectively for the reflectors so that for the low transmittance coatings associated with high finesse it is essential that the absorption should be kept even lower. The quarter wave stack is a considerable improvement over metal layers in this respect. In the visible and near infrared zinc sulphide and cryolite are the most popular materials because, apart from their excellent optical performance, they can be fairly easily removed without damage to the plates if the wavelength range of the interferometer is to be altered or should the coatings prove faulty.

The limited width of the high reflectance zone of the quarter wave stack is often a disadvantage and it is useful to be able to produce coatings having a wider high reflectance zone. Much work has been carried out on this topic and three principal techniques have emerged. The first consists simply of depositing a number of quarter wave stacks on top of each other with different reference wavelengths. The main precaution to be taken here is in avoiding the occurrence of transmission peaks in the high reflectance zone (Turner and Baumeister[23]). Unless special precautions are taken these always occur when two similar overlapping quarter wave stacks are deposited on top of each other because the assembly acts in a similar way to the Fabry Perot Filter discussed later. The cure (Turner and Baumeister[23]) is to insert a single coupling layer which has a thickness of one quarter wave at a wavelength between the centres of the two stacks. A typical design would be $H(LH)^n L''(H'L')^n H'$ where $H(LH)^n$ is centred on λ_1, $(H'L')^n H'$ on λ_2 and L'' on $(\lambda_1 + \lambda_2)/2$. The second technique consists of depositing a stack where the thicknesses of the layers are staggered in some way so that the effect is similar to a quarter wave stack where the reference wavelength varies throughout. Various different thickness progressions such as geometric or arithmetic have been tried with very similar results (Penselin and Steudel[24], and Heavens and Liddell[25]). A third approach uses a computer either to synthesise a suitable design from scratch or, more often, to refine a standing design which has performance only a little short of what is required (Baumeister and Stone[26]).

All these techniques yield much broader high reflectance zones than the basic quarter wave stack but there are a number of important disadvantages. The first is that the phase of the reflectance usually varies rapidly with wavelength unless special precautions have been taken in the design. This means that it can be dangerous to use broadband reflectors in Fabry Perot interferometers, (Ramsay and Ciddor[27]). Ciddor[28] has designed special coatings specifically for interferometric use. Broadband reflectors can of course be used without difficulty in many applications such as lasers which operate over an extended region. The second difficulty seems so far to have excaped mention in the literature. This is increased sensitivity to absorption losses in the materials which can be both disappointing and surprising when switching from a quarter wave stack to a staggered multilayer. The principle of operation of all the extended reflectance zone coatings is that at any wavelength only a limited group of layers contributes to the high reflectance, in much the same way as the quarter wave stack. For some wavelengths the layers giving high reflectance will be at the top of the stack and for these wavelengths the performance will be similar to the quarter wave stack. There will, however, be other wavelengths where the light must first penetrate to nearly the bottom of the stack before being reflected and because this light passes twice through a considerable thickness of material any residual absorption will have a major effect. At these wavelengths the reflection performance can be quite poor. When testing coatings it is much simpler to run a transmittance than a reflectance curve. The transmittance curve invariably shows the ideal theoretical shape for the coating and it is tempting to assume that all is well. When testing extended zone high reflectance coatings, the reflectance itself should always be measured. In the visible region the absorption losses are usually greater at the short wavelength end of the scale so that it is marginally better to have the thinnest layers at the top of the stack although much depends on the particular application.

Because of these difficulties and also because extended reflectance zone coatings involve large numbers of layers, for high reflectance over wide regions the simple metal layer is still the normal choice. Measurements of the reflectance of films of a number of the more useful metals together with notes on the technique of boosting the reflectance of metals over a limited region by adding several dielectric layers are given by Hass[29]. Aluminium layers are almost always used for reflectors in the visible region. It adheres well to glass and can be protected from abrasion by layers such as silicon oxide or aluminium oxide. Silver has the highest reflectance in the visible region with the great advantage that at high angles of incidence the polarisation effects are less than for aluminium and it is very easy to evaporate. However it tarnishes rapidly, is soft and protecting layers have never adhered well to it and so for these reasons silver has not been much used except in the laboratory. Recently the fact that silver sticks well to layers of Al_2O_3 has been used in the development of a new type of coating involving silver which has excellent resistance to abrasion and tarnishing together with high reflectance (Hass et al.[30]). The design is of the form: Substrate/Al_2O_3 (900 Å)/Ag (800–1000 Å)/Al_2O_3 (300 Å)/Silicon Oxide (1500 Å)/Air. The aluminium oxide layers act simply to improve the adhesion of the silver layer while the silicon oxide seals and protects the combination.

2.4 Edge Filters

Losses in the quarter wave stack are small provided the layers are reasonably free from absorption, so that the reflectance and transmittance characteristics are complimentary, consisting of pass bands, where transmittance is high and reflectance low, and stop bands, where transmittance is low and reflectance high. The quarter wave stack can therefore be used as a filter of several different types—long wave pass, short wave pass, band pass or band stop. The principal deficiency of the simple quarter wave stack is the prominent ripple in the pass bands. At an early stage it was found by trial and error that this could be reduced considerably by the addition of eight wave layers on either side of the stack. These were either of high or low index depending on the particular materials being used, and on whether the filter was intended to be long or short wave pass. Since then, design techniques have been established which make use of the properties of symmetrical assemblies already mentioned and remove the trial and error from the procedure.

The equivalent index, n_E, and equivalent phase thickness, ξ, of a symmetrical unit consisting of a quarter wave layer of one material n_B surrounded by two, eight wave layers of another, n_A, are given by

$$n_E = n_A \left\{ \frac{\cos \delta - \dfrac{r-1}{r+1}}{\cos \delta + \dfrac{r-1}{r+1}} \right\}^{\frac{1}{2}} \tag{18}$$

$$\cos \xi = \cos^2 \delta - \frac{1}{2}\left(r + \frac{1}{r}\right)\sin^2 \delta \tag{19}$$

where $\qquad r = n_A/n_B$

and $\qquad \delta =$ phase thickness of layer B

Figure 10 shows a plot of the equivalent index and thickness of a zinc sulphide and cryolite combination. The blank regions correspond to high reflectance zones where equivalent index and thickness are imaginary. In the pass regions the equivalent indices are real and the equivalent thicknesses are close to a half wave at the reference wavelength. A number of these combinations in series will therefore have the pass bands and the stop bands of the quarter wave stack, but, in the pass bands the multilayer will act as a single thick layer of material with index equal to the equivalent index. The reflectance of the stack in the pass band will therefore oscillate within a region with boundaries given by

$$R_1 = \left(\frac{n_0 - n_s}{n_0 + n_s}\right)^2$$

and

$$R_2 = \left(\frac{n_0 - n_E^2/n_s}{n_0 + n_E^2/n_s}\right)^2$$

The reason why the eight wave layers work is now clear. The combination

$$\left(\frac{H}{2} L \frac{H}{2}\right)$$

is a better match for glass as a long wave pass and

$$\left(\frac{L}{2} H \frac{L}{2}\right)$$

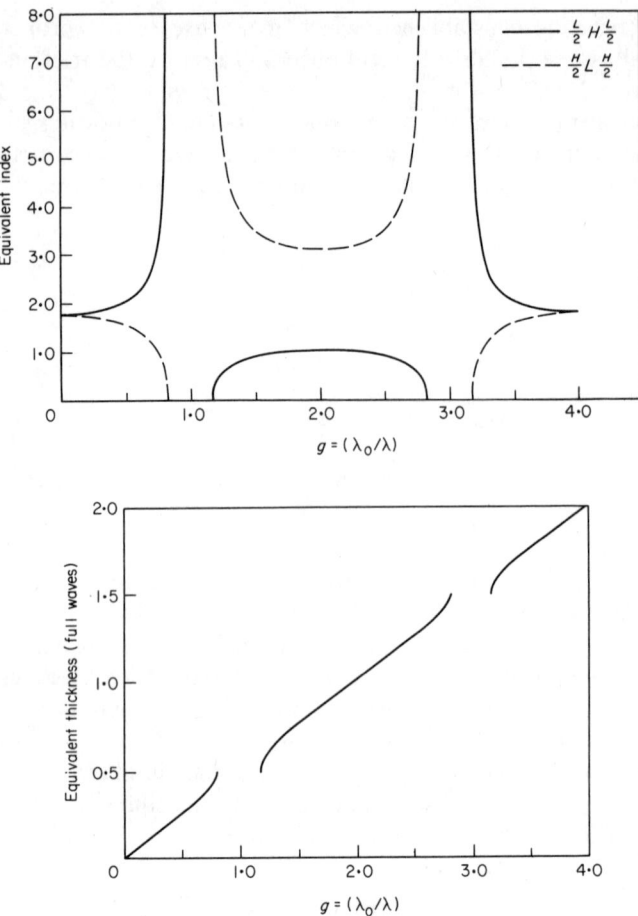

FIGURE 10. Equivalent index and thickness of $\left(\dfrac{L}{2}H\dfrac{L}{2}\right)$ and $\left(\dfrac{H}{2}L\dfrac{H}{2}\right)$ combinations with $n_L = 1.35$ and $n_H = 2.35$.

as a short wave pass filter. It is also possible to improve the performance still further. In the pass regions where the stack acts as a single thick slab of material two antireflection coatings are all that is required, one matching it to the substrate, and the other to the medium. Often the stack itself turns out to be a good match for either the substrate or the medium, or it can be made so by judicious choice of materials in which case just one matching layer is necessary. Figure 11 shows the design of such a coating. More information on the use of equivalent layers is given by Thelen[31]. Complex techniques exist for the elimination of ripple with greatly improved potential performance, see

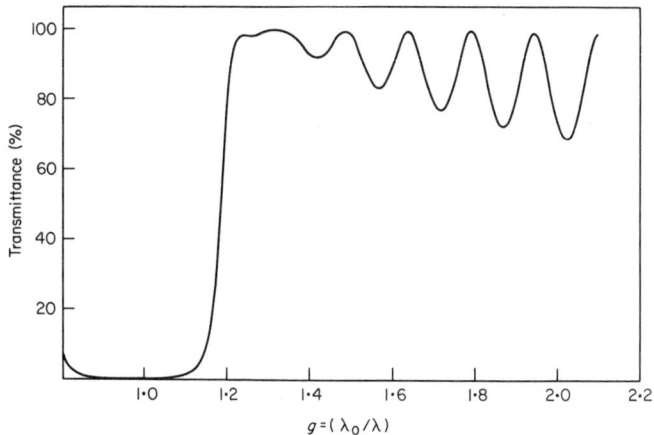

FIGURE 11. Short wave pass filter

$$\text{Ge} \left/ \frac{L}{1.25} \left(\frac{L}{2} H \frac{L}{2} \right)^6 \right/ \text{Air}$$

$L = \frac{1}{4}$ wave ZnS $n = 2.3$
$H = \frac{1}{4}$ wave Ge $n = 4.0$
$n_{\text{air}} = 1.0$.

for example Macleod[4] or Seeley et al.[32], but for the improvement to be realised (Evans et al.[33]) great accuracy is required in layer deposition.

When a quarter wave stack is used as a short wave pass filter the pass region has limited width because of the higher order reflectance zones at $\lambda_0/3$, $\lambda_0/5$, $\lambda_0/7$ and so on. Techniques have been developed by Thelen[34] for eliminating some of these higher orders. There is a potential high reflectance zone at every wavelength where the total phase thickness of the basic symmetrical period has a phase thickness of $m\pi$, m being an integer. If, however, the transmittance of the basic period can be made unity at any of these wavelengths then the zone of high reflectance will be suppressed. This happens automatically in the quarter wave stack for $m = 2, 4, 6, 8\ldots$, where all the layers are integral multiples of half waves. Thelen incorporates an antireflection coating in the basic period which at the correct wavelengths matches the period into a dummy medium. The antireflection coating normally consists of two layers and the index of the dummy medium is chosen to simplify the choice of antireflecting materials. When a number of similar periods are combined into a stack because they are matched to the

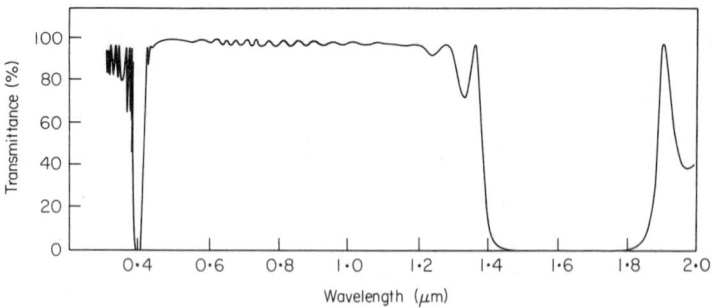

FIGURE 12. Short wave pass filter centred on 1.6 μm with the 2nd and 3rd reflectance zones suppressed. Design by Thelen[34]:
$$\text{Glass, } n = 1.52/\text{L}'/(\text{L M H M L})^{10}/\text{Air}$$
$$\text{L}' = 0.25\lambda_0 \; n = 1.38$$
$$\text{L, M, H} = 0.1\lambda_0 \; n = 1.38, 1.90, 2.30 \text{ respectively}$$
$$\lambda_0 = 1.6 \, \mu\text{m}.$$

dummy medium its thickness is unimportant and it can be allowed to shrink to zero so that it disappears altogether. Figure 12 shows the transmittance of a typical coating in which the second and third orders have been suppressed (Thelen[34]). The layers, arranged in the fashion ABCBA, are all of equal optical thickness, $\lambda_0/10$, while the refractive indices are connected by the relationship

$$\tan^2\left(\frac{2\pi}{5}\right) = \frac{n_A n_B - n_C^2}{n_B^2 - n_A n_C^2/n_B} \tag{20}$$

Of course it is also possible to reduce the width of the transmission zone by restoring some of the suppressed zones. A basic stack where layers are alternately $\lambda_0/6$ and $\lambda_0/3$ in optical thickness will have reflectance zones at λ_0 and $\lambda_0/2$. There will however be no reflectance peak at $\lambda_0/3$ because at that wavelength all the layers are multiples of half waves and the transmittance of the basic period will be unity.

Stop bands can be broadened simply by adding two stacks together. The centre wavelengths of each are chosen so that the reflectance zones just overlap. This procedure can be repeated a number of times but of course if simple quarter wave stacks are being used then, unless it is a long wave pass filter which is being constructed, higher order stop bands will eventually interfere with the pass region. If a short wave pass filter is required it will probably be necessary to use one of the suppressed zone filters. The quarter wave stack can also be used as a band stop filter. The stop band is the high reflectance region. Unfortunately, the width of the stop band cannot be made very narrow because it is determined by the ratio of the high and low refractive indices which in turn determines the number of layers required to achieve the necessary rejection. The narrower the band the greater the number of layers required and the tighter the layer thickness tolerances so that the chance of a serious reduction in performance due to errors is increased. Further since the number of layers traversed by the light is greater, absorption losses will be increased. All this limits achievable performance to some tens of nanometres which is unfortunate because there is a number of applications where rather narrower widths would be an advantage. Really narrow band stop filters using interference techniques are not available. The quarter wave stack band stop filter suffers as the other edge filters from ripple in the pass regions. Here the problem is more serious than in either short wave pass or long wave pass filters because the equivalent indices are different on either side of the pass band and the elimination of the ripple on one side always seems to be accompanied by an increase of the ripple on the other. Thelen[35] has devised an ingenious technique for removing ripple on both sides of the stop band simultaneously, which is based on the concept of equivalent index mentioned earlier. These band stop filters are often known as optical minus filters.

2.5 Band Pass Filters

The simplest type of band pass filter is a combination of long wave pass and short wave pass filter. This is a very useful and flexible arrangement for a broad but not a narrow pass band, because of the difficulty of producing either the long wave or short wave pass components with sufficiently steep edges. The minimum width which would normally be attempted by such a combination would be around one tenth of the centre wavelength.

2.5.1 Fabry Perot filters

For narrower filters, therefore, a completely different approach, based on the Fabry Perot interferometer, is used. In its classical form the Fabry Perot interferometer consists of two highly accurate optical flats, coated to have high reflectance, spaced apart and aligned to be parallel, so that multiple beam interference can take place in the layer of air between the plates. It can be shown (see for example Born and Wolf[1] or Knittl[2] or Macleod[4] that the transmittance of the Fabry Perot is given by

$$T_F = \frac{T_a T_b}{[1-(R_a R_b)^{\frac{1}{2}}]^2} \cdot \frac{1}{1+F\sin^2\left[\frac{1}{2}(\psi_a+\psi_b)-\delta\right]} \quad (21)$$

$$F = \frac{4(R_a R_b)^{\frac{1}{2}}}{[1-(R_a R_b)^{\frac{1}{2}}]^2}$$

$$\delta = \frac{2\pi\eta\, d\cos\theta}{\lambda} = \text{phase thickness of space layer}$$

R_a, R_b = reflectance of plates a and b respectively
T_a, T_b = transmittance of plates a and b respectively
ψ_a, ψ_b = phase shift on reflection from a and b respectively.

Usually the reflectors will be well matched and if we assume also that the phase shifts due to reflection is zero or 180° then the transmittance becomes

$$T_F = \frac{T^2}{(1-R)^2} \cdot \frac{1}{1+F\sin^2\delta}$$
$$F = 4R/(1-R)^2 \quad (22)$$

This function gives the well known series of narrow transmission bands centred on $\delta = m\pi$ where m is the order number. The bandwidth measured at half the peak height, or halfwidth $\Delta\lambda_H$ is given by

$$\Delta\lambda_H = \frac{(1-R)}{m\pi\sqrt{R}} \cdot \lambda_p \quad (23)$$

where λ_p is the peak wavelength.

Thus, keeping λ_p constant, the higher the reflectance R and the higher the order number m, the narrower is the pass band. Also, the higher the value of m, the narrower is the interval between neighbouring orders. A useful parameter is the finesse of the interferometer which is defined as the interval between fringes divided by the fringe width. Finesse is given by

$$\mathscr{F} = \frac{\pi\sqrt{R}}{(1-R)} \quad (24)$$

Then resolving power $= \dfrac{\lambda_p}{\Delta\lambda_H} = m\mathscr{F}.$ \quad (25)

A thin film Fabry Perot filter is similar in principle to the classical interferometer but the spacer is a transparent solid thin film and the two reflectors together with the spacer are all deposited as a single multilayer on one surface of the substrate. The order number m is low since the spacer is not usually thicker than some three or four wavelengths, because of the tendency to suffer from increasing scattering losses. Since the thickness distribution of the thin films depends primarily on the parameters of the deposition process, the dimensions of the plant, source position, and so on, but not on the flatness of the substrate, comparatively inexpensive substrates, flame polished microscope slide glass for instance, can be used.

In its simplest form, the thin film Fabry Perot filter consists of two metallic reflector layers on either side of a dielectric spacer and is known as a metal-dielectric filter. Absorption in the metal layers is the principal factor which limits the performance. Peak transmittance is given by

$$T_p = \frac{T^2}{(1-R)^2} = \frac{1}{\left(1+\frac{A}{T}\right)^2} \tag{26}$$

where A is the absorptance of each reflector. In the visible region, the most suitable metal is silver which has a residual absorption loss of around 4% which implies that the reflectance must be lower than 92% if the transmittance is to be greater than 25%. Typical first order metal-dielectric filters with bandwidths of 10–20 nm in the visible region have peak transmittances usually in the region of 20%–40%. Second order filters can be used to reduce the pass band width while retaining peak transmittance but they have the disadvantage of the unwanted first order peak at a longer wavelength. In a first order metal-dielectric Fabry Perot filter there is no peak of transmittance on the long wavelength side of the peak and so to use it as a filter we merely have to suppress the short wave second and higher order peaks. There is a wide range of glass absorption filters having long wave pass characteristics with sufficient edge steepness to use as suppression filters in this application but the position regarding short wave pass absorption filters is not nearly as good. In the visible and near infrared regions the usual combination of materials for these filters is silver and cryolite. A typical commercial filter will normally consist of a cemented sandwich of absorption glass and the thin film filter, with the absorption glass acting as a cover to protect the thin films from abrasion and moisture (Fig. 13). Because these filters are relatively straightforward to produce they are usually the cheapest which are available.

The optical performance of silver in the ultraviolet is very poor and aluminium is the most satisfactory metal, although not as good as silver in the visible and near infrared. Aluminium and magnesium fluoride are commonly used for ultraviolet filters. Aluminium reacts readily with any oxygen which

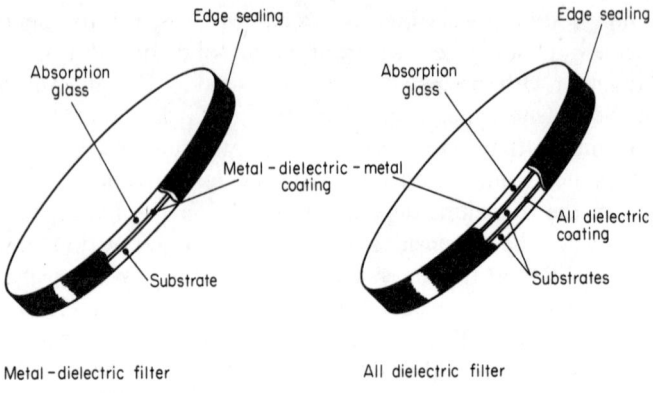

FIGURE 13. Construction of typical filters for the visible and near infrared.

is present and the resulting film of aluminium oxide reduces the ultraviolet reflectance of the coating appreciably especially beyond 300 nm. Thus the best technique for producing ultraviolet filters is to evaporate the aluminium rapidly at pressures as low as possible and certainly lower than 10^{-5} torr. Even after evaporation and while still under vacuum, the residual oxygen in the plant reacts with the aluminium films and so the deposition of the first aluminium layer should be followed immediately by the magnesium fluoride spacer and that of the second aluminium reflector by a magnesium fluoride protecting layer. This protecting layer can also be made to serve as a transmittance increasing layer since, if it is of the correct thickness, it will to some extent act as an antireflection coating on the aluminium. There are no really suitable cements in the ultraviolet beyond 300 nm and so the outer layer must continue to act as a protecting layer after production. A thin film of this sort never gives completely effective protection so that the life of uncemented ultraviolet filters is uncertain. Steps should always be taken to avoid exposure to high humidity and rapid changes of temperature. More information on metal-dielectric filters is given by Knittl[2] and Macleod[4].

The metal dielectric filter is attractive because of its simplicity and low cost but its performance is not adequate for a very large number of applications. Narrower filters can be constructed if, as in the case of the conventional Fabry Perot interferometer, the metal layers are replaced by all dielectric assemblies such as the quarter wave stack where the losses can be made very much smaller. Thus a peak transmittance in excess of 50% can be achieved even with bandwidths as low as 0.5 nm in the visible region. Zinc sulphide and cryolite are once again the most popular materials for filters operating in the visible and near infrared, while in the infrared beyond 2 μm combinations such as germanium ($n \simeq 4.0$) or lead telluride (transparent for $\lambda > 3.5$ μm, $n \simeq 5.6$) as the high index materials with silicon monoxide (transparent for $\lambda < 8$ μm, $n \simeq 1.7$) zinc sulphide ($n \simeq 2.2$) or zinc selenide ($n \simeq 2.3$) as the low index materials, are common.

Of course the production of these filters with their improved performance is a more exacting task than metal-dielectric filters since the number of

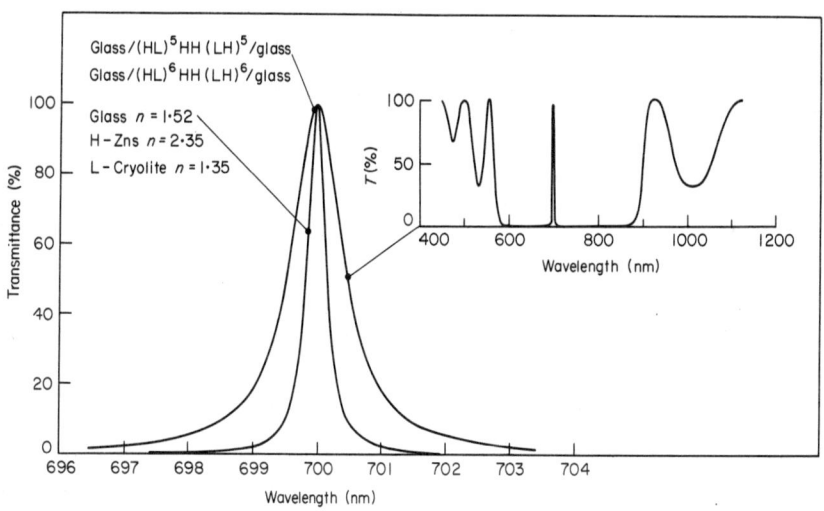

FIGURE 14. Fabry Perot filters.

layers and the precision required are both much greater and so the cost is inevitably much higher. Characteristic curves of typical all-dielectric narrow band filters are shown in Fig. 14. The quarter wave stack has a drop in reflectance at either side of the high reflectance zone and this is reproduced in the characteristic curves of the narrow band filters. The sidebands of transmission which result must normally be suppressed before the filter can be used, and this is usually accomplished by adding a metal-dielectric filter with its absence of long wave side bands, together with an absorption filter to eliminate the short wave remaining side bands. The construction of a typical narrow band filter is shown in Fig. 13.

Some of the problems in the production of narrow band filters are reviewed by Lissberger and Pearson[36] and by Macleod[37]. A major difficulty is in the exclusion from the layers of atmospheric moisture which is responsible for the drifts often observed after manufacture. The moisture penetrates in small patches rather than evenly over the filter surface and this causes irregular changes of filter profile during the drift (Macleod and Richmond[38]). The materials used for filters in the visible region are capable of halfwidths of less than 0.1 nm with useful peak transmission but there are difficulties in monitoring the layers for halfwidths as narrow as this as well as obtaining the necessary uniformity. Halfwidths of 0.3 to 0.5 nm are more readily achievable. For narrower filters use is made of what are called solid etalon filters or sometimes solid spacer filters.

Solid etalon filters are very high order Fabry Perot filters. For constant reflectance R the bandwidth is inversely proportional to the order number m, and thus a very high order seems a convenient way of narrowing a filter. It is difficult to increase the order of the conventional thin-film Fabry Perot filter beyond five or six and retain high peak transmittance apparently because scattering losses increase with layer thickness, although this is disputed by Lissberger and Pearson[36]. The solid etalon uses a slab of polished optical material as the spacer which allows high order without the scattering losses of a thick deposited film. Thin film reflectors are deposited on either side of the spacer in the normal way so that the spacer also acts as substrate. The solid etalon filter has the advantage of robustness and stability over the conventional Fabry Perot interferometer with comparable manufacturing difficulties.

An early worker in this field was Dobrowolski[39] who developed techniques for solid etalon filter production using mica as spacer. According to Dobrowolski, Billings was probably the first to use mica in this way, achieving halfwidths of 0.3 nm. Dobrowolski obtained substantially better halfwidths than this and also his is the first complete account of the technique. Mica can be cleaved readily to form thin sheets with very flat and parallel surfaces. There is a complication due to the anisotropy of the mica which has to be cleaved to form a half wave plate at the required wavelength otherwise the transmission peak depends on the plane of polarisation. This implies that there is a minimum thickness of mica which can be used and therefore a minimum order number. When the order number m is high, the separation between orders is given approximately by λ/m. For a peak wavelength of 546.1 nm, Dobrowolski found that the maximum order separation was 1.64 nm. Using this technique he did succeed in producing filters with halfwidths around 0.1 nm the narrowest being 0.085 nm. The filter must be used in conjunction with a conventional thin film narrow band filter to suppress the unwanted orders.

Another type of solid etalon filter which has recently become very popular uses as spacer a thin slice of appropriate material which has been optically worked so that the surfaces have the necessary degree of parallelism. This is no easy task. The most complete account so far of a method of producing such filters is by Austin[40]. Fused silica spacers as thin as 50 μm have been produced with the necessary parallelism for halfwidths as narrow as 0.1 nm in the visible region while thicker discs can give bandwidths as narrow as 0.005 nm. A 50 μm fused silica spacer gives an interval between orders of around 1.4 nm in the visible and again unwanted orders must be suppressed by subsidiary filters. The process of optical working is such that the absolute thickness error Δd over the spacer is independent of the thickness. This is true also of the conventional air-spaced Fabry Perot. Lack of parallelism in the spacer causes shifts in the position of the transmission peak over the area of the filter. If shifts are to be maintained at less than half the bandwidth then we can write

$$\frac{\Delta\lambda_0}{\lambda_0} = \frac{\Delta D}{D} \leqslant \frac{0.5\Delta\lambda_H}{\lambda_0}$$

where D is the optical thickness, nd, of the spacer. But

$$\frac{\Delta\lambda_H}{\lambda_0} = \frac{1}{m\mathscr{F}}$$

hence

$$\mathscr{F} \leqslant \frac{0.25\lambda_0}{\Delta D}$$

Now the achievable ΔD is of the order of $\lambda/100$ in the visible region which means limiting finesse is around 25. The required resolving power then has to be achieved by the order number m which determines both the spacer thickness $D = m(\lambda_0/2)$, and the interval between orders, λ_0/m. For a halfwidth of 0.01 nm at 500 nm resolving power is 50000 implying an order number of 2000, a spacer optical thickness of 500 μm and an interval between orders of 0.25 nm. For a halfwidth of 0.1 nm only 50 μm spacer optical thickness is required giving an interval between orders of 2.5 nm. The restricted free spectral range of the narrower filters means that a broader solid etalon filter is required in addition to a conventional thin film narrowband filter for sideband blocking. The temperature coefficient of peak wavelength of the solid etalon filters with fused silica spacers is 0.1 nm for 20°C at Hα = 656.3 nm. The solid etalon filter has also been used in the infrared. Smith and Pidgeon[41] described experimental results using a polished slab of germanium some 780 μm thick working at around 700 cm^{-1} in the 400th order coated on either face with a quarter wave of ZnS followed by a quarter wave of PbTe to give a reflectance of 0.82, a fringe halfwidth of 0.1 cm^{-1}, and a spacing between orders of 1.6 cm^{-1}. This particular arrangement was chosen so that a number of adjacent orders would match exactly the lines in the CO_2 R-branch at 14.5 μm which are spaced 1.6 cm^{-1} apart. Order sorting was not therefore a problem. Roche and Title[42] have recently constructed a range of solid etalon filters for the infrared. These filters are some 13 mm in diameter, have resolving powers of the order of 3×10^4 and the techniques used for their construction are those described by Austin[40]. For wavelengths out to near 3.5 μm, fused silica spacers are perfectly satisfactory. For longer wavelengths Roche and Title, after investigating a number of materials, finally decided on Yttralox which is a combination of yttrium and thorium oxide. With this material, solid etalon filters were produced which at 3.334 μm had halfwidths as low as 0.2 nm and at 4.62 μm, 0.8 nm halfwidths. The finesse achievable with this technique is 30 to 40 and the limit to the halfwidth achievable seems at present to be the interval between orders which determines the complexity of the additional sideband blocking filters.

2.5.2 Angular behaviour of the Fabry Perot filter

The form of the phase thickness of a thin film,

$$\delta = \frac{2\pi\, nd \cos\theta}{\lambda}$$

shows that its apparent optical thickness varies with the angle of incidence, and is greatest for normal incidence. For small tilts this effect predominates over the changes in effective index which also take place. For the same external angle of incidence, high index layers have smaller values of θ and are therefore less affected than those with a low index. In a Fabry Perot filter therefore the layers will all appear thinner when the filter is tilted, and although the effect in an individual layer depends on its refractive index, for tilts up to some 20–30° the overall result is simply a shift of the peak towards shorter wavelengths, peak transmittance and halfwidth remaining unchanged (see for example Lissberger and Wilcock[43], Pidgeon and Smith[44]). Pidgeon and Smith[44] show that this shift is similar to that which would be obtained from an ideal filter with spacer index n^*, intermediate between the high and low indices of the layers of the filter. In a simple first order Fabry Perot filter consisting of two materials n_H and n_L the effective index, n^*, is given by

$$n^* = \sqrt{n_H n_L} \tag{27}$$

for a high index spacer and

$$n^* = \left[\frac{n_L^2}{1 - \frac{n_L}{n_H} + \left(\frac{n_L}{n_H}\right)^2} \right]^{\frac{1}{2}} \tag{28}$$

for a low index spacer. Then

$$\lambda_{\text{peak}} = \lambda_0 \cos \theta$$

where

$$\theta = \sin^{-1}\left(\frac{n_i \sin \theta_i}{n^*}\right)$$

and θ_i, n_i refer to the incident medium. For small angles this is approximately

$$\frac{\Delta \lambda_0}{\lambda_0} = \frac{\theta_i^2}{2n^{*2}} \tag{29}$$

where

$$\lambda_{\text{peak}} = \lambda_0 - \Delta \lambda_0$$

and θ_i is given in radians. With θ_i in degrees this becomes

$$\frac{\Delta \lambda_0}{\lambda_0} = 1.5 \times 10^{-4} \frac{\theta_i^2}{n^{*2}} \tag{30}$$

This gives the shift in peak wavelength with angle θ_i. In practice a filter will be exposed simultaneously to a range of angles of incidence. The simplest case is a cone of semiangle Ψ. The analysis is simplest in terms of wavenumber $v(=(1/\lambda))$. The peak shifts to a new position

$$v_{\text{peak}} = \tfrac{1}{2}[v_0 + (v_0 + \Delta v')] = v_0 + \tfrac{1}{2}\Delta v' \tag{31}$$

where $\Delta v'$ is the shift corresponding to angle Ψ

$$\frac{\Delta v'}{v_0} = 1.5 \times 10^{-4} \frac{\Psi^2}{n^{*2}}$$

The new halfwidth is given by

$$V_\phi^2 = V_0^2 + (\Delta v')^2 \tag{32}$$

and the peak transmittance by

$$\frac{T'}{T_0} = \frac{V_0}{\Delta v'} \tan^{-1} \frac{\Delta v'}{V_0} \div \frac{1}{[1 + (\Delta v'/V_0)^2]} \tag{33}$$

where T_0 and V_0 refer to normal incidence.

We can extend this analysis of Pidgeon and Smith[44] to a cone of semi-angle Ψ incident at an angle other than normal provided we make some simplifying assumptions. Let the angle of incidence be Θ and the cone semiangle be Ψ. Then the range of angles encompassed by the filter will be $\Theta \pm \Psi$. If $\Theta < \Psi$ then the result is simply that for a filter illuminated normally by a cone of semiangle $\Theta + \Psi$. If $\Theta > \Psi$ then we have three frequencies, v_0 corresponding to normal incidence, v_1 the peak corresponding to angle of incidence $\Theta - \Psi$ and v_2 the peak corresponding to angle of incidence $\Theta + \Psi$. Then the new filter peak is simply

$$\tfrac{1}{2}(v_1 + v_2) \qquad (34)$$

The halfwidth is

$$V' = [W_0^2 + (v_2 - v_1)^2]^{\tfrac{1}{2}} \qquad (35)$$

and the peak transmittance

$$\frac{T'}{T} = \frac{V_0}{(v_2 - v_1)} \tan^{-1}\left\{\frac{(v_2 - v_1)}{V_0}\right\} \qquad (36)$$

$$\doteq \frac{1}{[1 + (v_2 - v_1)/V_0)^2]}$$

$(v_2 - v_1)$ is proportional to $\Theta \cdot \Psi$ and Hernandez[45] has found excellent agreement between measurements made on real filters with the results calculated from these assumptions for values of $\Theta \cdot \Psi$ up to 100 degree^2.

One particularly significant result which does emerge from the above analysis is that if a filter is to be used at maximum efficiency in a cone of light at normal incidence, then the peak wavelength in collimated light at normal incidence should be longer than that required, to allow for the shift in equation (31). (For example Dobrowolski[39], Linder[46], Lissberger and Wilcock[43] and Lissberger[47]). The extension of the concept of effective index n^* to metal-dielectric filters is described by Hemingway and Lissberger[48].

2.5.3 Multiple cavity filters

The shape of the Fabry Perot filter pass band is far from the ideal rectangle. The Fabry Perot can be thought of as a type of resonant cavity and it is possible to couple two or more such cavities together to give a combined response which is more rectangular in pass band shape and has greater rejection in the stop zones, in much the same way as the coupling of tuned circuits in electrical filters. The procedure is quite simple and consists of depositing a second Fabry Perot filter on top of the first with a single quarter wave coupling layer between. The design and performance of a typical two cavity filter or, as it is more often called, a double half wave filter (DHW filter) is shown in Fig. 15. This simple combination works very well for

Thin Film Optical Devices

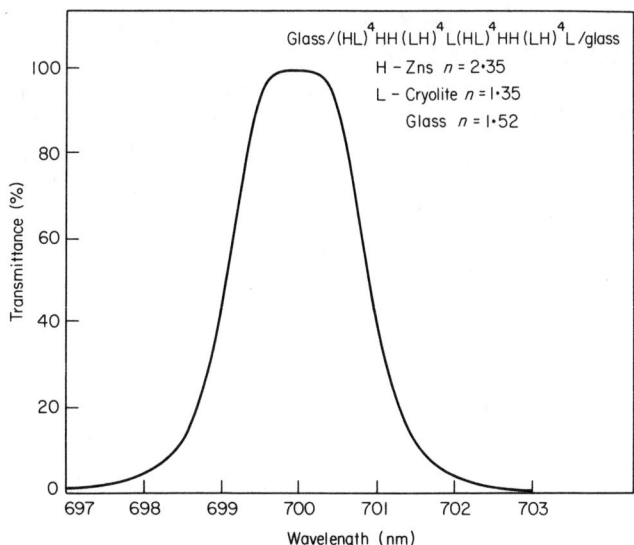

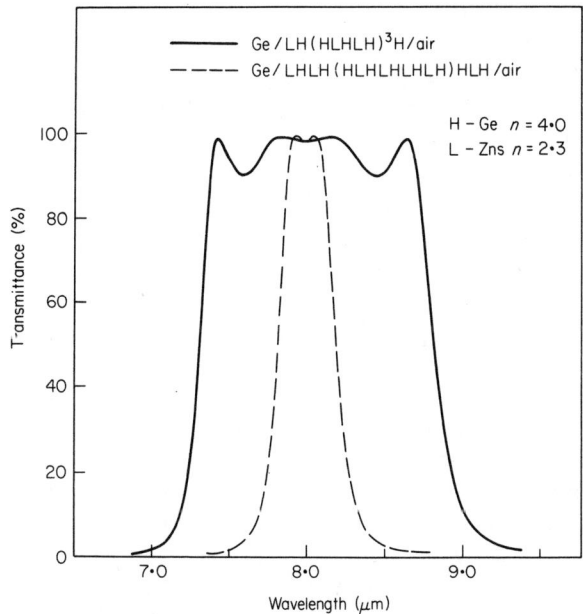

FIGURE 15. Multiple cavity filters.

virtually any pair of cavities but the addition of further cavities sometimes gives a pronounced ripple in the pass band rather than the expected improvement in shape. It turns out that empirical methods are not suitable for designing filters with three or greater number of cavities. The most successful approach is due to Thelen[31] and is based on symmetrical periods. Instead of dealing with cavities and coupling layers the filter is broken into a series of symmetrical periods with large numbers of layers, together with matching stacks at either end. The symmetrical periods can be shown to have real equivalent indices in the pass bands, that of the combination $(AB)^n A$, for instance, where A and B are quarter waves is $n_e = n_A(n_A/n_B)^n$, and then the filter can be looked on as a single thick slab with the equivalent index of the symmetrical periods which is matched by stacks to the substrate at one end and to the medium at the other. Figure 15 includes some filter curves and designs which were derived in this way.

The usefulness of multiple cavity filters is not limited to those of very narrow band widths. Broad band filters of the simple Fabry Perot type will have much less satisfactory shapes than narrow band filters because the lower reflectance of the mirrors means that the rejection outside the pass band is poor. Multiple cavity filters can avoid this problem.

2.5.4 Induced transmission filters

Another important type of filter is the induced transmission filter devised by Berning and Turner[47]. This filter is really very similar to a two-cavity all-dielectric filter in which the central reflectors and coupling layer are replaced by a single metal layer. The usual design technique, however, does not use this approach but concentrates instead on matching a single metal layer to its surroundings so that maximum transmittance is achieved. This is equivalent to designing an antireflection coating for each side of the metal layer. The antireflection coatings are, not surprisingly, similar to the spacer and second reflector of a Fabry Perot filter. The name of the filter is derived from this design approach which induces maximum transmittance through the metal. Various design techniques are described by Berning and Turner[49], Holloway and Lissberger[50], Landau and Lissberger[51] and Thetford[52]. The principal advantage of the induced transmission filter over the all-dielectric band pass filter is that the metal layer gives good rejection in the stop regions and, in particular, provided the filter is used in the first order, there are no sidebands on the long wavelength side of the peak. These filters are usually made with bandwidths several tens of nanometers in the visible and since with these widths they have excellent peak transmittance they are very suitable for suppressing the sidebands of narrower all-dielectric filters. A useful summary of the performance of current narrow band filters of various types is given by Baumeister[53]. A review paper which contains much useful information on narrow band filters as well as on other aspects of optical thin films has been written by Lissberger[54].

2.6 Polarisers

The reflectance of a single surface as the angle of incidence is changed is shown in Fig. 16. The reflectance for *p*-polarised light (electric vector in the plane of incidence) falls to zero at the Brewster angle, while that for

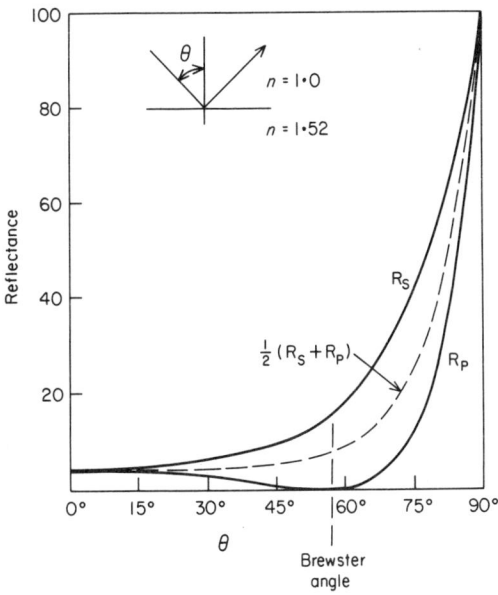

FIGURE 16. Reflectance of a single surface of glass ($n = 1.52$) in air ($n = 1.0$) as a function of angle of incidence in air.

s-polarised light increases gradually without passing through any minimum. Since the light reflected at the Brewster angle is entirely *s*-polarised while that transmitted is partially *p*-polarised, this effect can be used as the basis of a simple polariser. A pile of plates acts as a polariser and enhances the polarisation of the transmitted beam without loss of *p*-polarised intensity by tilting the stack to the Brewster angle. The principles of the pile of plates polariser and of the quarter wave stack can be combined in a thin film polarising beamsplitter (MacNeille[55], Banning[56]).

Consider the interface between two layers, one of high index, e.g. zinc sulphide $n = 2.35$, and one of low index, e.g. cryolite $n = 1.35$. For such a combination the Brewster angle, measured in the high index material, will be

$$\theta_B = \tan^{-1}(n_L/n_H)$$

A quarter wave stack for this angle of incidence will have high reflectance for the s-polarised light in the normal way but for the p-polarised light the reflectance will be zero. Thus the transmitted beam will be very highly p-polarised while the reflected beam will be highly s-polarised. Unfortunately when the Brewster angle is referred back to air as the incident

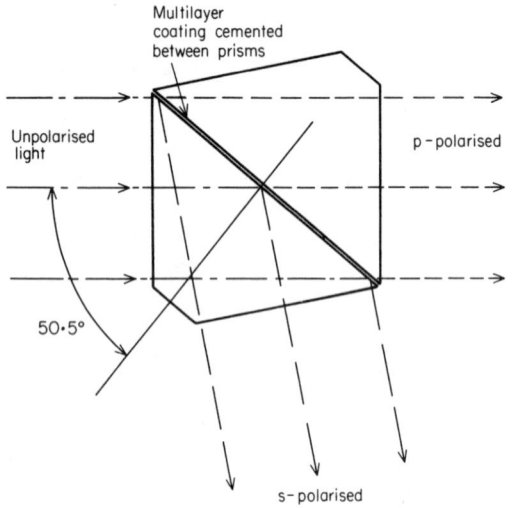

FIGURE 17. Construction of a polarising beam splitter.

medium using Snell's law, it is found to be greater than the critical angle. Thus a simple coating of this type can never permit the required angle of incidence in air. If, however, the incident medium is glass, then a real angle of incidence can be achieved. For the zinc sulphide and cryolite layers, glass of index 1.52 gives an angle of incidence of $50\frac{1}{2}°$. A simple arrangement of cemented prisms with a multilayer deposited on the diagonal surface of one of them is shown in Fig. 17. However, $50\frac{1}{2}°$ is not a particularly convenient angle for the prisms. With a glass of higher index it is possible to achieve an angle of incidence of $45°$ which implies a beam splitter in the form of a cube—a much more convenient optical component both for manufacture and use. Again for zinc sulphide and cryolite the glass index required is 1.66, corresponding to a fairly dense flint glass. An alternative approach is to use more suitable layer materials so that crown glass prisms with 45° angles can be used. This is the approach adopted by Clapham et al.[57], who give a very full account of the techniques for the production of these devices.

Recently there has been a renewal of interest in the use of polarisers to reduce dazzle from the headlamps of oncoming traffic. One of the problems is the polarising of the output of the headlamps themselves. Absorption type polarisers are completely unsatisfactory at the levels of power which are required because they reject over half the light output which causes an enormous temperature rise as well as destroying the efficiency of the lighting system. Zehender[58] describes how thin film polarising beam splitters can be used in this application. Both beams emerge from the headlamps, one first having its plane of polarisation rotated through 90°, so that the efficiency is retained and the cooling problems eliminated.

2.7 Dichroic Beam Splitters

The term dichroic beam splitter referred originally to coatings used for colour separation in the visible region but has now been extended to cover any situation where one wavelength band has to be separated from another. Normally, dichroic beam splitters consist simply of edge filters, or occasionally band pass filters, which are used tilted so that the reflected beam as well as the transmitted beam can be recovered. The dichroic beam splitter has the advantage of greatly increased efficiency over the more simple neutral beam splitter. The principal difficulty in the design of dichroic beam splitters occurs when the bands to be separated are adjacent and is that of maintaining edge steepness at the large angles of tilt, often 45°, which are required. The optical admittance of a material depends both on the refractive index and on the plane of polarisation and, for a quarter wave stack used as an edge filter, the result is that the width of the stop region for s-polarised light increases while that for p-polarised light falls, compared with that which existed at normal incidence. For unpolarised light, the characteristic is the mean of the two separate characteristics for each polarisation and, since the edges no longer coincide, the resultant edge is effectively less steep than either edge on its own. It has not been found possible to remove this effect entirely although there are a number of ways of improving the performance slightly. Instead of simple quarter wave stacks, stacks of construction (3H)L(3H)L(3H)...which contain an increased amount of high index material can be used (De Lang and Bouwhuis[59]). The higher index is less affected by tilting than the low index and so the polarisation splitting is slightly reduced. Rabinovich and Pagis[60] found that with a particular multilayer designed to have low ripple on one side of the high reflectance band, the splitting on the same side was a little less than in the simple quarter wave stack. The reason for this reduction in splitting is not entirely clear and it may be that there are other similar effects which are still to be discovered. The splitting increases with angle of incidence and so the smaller this can be made the better.

2.8 Factors which Limit Performance

The real performance of thin-film optical devices will always fall short of that theoretically possible for a number of reasons. The first is simply that the various layers have not been deposited with exactly the correct thicknesses and refractive indices. This is a question of manufacturing accuracy and allowable tolerances. And then there are defects in the layers—that is they depart in some way from the ideal homogeneous parallel-sided slab assumed in design calculations.

The field of manufacturing tolerances is a difficult one and not a great deal has been published on the subject. One can show that optical thickness is the most important parameter to be controlled. Assessing the effects of layer thickness errors is very involved because errors in different layers interact in a complex non-linear manner, unless they are exceedingly small, making straightforward analytical evaluations of multiple errors almost impossible except in special cases. The most fruitful general method has proved to be that of computer simulation. If a model of the appropriate monitoring process is set up on a digital computer it can be used in a Monte Carlo simulation of the production of a series of batches of coatings with appropriate random errors inserted. We are not able to go into this in detail here. Further information can be found in Macleod[37]. It has been shown that some systems of optical monitoring possess an error compensating mechanism which helps to maintain certain aspects of coating performance in spite of errors committed. In the case of narrow band filters, for example, the studies show that if all layers are monitored on the actual filter which is being produced by the technique known as turning value monitoring (where layers are terminated at extrema of reflection or transmittance) then the peak wavelength of the finished filter will be virtually the monitoring wavelength even with errors in layer thickness of perhaps 20 or 30 per cent. The bandwidth of the filter will be wider than theoretical however, and the peak transmittance a little reduced, the bandwidth effect being the greater. It is with simulation of this type that we can show that coating performance is a much more sensitive function of optical thickness than it is of refractive index. But it is important to take dispersion of refractive index into account in the design of coatings. Dispersion can cause problems because it affects optical thickness at wavelengths other than the monitoring wavelength.

Uniformity of layer thickness over the substrate, too, can be a problem. For any but the simplest coatings it is normal to rotate the substrates during deposition to achieve uniformity and special jigs are required for deeply curved substrates such as lenses. Narrow band filters which are usually on plane surfaces are the extreme case. Errors in uniformity should not cause shifts in peak wavelength over the surface of the filter exceeding 0.3 × halfwidth. This is straightforward to achieve for filters up to around 50mm in diameter with halfwidths of 2.0 nm or more in the visible region. For narrower or larger filters uniformity can cause problems and great attention must be paid to source and substrate positioning and smoothness of rotation.

Layers are often inhomogeneous, that is the refractive index varies across the thickness of the layer. This variation is often due to the columnar mode of growth of dielectric thin films. The columns vary in cross section as they grow causing a corresponding variation in film refractive index. Inhomogeneity is not a severe problem and indeed can sometimes be turned to advantage as Vermeulen has shown with antireflection coatings[61].

The penetration of atmospheric moisture into layers after deposition has already been mentioned in connection with narrow band filters. The effect occurs to a greater or lesser extent in virtually all thin film coatings including even tough oxide films deposited at high temperatures. The drifts towards longer wavelengths which is characteristic of this effect varies enormously with materials and technique but drifts in the visible region as high as 10 nm after manufacture are not uncommon. This is associated with a mottled appearance which gradually disappears as the coating stabilises. The behaviour of the coating is often referred to as ageing or settling.

For most purposes losses in the layers may be divided into scattering and absorption. In scattering the energy lost from the primary beams is emitted from the coating either at a different angle or different wavelength or both, although scattering processes in which the wavelength is changed are not normally of great significance in optical thin films. In absorption the light which is lost is trapped in the coating and raises its temperature. The scattering losses which are of importance are due to defects of some description in the coatings which may be classified into surface or volume defects. Surface defects are simply a departure from the smooth flat surface which ideally exists at the substrate surface and between the various layers. They may be due to a roughness of the substrate which is then reproduced throughout the multilayer or to the mode of growth of the films. Volume defects are any local variations in optical constants and are often simply dust particles or pinholes.

Recently much of the impetus for studying losses in thin film coatings has come from the laser field. Except in a few specialised applications such as frequency doubling, laser coatings normally are required to operate over a restricted range of wavelengths and so the basic designs are simple. For low power applications the most important factors are the scattering and absorption losses and the most severe requirements apply to the coatings which form the outer ends of the laser cavity where, because they are exposed to power levels which can be 50 to 100 times the output power of the laser, the absolute power loss is potentially greatest. For high power laser coatings additional factors are involved.

Barr[62] demonstrated the importance of dust particles in high reflectance laser coatings by introducing a rotating particle-filter which removed much of the dust from the stream of evaporant during deposition and Perry[20] showed how it is possible to improve laser mirror quality by taking careful precautions to eliminate dust from the process. Heitmann[19] too, achieved laser mirror coatings with especially low losses by similar means. In these studies, the actual scattering losses were not measured as such, and either they were included in an overall loss figure (Heitmann) or a simple assessment of quality based on a comparison of finished mirrors was made. Quantitative measurements of scattering have usually involved the measurement of the angular variation of scattered light intensity. Blazey[63] showed by analysis of scattering diagrams which he obtained that the results for a number of high quality laser mirrors were consistent with surface roughness as the major scattering source. Scattering losses in laser mirrors have also been studied by Günther et al.[64] who were able to identify both surface scattering, which predominated at angles greater than 20°, and volume scattering, which predominated at smaller angles. Total scattering losses measured for HeNe laser mirrors with multilayers of ZnS/ThF_4 or TiO_2/SiO_2 were in the range 0.1 to 0.25 per cent (Pulker[65]).

Since absorption losses result in the addition of heat to the coating, then the most direct way of measuring absorption is simply to measure the heat gained when the coating is illuminated under known conditions. Ahrens et al.[66] and Ahrens[67] have described a calorimetric method in which the temperature rise is measured by a thin film resistance thermometer deposited on the same substrate as the multilayer coating. They found at 1.06 μm absorption losses of 0.023 to 0.024 per cent in ZnS/THF_4 fully reflecting quarter wave stacks and 0.037 to 0.059 in TiO_2/SiO_2 stacks.

Laser mirrors which are to be used at very high power levels must withstand these levels without damage. The field of laser damage in optical components is growing rapidly and is now one of the more important areas of thin film research. Scattering losses will not in themselves give rise to coating damage because the scattered energy is emitted from the coating and the simplest potential source of laser damage is absorption. If the power absorbed in a coating is high enough, the temperature will rise to a level sufficient to cause damage, usually by thermal stress. The absorption may be continuous throughout the volume of the thin film or may be due to localised absorbing inclusions. Another potential source of damage, which has no counterpart at low power levels, is dielectric breakdown. If the power is high enough, the electric field of the wave can exceed the dielectric field strength of the material.

Bloembergen[68] has demonstrated the role of defects such as pores, cracks and absorbing inclusions on the damage threshold for dielectric materials. Cracks and pores, which are cavities in the dielectric, magnify the electric field both within them and also, since the field is continuous across the boundaries at the ends, within the dielectric material just at the ends of the cavities. The effect is especially severe in cracks parallel to the direction of propagation of the light where the magnification can cause reductions in the apparent breakdown intensity by factors as high as n^2. By reasoning based on the dimensions of the region over which the breakdown field operates, and on the thermal stresses due to the absorbing inclusions, Bloembergen came to the conclusion that defects can be ignored if their dimensions are less than 0.01 μm. Any defects larger than 1.0 μm can probably be avoided in the production process leaving those with size between 0.01 and 1.0 μm for which one can expect damage thresholds to be lower than for bulk material by up to a factor of five for low index materials and considerably greater for high index.

An important study of considerable significance was reported in 1973. Up till this time the spot size had not often been quoted in reports of damage measurements and DeShazer et al.[69] were able to show that the laser damage threshold increases with a decrease in the spot size of the beam illuminating the coating. Further, the manner in which the damage threshold varies is consistent with the existence in the coatings of defect centres. The larger the spot size the larger is the probability that the spot will illuminate a defect centre. For very small spot size, where the probability of including a defect is low, the damage threshold becomes the intrinsic threshold of the material. DeShazer and his colleagues found additional confirmation of this model through examination of the distribution and type of defects in their coatings using a scanning electron microscope. As a result of this work, plots of damage threshold with varying spot size and derived values of mean distance between defects, have become a regular feature of laser damage reports.

Further information on recent laser damage studies, including the role of phenomena such as stress, will be found in the report edited by Glass et al.[70].

3. Thin Film Light Guides

Communication systems using light propagating along glass fibres are currently the subject of considerable research and development effort. Such systems are attractive, partly because of the colossal information carrying capacity of a single optical channel and partly because of the very great simplicity of the glass fibre waveguide which, although it presents difficulties in manufacture, is exceedingly easy to install and use.

Of course the processing of the optical signals creates severe problems not the least being simply that of handling the enormous amount of information involved. It is expected that many operations will be carried out directly on the optical signals and the most promising techniques involve what are known as integrated optical circuits. These are similar to the more common integrated circuits in that they consist of single devices containing a large number of interconnected units, but most of the signals are in the form of light. A thin transparent film bounded on either surface by a medium of lower refractive index is capable of containing light which under certain conditions propagates along the film; the losses being simply due to scattering and absorption. Such a film is known as a light guide. In the integrated optical circuit simple thin film light guides form the connections between the various active sections. The active sections, too, are all visualised as thin film devices in which the light guiding property is used to contain the beams. As this chapter is being written, the integrated optical circuit is still very much in the future with current work being concentrated on the various components which might eventually be combined in such a device.

A second area in which the light guide may be expected to play a part and which may have applications in integrated optical circuits, is non-linear optics. As we shall see later, when a freely propagating light beam is coupled into a light guide the electric vector in the guide is many times larger than in the free beam. This very large magnification of the electric vector suggests that it should be possible readily to produce non-linear effects, such as frequency multiplication, in a thin film light guide.

A third area of interest is concerned with the measurement of the optical properties of materials. The propagation constants of the various modes of thin film light guides are very sensitive functions of film thickness and optical constants and so measurement of the modes can enable values of these quantities to be obtained with very high precision.

Yet another possible application is in the field of computing. The light distribution in the back focal plane of a lens is the Fourier transform of that in the front focal plane. Many applications of this property have been suggested including filtering and analogue computing. A thin film in the waveguiding mode acts as a simple two-dimensional space as far as the propagation of any one mode is concerned. Thus two-dimensional versions of the three-dimensional system could be constructed with the enormously improved robustness and stability of the integrated design.

Thus, although the major driving force behind the work being carried out on optical guides is undoubtedly the interest in integrated optics for communications, there is a number of other sufficiently important applications to justify research effort in the field of thin film optical guides. In what follows our first consideration will be the modes of propagation of energy in simple guides. Coupling is an important topic which is treated next and the remainder of the account is a very short description of some thin film devices, such as modulators, mode selectors and lasers.

For further information there are a number of reviews which are mentioned in the sections which follow but the most complete is probably that edited by Tamir[71]. Many of the most significant early papers including several reviews are collected by Marcuse[72].

3.1 The Simple Thin-film Light Guide

The most straightforward way of visualising the operation of a light guide, and the one which we shall use in the quantitative analysis which follows, is based on the phenomenon of total internal reflection. Light incident on a boundary between two dielectric media from the side of higher refractive index at an angle of incidence greater than the critical angle will be totally reflected. This occurs only in the higher refractive index material, which usually corresponds to the interior of an optical component, and so this phenomenon is known as total internal reflection. For the reflectance to be truly unity there must be no losses in either medium and no scattering at the boundary. If a single thin film is bounded on either side by media of lower refractive index then total internal reflection will be possible at both its surfaces. Light incident at a sufficiently large angle will be reflected backwards and forwards between the interfaces without loss, and the net propagation will be along the film rather than across it. In fact the film acts as a waveguide. There is a complication in that the situation is akin to the multiple beams found in the Fabry Perot interferometer and a similar interference between successive beams permits a steady state to be possible for certain well defined angles of incidence only. These angles of incidence depend on the optical constants of the guide and its surrounding media, on the thickness of the guide, and, because the phase shift on reflection varies with polarisation, on the state of polarisation of the wave. The effective velocity of propagation of phase in the guide depends on the angle of incidence; the lower this angle the higher the rate of propagation, and so, for each possible angle of incidence, we have a different velocity of propagation, that is a different waveguide mode. The properties of the modes can be shown to depend on the value of an integer, known as the order of the mode, and for the lowest order it is normal to assign a value of zero to the integer. The lowest permissible angle of incidence corresponds to the fastest propagation rate and the highest order mode. For s-polarised light we have the TE (or transverse electric) modes, written TE_j where j is the mode number, and for p-polarised we have the TM (or transverse magnetic) modes, TM_j.

The theory of light guide can be developed in a number of ways (e.g. Marcuse[73]; Kapany and Burke[74]; Tien[75]). It is possible for example to derive completely new solutions of Maxwell's equations with the appropriate boundary conditions. The treatment which follows relies very much on Tien[75] and is based on the ideas just outlined. We will need some of the results already mentioned in connection with thin film optical coatings.

First of all let us consider the reflectance of a single surface between two media when the angle is beyond the critical value. We choose axes as shown in Fig. 18. The amplitude reflection coefficient for either of the planes of

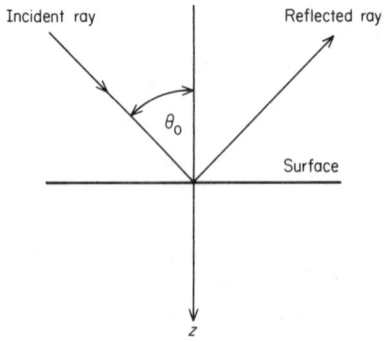

FIGURE 18. Ray reflected at single surface.

polarisation is given by equation (9) which we recall is

$$\rho = \frac{\eta_0 - \eta_1}{\eta_0 + \eta_1}$$

where the subscript 0 denotes the incident medium and 1 the medium of lower refractive index.

$$\left. \begin{array}{l} \eta_0 = n_0 \cos\theta_0 \\ \eta_1 = n_1 \cos\theta_1 \end{array} \right\} \quad \begin{array}{l} \text{TE waves} \\ (s\text{-polarisation}) \end{array}$$

$$\left. \begin{array}{l} \eta_0 = n_0/\cos\theta_0 \\ \eta_1 = n_1/\cos\theta_1 \end{array} \right\} \quad \begin{array}{l} \text{TM waves} \\ (p\text{-polarisation}) \end{array}$$

θ_0 and θ_1 being connected by Snell's law:

$$n_0 \sin\theta_0 = n_1 \sin\theta_1$$

In the following treatment we consider only dielectric lossless material, that is the refractive indices are real. If $n_1 < n_0$ then for values of θ_0 greater than a critical angle we find

$$\sin\theta_1 = \frac{n_0}{n_1}\sin\theta_0 > 1$$

In such cases we have

$$\cos\theta_1 = \sqrt{1-\sin^2\theta_1} = \sqrt{1-(n_0^2/n_1^2)\sin^2\theta_0}$$
$$= \pm i\sqrt{(n_0^2/n_1^2)\sin^2\theta_0 - 1}$$

Cos θ_1 is imaginary and, hence, so is η_1. Then, since η_0 is real

$$R = \rho\rho^* = 1$$

and we have the phenomenon of total reflection. The critical angle is given by

$$\theta_{\text{critical}} = \sin^{-1}\left(\frac{n_1}{n_0}\right)$$

We now investigate the phase of the reflectance which depends on the parameters θ_0, n_0, n_1. First of all we must decide on the sign of $\cos\theta_1$. The expression for the wave propagating into medium 1 is of the form

$$A\exp[i(\omega t - \mathbf{k}\cdot\hat{\mathbf{s}})]$$

with component along the z-direction

$$B\exp\left[i\left(\omega t - \frac{2\pi n_1 \cos\theta_1}{\lambda}\cdot z\right)\right]$$

If we replace $\cos\theta_1$ by $\pm i\alpha$ then the wave becomes

$$B\exp\left(\pm\frac{2\pi n_1\alpha z}{\lambda}\right)\exp(i\omega t) \tag{37}$$

This is a wave which is evanescent in space. For the solution to have reasonable physical reality we must choose the minus sign in the expression for $\cos\theta_1$ otherwise we have an amplitude which tends to infinity with z. Thus

$$\cos\theta_1 = -i\left(\frac{n_0^2}{n_1^2}\sin^2\theta_0 - 1\right)^{\frac{1}{2}} \tag{38}$$

The expressions for the reflection coefficients become

$$\rho = \frac{n_0\cos\theta_0 + i(n_0^2\sin^2\theta_0 - n_1^2)^{\frac{1}{2}}}{n_0\cos\theta_0 - i(n_0^2\sin^2\theta_0 - n_1^2)^{\frac{1}{2}}} \quad \text{TE waves}$$

$$\rho = \frac{\dfrac{n_0}{\cos\theta_0} - i\dfrac{n_1^2}{(n_0^2\sin^2\theta_0 - n_1^2)^{\frac{1}{2}}}}{\dfrac{n_0}{\cos\theta_0} + i\dfrac{n_1^2}{(n_0^2\sin^2\theta_0 - n_1^2)^{\frac{1}{2}}}} \quad \text{TM waves}$$

The phase shift on reflection is then given by

$$\psi = 2\Phi \tag{39}$$

where

$$\Phi_{TE} = \tan^{-1}\left[\frac{(n_0^2\sin^2\theta_0 - n_1^2)^{\frac{1}{2}}}{n_0\cos\theta_0}\right] \text{ in the first quadrant} \tag{40}$$

$$\Phi_{TM} = \tan^{-1}\left[-\frac{n_1^2\cos\theta_0}{n_0(n_0^2\sin^2\theta_0 - n_1^2)^{\frac{1}{2}}}\right] \text{ in the fourth quadrant} \tag{41}$$

These phases must be interpreted in the light of the sign convention given in Fig. 2. The expressions given by Born and Wolf[1] will at first sight appear different. This is simply because the sign conventions are different. Numerical results obtained from both sets of formulae and interpreted in terms of the appropriate sign convention will be identical.

Consider now a thin uniform film of index n_f bounded by n_0 on one side and n_s on the other as shown in Fig. 19. If $n_f > n_0$ and $n_f > n_s$ then for all values of θ_f above a limiting value there will be total internal reflection at both surfaces of the film. The resultant wave in the film can then be considered to consist of the sum of two component plane waves each inclined at θ_f to the surfaces, one travelling from b to a and the other from a to b. Now we want the resultant wave to have a stationary pattern propagating along the x-axis determined by a phase factor of the form $\exp[i(\omega t - \beta x)]$ where β is the propagation constant of the mode. This means that the components of the wave propagating along the z-axis must suffer a change of phase of $-2m\pi$ corresponding to the round trip from a starting point to one boundary followed by a reflection to the other boundary and then a reflection back to the starting point. This condition is

$$-2\cdot\frac{2\pi n_f d_f\cos\theta_f}{\lambda} + 2\Phi_a + 2\Phi_b = -2m\pi$$

i.e.

$$\frac{n_f d_f\cos\theta_f}{\lambda} = \frac{m}{2} + \frac{\Phi_a}{2\pi} + \frac{\Phi_b}{2\pi} \tag{42}$$

and then there will be a stable wave pattern in the film travelling along the x-axis with phase velocity ω/β where $\beta = k_f\sin\theta_f$ and $k_f = n_f k$, k being the wave vector with respect to free space, $2\pi/\lambda$.

Thin Film Optical Devices 373

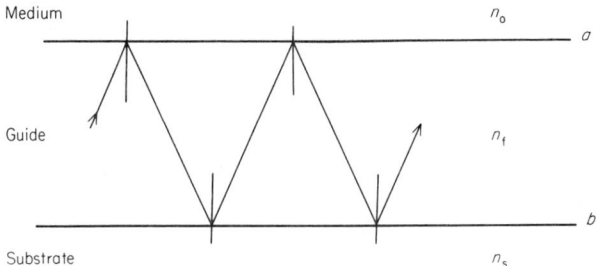

FIGURE 19. Path of ray within film guide.

The quantity

$$n_f \sin \theta_f = n_f \beta / k_f = \beta / k \quad (43)$$

can be considered to be the effective refractive index of the film for propagation in the particular mode. The magnitude of the effective wave vector is the propagation constant β, and the effective wavelength is given by

$$\lambda_{\text{eff}} = \frac{2\pi}{\beta} = \frac{\lambda}{n_f \sin \theta_f} \quad (44)$$

The field distribution in the guide can be obtained most easily if we use a device due to Tien[75]. We know that the distribution of the electric vector $\mathbf{E}_y$ or $\mathbf{E}_z$ will be of the form $A \cos(k_z z)$, k_z being given by $k_f \cos \theta_f$ where we have deliberately chosen the zero of z to coincide with the maximum of the distribution. We now fit the guide to this distribution rather than the other way round and derive an expression for the necessary thickness. This relationship can then be inverted to give the distribution in terms of guide thickness. The distribution can be split into a positive going and a negative going wave

$$\frac{A}{2} \exp(ik_z z) + \frac{A}{2} \exp(-ik_z z)$$

Now let the boundaries of the film be given by $z = -a$ and $z = b$ as shown in Fig. 20.

At $z = b$ we have the wave arriving at the boundary suffering a phase shift of $2\Phi_b$ to be transformed into the wave receding from the boundary. This relationship can be written

$$\frac{A}{2} \exp(-ik_z b + 2i\Phi_b) = \frac{A}{2} \exp(ik_z b)$$

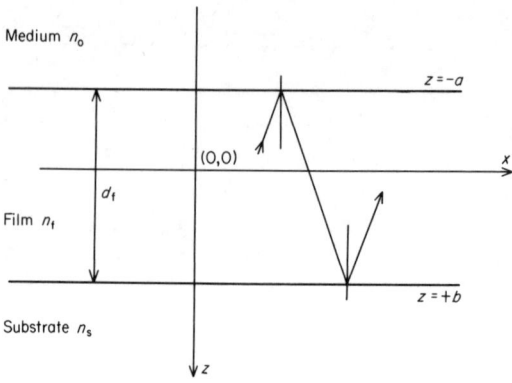

FIGURE 20. Coordinate system for deriving field distribution.

i.e.
$$2k_z b = 2\Phi_b + 2p\pi$$

i.e.
$$k_z b = \Phi_b + p\pi \tag{45}$$

Similarly at $z = -a$ we have

$$\frac{A}{2} \exp[ik_z(-a) + 2i\Phi a] = \frac{A}{2} \exp[-ik_z(-a)]$$

i.e.
$$k_z a = \Phi_a + q\pi \tag{46}$$

Then we write

$$d_f = a + b = \frac{\Phi_a}{k_z} + \frac{\Phi_b}{k_z} + \frac{m\pi}{k_z}$$

where we have written m for $(p+q)$. This is identical to (42). The electric field amplitudes at the boundary are then

$$E_a = \pm A \cos \Phi_a \quad \text{and} \quad E_b = \pm A \cos \Phi_b \tag{47}$$

depending on the value of m. The amplitude of the electric field component of the evanescent waves in the substrate and in the surrounding medium decays exponentially with the factor $\exp[-\gamma_b(z-b)]$ for the medium beyond the lower boundary and $\exp[\gamma_a(z+a)]$ for the upper, the particular forms of the factors being chosen to ensure that the wave tends to zero at very large distances from the boundary. Adapting equation (37) and (38) we have for γ

$$\gamma_a = \frac{2\pi n_0}{\lambda}\left(\frac{n_f^2}{n_0^2}\sin^2\theta_f - 1\right)^{\frac{1}{2}} \tag{48}$$

$$\gamma_b = \frac{2\pi n_s}{\lambda}\left(\frac{n_f^2}{n_s^2}\sin^2\theta_f - 1\right)^{\frac{1}{2}} \tag{49}$$

Some typical field distributions are shown in Fig. 21. Because of the signs

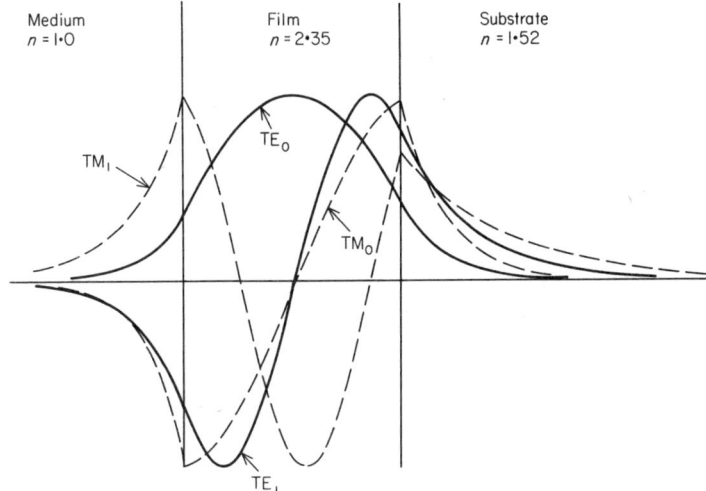

FIGURE 21. Typical electric field amplitude variation across thin film guide for TE and TM modes. The curves were calculated for a guide thickness of 0.5λ and $n_s = 1.52$, $n_f = 2.35$ and $n_0 = 1.0$.

of Φ, the lowest order mode corresponds to $m = 0$ for TE waves and $m = 1$ for TM waves. The mode number j is therefore given by m for TE waves and $m - 1$ for TM waves.

Clearly, a given thickness of film will support only a given number of modes. Once m becomes too large, the angle of incidence at the interfaces becomes too small and so the guide will not support the wave. When this happens the modes are known either as substrate modes, when there is still total reflection at the guide-air interface, or air modes when there is penetration at both interfaces. The thicker the guide, the greater the number of modes which will be supported. A symmetrical guide will always support the lowest order mode because of the variation of Φ with angle of incidence which permits equation (42) to be satisfied for $m = 0$ for TE waves or $m = 1$ for TM waves. In asymmetrical guides, however, one Φ will be zero when the other is non-zero and this determines a minimum thickness below which even the lowest order modes cannot propagate. This is of importance in the operation of the tapered coupler. Figure 22 shows the way in which the propagation of various modes varies with film thickness for a film of index 2.35 on glass of index 1.52; in air $n = 1.0$.

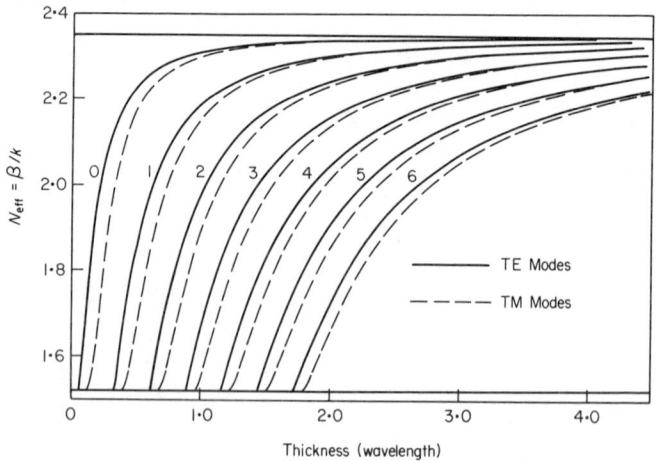

FIGURE 22. $N_{eff}(=\beta/k)$ plotted against thickness for a film guide with $n_s = 1.52$, $n_f = 2.35$, $n_0 = 1.0$. The horizontal line at the foot shows the cut-off at $N_{eff} = 1.52$.

3.2 Power Flow in a Thin-film Light Guide

The amplitude of the wave in the guide varies in quite a complicated manner across the thickness of the guide, and so calculation of the total power carried by it is somewhat tedious but of considerable importance. In this section we calculate the power carried by TE waves in a guide consisting of a film of index n_f bounded by a substrate of index n_s and a medium of index n_0. We use the same coordinate system as before.

Let the maximum electric field in the film be E_f which will be aligned along the y-axis. Again let us consider the distribution within the film as a combination of equal positive and negative going waves. Then we can write

Thin Film Optical Devices

for the electric vector (omitting the phase factor $\exp[i\omega t - \beta x]$ once again).

x-component   0
y-component $\frac{1}{2}E_f \exp(-k_z z) + \frac{1}{2}E_f \exp(+ik_z z)$
z-component   0

We know that the two waves are inclined at angle θ_f to the z-axis in the plane of the x- and z-axes. Using this fact together with equations (1) and (2) we have for the magnetic vectors

x-component $-\frac{n_f}{2} Y_0 E_f \cos\theta_f \exp(-ik_z z) + \frac{n_f}{2} Y_0 E_f \cos\theta_f \exp(+ik_z z)$

y-component   0

z-component $\frac{n_f}{2} Y_0 E_f \sin\theta_f \exp(-ik_z z) + \frac{n_f}{2} Y_0 E_f \sin\theta_f \exp(+ik_z z)$

Now the net power flow is along the x-axis and for this we need the y- and z-components of the vectors only. We use also that $\exp(i\theta) + \exp(-i\theta) = 2\cos\theta$. Whence

$$E_y = E_f \cos(k_z z) \qquad E_z = 0$$
$$H_z = n_f Y_0 E_f \sin\theta_f \cos(k_z z) \qquad H_y = 0$$

So that the power carried within the film by width l of the guide is

$$l \int_{-a}^{b} \tfrac{1}{2} R_e(\mathbf{E} \times \mathbf{H}^*) \, dz = l \int_{-a}^{b} \tfrac{1}{2} E_f \cos(k_z z) \cdot n_f Y_0 E_f \sin\theta_f \cdot \cos(k_z z) \, dz$$

$$= \tfrac{1}{2} l n_f Y_0 E_f^2 \sin\theta_f \int_{-a}^{b} \cos^2(k_z z) \, dz$$

Now

$$\int_{-a}^{b} \cos^2(k_z z) \, dz = \left[\frac{z}{2} + \frac{\sin 2k_z z}{4k_z}\right]_{-a}^{b}$$

and from equations (45) and (46) omitting $2p\pi$ and $2q\pi$

$$2k_z(-a) = -2\Phi_a$$
$$2k_z b = 2\Phi_b$$

So that the power in the film is

$$\tfrac{1}{2} l n_f Y_0 E_f^2 \sin\theta_f \left[\frac{W}{2} + \frac{\sin 2\Phi_b}{4k_z} + \frac{\sin 2\Phi_a}{4k_z}\right]$$

To this must be added the power carried in the n-direction by the evanescent waves in the substrate and the medium. Consider the substrate first. At the boundary we have from equation (47)

$$E_y = E_f \cos \Phi_b$$

and also we know that

$$E_z = 0$$

Thus the wave in the substrate is

$$E_y = E_f \cos \Phi_b \exp(-\gamma_b z') \exp[i(\omega t - \beta x)]$$
$$E_z = 0$$

γ_b being given by (49) and $z' = z - b$.

We need also the magnetic vector and to calculate that, we use one of Maxwell's equations

$$\text{curl } \mathbf{E} = -\frac{\partial \mathbf{B}}{\partial t}$$

At optical frequencies

$$\mathbf{B} = \mu_0 \mathbf{H}$$

Now

$$\mu_0 = \sqrt{\mu_0 \epsilon_0} \cdot \sqrt{\frac{\mu_0}{E_0}} = \frac{1}{cY_0}$$

where c is the velocity of light in free space.

Combining these various equations

$$\frac{\partial E_y}{\partial x} = -\frac{1}{cY_0} \frac{\partial H_z}{\partial t}$$

i.e.

$$-i\beta E_y = -\frac{1}{cY_0} i\omega H_z$$

i.e.

$$H_z = \frac{E_y}{\omega \beta c Y_0}$$

Now

$$\beta = k n_f \sin \theta_f$$

(equation 43) and

$$\omega/k = c$$

so that
$$H_z = n_f Y_0 \sin\theta_f \cdot E_y = n_f Y_0 \sin\theta_f E_f \cos\Phi_b \exp(-\gamma_b z')$$
Also $\quad H_y = 0.$

The power carried in the x-direction by a width l of the wave is then
$$l\int_0^\infty \tfrac{1}{2} R_e(E_y \cdot H_z^*)\,dz' = \tfrac{1}{2} l n_f Y_0 \sin\theta_f E_f^2 \cos^2\Phi_b \int_0^\infty \exp(-2\gamma_b z')\,dz'$$
$$= \tfrac{1}{2} l n_f Y_0 \sin\theta_f E_f^2 \frac{\cos^2\Phi_b}{2\gamma_b}$$

A similar expression holds for medium 0.

Then the total power flow is
$$\tfrac{1}{4} l n_f Y_0 \sin\theta_f E_f^2 \left[W + \frac{\sin 2\Phi_b}{2k_z} + \frac{\cos^2\Phi_b}{\gamma_b} + \frac{\sin 2\Phi_a}{2k_z} + \frac{\cos^2\Phi_a}{\gamma_a} \right]$$

We can simplify this by noting that
$$\frac{\sin 2\Phi_b}{2k_z} = \frac{\sin\Phi_b \cos\Phi_b}{k_z} = \frac{\sin^2\Phi_b}{k_z \tan\Phi_b} = \frac{\sin^2\Phi_b}{\gamma_b}$$

since from (40) and (49)
$$k_z \tan\Phi_b = k n_f \cos\theta_f \cdot \frac{\gamma_b/k}{n_f \cos\theta_f} = \gamma_b$$

Similarly
$$\frac{\sin 2\Phi_a}{2k_z} = \frac{\sin^2\Phi_a}{\gamma_a}$$

Then the power flow becomes (Tien[75])
$$\tfrac{1}{4} l n_f Y_0 E_f^2 \sin\theta_f \left[W + \frac{1}{\gamma_a} + \frac{1}{\gamma_b} \right] \qquad (50)$$

The term
$$\left[W + \frac{1}{\gamma_a} + \frac{1}{\gamma_b} \right]$$
is known as the effective thickness of the guide. Even if the film thickness W is made very small the effective thickness will still be much larger. This effect limits the magnitude of the field which can actually be achieved by feeding a given power into the guide in a particular mode.

3.3 Losses in Thin-film Light Guides

A detailed analysis of loss mechanisms in thin-film light guides is beyond the scope of this introduction although a brief outline of the general principles is given in this section.

The first source of loss in the guides is absorption in the material of the guide or of the media on either side of it. There are two aspects of this loss. The first, and most important is due to absorption and the second to leakage from the guide since complete total reflectance at a boundary does not occur when one of the media is absorbing.

When losses in guides are measured, usually by measuring the reduction in the power carried by the mode along the length of the guide, see for example Tien[75], it is found that their magnitude is much greater than can be accounted for by absorption alone. The additional losses are due to surface scattering.

An approximate analysis of surface scattering has been performed by Tien[75]. If σ is the variance of surface roughness then the specular reflectance of a surface at angle θ is given by

$$R = \exp\left\{-\left(\frac{4\pi\sigma}{\lambda}\cos\theta\right)^2\right\}$$

assuming that in the absence of roughness the surface would be perfectly reflecting. For small scattering this is approximately

$$1 - \left(\frac{4\pi\sigma}{\lambda}\cos\theta\right)^2$$

and the scattered fraction is $U^2 \cos^2\theta$ where

$$U^2 = \left(\frac{4\pi\sigma}{\lambda}\right)^2$$

U^2 can be identified as the scattering loss at normal incidence.

From the results of the previous section we know that the power in the guide is given by $I \cdot \sin\theta \cdot W_{\text{eff}}$ while the power in the component waves at angle θ is $I/2$. Thus the power loss per unit length of film is

$$U^2 \cos^2\theta \cdot \frac{I}{2} \cdot \cos\theta$$

and the attenuation per unit length is therefore

$$\frac{U^2 \cos^2\theta \cdot \frac{I}{2} \cos\theta}{I \cdot \sin\theta \cdot W_{\text{eff}}} = U^2 \frac{\cos^3\theta}{2\sin\theta} \cdot \frac{1}{W_{\text{eff}}} \tag{51}$$

This expression gives a much greater increase in scattering with mode number than simple absorption even allowing for the zig-zag path of the component beams. Tien[75] shows that the requirements for scattering loss are significantly more severe than those for high quality laser mirrors if the attenuation in the lowest order mode is to be as low as 1dB/cm.

More rigorous analyses of scattering losses are given by Marcuse[76,77].

3.4 Metal-clad Guides

It is important sometimes to have guides in which at least one of the boundary media is metallic. Metal electrodes are necessary for the operation of most active devices and can also act as useful buffer layers between light-guiding films and substrates of higher index (Tien et al.[78]).

Kaminow et al.[79] have examined, both theoretically and experimentally, metal-clad dielectric waveguides with either single sided or double sided metal cladding and found that the losses for the TE modes were much lower than for the TM modes. For polymer guides bounded on one side by glass and on the other by silver or gold films, the measured losses for the TE_0 modes at 632.8 nm were lower than 1dB/cm. The losses for the TM modes were much higher, the TM_0 mode having characteristics more like those of a surface wave along the metal-dielectric boundary than of a true waveguide mode. The study does make it clear that metal cladding for waveguides is a possibility, expecially for the lower order TE modes.

Rashleigh[80] has considered the case where a dielectric buffer layer of lower index is interposed between the metal cladding and the guide and has shown that the coupling between the TM_0 mode and the surface plasma wave along the metal is greatly reduced and can be much less than the other TM modes.

3.5 Strip Guides

The waveguides considered so far are unbounded within the plane of the guide so that two-dimensional propagation is possible. A guided wave can be further constrained by placing boundaries on the film area. This is the principle of the strip waveguide which has rectangular cross section and may consist of a high index core surrounded entirely with the same low index cladding or may be bounded by a medium of different index on all four sides. This type of guide is produced, for example, by depositing a thin film through a suitable mask or by etching it after deposition. It can be used as an interconnecting link between components in an integrated circuit or to direct the optical signal along a well defined path.

The mode characteristics of such a guide are more difficult to derive than are those of the two-dimensional guide. They have been analysed by Goell[81] and by Marcatili[82]. The distribution of field amplitude in the modes is qualitatively similar to a superposition of the two mode patterns for the corresponding two-dimensional guides of similar width but the quantitative details of the variation are different. Two mode numbers are required instead of just one.

When a strip guide is bent some of the guided energy will be lost by radiation. The losses in strip guides due to bends have been analysed by Marcatili[83] who has shown that if γ is the length over which the field of the evanescent wave on the outside of the curve decays to $1/e$ of its value then the radiation losses can be neglected provided

$$R > \frac{24\pi^2 |\gamma|^3}{\lambda_2}$$

γ can be reduced by increasing the index contrast between the guide and the surrounding medium, n_f/n_3, or, to a lesser extent, by increasing the width of the guide. For example, guides of width approaching the maximum which ensures a single guided mode will have losses at 632.8 nm of less than 0.1 dB/radian of bend for radii of curvature as small as 1 mm provided n_f/n_3 is greater than 1.01.

3.6 Coupling

So far we have considered the propagation of energy within a waveguide without considering how the energy is pumped into the guide in the first place. Usually the source of light will be a laser and unless the laser is part of the integrated optical device the output from the laser must be introduced into the guide in some way. The inverse process of removing energy from the guide is also important and these processes are known as coupling. The device for inserting energy is known as an input coupler and that for removing it as an output coupler.

Most of the published work on couplers to date is concerned with the coupling of free three-dimensional waves into the guides and vice versa. This is the type of coupling which is required in the laboratory when the properties of thin film waveguides are being investigated. The dimensions of an optical waveguide are such that in this application simply allowing the energy to enter or emerge from a squared-off end of the guide is not efficient and other techniques are necessary. Several types of couplers have been developed and used successfully in this application.

What is potentially the most important type of coupling for the future is that from an optical fibre into a guide and from a guide into a fibre since this is the most likely way in which integrated optical devices will be used. There appears still to be difficulties in this area.

We shall begin with the simplest type of coupler to understand, the prism output coupler.

3.6.1 Prism output coupler

The simplest technique for removing energy from an optical waveguide involves weakening the total internal reflection at one of the film boundaries so that energy can leak out of the guide. This can be done by what is known as frustrating the total internal reflection. Although the guide carries the energy, it is not confined strictly within the dimensional limits of the guide. As we have seen earlier, part of the energy is carried outside the guide in the form of an evanescent wave. The amplitude of this wave decays exponentially outside the guide. If a boundary of a medium which can support a wave is inserted into this evanescent wave it will remove energy. This is shown diagramatically in Fig. 23. The gap obviously depends on the rate of decay of

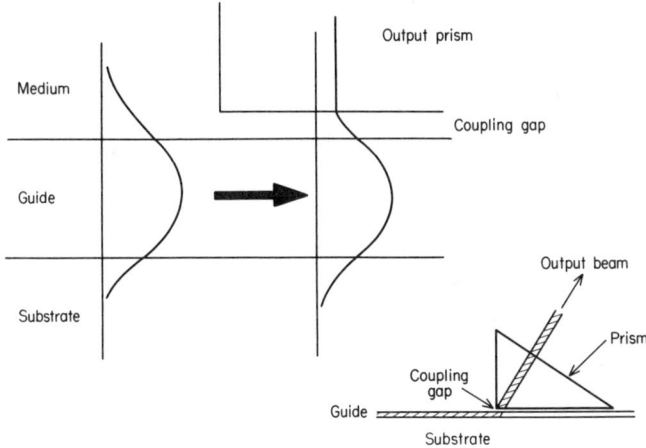

FIGURE 23. Representation of the electric field in a guide and in a prism output coupler.

the evanescent wave but normally it should be around or slightly less than $\lambda/4$. The requirement is simply that the refractive index of the coupling device must be of such a value that the mode of the guide will couple in with a real angle of incidence θ_c. This is

$$\sin \theta_c = \frac{n_f \sin \theta_f}{n_c} \leqslant 1$$

If $n_c = n_f$ then $\theta_c = \theta_f$.

As the wave progresses along the guide eventually all of its energy will couple into the wave in the output coupler. Since it is a linear process, the intensity of the wave in the output coupler will vary exponentially along its length. Finally, because of the angle θ_c at which the wave is propagated in the output coupler, it is convenient to make it in the shape of a prism as shown in Fig. 23.

It should be noted that the maximum field in the coupled out-wave occurs at the leading corner of the prism which must therefore be extremely sharp with a good optical finish if no losses are to be sustained. The length over which the wave couples out depends inversely on the transmission coefficient of the gap.

3.6.2 Prism input coupler

The prism input coupler is the reverse of the output coupler. If the directions of all the waves are reversed, retaining the amplitudes, then Maxwell's equations and boundary conditions remain satisfied. Thus by arranging a right angled prism and an incident beam which varies exponentially in the correct fashion along the coupling face of the prism up to the right angle, we would have completely efficient coupling of the light into the guide. To arrange this is extremely difficult and it is much more usual to use a simple uniform beam.

Many analyses of the prism coupler exist including those of Harris and Shubert[84], Harris et al.[85], Kapany and Burke[64], Midwinter[86], Tien and Ulrich[87], and Ulrich[88,89]. The following is a much simplified version.

The situation is sketched in Fig. 24. The guide has index n_f and thickness d_f and the angle of incidence within the guide is θ_f. We lose no generality if we suppose that the input medium has refractive index equal to that of the waveguide. If in a real case the index is different then the input beam can be referred to a notional input medium of index n_f by invoking Snell's law. We imagine that the input beam has unit amplitude and therefore intensity $\frac{1}{2}n_f Y_0$. We now slice the input beam into strips which are $d_f \sin \theta_f$ wide measured normal to the beam and we consider each slice as a separate beam. If we consider a width l of beam and guide then the power carried by each individual beam will be

$$\tfrac{1}{2}n_f\, Y_0 \cdot l d_f \sin \theta_f$$

We want to calculate the field within the guide at the point P in the diagram. The input beam to the right of P will have no effect at P and we imagine that the beam to the left holds $2m$ of the individual beams. Each set of alternate m beams will produce one of the two resultant beams at P as shown. The phase shift on transmission through the coupling gap is included in the amplitude transmission coefficient $t\, e^{i\psi t}$. The reflection coefficient at the lower surface b is unity and the phase shift on reflection is as before, $2\Phi_b$. The conditions at the upper surface, a, will be modified by the presence of the coupler and we

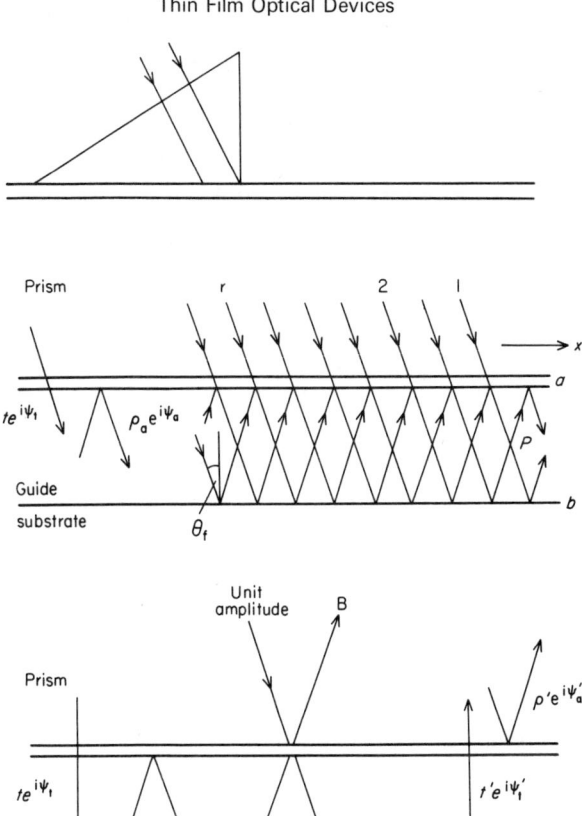

FIGURE 24. Rays in a prism input coupler.

will denote the amplitude reflection coefficient, including phase, as $\rho_a e^{i\psi_a}$. Then the phase shift between successive beams will be given by 2ϕ where

$$2\phi = 2\Phi_h + \psi_a - 2\delta$$

and

$$\delta = \frac{2\pi n_f d_f \cos\theta_f}{\lambda}$$

Consider now the beam labelled r which will suffer $(r-1)$ reflections at surface b and $(r-1)$ at a before reaching P. Thus the amplitude which this beam will have when it reaches P will be

$$t\, \mathrm{e}^{i\psi_t} \rho_a^{(r-1)} \mathrm{e}^{2i(r-1)\phi}$$

The m beams will therefore sum to produce a resultant amplitude $\mathscr{A}$ at P given by

$$\mathscr{A} = t\, \mathrm{e}^{i\psi_t} \sum_{r=1}^{m} \left\{ \rho^{r-1} \mathrm{e}^{2i(r-1)\phi} \right\}$$

$$= t\, \mathrm{e}^{i\psi_t} \left[\sum_{r=1}^{\infty} \left\{ \rho^{r-1} \mathrm{e}^{2i(r-1)\phi} \right\} - \rho^m \mathrm{e}^{2im\phi} \sum_{r=1}^{\infty} \left\{ \rho^r \mathrm{e}^{2i(r-1)\phi} \right\} \right]$$

i.e. $$\mathscr{A} = \frac{t\, \mathrm{e}^{i\psi_t} [1 - \rho^m \mathrm{e}^{2im\phi}]}{(1 - \rho \mathrm{e}^{2i\phi})} \tag{52}$$

There are two resultant beams corresponding to each set of m beams which carry the power in the guide at P. The power is given by

$$l \cdot d_f \sin \theta_f \times \text{Intensity of resultant}$$

in each beam

i.e. $$l d_f \sin \theta_f \cdot \tfrac{1}{2} n_f Y_0 \mathscr{A} \mathscr{A}^*$$

so that the total power in the guide is

$$l d_f \sin \theta_f n_f Y_0 \mathscr{A} \mathscr{A}^*$$

Now

$$\mathscr{A}\mathscr{A}^* = \frac{t^2 [1 - \rho^m \mathrm{e}^{2im\phi}][1 - \rho^m \mathrm{e}^{-2im\phi}]}{[1 - \rho \mathrm{e}^{2i\phi}][1 - \rho \mathrm{e}^{-2i\phi}]}$$

$$= t^2 \frac{(1-\rho^m)^2}{(1-\rho)^2} \cdot \frac{1 + \dfrac{4\rho^m}{(1-\rho^m)^2} \sin^2 m\phi}{1 + \dfrac{4\rho}{(1-\rho)^2} \sin^2 \phi} \tag{53}$$

which is similar in many respects to the expression for fringes in the Fabry Perot interferometer.

For efficient coupling this expression shows that ϕ must be made sufficiently near $s\pi$, where s is an integer, to ensure that the factor on the right hand side of the expression (53) is as near unity as possible. If this is so then the expression becomes

$$\mathscr{A}\mathscr{A}^* = t^2 \frac{(1-\rho^m)^2}{(1-\rho)^2}$$

Now
$$t^2 = 1 - \rho^2$$
so that multiplying above and below by $(1+\rho)^2$
$$\mathscr{A}\mathscr{A}^* = \frac{t^2(1+\rho)^2(1-\rho^m)^2}{(1-\rho^2)^2}$$
$$= \frac{(1+\rho)^2(1-\rho^m)^2}{t^2}$$
and if $\rho \simeq 1$ then we have
$$\mathscr{A}\mathscr{A}^* = \frac{4}{t^2}(1-\rho^m)^2 \tag{54}$$

We see immediately that as $m \to \infty$
$$\mathscr{A}\mathscr{A}^* \to \frac{4}{t^2}$$
which shows that the power in the guide will build up to a maximum value regardless of the length over which coupling takes place. To examine more closely the manner in which the power builds up to its maximum we make use of a device due to Kapany and Burke[74]
$$\rho^m = \exp(m \log_e \rho)$$
and
$$\log_e \rho = \log_e (1-t^2)^{\frac{1}{2}} \doteqdot \frac{t^2}{2}$$
so that from (54)
$$\mathscr{A}\mathscr{A}^* = \frac{4}{t^2}\left(1 - \exp\left[-m\frac{t^2}{2}\right]\right)^2 \tag{55}$$

Now let x be the coordinate along the direction of the guide, parallel to the surface and let the edge of the incident wave coincide with $x = 0$. Then
$$m = \frac{x}{2d_f \tan \theta_f} \tag{56}$$
where we are now allowing m to take on non-integral values. For the power in the guide to approach its maximum value the length over which the coupling takes place will be at least of the order given by
$$m = \frac{2}{t^2} = \frac{x}{2d_f \tan \theta_f}$$

i.e.
$$x = \frac{4d_f \tan \theta_f}{t^2} \tag{57}$$

Any power incident on the coupler beyond the point at which the power in the guide approaches its maximum will simply be reflected from the coupling gap and will not contribute to the useful power in the guide. If however the input beam were simply to be cut off then beyond the cut-off point the coupling gap would serve to couple power back out of the guide once again just as a normal output coupler. (This effect is similar to the well known Goos Haenchen shift). Thus at the cut-off point of the incident beam we must have the sharp right angle of the coupling prism so that beyond that point the power is contained within the guide.

We now calculate the efficiency of the coupler which we write as

$$\text{Efficiency} = \frac{\text{power in guide}}{\text{incident power}}$$

where the power in the guide is measured at the prism corner and the incident power includes only that which is actually incident on the coupling gap.

The incident power is

$$l \cdot x \cdot \tfrac{1}{2} n_f Y_0 \cos \theta_f$$

and so the efficiency can be written as

$$\frac{l d_f \sin \theta_f n_f Y_0 \mathscr{A}\mathscr{A}^*}{l x \cdot \tfrac{1}{2} n_f Y \cos \theta_f} = \frac{2 d_f \tan \theta_f}{x} \mathscr{A}\mathscr{A}^*$$

and using (55) and (56) this gives

$$\text{Efficiency} = \frac{2(1 - e^{-\alpha})^2}{\alpha} \tag{58}$$

where

$$\alpha = \frac{x t^2}{4 d_f \tan \theta_f}$$

The maximum corresponds to

$$2\alpha e^{-\alpha} = 1 - e^{-\alpha}$$

i.e.
$$\alpha = 1.256$$

and the corresponding efficiency is 82%. x is given by

$$x = \frac{4 d_f \tan \theta_f \cdot \alpha}{t^2} = \frac{5.024 \, d_f \tan \theta_f}{t^2} \tag{59}$$

t will usually be sufficiently small for x to be several millimetres.

The power distribution in the reflected beam is also of interest. At each point along the coupler there are four beams involved as shown in the lower part of Fig. 24. The amplitude of the wave leaving the point within the guide is just $\mathscr{A}$, already determined. The wave within the guide arriving at the point then has amplitude

$$\frac{\mathscr{A} - t\, e^{i\psi_t}}{\rho\, e^{i\psi_a}}$$

The amplitude of the resultant reflected wave in the incident medium can then be written

$$\mathscr{B} = \frac{(\mathscr{A} - t\, e^{i\psi_t})}{\rho\, e^{i\psi_a}} t'\, e^{i\psi'_t} + \rho'\, e^{i\psi'_a} \tag{60}$$

Since the incident medium and the guide are both supposed to have index n_f we can write

$$t'\, e^{i\psi'_t} = t\, e^{i\psi_t}$$

$$\rho'\, e^{i\psi'_a} = \rho\, e^{i\psi_a}$$

We assume that the resonance condition $\phi = s\pi$ is satisfied so that $\mathscr{A}$ is given by (52) as

$$\mathscr{A} = \frac{t\, e^{i\psi_t}(1 - \rho^m)}{(1 - \rho)}$$

Then, substituting in (60)

$$\mathscr{B} = \frac{t^2\, e^{2i\psi_t}}{\rho\, e^{i\psi_a}} \cdot \frac{(1 - \rho^m) - (1 - \rho)}{(1 - \rho)} + \rho\, e^{i\psi_a}$$

$$= \rho\, e^{i\psi_a}\left[1 + \frac{t^2\, e^{2i\psi_t}}{\rho^2\, e^{2i\psi_a}} \cdot \frac{\rho(1 - \rho^{m-1})}{(1 - \rho)}\right]$$

but

$$t^2 = (1 - \rho^2)$$

so that

$$\mathscr{B} = \rho\, e^{i\psi_a}\left[1 + \frac{(1 + \rho)(1 - \rho^{m-1})}{\rho} e^{2i(\psi_t - \psi_a)}\right]$$

In the steady state as $m \to \infty$ we must have a reflected amplitude equal to the incident except for a change of phase. This implies that

$$e^{2i(\psi_t - \psi_a)} = -1$$

The reflected intensity is then

$$\frac{n_f}{2} Y_0 \mathscr{B}\mathscr{B}^* = \frac{n_f}{2} Y_0 [1 - 2(1 - \rho^{m-1})]^2$$

and for large m we can write $m = (m-1)$ so that

$$\text{Intensity} = \frac{n_f}{2} Y_0 \left[1 - 2\left\{1 - \exp\left(-m\frac{t^2}{2}\right)\right\}\right]^2$$

This falls to zero when

$$\exp\left(-m\frac{t^2}{2}\right) = \tfrac{1}{2}$$

The existence of a dark band in the reflected beam has been demonstrated by Tien et al.[90].

An experimental arrangement of a prism film coupler is described briefly by Tien[75]. The film side of the substrate is pressed against the base of the prism by a knife edge situated only a millimetre or less from the corner and the gap is maintained by the dust particles which are inevitably present on the surfaces. The input beam is directed at the correct angle on the region between knife edge and prism corner and for best efficiency it should just fill it. The pressure of the knife edge bends the substrate very slightly so that the air gap is less at the pressure point than at the prism corner making the coupling vary from stronger to weaker and improving slightly the efficiency of the device.

The theory of the prism coupler developed so far applies equally well to any coupler in which an evanescent wave penetrates through a thin coupling layer. Any material of index lower than that of the prism and guide can be used as the coupling layer. A thin solid film gives greater stability and robustness than an air gap. Ulrich[88] has suggested two arrangements in which the prism has the dual role of coupling prism and substrate and where the gap is formed by a thin film of index lower than substrate or waveguide. These are shown in Fig. 25. In case (a) the gap is simply increased at the end of the coupling length so that the leakage through it is reduced virtually to zero while in case (b) the guide thickness is increased to the extent that the effective index associated with the mode is increased above the index of the prism. This implies that θ_f is now greater than the critical angle between guide and prism material and allows the coupling gap to be reduced to zero without causing leakage from the guide.

Harris and Shubert[84] pointed out that the efficiency of the prism-film coupler could be increased by allowing the thickness of the coupling gap to vary in the direction of propagation. In particular, for gaussian input beams and coupling gaps which increased linearly an efficiency of over 90% could be obtained. Ulrich[89] has generalised the theory of the prism-film

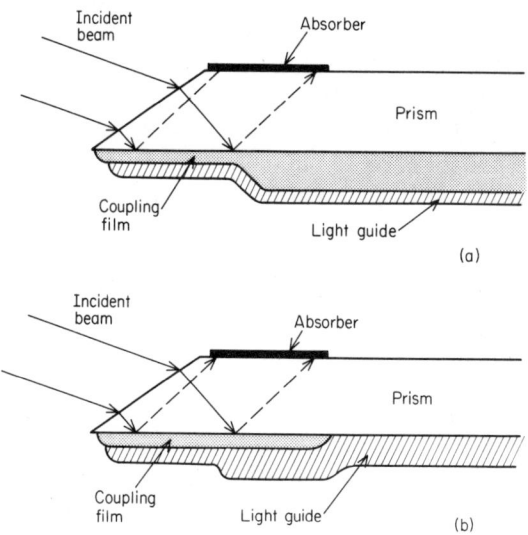

FIGURE 25. Two possible arrangements of prism input couplers where the films are deposited on the prism and the coupling gap consists of a thin film of index lower than prism or guide. (After Ulrich[88].)

coupler so that the thicknesses of both the coupling gap and the film guide can vary in the direction of propagation. In this way it is possible to arrange that the portion of beam C transmitted through the gap in the lower half of Fig. 24, exactly cancels the reflected portion of the incident beam so that beam B is zero and hence the efficiency of the coupler becomes 100%. The film thickness adjusts the phase of beam C and the thickness of the gap adjusts the transmittance and reflectance.

This approach has the great advantage that for beams such as gaussian where the amplitude shows only smooth variations the gap also varies smoothly and, especially useful, does not require the abrupt termination necessary with the uniform gap and uniform beam case. For linear gaps and gaussian beams the maximum efficiency calculated was 96%. Experimental measurements of 88% were made which is appreciably greater than the uniform beam and gap figure of 82%. This type of coupler lends itself to configurations similar to (a) in Fig. 25.

3.6.3 Grating couplers

Gratings form another important class of coupler. As we have already seen, the propagation constant of the wave along the guide is

$$\beta = kn_f \sin\theta_f$$

and because of the total internal reflection we are unable to match this with an incident wave outside the guide in medium n_0 or n_s because this would demand a sine of angle of incidence greater than unity. However there is a way of using a diffraction grating to obtain the correct angle of incidence θ_f in the guide. Imagine a wave incident at a fairly small angle on one surface of a guide. This wave will penetrate into the film with θ_f much less than critical and the guide will be unable to contain the wave. Now consider what happens if the opposite surface of the guide instead of being just a simple surface is a diffraction grating with grooves perpendicular to the desired direction of propagation in the guide. The grating will produce a series of diffracted orders according to the grating equation

$$\Lambda(\sin\theta_0 + \sin\theta_d) = m\lambda \qquad (61)$$

where Λ is the groove spacing, θ_0 and θ_d are the angles of incidence and diffraction respectively with the sign convention that both angles have the same sign when on the same side of the grating normal, m is the order number and λ is the wavelength in the medium. Provided one of the possible θ_d's corresponds to the required θ_f then power will be fed into that mode of the guide. Similarly, a guided wave incident on a grating will produce a range of orders some of which are able to leave the waveguide.

The efficiency of the device will not be large if energy is being shared amongst a number of diffraction orders and there are two principal ways of ensuring that a reasonable proportion of the incident light is directed into the correct order. The first is simply to limit the possible number of orders by making Λ sufficiently small. The second is to make the grating thick and to arrange that the Bragg condition is satisfied for the input and required output directions when the energy will be preferentially coupled into the correct order.

The interaction between the input or output beam and the mode of the guide is similar to that in the prism coupler, expecially if the grating has been designed so that only two beams are involved. The operation as an output coupler is then straightforward, all the energy eventually being removed from the guide. As an input coupler, once again there is the complication that as an input beam is coupled into the guide, the guided wave is being coupled out. Thus the power builds up along the grating in much the same way as in the prism coupler and the grating must terminate abruptly to prevent the guided wave from leaking out where the input beam intensity drops away.

We have mentioned reflection gratings in this brief description because they are a more familiar type and their operation is easier to visualise. Transmission gratings can also be used. The production of a grating on the surface of a film guide is not easy and the technique which has been most successful is to expose photoresist to an optical interference pattern. The gratings in this case are phase gratings and it is worthwhile examining the operation of phase gratings in a little more detail.

3.6.4 Thin phase-grating

The phase grating has a refractive index which varies along the direction of propagation, i.e. the x-axis, as

$$n_g + n_h \sin \frac{2\pi x}{\Lambda} \tag{62}$$

We consider a grating of thickness g along the z-axis and an incident medium of index n_g. Now, neglecting the time variation of the wave, a wave incident at angle θ_i will have phase factor given by

$$\exp[-i(kn_g \sin \theta_i x + kn_g \cos \theta_i z)] \tag{63}$$

Let us take the zero of the z-axis as the exit face of the grating so that the entrance face is given by $z = -g$. Then the phase factor of the incident wave at the entrance face of the grating will be given by

$$\exp -i(kn_g \sin \theta_i x - kn_g \cos \theta_i \cdot g)$$

Inside the grating the refractive index varies with x according to equation (62) and the term $kn_g \cos \theta_i z$ in equation (63) must be replaced by

$$k\left(n_g + n_h \sin \frac{2\pi x}{\Lambda}\right) \cos \theta_i \cdot z$$

Thus the phase factor of the wave which emerges from the grating becomes

$$\exp\left\{-i\left(kn_g \sin \theta_i x - kn_g \cos \theta_i g + k\left[n_g + n_h \sin \frac{2\pi x}{\Lambda}\right] \cos \theta_i \cdot g\right)\right\}$$

$$= \exp\left\{-i\left(kn_g \sin \theta_i x + kn_h \sin \frac{2\pi x}{\Lambda} \cdot \cos \theta_i \cdot g\right)\right\}$$

This can be written as

$$\exp\left[-i\left(px + \alpha \sin \frac{2\pi x}{\Lambda}\right)\right] \tag{64}$$

Expressions of this type have been studied in the field of frequency modulation and it can be shown that expression (64) can be replaced by

$$J_0(\alpha)\exp(-ipx) + J_1(\alpha)\exp\left[-i\left(px + \frac{2\pi}{\Lambda}x\right)\right] - J_1(\alpha)\exp\left[-i\left(px - \frac{2\pi}{\Lambda}x\right)\right]$$

$$+ J_2(\alpha)\exp\left[-i\left(px+\frac{4\pi}{\Lambda}x\right)\right] + J_2(\alpha)\exp\left[-i\left(px-\frac{4\pi}{\Lambda}x\right)\right] + \ldots$$

$$\ldots + J_m(\alpha)\exp\left[-i\left(px+m\frac{2\pi}{\Lambda}x\right)\right] + (-1)^m J_m(\alpha)\exp\left[-i\left(px-m\frac{2\pi}{\Lambda}x\right)\right]$$

$$+ \ldots \tag{65}$$

where the J's are Bessel functions of the 1st kind. This represents a series of waves with propagation constants

$$\beta = p + m\frac{2\pi}{\Lambda} \tag{66}$$

where

$$p = kn_g \sin\theta_i$$

These are identical with those from a conventional grating.

The amplitudes of the various orders are given by

$$J_m(\alpha) \tag{67}$$

where

$$\alpha = kgn_h \cos\theta_i$$

q is sometimes written $\Delta\phi$ and known as the phase depth of the grating. Thus a thin phase-grating gives exactly the same diffracted orders as a conventional grating. In the conventional grating the intensities of the various orders are given by the diffraction pattern of a single slit or groove and so the intensity distribution of the phase grating is rather different.

All the diffracted waves which have propagation constants greater than the cut-off for the guide will couple into the modes of the guide. Efficient coupling into only one mode is best arranged by ensuring that the first order diffracted beam has a propagation constant coinciding with that of the lowest order guide mode. This will mean that $2\pi/\Lambda$ must be large enough and hence Λ will be small and of the order of a wavelength.

This type of coupling has been described by Dakes et al.[91]. Their grating consisted of exposed Shipley AZ-1350 photoresist with $\Lambda = 1.2\lambda$ and a measured efficiency of around 40%.

The difficulty of ensuring that the coupling is into one single mode with reasonable efficiency makes the thin grating less attractive than the thick grating to be considered now, where it is possible to arrange the coupling to be very strongly into just one diffracted order.

3.6.5 Bragg grating couplers

The very fine groove spacing necessary for efficient coupling in a thin grating renders the device not easy to manufacture. An alternative approach is a thick phase grating in which what is known as the Bragg condition will ensure that there will be strong coupling between the input beam and just one diffracted beam. In x-ray diffraction where the diffraction is arising in a 3-dimensional crystal, it can be shown that the relationship between incident and diffracted waves is given by

$$K - K' = Q \qquad (68)$$

where $\mathbf{Q}$ is a reciprocal-lattice vector. A similar relationship applies for the thick diffraction grating. In the thin diffraction grating the relationship is similar but less restrictive than (68) and more diffracted orders can appear with appreciable intensity. In fact the thin grating is similar to x-ray powder diffraction. It is worthwhile examining the conventional grating equation to see how it compares with (68).

The conventional optical arrangement is shown in Fig. 26. It is usual to assume that θ_0 and θ_d have the same sign when, as at (a), they are on the

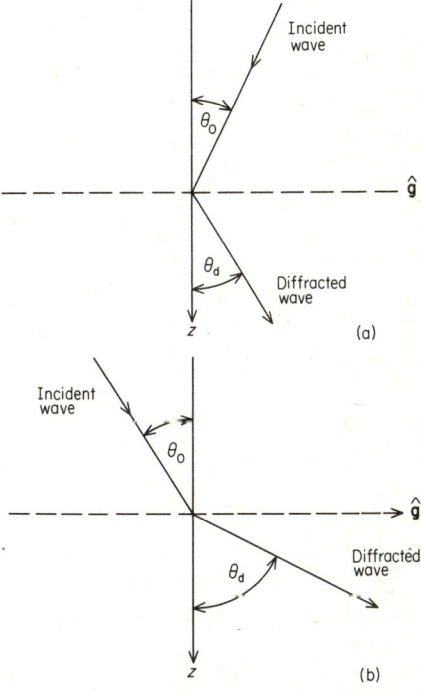

FIGURE 26. Sign convention for angles in a diffraction grating. The positive sense of the angles is shown. (a) Conventional arrangement. (b) Modified arrangement.

same side of the grating normal and then we have the grating equation

$$\Lambda(\sin\theta_0 + \sin\theta_d) = m\lambda$$

If λ is the free space wavelength and the grating is immersed in a medium of refractive index n_g the grating equation can be written

$$n_g\Lambda(\sin\theta_0 + \sin\theta_d) = m\lambda$$

which can be manipulated into

$$\frac{2\pi n_g}{\lambda}(\sin\theta_0 + \sin\theta_d) = \frac{2\pi m}{\Lambda}$$

Now the sign convention means that the angles have opposite sign when the direction of incident and diffracted beams are identical. It is more convenient in the following to change this convention so that they have the same sign when in the same direction. We therefore replace θ_0 by θ_i with positive direction as shown in Fig. 26(b). Then the grating equation becomes

$$\frac{2\pi n_g}{\lambda}\sin\theta_d = \frac{2\pi n_g}{\lambda}\sin\theta_i + m\cdot\frac{2\pi}{\Lambda} \tag{69}$$

We can write

$$\frac{2\pi}{\lambda}n_g = kn_g = k_g = |\mathbf{k}_g| \tag{70}$$

so that we can denote the incident beam by a propagation vector $\mathbf{k}_g$ and the diffracted beam similarly by $\mathbf{k}_g'$ where the magnitudes $|\mathbf{k}_g|$ and $|\mathbf{k}_g'|$ are identical but the directions are given by θ_i and θ_d. $2\pi/\Lambda$ we can denote by G and we can introduce the vector $\mathbf{G} = G\hat{\mathbf{g}}$ where $\hat{\mathbf{g}}$ denotes the unit vector along the surface of the grating perpendicular to the grooves. The grating equation can then be written

$$\mathbf{k}_g'\cdot\hat{\mathbf{g}} = \mathbf{k}_g\cdot\hat{\mathbf{g}} + m\mathbf{G}\cdot\hat{\mathbf{g}}$$

and for the first order only where $m = 1$

$$\mathbf{k}_g'\cdot\hat{\mathbf{g}} = (\mathbf{k}_g + \mathbf{G})\cdot\hat{\mathbf{g}} \tag{71}$$

Now in thick gratings the coupling between input and diffracted beams is strong only if the condition given in equation (68) is satisfied. Written in terms of the parameters of the diffraction grating this is

$$\mathbf{k}_g' = \mathbf{k}_g + \mathbf{G} \tag{72}$$

Clearly this is a special case of equation (71). It is known as the Bragg relationship and a grating aligned so that (72) is satisfied is said to be in the Bragg condition. Since the magnitudes of the propagation vectors are identical it is only the directions that can be changed by the grating and this means that only one order can be coupled to the input wave at any one time as the diagram of Fig. 27(a) will make clear.

Kogelnik[92] has considered the coupling in gratings of this type and Kogelnik and Sosnowski[93] describe experiments on such a coupler. Because

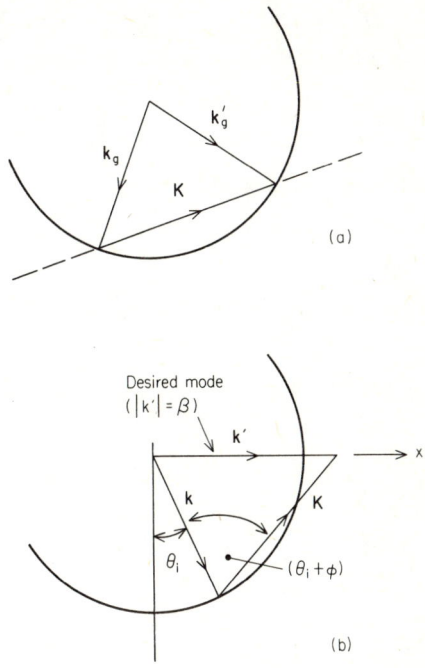

FIGURE 27. Vectors in Bragg gratings. (a) Simple Bragg grating. (b) Bragg grating used as coupler. (After Kogelink and Sosnowski[93].)

n_f will usually be greater than n_g we cannot quite reach the optimum coupling because the maximum magnitude of k' is $n_g k_0$ and this must be less than β which is approximately $n_f k_0$. This seems to mean that the Bragg condition must after all be violated if the input wave is to couple into the desired waveguide mode. The vector diagram is shown in Fig. 27(b).

The phase matching condition is given by

$$\beta^2 = k^2 + G^2 - 2kG \cos(\theta_i + \phi) \tag{73}$$

while the Bragg coupling can be written

$$\cos(\theta_i + \phi) = \frac{G}{2k} \tag{74}$$

We can express the difference between these two conditions as a difference in θ_i assuming all other parameters are constant. Equation (73) becomes

$$\beta^2 = k^2 + G^2 - 2kG\cos(\theta_i + \Delta\theta_i + \phi)$$

while (74) remains unchanged. Then, if $\Delta\theta_i$ is small,

$$\Delta\theta_i = \frac{\beta^2 - k^2}{2kG} \cdot \frac{1}{\sin(\theta_i + \phi)} = \frac{\beta^2 - k^2}{2kG(1 - [G^2/4k^2])^{\frac{1}{2}}}$$

The smallest value of $\Delta\theta_i$ is given when

$$G^2 = 2k^2$$

that is when

$$\Delta\theta_i = \frac{\beta^2 - k^2}{2k^2} = \frac{(\beta - k)(\beta + k)}{2k^2} \doteq \frac{\beta - k}{k}$$

since $(\beta - k)$ is small. This is valid if β is approximately $n_f k_0$

$$\Delta\theta_i \doteq \frac{n_f - n_g}{n_g} \tag{75}$$

Now Kogelnik has shown that the coupling will still be strong in spite of the violation of the Bragg condition provided that for transmission gratings

$$\Delta\theta_i < \frac{\Lambda}{2d}$$

and for reflection gratings that

$$\Delta\theta_i < \frac{\Lambda}{2d}\cot\theta_i$$

The assumption made in deriving these expressions was that the parameter

$$Q = \frac{2\pi\lambda d}{n_g \Lambda^2}$$

is greater than 10, where d is the grating thickness. Normally this relationship is readily satisfied. Kogelnik's expressions can be combined as

$$\frac{\pi\lambda}{10 n_g \Lambda} > \frac{\Lambda}{2d} > \frac{n_f - n_g}{n_g} \tag{76}$$

Provided this inequality is satisfied then the efficiency of the grating coupler will be high.

Useful theoretical analyses and accounts of experimental work on grating couplers have been described by Dalgoutte and Wilkinson[94], Harris et al.[95] and Ulrich[96]. Tamir[97] has surveyed many types of couplers in terms of leaky waves and the review by Taylor and Yariv [98] contains much useful information on couplers.

3.6.6 Tapered film coupler

Originally described by Tien and Martin[99] this is the simplest of couplers consisting of a tapered edge to the film normal to the propagation beam (Fig. 28). As already noted, even in the lowest order modes of an asymmetrical guide there is a minimum film thickness below which the mode cannot

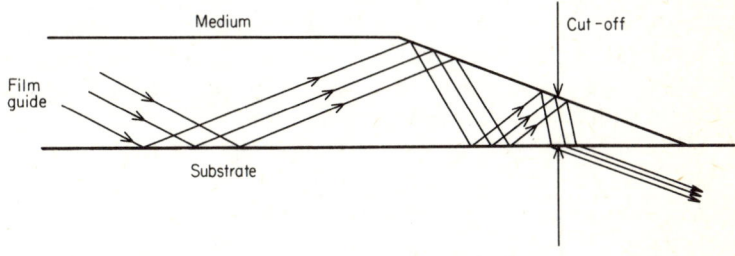

FIGURE 28. Tapered film coupler. (After Tien and Martin[99].)

propagate. The taper gradually reduces the thickness of the film below this minimum at which the reflectance at the film-substrate boundary becomes less than total and the energy escapes as a free wave in the substrate. An alternative way of visualising the operation is that the wave is reflected backwards and forwards in the film but with each reflection at the tapered surface the angle of incidence is reduced by $2\theta_t$, where θ_t is the taper angle, so that eventually it drops below the critical angle and the light can escape. Since the critical angle between film and substrate is less than that between film and air the light always enters the substrate rather than the air.

As with the other forms of coupler, energy can be coupled in as well as out. For maximum coupling efficiency the input beam should possess the same intensity distribution as occurs with an output wave. With straightforward beams, Tien and Martin[99] achieved efficiencies up to 40%.

3.6.7 Interconnection in integrated optical circuits

The integrated optical circuit is likely to consist of a number of devices which must be connected together and which may well be of different materials having different refractive indices. The simplest and most useful solution to this problem has been proposed by Tien et al.[100]. The two sections of guide that are to be interconnected are tapered at the edges. The interconnection consists of a short length of film, itself with tapered edges, deposited across the gap and overlapping the guides. The tapers cause smooth transitions, with complete coupling without mode changes, between the various sections— film guide to composite sandwich to coupling guide to second composite sandwich to second film guide. No tight tolerances are required. The tapers must simply be many wavelengths long but the extent of the overlap regions is completely arbitrary. The other requirement is that the waveguide mode must not cut off in the coupling guide which implies a minimum thickness of coupler.

When the retractive index of the substrate is substantially lower than that of the guides another form of two-layered construction described by Tien et al.[101] becomes possible. The devices are deposited as usual on the substrate with tapered edges where coupling is to take place. The entire system is then covered with a single film of index much lower than that of the guides but higher than that of the substrate. The presence of this film over the devices does not affect their optical guiding properties and they contain the light in the normal way. At the tapers, however, the wave is coupled out into the covering film in which it is guided across the gap and then coupled into the neighbouring device through the taper. The devices may take the form of lasers, modulators, thin-film lenses, prisms and the like (see Fig. 29). Lenses may be added in the gaps to focus the light beams between devices.

3.6.8 Fibre-waveguide coupling

Possibly the most important types of coupler are those which involve coupling into or out of optical fibres. Efficient couplers of this type are essential for optical communications. At the time of writing there still appear to be some difficulties in this area. The major problem seems to be that the modes of the thin film light guide are not a good match for those of the optical fibre.

A straightforward coupler, described by Boivin[102], was simply a butt-joint between the squared-off ends of the film and fibre. To improve the matching between film and fibre a thin film lens (see Fig. 29) was included in the film to focus the mode of the guide onto the end of the fibre. Very accurate alignment between fibre and guide is essential and a technique of etching a v-shaped groove in a silicon substrate to hold the fibre was developed. This guide-fibre coupler was developed as part of a complete laser-fibre coupling system, the laser output first of all being coupled into the guide through a grating. The overall efficiency of the system did not exceed

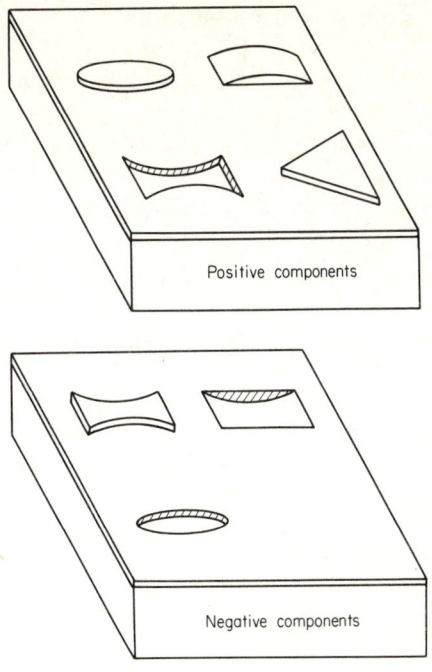

FIGURE 29. Typical two-dimensional optical components.

10% with grating efficiencies of up to 40%, indicating film-fibre coupling efficiencies of a maximum of 25%.

Kersten (in Pole et al.[103]) has devised a technique for coupling from a guide to a multimode fibre. A droplet of high index liquid on the surface of the guide couples the wave out of the guide into a fibre that is inserted into the droplet at the most favourable angle. Maximum coupling efficiency is around 50% for multimode film to multimode fibre.

Smolinsky et al. (in Pole et al.[103]) have devised an arrangement of tapered-film coupler for coupling to a fibre. The substrate has a hole drilled in it with a hemispherical end which is filled with high index liquid so that the light leaving the film is focussed into the end of a fibre introduced into the liquid. Coupling efficiencies of 50–60% have been achieved. Unfortunately the reverse coupling from fibre to guide is much less efficient.

Guttmann et al.[104] have developed a device for accurate alignment of an optical fibre which can be used in a variety of applications, including the butt-joint-film fibre arrangement. The unit consists basically of a rotatable cylinder itself carrying a second eccentrically mounted rotatable cylinder with axis parallel to that of the first and carrying the fibre mounted eccentrically. Rotation of either or both cylinders moves the fibre in translation to any point within a large area of circular shape with a small forbidden area in the centre, the existence of which is inevitable without impossibly accurate dimensions. In single mode fibre–fibre coupling applications the device has achieved coupling efficiencies of up to 95%.

In order to match the modes of fibres more nearly to that of thin film devices it has been suggested (Smolinski *et al.* in Pole *et al.*[103]) that fibres should be flattened. Dalgoutte *et al.*[105] have developed a technique for flattening the end of a circular fibre so that the aspect ratio is similar to that of a semiconductor laser. This improves considerably laser-fibre coupling efficiency. The gradual transition from flattened end to circular fibre shifts the characteristics of the modes from one type to another without loss, and it is likely that such an arrangement would have useful application in forms of coupling other than laser-fibre.

3.7 Two-dimensional Optics

As far as the behaviour parallel to the surfaces of the film is concerned, the propagation of a single guided mode is of exactly the same form as that of a three-dimensional plane wave. The variation normal to the film boundaries is somewhat different, but it is stationary, and for power flow calculations can be neglected if the effective thickness of the film is used. Thus we can treat the modes as straightforward two-dimensional beams which follow the same rules as normal three-dimensional free beams. Two-dimensional optics is the term which is used to describe this aspect of thin film waveguides. The propagation is governed by what we have called the propagation constant β which, as we have seen, is given by $\beta = k n_f \sin \theta_f$ and the angular frequency of the wave, ω, which always retains its free space value. Thus the wave velocity is given by

$$\frac{\omega}{\beta} = \frac{c}{n_f \sin \theta_f}$$

and we can think of $n_f \sin \theta_f = \beta/k$ as the effective refractive index associated with the mode. We will denote β/k by N in this context. N can take any value between the index of the substrate and that of the guide. Of course the effective index, N, has a different value for each of the modes of the film and so the correct operation of two-dimensional optical devices usually depends on there being just one normal mode present which is usually the lowest order. For this reason, the thickness of the films is often kept below the cut-off for the higher order modes. Because it is easy to change the refractive index of a guide simply by varying its thickness, the construction of all the usual three-dimensional refracting optics in two-dimensional form becomes simply a matter of varying the thickness of the guiding film in a prescribed manner over the area concerned. In fact it becomes possible to produce two-dimensional components the three-dimensional form of which would be exceedingly difficult if not impossible to construct and components with graded indices are examples of this facility.

Shubert and Harris[106] describe a number of components, including diffraction gratings and lenses. Ulrich and Martin[107] have published a treatment of two-dimensional optics which is essentially the normal three-dimensional treatment based on Snell's law adapted for two dimensions. They show formally that the propagation of two-dimensional waves is governed by equations which are identical to those of free three-dimensional waves. Their treatment involves manipulating the three-dimensional wave equation into a two-dimensional wave equation containing the index N which is obtained as an eigenvalue of a second differential equation relating to the z-variation of the fields.

Typical optical elements are shown diagrammatically in Fig. 29. These can be produced quite simply by depositing extra identical film material on top of the existing guide or by masking off part of the film so that there is an indentation. Extra material gives a component with an effective index larger than that of the guide, while an indentation gives a lower effective index. The components may alternatively be made of high index material embedded in the substrate or in the layer, or deposited on top of the guide (Shubert and Harris[105]).

Tien et al.[101] describe a useful technique, mentioned already in the section on coupling, which involves simply depositing high index components with tapered edges and then covering the entire surface with a single guiding film of index between that of the components and the substrate. The advantage of being able to use a different material for the component is that lossy materials, unsuitable for extended lengths of guide but useful optically, can be employed.

It is possible to distort a waveguide in the third dimension without altering its thickness and to produce novel components in this way. Righini et al.[108] have described a geodesic lens consisting of a spherical identation of a constant thickness guide such as might be produced by coating a substrate with an indentation with a uniform film. Around the distorted surface the minimum path lengths are given by geodesics and the overall effect is a focussing of the beams in the manner of a positive lens. Such a component depending on the geometry of the distorted surface shows no dispersion or dependence on the mode of the guide but it does possess spherical aberration. Harper and Spiller (in Pole et al.[103]) have recently described a technique for compensating the spherical aberration of this type of lens which involves a thickening of the guide in the centre of the depression.

We can follow Shubert and Harris[105] and take the useful range of refractive indices possible from a single thin film material to be those corresponding to thicknesses below the cut-off for the second normal mode of the guide. These can simply be found from a diagram such as Fig. 22. Shubert and Harris [106] give curves showing the range of indices which can be achieved with films of index up to 1.75 on a substrate of index 1.52 in a medium of index 1.0. The variation in the effective indices which can be achieved is a little less than in three dimensions with available optical glasses. With high film indices, like the 2.35 of zinc sulphide, a rather greater range is possible. From Fig. 22 we can see that the refractive index varies almost linearly with film thickness over the greater part of the available range which simplifies the construction of graded index devices.

With a normal three-dimensional optical component there is a reflection loss at the surface unless an antireflection coating has been applied. The situation is much the same with two-dimensional components and Marcuse[77] has shown that for zero order TE waves the reflectance is simply

$$\left[\frac{N_1 - N_2}{N_1 + N_2}\right]^2$$

just as in the three-dimensional case, although expressions for other cases can be more complicated. However, as Ulrich and Martin[107] explain, an antireflection coating can simply consist of a tapered edge to the thin film component. Provided the taper is many wavelengths long then the reflection loss is negligible.

The dispersion of thin film guides has been studied by Ulrich and Martin[107] who have shown that the dispersion can often be greater than that of the material used for the guides. Figure 30 shows the dispersion of a ZnS film on BK7 glass at 632.8 nm against the effective index N. It is possible by choosing two points such as I and II which have different effective indices but identical dispersion to construct achromatic components since the ratio of the refractive indices will remain constant.

Ulrich[109] has recently surveyed two-dimensional optical components.

3.8 Waveguide Filters and Beam Splitters

3.8.1 Narrow-band grating filters

The grating filter has received more attention than others and consists of a waveguide which has a periodic disturbance in its characteristics: this is usually a corrugation of the profile of the guide running normal to the direction of propagation. The principal characteristic of the filter is a reflection peak of very narrow bandwidth.

There are several ways of understanding the operation of such a device.

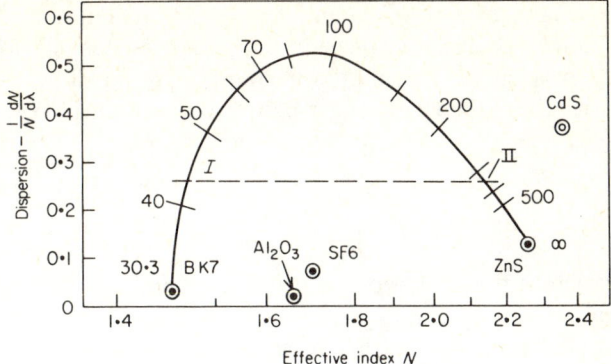

FIGURE 30. Plot of dispersion against index for the TE_0 mode of a ZnS ($n = 2.35$) guide on Schott BK7 glass in air at a wavelength of 632.8 nm. The thickness of the guide in nm is shown along the curve. The dispersions of several bulk optical materials is also shown. SF6 is the designation of another type of Schott optical glass. (After Ulrich and Martin[107].)

The effect of the corrugations is to cause the effective index of the guide to vary in a cyclic manner. This is not unlike the variation of refractive index through a quarter wave stack (described in part I). If the period of the corrugations is Λ and the mean effective index N then $N\Lambda$ will be the optical thickness of the equivalent of a high index plus a low index layer and therefore is half the wavelength corresponding to peak reflectance. The peak of the filter is therefore given by $\lambda = 2N\Lambda$ where λ is the free-space wavelength. We can see qualitatively that the bandwidth of the reflection peak, as for the quarter wave stack, will be narrower as the corrugation depth decreases but this must be accompanied by an increase in the length of the grating filter in order to maintain the peak reflectance.

Another way of visualising the operation of the filter is to consider it simply as a thick diffraction grating and to apply the coupled mode approach to performance calculations as in the thick grating coupler. The coupling will be especially strong when the Bragg condition is satisfied. In the case of an input wave normal to the grating and an output wave of identical frequency also normal to the grating but travelling in the reverse direction, we have

$$\frac{2\pi N}{\lambda} = \frac{1}{2} \cdot \frac{2\pi}{\Lambda}$$

which gives, as before,

$$\lambda = 2N\Lambda$$

for the wavelengths of peak reflection. This is the normal incidence value but the Bragg condition can be satisfied for any angle of incidence and the appropriate wavelength. Thus the grating filter can be tuned by tilting. If θ is the angle of incidence then the Bragg condition gives for the peak wavelength:

$$\frac{2\pi N}{\lambda} \cos \theta = \tfrac{1}{2} \cdot \frac{2\pi}{\Lambda}$$

i.e. $\qquad \lambda = 2N\Lambda \cos \theta$

which is similar to the expression for the effect of tilts in thin film filters.

The bandwidth is inversely proportional to the effective length of the grating, i.e. to the length which contributes to the peak reflectance. The weaker the perturbations due to the corrugations, the longer the grating has to be and the narrower is the bandwidth. Flanders et al.[110] describe grating filters with reflectances greater than 75% and bandwidths less than 0.2 nm. These were constructed from sputtered glass guides by ion beam machining through a layer of photoresist which had been exposed to an interferogram. The performance of one of the filters is shown in Fig. 31.

In the quarter wave stack, the magnitude of the sidelobes is a function of the refractive index contrast. Similarly in the grating filter the magnitude of the sidelobes is a function of the coupling coefficient between the incident and diffracted beams and the effective length of the grating. Matsuhara and Hill[111] show how the sidelobes may be reduced in magnitude by profiling the grating so that the groove depth varies slowly along the length of the grating; being lowest at the ends and highest in the middle.

Schmidt et al.[112] have produced still narrower filters. Their gratings were produced in the same way as those of Flanders et al. and in a length of around 10 nm. The coupling coefficient was then reduced by immersing the grating in a fluid of index near that of the sputtered glass. Bandwidths of 0.015 nm were obtained and, using this technique, grating filters with bandwidths of 0.01 nm are feasible.

3.8.2 Beam splitters

Beam splitters are certain to be essential components in integrated optical devices. The simplest type of beam splitter involves some perturbation of the guide causing a local discontinuity in effective index similar to a single thin film used in three-dimensional optics. The perturbation must not be such that light can be scattered into higher modes and lost. Tsang and Wang[112] have described a simple beam splitter of this type which consists either of a local thinning or a local thickening of a guide. The rate of etching of the (100) face of silicon is much faster than that of the (111) faces inclined at 54.74° to it. Thus preferential etching of a (100) face can be used to produce v-shaped

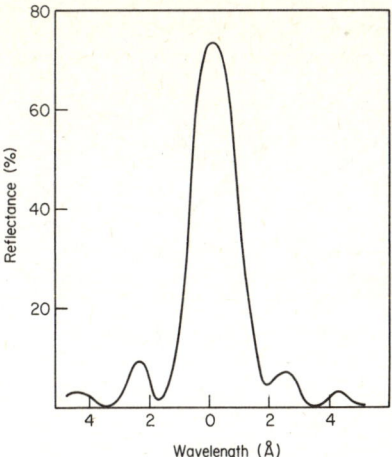

FIGURE 31. Measured performance of a narrow-band grating filter. (After Flanders et al.[108].)

grooves or protruding ridges with perfectly smooth walls consisting of (111) planes. The width of the mask defining the groove determines its depth. The substrate can then be coated with organic light guiding films by spinning on a solution of the appropriate material. The upper surface of the guide is completely flat so that the local thickness is either increased at a groove or decreased at a protrusion. Of course, silicon is a high index substrate and the organic light guiding films are very much lower in index and so a thick, low index coating, which can most conveniently be silicon dioxide grown by thermal oxidation, has first of all to be applied to the substrate before the light guide to isolate it from the substrate.

A particularly interesting device has been described by Mahlein et al.[114]. This is a TE-TM mode splitter which is based on the thin film polarising beam splitter described earlier but the method could readily be extended to make use of any thin film optical filter in an integrated optical circuit. The basis of the technique is quite simply to insert in the waveguide a conventional thin film multilayer so that it forms a right angle with the plane of the guide. The steps are shown in Fig. 32. The cut faces of the substrates are polished before the glass plates in (b) are optically contacted so that they protrude only a few tenths of a millimetre on the side to be coated. Once in place the glass plates are held by PTFE clamps and then the waveguide, which in this case is a thermally evaporated glass film with index greater than that of the substrate, is deposited. The remainder of the steps are as illustrated

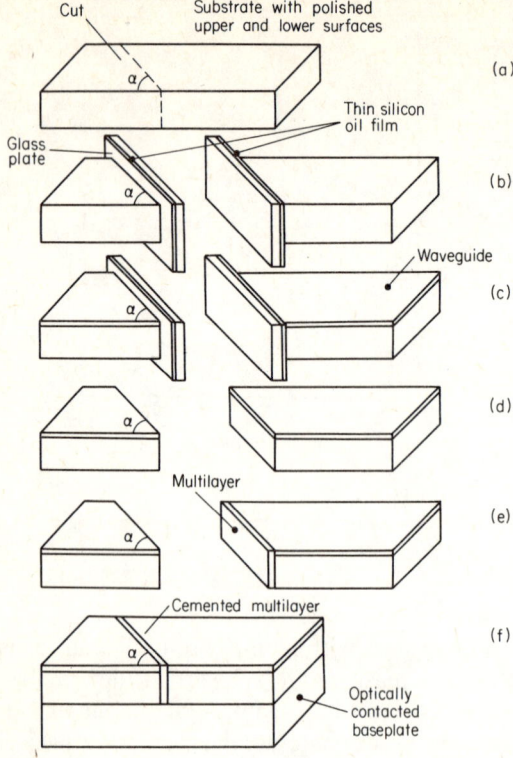

FIGURE 32. Steps in the construction of an integrated optical filter. (After Mahlein et al.[112].)

with the final assembly being optically contacted to the baseplate before the cement is cured to ensure correct alignment. The particular thin film component which Mahlein et al.[114] constructed was a polarising beam splitter consisting of a quarter wave stack of zinc sulphide and cryolite layers the TE waves corresponding to the p-waves and the TM to the s-waves so that the beam splitter transmitted the TE waves and reflected the TM. The success of this technique means that virtually the whole range of conventional thin film filters is available for use in integrated optical circuits.

3.9 Second Harmonic Generation

For small fields the polarisation of a dielectric is a linear function of electric field. For higher fields non-linearities cannot be neglected and the expression for polarisation becomes of a form similar to

$$P = \chi_1 E + \chi_2 E^2 + \chi_3 E^3 \qquad (77)$$

The series converges rapidly and (77) is a very much simplified form of the relationship. Polarisation and electric field are vectors rather than the scalars in (77) and χ_1, χ_2 and χ_3 are usually tensors. The first term in (77) leads simply to the normal linear wave propagation in crystals. Even in crystals having a comparatively large non-linearity the second term is small unless E is very large.

An incident wave of frequency produces a polarisation

$$P = \chi_1 \mathscr{E} \cos(\omega t - kx) + \chi_2 \mathscr{E}^2 \cos^2(\omega t - kx) + \ldots$$
$$= \chi_1 \mathscr{E} \cos(\omega t - kx) + \tfrac{1}{2}\chi_2 \mathscr{E}^2 [1 + \cos(2\omega t - 2kx)]$$

The terms of frequency 2ω will radiate an electromagnetic field of the same frequency. This field will be propagated through the crystal with appropriate value k' of k corresponding to frequency 2ω. The polarisation term is propagating with velocity $(2\omega/2k) = (\omega/k)$ while the electromagnetic field has velocity $2\omega/k'$. If these velocities are unequal then the radiated field at any instant will pass in and out of phase with the already existing field so that electromagnetic intensity, instead of steadily increasing throughout the crystal, will simply fluctuate. The total relative change of phase between the polarisation and the electromagnetic field through the length 1 of the crystal should therefore be not greater than $\pi/2$.

i.e.
$$(2k - k')l \leqslant \frac{\pi}{2} \qquad (78)$$

When the two k values are sufficiently close the crystal is said to be phase matched.

There are difficulties in arranging phase matching in a bulk crystal. The normal technique is for a direction to be chosen in a birefringent crystal so that the refractive index for one plane of polarisation at ω is equal to that for the other plane of polarisation at 2ω. Such an approach cannot be used in isotropic crystals.

The conditions, therefore, for efficient second harmonic generation are as intense a field as possible within the crystal to ensure a sufficiently large second harmonic polarisation, and phase matching to ensure the efficient coupling of the electromagnetic field due to the polarisation into the second harmonic wave. Both of these conditions are more straightforward to establish in a thin film light guide than in a bulk crystal. The maximum intensity within the thin film guide is many times that of the free incident wave. Further, the dispersion of the guide is a sensitive function of its dimensions rather than of the dispersion of the material from which it is made. Thus a TE mode fundamental may be non-linearly coupled to a second harmonic TM mode and the dimensions chosen so that accurate phase matching is obtained over considerable lengths of guide. A useful discussion of the technique is given by Tien[75] and a detailed theoretical analysis of the problem has been performed by Conwell[115].

Anderson and Boyd[116] have used GaAs waveguides to produce phase matched second harmonic radiation from a primary beam of CO_2 laser radiation. GaAs is isotropic and so the normal bulk technique based on birefringence cannot be used for phase matching. In thin film form however, phase matching is readily arranged by choosing appropriate dimensions for the guide. The range of wavelengths over which phase matching was achieved was sufficiently large for CO_2 wavelengths from 9.2 µm to 10.8 µm to produce phase matched second harmonics from the same GaAs guide. Dimensions of guides used were of the order of 3.25 µm thick and around 4000 µm long. To increase the wavelength range over which phase matching could be obtained in the same guide, some were tapered in thickness.

The second harmonic need not necessarily be generated in the thin film guide itself. A linear guide on a non-linear substrate can also be used. This was the approach used by Tien et al.[117] to produce second harmonic light in the zinc oxide substrate of a zinc sulphide guide. In this case the waves were not phase matched but the guide was arranged so that the phase velocity of the second harmonic polarisation in the substrate was faster than the phase velocity of a free wave in the substrate at the same frequency; just the situation with Cerenkov radiation where the direction of propagation of the light forms a cone of angle α around the track of a very fast particle. Here, the free wave is a plane wave at an angle to the substrate surface where

$$\cos \alpha = \frac{\text{free wave phase velocity}}{\text{polarisation phase velocity}}$$

The c-axis of the ZnO substrate was arranged parallel to the substrate surface and at right angles to the direction of propagation in the guide. Then, when the light of wavelength 1.06 µm from a Nd-doped YAG laser was coupled into a TE mode, the evanescent wave in the substrate produced non-linear polarisation along the c-axis which radiated into a free plane wave inclined at an angle to the surface. In one case the guide had a thickness of 0.204 µm and the Cerenkov angle was 13.5°.

Second harmonic generation in ZnO waveguides has been achieved by Zemon et al.[118]. A TE_0 mode at 1.06 µm was converted into a TM_1 mode at 0.53 µm. Van der Ziel et al.[119] obtained second harmonic generation in a GaP waveguide.

Zinc sulphide itself is non-linear at sufficiently high fields and an early experiment on second harmonic generation in polycrystalline zinc sulphide films was performed by V. E. Sotin in 1969. This is described in a review by Zolotov et al.[120]. The zinc sulphide guide was deposited over a lithium fluoride coupling layer on the base of a prism and the exciting light was coupled into the zinc sulphide layer and then directly out again by illuminating internally the centre of the prism base with a Nd laser beam. In one arrangement the outer surface of the ZnS thin film guide was immersed in nitrobenzene with a dispersion of index such that the second harmonic light was not contained within the guide but escaped through the nitrobenzene as a coherent plane wave. More recently, Uesugi (in Pole et al.[103]) has described phase-matched second and third harmonic generation in polycrystalline zinc sulphide guides.

Dabby et al.[121] give a theoretical analysis of waveguides containing a periodic structure and show how such an arrangement can be used to achieve phase matching.

3.10 Materials

For a successful thin film waveguide, the refractive index of the film must be greater than those of the adjoining media, and the film should be as free from absorption as possible. The first requirement is relatively easy to satisfy but the second is much more difficult to achieve.

The simplest way of producing thin film waveguides is just to take the materials and techniques which have been long established in the field of thin film optical coatings. Unfortunately the conditions of use in thin film guides are quite different and in particular the modes of propagation along rather than across the films mean that even in what is a relatively short length of thin film guide, the total path length traversed in the film is several orders of magnitude greater than it would be in an optical coating. Thus, many materials having acceptable performances for optical coatings are quite unsuitable for thin film waveguides. This has been well understood in the early stages of development of the field, especially since optical fibres with their even more severe absorption problem have been extensively studied. Right from the beginning a range of techniques has been used some of which are quite different from those more familiar to the optical thin film coating worker.

An early technique, which did not involve vacuum deposition, was the irradiation of fused silica by protons (described by Schineller et al.[122]) which increases the refractive index of the fused silica in a localised region. The index change is not large and varies linearly with the proton dose up to a maximum of around 0.01. The change appears to be due to displacements of the atoms of the fused silica caused by collisions with the protons. Initially when the protons enter the fused silica the primary loss mechanism is by ionisation and it is only when the proton has lost some energy and is below the surface that collisions become important. For 1.5 MeV protons, calculation and experiment show that the region of increased index should be some 27 µm below the surface, around 1 µm wide, and that the 0.01 index change should occur with a total dose of 10^{15} protons per cm^2. Optical waveguides have been produced using this technique but no measurements have been made of losses in the guides mentioned.

Standley et al.[123] used a similar technique in which fused quartz was bombarded with a wide variety of ions the best results being obtained using Li^{+7} ions. The refractive index change can be as high as 0.05 and it is proportional to the total dose of radiation. However it again appears that it is the displacement of the atoms of the fused quartz, rather than the implantation of lithium, which causes the index increase since most of the change can be annealed out. The losses appear to be less if bombardment is carried out at an elevated temperature. Losses below 0.2 dB/cm with refractive index changes of 0.01 were measured after a dose of 10^{15} particles/cm^2 at 220°C followed by annealing for 1 hour at 300°C. The annealing reduced both the index and the losses immediately after bombardment.

Sputtering is another technique which has been used in the production of thin film light guides. Tien[73] describes measurements of losses in sputtered ZnO films. Losses of more than 60 dB/cm were found initially primarily due to the rough surface of the films. After polishing the films, the loss dropped towards 20 dB/cm. Tantalum oxide films were then produced in an attempt to reduce the losses. The procedure is described by Hensler et al.[124] and summarised by Tien[75]. High purity tantalum metal was first sputtered in an argon atmosphere and then the β-tantalum which was deposited on the substrate was converted to Ta_2O_5 by heating in pure oxygen at 500°C. This gave films of refractive index 2.2136 at 632.8 nm with measured losses of 0.9 dB/cm. In the blue region the losses rose, becoming 4.1 dB/cm for a wavelength of 488.0 nm.

Goell and Standley[125] used R.F. sputtering in oxygen to prepare glass waveguiding films. Corning 7059 glass gave the best results and gave high quality films of refractive index 1.62. The absorption losses of the films are not quoted but they must have been very small. Scattering losses were measured as around 1 dB/cm.

Aagard[126] has formed niobium pentoxide films with losses of 1–2 dB/cm, occasionally 0.5 dB/cm, by reactively sputtering niobium metal in oxygen.

A useful report on the R.F. sputtering of materials for thin film light guides has been published by Pitt et al.[127] which describes the authors' own results on glass, ZnO, Ta_2O_5, and TiO_2 as well as discussing in general terms the sources of loss in optical guides and the criteria for selecting guide materials. A very full account of their sputtering technique is included.

R.F. sputtering was also used by Watts et al.[128] to prepare films of certain chalcogenide glasses, As_2S_3, $Ge_{28}Sb_{12}Se_{60}$ and $Ge_{33}As_{12}Se_{55}$. The properties were measured at 1.064 nm and the As_2S_3 glass had the lowest loss of around 0.4 dB/cm for a TE_0 mode. The $Ge_{33}As_{12}Se_{55}$ glasses had the highest losses at 22 dB/cm. These glasses have high refractive index, 2.3 to 2.8 and appear potentially useful in acousto-optic devices.

Diffusion techniques have been used by a number of authors. Carruthers et al.[129] produced Li_2O guides on the surface of lithium niobate and lithium tantalate by heating the crystals in a vacuum at temperature in the region of 1000°C causing out-diffusion of lithium oxide. The removal of lithium oxide increases the extraordinary refractive index, although the ordinary index is not affected, producing very low loss guides with an exponential refractive index profile. Similar out-diffused guides have been produced by Noda et al.[130].

The indiffusion of material has also been successfully applied. Most of this work has been aimed at fabricating guides in which large electro-optic effects could be produced. Taylor et al.[131] formed single crystal optical waveguides by diffusing Se into single crystal CdS. Losses measured were 10–15 dB/cm at 632.8 nm. Martin and Hall[132] reported the diffusion of Cd and Se into ZnS and of Cd into ZnSe which gave losses as low as 3 dB/cm at 632.8 nm. $LiTaO_3$ and $LiNbO_3$ were used as substrates by a number of workers. Noda et al.[133] diffused copper into $LiTaO_3$ forming satisfactory guides at 632.8 nm. Schmidt and Kaminow[134] formed guides by diffusing titanium, vanadium or nickel into $LiNbO_3$ with losses of around 1 dB/cm at 632.8 nm. Hammer and Phillips[135] and Standley and Ramaswamy[135] diffused Nb into $LiTaO_3$ to give losses as low as 1 dB/cm, again at 632.8 nm. These diffused waveguides present problems in thickness control which are more severe than in many of the other techniques and this has prompted Weller and Giallorenzi[137] to propose the use of thin-film overlays to adjust the characteristics of the diffused waveguides.

Vacuum evaporation, the usual technique for optical coating manufacture, has also been used to produce waveguides and the low loss of ZnS in optical coatings together with its high index has made it an obvious choice. Unfortunately even when deposited on substrates at low temperatures to reduce the scattering losses, it has been found that the intrinsic absorption is sufficiently high to give losses of around 5 dB/cm at 632.8 nm (Tien[75]).

One of the best results of vacuum evaporated films is that of Kersten and Rauscher[138] who used an electron beam source to evaporate G.V. Planar CAS 10 glass. The losses measured in these films at 676.4 was approximately 1.2 dB/cm. The corresponding refractive index of the glass is 1.4675 and so low index substrates, in this case fused quartz, must be used. The losses are greater for shorter wavelengths reaching approximately 4 dB/cm at 470 nm.

These figures compared well with those of Goell and Standley[124] for sputtered glass.

Epitaxially-grown ZnO films on sapphire substrates have been studied by Channin et al.[139]. The films were subjected to various treatments after deposition including annealing and polishing and the best gave losses of less than 1 dB/cm at 632.8 nm. This can be compared with the results for sputtered ZnO films quoted by Tien[75] and mentioned above.

Epitaxial deposition of single crystal GaAs onto N-type GaAs substrates is reported by Cheo et al.[140]: at 10.6 μm the substrate index is 2.975 and that of the film 3.275. The losses in the waveguides are very low and are due not to the guide material itself but to the substrates. The losses vary with film thickness and mode but can be considerably less than 1 dB/cm.

Certain solution deposited films have been found to have a high performance as light guides. Harris et al.[85] have reported the use of polyester and polyurethane films deposited from solution. Ulrich and Weber[141] give a full account of the preparation procedures for a number of solution deposited films. The materials are dissolved in a suitable solvent, applied to the substrate and the solvent is permitted to evaporate very slowly. This gives very thin films of uniform thickness and was used to produce films of polyurethane ($n = 1.555$ for $\lambda = 632.8$ nm) and epoxy ($n = 1.581$), both polymerised after deposition by baking in air, lead-silica ($n = 1.664$), slightly baked after deposition, and photoresist ($n = 1.615$), which were exposed to UV radiation and developed. The losses were almost all less than 1 dB/cm at both 632.8 nm and 1.064 μm except for the photoresist which had rather higher losses of 7 dB/cm at 632.8 nm and 1 dB/cm at 1.064 μm. Ramaswamy and Weber[142] have reported losses of less than 0.2 dB/cm from films formed from a mixture of polymethyl methacrylate and styreneacrylonitrite (3:1) copolymer dissolved in methylisobutylketone with a very small addition of silicone oil. The silicone oil inhibits the orange peel effect which otherwise sometimes appears. The deposited films were baked for 70 hours at 42°C. The advantage of using this mixture of polymers is that a range of refractive indices from 1.489 to 1.566 can be produced simply by altering the composition. Baking at 80°C slightly increases the indices by up to approximately 0.007 but the losses are also very slightly increased.

Tomlinson et al.[143] describe a method of forming waveguides and couplers, including gratings, in a polymer film by a photolocking process in which a dopant is dispersed in a polymer and locked in the film by a photochemical reaction on exposure to ultraviolet light. Unreacted dopant is simply removed by heating. Losses of 0.3 dB/cm were measured on the doped polymer at 632.8 nm.

Possibly the lowest losses reported so far have been achieved with plasma polymerised films. Tien et al.[144] describe the production of organosilicon films from vinyltrimethylsilane and hexamethyldisiloxane monomers in an argon R.F. discharge. The polymerised vinyltrimethylsilane films have an index of 1.531 which makes it satisfactory for use on microscope slide glass while the hexamethyldisiloxane films have an index of 1.488 suitable for use on fused silica or certain lower index glasses. Mixtures could be used to produce intermediate indices. The best films produced losses of less than 0.04 dB/cm at 632.8 nm.

Szentesi and Noga[145] have deposited polymonochloropara-xylylene films from a reactive vapour of the monomer formed by pyrolising the dimer. Films have an index of 1.659 at 632.8 nm and losses measured were of the order of 0.8 dB/cm.

Very little definite pattern emerges from the various published techniques. In the end it is likely that the chosen method in any particular case will be adopted because of compatability with the remainder of the process. Certainly every possible technique for the production of thin films is being investigated in what is a fast moving field.

3.11 Active Devices

The devices we have considered so far can be classified as passive, but integrated optical circuits will need also to include active devices. The difficulties of coupling laser beams into thin film waveguides have been mentioned already and a better arrangement in many respects is to have the laser built into the integrated optical circuit and so the development of thin film lasers is now receiving much attention. Once the light signal is generated by the laser, modulators are required to impress on it the information to be carried and switches to direct it to the appropriate channel. Ideally, the devices should act on the optical beam directly rather than involve any form of conversion and then reconversion. In the integrated optical circuit, the light will be propagating along a thin film waveguide and the simplest way in which the desired interaction can take place is through some perturbation of the propagation characteristics of the guide. The propagation constant of the guide depends on the mode, the guide dimensions, and the optical constants of the thin film material. Assuming that the mode remains the same this leaves the dimensions and the optical constants as possible parameters. Those effects which appear specially suited for application in this area are magneto-optical, acousto-optical and electro-optical effects. Thin film devices utilising these effects can be made rather more efficient than the corresponding bulk devices because of the greatly increased power density which characterises propagation in thin films.

3.11.1 *Electro-optic devices*

The phase change induced in a beam of light on passing through a given length of guide depends only on the propagation constant or the effective index. The phase change induced by a length of guide l is given by βl where β is the propagation constant appropriate to the mode. If the index n_f is changed by Δn_f then the corresponding change in phase is given by

$$kl \frac{d(\beta/k)}{dn_f} \cdot \Delta n_f$$

The important quantity is $d(\beta/k)/dn_f$ and we can follow Tien[75] and derive an expression from equations (40) to (49). We limit this to the TE modes.

We have

$$\frac{\beta}{k} = n_f \sin\theta_f \tag{43}$$

and k_z within the film $= kn_f \cos\theta_f$. If we use the symbol α for this quantity then

$$\frac{\alpha}{k} = n_f \cos\theta_f$$

Further, from (40) and (41),

$$\Phi_a = \tan^{-1}\left[\frac{(n_f^2 \sin^2\theta_f - n_0^2)^{\frac{1}{2}}}{n_f \cos\theta_f}\right] = \tan^{-1}\frac{\gamma_a}{\alpha}$$

$$\Phi_b = \tan^{-1}\left[\frac{(n_f^2 \sin^2\theta_f - n_s^2)^{\frac{1}{2}}}{n_f \cos\theta_f}\right] = \tan^{-1}\frac{\gamma_b}{\alpha}$$

where, from (48) and (49),

$$\gamma_a = k(n_f^2 \sin^2\theta_f - n_0^2)^{\frac{1}{2}}$$

and

$$\gamma_b = k(n_f^2 \sin^2\theta_f - n_s^2)^{\frac{1}{2}}$$

Also, from (42),

$$\frac{2\pi n_f d_f \cos\theta_f}{\lambda} = d_f \alpha = m\pi + \Phi_a + \Phi_b$$

$$= m\pi + \tan^{-1}\left(\frac{\gamma_a}{\alpha}\right) + \tan^{-1}\left(\frac{\gamma_b}{\alpha}\right) \tag{79}$$

$$\frac{d\alpha}{dn_f} = \frac{k^2\left[n_f - \left(\frac{\beta}{k}\right)\frac{d(\beta/k)}{dn_f}\right]}{\alpha} \tag{80}$$

$$\frac{d\gamma_a}{dn_f} = \frac{k^2(\beta/k)d(\beta k)/dn_f}{\gamma_a} \qquad (81)$$

and similarly for

$$\frac{d\gamma_b}{dn_f}$$

Differentiating (79) we find

$$d_f \frac{d\alpha}{dn_f} = \frac{1}{1+\frac{\gamma_a^2}{\alpha^2}} \left\{ \frac{1}{\alpha} \frac{d\gamma_a}{dn_f} - \frac{\gamma_a}{\alpha^2} \cdot \frac{d\alpha}{dn_f} \right\} + \frac{1}{1+\frac{\gamma_b^2}{\alpha^2}} \left\{ \frac{1}{\alpha} \frac{d\gamma_b}{dn_f} - \frac{\gamma_b}{\alpha^2} \cdot \frac{d\alpha}{dn_f} \right\}$$

Substituting from (80) and (81), and collecting terms we find

$$\frac{d(\beta/k)}{dn_f} = \frac{kn_f}{\beta} \cdot \frac{d_f + \frac{\gamma_a}{\alpha^2+\gamma_a^2} + \frac{\gamma_b}{\alpha^2+\gamma_b^2}}{d_f + \frac{1}{\gamma_a} + \frac{1}{\gamma_b}}$$

which is exactly the expression obtained by Tien[75].

Now for d_f equal to the cut-off thickness we have that $\beta/k = n_s$ and $\gamma_b = 0$, so that

$$\frac{d(\beta/k)}{dn_f} \to 0$$

as d_f is reduced to the cut-off thickness.

Increasing the thickness we find that $(\beta/k) \to n_f$ and $\alpha \to 0$ so that

$$\frac{d(\beta/k)}{dn_f} \to 1$$

A typical curve showing the way in which

$$\frac{d(\beta/k)}{dn_f}$$

varies is shown in Fig. 33. Clearly the efficiency of a modulator is going to be low if the thickness of the guide is approaching cut-off.

For phase modulation of an input beam all that is required is an electro-optic waveguide with electrodes attached. For intensity modulation the basic phase modulation is retained but the phase is converted into intensity either by interference with an unmodulated reference beam or by

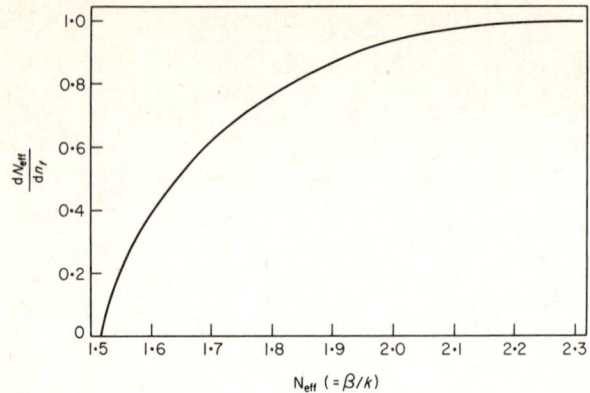

FIGURE 33. Plot of dN_{eff}/dn_f against $N_{eff}(=\beta/k)$ for the TE_0 mode of a guide of index 2.35 on a substrate of index 1.52 in a medium of index 1.0.

rotation or by any other suitable change in the state of polarisation of an input beam detectable by an analyser.

There are three principal ways in which electro-optic phase modulators have been constructed and most of the published designs use arrangements of one or other of the basic types. The first is simply a slab of electro-optic material with electrodes applied. The electro-optic material must normally be oriented and so the film is usually grown epitaxially on a single crystal substrate. This makes it difficult to deposit electrode films before and after the electro-optically active film and so a frequent approach is to etch away a film which has already been deposited leaving a narrow strip with rectangular cross-section. Electrodes are then applied to either side of the slab. The substrate must be insulating. The process is shown diagrammatically in Fig. 34. Modulators of this type have been described by a number of workers. A good description is given by Noda et al.[146]. Copper was diffused into a $LiTaO_3$ crystal to form a light guiding layer which was then etched so that the approximate dimensions of the waveguide were 5 μm thickness, 25 μm wide and 5.4 mm long. This gave excellent modulation performance stretching virtually unchanged from D.C. to 1GHz.

The second basic type of modulator is similar but the electrodes are deposited over the light guiding electro-optic film with a small gap between (Fig. 35). The index of the guide is reduced by the metal cladding of the electrode so that the mode is confined to the space between the electrodes. Uehara et al.[147] describe the formation of such a modulator based on a film

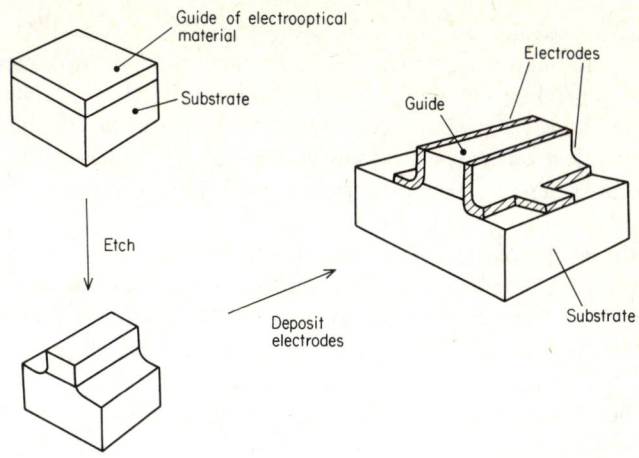

FIGURE 34. Steps in the construction of an electro-optical component.

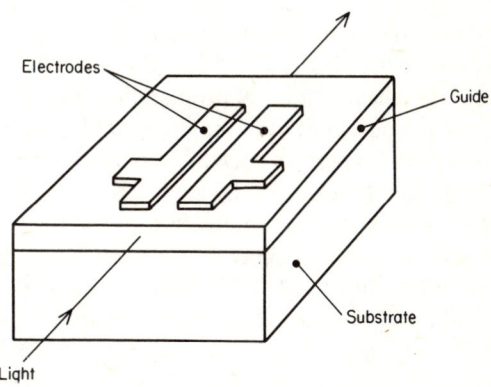

FIGURE 35. Diagram of electro-optic modulator with electrodes deposited over the active film.

of $LiTaO_3$. First of all a lithium tantalate plate was coated with a sputtered Al_2O_3 film as cladding and then attached to an $LiTaO_3$ substrate by epoxy adhesive, the Al_2O_3 film being adjacent to the epoxy. The outer surface of the plate was then polished down to a thickness of 20 μm followed by ion etching to 2.5 μm. Aluminium electrodes 7.5 mm long were added with a 22 μm gap between. Kaminow et al.[148] used a similar technique on an out-diffused waveguide on an $LiNbO_3$ substrate. Webster and Zernike[149] describe a modulator in which a similar arrangement of electrodes is arranged to produce an electro-optic effect in a slab of $LiNbO_3$. The waveguide is, however, a passive Nb_2O_5 strip guide, deposited on the $LiNbO_3$ between the electrodes. The device was fabricated by etching the Nb_2O_5 which had been deposited by sputtering on the $LiNbO_3$, and then depositing the electrodes.

In the third type of modulator the electric field is applied normal to the plane of the film. The substrate is a semiconductor and the light guide is formed by reverse biasing a junction or Shottky barrier. Hall et al.[150] deposited on a low resistivity GaAs substrate, first of all a 12 μm thick epitaxial layer of high resistivity GaAs followed by an aluminium electrode to form a Schottky barrier. With no voltage applied to the device the change in index between substrate and film was insufficient for light guiding. With a reverse bias voltage applied to the metal-semiconductor junction the electro-optic increase in index of the high resistivity layer was sufficient to constrain a light beam. Campbell et al.[151] describe a modulator, again using an expitaxial layer of high resistivity GaAs on a low resistivity substrate but with an electrode of gold platinum alloy deposited in the form of a narrow strip. When the Schottky barrier was reverse-biased the light was confined beneath the barrier strip because of the discontinuity in index in the y-direction; i.e. across the direction of propagation. This gives, in addition to phase modulation, the possibility of amplitude modulation by coupling into a receiver with very small aperture such as an optical fibre. A large change in the intensity of the signal picked up by such a receiver can result when the beam is laterally confined by the application of the reverse-bias voltage. A double heterostructure $GaAs-Al_xGa_{1-x}As$ system, where greater efficiencies are possible, has been described by Reinhart and Miller[152]. Their particular system had cut-off frequencies, based on capacitance measurements, of around 4GHz but with potentially much higher values by improving the contacts.

Phase modulators have been converted to intensity modulators by a number of workers. Webster and Zernike[149] constructed a push-pull modulator by placing a phase modulator in each arm of a Jamin two-beam interferometer. The interferometer itself was a conventional one with the light being coupled in and out of the modulators by prism couplers. Martin[153] has constructed a complete thin film Mach-Zehnder interferometer using strip guides of single crystal ZnSe. Phase modulators can also be fed with light plane polarised at 45° to the guide normal. The differential phase shift between TE and TM modes rotates the plane of polarisation and causes intensity modulation of the light transmitted by an analyser (Uehara et al.[147], and Noda et al.[146]).

Hammer et al.[154] have constructed a modulator/deflector of a completely different type. Electrodes are arranged to produce an electric field with a spatial period so that the induced change in refractive index forms a phase grating.

Simple switches can also be made using electro-optic effects and the two beam interferometer of Martin[153] can be used for this purpose. A particularly simple switch can be constructed by arranging for a waveguide to be near the cut-off thickness. An electric field can then be applied to reduce the appropriate refractive index and cut-off of the guide (Hall et al.[150]). Alternatively a film can be arranged to become guiding by the application of an electric field (Chanin[155]). Electro-optic deflectors have been constructed by Cheo[156] which use triangularly shaped electrodes to produce a two-dimensional prism in the guide which simply deflects the beam by two-dimension refraction.

3.11.2 Magneto-optical devices

The operation of magneto-optical modulators is a little different. A magnetic field applied in the direction of propagation rotates the plane of polarisation. This is equivalent to coupling out of a TM mode, say, onto a TE mode. Unfortunately, the phase velocities of these modes are not matched and to ensure coupling the magnetic field must be spatially modulated to form a periodic disturbance with the correct spatial period for strong Bragg coupling between the modes. This has been achieved with iron garnet films deposited epitaxially on gadolinium garnet substrates by Tien et al.[157].

Wang et al.[158] have analysed the propagation of waves in optical waveguides deposited on a substrate which is either anisotropic or gyrotropic and have suggested the construction of mode converters using electro-optic or magneto-optic substrates.

3.11.3 Acousto-optical devices

Acousto-optic modulators and deflectors work on a similar principle. The acoustic wave sets up a periodic variation in refractive index which forms a phase grating which may be used as a simple deflector if the light wave propagation is across the path of the acoustic wave (Kuhn et al.[159]; Chubachi et al.[160]).

3.11.4 Thin film lasers

Perhaps the most promising type of laser for use in integrated optical systems is the semiconductor injection laser. A typical laser will consist (see Taylor and Yariv[98]) of a thin film of GaAs which is the active layer bounded by lower index $Ga_{1-x}Al_xAs$, p-type on one side and n-type on the other. The advantages are firstly that the lasing action takes place in a thin film guide and second that the input to the laser is electrical. The conventional injection laser however must have polished faces to define the cavity which makes it more difficult to couple them into other waveguides and the short length of the laser cavity means that the output will contain many modes.

To avoid these difficulties and also to make lasers which can more easily be built into the optical circuit Kogelnik and Shank[161] devised the distributed feedback laser. What distinguishes this from the conventional laser is that the coupling between the forward and backward waves is distributed along the length of the cavity instead of being concentrated at the ends in the form of two reflectors. The coupling is derived from a periodic disturbance which can take the form of a modulation of gain, index or thickness running along the guide in the direction of propagation of the laser modes. The modulation acts as a thick Bragg grating just as in the grating filters already discussed so that the condition for feedback is $(2\pi/\Lambda) = 2\beta$ where Λ is the period of the modulation. The theory of distributed feedback lasers has been treated by Kogelink and Shank[162] and by Wang[163]. The earliest lasers of this type utilised dyes such as rhodamine 6G in gelatine (Kogelink and Shank[161]), polymethyl methacrylate (Cheremiskin and Chekhlova[164], and polymethane (Shinke et al.[165] and Wang and Sheem[166]). Optically pumped solid state lasers using surface corrugation for example in GaAs (Nakamura et al.[167] and Shank et al.[168]) have also been developed. These have the disadvantage from the point of view of integrated optics that they require external optical pumping sources.

There are great difficulties in the realisation of distributed feedback injection lasers although Scifres et al.[169] have succeeded in constructing such a laser which can be operated at 77 K. The difficulties are summarised by Wang[163] who proposes a modification from distributed feedback to distributed Bragg reflector laser simply by moving the reflectors out of the active part of the cavity. This would retain the mode selection and other advantages of the distributed feedback laser but remove the difficulties associated with the fabrication of corrugations without introducing too many defects.

References

1. M. Born and E. Wolf, "Principles of Optics", 4th Ed., Pergamon Press, Oxford, New York, (1970).
2. Z. Knittl, "Optics of thin films", John Wiley and Sons, London and New York, and SNTL, Prague, (1976).
3. O. S. Heavens, "Optical Properties of Thin Solid Films", Butterworths Scientific Publications, London 1955. (Reprinted by Dover Publications, 1965.)
4. H. A. Macleod, "Thin Film Optical Filters", Adam Hilger, London, and American Elsevier, New York, (1969).

5. J. T. Cox and G. Hass, *J. Opt. Soc. Am.*, **48**, 677, (1958).
6. J. T. Cox, G. Hass and G. F. Jacobus, *J. Opt. Soc. Am.*, **51**, 714, (1961).
7. R. Jacobsson and J. O. Martensson, *Applied Optics*, **5**, 29, (1966).
8. R. Jacobsson, in "Physics of Thin Films," Vol. 8, eds., G. Hass, M. H. Francombe and R. W. Hoffman, Academic Press, New York and London, (1975).
9. P. B. Clapham and M. C. Hutley, *Nature*, **244**, 281, (1973).
10. L. Young, *J. Opt. Soc. Am.*, **51**, 967, (1961).
11. A. Musset and A. Thelen, in "Progress in Optics," Vol. 8, 201, ed., E. Wolf, North Holland, Amsterdam and London, (1970).
12. J. T. Cox, G. Hass and A Thelen, *J. Opt. Soc. Am.*, **52**, 965, (1962).
13. P. H. Berning, *J. Opt. Soc. Am.*, **52**, 431, (1962).
14. A. J. Vermeulen, *Optica Acta*, **18**, 531, (1971).
15. J. Ward, *Vacuum*, **22**, 369, (1972).
16. J. T. Cox and G. Hass, in "Physics of Thin Films," Vol. 2, eds., G. Hass and R. E. Thun, Academic Press, New York and London, (1964).
17. E. Ritter, in "Physics of Thin Films," Vol. 8, eds., G. Hass, M. H. Francombe and R. W. Hoffman, Academic Press, New York and London, (1975).
18. K. H. Behrndt and D. W. Doughty, *J. Vac. Sci. Technol.*, **3**, 264, (1966).
19. W. Heitmann, *Z. Angew. Phys.*, **21**, 503, (1966).
20. D. L. Perry, *Applied Optics*, **4**, 987, (1965).
21. H. E. Bennet and J. M. Bennet, in "Physics of Thin Films," Vol. 4, eds., G. Hass and R. E. Thun, Academic Press, New York and London, (1967).
22. D. J. Hemingway and P. H. Lissberger, *Optics Acta*, **20**, 85, (1973).
23. A. F. Turner and P. W. Baumeister, *Applied Optics*, **5**, 69, (1966).
24. S. Penselin and A. Steudel, *Z. Phys.*, **142**, 21, (1955).
25. O. S. Heavens and H. M. Liddell, *Applied Optics*, **5**, 373, (1966).
26. P. W. Baumeister and J. M. Stone, *J. Opt. Soc. Am.*, **46**, 228, (1956).
27. J. V. Ramsay and P. E. Ciddor, *Applied Optics*, **6**, 2003, (1967).
28. P. E. Ciddor, *Applied Optics*, **7**, 2328, (1968).
29. G. Hass, *J. Opt. Soc. Am.*, **45**, 945, (1955).
30. G. Hass, J. B. Heaney, H. Herzig, J. F. Osantowski and J. J. Triolo, *Applied Optics*, **14**, 2639, (1975).
31. A. Thelen, *J. Opt. Soc. Am.*, **56**, 1533, (1966).
32. J. S. Seeley, H. M. Liddell and T C. Chen, *Optica Acta*, **20**, 641, (1973).
33. C. S. Evans, R. Hunneman and J. S. Seeley, *J. Phys. D.*, (*Appl. Phys.*), **9**, 309, (1976).
34. A. Thelen, *J. Opt. Soc. Am.*, **53**, 1266, (1963).
35. A. Thelen, *J. Opt. Soc. Am.*, **61**, 365, (1971).
36. P. H. Lissberger and J. M. Pearson, *Thin Solid Films*, **34**, 349, (1976).
37. H. A. Macleod, *Thin Solid Films*, **34**, 335, (1976).
38. H. A. Macleod and D. Richmond, *Thin Solid Films*, **37**, 163, (1976).
39. J. A. Doborowolski, *J. Opt. Soc. Am.*, **49**, 794, (1959).
40. R. R. Austin, *Optical Engineering*, **11**, 65, (1972).
41. S. D. Smith and C. R. Pidgeon, *Memoires Soc. R. Sc. Liege, 5ième serie*, **9**, 336, (1963).
42. A. E. Roche and A. M. Title, *Applied Optics*, **14**, 765, (1974).

43. P. H. Lissberger and W. L. Wilcock, *J. Opt. Soc. Am.*, **49**, 126, (1959).
44. C. R. Pidgeon and S. D. Smith, *J. Opt. Soc. Am.*, **54**, 1459, (1964).
45. G. Hernandez, *Applied Optics*, **13**, 2654, (1974).
46. S. L. Linder, *Applied Optics*, **6**, 1201, (1967).
47. P. H. Lissberger, *J. Opt. Soc. Am.*, **58**, 1586, (1968).
48. D. J. Hemingway and P. H. Lissberger, *Applied Optics*, **6**, 471, (1967).
49. P. H. Berning and A. F. Turner, *J. Opt. Soc. Am.*, **47**, 230, (1957).
50. R. J. Holloway and P. H. Lissberger, *Applied Optics*, **8**, 653, (1969).
51. B. V. Landau and P. H. Lissberger, *J. Opt. Soc. Am.*, **62**, 1258, (1972).
52. A. Thetford, *Optica Acta*, **19**, 533, (1972).
53. P. Baumeister, *Applied Optics*, **12**, 1993, (1973).
54. P. H. Lissberger, *Rep. Prog. Phys.*, **33**, 197, (1970).
55. S. M. MacNeille, U.S. Patent 2403731, 9 July, (1946).
56. M. Banning, *J. Opt. Soc. Am.*, **37**, 792, (1947).
57. P. B. Clapham, M. J. Downs and R. J. King, *Applied Optics*, **8**, 1965, (1969).
58. E. Zehender, *Lichttechnik*, **3**, 100, (1973).
59. H. De Lang and G. Bouwhuis, *Philips Technical Review*, **24**, 263, (1963).
60. K. Rabinovitch and A. Pagis, *Optica Acta*, **21**, 963, (1974).
61. A. J. Vermeulen, *Optica Acta*, **23**, 71, (1976).
62. W. P. Barr, *J. Phys. E., Sci. Inst.*, **2**, 1112, (1969).
63. R. Blazey, *Applied Optics*, **6**, 831, (1967).
64. K. Guenther, H. L. Gruber and H. K. Pulker, *Thin Solid Films*, **34**, 363, (1976).
65. H. K. Pulker, *Thin Solid Films*, **34**, 343, (1976).
66. H. Ahrens, H. Welling and H. E. Scheel, *Applied Physics*, **1**, 69, (1973).
67. H. Ahrens, *Vakum-Technik*, **24**, 33, (1975).
68. N. Bloembergen, *Applied Optics*, **12**, 661, (1973).
69. L. G. DeShazer, B. E. Newnam and K. M. Leung, *App. Phys. Letts.*, **23**, 607, (1973).
70. A. J. Glass and A. H. Guenther, "Laser Induced Damage in Optical Materials 1975," NBS Special Publication No. 435, (1976). See also A. J. Glass and A. H. Guenther, *Applied Optics*, **15**, 1510, (1976).
71. T. Tamir (ed.), "Integrated Optics", Springer-Verlag, Heidelberg, (1975).
72. D. Marcuse (ed.), "Integrated Optics", IEEE Press, The Institute of Electrical and Electronics Engineers, New York, (1973).
73. D. Marcuse, "Theory of Dielectric Optical Waveguides", Academic Press, New York and London, (1974).
74. N. S. Kapany and J. J. Burke, "Optical Waveguides", Academic Press, New York and London, (1972).
75. P. K. Tien, *Applied Optics*, **10**, 2395, (1971).
76. D. Marcuse, *Bell Systems Technical Journal*, **48**, 3187, (1969).
77. D. Marcuse, *Bell Systems Technical Journal*, **49**, 273, (1970).
78. P. K. Tien, R. J. Martin and S. Riva-Sanseverino, *App. Phys. Letts.*, **27**, 251, (1975).
79. I. P. Kaminow, W. L. Mammel and H. P. Weber, *Applied Optics*, **13**, 396, (1974).
80. S. C. Rashleigh, *Optical and Quantum Electronics*, **8**, 49, (1976).
81. J. E. Goell, *Bell Systems Technical Journal*, **48**, 2133, (1969).

Thin Film Optical Devices   425

82. E. A. J. Marcatili, *Bell Systems Technical Journal*, **48**, 2071, (1969).
83. E. A. J. Marcatili, *Bell Systems Technical Journal*, **48**, 2103, (1969).
84. J. H. Harris and R. Schubert, *IEEE Trans. Microwave Theory Tech.*, **MTT-19**, 269, (1971).
85. J. H. Harris, R. Schubert and J. Polky, *J. Opt. Soc. Am.*, **60**, 1007, (1970).
86. J. E. Midwinter, *IEEE J. Quantum Electronics*, **QE-6**, 583, (1970).
87. P. K. Tien and R. Ulrich, *J. Opt. Soc. Am.*, **60**, 1325, (1970).
88. R. Ulrich, *J. Opt. Soc. Am.*, **60**, 1337, (1970).
89. R. Ulrich, *J. Opt. Soc. Am.*, **61**, 1467, (1971).
90. P. K. Tien, R. Ulrich and R. J. Martin, *App. Phys. Letts.*, **14**, 291, (1969).
91. M. L. Dakss, L. Kuhn, P. F. Heidrich and B. A. Scott, *App. Phys. Letts.*, **16**, 523, (1970).
92. H. Kogelnik, *Bell Systems Technical Journal*, **48**, 2909, (1969).
93. H. Kogelnik and T. P. Sosnowski, *Bell Systems Technical Journal*, **49**, 1602, (1970).
94. D. G. Dalgoutte and C. D. W. Wilkinson, *Applied Optics*, **14**, 2983, (1975).
95. J. H. Harris, R. K. Winn and D. G. Dalgoutte, *Applied Optics*, **11**, 2234, (1972).
96. R. Ulrich, *J. Opt. Soc. Am.*, **63**, 1419, (1973).
97. T. Tamir, *Nouv. Rev. Optique*, **6**, 273, (1975).
98. H. F. Taylor and A. Yariv, *Proc. IEEE*, **62**, 1044, (1974).
99. P. K. Tien and R. J. Martin, *App. Phys. Letts.*, **18**, 398, (1971).
100. P. K. Tien, R. J. Martin and G. Smolinsky, *Applied Optics*, **12**, 1909, (1973).
101. P. K. Tien, S. Riva-Sanseverino, R. J. Martin and G. Smolinsky, *App. Phys. Letts.*, **24**, 547, (1974).
102. L. P. Boivin, *Applied Optics*, **13**, 391, (1974).
103. R. V. Pole, E. M. Conwell, H. Kogelnik, P. K. Tien, J. R. Whinnery, A. Yariv and A. J. De Maria, *Applied Optics*, **14**, 569, (1975).
104. J. Guttman, O. Krumpholz and E. Pfeiffer, *Applied Optics*, **14**, 1225, (1975).
105. D. G. Dalgoutte, G. L. Mitchell, R. L. K. Matsumoto and W. D. Scott, *App. Phys. Letts.*, **27**, 125, (1975).
106. R. Shubert and J. H. Harris, *J. Opt. Soc. Am.*, **61**, 154, (1971).
107. R. Ulrich and R. J. Martin, *Applied Optics*, **10**, 2077, (1971).
108. G. C. Righini, V. Russo, S. Sattini and G. Toraldo di Francia, *Applied Optics*, **11**, 1442, (1972).
109. R. Ulrich, *Nouv. Rev. Optique*, **6**, 253, (1975).
110. D. C. Flanders, H. Kogelnik, R. V. Schmidt and C. V. Shank, *App. Phys. Letts.*, **24**, 194, (1974).
111. M. Matsuhara and K. O. Hill, *Applied Optics*, **13**, 2886, (1974).
112. R. V. Schmidt, D. C. Flanders, C. V. Shank and R. D. Standley, *App. Phys. Letts.*, **25**, 651, (1974).
113. W-T. Tsang and S. Wang, *App. Phys. Letts.*, **27**, 588, (1975).
114. H. F. Mahlein, R. Oberbacher and W. Rauscher, *Applied Physics*, **7**, 15, (1975).
115. E. M. Conwell, *IEEE Trans. Quantum Electronics*, **QE-9**, 867, (1973).
116. D. B. Andersen and J. T. Boyd, *App. Phys. Letts.*, **19**, 266, (1971).
117. P. K. Tien, R. Ulrich and R. J. Martin, *App. Phys. Letts.*, **17**, 447, (1970).

118. S. Zemon, R. R. Alfano, S. L. Shapiro and E. Conwell, *App. Phys. Letts.*, **21**, 327, (1972).
119. J. P. van der Ziel, R. M. Mikulyak and A. Y. Cho, *App. Phys. Letts.*, **27**, 71, (1975).
120. E. M. Zolotov, V. A. Kiselev and V. A. Sychugov, *Soviet Physics—Uspektu*, **17**, 64, (1974).
121. F. W. Dabby, A. Kestenbaum and U. C. Paek, *Optics Communications*, **6**, 125, (1972).
122. E. R. Schineller, R. P. Flam and D. W. Wilmot, *J. Opt. Soc. Am.*, **58**, 1171, (1968).
123. R. D. Standley, W. M. Gibson and J. W. Rodgers, *Applied Optics*, **11**, 1313, (1972).
124. D. H. Hensler, J. D. Cuthbert, R. J. Martin and P. K. Tien, *Applied Optics*, **10**, 1037, (1971).
125. J. E. Goell and R. D. Standley, *Bell Systems Technical Journal*, **48**, 3445, (1969).
126. R. L. Aagard, *App. Phys. Letts.*, **27**, 605, (1975).
127. C. W. Pitt, F. R. Gfeller and R. J. Stevens, *Thin Solid Films*, **26**, 25, (1975).
128. R. K. Watts, M. de Wit and W. C. Holton, *Applied Optics*, **13**, 2329, (1974).
129. J. R. Carruthers, I. P. Kaminow and L. W. Stulz, *Applied Optics*, **13**, 2333, (1974).
130. J. Noda, N. Uchida and T. Saku, *App. Phys. Letts.*, **25**, 131, (1974).
131. H. F. Taylor, W. E. Martin, D. B. Hall and V. N. Smiley, *App. Phys. Letts.*, **21**, 95, (1972).
132. W. E. Martin and D. B. Hall, *App. Phys. Letts.*, **21**, 325, (1972).
133. J. Noda, T. Saku and N. Uchida, *App. Phys. Letts.*, **25**, 308, (1974).
134. R. V. Schmidt and I. P. Kaminow, *App. Phys. Letts.*, **25**, 458, (1974).
135. J. M. Hammer and W. Phillips, *App. Phys. Letts.*, **24**, 545, (1974).
136. R. D. Standley and V. Ramaswamy, *App. Phys. Letts.*, **25**, 711, (1974).
137. J. F. Weller and T. G. Giallorenzi, *Applied Optics*, **14**, 2329, (1975).
138. R. Th. Kersten and W. Rauscher, *Optics Communications*, **13**, 189, (1975).
139. D. J. Channin, J. M. Hammer and M. T. Duffy, *Applied Optics*, **14**, 923, (1975).
140. P. K. Cheo, J. M. Berak, W. Oshinsky and J. L. Swindal, *Applied Optics*, **12**, 500, (1973).
141. R. Ulrich and H. P. Weber, *Applied Optics*, **11**, 428, (1972).
142. V. Ramaswamy and H. P. Weber, *Applied Optics*, **12**, 1581, (1973).
143. W. J. Tomlinson, H. P. Weber, C. A. Pryde and E. A. Chandross, *App. Phys. Letts.*, **26**, 303, (1975).
144. P. K. Tien, G. Smolinski and R. J. Martin, *Applied Optics*, **11**, 637, (1972).
145. O. I. Szentesi and E. A. Noga, *Applied Optics*, **13**, 2458, (1974).
146. J. Noda, N. Uchida, M. Minakata, T. Saku, S. Saito and Y. Ohmachi, *App. Phys. Letts.*, **26**, 298, (1975).
147. S. Uehara, K. Takamoto, S. Matsuo and Y. Yamuchi, *App. Phys. Letts.*, **26**, 296, (1975).
148. I. P. Kaminow, J. R. Carruthers, E. H. Turner and L. W. Stulz, *App. Phys. Letts.*, **22**, 540, (1973).
149. J. C. Webster and F. Zernike, *App. Phys. Letts.*, **26**, 465, (1975).
150. D. Hall, A. Yariv and E. Garmire, *App. Phys. Letts.*, **17**, 127, (1970).

151. J. C. Campbell, F. A. Blum and D. W. Shaw, *App. Phys. Letts.*, **26**, 640, (1975).
152. F. K. Reinhart and B. I. Miller, *App. Phys. Letts.*, **20**, 36, (1972).
153. W. E. Martin, *App. Phys. Letts.*, **26**, 562, (1975).
154. J. M. Hammer, D. J. Channin and M. T. Duffy, *App. Phys. Letts.*, **23**, 176, (1973).
155. D. J. Channin, *App. Phys. Letts.*, **19**, 128, (1971).
156. P. K. Cheo, *Applied Physics*, **6**, 1, (1975).
157. P. K. Tien, R. J. Martin, R. Wolfe, R. C. LeCraw and S. L. Blank, *App. Phys. Letts.*, **21**, 394, (1972).
158. S. Wang, M. L. Shah and J. D. Crow, *IEEE Trans. Quantum Electronics*, **QE-8**, 212, (1972).
159. L. Kuhn, M. L. Dakss, P. F. Heidrich and B. A. Scott, *App. Phys. Letts.*, **17**, 265, (1970).
160. N. Chubachi, J. Kushibiki, H. Sasaki and Y. Kikuchi, Oyo Buturi, **43**, *Suppl.*, 199, (1974).
161. H. Kogelnik and C. V. Shank, *App. Phys. Letts.*, **18**, 152, (1971).
162. H. Kogelnik and C. V. Shank, *J. App. Phys.*, **43**, 2327, (1972).
163. S. Wang, *Wave Electronics*, **1**, 31, (1974).
164. I. V. Cheremiskin and T. K. Chekhlova, *Sov. J. Quantum Electronics*, **4**, 387, (1974).
165. D. P. Shinke, R. G. Smith, E. G. Spencer and M. F. Galvin, *App. Phys. Letts.*, **21**, 494, (1972).
166. S. Wang and S. Sheem, *App. Phys. Letts.*, **22**, 460, (1973).
167. M. Nakamura, H. W. Yen, A. Yariv, E. Garmire, S. Somekh and H. L. Garvin, *App. Phys. Letts.*, **23**, 224, (1973).
168. C. V. Shank, R. V. Schmidt and B. I. Miller, *App. Phys. Letts.*, **25**, 200, (1974).
169. D. R. Scifres, R. D. Burnham and W. Streifer, *App. Phys. Letts.*, **25**, 200, (1974).

Note Added in Proof

Since this chapter was completed early in 1976, a process known as molecular beam epitaxy has emerged as likely to be of considerable importance for the production of components for integrated optics. It is a powerful technique for the production of epitaxial films of very high quality of materials such as gallium arsenide and gallium aluminium arsenide. The process consists essentially of closely controlled deposition under ultra high vacuum conditions from collision-free evaporant beams. Reflection of evaporant from chamber walls and other parts of the structure is prevented by liquid nitrogen cooled screens. Further information on the process will be found in A. Y. Cho and J. R. Arthur, "Progress in Solid State Chemistry", Vol 10, p. 157, eds., J. McCaldin and G. Somorjai, Pergamon, Oxford, (1975).

Chapter 9†

Thin Film Photoconductors

V. Vincent

The General Electric Co. Ltd.,
Hirst Research Centre,
Wembley, London,
England

1. Introduction 430
   1.1 Thin Films 431
2. Photoconductivity Theory 432
   2.1 The Effect of Traps 437
   2.2 The Effect of Space Charge Limiting Currents 441
   2.3 The Non-crystalline Lattice 441
3. Photoconductivity in Thin Films 443
   3.1 Elemental Photoconductors 443
       (a) Selenium 443
       (b) Silicon and germanium 446
   3.2 Inorganic Compounds 447
       (a) Cadmium sulphide and cadmium selenide 447
       (b) Antimony trisulphide 454
       (c) Indium antimonide 454
       (d) Zinc oxide 456
       (e) Mixed photoconductive films 457
   3.3 Organic Photoconductors 463
4. Applications 470
   4.1 Radiation Detectors 470
   4.2 The Vidicon 472
   4.3 Electrophotography 475
   4.4 Display Devices 477
       (a) Addressing displays 477
5. Conclusion 482
   Acknowledgements 483
   References 483

† For a list of symbols used in this chapter and their definitions see p. xvi.

1. Introduction

Devices based on thin film photoconductors have made a large impact on our everyday lives both at home and at work; the most obvious examples being the vidicon or T.V. camera tube and the electrophotographic plate or photocopying machine. Both have been so well assimilated into our way of life that they are taken for granted. Others, perhaps more familiar examples of photoconductor-based devices form part of the range of radiation detectors which are used variously as light meters in photographic work and photocells switching on street lights, burglar alarms and smoke detectors. More sophisticated applications include gamma ray detectors and photoconducting sensors in optical communication systems, while research and development work has been directed towards producing solid state display panels complete with photoconductively operated circuitry integrated onto the panel itself.

These devices all rely on the ability of a photoconductor to increase its conductivity in proportion to the amount of radiation it absorbs. In radiation detectors, this effect is exploited by applying an electric field across a photoconducting element and measuring the current flowing. Used in this mode, some background level of signal will normally be present, denoting the resistance of the material in the dark, and so the effect of the light must be sufficiently great to produce a signal which is large in relation to the noise level. Happily, photoconductors can behave as solid state analogues of the photomultiplier tube, producing many more than one carrier per photon absorbed. This internal amplification factor or gain can be as high as 10^2 or 10^3 in some cases. Once generated, the carriers contribute to the current until they terminate by recombining. The length of the carrier lifetime is the key to high gain and thus to high photosensitivity. Particularly good photoconductors can have carrier lifetimes in the region of milliseconds.

In sharp contrast, both the vidicon and electrophotographic plates use the change in conductivity to allow charge to flow across the photoconducting film (which in practice needs to have a very high dark resistance). Initially, the photoconductor surface is charged up and this charge is allowed to leak through the film in the illuminated regions; then, the final surface potential is "read". In the vidicon, this is done by scanning an electron beam across the surface in typical T.V. raster fashion (line by line) and the signal from the beam can be transmitted for translation back on a conventional CRT screen. In electrophotography, small black dielectric particles are attracted electrostatically to the undischarged portions of the photoconducting plate to form a black, grey and white image which is subsequently fused onto a sheet of paper. In these devices, a carrier can only live for the time it takes to cross the film and so gain is limited to unity. The ability to retain surface charge is a more important photoconductor property than gain for these applications.

Table 1

Device	Application	Required photoconductor properties
Electrophotographic plate	Photocopying	Dark resistivity $10^{12}\,\Omega\,cm$ Response time $\sim 10^{-2}$ sec
Vidicon	T.V. camera tubes especially closed circuit T.V.	Frequency response to match illumination source
Radiation detectors	(1) light meters	Dark resistivity $10^{6}\,\Omega\,cm$ Gain > 10 Response time 10^{-3} to 10^{-1} sec Frequency response to correspond with film response
	(2) photocells for smoke detection, etc.	Ratio between dark and light resistance $> 10^{2}$ Response time $\sim 10^{-3}$ sec Frequency response to correspond with source, often IR
	(3) sensors in optical communication	Dark resistivity $10^{10}\,\Omega\,cm$ Gain $\sim 10^{2}$ Response time 10^{-6} sec

Table 1 compares the characteristics required of photoconductors for their various applications.

1.1 Thin Films

In one sense, almost all photoconductor devices are thin film devices by virtue of their basic design. For any given device a material will have been chosen to have maximum sensitivity at a certain frequency, hence it will strongly absorb that radiation incident upon it. An approximate calculation based on the Beer-Lambert law can be used to give an indication of typical light penetration depths. (This will later be shown to be an oversimplification, although its usefulness makes it worth pursuing.) The Beer-Lambert law can be stated as $I = I_0 \times 10^{-\varepsilon Cl}$ where I_0 and I are respectively the incident and transmitted light intensities passing through a sample l cm thick, where the concentration of absorbing species is C moles/litre and the extinction coefficient of the species is

ε. In this general form, the law can be applied to crystals, solid mixtures or liquids. If we assume the transmitted flux to be 10% of the incident flux then $\log_{10} I/I_0 = -1$ and, equating indices, $-1 = -\varepsilon Cl$. Typically ε is of the order of 10^4 to 10^5 in intrinsic photoconductors† or in materials where band to band transitions occur, and for a solid C lies typically between 1 and 10 moles/litre

$$\left(C \text{ is given by } C = \frac{\text{density in g/cm}^3 \times \text{mol. wt}}{1000} \right)$$

Thus, l can be calculated to be in the region of a micron. No significant advantage in sensitivity can be gained from using thicker photoconducting layers because the light will not penetrate to the back of the sample.

This provides a strong case for using thin film photoconductors although in no material so far has a thin film been found to be superior, or even equal, in performance to a good single crystal. However, the thin film offers the commercial advantage of ease of large scale production at relatively low cost and so where its performance is acceptable, it is to be preferred to single crystal material. It is probably true to say that there is no one property of a thin film photoconductor which is attributable just to its thickness. All its properties are related to those found in the single crystal.

2. Photoconductivity Theory

The theories of photoconductivity can be most easily understood by following the phenomenological approach of Albert Rose[1]. He has developed a clear understanding of the processes involved by taking the mathematical models and giving physical insight to their interpretation. Using this approach, consider first an intrinsic, insulating photoconductor (i.e. a material with no defect levels in the forbidden band) under conditions of steady state illumination. The number of photogenerated carriers, n, will be given by the product of the number of electron hole pairs per second per unit volume, f, produced by the light and the free lifetime of the carriers, τ

i.e. $\qquad\qquad\qquad n = f\tau \qquad\qquad\qquad (1)$

This equation simply states that the generated carriers are in equilibrium with those lost, but both these processes are complex and need further consideration.

Firstly, the generation process; two events need to take place for conversion of photons into charge carriers to occur. Light, or any radiation, must be absorbed by the material so as to produce excitons and the excitons must decay into charge carriers. Excitons can be regarded as electron-hole pairs

† A small group of photoconductors, including some extrinsic materials, where ε is significantly lower, will clearly need to be proportionately thicker.

bound together. Two main types of exciton can be identified; the Frenkel exciton is tightly bound, has a radius comparable with that of a lattice site and is dominant in ionic lattices, while the Wannier exciton is loosely bound, with a radius extending over several lattice sites and is found in molecular crystals. It is common to find the spectral dependence of photoconductivity follows the absorption spectrum for a given material, such that a photoconduction peak occurs at the absorption edge. There are cases, however, where the peak of photoconduction with respect to wavelength is offset from the absorption edge and this can be attributed to a change in the excitonic decay process with energy (an example of this is shown for selenium in Fig. 3).

A range of decay processes are open to the exciton, not at all of them resulting in carrier formation. For example, a singlet exciton (Wannier or Frenkel) may decay in any one of the following ways:

It may dissociate into charge carriers due to the influence of
  (i) thermal or lattice effects (autoionisation)
  (ii) a photon collision (photoionisation)
  (iii) an applied field (the Poole Frenkel effect)
  (iv) the presence of a surface, an impurity or an electrode

or, it may revert to the ground state via
  (v) a radiationless transition
  (vi) a fluorescence emission, or
  (vii) as a result of a two exciton collision one may drop to the ground state while the other forms carriers.

Similar options are open to triplet excitons. One of the features common in a "good photoconductor" is an efficiency approaching unity for the conversion of excitons to carriers.

Once the carrier is formed, its "history" is lost and "dark" and "photo" carriers alike then live out their free lifetime, τ before recombining. This can occur in several ways, free carriers can recombine directly or one carrier can be held at a recombination centre until the opposite carrier becomes attracted to it. Such centres are frequently introduced by deliberately doping the specimen but may also be due to vacancies, defects, surfaces, etc. The free carrier lifetime then, is inversely proportional to the thermal velocity of the carrier, v, $[\sim 10^7 \times (T/300)^{\frac{1}{2}} \text{ cm/sec}]^{(2)}$ the concentration of available recombination centres, N, and their capture cross section, S.

Thus, $\qquad\qquad\qquad \tau = 1/vSN \qquad\qquad\qquad (2)$

The value of N can vary from 10^{12} to $10^{19}/\text{cm}^3$, while S can vary between 10^{-12} to 10^{-15} cm^2 for attractive coulombic centres but can be as low as 10^{-22} cm^2 for repulsive coulombic centres (which are the type needed for the most sensitive photoconductors). The energy profiles of such centres are shown in Fig. 1.

The extremes of these ranges give a lifetime range of 10^{-14} to 10^3 second. So, by changing the concentration and type of centres in the material, by varying conditions of growth or doping concentrations, the lifetime and hence the

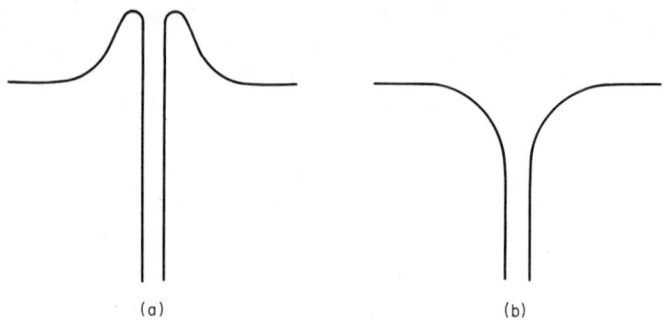

FIGURE 1. Energy profiles for (a) a repulsive and (b) an attractive coulombic centre.

magnitude of the photocurrent can be changed drastically, or, in other words, the material can be sensitised.

The dependence of photocurrent upon the light intensity (the lux-ampere characteristic) follows directly. Combining equations (1) and (2) the number of "photogenerated" carriers is given by

$$n = \frac{f}{vSN} \qquad (3)$$

The equivalent expression for "dark" carriers is

$$n_0 = \frac{g}{vSN} \qquad (4)$$

where n_0 is the number of thermally generated carriers produced by a rate of thermal generation g.†

In the dark, equation (4) applies.

† Assuming that the recombination centres are at depth E_1, below the bottom of the conduction band it can be shown that $g = nvS_nN_c(-E_1/kT)$, where n is the density of centres occupied by an electron, S_n is the capture cross section for an electron by a centre occupied by a hole, and N_c is the effective density of states at the conduction band. An essentially similar equation applies for direct thermal excitation from the valence band itself although clearly this rate will normally be significantly smaller.

In the light,

$$n_0 + n = \frac{g+f}{vSN} \quad (5)$$

From these equations we can consider two cases,
(a) where $n_0 + n \ll N$; which is applicable for intrinsic photoconduction, here f is proportional to n, giving rise to an essentially monomolecular dependence.
(b) where $n_0 + n = N$; rewriting equation (5),

$$f + g = (n_0 + n)^2 vS \quad (6)$$

Thus $n \propto f^{\frac{1}{2}}$ and a half law dependence or bimolecular dependence is seen. In an insulator where $n_0 \ll n$ and $g \ll f$

$$n \propto f^{\frac{1}{2}}$$

In a semiconductor where $n_0 \gg n$

$$n \propto f$$

Thus for materials containing a density of recombination centres comparable with the carrier density and with photocurrents large compared with their dark currents, a simple model will predict an index of 0.5 in the lux-ampere characteristic. For materials with relatively small photocurrents compared to their dark current or a lower density of carriers compared with the recombination centre density, the index will be unity. The introduction of additional types of centre can produce intermediate values of the index. Using this model the response time of the material will be that of the free carrier lifetime.

Suppose we wish to apply a field across the photoconductor. In order to do so we must attach contacts, which will themselves affect device properties. By choosing an electrode material to have a work function relatively greater, less than or equal to that of the insulator then respectively either a blocking, neutral or ohmic contact will be formed with an n-type material. In practice, contamination often causes surface barrier layers and mechanical damage (for example ion bombardment) can then be used to disturb the chemical potential of both surfaces sufficiently to ensure an intimate contact is formed. The damaged surface causes the work functions to be equalised (via the chemical potential) and ohmic contacts can result.

With an ohmic contact, the classical conductivity equations can be written

$$\sigma = ne\mu_n + pe\mu_p \quad (7)$$

where p is the hole concentration and μ_n and μ_p are electron and hole mobility respectively.

Also
$$\sigma = \sigma_0 \exp(-E/kT) \qquad (8)$$
where E represents the energy difference between the bottom of the conduction band (or top of the valence band) and the level from which the excitation has been stimulated. In the intrinsic case
$$\sigma = \sigma_0 \exp(-E_{gap}/2kT)$$
A parameter for sensitisation, known as gain, G, is quantified as the ratio of free carrier lifetime to the transit time, T_r, of the carrier between the electrodes
$$G = \frac{\tau}{T_r} \qquad (9)$$
Transit time is given by
$$T_r = l^2/\mu V$$
where l is electrode separation and V the applied voltage. The shorter the transit time, the higher the sensitivity, with the magnitude of the photocurrent flowing given by
$$J = \frac{ne}{T_r} = \frac{F\tau e}{T_r}$$
As has already been pointed out, devices such as the vidicon and xerophotographic plates have gains limited to unity as the carriers can only traverse the electrode separation once and then effectively recombine at the electrodes (i.e. the observed decay time cannot exceed the relaxation time). Photocells on the other hand can have gains very much greater than unity. For a device with given electrode separation and applied voltage, the value of gain is determined by the carrier lifetime and mobility $G = \tau \mu V/l^2$. It follows that the longer the lifetime (which is also the response time of the ideal device) the higher the gain. The gain is eventually limited when the transit time becomes limited to the dielectric relaxation time τ_{rel}, where τ_{rel} can be expressed in terms of an rc equivalent circuit in which r is the resistivity, ρ, and c is the dielectric constant of the photoconductor, κ; thus
$$\tau_{rel} = \rho \kappa$$
The trade-off between response time, τ_0, and gain can be examined in the form of the gain bandwidth product, G/τ_0, where
$$G_{max} = \tau_0/T_{rmax} = \tau_0/\tau_{rel}$$
thus
$$(G/\tau_0)_{max} = 1/\tau_{rel} = 1/\rho\kappa \qquad (10)$$

Other factors can further limit the behaviour of an ideal photoconductor and these will now be considered.

2.1 The Effect of Traps

In photoconductivity theory all defect levels in the forbidden band may be divided into two main classes, namely recombination centres and trapping centres. Free electrons (or holes) have a finite probability of being captured by both types of centres. It is the subsequent transition that decides the role of a particular defect level under given conditions. Thus, as has already been implied,† in a recombination centre the electron will subsequently transfer to the valence band (which is equivalent to saying that a hole is captured by that centre). In a trapping level, the electron will have a greater probability of being thermally excited into the conduction band. Electron traps are frequently sited close to the conduction band and hole traps close to the valence band but this must not be regarded as a general statement. For a given concentration of one type of defect, levels may act as, say, trapping levels under one condition of illumination and as recombination levels under a second condition of illumination. The relative occupancy of the level being the deciding factor.

This can be better understood by making use of the Fermi equation

$$n = N_c \exp(-E_f/kT) \tag{11}$$

where n is the number of free electrons and N_c is the effective density of states near the bottom of the conduction band. Thus, knowing the concentration of electrons in the conduction band, we can deduce the depths of the Fermi level and, as is well known, this defines the occupancy of all levels in the forbidden band as a function of depth, provided the specimen is in thermal equilibrium and in the dark; (at the Fermi level itself, the occupancy is $\frac{1}{2}$).

In a photoconductor, light changes the density of free carriers (both electrons and holes) and in turn, the occupancy of the levels in the forbidden band. By introducing the concept of quasi Fermi levels—one for electrons, E_{fn} and one for holes E_{fp}—each being governed by the concentration of free electrons and holes respectively, a new equilibrium situation can be set up for a given condition of illumination such that the same basic equations can be used, as were appropriate in the dark

Consider for example an electron trapping level at a depth E_t below the conduction band, then the occupancy is represented by the equation

$$n_t/n = (N_t/N_c) \exp(-E_t/kT) = \theta \tag{12}$$

where n_t represents the number of trapped electrons and N_t represents the density of trapping levels at a depth E_t. θ is the ratio of trapped to free charge. Using typical values of $N_c = 10^{19}/cm^3$ and a trap depth of say 0.1 eV for a low trap level of $10^{15}/cm^2$, θ is about 10^{-2}, while for a trap density of $10^{17}/cm^3$

† The case considered previously actually involved the transition of a hole from the valence band into a centre but in this context, the effect is equivalent.

(more probable in a thin film) θ is around 10^{-4}. It can be seen that under illumination in order to increase n by ∂n, $\theta \partial n$ electrons will have to be excited into the trapping levels via the conduction band. Thus θ times as many photons must be absorbed to raise the extra electrons into the conduction band and so, to a first approximation, the measured response time of the detector will be θ times the free carrier lifetime. Note, however, that from this model we can deduce the photosensitivity, which is determined by n, is not affected by traps.

However, a significant change in the occupancy of the electrons in the defect levels will affect photoconductivity behaviour. On one hand traps can become saturated, i.e. the occupancy for electron traps at a given level increases from near zero to near unity (as the quasi Fermi level moves through that level with increasing light levels incident on the sample). Then further increases in light intensity generate electrons which are excited into the conduction band as if the material were trap-free and the response time returns to that of the free carrier lifetime. The opposite situation can arise when θ is large (e.g. with a deep trapping level) such that the occupancy of recombination centres changes from near unity to near zero, (it will be apparent that θ times as many holes will be found in the recombination centres). When this situation occurs, the free carrier lifetime will start to be reduced, as will the photosensitivity or gain factor. From previous discussion the onset of this condition will change the lux-ampere index from unity to 0.5. However the value of the index has a complex behaviour according to the distribution of traps present in the forbidden band, varying in practice from less than 0.5 to greater than unity. This behaviour is worth investigating as the interpretation of the index can be misleading unless fully understood.

The useful concept of a demarcation level helps to illustrate a concise, physical model. It denotes the level at which an electron (or hole) is equally likely to recombine or escape from a trap.

Thus for an electron it can be shown that

$$N_c v S_n \exp(-|D_n - E_c|/kT) = pvS_p \qquad (13)$$

where $|D_n - E_c|$ represents the modulus of the energy difference between the electron demarcation level and the bottom of the conduction band and p is the hole concentration.

Re-arranging equation (13) we have

$$|D_n - E_c| = kT \ln(N_c S_n / p S_p) \qquad (14)$$

Using $n = f/vS_n P_r$ (equation 3) and by analogy $p = f/vS_p N_r$, and also using

$$n = N_c \exp(-|E_{fn} - E_c|/kT)$$

we can express equation (14) as

$$|D_n - E_c| = |E_{fn} - E_c| + kT \ln(N_r/P_r) \qquad (15)$$

A similar derivation for holes gives

$$|D_p - E_v| = |E_{fp} - E_c| + kT \ln(N_r/P_r) \qquad (16)$$

Comparing equations (15) and (16) we can deduce that the demarcation levels lie close to the quasi Fermi levels, the separation between demarcation levels is the same as that between Fermi levels and so the small displacement between a Fermi and demarcation level will be in the same direction for either carrier. This information is summarised in Fig. 2.

Taking an exponential trap distribution

$$N_t(E) = A \exp\left(\frac{|E_c - E_t|}{kT_1}\right) \qquad (17)$$

where T_1, is a formal parameter such that the density of states can be varied with respect to energy. Let N_r and N_t respectively be the number of recombination centres and trap states. We will assume, for simplicity, that all capture cross sections for a given carrier are equal (but not equal for opposite carriers).

Let

$$N_r > \int_{E_f}^{E_c} N_t(E)\,dE \quad \text{and} \quad S_n \ll S_p \quad \text{i.e.} \quad n \gg p$$

As the light intensity is increased, the quasi Fermi level will move, taking its demarcation level with it, such that some of the states that were initially N_t states will become N_r states. Thus P_r, the density of recombination states for electrons, increases and so causes a decrease in the free electron lifetime. The density of empty P_r states is given approximately by the number of N_t states between the initial and final quasi Fermi levels. So

$$P_r = \int_{E_f}^{E_{fn}} A \exp(-|E_c - E_t|/kT_1)\,dE \qquad (18)$$

We know

$$n = f/P_r v S_n = N_c \exp(-|E_c - E_{fn}|/kT)$$

so, substituting for P_r we have

$$n = \left\{\frac{fN_c^{T_1/T_2}}{kT_1 A v S_n}\right\}^{\frac{T_1}{T+T_1}} \qquad (19)$$

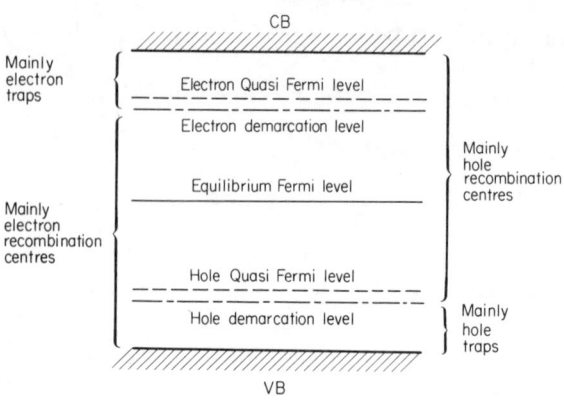

FIGURE 2.   Demarcation and quasi Fermi levels for carriers in an insulator.

If $T_1 \geqslant T$ the exponent

$$\frac{T_1}{T_1 + T}$$

lies between 0.5 and 1.

A change in the distribution to more nearly uniform causes T_1 to tend to infinity and the index approaches 1. If $T_1 < T$ the index can be less than 0.5. The introduction of levels with different capture cross sections can produce an index greater than unity.

So far, traps can be seen to complicate the behaviour of photoconduction. For the materials scientist attempting to fabricate devices, the foregoing theory can be broadly summarised in three statements.

1. The presence of recombination centres in a photoconductor forms a means of tailoring and sensitising the material.

2. The presence of traps can limit device performance, at best, leaving it relatively unchanged as compared with "trap free" performance or at worst, drastically impairing sensitivity and speed of response.

3. Recombination centres and traps can interchange roles, according to the prevailing conditions (such as applied voltage and incident light).

This set of circumstances frequently seems analogous to the three laws of thermodynamics but because of the high degree of control achieved during growth, doping and handling stages of device fabrication, many good photoconductor based devices are produced. The constraints are severe and "theoretical best" performance is rare even in single crystal material. In thin film work, it is probably unachievable but a fairly close approach is more than adequate for many applications, as will be seen in Sections 3 and 4.

2.2 The Effect of Space Charge Limiting Currents

Space charge limited currents (SCLC) in solids form a direct solid state analogue of the behaviour of electrons in vacuum tubes. In the latter case, the Child-Langmuir space charge law applies giving $J \propto V^{3/2}$. In insulators it will be seen that $J \propto V^2$ in the space charge region. An insulator or semiconductor with ohmic contacts can, under the influence of an applied voltage, carry a current of injected charge which is limited only by space charge considerations. It is this mechanism which allows gains exceeding unity to be measured. A value for the theoretical upper limit of the SCL current, J_L can easily be calculated by assuming the photoconductor to be a dielectric layer in a parallel plate condenser. Then $Q = cV \approx \kappa V/4\pi l \times 10^{-12}$ coulombs/cm. Where, κ is the dielectric constant, V the applied voltage, l the electrode separation and the factor 10^{-12} arises from approximating

$$1 \text{ farad} = 8.99 \times 10^{-11} \sim 10^{-12} \text{ esu}$$

We already know that the transit time, T_r, of carriers between electrodes is given by $T_r = l^2/\mu V$ thus J_L can be given by

$$J_L = \frac{Q}{T_r} = \frac{kV^2\mu}{4\pi l^3} \times 10^{-12} \text{ amp/cm}^2$$

(showing the square law dependence and inverse cube dependence on distance between electrodes). The order of magnitude of such currents can be indicated by substituting values for CdS, say, into the equation; thus for $\kappa \sim 10$, $\mu \sim 300 \text{ cm}^2/V$ sec with $V = 10$ volts and $l = 10^{-3}$ cm, J is approximately 30 amp/cm².

Such currents are occasionally found in high quality thin single crystals[2]. However, the presence of traps can drastically reduce the Hall mobility of the carriers by the factor of θ. Thus J_L becomes

$$J_L(\text{trap}) = \frac{kV^2\theta\mu}{4\pi l^3} \times 10^{-12} \text{ amp/cm}^2$$

The very small values of θ frequently found in photoconductors can easily reduce the SCLC flow to a value less than that of the ohmic dark current and this is one basic reason why they are seldom observed in practice (non-ohmic contacts being another common limitation).

2.3 The Non-crystalline Lattice

Use of band theory in an "amorphous" material leads to problems of interpretation. While any solid structure has a band structure, it need bear no relationship with that found in the single crystal material and the bands may be so narrow that the concept is meaningless. There is great reluctance at this point to abandon the security of such a familiar theory for other models but fortunately, other models with simple physical interpretations do exist. As a first step from the familiar, we can superimpose Petritz's barrier theory onto band theory to find a model applicable to semiconductors of a polycrystalline nature which provide intergrain boundary barriers for the carriers to negotiate. This theory has already been presented by Professor Anderson in his chapter on thin film transistors and so will not be presented here. However, we can usefully compare the expression he derives for current flow in the presence of barriers with the familiar ohmic expression. His notation is slightly different from that of the present author and to avoid confusion the equation has been re-written in the notation used throughout this chapter.

$$J \propto ne^2 \bar{\mu} V / kT \cdot \exp(-e\phi_0/kT)$$

where ϕ_0 represents the intergrain barrier height.

Ohmic current can be expressed as

$$J \propto ne\mu V$$

The essential difference lies with the mobility term which has acquired the characteristics of an activated process when applied to polycrystalline films. Mobility will be low until ϕ_0 becomes smaller than kT, but once ϕ_0 is less than kT, "band theory" behaviour will take over. At low temperatures or high barrier conditions, conductivity will be limited by this mobility constraint.

Another less quantifiable effect of grain boundaries is their ability to provide energetically favourable sites for impurities causing them to concentrate along boundaries, while the boundary itself can form a characteristic defect level in the forbidden band gap.

To explain organic photoconduction, we find band theory is not applicable to even the single crystals. Instead, molecular crystals are based on discrete levels. The mechanisms of absorption of light to produce excitons remain the same, as does the decay of excitons into carriers. Although now the factors governing the likelihood of the exciton decay processes become more important than in inorganic materials, in that, if there is a very favourable decay to the ground state, carrier production will be low. If decay to the ground state can be made unfavourable say, by promoting weak overlap integrals between excited and ground states, carrier generation will be more probable.

The chief differences arise in carrier separation and migration where a hopping model is more appropriate in molecular crystals. The necessity for an effective separation of carrier mechanism marks the main distinction between organic and inorganic photoconductivity. The types of mechanism involved in carrier separation are better understood by experimental illustrations and so will be considered in Section 3.

It remains only to point out the close similarity between polycrystalline and organic photoconductivity mechanisms by noting that a polycrystalline material is composed of a set of crystallites separated by energy barriers while an organic material is composed of a set of molecules separated by energy barriers. As we move to polymeric materials, the molecular dimensions can approach that of a small crystallite and so it can be anticipated that the photoconductivity behaviour of these two systems may be very similar.

3. Photoconductivity in Thin Films

Some of the most useful photoconducting materials will be considered in detail in this section. Many explanations of "anomalous" behaviour still exist in the literature, indicating, in part, the complexity of the photoconducting mechanisms and in part, reflecting the very limited understanding of many of the materials only studied recently. First we consider the simplest materials, the elements.

3.1 Elemental Photoconductors

(a) Selenium

Commercially, the most important thin film photoconductor is amorphous selenium. It is an appropriate beginning as photoconductivity itself was first noted in selenium just over a century ago[3]. To put the subject in perspective, an historical approach has been adopted for "the oldest photoconductor" to illustrate just how recently its behaviour has been explained. Selenium occurs naturally in two crystalline phases. The normal hexagonal phase is grey or metallic selenium and, with respect to the other semiconducting phases, has a relatively high conductivity (in the region of 10^{-5} mho cm). Red or monoclinic selenium has a low conductivity (around 10^{-11} to 10^{-12} mho cm) and is a less stable phase, reverting to the grey form on heating. This red phase is thought to be based on a structure of Se_8 rings whilst grey or metallic selenium consists of long chains of atoms, allowing a path for carrier migration. Although the correlation between the properties and structure of selenium has still not been fully confirmed, the original paper by Von Hippel[4] provides a good discussion of these aspects. A review of the electrical properties of these two phases can be found in the book by Moss[5], *Photoconductivity in the Elements.*

A third phase, amorphous selenium, can be prepared by rapid cooling a melt or by vacuum deposition both of which result in the formation of a red, glassy phase which has commercial application as a photoconductor. The "structure" of vacuum deposited layers is mainly determined by the nature of substrate[6,7] and this is also true for amorphous silicon and germanium although there is a continuing debate as to why the dependence arises[8,9].

Above a critical substrate temperature selenium recrystallises, impairing the photoconductive properties of the amorphous film. The first major paper to report on the photoconductive properties of evaporated amorphous selenium thin films was that of Weimer and Cope[10] in 1951. They reported dark resistivities of about $10^{12}\,\Omega\,\text{cm}$ ($10^{14}\,\Omega\,\text{cm}$ is more representative, their value has been lowered by injection of carriers, probably due to space charge current flowing) and found photoconduction was due to holes, also, under conditions of high fields and strong illumination at the positive electrode, quantum yields approaching unity were measured. A particularly noteworthy feature of the spectral response of the photocurrent was its displacement by about 0.6 eV to the high energy side of the optical absorption spectrum, as shown in Fig. 3. From samples ranging in thickness from 0.1 to 10 micron, they noted that photocurrents flowed through the films even when the thickness was up to an order of magnitude greater than the penetration depth. This is now explained in terms of exciton diffusion prior to dissociation[11].

Weimer and Cope[10] made their measurements using a sandwich cell configuration comprising a selenium layer deposited onto a conducting glass electrode with either a scanning electron beam or an evaporated metal film forming the counter-electrode. They found the response times of the films were better than the 50 microsecond response time of their equipment but they noted that photocurrent dropped slowly with time under steady state conditions. They called this phenomenon fatigue and attributed it to space charge effects, although traps can produce the same result. They investigated the effect of doping at a "few per cent" level (using Sb, Cd, CdS), in all cases there were increases in sensitivity, dark current and response time of between two and three orders of magnitude. In some cases a low energy response in the red was produced.

Li and Regensburger[12] attempted to explain the behaviour of photocurrents in selenium in terms of a range limited model, that is the photocurrent is limited by the distance carriers travel before falling into traps. But this was disproved by Tabak[11] when he measured photoconductivity as a function of thickness and found photocurrents were thickness independent over a range of samples varying in thickness from 1 to 50 micron. Instead, he supported the view that photogeneration was a field assisted process. Later work by Tabak and Warter[13] and by Pai and Ing[14] corroborated this view. The light is absorbed to produce mobile excitons which require a given field strength for them to dissociate. This would account for non-linearity of response with respect to applied field. It could also explain an apparent thickness dependence at constant applied field and accounts for the displacement of the photoresponse with respect to absorption.

Photocurrent can be determined by measuring the photoinduced decay of surface potential with respect to time. This is shown in Fig. 4 where a thick

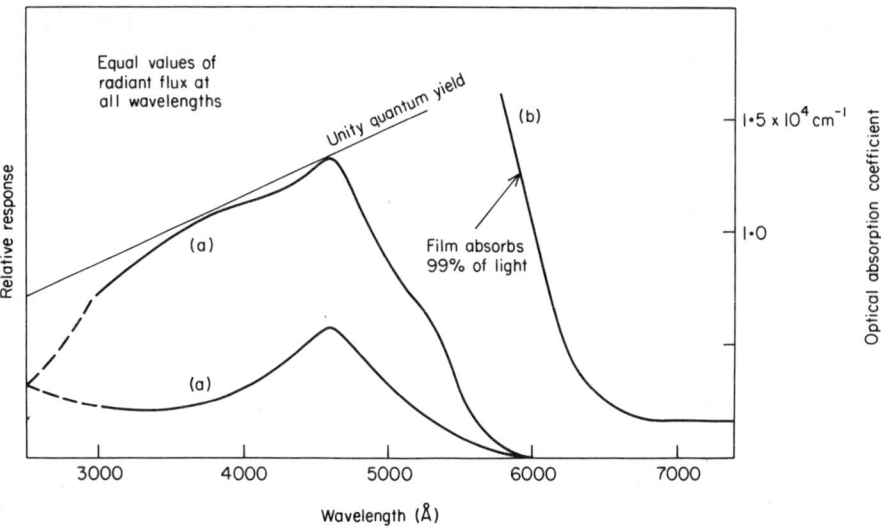

FIGURE 3. (a) Photocurrent and (b) optical absorption as a function of wavelength for amorphous selenium after Weimer and Cope[10].

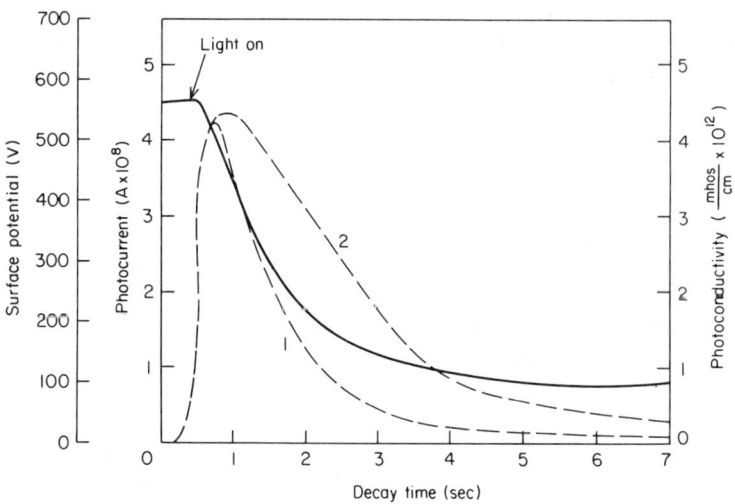

FIGURE 4. Representative photo-induced discharge of a xerographic plate, showing the decay of surface potential (solid line), photocurrent (curve 1) and photoconductivity (curve 2) for a 50 μm thick film of amorphous selenium on an aluminium substrate (positive surface charge) after Schaffert[15].

selenium film has been deposited onto an aluminium substrate. A study of the physical processes involved in the photodischarge of a homogenous photoreceptor has been made by H. Seki[16] for those interested in a deeper insight.

To summarise, amorphous selenium is an insulator in the dark, with a spectral response in the visible. It can hold a high surface charge when in the dark, has fast light discharge characteristics and good resolution. Its properties can be altered by controlling the deposition parameters; dopants can enhance sensitivity and spectral range but only at the expense of longer response times. This combination of properties makes it very suitable for vidicon or xerophotographic use. An additional feature is its sensitivity to x-rays, although somewhat thicker films are needed to absorb sufficient of the higher energy radiation. This enables x-rays, cameras and photocopiers to be considered.

(b) Silicon and germanium

Silicon and germanium must be briefly considered as they also are elemental photoconductors. Both form good junction devices and as a result research has concentrated on these devices more fully. Their technology has been developed so as to produce large area single crystal slices rather than thin films. Both can be used as extrinsic, IR detectors but only germanium single crystal detectors (typical element dimensions $1 \times 6 \times 2 \, mm^3$) are a commercial venture.

Unlike selenium, both silicon and germanium have a high conductivity and so to use them as photoconductors involves removing this level of "background noise" by either backing it off electrically or cooling the device. Two excellent reviews on IR detectors[17,18] detail the merits of these and many other types of detector.

An intrinsic photoconduction lies in the IR, it can be argued that a doped sample may provide a response even at microwave frequencies. An example of recent work aimed at producing thin film germanium detectors is that of Taylor et al.[19] who have investigated the response of evaporated thin films (around $10 \, \mu m$) of amorphous germanium to far IR and microwave frequencies. (Thicker films tend to be used for extrinsic photoconductors because the impurity absorption coefficient is smaller than for intrinsic photoconductors.) They found that the conductivity depended heavily upon preparation conditions and in particular on the rate of deposition—a high deposition rate giving high residual conductivity. They ascribe this dependence to dangling bonds† and after comparing their results with that of other workers on sputtered germanium films[20,21] conclude that the dependence on dangling bonds is present in all vacuum deposited germanium films.

† It can be formally pictured that the bonding of "outside" or surface atoms are not satisfied and just "dangle".

3.2 Inorganic Compounds

The spectral response and "optical context" of a range of inorganic photoconductors are shown in Fig. 5.

Most of these materials are used as intrinsic photoconductors, the major difference between them being the frequency of their peak response. Since it would be impractical to consider them all, cadmium sulphide and cadmium selenide will be considered as typical. The widespread use of CdS in camera

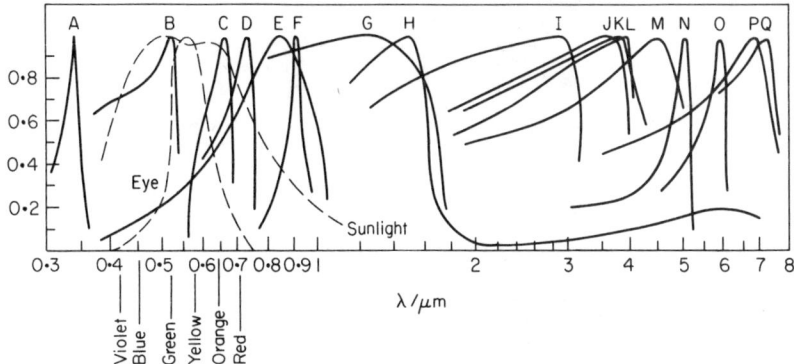

FIGURE 5. Normalised spectral response for a range of semiconductors. Also shown are the eye response and the sunlight spectrum curves after Tang[22].

A ZnS (300K)
B CdS (300K)
C Se (300K)
D CdSe (300K)
E Si (300K)
F GaAs (300K)

G He (300K)
H Au-doped Ge (300K)
I PbS (300K)
J PbS (90K)

K PbSe (300K)
L PbTe (300K)
M PbTe (90K)
N InSb (4.2K)
O InSb (77K)
P PbSe (90K)
Q InSb (300K)

light meters and in switching on street lighting merits this treatment. Then brief mention will be made of antimony trisulphide for its importance in the vidicon, indium antimonide as an intrinsic IR detector and zinc oxide for its importance in electrofax.

(a) Cadmium sulphide and cadmium selenide

Figure 5 shows both the materials to be sensitive in the visible region of the spectrum, CdS in the green, CdSe in the red. Their major commercial application uses sintered films between 5 and 25 μm thick but perhaps more interesting are the potential applications in optical addressing techniques.

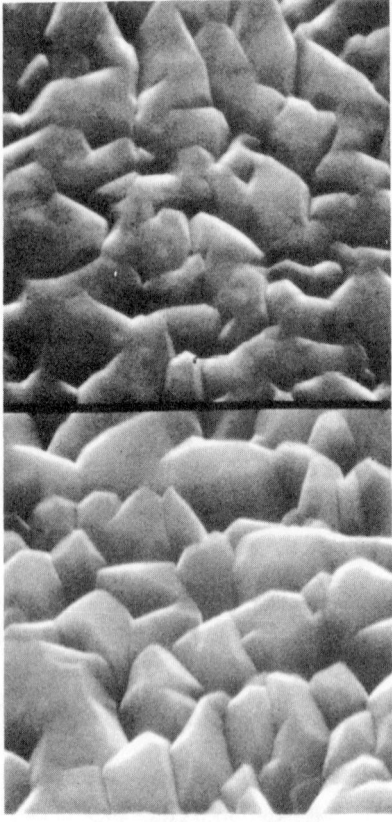

FIGURE 6. Scanning electron micrograph of "as sputtered" CdS films. The upper view shows CdS on 7059 glass and the lower shows CdS on indium tin oxide coated glass. Reproduced by kind permission of the American Institute of Physics from Fraser and Melchior[23].

Evaporated and sputtered layers of any II-VI compound show oriented growth with the c-axis normal, or within a solid angle of 30° of normal to the substrate. The nature of the substrate is important in determining crystal habit and crystallite size as can be seen in Fig. 6 where films of CdS have been sputtered on glass and on indium tin oxide coated glass; although it is not clear from the photograph, the c-axis is normal to the substrate in both cases.

Conditions of deposition are crucial in establishing device properties, in particular variation of substrate temperature, rate of deposition and the partial pressures of components during growth effect the stoichiometry and crystal habit of the film[23-26]. Films can be either hexagonal or cubic or both; the hexagonal form being preferred for photoconductivity. Evaporated film work[24,25] indicates that variation of substrate temperature affects the ratio of hexagonal to cubic phase present but that some cubic is very often found. (It must be borne in mind that x-ray glancing angle diffraction plates show up the cubic phase very readily while the hexagonal phase is not quite so easily seen because of the relative intensities of their characterising lines.) Sputtered films can be deposited in a hexagonal phase either with or without substrate heating by both D.C. and R.F. sputtering techniques[23,25]. The author has found that changing the sputtering gas affected habit, with D.C. sputtering in argon producing films with both phases present while D.C. sputtering in nitrogen produced the hexagonal phase only. Changing the stoichiometry by a small amount has a large effect on electrical properties. At the stoichiometric point II–VIs are insulators, increasing the group VI component maintains the insulating properties in the dark, while increasing the group II component increases dark conduction. In general, anion vacancies behave as donors (as do group III and VII impurities), cation vacancies act as acceptors (together with group I and V impurities). Crystalline defects behave much as impurities and while stoichiometric material is not necessarily defect free, non-stoichiometric material must contain some defects, thus stoichiometric films need not be good photoconductors but there is a better chance that they can be doped into a suitable material. The wide difference between the vapour pressure of elemental cadmium and either sulphur or selenium makes the formation of stoichiometric films difficult to achieve. The different sticking coefficients of the elements causes further problems as re-evaporation of the group VI element from the compound readily occurs. This effect is enhanced when a heated substrate is used but various growth techniques have been employed to overcome these problems. Dual source evaporation[25] or direct evaporation of the compound in a flux of excess S or Se vapour[24] appear to give the best results whilst good quality films can also be prepared by sputtering, either from a target containing an excess of sulphur or selenium[26] (by molar ratio) or by introducing a quantity of selenious or sulphurous gas with the sputtering gas. Fraser and Melchior[23] and Snelling[27] have successfully sputtered CdS in a mixture of argon and a few per cent of H_2S, the author has sputtered CdSe in argon with H_2Se added.

Chemical vapour deposition can be accomplished by causing an organo-cadmium compound to decompose at a hot surface in the presence of sulphur or selenium species to produce high dark resistance, high photosensitivity films[28].

With all these deposition methods, response times of milliseconds or greater are typical, due to the presence of traps but provided the device can accommodate these, there is no need to seek to improve upon them.

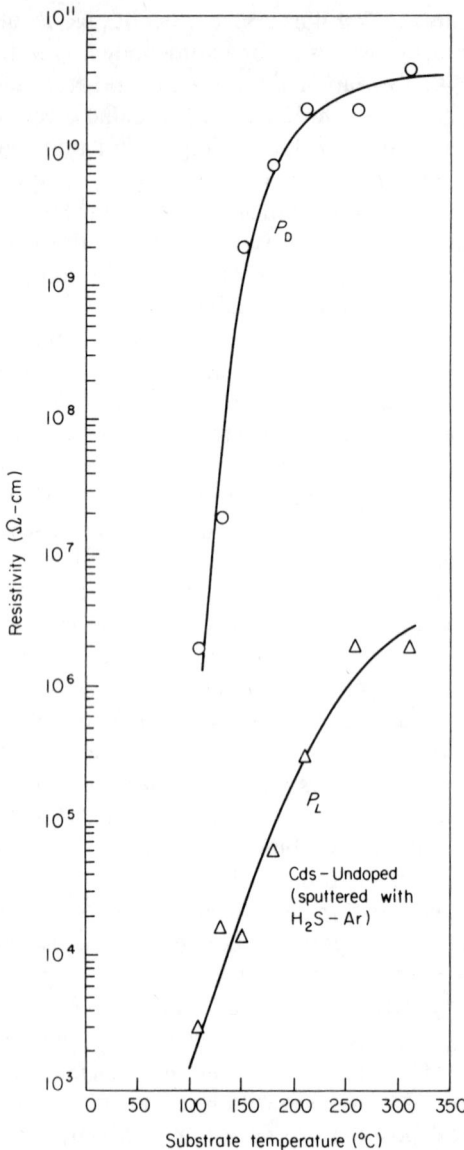

FIGURE 7. Resistivity values in the dark and light for CdS films deposited on substrates of various temperatures. Reproduced by kind permission of the American Institute of Physics from Fraser and Melchior[23].

The light to dark resistance ratio can be optimised by varying substrate temperature (see Fig. 7) and the latter also governs the mean surface scattering length (i.e. the distance a carrier travels before being scattered by an imperfection). In films thin enough for the half thickness, d, to be comparable with the mean surface scattering length, λ, mobility is affected. This effect has been studied by Kazmerski[29] in evaporated thin film of CdS. He shows the thickness dependence of the Hall constant as a function of temperature, see Fig. 8, and empirically establishes the relationship

$$\mu = \mu_b(-q\phi/kT)(1+\lambda/d)^{-1}$$

where μ is actual measured mobility, μ_b is bulk mobility, q is the charge on the electron and ϕ is the Petritz barrier energy.

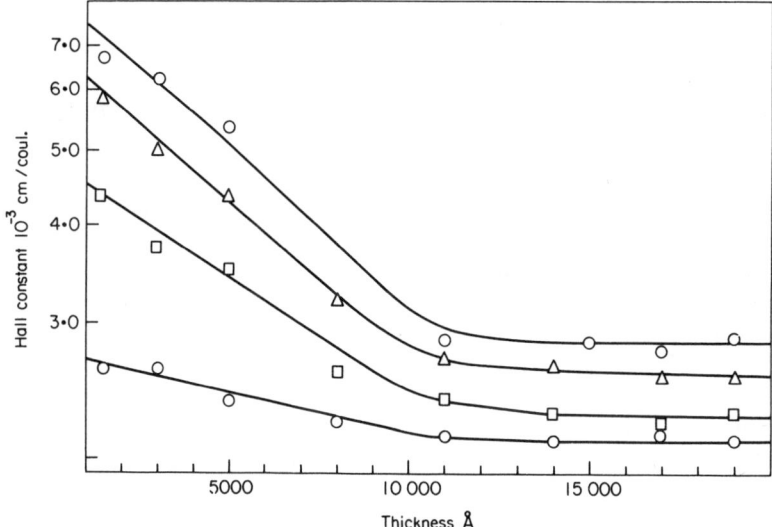

FIGURE 8. Hall data dependence on thickness for CdS films grown at various substrate temperatures (○—220°C, △—180°C, □—150°C, ●—100°C), after Kazmerski[29].

These findings are explained in terms of the dependence of λ directly on substrate temperature and inversely on deposition rate. They also note the correlation between resistivity and deposition rate, which arises from the general property that films with high carrier concentration (low resistivity) have low λ values, λ being proportional to relaxation time[30]. Finally, they show mobility to increase with surface potential which they explain in terms of a modified Petritz theory.

Light to dark resistance ratio can also be affected by changing sample thickness as shown in Fig. 9.

The work by Takeuchi et al.[26] and the paper by Leighton[25] provide the only systematic comparisons of gap and sandwich cell properties for sputtered films. Takeuchi et al.[26] find the light to dark resistance ratio for a sandwich cell to be 10^4 or 10^5 while for a gap cell configuration, it falls to 50. The decay time in a sandwich is sharper than in a gap cell as shown in Fig. 10. Leighton[25] notes that the dark activation energy is the same perpendicular or parallel to the substrate (0.7 eV) but the photo activation energy is 0.03 eV "across" the film (gap configuration) while the "through" thickness photocurrent is temperature independent. The difference between the configurations is thought to be due to the effects of oxygen or other contaminant gases enhancing surface effects and inhibiting photoconduction. The nature of these interactions has been investigated by South and Hughes[31]. A more serious form of surface damage affecting photoconductivity is the bombardment of films by electrons during sputtering; this damage is known to reduce mobility by increasing scattering. There are indications that better quality films are obtained for this and other reasons, if a sputter gun is used[32].

A neat way of improving light to dark resistance ratio further is to employ contacts such as gold which form blocking contacts under dark conditions, but in the light the semiconductor Fermi level moves and so forms an ohmic contact allowing SCLC to flow[33]. Thus a simple on-off switch is produced.

It must be noted that CdS and CdSe can only form n-type semiconductors. The increasing ionic content of the lattice binding energy with decreasing atomic weight renders it difficult to produce both a p- and n-type condition. When an acceptor is added into the lattice it is automatically compensated by electrically active lattice sites. The only means by which a p-type condition is claimed to have been induced is by ion bombardment with say bismuth[34].

Table 2

Device Data for CdS and CdSe at Room Temperature

Material	Energy gap E_g(eV)	Majority mobility in cm^2/V sec (single crystal values)	Dark conductivity Ω-cm	Light:Dark resistance ratio		Gain		Response time τ_0 in seconds	
				Best	Typical	Best	Typical	Best	Typical
CdS	2.4	300	10^{8-10}	10^7	10^5	10^4	10^{1-2}	10^{-6}	10^{-3}
CdSe	1.7	500	10^{8-10}	10^7	10^5	10^4	10^{1-2}	10^{-7}	10^{-3}

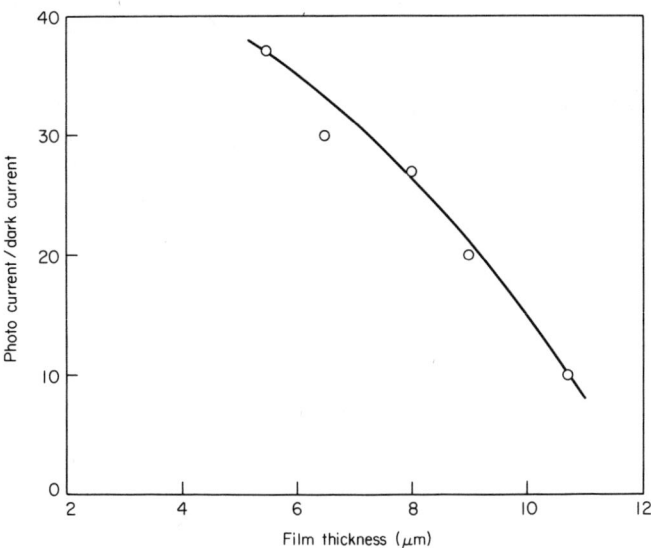

FIGURE 9. Light to dark conductivity ratio for CdS tells in a gap configuration as a function of film thickness. Illumination consisted of monochromatic light of 500 nm at an intensity of 1 mW/cm^2, after Takeuchi et al.[26].

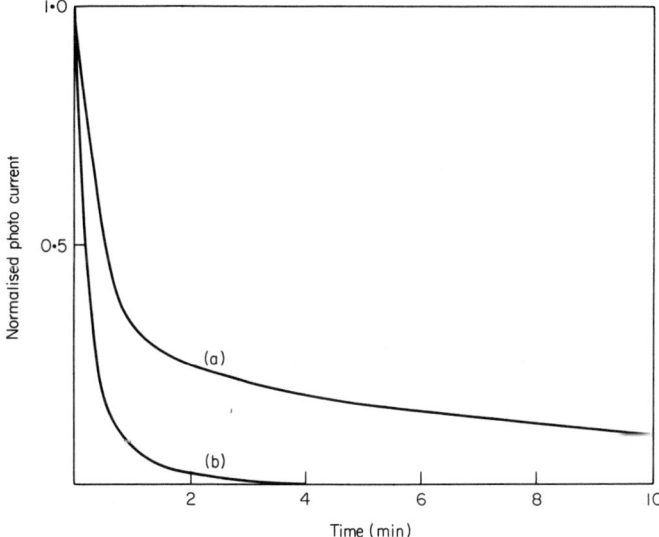

FIGURE 10. Decay of photocurrent in CdS on cessation of illumination for (a) a gap cell; (b) a sandwich cell configuration, after Takeuchi et al.[26].

(b) Antimony trisulphide

Although much work has been done on antimony trisulphide, very little of it has been published, mainly by reason of commercial confidentiality. As a thin film photoconductor it is p-type, has a dark resistance in the region 10^{11} to $10^{13} \, \Omega$ cm, a good sensitivity of up to 1 amp/watt ($10^3 \, \mu$A/lumen)[35] and a response time which is short with respect to the human eye ($\frac{1}{30}$ sec). The feature that has made it so suitable in imaging applications is its spectral response, shown in Fig. 11, which nearly matches that of the human eye. The lux-ampere

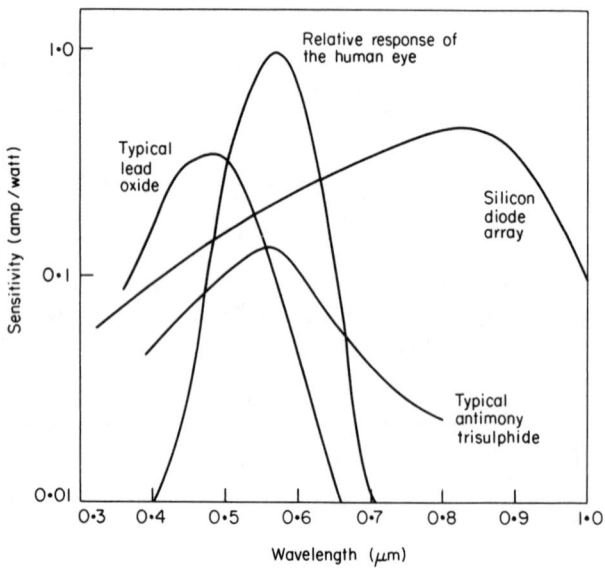

FIGURE 11. Spectral response of (a) a typical antimony trisulphide vidicon target; (b) a typical lead oxide target and the relative eye response. For comparison a silicon diode array target is included, after Longhurst and Woolgar[36].

characteristic of evaporated films is always about 0.7, indicating the presence of a range of trap levels. This may be because evaporation can only take place in a poor vacuum, as high vacuum conditions promote crystalline growth. To some extent, the spectral photoresponse peak can be varied as a function of film thickness; a "thick" layer ($>2 \, \mu$m) peaks in the red while a "thin" layer (unspecified) peaks in the blue. This effect is explained in terms of the short carrier range[35].

(c) Indium antimonide

InSb is a typical, direct narrow bandgap IR detecting material and as such its dark conductivity is high compared with the induced photoconductivity. Devices are made in thin single crystal form. First, single crystals are pulled by, either the crzochalski, or zone-levelling techniques from polycrystalline, zone-refined and etched starting ingots. These are cut into slabs and lapped down. They are mounted and diced into device "chips" and finally etched down to their optimum thickness (5 to 10 μm for 300 K operation or 25 μm for 77 K operation). Etching leaves the surface with a high recombination velocity, so that surface conductivity is minimised. Investigations have been made of the temperature dependence of conductivity, mobility and Hall coefficient for both n and p material[37]. Studies of these properties as a function of purity show that the resistivity maximum occurs in slightly p-type material and not in intrinsic material, this is accounted for by electron mobility being about a hundred times greater than the hole mobility. An account of the band model structure applicable to the electrical properties of InSb can be found (together with much other data and references) in a review by P. W. Kruse[38].

The spectral responsivity and detectivity are shown in Fig. 12 as a function of majority carrier concentration and can be seen to broadly follow the profile of the resistivity against majority carrier concentration dependence. There is a slightly higher sensitivity of the p-type which is surprising in view of the much lower mobility in p-type material. A summary of the major parameters is given in Table 3.

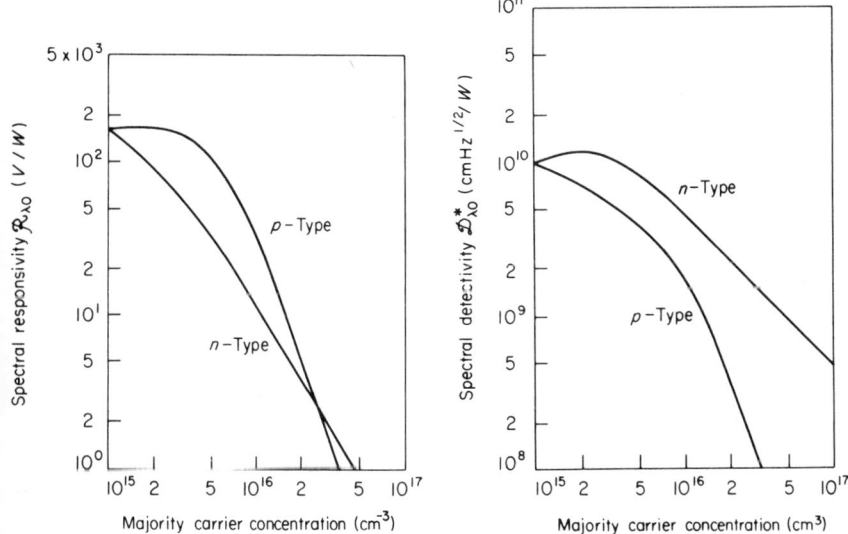

FIGURE 12. Dependence of (a) spectral responsivity and (b) spectral detectivity for photoconductive operation at 195 K upon majority carrier concentration in n- and p-type InSb. The sample thickness is 10 μm, after Kruse[38].

Table 3
InSb Device Parameters

Parameter	300 K pcond	195 K pcond	77 K pcond	300 K Photel
Peak wavelength in μm	5.5	5.0	5.3	5.4
Detectivity $\mathscr{D}^*$ (cm Hz$^{\frac{1}{2}}$/W)	1.9×10^8	8.6×10^9	6.5×10^{10}	1.7×10^8
f mod (Hz)	flat	1×10^4	1×10^4	> 75
Responsivity $\mathscr{D}_{\lambda p}$ (V/W)	1.96	140	1.1×10^5	1.1
R (ohms)	117	60	2.9×10^4	41
Response time (μ sec)	<1	<1	<2	<1

Data from NOLC reports data sheets 708, 710 and 716 of NOLC report 557 and sheet 704 of NOLC report 551.

(d) Zinc oxide

Zinc oxide in the form of a powder mixed with a binder is spread as a film, 10 to 15 μm thick, onto paper, for use in the electrofax process. This powder is mixed with the binder, typically a silicone based resin, in roughly a one to one weight ratio. The suspension is then used to coat paper by standard photographic production methods. It can be processed in this form, apparently by a fluke of nature; provided the zinc oxide is sulphur free, the otherwise relatively impure powder exhibits the unique property of behaving as a semiconductor material undegenerated in performance by traps, while remaining amenable to cheap and large scale deposition. Furthermore, when in a finely divided particulate form (0.25 μm diameter particles seem optimum) surface absorption of oxygen (say) is high, producing a very low dark current. Discharge characteristics for paper mounted ZnO/binder layers by various manufacturers are shown in Fig. 13.

With a band gap of 3.1 eV, the photoresponse peak is around 370 nm (UV) but if excess zinc is present then the response occurs at lower energy and is always n type. Light penetration is only about a micron. The mechanism has, until recently, been widely accepted as a recombination by photogenerated holes in two distinct steps. Firstly a fast process involves the optical ionisation of excess Zn^{2+} ions, that is, holes are optically generated. The second, slow step is the optical release of oxygen from the surface, thus leaving an excess of zinc ready for step one. In the dark, oxygen is adsorbed and once on the surface, forms traps for conduction band electrons and so reduces the dark current. The mechanism involved in adsorption and desorption are discussed fully by Eger et al.[40] Schaffert[15] and Dessauer and Clark[7]. Recent work by Shapira et al.[41] indicates that this mechanism is slightly more subtle than previously realised, in that the second step is shown to be due to the photo desorption of carbon dioxide, not oxygen, and that this process only takes place when the incident photon energy is greater than the ZnO band gap energy. They point out that in practice the major difference between the two models is the presence of carbon impurity at the surface. The frequency response of ZnO can be altered by sensitising the films with dyes[42]. The sensitivity is found to have a maximum at a given concentration of additive. The introduction of an electron accepting compound (e.g. chloranil) enhances the ZnO "intrinsic" response. The combination of dyes and acceptors enhances both "intrinsic" and "impurity" responses. Thus ambition will lead us ultimately towards a mixture of dyes and acceptors to produce a sensitive film with panchromatic response. Sensitivity is also affected by particle size, where too small a diameter particle results in a loss of sensitivity. Response times are slow in these layers probably due to shallow traps severely reducing electron mobility. A typical discharge characteristic of an electrofax film is shown in Fig. 14.

Much data is available on the performance of given commercially available films which are not directly relevant to the thin film aspects of photoconductivity, however both the texts by Dessauer and Clark[7] and by Schaffert[15] provide good treatises of ZnO for those interested in a wider study. The impact of thickness in the electrofax process is cost related, the thinner the layer the lower the cost of materials.

(e) Mixed photoconductive films

Producing photoconductors with a predetermined set of properties has always been a goal of scientists working in this field. The opportunities presented by mixing photoconductors together to form a tailored balance of properties looks too promising to ignore, but how realistic this object is remains to be seen.

Mixed photoconductors can be divided conveniently, if arbitrarily, into two categories. Type 1—both components are photoconductors and combining them gives a linear correlation of composition with intermediate property. This can be exemplified by $Cd_{1-x}Zn_xS$. Type 2—the chalcogenides where, perhaps, only one component need be photoconductive.

(i) Type 1; $Cd_{1-x}Zn_xS$

The band gap of the mixture varies linearly with composition[43] and from this, the remaining properties follow. The peak of the photoconductive response with respect to wavelength varies continuously with composition between the band gap energies of ZnS and CdS[44]. Maximum effective gain also varies with composition; the highest gain being in mixtures containing least zinc as these have longer electron lifetimes.

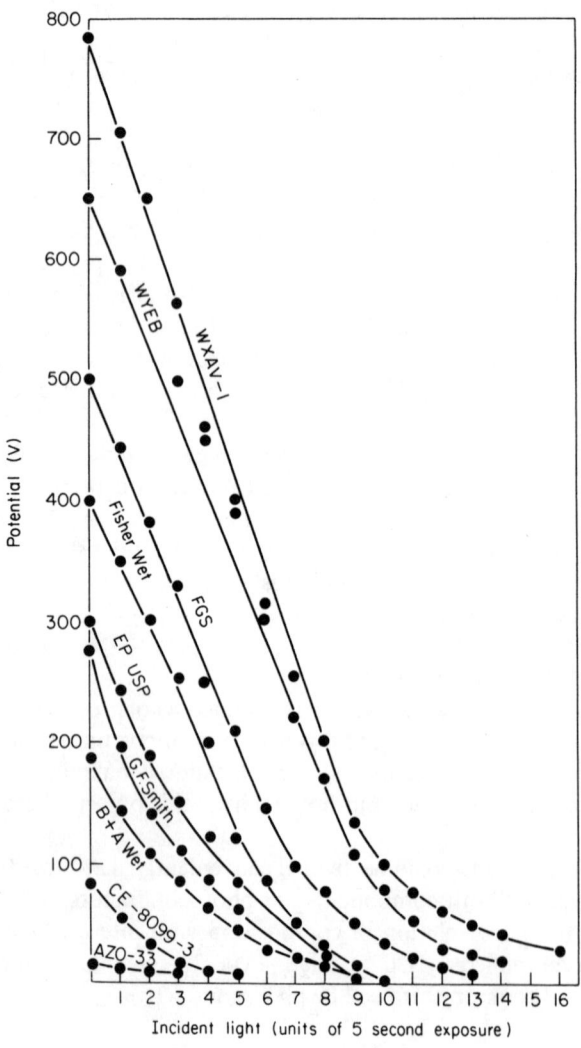

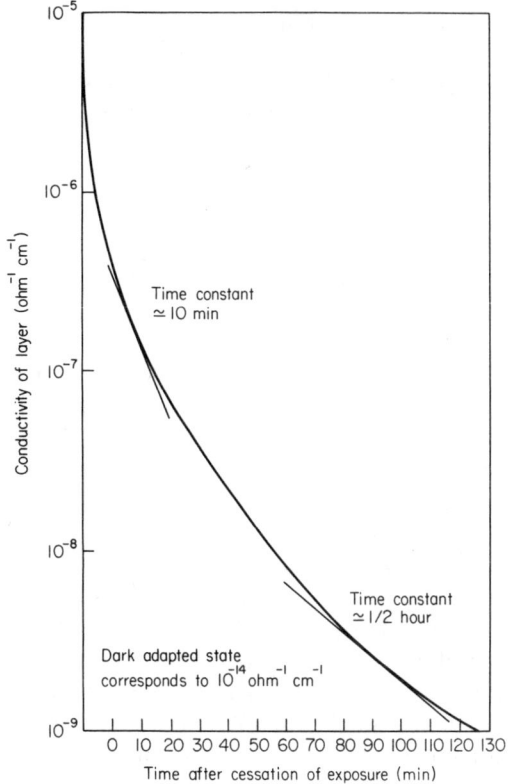

FIGURE 14. Typical photoconductivity decay characteristic of an electrofax $\frac{1}{2}$ mil layer coated on mica with electrodes 4 cm long and $\frac{1}{2}$ cm apart, after Amick[39].

FIGURE 13. Light discharge characteristics of electrofax layers. The standard layer thickness is 14.5×10^{-4} cm, and each 5 second light pulse provides 1.06×10^{11} photons/cm² at the electrofax surface, after Amick[39].

Key to Source of Layer

WXAV-1	Mallinckrodt Analytical Reagent WXAV-1
WYEB	Mallinckrodt Analytical Reagent WYEB
FGS	Florence Green Seal
Fisher Wet	Eimer and Amend Wet Process
EP USP	Eagle Picher US Pharmacopoeia
G. F. Smith	G. F. Smith
B & A Wet	Baker and Adamson Wet Process
CE-8099-3	CE-8099-3 (New Jersey Zinc Co.)
AZO-33	AZO-222-33 (American Zinc Sales Co.)

Table 4

A Comparison of the Properties of CdS with Cd ZnS, data from Fraser[44]

Material	Dark resistivity Ω-cm	Gain	Response time τ_0
CdS	10^7–10^{10}	10^3	microsec
$Cd_{1-x}Zn_xS$ for $x = 0.3$	$\sim 10^{12}$–10^{14}	10	microsec

The properties of such films are similar to those of CdS films; $Cd_{1-x}Zn_xS$ films being mechanically harder, having higher dark resistance, and so are not suitable for D.C. sputtering, but they have lower gain. These materials have been successfully sputtered using a magnetron system and a few percent of H_2S gas mixed with the argon.

(ii) Type 2; Chalcogenides

The class of glasses known as chalcogenides (i.e. compounds containing a group VI element) have really only been of interest since Ovshinsky[45] reported on their curious electrical switching behaviour (see Chapter 7 for a review of the electrical properties of these materials). Now investigations are under way to establish what other properties they may possess, including their electrical properties under illumination. With the limited data available there have been recent attempts to unify the experimental observations into a simple theory. At this stage little more is possible than fitting a concept to a set of observations. Weiser [46] has put forward a good model applicable to bulk effects, involving carriers excited above the mobility edges[47]. (This corresponds approximately to the position where a band edge might be if a true band structure were applicable.) It is assumed that the mobility edge exists in a continuum of states whose density increases with increasing energy and that the rate of relaxation of a hot carrier decreases as it falls into lower energy states. Finally, it is assumed that the recombination rate between electrons and holes depends only on their respective energies. Weiser infers that due to the low carrier mobilities characteristic in these materials free carrier recombination will be the dominant loss process. From this model, it follows that a carrier initially above the mobility edge will gradually drop down through the energy states as it uses up kinetic energy, but as it drops it encounters fewer and fewer states and so further energy loss becomes difficult. At some point its probability of dropping to a lower energy state balances with the probability of recombination. Using this model, he can account for the very strong temperature depence of photoconductivity characteristic in these materials. Another curious feature is the changeover of lux-ampere indices; they can be unity at low light levels and revert to 0.5 at higher levels. This is explained by saying that at low light levels a "photogenerated" carrier has a high probability of recombining with a "dark" carrier thus giving the impression of a monomolecular decay process. At higher light levels normal intrinsic recombination kinetics become apparent. (*Comment*: This could be explained in terms of a changeover from $n_0 > n$ to $n > n_0$ in an intrinsic recombination dominated material—see equation (4).)

Photoconductivity is predicted to be an activated process with the activation energy for electrons at high light levels being the energy difference between the mobility edge and the level at which the probability cross-over occurs. For low light levels the activation energy, E_a, is more complex.

$$-E_a = -(2\Delta E_R - \tfrac{1}{2}E_g)$$

where ΔE_R is the energy difference between the mobility edge and the energy level at which the probability cross-over sets in for holes. E_g represents the band gap. Significantly, this could be either a positive or negative quantity, thus allowing both positive and negative activation energies to exist. Comparison with experimental results for As_2Te_2Se shows very good agreement with activation switching from a negative to a positive quantity as light intensity is increased, but for As_2Se_3 agreement is poor.

Other noteworthy phenomena found in chalcogenides still appear with individual rationale for their behaviour. For example $As_2Se_3 - CdO$ changes its photoproperties as the CdO concentration is varied[48]. Small concentrations of CdO impair the photoconductive properties of the glass but as further additions are made, a strong photoresponse sets in. This is attributed to the reaction

$$As_2Se_3 + 3\,CdO \rightarrow As_2O_3 + 3\,CdSe$$

The response is characteristic of CdSe, as can be seen from Figs 15 and 16. The intensity dependence is not the 0.5 value, characteristic of chalcogenides, but is 0.8 and the photoconductive activation energy is nearly temperature independent compared with an expected 0.23 eV activation energy of the parent material. The low wavelength response is supposed to be due to arsenic doped CdSe. Other similar systems do not behave like this; in particular it may be noted $As_2S_3 - CdO$ does not show this effect.

Another curious effect is that of photodoping observed by Kokado et al.[49] in chalcogenide glasses, whereby silver from the metal contact diffuses into the glass when the sample is illuminated. It is explained by supposing that a junction barrier exists at the metal-glass interface, which, in the dark, prevents any reaction. In the light, carriers are separated, the silver capturing holes and electrons migrating through the chalcogenide bulk, eventually being lost by recombination or trapping. The silver ion now can migrate into the

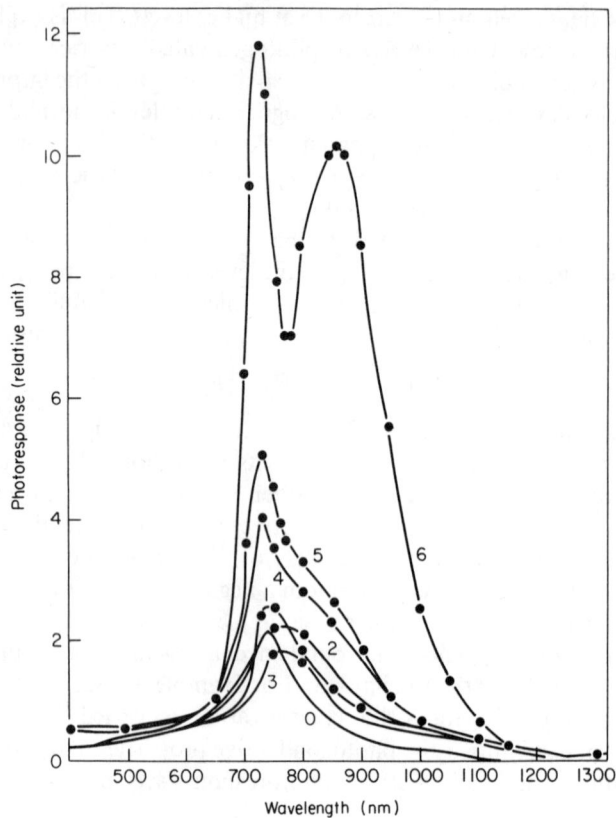

FIGURE 15. Photocurrent as a function of wavelength for samples of As_2Se_3 glass containing different amounts of CdO, after Minami et al.[48]. Glass composition is as follows: 0 is the parent glass, glasses 1 to 5 contain respectively 1, 2, 3, 4 and 5 molar % CdO, glass 6 contains 10 molar % CdO.

chalcogenide region. By measuring spectral sensitivity as a function of thickness, the rate of diffusion and hence rate of doping was established by Kokado et al. Two mechanisms, a fast step followed by a slow step were found, indicating some (as yet unknown) change in the rate determining step.

Tao and Wang[50] have investigated the photo-properties of $Pb_{1-x}Sn_xTe$, a narrow energy gap, band inversion alloy semiconductor. The change in spectral response as a function of composition can be seen from Fig. 17, the graph of detectivity against composition. As x is increased, response time lengthens. Rise and decay times lie in the 10 nsec region.

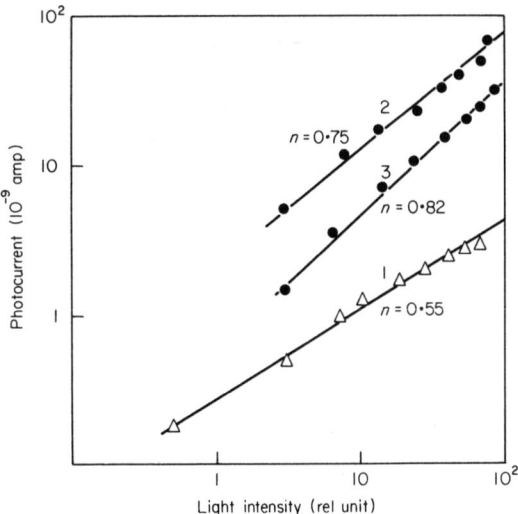

FIGURE 16. Lux-ampere characteristics for samples of As_2Se_3 glass containing different amounts of CdO. The applied voltage is 10 V, after Minami et al.[48]. Curve 1 is for the parent glass at 740 nm. Curve 2 is for 10% Molar content CdO at 720 nm illumination. Curve 3 is again 10% Molar content but at 860 nm.

A comparison of thin films of different orientation indicate that the polycrystalline samples with 100 orientation have higher photosensitivity than the epitaxial 111 orientation thin film samples. This is interesting, as other work by Tao and Wang[51] on $Pb_{1-x}Sn_xSe$ and by Logothetis and Holloway[52] on $Pb_{1-x}Sn_xTe$ both find mobilities in the 111 orientated layers higher than in the 100 layers. This may be evidence for longer carrier lifetime but lower mobility in the 100 orientated layers.

3.3 Organic Photoconductors

Photoconductivity was first observed in single molecular crystals such as anthracene and violanthrene[53]. As interest in organic photoconductivity continued, it proved difficult to grow some of the compounds in crystal form and consequently a spate of work on polycrystalline materials was published[54,55,56] using pressed pellet, evaporated layers, melts squashed between glass slides and cast or dipped polymeric films. These marked the turning point, for once polymers had been found to exhibit photoconductivity, great commercial interest was generated. (At this point the interest was enhanced by the discovery of conducting salts of TCNQ (tetracyanodiquinomethan), which generated much speculation about room temperature superconductors.)

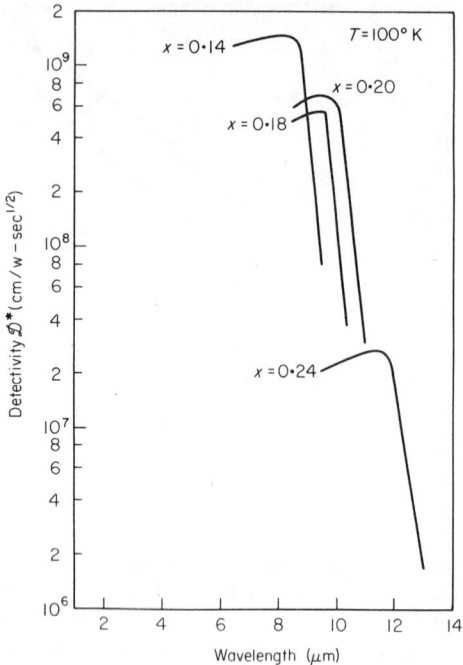

FIGURE 17. Spectral photoconductive detectivity at 100 K of four polycrystalline (100) films of $Pb_{1-x}Sn_xTe$ ($x = 0.14, 0.18, 0.20$ and 0.24) after Tao and Wang[50].

To understand the mechanism in organic materials, a simple model comprises a structure of charge transfer complexes which contain the essential features of most polymeric materials. The photoconducting polymer is typically composed of a central chain of an electron acceptor (or donor) which has side chains with electron donating (or accepting) properties. The transfer of charge inherent in this type of structure provides paths for conductivity. A charge-transfer complex is bound together by the "donated" electron being taken by the acceptor and it returning to its ground state on the donor at a fast enough rate to form a weak bond. Characteristically, the complexes have a broad band absorption with very high extinction coefficient. They can be seen to behave as inorganic photoconductors in their J-V characteristics and lux-ampere characteristics (for specific information on particular materials, the major text on organic semiconductors to consult is Gutmann and Lyons[57]).

Although dark conductivity is affected by impurities the spectral response of complexes are relatively insensitive to the presence of guest molecules (this is also characteristic of inorganic semiconductors). Taking a typical electron acceptor which has caused much interest, see Fig. 18, the spectral response of TCNQ itself is strongly dependent on impurities, particularly metals, which it treats as donors, but when it is complexed with, say, organic donors, impurities play a much less important role. The photocurrent peak for a donor or acceptor alone occurs as expected at the band edge. In a charge-transfer

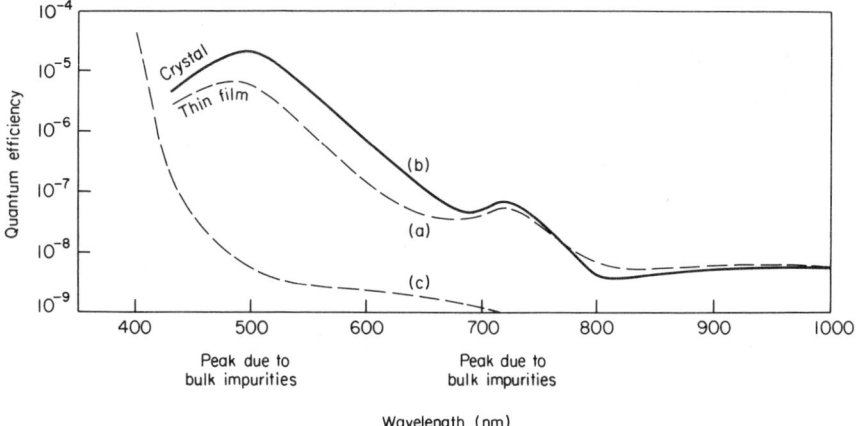

FIGURE 18. The spectral dependence of photoconductivity in TCNQ for (a) a thin film[58]; (b) a single crystal[58] and (c) an approximate curve for a very high quality single crystal[59].

complex, however, the peak can be found in one of two regions as shown in Figs 19 and 20. Either (1), the photocurrent follows the charge-transfer spectrum, the principal mechanism being the field or thermal dissociation of Wannier excitons into carriers or, (2), the photocurrent peak occurs between a singlet absorption of one component and the charge-transfer peak. In the latter case the following mechanism is proposed; (for example), let us assume the donor singlet overlaps with the charge transfer band. Then the interaction of the light will cause two processes to take place, production of a Wannier exciton and an excited singlet state. Using D and A to represent the donor and acceptor ground states and denoting the excited state by an asterisk

$$DA \xrightarrow{hv} D^*A$$
$$DA \xrightarrow{hv} (D^+A^-)$$

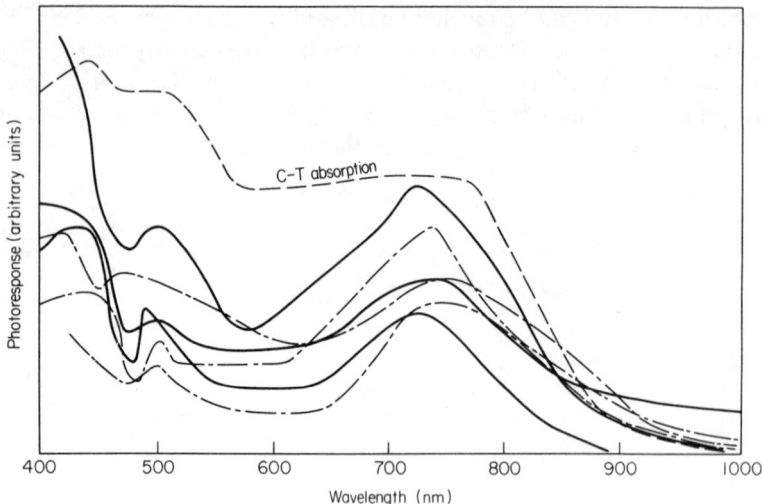

FIGURE 19. Photocurrent as a function of wavelength for Pyrene/TCNQ 1:1 charge transfer complex for several crystals. The response follows the absorption curve shown as a dotted line[60].

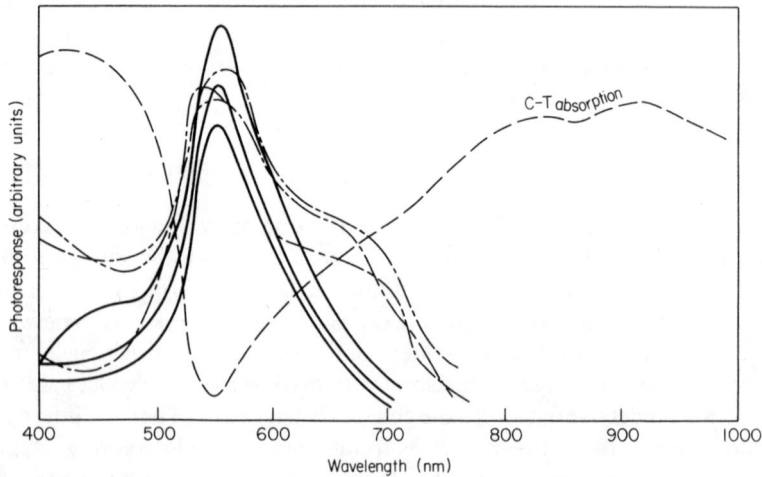

FIGURE 20. Photocurrent as a function of wavelength for several crystals of perylene/TCNQ 1:1 charge transfer complex. The peak occurs in the region of overlap of the perylene singlet and charge transfer absorption curves, shown as one dotted line, the higher energy peak being the singlet absorption[60].

When these two states are on adjacent molecules then

$$D^*A(D^+A^-) \xrightarrow{1} (D^+A)(DA^-) \xrightarrow[\text{activated process}]{2} 2DA + e + h$$

in step 1, charge separation is achieved, in step 2 the free carriers are produced. This mechanism can be seen more clearly by considering the donor energy levels involved pictorially as in Fig. 21.

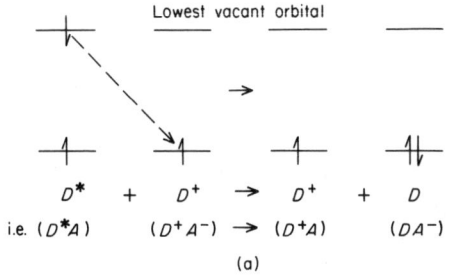

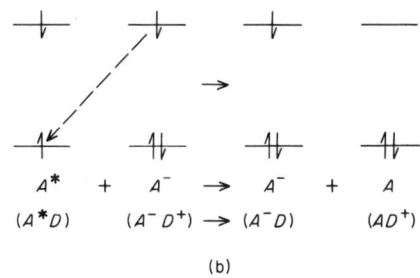

FIGURE 21. Diagrammatic representation of the energy transfer mechanism involved in carrier production where both singlet and charge transfer absorption is present[60]. (a) Donor singlet overlap. (b) Acceptor singlet overlap.

Since a hopping model of conduction applies to these materials, the migration process for carriers must be considered; the activation process for electrons and holes may be different, in which case one carrier may become trapped while the other is free to migrate. In fact, there is evidence to show that a difference in barrier heights must exist in order to separate the carriers effectively. If both barriers (measured in terms of the π orbital overlap between adjacent molecules) are comparable, the Wannier exciton can migrate without bothering to dissociate. Too high a barrier results in no carriers migration. It follows that the majority carriers will be holes, if donors overlap and electrons, if acceptors overlap. The exciton production step can be considered to be governed by thermodynamic laws while the exciton dissociation step is governed by kinetic theory. In general, organic photoconductors, particularly complexes, have very high dark resistivity ($\sim 10^{10}\,\Omega\,\text{cm}$), small light to dark resistance ratios (10:1 at best) and slow response times (10^{-3} to 1 sec) with either carrier being the majority species. The only difference between single crystal and squashed oriented films is the loss of anisotropy in the melted layers.

Using the models developed on complexes, the field of great interest, the polymeric films, can be investigated. A long chain molecule provides an easy path for migration, the ends forming the only problems. The best known photoconducting polymer is polyvinylcarbazole known variously as PVC_2, PVCz or PVK. Its photosensitivity can be increased by additions of various compounds. But perhaps as a result of influence from inorganic material technology, in early doping experiments only low concentrations of "dopants" were added, although a 1:1 complexing ratio would seem more appropriate. In 1971 Schaffert et al.[61] added an equimolar quantity of TFN (trinitro→9-fluorenone) to PVK and found it to give a promising xerographic material. Its negative charge acceptance is a function of film thickness, as seen in Fig. 22 but Schaffert et al.[61] calculate that when capacitance is taken into account, the maximum surface charge acquired by a 25 μm layer is 1.7×10^{-7} c/cm^2 and for a 10 μm layer is 1.96×10^7 c/cm^2 from which they infer thinner samples can sustain higher charge levels. Sensitivity as a function of field strength also shows a thickness effect and it seems field assisted photogeneration mechanisms, similar to those in selenium, are responsible. This is corroborated by the activation energy being a function of applied voltage and quantum yield depending on the field strength.

A whole range of related polymeric films are now found in commercial use and are being developed[62]. Their chief failing appears to be a build up of deep traps which ultimately retain a memory of previous signals and gradually cause the film properties of the film to fatigue.

Before moving on from organic photoconducting thin films, mention must be made of the work by Rosenberg[63] which has, at present, no application at all but must be of great potential in biological and other fields. He made a study of bimolecular lipid membranes which can be used as model systems for biological membranes. The membrane consisted of an oxidised cholesterol bimolecular layer assumed to have a structure of two molecules end to end as shown schematically in Fig. 23.

The material was known to be a semiconductor in its solid state and experiments on the bilayer established that it too was a semiconductor. The

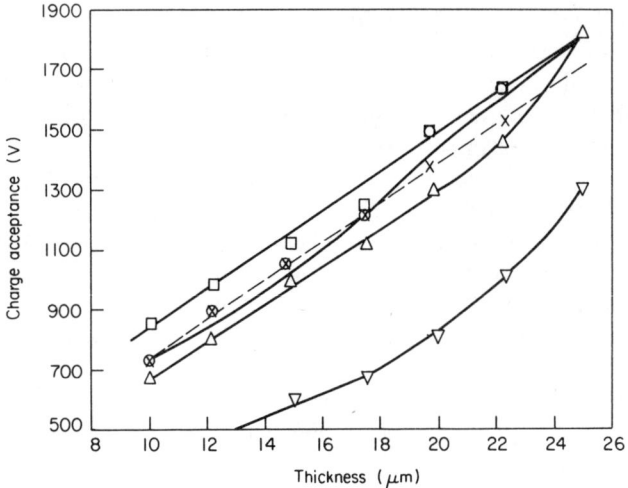

FIGURE 22. Negative charge acceptance as a function of thickness for the 1:1 trinitrofluorenone/vinylcarbazole complex for various values of charge acceptance; ▽—5000 V, △—6000 V, ○—7000 V, × —8000 V, □—Max^m charge acceptance line, after Schaffert[61].

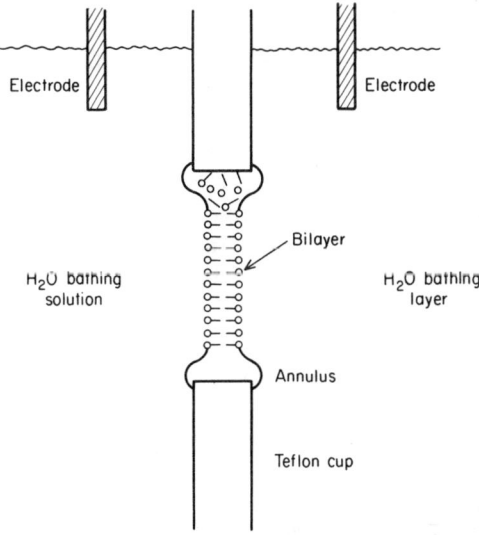

FIGURE 23. Schematic diagram of the assumed structure of the bimolecular lipid membrane. The circles represent polar groups and lines represents hydrophobic components, after Rosenberg[63].

addition of donors or acceptors to form complexes greatly enhanced the conductivity. Electronic conduction was demonstrated by passing electrons into one side of the layer while a solution containing Hg^{2+} ions was in contact with the other side of the layer. On passage of the current a mercury mirror formed on the layer. Asymmetric solutions on each side of the layer were found to be able to give rise to rectifying characteristics being generated by the layer. The layers absorb in the far infrared but addition of pigments or certain ions such as Fe^{3+} on the non-illuminated side of the layer produce photovoltages and currents up to two orders of magnitude higher than the "intrinsic" layer when illuminated with visible light. Saturation sets in eventually. It is thought the Fe^{3+} ion must inject holes into the bilayer and water is then oxidised at the other side of the layer.

The insight that biological membranes can exhibit all the standard semiconducting and photoconducting behaviour provides perspective on an almost unknown field for study, which must surely yield some exciting new devices whether their applications be in physics or medicine. If, for example, it can be shown that the body works on networks of electronic circuitry already known to us, then in one sphere we have the potential to construct prosthetics very closely approximating to a normal limb or the means of repairing, say, severe nerve damage. At the other extreme the certainty that a human brain functions by "standard electronics" would give us the insight to build (or grow) far more advanced but very compact computers.

4. APPLICATIONS

4.1 Radiation Detectors

A typical radiation detector consists of a photosensitive element attached to a substrate. Electrodes are evaporated onto the element, often in a comb configuration, so that electrode separation is small, while sensitive area exposed remains large. This unit is then encapsulated under vacuum or inert gas to prevent random changes in surface conductivity taking place; a window with suitable transmission providing the necessary "optical access". The element itself is usually a sintered layer about 10–15 μm thick as in CdS detectors but can be an evaporated layer (also in some CdS detectors) or lapped down single crystal material as in InSb detectors. Figure 24 shows a CdS photoconductive cell with this comb electrode configuration described above.

The advantages of this design can be appreciated by recalling the theory developed in Section 2. The area exposed to radiation is maximised while the electrode separation and so transit time is minimised, thus gain will be optimised. Gain could be further improved if a sandwich configuration were used with a top transparent electrode, as area and transit time then need not be

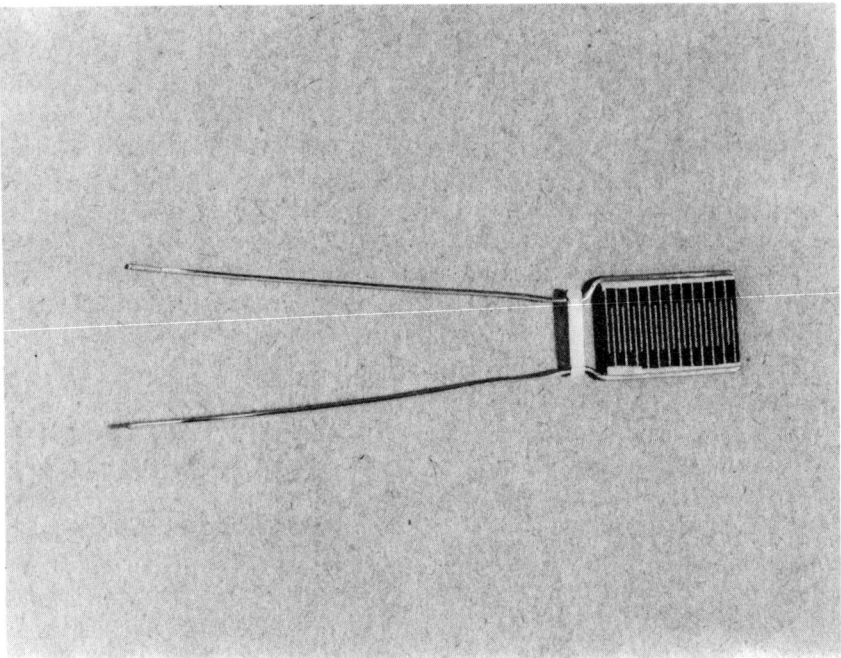

FIGURE 24. Photograph of a CdS photoconducting cell (Mullard, type RPY82). The photograph is supplied and reproduced by kind permission of Philips Industries.

traded against one another. However, the sandwich design would be more costly as more operations would be involved but perhaps more serious, the electrodes may be unsymmetric. Even if perfectly ohmic contacts cannot be achieved, the single evaporation to deposit both contacts provides symmetry, although even in this case a large deviation from ohmicity could not be tolerated.

A voltage is applied to the device and for that voltage the resultant photocurrent will be directly proportional to the intensity of illumination. Using a standard light source, detectors can be calibrated, but normally they are used for measuring relative intensities.

The detection of radiation by photoconductive methods begins to present difficulties as wavelength increases. Materials with band gaps narrow enough to produce intrinsic photoconductivity in the IR and microwave regions must be semiconductors with significant dark currents at room temperature. As mentioned earlier two approaches are used to avoid losing the photogenerated signal in the dark noise, the dark current can be backed off electrically or the detector can be cooled. The resultant sensitivity can then be discussed in terms of detectivity and responsivity as in Section 3. In addition, it is common practice to drive such elements so as to minimise power consumption by using pulsed power supplies and chopped radiation. Low energy response can also be produced by extrinsic or impurity photoconduction (as illustrated by the germanium thin films discussed in Section 3 responding at microwave wavelengths instead of their intrinsic IR value). Doped films generally have longer response times than intrinsic material but provided this can be tolerated by the application this need not be a problem.

The major use of detectors have already been mentioned in the introduction. Cadmium sulphide is used mainly for light meters and in street lighting switches. CdSe or PbS are commonly used in alarms or mechanisms involving light beams being interrupted because red or very near IR light emitting diodes are convenient light sources and undetectable by eye; ideal in burglar alarms, unobtrusive when opening lift doors. Elements in smoke detectors can be CdS or ZnS. They too rely on the interruption of a light beam, with white or blue light (tungsten or mercury lamps) being used, to avoid confusion with thermal radiation generated in the case of a fire. The smoke itself forms the interruption and the device can then activate sprinklers.

The possible use of photoconductors in optical communications applies far more stringent requirements on the photoconductor, as a fast, high gain IR detector. Light emitting diodes or lasers are used to feed optical signals down fibre optic light pipes which can be interfaced with computers to provide a rapid delivery of information. The photoconductor detectors therefore need nanosecond response times to be usable in this application. Photoconductive mode germanium cells are used in this way and research is being done with a view to developing CdS or CdSe cells.

4.2 The Vidicon

The vidicon camera tube was developed by RCA around 1950[35] but since then the name has come to represent any T.V. camera tube using electron beam scanning and direct readout from a photoconductive target. A typical vidicon tube is shown schematically in Fig. 25 together with its equivalent circuit. It consists of a glass cylinder with an optically polished end plate coated with a transparent electrode and having a photoconductive layer evaporated onto it. The photoconducting target can be either an antimony trisulphide layer about $5\,\mu m$ thick or a lead oxide layer of comparable thickness. The target is scanned by an electron beam focused down the axis of the cylinder so that the beam can charge up the free side of the photoconductor to the point where the surface will not accept further charge. Meanwhile, the transparent glass signal plate is connected to a fixed positive potential (between 10 and 100 volts). When an image falls on the sensitive area, charge leaks across the film, the charge transported being proportional to the light intensity. The free surface potential at an illuminated point is then lowered,

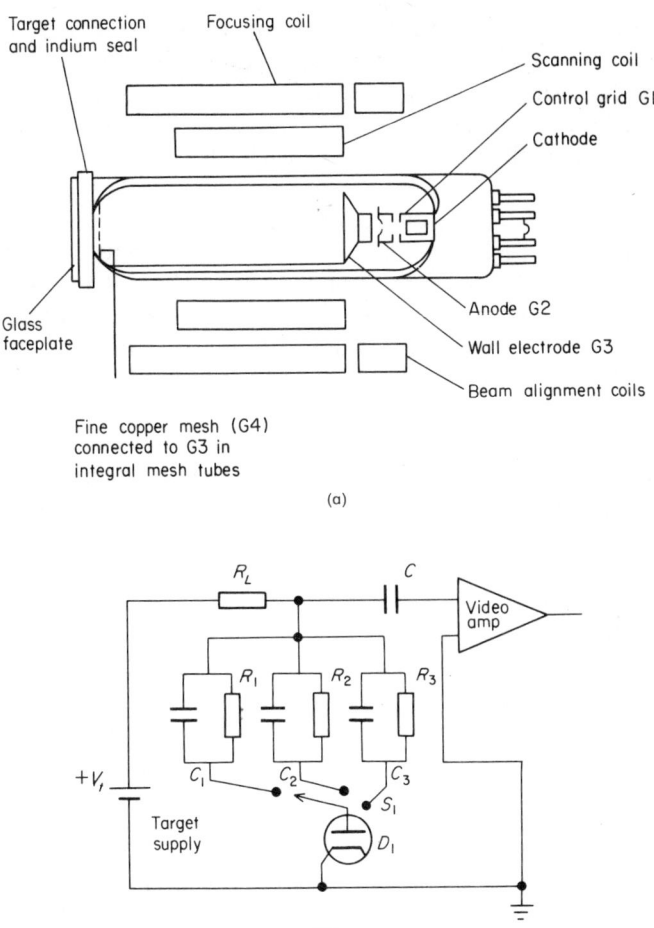

FIGURE 25. Schematic diagram of (a) a vidicon camera tube with (b) its equivalent circuit, after Longhurst and Woolgar[36].

only to be compensated by the beam as it rescans that point; the gun provides the means of converting the picture into electrical signals which can be read sequentially as an output.

Electrically then, the photoconductor must behave as a good insulator except in the areas of illumination so that contrast is achieved. The spectral sensitivity of the photoconductor should be centred over the visible spectrum for normal camera use, or in the IR for infrared cameras. The decay time should be long enough for the charge to leak "evenly" or information on relative brightness will be lost. Against memory, the element must not retain the previous scene when a new scene is projected. Thus ideally, charge at the illuminated element can leak away slowly so long as it takes less than a scan interval ($\frac{1}{30}$ sec). Detrapping is the major material cause of lag, but the action of the electron gun itself gives rise to a residual lag because the surface potential state does not have a sharp cut off. So long as the carriers can move right through the film without becoming trapped, their transit time will be equal to the dielectric relaxation time, thus the restriction imposed is that the mean distance a carrier can move before being trapped (known as a schubweg) must be greater than the thickness of the film. The lux-ampere characteristic of the device, known as the contrast or gamma factor, determines contrast. Low gamma values produce poor contrast but moving to high gamma values results in poor dynamic range.

Antimony trisulphide with its "human eye response" spectrum, dark resistivity of 10^{13} Ω cm, lag of $\frac{1}{20}$ second to fall to 10% of the original signal value and gamma of 0.6 makes it an eminently suitable material. PbO with a similar high resistivity, a response slightly further to the red, shorter lag and higher gamma than Sb_2S_3 provides a faster, higher contrast vidicon. Doping the photoconductors extends the range of vidicons. Antimony trisulphide can be doped to produce vidicons with high or low, red or blue sensitivity and in addition, lag can be increased to provide partial memory (useful in slow scanning modes).

Using lead oxide, UV, IR and x-ray sensitivity can be produced and high sensitivity can be combined with low lag. The lead oxide vidicon is more expensive because a cleaner, more controlled process is necessary to form the polycrystalline tetragonal lead oxide layer and so the evaporation must be carried out directly onto the vidicon tube. The less stringent purity conditions needed for antimony trisulphide enable it to be deposited onto the face plate which can subsequently be sealed into place. Thus rejection of unacceptable lead oxide vidicon occurs much later in the fabrication process than in the case of Sb_2S_3 devices.

Vidicons are cheap, established devices giving good "black and white" response and fairly good colour response. It seems likely that silicon photodiode arrays will eventually supercede vidicons because of their superior response, to colour particularly. However, vidicons will still have a role in closed circuit television where they can be applied to a range of activities including monitoring industrial processes, where inspection in a hostile environment or an inaccessible location is necessary, providing closed circuit T.V. for "overflow" lectures in education and in medical teaching where operations can be transmitted for viewing by a whole class of students. Surveillance of public areas is also becoming a more common practice, for the purposes of monitoring security, "spying", or for traffic control!

4.3 Electrophotography

This general term covers any process involving electrostatic reproduction of images—a convection has arisen whereby xerography is taken to mean the process in which an image is first produced on a photoconducting plate and then transferred to a piece of ordinary paper (known in the field as a dielectric sheet). The second main process, electrofax, is taken to mean the process where a photoconductive layer coated onto a sheet of paper produces an image directly. In the first process the same photoconducting plate is recycled, in the latter the film is used once only.

The principle of operation is the same in both cases. The plate can be considered to be made of a photoconductor layer and an insulating layer sandwiched between electrodes as shown in Fig. 26. Using a corona discharge the surface of the insulator is charged up (in an analogous manner to the vidicon) and by induction the interface between photoconductor and insulator acquires the opposite charge. Projecting an image onto the plate allows the induced charge to be neutralised at the illuminated portions of the plate, by charge flowing through the photoconductor. The plate is then dusted with black particles known as toner particles which fulfil the role of an ink. These particles are held by electrostatic forces at "dark" regions. In practice, this step is highly sophisticated and involves mixing toner particles with heavier carrier particles and the mixture being drawn across the plate.

At this point electrofax and xerography develop the copy differently. In the electrofax process the plate is heated up so that the toner particles fuse onto the plate to produce a fixed image, whereas in the case of xerography, a sheet of paper held just above the plate is charged by the corona discharge and the toner particles are electrostatically attracted onto the paper. Finally the image is fixed by heating. The plate is then recycled by brushing off any excess toner particles and flooding the whole plate with light to remove any residual charges. A vast amount of research has been devoted to the whole photocopying process; the nature of the corona, the toner particles, the carrier particles and modes of "sweeping up" excess toner particles have all been areas of great interest. Good expositions of the whole electrophotographic process are given by Dessauer and Clark[7] and by Schaffert[15]. Of more direct interest here is the nature of the photoconductor used. (Traditionally, selenium is the photoconductor used in xerography and indeed, it is highly suitable for such an application.) Electrofax is only possible because mass-produced thin layers of zinc oxide behave as photosensitive insulators and it is unlikely that any other material will be found which offers serious competition. Selenium however can be challenged by other materials and to some extent CdS has always been a threat. More recently organic photoconductors have proved themselves to have many advantages for xerography. They are readily formed into thin, low density films. They are nearly transparent and offer prospects of

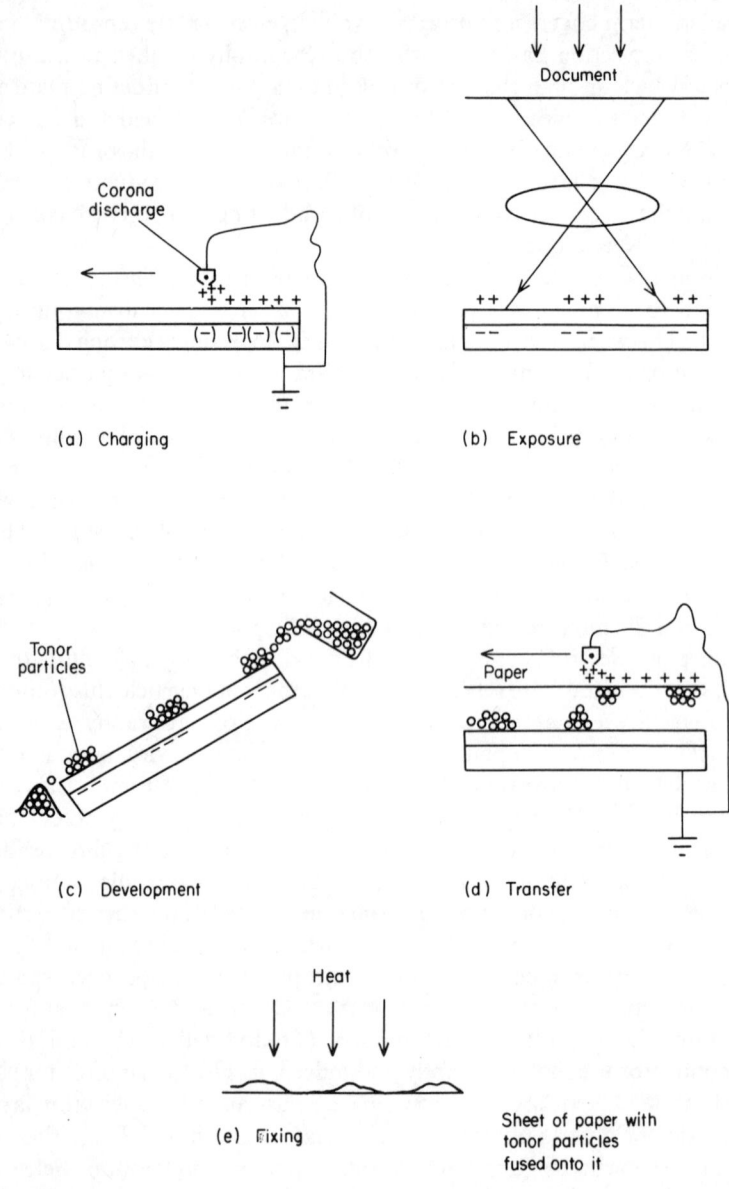

FIGURE 26. Schematic representation of the electrophotographic process.

spectral and chemical sensitisation, high resolution and often the possibility of both positive and negative charging. Their present limitation is their tendency to fatigue by retaining charge in deep traps.

The major advance which is sought by ambitious electrophotographers is the colour photocopy! This would require three plates each sensitive to a primary colour (for example anthracene, cadmium sulphide and selenium). In principle, it involves three consecutive exposures, suitable registered with the relative sensitivity of each plate chosen to give a resultant colour balance. Perhaps electrofax will provide an easier approach using various dyes or perhaps we shall return enthusiastically to the promise that organic materials offer.

4.4 Display Devices

(a) Addressing displays

Thin film photoconductors may in principle be used to solve two of the major problems encountered in "$x-y$" addressing display systems such as D.C. electroluminescent panels. Such panels are based on ZnS powder and they offer the technical possibility of producing large area panels at reasonable costs.

(i) The first problem is that of addressing a very large panel with perhaps more than 1000 characters. A simple $x-y$ addressing scheme can be employed to scan the display row by row and sequentially apply half the required driving voltage to each appropriate row and column so as to activate a given crossover; each cross-over sandwiches an element of luminescent material. The element at the cross-over site will light up but in addition light of a lower intensity will be emitted from other elements along x and y lines which will be essentially half selected; this gives rise to reduced contrast and in some cases complete loss of image.

A second factor in the $x-y$ addressing of large panels is the duration of the address pulse on any given element compared with the total cycle time of the display. The luminance of an element during the short period it is being addressed must be greater than its luminance for the remaining time period. Also, the constrast ratio will be further impaired for the following reason; if an OFF element occurs in a row of predominantly ON elements then that element will experience half-select pulses almost continuously. Conversely if an ON element occurs in a predominantly OFF row then that element will only experience one full select pulse. A worst case contrast ratio, c, is given by $c = (1/N)[L_v/(L_v/2)]$ where L_v and $L_v/2$ are the luminance at full and half voltage respectively and N is the number of columns. For a more typical situation with about 50% of elements in a row being selected, the contrast ratio becomes $c = (2/N)[L_v/(L_v/2)] + 1$.

A solution could be provided by the cross-point latch or on-panel memory, a photoconducting element in series with the active element would "see" the element was alight and so would be in a low resistance state. The latch "looking" at an element intended to be off, would remain in a high resistance state, thus power could be applied to the whole panel all of the time and only the required elements would be on (initial lighting could be effected by a voltage spike applied to the relevant elements). A similar principle applies to the liquid crystal displays where photoconducting elements can be used to sense light transmitted through a liquid crystal and effect feedback. In general, displays are designed for use in the visible and so the appropriate photoconductors are those with visible response, high sensitivity (the light emitting areas are small and so the number of photons emitted per element is low) and fast response (to be activated by the light emitted due to a voltage spike). Alternately a design could be envisaged where response time is low and the display is refreshed; in effect a lag is being introduced. Cadmium sulphide or selenide would be suitable for this type of application with suitable tailoring, for either design.

In order to be a good on/off switch, a high dark resistance is essential and the element geometry must be optimised for sensitivity. Thus the film thickness must be comparable with the absorption depth of visible light, i.e. 2–5 μm, the electrode separation would need to be small so that transit time is short and the area must be large enough to collect the maximum amount of light without obscuring too much of the active element. Figure 27 shows a typical design for a single element.

The trap concentration must be low as discussed in Section 3. Material suitable for use in this application has been produced, but this mode of operation means elements are fully on, or off completely, that is grey scale facility has been sacrificed. However, such large panels could not be addressed directly at all by the standard techniques used for smaller panels. This approach enables the panel to be addressed, but how great a demand exists for such a panel is not yet known.

(ii) The second addressing problem that can be tackled using photoconductors is that of finding reliable inter-connections. The connectors available are basically strips of spring-loaded metal clips which push-fit onto the side of a sheet of glass and are aligned with the tin oxide contact pads. A bent clip or uneven piece of glass leads to poor contact, while repeatedly taking the clip on and off causes the glass to chip at the edges.

The mechanical approach may be superceded by using an optical link. The operation of the photoconductor as part of an addressing system can be seen most easily from Fig. 28. The display is constructed with the x-lines connected onto (say) a positive bussbar and the y-lines onto a negative bussbar interposed between any one row or column and the supply voltage is a

Thin Film Photoconductors

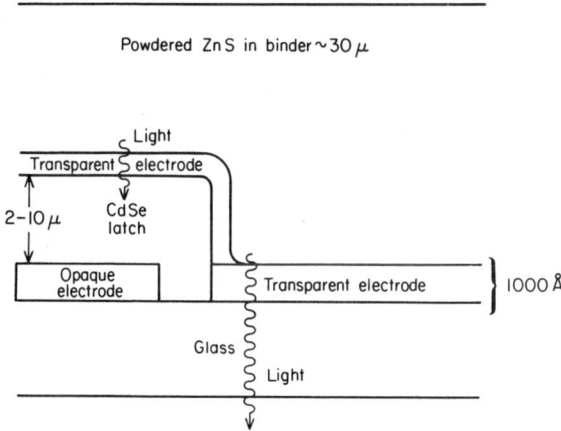

FIGURE 27. Schematic diagram of a photoconductive cross-point latch in a dcel panel, sectional view.

photoconducting element behaving as a switch. Thus a given cross-point can be lit by directing light on the elements of the required lines. Thus the panel itself requires only a pair of bussbars (and, in practice, an earth line to return all "off" elements to a zero potential; this is effected by use of a second independent array of photoconductor switches). The drive electronics can then be used to operate a set of light sources (e.g. light emitting diodes) in conventional multiplexing sequences, remote from the panel but connected optically to the photoconductors either by mechanical alignment or light pipes. The requirements of such a photoconductor are high sensitivity, very short response time and the capability of withstanding high power densities. There is more flexibility on the "shape" of the element. Electrode separation must be kept small and trap concentrations low in order to maintain a short response time. Such a connector should be cheaper and more reliable but so far no material has been produced which has such a high specification.

The same basic principle as the latch is employed in the use of x-ray image intensifiers where a photoconductor sensitive to x-rays (usually CdS) can convert an x-ray image into a visible image, using say, an electroluminescent ZnS panel. In this case, the technical problems are less severe as the conversion is from radiation to an electrical signal, whereas the latch involves the conversion of an electrical signal to radiation and then back to an electrical signal. All these and other similar applications rely on the photoconductor film thickness being just great enough to absorb all radiation without losing sensitivity.

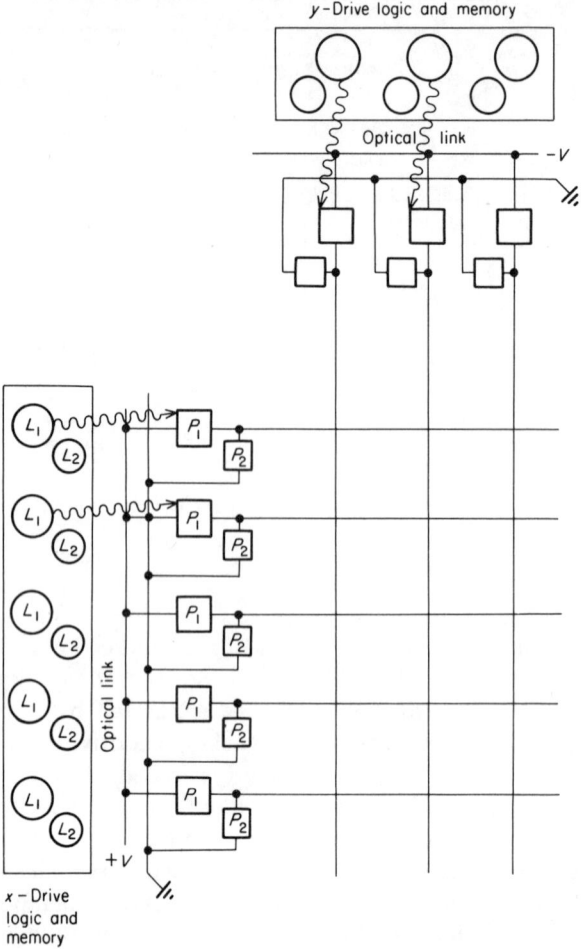

FIGURE 28. Schematic representation of a photoconductivity operated line drive mechanism suitable for operating an x-y display panel. L_1 elements are light emitting elements activating P_1 switches. L_2 elements activate P_2 switches to "switch the line off" by earthing.

The final type of display application considered is the use of photoconductors in the so-called lead lanthanum zirconium titanate (**PLZT**) ceramics to produce light valve devices, such as the ferpic (ferroelectric-photoconductor picture image converter). These are used in computer interfacing and are addressed by scanning lasers. The device is shown schematically in Fig. 29.

Thin Film Photoconductors   481

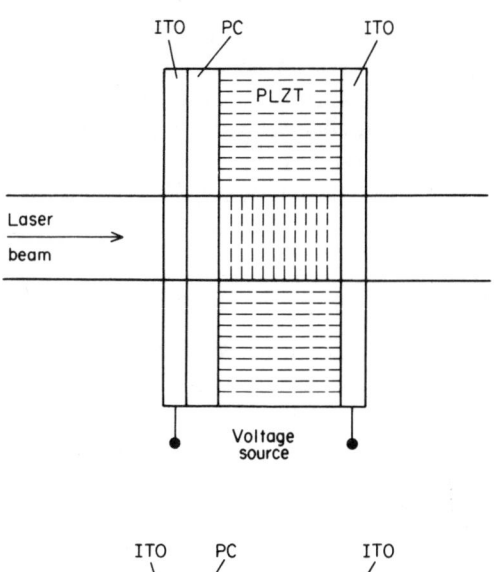

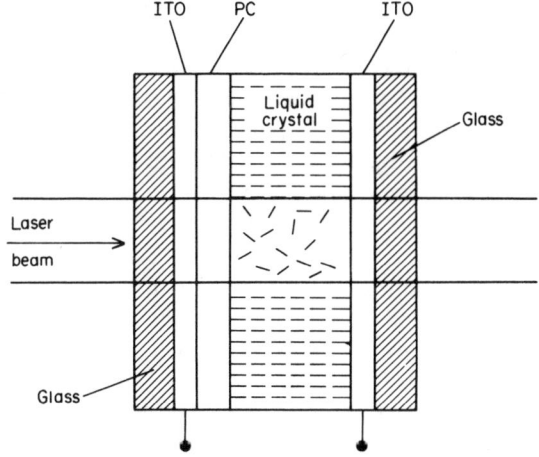

FIGURE 29. Sectional view of basic light valve structures, using laser light to modify light scattering in a PLZT plate or liquid crystal layer, after Fraser[44].

A voltage is applied across the layer but is only transmitted to the ferroelectric when the photoconductor has been illuminated. The effect of voltage on the strain-biassed ferroelectric ceramic, typically bismuth titanate, is to cause it to switch from one magnetic polarity to the other. Thus the information has a binary positive or negative polarity across the ferroelectric. The advantage of this arrangement is that information remains in this stored state (with no voltage applied) and can be randomly accessed. Holographic plates can be made from this type of device. They are a solid state analogue of a liquid crystal in that they switch from an ordered to disordered state and in fact light valve devices work equally well using liquid crystals. Photoconductors need high impedance for this type of application and it is hoped that a suitable organic material will be found. So far $Cd_{1-x}Zn_xS$ seems the best compromise.

5. Conclusion

A wide range of devices employ photoconductors in thin film form. The specific properties of the photoconducting material are important in all the applications but the variation of dimensions enables given parameters, mainly sensitivity and response time, to be optimised. This compensates largely for the loss in performance found in moving from a single crystal to a polycrystalline film. From a production viewpoint thin film devices involve less complications than do single crystal elements and so devices are oriented mainly towards sintered, sputtered or evaporated films. Experience from the organic materials suggests that cast, dipped and sprayed films will also be widely used. So far as advances in the field are concerned, even the so-called classical photoconductors selenium and cadmium sulphide are not yet fully understood, while the new materials have barely been characterised. The scope is immense. Imaging and displays are the present métière of photoconductors but they may easily move next into biology and energy conversion along the lines described in Section 3.

As previously mentioned, the integration of electronics with biological systems holds the promise of highly sophisticated surgery becoming an achievable goal (thus posing the question, will this produce a truly bionic man?).

On a more practical level, plants have been using photoconductivity in their energy conversion processes since evolution. The efficiency is low, four photons are required to generate one electron, but sunlight is abundant. As our fossil fuel resources become depleted and our fears of plutonium grow, solar energy looks a very attractive alternative. We already have photovoltaic cells but these are very costly (at least £1000/watt) and produce only a few volts output.

Indirect means of harnessing energy produced by photosynthesis are in use in prototype plants; blue-green algae is grown in sunlight and then under conditions causing anaerobic respiration, generates methane gas from, say, sewage sludge. The dried algae can subsequently be used as cattle fodder. Direct means are being investigated[64] by setting up electrochemical cells containing organically based solutions which generate photovoltaic responses. To date, no sign of a cheap commercial cell has emerged but with the range of organic materials available the chance of such a system existing is high but its efficiency would be greatly enhanced by a photoconducting component. If such a cell were available, the market would be worldwide.

Acknowledgements

The author would like to express her thanks to Dr. P. D. Fochs for his encouragement and many invaluable discussions on photoconductors and related topics; to Dr. D. K. Wickenden for all his helpful advice; to Professor J. C. Anderson for providing her with a draft of his chapter and to Dr. J. D. Wright for making available useful material. Finally thanks are due to Dr. T. Coutts for patience and understanding.

References

1. A. Rose, "Concepts in Photoconductivity and Allied Problems", Wiley, New York, (1963).
2. R. H. Bube, "Photoconductivity in Solids", Wiley, New York, (1960).
3. W. Smith, *Nature*, **7**, 303, (1873).
4. A. Von Hippel, *J. Chem. Phys.*, **16**, 372, (1948).
5. T. S. Moss, "Photoconductivity in the Elements", Academic Press, London and New York, (1952).
6. B. Petretis, H. Rogass and A. Satas, *Phys. Stat. Sol.*, **A30**, K105, (1975).
7. J. H. Dessauer and H. E. Clark, eds., "Xerophotography and Related Processes", Focal Press, London and New York, (1965).
8. W. Beyer and J. Stuke, *Phys. Stat. Sol.*, **A30**, 511, (1965).
9. P. Thomas, A. Barna, P. B. Barna and G. Radnoozi, *Phys. Stat. Sol.*, **A30**, 637, (1975).
10. P. K. Weimer and A. D. Cope, *RCA Review*, **12**, 314, (1951).
11. M. D. Tabak, *Trans. Met. Soc. A.I.M.E.*, **239**, 330, (1967).
12. H. T. Li and P. J. Regensburger, *J. Appl. Phys.*, **34**, 1730, (1963).
13. M. D. Tabak and P. J. Warter, *Phys. Rev.*, **173**, 899, (1968).
14. D. M. Pai and S. W. Ing Jnr., *Phys. Rev.*, **173**, 729, (1968).
15. R. M. Schaffert, "Electrophotography", Focal Press, London and New York, (1965).
16. H. Seki, *IEEE Trans. Electron Devices*, **ED-19**, 412, (1972).
17. E. H. Putley, *Phys. Stat. Sol.*, **6**, 571, (1964).
18. H. Melchior, M. B. Fisher and F. R. Arams, *Proc. IEEE*, **58**, 1466, (1970).
19. P. C. Taylor, U. Strom, J. R. Hendrickson and S. K. Bahl, *Phys. Rev.*, **B-13**, 1711, (1976).
20. R. W. Stimets, J. Waldman, J. Lin, T. S. Chang, R. J. Tempkin and G. A. N. Connell, *Sol. St. Comms.*, **13**, 1485, (1973).

21. M. H. Brodsky and A. Lurio, *Phys. Rev.*, **B-9**, 1646, (1974).
22. T. B. Tang, *J. Electron Mat.*, **4**, 1229, (1975).
23. D. B. Fraser and H. Melchior, *J. Appl. Phys.*, **43**, 3120, (1972).
24. R. M. Moore, J. T. Fischer and F. Kozielec Jnr., *Thin Solid Films*, **26**, 363, (1975).
25. W. H. Leighton, *J. Appl. Phys.*, **44**, 5011, (1973).
26. M. Takeuchi, J. Sakagawa and H. Nagasaka, *Thin Solid Films*, **33**, 89, (1976).
27. J. B. Snelling, Technique for Sputtering Transparent CdS and CdZnS Photoconductive Films. Report No. SAND. 75-0466, (1975). (Internal report—unlimited release, Sandia Labs, Albuquerque.)
28. R. S. Feigelson, Ph.D. Thesis. Univ. Stanford. Photoelectric Properties of Chemically Deposited Cadmium Sulpho-selenide films, (1974).
29. L. L. Kazmerski, *Thin Solid Films*, **21**, 273, (1974).
30. A. Waxman, V. E. Henrich, F. V. Shallcross, H. Borkan and P. K. Weimer, *J. Appl. Phys.*, **36**, 168, (1965).
31. G. South and D. M. Hughes, *Thin Solid Films*, **20**, 135, (1974).
32. D. B. Fraser, *Proc. IEEE*, **61**, 1013, (1973).
33. B. S. Sharma and R. R. Metha, *IEEE Trans. Sonics & U/S*, **SU-19**, 225, (1972); (same article in *Ferroelectrics*, **3**, 225, (1972)).
34. T. N. Bhar and J. S. Linder, *Thin Solid Films*, **21**, 267, (1974).
35. S. V. Forgue, R. R. Goodrich and A. D. Cope, *RCA Review*, **12**, 335, (1951).
36. W. P. Longhurst and A. J. Woolgar, *Electronic and Radio Tech.*, **4**, 15, (1970).
37. K. C. Hilsum and A. C. Rose-Innes, "Semi-conducting III-V Compounds", MacMillan, (Pergamon), New York, (1961).
38. P. W. Kruse, in "Indium Antimonide Detectors from Semi-conductors and Semimetals," Vol. 5, eds., R. K. Willardson and A. C. Beer, Academic Press, London and New York, (1970).
39. J. Amick, *RCA Review*, **20**, 753, 770, (1959).
40. D. Eger, Y. Goldstein and A. Many, *RCA Review*, **36**, 508, (1975).
41. Y. Shapira, S. M. Cox and D. Lichtman, *Surf. Sci.*, **54**, 43, (1976).
42. See for example L. R. Weisberg, in "Solid State Devices", IOP Conf. Series No. 25, p. 15, (1974).
43. S. Larach, R. E. Schrader and C. F. Stocker, *Phys. Rev.*, **108**, 587, (1957).
44. D. B. Fraser, *Proc. IEEE*, **61**, 1013, (1973).
45. S. R. Ovshinsky, *Phys. Rev. Letts.*, **21**, 1450, (1968).
46. K. Weiser, *J. Non-Cryst. Sol.*, **8-10**, 922, (1972).
47. M. H. Cohen, H. Fritzsche and S. R. Ovshinsky, *Phys. Rev. Letts.*, **22**, 1065, (1969)
48. T. Minami, M. Hibino and M. Tanaka, *J. Non-Cryst. Sol.*, **15**, 141, (1974).
49. H. Kokado, I. Shimizu and E. Inoue, *J. Non-Cryst. Sol.*, **20**, 131, (1976).
50. T. F. Tao and C. C. Wang, in "Physics of IV-VI Compounds and Alloys", ed., S Rabii, Gordon and Breach, New York, London and Paris, (1974).
51. T. F. Tao and C. C. Wang, *J. Appl. Phys.*, **43**, 1313, (1972).
52. E. M. Logothetis and H. Holloway, *J. Appl. Phys.*, **43**, 256, (1972).
53. H. Akamatu and H. Kuroda, *J. Chem. Phys.*, **39**, 3364, (1963).
54. H. Kokado, K. Hasegawa and W. G. Schneider, *Organic Crystal Symposium*, NRC Ottowa, (1962).
55. C. K. Prout, R. J. P. Williams and J. D. Wright, *J. Chem. Soc. A.*, 747, (1966).

56. P. J. Reucroft and W. H. Simpson, *Photochem. and Photobiol.*, **10**, 79, (1969).
57. F. Gutmann and L. E. Lyons, "Organic Semi-conductors", Wiley Interscience, (1967).
58. A. A. Bright, P. M. Chaiken and A. R. McGhie, *Phys. Rev. B*, **10**, 3560, (1974).
59. R. J. Hurditch, V. M. Vincent and J. D. Wright, *J. Chem. Soc., Farad. Trans.*, *I*, **68**, 465, (1972).
60. V. M. Vincent and J. D. Wright, *J. Chem. Soc., Farad. Trans.*, *I*, **70**, 58, (1974).
61. R. M. Schaffert, *IBM. J. Res. Develop.*, **15**, 75, (1971).
62. K. Okamoto, S. Kusabayashi, M. Yokoyama, K. Kato and H. Mikawa, Soc. of Photographic Scientists and Engineers, *2nd Int. Conf. on Electrophotography*, (1974).
63. B. Rosenberg, *Disc. Farad. Soc.*, No. 51, 190, (1971).
64. See for example work begun by Dr. M. Archer and Professor Page at the Royal Institution in 1973.

Chapter 10†

Thin Film Solar Cells

R. Hill

*Department of Physics and Physical Electronics,
Newcastle upon Tyne Polytechnic,
Newcastle upon Tyne,
England*

1. Introduction 488
2. Theoretical Treatment of Solar Cells. 491
    2.1 The Ideal Solar Cell 491
        2.1.1 The efficiency of ideal solar cells 500
    2.2 Practical Solar Cells 501
        2.2.1 Current collection in p–n junction devices 502
        2.2.2 Current/voltage characteristics of homojunctions 505
        2.2.3 Semiconductor heterojunctions 508
        2.2.4 Schottky barriers 512
    2.3 The Characteristics of Practical Cells. 513
        2.3.1 Equivalent circuit analysis 513
        2.3.2 Origin of shunt and series resistance 516
3. Structure of Thin Films 517
    3.1 Influence of the Deposition on the Properties of Thin Films . . . 518
    3.2 The Influence of Polycrystallinity on the Electrical Properties of Thin Films. 523
4. Cadmium Sulphide–Copper Sulphide Solar Cells 526
    4.1 Introduction 526
    4.2 The Formation of Cadmium Sulphide Layers 527
        4.2.1 Thermal evaporation 527
        4.2.2 Sputtering 532
        4.2.3 Vapour deposition 534
        4.2.4 Spraying techniques 534
        4.2.5 Heterogeneous conversion of cadmium 535
        4.2.6 Sintered and screen-printed layers 535
    4.3 Formation of the Copper Sulphide–Cadmium Sulphide Junction . . 535
        4.3.1 Chemiplating processes 536
        4.3.2 Evaporation techniques 539
    4.4 Properties of Cuprous Sulphide 544

† For a list of symbols used in this chapter and their definitions see p. xvii.

4.5 Analysis of the Cadmium Sulphide–Copper Sulphide Junction . . . 553
 4.6 The Characteristics of Copper Sulphide–Cadmium Solar Cells . . . 569
 4.7 Degradation of Power Conversion Efficiency 574
5. Other Thin Film Solar Cells 577
 5.1 Thin Film Homojunctions 578
   5.1.1 Cadmium telluride homojunctions 578
   5.1.2 Silicon homojunctions 578
 5.2 Schottky Barrier Cells. 579
   5.2.1 Silicon Schottky barrier cells 579
   5.2.2 Gallium arsenide Schottky barrier cells 580
 5.3 Thin Film Heterojunction Cells 581
   5.3.1 Cadmium telluride–copper telluride 581
   5.3.2 Cadmium sulphide–cadmium telluride solar cells 582
   5.3.3. Indium phosphide and ternary indium compound cells . . . 583
 5.4 Semiconductor–Electrolyte Junction Solar Cells. 585
6. Future Uses of Thin Film Solar Cells 587
7. Conclusions 593
 References. 594

1. Introduction

A solar cell is a device which converts sunlight into electrical power. Although these have only recently become of serious commercial interest, the principles on which they are based have been known for a long time.

In 1839 Becquerel[1] studied the current produced between two metal electrodes in an electrolyte whilst one electrode was illuminated by sunlight. Becquerel used electrodes of platinum, brass and silver, and managed to estimate the spectral dependence of the photocurrent. Becquerel's work was confirmed and extended by a number of workers in the following years[2], and has been developed in the past fifteen years to the stage where it deserves serious consideration as a power source, mainly as a result of the work of Gerischer[3,4,5].

Photoconductivity in selenium seems to have been discovered by Smith[6] in 1873. In an elegant series of experiments, Sabine[7] investigated the effect of light on the resistance of selenium and its temperature coefficient in the dark and under illumination. He also made what appears to be the first measurement of contact resistance, and showed that this could be significantly higher than that of the selenium samples. Adams and Day[8] also investigated the effect of light on the currents flowing in selenium with platinum contacts and discovered that the action of light falling on the positively biased end of the sample was to reduce the current whilst light falling on the negatively biased end caused an increase. They found further that an E.M.F. could be induced across the samples when they were illuminated. This latter point was disputed by Sabine[7] who tried to distinguish between effects due to a photo-

E.M.F. and a simple reduction in resistance and found that his experimental results supported the latter. The existence of an E.M.F. induced by illumination was confirmed by Fritts[9] who produced a selenium cell with a transparent gold counter electrode, and Minchin[10] whose cells used selenium with an aluminium back contact and an electrolyte plus aluminium electrode as the front contact and produced photovoltages of over 0.5 volts in daylight. Minchin also developed what was probably the first thin film cell when he chemically oxidised tin foil to give a tin–tin oxide cell, again with an electrolyte and counter electrode as a front contact. These cells he called "impulsion cells"[11] since they needed to be flicked with the finger in order to make them work. Minchin also seems to have been the first person to suggest that photovoltaic cells could be used to produce useful electrical power from sunlight. This idea was supported by Appleyard[12] who "beheld the blessed vision of the Sun, no longer pouring his energies unrequited into space, but, by means of photo-electric cells and thermo-piles, these powers gathered into electric storehouses to the total extinction of steam engines, and the utter repression of smoke".

The first observation of rectification in metal systems appears to have been made by Munck[13] in 1835. Rectification in metal sulphide systems was studied by Braun[14] and in copper–copper oxide by Schuster[15] in 1874, but it was not until the turn of the century that practical uses were found[16].

The photo-electric activity of the metal–cuprous oxide system was first noticed by Hallwachs[17] in 1904. The cuprous oxide cell was developed by Grondahl[18], and made commercially viable by Lange[19].

Until the 1950's the selenium cell and the copper oxide or copper sulphide cells were the types used in the large majority of cases[20], although both thallium sulphide[21] and bismuth sulphide[22] were also used in photovoltaic cells.

All of these early cells were rendered obsolete by the discovery[23] of the silicon solar cell by Chapin, Fuller and Pearson[24]. The early cells, developed from large area power rectifiers, had an efficiency of about 4%[23] but due to being lithium doped, were rather unstable. The development of boron diffusion technology produced stable cells with an efficiency of 6%[24] in 1954.

The first analytical treatment of solar cell efficiency was published in 1955 by Prince[25], and then Loferski[26] showed that there is an optimum energy-gap of about 1.5 eV for a p–n homojunction solar cell. This stimulated investigation into other materials such as gallium arsenide, cadmium telluride and indium phosphide which all have energy-gaps close to the optimum[27].

In 1954 D.C. Reynolds et al.[28] announced the observation of a photovoltaic effect in cadmium sulphide single crystals with an indium base electrode and silver, gold, copper and platinum counter electrodes. They found open circuit voltages of 0.4 volts and short circuit currents of 15 mA cm^{-2} in sunlight and a spectral response with a pronounced peak in the red. The power conversion efficiency of Reynolds' cells with a copper counter electrode reached 1.5%. Single crystal CdS cells were developed by Hammond and Shirland[29,30] and efficiencies of over 5% were achieved by 1959, and have been investigated by many groups[31] since 1959, mainly to gain understanding of the heterojunction.

A thin film CdS cell was developed by Carlson[32] starting in 1954, but it was not until 1960 that power conversion efficiencies of over 1% were reached[31]. Thin film cadmium sulphide–copper sulphide solar cells were subject to intensive development during the 1960's and by 1965, cells up to 55 cm^2 in area were being produced with efficiencies of 6%, with some cells having efficiencies of over 8%[33,34]. Despite a considerable research effort throughout the world and an enormous increase in understanding of the mechanisms in the $CdS:Cu_2S$ cells, very little improvement in power conversion efficiency has been obtained since that time.

In the 1960's many groups produced solar cells using a variety of different materials and techniques[23-38]. All of the cadmium chalcogenides were investigated, and amongst the most significant developments in this period were the $CdTe-Cu_2Te$[39] cells which reached efficiencies of up to 6%, $pCdTe-nCdS$[40] and graded band-gap CdS_xTe_{1-x} anisotype heterojunctions[41]. These latter devices are interesting since Tauc[42] has shown that graded band-gap solar cells should be more efficient than abrupt-junction devices, although these CdS_xTe_{1-x}[41] cells were not subjected to more than a preliminary study, and their efficiency was low.

The most efficient solar cells produced so far are those using single crystal gallium arsenide, which has an energy-gap equal to the optimum for $p-n$ homojunctions. As the material properties of GaAs improved, the conversion efficiency of $p-n$ GaAs solar cells improved up to about 6%, where it was limited by surface recombination and low minority carrier lifetimes. These limitations were overcome by using a window of $Ga_xAl_{1-x}As$ or by forming $Ga_xAl_{1-x}As:GaAs$ heterojunctions, and these have now been developed to give power conversion efficiencies of over 20%[43].

The driving force behind the intensive development of solar cells in the years up to 1970 was the space programme. The first solar cells were flown in Vanguard 1 in March 1958, and since that time, solar cells have been essential for providing electrical power on space missions. The development of thin film solar cells was not as intensive as that of the more efficient silicon cells, but they did offer advantages of flexibility, greater resistance to radiation damage and a higher power/mass ratio, and both United States and French $CdS:Cu_2S$ solar cells have worked satisfactorily in space.

As the space programme languished, funds for work on solar cells diminished until 1973 when it became much more generally recognised that renewable energy sources would be essential for the next century, and development of solar cells for terrestrial use was greatly expanded. Thin film solar cells would seem to have great advantages due to their ability, in principle at least, of being deposited easily and cheaply as large area devices with minimum use of materials and energy. Apart from the cost of the cells, which is orders of magnitude too high at present to compete with conventional electricity generating stations, the storage of electricity produced is the major problem in designing systems for widespread use. The uses of thin solar cells will be discussed in the final section of this article.

2. Theoretical Treatment of Solar Cells

The first step in the conversion of sunlight into electrical power is the absorption of the light by the semiconductor. When a photon of energy greater than the energy-gap of the semiconductor is absorbed, an electron is excited from the valence band to the conduction band. Absorption of such photons thus produces an increase in the density of electrons in the conduction band and an exactly equal increase in the density of holes in the valence band. The electron and hole concentrations are both increased above the thermal equilibrium concentrations, and the excess electrons and holes will tend to recombine. The presence of these excess carriers will merely reduce the resistivity of the material and can not give rise to a voltage, and hence to the output of electrical power, unless there is some inhomogeneity in the system[42]. This may take the form of non-uniform illumination, inhomogeneous material or non-uniform doping of homogeneous material. The inhomogeneity leads to the separation of the photogenerated charge carriers, and hence produces a photovoltage whilst the motion of the photogenerated carriers and their recombination after separation produces the photocurrent.

A precise description of the operation of a solar cell is quite complex and it is useful to consider first the simplest case.

2.1 The Ideal Solar Cell

Consider a thin slice of semiconductor of thickness x and optical absorption coefficient α, with radiation of intensity I_0 and energy hv incident normally on one face. The absorbed intensity is $I_0 - I_0 e^{-\alpha x} = I_0 \alpha x$ if $\alpha x \ll 1$, and the power absorbed per unit volume is thus $I_0 \alpha$. The number of photons absorbed per unit time per unit volume is then $I_0 \alpha / hv$ and the absorption of these causes the electron and hole concentrations to increase at a rate $R_G = I_0 \alpha q / hv$ where q is the quantum efficiency.

$$q = \frac{\text{Number of electron-hole pairs produced}}{\text{Number of photons absorbed}}$$

In equilibrium, the rate of generation of electron-hole pairs is equal to their rate of recombination, which can be written as $\Delta n/\tau_e$ in p-type semiconductors and as $\Delta p/\tau_h$ in n-type semiconductors, where τ_e and τ_h are the recombination lifetimes of excess minority electrons and holes. Thus

$$\Delta p = \frac{I_0 \alpha q \tau_h}{h\nu}$$

In an ideal p–n junction, the forward current density in the dark is given by

$$J = J_0\left(e^{\frac{eV}{kT}} - 1\right) = J_0(e^{\lambda V} - 1)$$

where J_0 is the reverse saturation current density, V is the applied bias, T is the temperature and $\lambda = e/kT$. The reverse saturation current density can be written

$$J_0 = n_p e v_e + p_n e v_h$$

where n_p and v_e are the concentration and velocity of electrons in the p-region of the junction and p_n and v_h are the concentration and velocity of the holes in the n-region. In the recombination times τ_e and τ_h the electrons and holes drift over distances of L_e and L_h, the electron and hole diffusion lengths. Thus

$$J_0 = n_p e L_e/\tau_e + p_n e L_n/\tau_h \tag{1}$$

This is obviously not an exact derivation of this equation, but the result is correct[44] for semi-infinite n- and p-regions.

When the solar cell is illuminated, the carrier densities are increased due to the absorption of photons and $n_p \to n_p + \Delta n$ whilst $p_n \to p_n + \Delta p$ where

$$\Delta n = \Delta p = \frac{I_0 \alpha q \tau_R}{h\nu} = R_G \tau_R \tag{2}$$

where τ_R is the appropriate recombination time.

The saturation current density increases when the junction is illuminated and becomes

$$J_0(\text{Ill}) = (n_p + \Delta n)eL_e/\tau_e + (p_n + \Delta p)eL_h/\tau_h$$

$$J_0(\text{Ill}) = J_0 + J_L$$

where J_L is the photogenerated current and is given by

$$J_L = \Delta p e \left(\frac{L_e}{\tau_e} + \frac{L_h}{\tau_h}\right) = R_G e(L_e + L_h)$$

$$J_L = \frac{I_0 \alpha q}{h\nu} e(L_e + L_h) \tag{3}$$

The ideal solar cell has electrical characteristics which are well represented by the equivalent circuit shown in Fig. 1. The photogeneration of current in the solar cell is represented by an ideal current generator G, which gives a current of density J_L. The diode characteristics of the cell are represented by the ideal

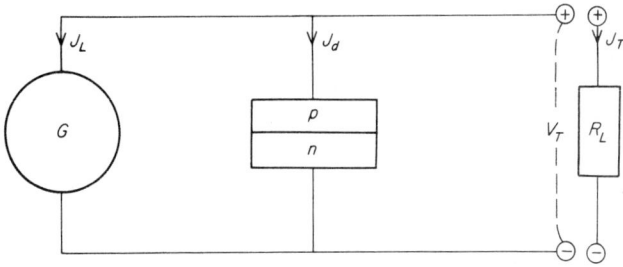

FIGURE 1. Equivalent circuit diagram of an ideal solar cell.

junction which gives a forward current of density J_d due to the forward voltage V_d across the diode. The load resistor R_L carries a current J_T due to the voltage V_T across the terminals.

From Kirchoff's law of junctions,

$$-J_T = J_d - J_L \text{ (forward currents positive)}$$

i.e.
$$-J_T = J_0(e^{\lambda V_d} - 1) - J_L$$

As the forward voltage across the diode increases, the forward current must increase until at the open circuit voltage V_{OC}, the forward diode current is equal to the reverse photocurrent and the current through the load is zero. Then

$$J_L = J_0(e^{\lambda V_{OC}} - 1)$$

i.e.
$$V_{OC} = \frac{1}{\lambda} \ln\left(1 + \frac{J_L}{J_0}\right) \qquad (4)$$

Since $J_L/J_0 \gg 1$

$$V_{OC} = \frac{kT}{e} \ln\left[\frac{I_0 \alpha q}{h\nu} \frac{e}{J_0}(L_e + L_h)\right] \qquad (5)$$

Figure 2 shows a band diagram for a p–n homojunction cell in the dark and shows the effect of illumination on the electron energy levels. Figure 3 shows

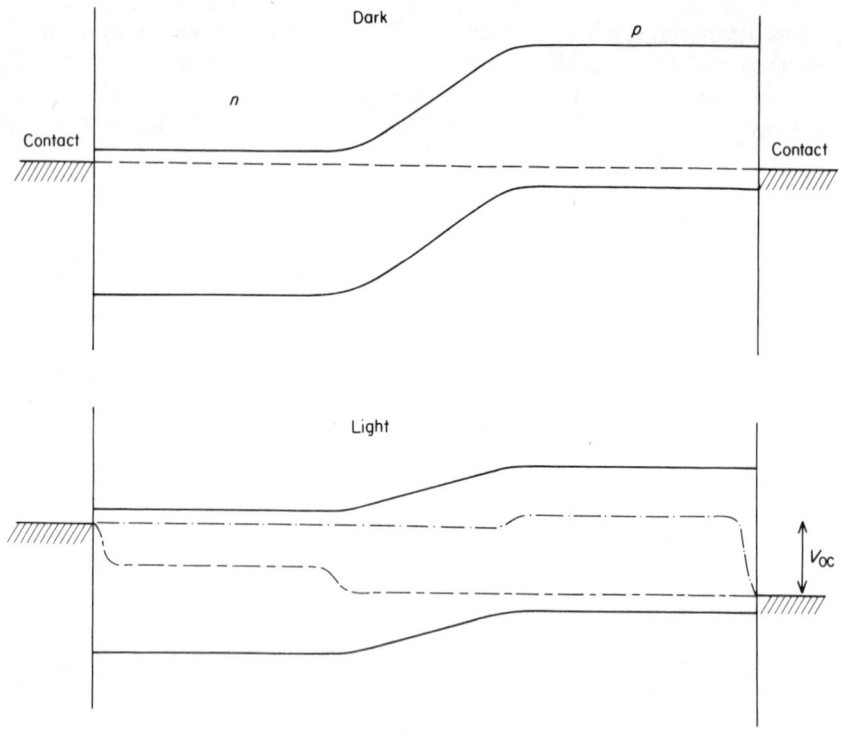

FIGURE 2. Energy-band diagram of a p–n homojunction in the dark and under illumination showing the electron (—·—·—·) and hole (—————) quasi Fermi levels.

the current/voltage characteristics of an ideal solar cell in the dark and under illumination.

The condition for maximum power output is

$$\frac{d}{dV_T}(J_T V_T) = 0$$

i.e.

$$\frac{d}{dV_T}[J_0 V_T(e^{\lambda V_T}-1) - J_L V_T] = 0$$

The voltage V_m at the maximum power point is thus given by

$$(1+\lambda V_m)e^{\lambda V_m} = \frac{J_L}{J_0} \qquad (6)$$

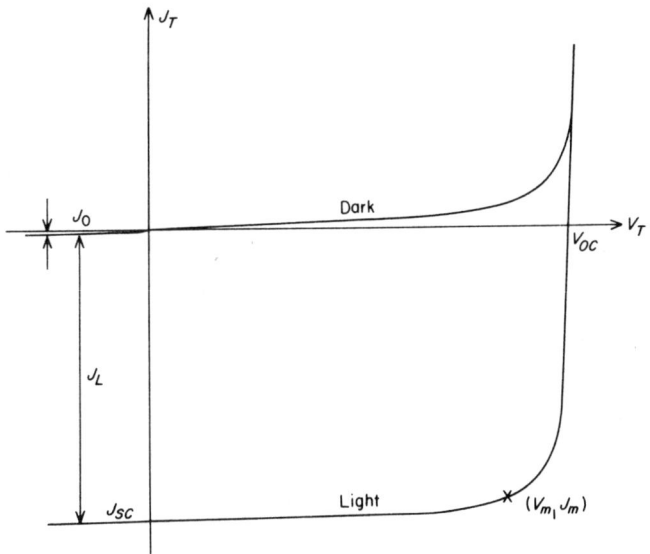

FIGURE 3. Current/voltage characteristics of an ideal solar cell.

Similarly the current density J_m for maximum power is

$$J_m = (J_L + J_0)\frac{\lambda V_m}{1 + \lambda V_m} \qquad (7)$$

It is usual to define a parameter called the fill-factor which relates the maximum power which a cell can deliver to the short circuit current and the open circuit voltage, and the fill-factor thus is given by

$$F.F. = \frac{J_m V_m}{J_{SC} V_{OC}} \qquad (8a)$$

Now $J_m = J_{SC} - J_d(V_m)$ (considering only the magnitude of the terminal current)

$$J_m = J_{SC} - J_0(e^{\lambda V_m} - 1)$$

$$\therefore \qquad F.F. = \frac{V_m}{V_{OC}}\left[1 - \frac{J_0}{J_{SC}}(e^{\lambda V_m} - 1)\right] \qquad (8b)$$

but $\qquad J_{SC} = J_0(e^{\lambda V_{OC}} - 1)$

Thus

$$F.F. = \frac{V_m}{V_{OC}}\left[1 - \frac{e^{\lambda V_m} - 1}{e^{\lambda V_{OC}} - 1}\right] \qquad (9)$$

The power conversion efficiency is defined as

$$\eta = \frac{J_m V_m}{I_0} = \frac{J_{SC} V_{OC} F.F.}{I_0} \qquad (10)$$

Table 1

Parameter	Cell A	Cell B	Derived from equation
I_0 mW cm^{-2}	100	100	
J_L mA cm^{-2}	30	20	
J_0 A cm^{-2}	10^{-12}	10^{-10}	
J_L/J_0	3×10^{10}	2×10^8	
V_{OC} volts	0.602	0.481	(4)
V_m volts	0.53	0.41	(6)
J_m mA cm^{-2}	28.6	18.9	(7)
F.F.	0.84	0.80	(8a)
η	15.2	7.7	(9)

It is notable that all the parameters of an ideal solar cell can be deduced from a knowledge of J_0 and J_L, and two examples are given in Table 1. Example A shows data which are similar to those for silicon single crystal cells, whilst example B uses data similar to those for a high efficiency thin film cell. It can be seen from these examples and from equations (4) to (10) that the power conversion efficiency of a solar cell depends on the magnitude of the photogenerated current J_L and on the ratio J_L/J_0, both of which should be as large as possible to ensure maximum efficiency. Equation (1) shows that the reverse saturation current J_0 depends on the dark minority carrier densities and on the ratio of minority carrier diffusion length and recombination time L/τ which is equal to $\mu kT/eL$ where μ is the minority carrier mobility. Equation (3) shows that the photogenerated current is maximised when $I_0 \alpha q/hv$ is as large as possible and when the minority carrier diffusion lengths are as long as possible. Three requirements are thus apparent from this

analysis of the ideal solar cell if the power conversion efficiency is to be minimised:
(1) the minority carrier densities should be as small as possible;
(2) the minority carrier diffusion lengths must be as long as possible;
(3) the factor $I_0 \alpha q/hv$ must be as large as possible.

The magnitude of the factor $I_0 \alpha q/hv$ depends on the semiconductor material but largely on the solar radiation. The simple theory given on page 491 takes no account of the spectral distribution of solar insolation, and must obviously be modified. The intensity I_0 must be replaced by a spectral distribution function $G(v)$, and it must be remembered that both the absorption coefficient α and the quantum efficiency q are functions of photon energy. The rate of generation of electron-hole pairs is then

$$R_G = \int_0^\infty \frac{G(v)\alpha(v)q(v)}{hv} dv$$

The spectral distribution function has been measured in a number of locations under a variety of conditions[45]. The total intensity $\int_0^\infty G(v) dv$ and the spectral distribution depend on the location, the time of year and time of day and on atmospheric conditions. It is usual to take as a standard the radiation incident on a horizontal surface when the sun is directly overhead and the atmosphere is as free as possible from scattering centres. The solar radiation is then subjected to absorption by unit thickness of the Earth's atmosphere and is designated Air Mass 1 or AM1. Under these conditions the intensity is 1 kwatt m^{-2}. For the purpose of calculating the electron-hole pair generation rate, it is more convenient to plot $G(v)/hv$ against hv and this is shown in Fig. 4.

The energy dependence of the absorption coefficient has been measured for a number of semiconductors, and some of those important in solar cells are shown in Fig. 5.

The absorption coefficient can be successfully calculated for allowed direct transitions[46] and it is found that

$$\alpha = 2.7 \times 10^5 \frac{f}{N}\left(\frac{2m_r}{m_0}\right)^{3/2} (hv - E_g)^{1/2} \text{ cm}^{-1}$$

where N is the refractive index, m_r is the reduced effective mass, i.e. $m_r = m_e m_h/(m_e + m_h)$, m_0 is the free electron mass and f is the oscillator strength $f \approx 1 + (m_0/m_h)$.

For indirect transitions, there are three processes occurring, photon absorption and phonon absorption or emission. The absorption coefficient

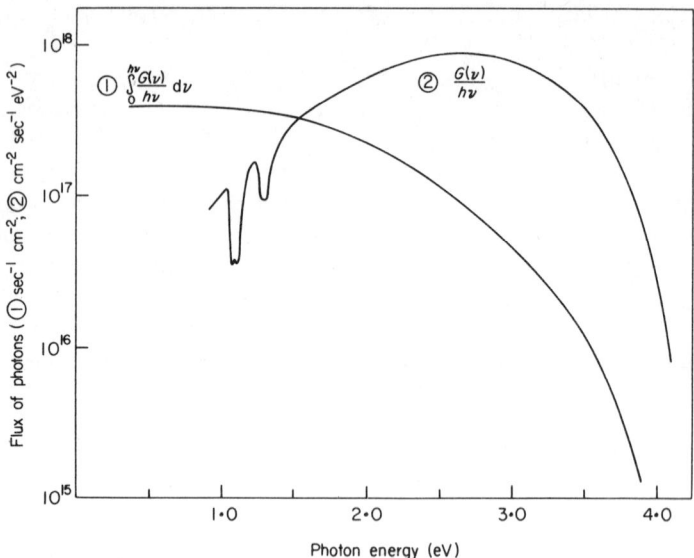

FIGURE 4. The number of photons per electron volt, and the number of photons of energy less than $h\nu$ as a function of photon energy $h\nu$ for the AM1 solar spectrum.

can be written[44]

$$\alpha = \frac{C}{E_g^2}\left[\frac{(h\nu-E_g-E_p)^2}{1-\exp(-E_p/kT)} + \frac{(h\nu-E_g+E_p)^2}{\exp(-E_p/kT)-1}\right] \text{ for } h\nu > (E_g+E_p)$$

where C is a constant and E_p is the phonon energy.

Except in very high purity single crystal semiconductors, there is a significant amount of absorption for photons of energy less than the energy gap of the semiconductor. This is partly due to free carrier absorption, but is mostly due to impurities and defects.

Many impurity and defect states exist close to the conduction band and valence band edges, particularly on thin films. A photon of energy $h\nu = E_g - E_{ID}$ can excite an electron from an impurity or defect state which has an energy E_{ID} above the valence band edge into the conduction band. An electron can be subsequently captured from the valence band into the unoccupied impurity or defect state, leaving a hole in the valence band. It is thus possible for photons of energy less than E_g to create an electron-hole pair. The magnitude of this effect depends on the density of defect and impurity

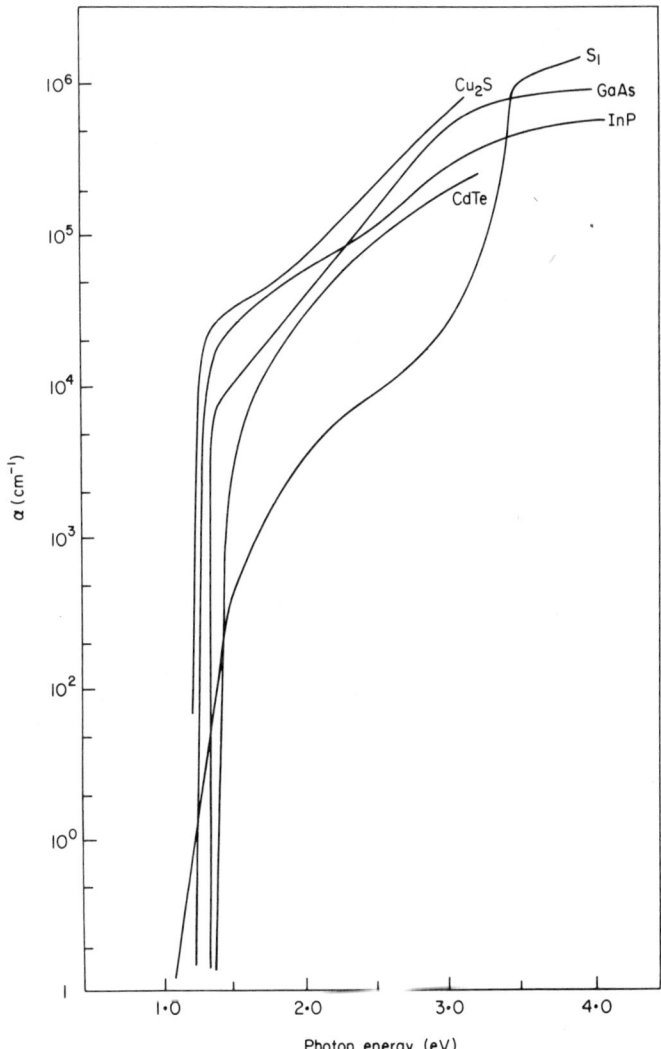

FIGURE 5. The absorption coefficient of a number of semiconductors as a function of photon energy.

states, which is not usually small in thin films. However, the total density of states in the band tails is small compared to the density of states in the bands and the contribution to total electron-hole generation rate is small.

The absorption coefficient can also be affected by the doping levels of the semiconductor. If doping levels are high the lower levels of the conduction band may be substantially occupied or the upper levels of the valence band substantially empty. The probability of photon absorption by these levels is thus reduced, and a shift in the optical absorption edge, known as the Moss-Burstein shift[47], can occur. In order for such a shift to occur, it is not essential that the lowest conduction band states are completely occupied or that the upper valence band states be completely empty, but only that the density of available states corresponding to a particular absorption coefficient should occur at higher energy[47]. We have seen above that solar cells should be highly doped in order to reduce the minority carrier densities and hence attain high efficiency, and the Moss-Burstein shift can occur in some types of cells.

The quantum efficiency is also a function of photon energy. We have seen above that some electron-hole pair generation can occur for photons of energy below the energy gap of the semiconductor, due to excitation from band tails with subsequent thermal activation into the bands, but this effect is small and usually neglected. It is usual to take $q = 0$ for $h\nu < E_g$ and $q = 1$ for $h\nu \geq E_g$.

2.1.1 The efficiency of ideal solar cells

The Carnot efficiency of a cell at 300 K illuminated by the Sun, assumed to be a black body at 6000 K is

$$1 - \frac{300}{6000} = 95\%$$

Landsberg[48] has shown that the constraint on efficiency due to the second law of thermodynamics is also irrelevant since the maximum permitted efficiency approaches 100%.

The number of photons of energy greater than E_g incident on unit area in unit time radiated from a black body at temperature T_s is

$$Q = \frac{2\pi k^3}{h^3 C^2} T_s^3 \int_{E_g/kT_s}^{\infty} \frac{x^2 \, dx}{e^x - 1}$$

Landsberg[48] shows that the efficiency of a solar cell composed of material with energy-gap E_g is

$$\eta_u = \frac{E_g}{kT_s} \int_{E_g/kT_s}^{\infty} \frac{x^2 \, dx}{e^x - 1} \bigg/ \int_0^{\infty} \frac{x^3 \, dx}{e^x - 1}$$

The variation of this efficiency η_u with E_g is shown in Fig. 6. This curve is nearly identical to that derived by Shockley and Queisser[49] using a very similar approach. Since this approach specifically excludes any consideration of loss mechanisms, either intrinsic or extrinsic, it must be regarded as the

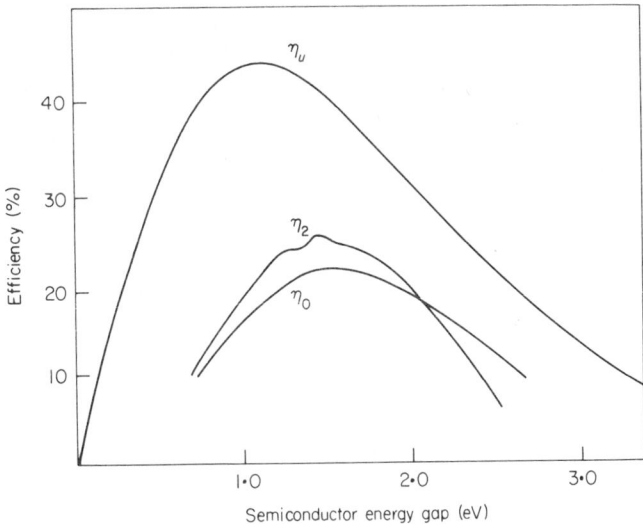

FIGURE 6. The ultimate efficiency η_u, and practical efficiencies for AM0 (η_0) and AM2 (η_2) solar spectra for p–n homojunctions.

ultimate efficiency which could be reached by a perfect homojunction cell. It is not possible for a cell composed of any material presently available to approach this efficiency.

Semi-empirical calculations of efficiency vs. energy-gap have been made by a number of authors[50,51,52] and their results for ideal cells are also shown in Fig. 6. These calculations use values for the charge transport parameters which are attainable in practice, and thus establish a maximum attainable efficiency.

De Vos[53] has shown that the power conversion efficiency which can be achieved by a heterojunction solar cell is higher than that possible for a p–n homojunction or Schottky barrier cell. A practical attainable maximum efficiency of about 32% could be achieved with a material of energy-gap 1.1 eV forming a junction with another semiconductor whose energy-gap was greater than 1.3 eV.

2.2 Practical Solar Cells

In practice, solar cells differ considerably from the ideal case considered in the last section. In order to absorb as much light as possible, the cells are much thicker than the inverse absorption coefficient, i.e. $\alpha x > 1$. Not all of the photogenerated carriers reach the junction, and it is necessary to consider the diffusion and recombination of these. Furthermore junctions are rarely ideal, so the junction equation must be modified. The equivalent circuit of the solar cell must also be modified to include series and shunt resistances which can affect the observed current/voltage characteristics. The effect of these considerations on solar cell efficiency will be briefly discussed in this section. More detailed discussions can be found in recent reviews[54,55,37].

2.2.1 Current collection in p–n junction devices

The minority carrier diffusion equations are

$$\frac{\partial p}{\partial t} = D_p \nabla^2 p - \mu_p \varepsilon \cdot \nabla p - \mu_p p \nabla \cdot \varepsilon + R_{G,p} - \left(\frac{p}{\tau_p} - \frac{p_0}{\tau_{p_0}} \right) \quad (11)$$

and

$$\frac{\partial n}{\partial t} = D_n \nabla^2 n + \mu_n \varepsilon \cdot \nabla n + \mu_n n \nabla \cdot \varepsilon + R_{G,n} - \left(\frac{n}{\tau_n} - \frac{n_0}{\tau_{n_0}} \right) \quad (12)$$

The first term in equations (11) and (12) represents the diffusion of carriers whilst the second term represents transport due mainly to an external applied electric field. The third term describes transport under the influence of the photogenerated internal electric field or Dember field which is the result of the different mobilities of electrons and holes. The fourth term is the rate of generation of electron-hole pairs, whilst the final term describes the recombination rate of the photogenerated pairs.

For a junction with a p-type outer layer, the solution to equations (11) and (12), neglecting electric fields, in the p-region is

$$n - n_0 = \Delta n = A \cosh\left(\frac{x}{L_n}\right) + B \sinh\left(\frac{x}{L_n}\right) - \frac{R_{G,n}\tau_n}{\alpha^2 L_n^2 - 1} \exp(-\alpha x) \quad (13)$$

The two constants, A and B can be found from the two boundary conditions. The first boundary condition takes account of the recombination at the outer surface of the p-layer, which determines the electron distribution near this surface since

$$D_n \frac{\partial (\Delta n)}{\partial x} \bigg|_{x=0} = s_n \Delta n$$

where s_n is the surface recombination velocity.

The second boundary condition concerns the excess electron density at the edge of the depletion layer. The electrons which arrive at the edge of the depletion layer are swept across the junction by the junction field, and the excess electron density can be taken as zero, i.e. $n = 0$ when $x = x_j$, the junction depth.

The excess electron density in the p-region is then

$$\Delta n = \frac{R_{G,n}\tau_n}{\alpha^2 L_n^2 - 1}$$

$$\times \left[\frac{\left(\dfrac{s_n\tau_n}{L_n} + \alpha L_n\right) \sinh\left(\dfrac{x_j - x}{L_n}\right) + \exp(-\alpha x_j)\left(\dfrac{s_n\tau_n}{L_n}\sinh\dfrac{x}{L_n} + \cosh\dfrac{x}{L_n}\right)}{\dfrac{s_n\tau_n}{L_n}\sinh\dfrac{x_j}{L_n} + \cosh\dfrac{x_j}{L_n}} - e^{-\alpha x} \right]$$

and the resulting photocurrent density per unit bandwidth of the incident radiation is

$$J_n = \frac{R_{G,n} e L_n}{\alpha^2 L_n^2 - 1}$$

$$\times \left[\frac{\left(\dfrac{s_n\tau_n}{L_n} + \alpha L_n\right) - e^{-\alpha x_j}\left(\dfrac{s_n\tau_n}{L_n}\cosh\dfrac{x_j}{L_n} + \sinh\dfrac{x_j}{L_n}\right)}{\dfrac{s_n\tau_n}{L_n}\sinh\dfrac{x_j}{L_n} + \cosh\dfrac{x_j}{L_n}} - \alpha L_n e^{-\alpha x_j} \right] \quad (14)$$

This current flows across the junction from the p-layer and contributes to the total photogenerated current.

The boundary conditions for the n-region are

$$\Delta p = 0 \quad \text{when} \quad x = x_j + W$$

and

$$-D_p \frac{\partial(\Delta p)}{\partial x}\bigg|_{x=H} = s_p \Delta p$$

where W is the width of the depletion layers and H is the thickness of the cell. If the back contact of the cell is ohmic then $s_p \to \infty$.

The photocurrent per unit bandwidth due to holes collected at the edge of the depletion layer is then

$$J_p = \frac{R_{GP} e L_p}{\alpha^2 L_p^2 - 1} e^{-\alpha(x_j + W)}$$

$$\times \left[\alpha L_p - \frac{\dfrac{s_p\tau_p}{L_p}\left(\cosh\dfrac{x_n}{L_p} - e^{-\alpha x_n}\right) + \sinh\dfrac{x_n}{L_p} + \alpha L_p e^{-\alpha x_n}}{\dfrac{s_p\tau_p}{L_p}\sinh\dfrac{x_n}{L_p} + \cosh\dfrac{x_n}{L_p}} \right] \quad (15)$$

where $x_n = H - (x_j + W)$ = thickness of base n-region.

For a heterojunction device, the generation rate becomes a little more complicated than for a homojunction, since the two materials have different optical absorption properties, and light of frequency v incident on the n-type material beyond the depletion region is now of intensity

$$G(v)[1-\mathcal{R}(v)]\exp(\alpha_p(v)\{x_j+W_p\})\exp[-\alpha_n(v)W_n]$$

where W_p and W_n are depletion layer widths in p-type and n-type materials.

Absorption of light also takes place within the depletion region. The electric fields in the depletion region are quite large, and it is usually assumed that electrons and holes photogenerated in this region will be swept across the junction before they can recombine. The photocurrent/unit bandwidth is thus

$$J_{DR} = eG(v)[1-\mathcal{R}(v)]e^{-\alpha x_j}[1-e^{-\alpha W}] \qquad (16)$$

The total photocurrent is the sum of these three contributions

$$J_{PC} = J_n + J_p + J_{DR} \qquad (17)$$

This analysis ignores the optical effects which can occur in thin film solar cells. Milnes and Feuch[56] have considered the reflection which occurs at the interface in a semiconductor heterojunction when the outer layer is optically very thick. A complete analysis has been carried out in the author's group by Mr. B. Gandham using techniques similar to those used in the analysis of thin film multi-layer devices. The solar cell is considered to be divided into three layers as before, and the photocurrent from the 1st layer due to absorption in that layer (assumed to be p-type) is

$$J_1 = \frac{e\sigma_1}{2}\left[2\frac{C_2}{L_n}e^{(x_j/L_n)} + \frac{\omega_1}{L_n} - \frac{2P_4 S}{S^2+L_n^{-2}} + \frac{2\alpha P_5}{\alpha^2-L_n^{-2}}\right]F(\lambda) \qquad (18)$$

where $F(\lambda)$ = intensity distribution of light incident on outer surface of cell

$$\sigma_1 = 4\pi\varepsilon_0 n_1 k_1 v = \text{conductivity of 1st layer}$$

Refractive index of 1st layer $N_1 = n_1 - ik_1$

$$S = \frac{4\pi}{\lambda}n_1 \qquad \alpha = \frac{4\pi}{\lambda}k_1$$

$$C_2 = \frac{\omega_1 e^{x_j/L_n} - \left(\dfrac{S_n}{D_n}\omega_3 - \omega_2\right)\bigg/\left(\dfrac{S_n}{D_n}+\dfrac{1}{L_n}\right)}{\left[\left(\dfrac{S_n}{D_n}-\dfrac{1}{L_n}\right)\bigg/\left(\dfrac{S_n}{D_n}+\dfrac{1}{L_n}\right)\right]-e^{2x_j/L_n}}$$

s_n = recombination velocity of outer surface of 1st layer

$$\omega_1 = \frac{P_3}{S^2 + L_n^{-2}} - \frac{P_2}{\alpha^2 - L_n^{-2}}$$

$$\omega_2 = \frac{P_3 \sin(Sx_j)}{S^2 + L_n^{-2}} - \frac{2P_4 S \cos(Sx_j)}{S^2 + L_n^{-2}} + \frac{P_2 \alpha \sinh(\alpha x_j)}{\alpha^2 - L_n^{-2}} + \frac{2P_5 \alpha \cosh(\alpha x_j)}{\alpha^2 - L_n^{-2}}$$

$$\omega_3 = \frac{P_3 \cos(Sx_j)}{S^2 + L_n^{-2}} + \frac{2P_4}{S^2 + L_n^{-2}} - \frac{P_2 \cosh(\alpha x_j)}{\alpha^2 - L_n^{-2}} - \frac{2P_5 \sinh(\alpha x_j)}{\alpha^2 - L_n^{-2}}$$

$$P_1 = [(X_1^2 + Y_1^2)(n_1^2 + k_1^2)]\varepsilon\varepsilon^*$$

The effective refractive index of the assembly under the 1st layer $Z_1 = X_1 - iY_1$

$$P_2 = (1 + P_1) \quad \text{and} \quad P_3 = (1 - P_1)\varepsilon\varepsilon^*$$
$$\cdot P_4 = [(X_1 n_1 - Y_1 k_1)/(n_1^2 + k_1^2)]$$
$$P_5 = [(X_1 n_1 + Y_1 k_1)/(n_1^2 + k_1^2)]\varepsilon\varepsilon^*$$

Similar expressions can be derived for the depletion region and the n-type base layer, and the total photocurrent found.

The results obtained by Gandham will be discussed in Section 4.5 for copper sulphide–cadmiun sulphide solar cells.

The current collection efficiency of a cell can be significantly improved if the carrier diffusion is aided by built-in electric fields due to dopant concentration gradients. A number of authors have considered this case theoretically[57,58,59,60], and silicon cells containing these built-in fields have shown the expected increase in efficiency[37]. In thin film solar cells, however, the control over the doping levels is not adequate to allow a well-defined dopant concentration gradient to be established. It is possible to produce CdS layers which are heavily doped near the substrate and lightly doped near the junction, giving a "back surface field". This does effectively reduce the recombination velocity of the back contact and can increase the current collection efficiency. However, since the diffusion length is usually much smaller than the thickness of the cadmium sulphide layer, the effect must be mainly due to the improved contact to the conducting substrate.

2.2.2 Current/voltage characteristics of homojunctions

The forward current/voltage characteristics of a cell are of great importance, since they largely determine the open circuit voltage and fill-factor of the cell under illumination.

The forward current of a junction in the dark can always be described by a relationship of the form

$$J = \sum_i J_{0i} \left[\exp\left(\frac{eV}{A_i kT}\right) - 1 \right] \tag{19}$$

where J_{0i} is equivalent to a "reverse saturation current" and A_i is a diode factor or ideallity factor. It is usually found that the forward current is the sum of forward currents due to two mechanisms, and equation (19) becomes

$$J = J_{01}\left[\exp\left(\frac{eV}{A_1 kT}\right) - 1\right] + J_{02}\left[\exp\left(\frac{eV}{A_2 kT}\right) - 1\right] \quad (20)$$

Three forward current mechanisms have been identified, and these are usually labelled injection, recombination and tunnelling.

a. Injection currents. The injected current component consists of electrons injected from the *n*-type region over the potential barrier at the junction into the *p*-type region. These excess minority carriers then diffuse and drift away from the junction and eventually recombine either at a surface or in the bulk. There is also a similar injected current of holes injected from the *p*-type region over the junction into the *n*-region. This is the standard situation discussed in most text-books on *p*–*n* homojunctions. Assuming that the doping levels are uniform and that all donors and acceptors are ionised, the forward current is

$$J = J_0\left[\exp\left(\frac{eV}{kT}\right) - 1\right] \quad (21)$$

where

$$J_0 = en_i^2 \left\{ \frac{L_p}{\tau_p N_d} \left[\frac{\frac{S_p \tau_p}{L_p} \cosh\left(\frac{x_n}{L_p}\right) + \sinh\left(\frac{x_n}{L_p}\right)}{\frac{S_p \tau_p}{L_p} \sinh\left(\frac{x_n}{L_p}\right) + \cosh\left(\frac{x_n}{L_p}\right)} \right] \right.$$

$$\left. + \frac{L_n}{\tau_n N_a} \left[\frac{\frac{S_n \tau_n}{L_n} \cosh\left(\frac{x_p}{L_n}\right) + \sinh\left(\frac{x_p}{L_n}\right)}{\frac{S_n \tau_n}{L_n} \sinh\left(\frac{x_p}{L_n}\right) + \cosh\left(\frac{x_p}{L_n}\right)} \right] \right\} \quad (22)$$

x_n and x_p are the thicknesses of *n*- and *p*-layers outside depletion region, N_d and N_a are donor and acceptor concentrations in *n*- and *p*-regions respectively, and n_i is the intrinsic carrier concentration.

The maximum photovoltage which is theoretically possible is equal to the junction barrier height V_D at zero bias where

$$V_D = \frac{kT}{e} \ln\left(\frac{N_d N_a}{n_i^2}\right) \quad (23)$$

If the dopant concentrations are not uniform, then built-in electric fields will exist. Ellis and Moss[57] have derived an expression for the reverse saturation current assuming uniform electric fields in the n- and p-regions, but this is not a situation which can be guaranteed in thin film solar cells.

When the base region is highly doped near the back surface contact, the dark current can be reduced due to an effective reduction in the surface recombination velocity at this contact. The expression for J_0 has been shown[61] to be identical to equation (22) except that the surface recombination velocity s_p for the n-type base region is replaced by an effective velocity s_e where

$$s_e = \frac{N_d D_p^*}{N_d^* D_p} \left[\frac{\frac{S_p L_p^*}{D_p^*} \cosh\left(\frac{x_n^*}{L_p^*}\right) + \sinh\left(\frac{x_n^*}{L_p^*}\right)}{\cosh\left(\frac{x_n^*}{L_p^*}\right) + \frac{S_p L_p^*}{D_p} \sinh\left(\frac{x_n^*}{L_p^*}\right)} \right] \quad (24)$$

and the starred parameters refer to the highly doped region.

b. *Recombination currents.* Impurity levels within the energy-gap can promote electron-hole recombination, and this can significantly affect both the reverse saturation current and the diode factor. The recombination rate depends on the carrier densities, and in the centre of the depletion region of a p–n homojunction.

$$n = p = n_i \exp\left(\frac{eV}{2kT}\right)$$

The recombination current density J_{rg} can thus be expected also to be proportional to $\exp(eV/2kT)$. Sah et al.[62] derived expressions for the forward current of a symmetrical p–n homojunction with recombination centres in the centre of the energy-gap having approximately equal capture cross-sections for electrons and holes. Evans and Landsberg[63] extended this treatment to include Auger recombination. Choo[64] considered the general case of non-symmetrical junctions with recombination centres away from the centre of the energy-gap and with non-equal capture cross-sections. Both Sah et al.[62] and Choo[64] find that the forward dark current due to recombination is

$$J_{rg} = \frac{2 n_i W}{(\tau_{po} \tau_{no})^{1/2}} \frac{\sinh(eV/2kT)}{(V_D - V)/kT} f(b) \quad (25)$$

where V_D is the built-in voltage, W is the thickness of the depletion layer and τ_{po} and τ_{no} are the minority carrier lifetimes on the n and p sides of the junction. The function $f(b)$ is a complicated function of bias, trapping levels, dopant concentrations and carrier lifetimes. Choo[64] shows that the recombination current at moderate forward bias shows a diode factor of two, but this current saturates at forward bias values of about 5 to $15kT/e$, i.e. 0.1 to 0.3 volt depending on dopant concentrations, trap depths and carrier lifetimes.

These analyses are valuable in showing how recombination in the junction region can give rise to a diode factor of two, and larger dark currents than would be expected from diffusion theory. The dark current/voltage characteristics of actual $p-n$ homojunction solar cells are usually much more complex than would be predicted by assuming the dark currents to be the sum of injected and recombination currents given in equations (22) and (25). This is probably due to the simplifying assumptions made in both derivations such as linear potential distribution in a well-defined depletion layer, which are necessary if analytical solutions are to be found.

c. Tunnelling currents. Quantum mechanical tunnelling can occur when an empty electron state is separated from a filled electron state by a short distance and a relatively small potential barrier. Tunnelling occurs in highly doped $p-n$ homojunctions, particularly in reverse bias, but is usually negligible in the operation of conventional single crystal solar cells.

Tunnelling is a very important, and often dominant, current mechanism in heterojunctions and Schottky barrier cells. Since these junctions have characteristics which are somewhat different from those of $p-n$ homojunctions, and are used in the majority of thin film solar cells, a brief description of them will be useful at this point.

2.2.3 Semiconductor heterojunctions

A semiconductor heterojunction is one formed by two dissimilar semiconducting materials. If the materials are both n-type or both p-type they form an isotype heterojunction, whilst materials of different conductivity types form an anisotype heterojunction. The junctions may be either graded or abrupt, i.e. widths of many diffusion lengths or a few atomic spacings.

In a heterojunction, the electron affinities and photoelectric thresholds are different for the two materials. This leads to discontinuities in the conduction bands and valence bands at the junction. The first model of a heterojunction was proposed by Anderson[65] and is shown in Fig. 7. The height of the "spike" in the conduction band ΔE_c is equal to the difference in electron affinities $\chi_2 - \chi_1$ whilst the height of the "notch" in the valence band is equal to the difference in photoelectric thresholds $\Phi_2 - \Phi_1$.

Anderson derived the current/voltage characteristics of this junction in a manner similar to that used to derive the injection current of an ideal $p-n$ homojunction, and showed that the forward current density is given by

$$J = eXN_{d2}\left(\frac{D_{n1}}{\tau_{n1}}\right)^{1/2} \exp\left(-\frac{eV_{p2}}{kT}\right)\left[\exp\left(\frac{eV_2}{kT}\right) - \exp\left(\frac{eV_1}{kT}\right)\right] \quad (26)$$

where V_1 and V_2 are the fractions of the applied voltage, $V = V_1 + V_2$, dropped across the p and n sides of the junction, N_{d2} is the donor concentration in the n-region, D_{n1} and τ_{n1} are the electron diffusion coefficient and lifetime in the p-

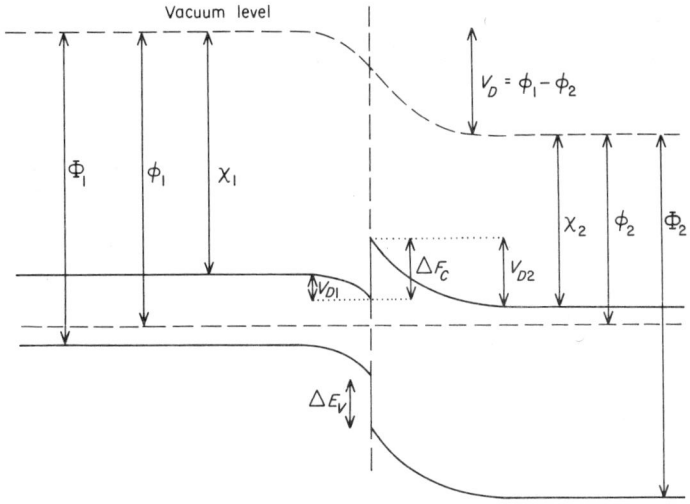

FIGURE 7. The Anderson[65] model of a heterojunction showing the electron affinity χ, the work function ϕ and photo-electric threshold Φ. The diffusion potentials V_{D1} and V_{D2} and the discontinuities in the conduction band ($\Delta E_c = \chi_1 - \chi_2$) and valence band ($\Delta E_v = \Phi_1 - \Phi_2$) are also shown.

region, and X is the transmission coefficient for electrons across the junction.

This model ignores the states which inevitably occur at the interface which give rise to recombination and tunnelling currents.

Dolega[66] considered heterojunctions formed between semiconductors having different crystal lattices. It is possible to consider the junction as the interface between two surfaces, as shown in Fig. 8, each surface having its own surface state density distribution. The trapped charges normally associated with surfaces can recombine due to the overlap of the surface state distributions. Thus fast recombination occurs at the interface, of electrons and holes which reach the recombination centres by thermal activation.

Dolega's expression for forward current has been re-written by Van Opdorp[67] in the form

$$J = B \exp\left(-\frac{eV_D}{kT}\right)\left[\exp\left(\frac{eV}{AkT}\right) - 1\right] \quad (27)$$

B is a weak function of temperature and the value of A depends on the ratio of the density of imperfections in the two materials, but $1 < A < 2$. The simplest

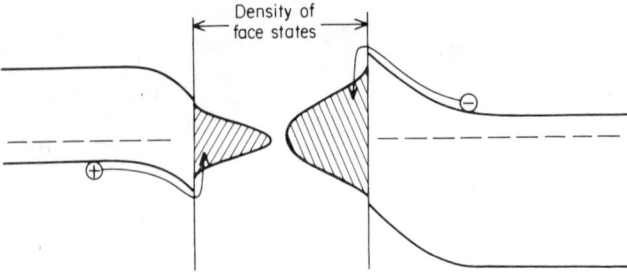

FIGURE 8. The Dolega[66] model of a heterojunction assuming emission of carriers into interface states, with subsequent recombination.

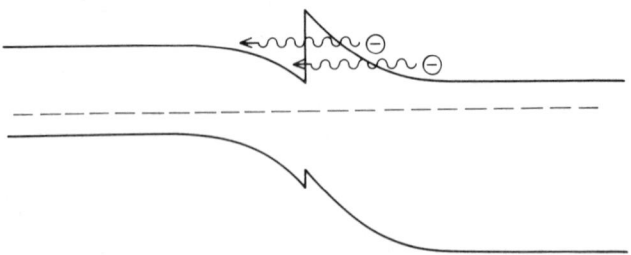

FIGURE 9. A tunnelling model of a heterojunction.

case of tunnelling in a heterojunction is shown in Fig. 9. Electrons tunnel through the spike to the conduction band of the *p*-type material, where they subsequently recombine either close to the interface or after thermal excitation out of the accumulation region[68]. The tunnelling current has been found to be[69]

$$J = J_{s0} \exp\left(\frac{T}{T_0}\right) \exp\left(\frac{V}{V_0}\right) \qquad (28)$$

The current/voltage characteristics of some heterojunctions are quite well described by equation (28), but for the majority of heterojunctions, the current seems to be due to a mixture of tunnelling and recombination.

Riben and Feucht[70] proposed a model containing both tunnelling and recombination at the junction, as shown in Fig. 10. Electrons from the *n*-type

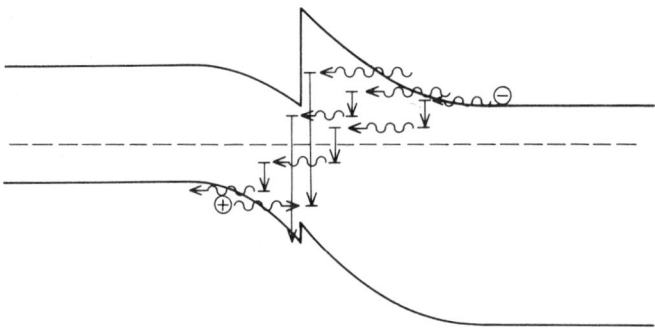

FIGURE 10. Tunnelling–recombination in a heterojunction.

material can tunnel through the spike to states within the energy-gap of the p-region and there recombine, or electrons can tunnel through to interface states and by a series of thermal de-excitation and tunnelling steps can reach the valence band of the p-region. If tunnelling originates from the bottom of the conduction band or the top of the valence band, then the dark forward current is of the form[71]

$$J = B\exp[-\alpha_T(V_D - V)] \tag{29}$$

where

$$\alpha_T = \frac{4}{3\hbar}\left(\frac{m_e\varepsilon}{N_{d2}}\right)^{1/2} \tag{30}$$

This expression for α_T assumes a single step process and a linear barrier, and agreement with experiment is not usually good[72]. More satisfactory agreement with experimental results is obtained[70,73] if a multi-step tunnelling process is considered. If the tunnelling probability of Franz[74] is assumed then the current can be written

$$J = C\exp[-\alpha_{MT}R^{-1/2}(V_D - V)] \tag{31}$$

where

$$\alpha_{MT} = \frac{8}{3\hbar}\left(\frac{m_e\varepsilon}{N_{d2}}\right)^{1/2} \tag{32}$$

and R is the number of tunnelling steps.

For a barrier of general shape and a multi-step tunnelling process, the value of α_{MT} depends on the shape of the barrier and the interface state density distribution. Experimentally, α_{MT} is often found to be between 20 and 30, which would give semilog current/voltage curves with a slope equivalent to a diode factor of between 2.0 and 1.3. However, the temperature dependence of the forward current is very weak in the tunnelling recombination process, so these currents can be distinguished from thermal currents with diode factors in this range by measuring the temperature dependence.

2.2.4 Schottky barriers

The standard analysis of the forward current in a Schottky barrier leads to

$$J = A^* T^2 \exp\left(-\frac{e\phi_B}{kT}\right)\left[\exp\left(\frac{eV}{AkT}\right) - 1\right] \tag{33}$$

where A^* is a modified Richardson constant and the Schottky barrier height ϕ_B is approximately equal to the difference between the metal function and the electron affinity of the semiconductor. It is found experimentally that the less ionic semiconductors have barrier heights which are almost independent of the metal work function. The barrier height ϕ_B can be written as

$$\phi_B = \mathscr{S}\phi_M + \text{constant} \tag{34}$$

The constant $\mathscr{S}$ depends on the electronegativity difference of the elements in the semiconductor, and has been shown[75] to be given by

$$\mathscr{S} = (1 + 4\pi e^2 D_{FL} \delta)^{-1} \tag{35}$$

where D_{FL} is the density of surface states near the Fermi level and δ is the thickness of the interfacial layer. For semiconductors more covalent than CdTe or GaAs, the value of $\mathscr{S}$ is less than 0.2[75], so that the barrier height depends only weakly on the metal work function.

A careful analysis of ideal Schottky barrier cells has been carried out by McQuat and Pulfrey[76]. Their results show the great importance of providing as large a barrier height as possible to reduce J_0 and hence increase V_{OC}. This is difficult for semiconductors with an energy-gap of around 1.4 eV since these are materials of low ionicity, and the barrier heights of surface state dominated materials tend to be about 2/3 of the energy-gap[78]. This difficulty can be overcome by using non-ideal Schottky barriers with a very thin insulating layer between the metal and the semiconductor. This can produce a junction with a high diode factor which leads to higher open circuit voltages, but lower fill-factors. Schottky barrier cells which have given the highest efficiencies have all used this type of MIS structure[78,79].

Card and Rhoderick[80] have analysed the metal-insulator-semiconductor junction where surface states at the insulator-semiconductor interface do not necessarily interact with the electrons in the metal. This analysis has been extended by Card and Yang[81] to illuminated MIS junctions and they point out that the charges on the interface states are not the same in the dark junction under forward bias as in an illuminated junction. The open circuit voltage under illumination does not depend on the diode factor found in the dark I/V characteristics, but is the sum of an ideal term and a term containing the tunnelling transmission coefficient.

A general analysis of MIS junctions has been made by Fonash[82] for semiconductors with different oxide layer thickness and metal work functions. This analysis by Fonash[82] agrees with the results obtained by Card and Yang[81] and Shewchun et al.[83] for the cases which they studied, but he also pointed out that by having interface states which have a high capture cross-section for electrons and a low cross-section for holes, it is possible to produce open circuit voltages of larger magnitude than the built-in diffusion voltage.

2.3 The Characteristics of Practical Cells

2.3.1 Equivalent circuit analysis

We have seen that the simple theory of solar cells put forward in Section 2.1 which assumed that the cell was an ideal p–n homojunction must be modified. The photocurrent is a complicated function of carrier transport properties, layer thicknesses, material properties and dopant concentrations as shown in equations (14) to (18). The dark characteristics are complicated by the existence of three possible current carrying processes, injection, recombination and tunnelling, at least two of which are present in almost all solar cells. There are also technical problems in cell manufacture in that the photocurrent must be extracted from the upper layer of the cell whilst causing minimum obstruction to the incident light. The sheet resistance of the upper layer combined with the bulk resistance of the base layer gives rise to a resistance in series with the cell. Some leakage currents are inevitable and these can be represented by a shunt resistance. The solar cell is then more closely represented by an equivalent circuit such as that shown in Fig. 11.

Taking the convention of forward currents as positive, Kirchoff's law gives

$$J_T = J_d - J_L + J_{SH}$$

Representing the forward diode current J_d by just one characteristic

$$J_d = J_0 \left[\exp\left(\frac{eV_d}{AkT}\right) - 1 \right] = J_0 [\exp(\lambda' V_d) - 1]$$

The voltage V_d across the diode is greater than the voltage V_T at the terminal since

$$V_d = V_T + J_S R_S + \frac{R_S}{R_{SH}} V_T$$

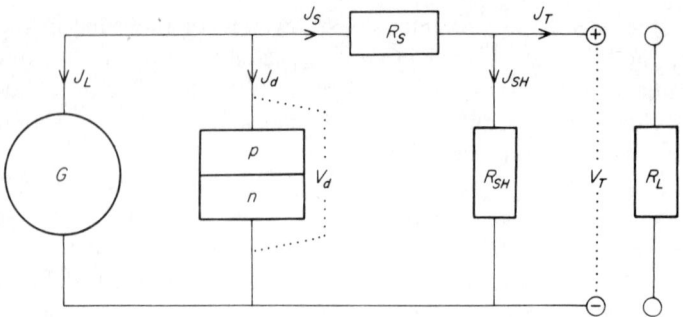

FIGURE 11. Equivalent circuit of a solar cell including series and shunt resistances.

Thus

$$J_T = J_0\left[\exp \lambda'\left(V_T + J_S R_S + \frac{R_S}{R_{SH}} V_T\right) - 1\right] + \frac{V_T}{R_{SH}} - J_L \qquad (36)$$

When the terminal voltage is zero,

$$J_T(V_T = 0) = J_0[\exp(\lambda' J_S R_S) - 1] - J_L$$

It can be seen that the short circuit current is reduced by the series resistance, but is not affected by the shunt resistance.

When the terminal current is zero,

$$J_0\left[\exp \lambda'\left(V_{OC} - J_S R_S + \frac{R_S}{R_{SH}} V_{OC}\right) - 1\right] + \frac{V_{OC}}{R_{SH}} = J_L$$

Since $J_S \approx 0$ when $J_T = 0$

$$V_{OC} = \frac{AkT}{e} \frac{R_{SH}}{R_S + R_{SH}} \ln\left[\frac{1}{J_0}\left(J_L - \frac{V_{OC}}{R_{SH}}\right) + 1\right] \qquad (37)$$

The open circuit voltage is little affected by the series resistance since $R_{SH} \gg R_S$, but is reduced when the shunt resistance falls to values comparable to V_{OC}/J_L.

The shunt and series resistances also affect the shape of the current/voltage characteristics (see Fig. 12). Differentiating equation (36) we find

$$dJ_T = \lambda' J_d\left[dV_d\left(1 + \frac{R_S}{R_{SH}}\right) - dJ_S R_S\right] + \frac{dV_T}{R_{SH}} \qquad (38)$$

In the short circuit condition $J_d \approx 0$

$$\left(\frac{dJ_T}{dV_T}\right)_{V_T = 0} = \frac{1}{R_{SH}} \qquad (39)$$

Re-writing equation (38), and putting $J_S = J_T$

$$dJ_T(1+\lambda'J_dR_S) = dV_T\left[\frac{1}{R_{SH}} + \lambda'J_d\left(1+\frac{R_S}{R_{SH}}\right)\right] \quad (40)$$

In open circuit conditions $\lambda'J_d \gg 1$ or $1/R_{SH}$ and so

$$\left(\frac{dJ_T}{dV_T}\right)_{J_T=0} = \frac{1}{R_S}\left(1+\frac{R_S}{R_{SH}}\right) \approx \frac{1}{R_S} \quad (41)$$

Using an equivalent circuit with the series resistance between the terminal and the shunt resistance, Bryant and Glew[84] have shown that

$$R_S = \left(\frac{dV_T}{dJ_T}\right)_{J_T=0} - \left(\lambda'J_d(V_{OC}) + \frac{1}{R_{SH}}\right)^{-1} \quad (42)$$

and that this series resistance can be negative under conditions of low illumination in copper sulphide–cadmium sulphide solar cells[84]. Equivalent circuits of the types shown in Fig. 11 can not therefore represent the operation of a solar cell under all conditions.

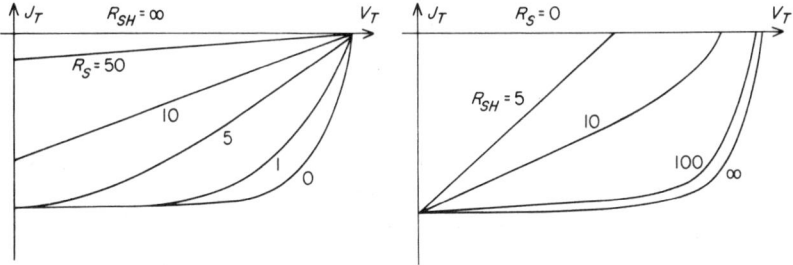

FIGURE 12. Current/voltage characteristics showing the effects of increasing series resistance or decreasing shunt resistance.

Equation (37) suggests that the open circuit voltage should increase linearly with the diode factor. Such an effect can be observed in some Schottky barrier cells[79], but for p–n homojunctions and to a lesser extent for heterojunctions, the current mechanisms which give rise to diode factors greater than unity also give rise to larger values of J_0, thus reducing the ratio J_L/J_0. The nett result is that the open circuit voltage alters little with diode factor. The fill-factor, and hence the current and voltage at maximum power output are all reduced as the diode factor increases. Figure 13 shows the variation of maximum power parameters and fill-factor as a function of $\lambda'V_{OC}$ assuming that the effects of shunt and series resistances are negligible. The advantages of high open circuit voltage and low diode factor can be clearly seen.

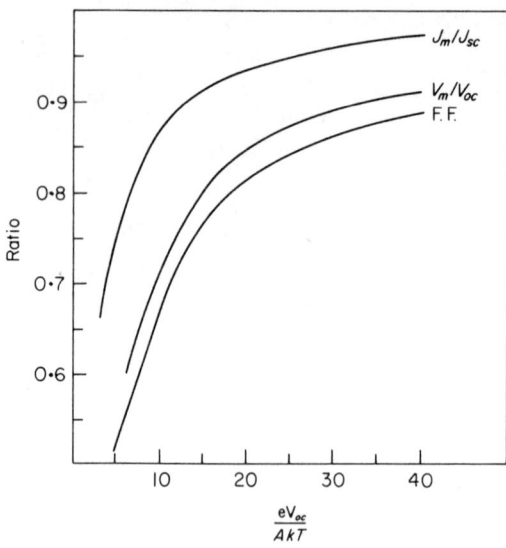

FIGURE 13. The power output parameters as a function of the normalised open-circuit voltage.

2.3.2 Origin of shunt and series resistance

The shunt resistance in the equivalent circuit arises because of current leakage paths within the solar cell. The outer boundaries of a junction device will inevitably permit some leakage of current, but even in single crystal cells, some areas of the cell are found to have much higher leakage currents than average[85]. Surface damage during manufacture seems to play a significant role in reducing the shunt resistance in these small areas[85].

In thin film devices the possibilities for current leakage are significantly increased due to the structure of the films. As discussed in the next section, thin films usually have a columnar structure with a junction formed on each crystallite. Unless great care is taken in the deposition of the base film to ensure high packing density, and in the formation of the junction, leakage paths can exist down the crystallite boundaries.

Surface damage can also play an important role in reducing the shunt resistance in thin film cells, and the presence of pin-holes in the outer junction layer of cells formed by sequential evaporation can lead to shorting between the collecting grid and the base semiconductor layer.

The series resistance has a number of components. The back surface contact can introduce a resistance in series with the photocurrent. This effect can be reduced to negligible proportions by correct contact production techniques, although it remains as a possible source of reduced efficiency in a small fraction of cells during a production run. The current must be conducted through the base semiconducting layer, which is usually of 1 kΩ cm to 10 Ω cm volume resistivity, and this introduces a series resistance of 1.0 to 0.01 Ω cm^2 for a cell of 1 cm^2 area and thickness of 10 μm. This resistance decreases as cell area increases, and for cells of typical production sizes—20 to 100 cm^2—the series resistance due to this effect will also be very small.

The major contribution to series resistance is due to the collection of the photocurrent at the outer surface. The current collecting grid must be of the highest possible optical transparency consistent with the lowest possible resistance to the passage of current from the outer semiconductor layer to the current output lead.

Wolf[86] considered the resistance of the outer semiconductor layer of a cell feeding into metal contacts in the form of "fingers". He derived self-consistent equations relating the sheet resistances of the semiconductor layer and metal contacts, and the width, height and spacing of the fingers, giving maximum collected photocurrent. Handy[87] has produced an equivalent circuit containing all sources of series resistance, and derived equations for the lumped series resistance as a function of all these components. He derived equations for the resistence of the upper layer for carriers flowing to the grid and contact strips of an upper electrode with fingers, and calculated the lumped series resistances for some practical solar cells. The results compared extremely well with the series resistances determined experimentally for silicon cells[87].

The interpretation of the current/voltage characteristics of thin film solar cells in terms of equivalent circuits of the type shown in Fig. 11 is not always appropriate, since other factors can give rise to a slope in the characteristics in short circuit conditions, which could be wrongly attributed to a low shunt resistance. The series resistance derived exactly from equation (40) is

$$R_S = \frac{1 + \lambda' R_{SH} J_d(V_{OC})}{\lambda' R_{SH} J_d(V_{OC})} \left[\left(\frac{dV_T}{dJ_1}\right)_{J_T=0} - \left(\lambda' J_d(V_{OC}) + \frac{1}{R_{SH}}\right)^{-1} \right] \quad (43)$$

which is nearly identical to the expression derived by Bryant and Glew[84]. The series resistance is thus not simply related to the slope of the current/voltage characteristic and the work of Bryant and Glew[84], in which negative series resistances at low levels of illumination were found, shows that real cells are more complex than the equivalent circuit would suggest.

3. Structure of Thin Films

The deposition of thin films has been discussed in Chapter 2 of this book, and deposition techniques will not be considered in detail except where the technical details influence the properties of the deposited films.

The techniques of thin film deposition used in producing thin film solar cells divide broadly into two areas: physical vapour deposition (PVD) and chemical vapour deposition (CVD). Physical vapour deposition uses mainly the techniques of thermal or electron-beam evaporation or of sputtering to produce a vapour of the material to be deposited which subsequently condenses on a nearby substrate[88]. Chemical vapour deposition occurs when a gas containing the elements composing the material to be deposited is introduced into a reaction chamber, and deposition occurs onto a substrate placed at the point where the reaction producing the required material takes place[89].

Any semiconductor can in principal be deposited by either PVD or CVD and the choice of technique is governed by the technical difficulties and the cost of producing films with the required properties. In general, the compound semiconductors of groups II/VIb are deposited by PVD whilst those of III/Vb and the group IV element semiconductors are deposited by CVD, but this merely reflects the techniques used at present by the majority of workers and exceptions often occur.

The electrical and optical properties of thin films of a particular semiconductor can vary very widely depending on the mode of deposition, the deposition parameters and the type of substrate used. It is usually, though not always[88], the case that PVD films are polycrystalline, and the existence of the large area of grain boundaries can significantly affect their optical and electrical properties. CVD films can also be polycrystalline, although this technique is usually used with a single crystal substrate to attain epitaxial growth of the film in the form of a single crystal. If thin film solar cells are to produce power at reasonably low cost, it is almost inevitable that the substrates will not be single crystals, and the films deposited by either PVD or CVD will be polycrystalline. It is useful at this stage therefore to consider the effect on thin film properties of their polycrystalline nature, and since the majority of solar cells are at present produced by physical vapour deposition, this technique will be assumed throughout this discussion. The special problems which occur in chemical vapour deposition will be considered during the discussion of those thin film solar cells, mainly of silicon, where this technique is used.

3.1 Influence of the Deposition on the Properties of Thin Films

There are many factors which influence the opto-electronic properties of thin films, in fact every aspect of the deposition process has some influence on the quality of the film, but some are of paramount importance.

The substrate plays an important role in thin film solar cells. It provides the back-surface contact and in some heterojunction cells its optical properties can influence the overall efficiency of the cell, and it also plays an important role in determining the structure of the thin film. The substrate thus needs to be a good electrical conductor, and it is usual to use a metallic or a metallised substrate, whilst the substrate surface should be such as to promote the growth of high quality films. Since most films are deposited onto a heated substrate, it is important that the thermal expansion coefficients of the substrate and the film should be as nearly equal as possible. It is also important that diffusion of material from the substrate into the growing film should not cause any degradation of either the bulk properties of the film or of the back-surface contact.

The surface structure and temperature of the substrate are very important in the initial stages of the growth of a thin film[90]. When vapour evaporated from a source impinges on the substrate, some will condense on the substrate, whilst some will remain free. The probability that a molecule will remain permanently on the substrate surface is the sticking coefficient (see Chapter 2). After condensation on the substrate, the molecule moves around the surface until it finally arrives at its lowest energy configuration. This usually occurs at specific sites on the substrate surface, known as nucleation sites, at which molecules of the condensate are most tightly bound to the surface. As further molecules of condensate arrive at the nucleation site, a monomolecular layer occurs in which the atoms are arranged in the lowest possible energy configuration. Unless the substrate surface structure closely matches the structure of a single crystal of the deposited material, the lowest energy configuration of atoms arriving at the nucleating layer will not in general be the same as their crystalline configuration, and layers from adjacent nucleation sites will not have identical orientation. Kirk and Raven[91] have shown that the structure of zinc selenide films on passivated single crystal germanium substrates depends on the free energy of formation of the zinc selenide, and have shown generally that the free energy of formation depends on the source and substrate temperatures and on the impingement rate. They found that epitaxial films of zinc selenide would form only for a well-defined range of free energy values, with polycrystalline films being formed for free energy values outside this range. With correct choice of source and substrate temperatures, Kirk and Raven[91] deposited single crystal films of zinc selenide at quite high deposition rates.

With careful deposition on non-single crystal substrates such as mica, it is possible to prepare films in which the diameter of the crystallites is much larger than the film thickness[88,92], but with the substrates used in thin film solar cells it is usually found that the crystallite size is less than the thickness of the film. Figure 14 shows cadmium sulphide films of 10 μm thickness on molybdenum substrates. The columnar structure is quite apparent and the average size of the crystallites is between 1 μm and 2 μm. These films consist of columnar crystallites which are quite tightly packed, i.e. the boundaries of each

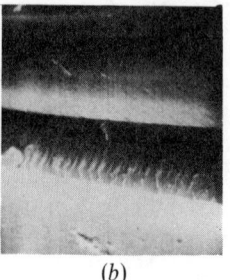

(a)   (b)

FIGURE 14. Micrographs of cadmium sulphide: (a) an evaporated, 15 μm thick film and (b) 2 μm thick sputtered film.

crystallite are almost everywhere in contact with the boundaries of neighbouring crystallites. The degree of packing of a film can be expressed by the packing density which is the total volume of the crystallites divided by (area of film × thickness of film), or by the porosity which is equal to 1-packing density. The packing densities of a number of materials on glass substrates have been measured as a function of substrate temperature[93]. It is found that the packing densities of many optically useful materials increase from about 0.7–0.9 to unity as the substrate temperature increases from about 30°C to 300°C. Most materials, except some oxides, exhibit a packing density of unity when deposited on substrates at a temperature of 300°C or higher[93].

A thin film which has pores reaching from the outer to the inner surface presents many pathways for shorting out the junction and for the diffusion of impurities into the bulk of the film, leading to degradation of the optical and electrical properties of the final solar cells. It is important therefore to deposit films on substrates held at temperatures near to 300°C, to reduce the porosity of the films if this is possible without causing degradation of the crystallinity of the material within the grains.

Both the electrical and optical properties of thin films improve as the temperature of the substrate on which they are deposited is increased. Most films have very small crystallites and are frequently non-stoichiometric when deposited at room temperature, but properties such as the sharpness of the optical absorption edge, the charge carrier mobilities and diffusion lengths all increase with increasing substrate temperature, whilst charge carrier densities decrease[88,92,94]. There is an optimum substrate temperature for these properties, since too high a deposition temperature often results in non-stoichiometric films, and for vacuum deposited films, this optimum is frequently between 200°C and 300°C[95].

The condition of the substrate surface determines the initial nucleation of the deposited film. It is obviously essential that the substrate be uniformly clean and free from any extraneous matter, and this is not always easy to arrange in vacuum deposition. It is not difficult to ensure that the substrate is clean and dust free when it is put into the vacuum chamber, but the initial pump down causes air currents in the vacuum chamber which deposit dust on the substrate. The presence of dust particles on the substrate during deposition leads to pin-holes in the film which can result in shorting of the junction in the completed solar cell, or at best a reduced shunt resistance. This can be avoided if the initial pump down is very slow, taking say one or two hours to reach a pressure of 1 torr. It is then possible to produce pin-hole-free films of thickness down to 0.1 μm. Active cleaning of the substrate after pump-down is also satisfactory, using for instance ion-beam cleaning or sputter etching. Another cause of pin-holes is spattering of particles from the evaporation source, which can be reduced, but not completely eliminated by careful design of the source[31]. Using standard pumping rates, it is found that films must be of thickness greater than a few microns before the density of pin-holes falls below 0.1 cm^{-2}[96].

The uniform cleanliness of the substrate is of vital importance if uniform, high quality thin films are to be deposited.

The effect of even slight contamination can be quite dramatic, as shown in Fig. 15. This shows the effect on the resistivity of a copper sulphide layer plated on a CdS film which was deposited on a zinc coated copper substrate. The high resistance regions are the result of the contact of gloved fingers with the cleaned copper substrate before insertion into the vacuum chamber. The reason why the finger prints cause the high sheet resistance is not certain, but it seems likely that it is a result of extra nucleation sites due to etching of the substrate by finger grease, resulting in smaller cadmium sulphide crystallites and smaller copper sulphide grains. This experiment provides a dramatic illustration of the importance of correct substrate preparation if the deposited films are to have uniform electrical properties.

The columnar film structure shown in Fig. 14 permits electric current to be conducted from the top to bottom surface of the film without having to cross crystallite boundaries. It is important that the crystallites should extend throughout the film so as to avoid the potential barriers which would be introduced by the presence of crystallite boundaries, since these crystallite boundaries give rise to diode-like behaviour[97].

Figure 16 shows an example of a film of cadmium sulphide deposited at high rate on to a silver/zinc substrate at 200°C in which the formation of intermediate grain boundaries and of voids can be clearly seen. It is possible to anneal thin films to promote crystallite growth[88,98] and filling of voids[99], but quite high temperatures are usually required to promote significant growth or filling of voids, since these processes tend to rely on diffusion of interstitials.

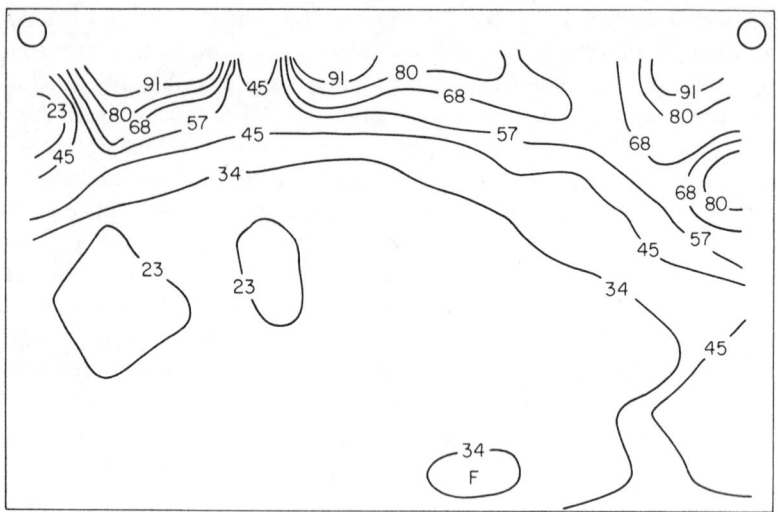

FIGURE 15. The variation of sheet resistivity with position for a cuprous sulphide layer on a cadmium sulphide film, as found by the Institute of Energy Conversion, University of Delaware[96].

FIGURE 16. Micrograph of a 10 μm thick film of cadmium sulphide evaporated at a high rate onto a cool substrate. The voids in the film can be clearly seen.

It is important for the efficient operation of thin film solar cells that the crystallographic axes be correctly aligned with respect to the substrate. Most thin films have a preferred axis along which the crystallites grow, and if deposition takes place at non-normal incidence, this axis, and the direction of crystallite growth, are at an angle which is intermediate between the substrate normal and the angle of incidence and which is dependent on the film thickness[100]. The anomalous photovoltaic effect in CdTe is a striking illustration of this effect. If cadmium telluride is deposited on to a substrate which is tilted so that the normal to the substrate is at an angle θ to the line joining the source and the centre of the substrate, the photovoltage is found to vary with θ and can reverse in sign. The efficiency of cadmium sulphide–copper sulphide cells is also adversely affected by deposition of the cadmium sulphide at even quite small angles of normal[101], and this raises serious problems in the deposition of large area devices.

3.2 The Influence of Polycrystallinity on the Electrical Properties of Thin Films

The effects of the polycrystalline nature of most thin films on their optoelectronic properties have been investigated by a number of workers. Bednarczyk et al.[102] have considered the effect of surface states at the crystallite boundaries and find that the band-bending caused by the filling of these traps can extend for a considerable distance into the crystallite. For crystallites at 300 K, donor concentrations of 10^{18} cm^{-3} and surface state densities less than about 10^{12} cm^{-2}, the barrier height at the crystallite boundaries is only about $5kT$ (~ 0.1 eV) but for surface state densities above 10^{13} cm^{-2}, the barrier height is greater than $20kT$ (~ 0.5 eV). Bednarczyk et al. showed that the effect of the conduction band bending is to reduce the electron concentration near the surface by many orders of magnitude below that in the centre of the crystallite.

Snejdar and Jerhot[103] have considered the currents which could flow across the intergrain boundary by either tunnelling, thermionic emission or by ohmic conduction. They assumed that the grain boundary could be treated as a heterojunction and related the physical parameters affecting the conductivity to elements in an equivalent electrical resistance network. They also showed that earlier models of conduction in polycrystalline films could be reduced to special cases of their general model. In particular thermionic emission over grain boundaries in materials such as the cadmium chalcogenides leads to an expression for the conductivity σ where

$$\sigma = \frac{e^2 n_G \langle v_{thG} \rangle}{4 N_G kT} \exp\left(-\frac{\phi_{GB}}{kT}\right) \quad (44)$$

where $\langle v_{thG} \rangle$ is the mean thermal velocity of carriers in the grains and N_G is the number of grains encountered by the current path.

The grain boundary barrier height is given by

$$\phi_{GB} = kT \ln\left(\frac{n_G}{n_B}\right) \quad (45)$$

where n_G and n_B are the electron densities in the grain and between grains. These expressions are identical to the expressions derived by Petritz[104] and Waxman et al.[105].

Kuznicki[106] has developed the electrical network model of polycrystalline films based on the existence of resistance within the grain, a different resistance within the intergrain material, and a diode between the two. An equivalent circuit can thus be constructed and analysed to find the current/voltage relationships for cross-film current flow. Kuznicki performed this analysis for cadmium selenide films and compared the derived current/voltage characteristics and the voltage distributions with those which he found experimentally. The forms of the distributions and I/V characteristics were well reproduced by the network model although discrepancies were present as might be expected. Kuznicki[106] included in his network a series resistance due to states on the upper surface of the grains, and Anderson[107] has shown that these surface states play an important role in conduction in thin film transistors. The important role of grain boundary scattering has been established by a number of workers in cadmium selenide[106,107,108], indium antimonide[109] and cadmium sulphide[104,105,110] and this scattering mechanism is often dominant.

The conductivity through the film is much less affected by grain boundary scattering than the cross-film conductivity if the crystallites extend without break from the substrate to the outer surface of the film. The existence of a depletion layer at the grain surfaces will affect the conductivity through the film not only because the effective grain area for conduction is reduced, but because of the high electric fields associated with the depletion layers. These electric fields can be regarded as approximately cylindrically symmetrical and a charge outside the depletion layer would experience a field towards the centre of the grain. Electrons would thus tend to be forced towards the grain boundary whilst holes would be forced away from the boundary. Electrons would in general be unlikely to penetrate far into the depletion layer, but there will be an increased probability of their being captured by a recombination centre, and hence their diffusion length will be reduced. The minority holes, on the other hand would tend to be forced away from the grain boundary and their effective grain boundary recombination velocity would be reduced. The effect on hole recombination at the grain boundaries would be similar to that due to the back-surface field on back-surface recombination velocity as discussed in Section 2.2. If the crystallites were not normal to the substrate, there would be a component of the field due to the photovoltage forcing the minority carriers towards a grain boundary. For relatively small angles ($\sim 10°$) of deviation of the crystallite axes from normal, this component of the field would not be large (10 to 100 volts/cm) but would probably be sufficient to cause some increase in the rate of hole recombination at the edge of the depletion layer with a reduction in hole diffusion length and consequent reduction in solar cell efficiency.

The effect of surface accumulation or depletion layers on majority carrier mobility has been considered by Many et al.[111] who showed that the surface mobility was less than the mobility in the centre of the crystallite, and increased as the barrier potential increased up to a maximum value equal to the crystallite mobility.

The effect on mobility of scattering from surfaces in the case when the barrier potential is zero has been reviewed by Chopra[112], but for typical crystallite sizes in thin film solar cells, the effect is negligible.

The height of the potential barrier and the shape of the depletion layer at grain boundaries depends on the carrier densities in the grain and the intergrain region [through equation (45)] and on the distribution of donor states and impurities and defects near the grain boundary, and will thus depend critically on the deposition parameters. As discussed earlier Kirk and Raven[91] have shown the close inter-relationship between the structure of a deposited thin film with source and substrate temperature and impingement rate. It seems likely that the commonly found lack of reproducibility of the carrier transport parameters in evaporated thin films is due to variations below the normal control levels in the deposition parameters. This emphasises the importance of ensuring close control over both the magnitude and homogeneity of substrate and source temperatures and the impingement rate of the evaporant.

It is frequently found that evaporated thin films have a high degree of stress, with magnitudes up to 10^8 or 10^9 Nm^{-2} being quite common. Matsuura and Tsurumi[113] have measured the effect of uniaxial stress on the electron mobility in single crystals of cadmium sulphide and found that the "intrinsic" mobility had a stress coefficient $d(\ln \mu)/dp$ of 2.9×10^{-10} N^{-1} m^2 whilst the "extrinsic" mobility had a stress coefficient of 2.7×10^{-9} N^{-1} m^2. The stress coefficient of electron concentration was found to be approximately zero $d(\ln n)/dp = (6.5 \pm 7.8) \times 10^{-12}$ N^{-1} m^2.

Stress can arise in thin films from two mechanisms. The most obvious is that due to differences in thermal expansion coefficients of film and substrate, leading to stress when a film deposited on a hot substrate is cooled to room temperature. This thermal stress is given by

$$S_{Th} = \Delta\alpha Y_F \Delta T \qquad (46)$$

where $\Delta\alpha$ is the difference between the thermal expansion coefficients of film and substrate, Y_F is the Young's modulus of the film and ΔT is the difference between deposition temperature and the temperature at which the stress is measured. Typical stresses due to mis-match in the thermal expansion coefficients are 10^7–10^8 N m^{-2} for substrate temperatures of about 250°C.

There are also intrinsic stress-producing mechanisms[112] which give rise to stresses whose magnitude depends on the deposition rate, substrate temperature, film thickness, angle of deposition, film and substrate material and surface condition. The intrinsic stresses are often[112] of the order of 10^8 Nm^{-2}, which, on the basis of the stress coefficients of Matsuura and Tsutumi[113] would cause changes of -7% in "intrinsic" electron mobility and $+19\%$ in "extrinsic" mobility in CdS. In the author's laboratory, changes in room temperature Hall mobility have been observed in cadmium sulphide thin films after annealing at temperatures of up to 350°C. The crystallite sizes increased after annealing by factors of up to ten, whilst the Hall mobility sometimes increased, but usually decreased by up to a factor of two, although no correlation could be found between the changes in mobility and the film deposition parameters or the annealing process. It seems possible that intrinsic stress and its changes after heat treatment are further factors which can cause films with nominally identical deposition parameters to have different carrier transport properties.

4. Cadmium Sulphide–Copper Sulphide Solar Cells

4.1 Introduction

Cadmium sulphide–copper sulphide anisotype heterojunction solar cells are the most important of the thin film solar cells. More work has been done throughout the world on these cells than on any other thin film cell, and the power conversion efficiencies are now approaching 10%; higher than any other thin film cell. A very extensive review of the work on these cells has been given by Stanley[31], and the development of thin film cadmium sulphide–copper sulphide cells has also been reviewed by Komons[114], Massie[115], Shirland[116] and Perkins[117].

After Reynolds[28,118] original discovery of the photovoltaic effect in rectifiers formed from single crystal cadmium sulphide with an electroplated copper layer, the single crystal cell was developed by Shirland and co-workers[29,30] up to efficiencies of over 5%.

Stanley[31], in his Table III, lists the development of the thin film cell from the original attempts by Carlson[32] in 1956 up to 1973 giving information on substrate, CdS layer, barrier formation, counter electrode, encapsulation and electrical parameters.

The substrates used have included transparent conducting glass, metallised polyimide film, and metals such as molybdenum and copper, usually with a zinc plating. The cadmium sulphide layer was usually between 10 μm and 50 μm thick, and was deposited on to the hot substrate by thermal evaporation in the large majority of cases, although other techniques have been used successfully.

The most successful barrier formation technique at the present time uses a chemical plating bath containing cuprous chloride, and this has been used in the majority of cases. The cuprous sulphide layer has also been formed by thermal or flash evaporation of cuprous sulphide; evaporation of copper, with subsequent heat treatment; spraying of cuprous sulphide; and evaporation of cuprous chloride with subsequent heat treatment.

The counter electrode, which makes contact to the cuprous sulphide has usually been in the form of a copper or gold mesh. To prevent degradation of the cells due to copper diffusion, and to reduce the cost, it has become standard practice to use gold coated copper grids for this electrode.

The efficiencies of cells constructed by the Clevite Corp. reached 8% in 1965 with short circuit currents of $20\,\text{mA}\,\text{cm}^{-2}$ and open circuit voltages of 0.5 volts, for cells up to 3×3 inches in size. The power conversion efficiencies achieved by other workers ranged from 0.1% to 9% with values of about 3% to 5% being typical. Open circuit voltages were typically in the range 0.46 V to 0.52 V with short circuit currents between $10\,\text{mA}\,\text{cm}^{-2}$ to $20\,\text{mA}\,\text{cm}^{-2}$ whilst fill-factors were usually in the range of 0.6 to 0.7.

4.2 The Formation of Cadmium Sulphide Layers

4.2.1 Thermal evaporation

The thermal evaporation of cadmium sulphide single crystals has been studied in some depth, most notably by Somorjai and co-workers[119,129,121]. The process of evaporation has four main steps within the overall reaction[120].

$$\text{CdS(solid)} \rightleftharpoons \text{Cd(gas)} + \tfrac{1}{2}\text{S}_2(\text{gas}) \qquad (47)$$

The first step is the dissociation of bulk CdS to cadmium and sulphur on the surface

$$\text{CdS(solid)} \rightleftharpoons \text{Cd(surface)} + \text{S(surface)} \qquad (48)$$

The loosely bound cadmium atoms on the surface can evaporate directly

$$\text{Cd(surface)} \rightleftharpoons \text{Cd(gas)} \qquad (49)$$

but the single sulphur atoms are more likely to associate

$$2\text{S(surface)} \rightleftharpoons \text{S}_2(\text{surface}) \qquad (50)$$

before evaporating

$$\text{S}_2(\text{surface}) \rightleftharpoons \text{S}_2(\text{gas}) \qquad (51)$$

The first step [equation (48)] in the sequence is the rate limiting step[120] and the evaporation rate from a (0001) or c face is about one order of magnitude less than would be expected from the equilibrium vapour pressure.

The rate of evaporation depends on the crystal face from which the evaporation takes place[121]. Evaporation from (0001), (000$\bar{1}$) and (11$\bar{2}$0) faces give equal evaporation rates which are not dependent on the duration of the evaporation, whilst (10$\bar{1}$0) faces have an evaporation rate about half that of the other faces and the rate increases with time. Excess sulphur or cadmium on or near the surface depresses the evaporation rate of cadmium sulphide[120] and the element in excess appears to evaporate preferentially until the region near the surface is approximately stoichiometric after which the evaporation rate approaches that of the pure single crystal.

If cadmium sulphide is evaporated in the form of powders or finely ground flow crystals, one would expect evaporation to take place from all crystal faces simultaneously. The rate of evaporation would be an average of the rates from these faces, and should thus be stable. The rate of evaporation from powders or ground flow crystals is higher than that from a single crystal of the same volume, partly because the surface area from which evaporation can take place is much larger, and partly because the surfaces of powders or ground flow crystals are damaged thus reducing the energy required for the rate limiting reaction [equation (48)].

In practice cadmium sulphide in the form of powder or ground flow crystals is used for evaporation, although this leads to problems of spattering. It is usual to contain the cadmium sulphide in a quartz or carbon ampoule to prevent chemical attack of the heated metal source, and to reduce spattering by use of a quartz wool or sintered glass plug at the neck of the ampoule. A comprehensive review of the deposition of cadmium sulphide has been given by Stanley[31].

A wide variety of substrates have been used, the most successful to date being molybdenum, zinc-coated copper, and metallised Kapton. The substrate must have a coefficient of thermal expansion which is close to that of cadmium sulphide, which is quite temperature dependent[122] but averages about $5.10^{-6} K^{-1}$ between room temperature and 300°C. Molybdenum is the best match to this value, but tends to oxidise at the higher substrate temperatures leading to poor adhesion of the cadmium sulphide film and high contact resistance. These problems can be overcome if the molybdenum is etched just prior to its being used and then baked whilst under high vacuum. The best cells to date have used either zinc-plated copper or Kapton with a zinc-coated screen-printed silver metallisation layer. The temperature of the substrate is usually about 220°C–250°C[95], but the optimum temperature depends to some extent on the evaporation system used and on the other deposition parameters and it has been found at the I.E.C.[123] and by the author that a substrate temperature close to 300°C gives rather better results.

The cadmium sulphide is deposited as a polycrystalline layer, and the electron micrograph shown in Fig. 14 is typical. The thickness of the cadmium sulphide layers used by different workers has varied between a few microns to a few tens of microns. The thickness is determined by the need to avoid short circuits in the subsequent junction formation steps, and a lower thickness limit is determined by the increasing pin-hole density as film thickness is reduced. The average crystallite size in these films is typically a few microns, although a subsequent annealing step can increase the crystallite size considerably. The crystallite size depends quite critically on the substrate temperature and deposition rate with very small crystallites being found for substrate temperatures less than about 150°C and for high deposition rates. The crystallite size increases as the film thickness increases up to about one micron, and this increase in crystallite size can be seen in Fig. 14.

The fast growth direction in cadmium sulphide films is parallel to the c-axis, so one would expect from the discussion in Section 3 that the films would be oriented with the c-axis normal to the substrate. For films thicker than a micron or so, this is observed, although it is usual to find that individual crystallites have a c-axis orientation within a cone whose axis is normal to the substrate and whose apex angle depends on the deposition conditions, but is typically about 10° to 20°. Vecht et al.[124,125] have shown that, at least in recrystallised films, the (000$\bar{1}$) sulphur surface is bonded to the substrate and the cadmium (0001) plane forms the outer crystal surface.

If the beam of evaporant is not normal to the substrate, the c-axis of the deposited crystallites are not longer normal to the substrate on average. The angle between the substrate normal and the crystallite c-axis increases with film thickness and tends to an angle equal to about half the angle of incidence, up to incidence angles of 55°[126]. Above angles of incidence of 55°, the c-axes of the CdS crystallites become more nearly normal to the substrate[126]. Problems of non-uniform c-axis orientation thus arise when large area films are being deposited from a single source, since the angle of incidence varies over the substrate. An attempt has been made[101] to reduce this problem by using four sources placed at the corners of a square. The uniformity in film thickness improved substantially but the c-axes of the crystallites were oriented as if there had been just one source at the centre of the square[101]. Since the efficiency of these solar cells is reduced if the cadmium sulphide c-axis is not normal to the substrate, this variation in c-axis orientation is a serious problem in the thermal evaporation of large area cells.

The surface topography of the cadmium sulphide films is important since it determines the optical reflectance of the outer surface of the cell. A rough surface, with roughness of the order of a micron, admits significantly more light than an optically smooth surface. Since the refractive index of cuprous sulphide is about 3.2, giving a normal incidence reflection coefficient of about 27%, this can lead to useful improvements in power conversion efficiency.

Evaporation of cadmium sulphide in an open system at low rates on to smooth substrates at high temperatures usually produces a film which is optically flat. The surface topography of single crystal cadmium sulphide films grown on to single crystal strontium fluoride has been investigated by Christmann et al.[127]. At low deposition rates and high substrate temperature they found that their films were optically flat, whilst at higher deposition rates and/or lower substrate temperatures flat-topped hexagonal truncated pyramids were formed with dimensions of the order of a few microns. For thicker films and the highest deposition rates, complete hexagonal pyramids were formed, which they considered to have nucleated at cadmium droplets. When the substrate surface was roughened, a high density of hexagonal pyramids were formed.

The surface topography of polycrystalline cadmium sulphide films is much less regular than that of the single crystal films, but the same considerations apply. A rough outer film surface can be produced by evaporating onto a roughened substrate or by ensuring a high degree of supersaturation at the growth surface of the film. Roughening of the substrate surface is the easiest method to adopt but care must be taken not to adversely affect the nucleation properties. A high degree of supersaturation can be produced near substrates even at high temperatures ($\sim 300°C$) by the use of hot wall techniques, in which the substrate and source are at opposite ends of a heated cylinder. The degree of supersaturation can be varied by altering the temperature of the walls of the cylinder, and this technique also leads to improved control over the stoichiometry of the cadmium sulphide film.

Both methods of ensuring a rough surface are successful. Figure 17 shows an electron micrograph of a typical surface produced by the hot wall technique, and Fig. 18 shows a scattering polar diagram[128] from the sample shown in Fig. 17.

Cadmium sulphide films with nearly perfect stoichiometry have a volume resistivity above $10^6 \,\Omega$ cm. For a solar cell with a 10 μm thick CdS layer and an area of 1 cm^2, this would give a series resistance of 100 Ω, which would render the cell useless. The volume resistivity can be reduced by allowing sulphur vacancies to occur or by doping. The easiest way of reducing the resistivity is to adjust the deposition parameters so as to introduce sulphur vacancies, but this will probably also result in films with reduced mobilities and recombination times. However, since the cadmium sulphide layer plays a minor role in photocurrent generation (see Section 4.5) this is not a major disadvantage.

Reduction of the resistivity by doping during the evaporation has been studied by a number of workers, and reviewed by Stanley[31]. Doping with a metal such as indium will give the required resistivity of about 10 Ω cm, but there is a danger that the indium will condense out at the grain boundaries, particularly in films below 5 μm thickness, leading to very low shunt resistances. David et al.[129] have used $CdCl_2$ as a dopant and found that the

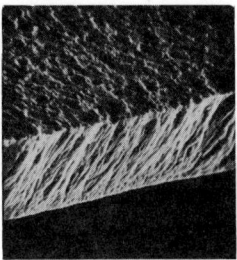

FIGURE 17. Micrograph of the surface of a cadmium sulphide film. The microroughness of the surface is about 0.5 μm.

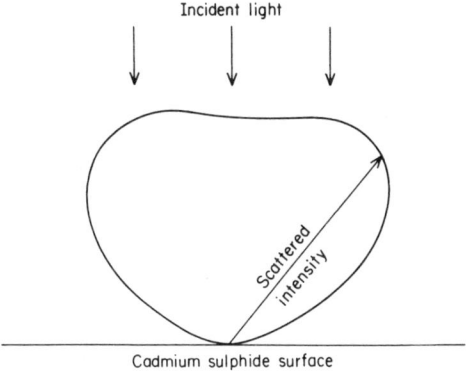

FIGURE 18. Scattering polar diagram from the surface of the film shown in Fig. 17.

series resistance decreased and the photovoltaic response improved. It has been found in the author's laboratory that if the CdS charge to be evaporated is composed of pure CdS in the bottom half of the ampoule and CdS + 0.01% $CdCl_2$ in the top half of the ampoule, the resistivity of the CdS film is low at the back surface, leading to improved contact with the molybdenum substrate, and increased towards the outer surface. The conversion efficiencies of these cells were higher than those of uniformly doped cells, mainly due to an increase in photogenerated current.

The electron transport parameters measured in thin films are usually dependent on the deposition parameters. The electron mobilities reported for good quality films vary from $40 \, cm^2 \, V^{-1} \, sec^{-1}$ to over $100 \, cm^2 \, V^{-1} \, sec^{-1}$, and it should be noted that these mobilities are almost invariably measured across the film, and therefore are influenced by the thermal activation of electrons over the intergrain potential barriers. Mobilities equal to the best single crystal mobilities ($\sim 300\,\text{cm}^2\,\text{V}^{-1}\,\text{sec}^{-1}$) have been reported[98] for recrystallised films under intense illumination, although these results have not been repeated.

The minority carrier diffusion length has been found to be a few tenths of a micron[96], and this would suggest that the recombination time is about 10^{-8} sec.

Evaporation of cadmium sulphide as a means of depositing the base layer has a number of disadvantages when considering the production of large area, low cost devices. It is inefficient in its use of the evaporated material, since at best only a few percent of the initial charge is deposited on the substrate. With careful design of the evaporation system, it is possible to recover most of the evaporated material which is not deposited on the substrate, but the stoichiometry of this recovered cadmium sulphide is usually such that it must be reprocessed before it can be used again. As discussed earlier it is important to ensure that the evaporant is incident on the substrate at nearly normal angles. For large area devices, this requirement will mean that the distance between the source and the substrate must be large, resulting in a large, and therefore costly, vacuum system. This problem can be alleviated to some extent by the use of "hot wall" techniques, which also improve the thickness uniformity of the deposited films. The thickness uniformity can be ensured by the use of multiple sources, but, as discussed previously, this does not ensure that the crystallites are normal to the substrate[101]. It is usually assumed that large area cells will be produced by techniques similar to those at present used for roll-coating. The jump in technology needed to move from the present laboratory sized "one-off" systems to a roll-coating system is rather large, and a great deal of work must first be done to ensure that roll-coating techniques are capable of producing cadmium sulphide layers of a quality suitable for solar cells.

Another serious problem with the use of evaporation techniques for producing commercially viable solar cells is the capital intensive character of the process. The cost of the cells would be governed to a significant extent by fixed capital charges, and the cells would be very expensive if produced in small numbers. The problems associated with the introduction of solar cells for general use will be discussed in the final chapter.

Because of these problems, a number of other techniques for the production of the base cadmium sulphide layer have been investigated.

4.2.2 Sputtering

Cadmium sulphide thin films have been deposited successfully using both D.C. and R.F. sputtering[31] at rates of up to $1\,\mu\text{m/min}$. The target is usually compressed or sintered cadmium sulphide, and dopants such as cadmium chloride can be introduced into the target.

The structure and properties of sputter-deposited cadmium sulphide films have been studied by a number of workers[130-136]. The films produced by Lagnado and Lichtensteiger[130] were typically a few microns thick and had crystallites about a micron in size. The crystallites were oriented with the c-axis normal to the substrate with a much smaller variation about the normal than is found for evaporated films. The film surfaces were not flat, but had a surface roughness of 0.1–0.5 μm which caused scattering of light incident on the surface.

The resistivity of the films prepared by different groups varied from about 10 Ωcm to 10^{10} Ωcm depending on the deposition parameters. The dark resistivity of films sputtered in Ar + H_2S increased by about four orders of magnitude as the substrate temperature increased from 100°C to 300°C[131]. The resistivity under illumination increased by about three orders of magnitude over the same substrate temperature range, with an optimum substrate temperature of about 200–250°C[131]. Above this temperature, the deposition decreased, and dark resistivity increased only slightly.

In pure argon, the orientation of the films improved with increasing substrate temperature and deposition rate[130] and orientation of the crystallites within the basal plane was achieved with substrate temperatures above 90°C and growth rates above 5 Å/sec. The resistivity of films sputtered in pure argon is much lower than that of films sputtered in a mixture of argon and H_2S due to sulphur vacancies. The resistivity was found to have an activation energy of about 0.3 eV at room temperature whilst the mobility had an activation energy of about 0.02 eV[130].

Leighton[132] found that the light conductivity of D.C.-sputtered films increased linearly with the intensity of illumination up to a certain value and then was proportional to the square root of intensity, whilst the current/voltage characteristics were ohmic up to a critical applied field and then showed current saturation, probably because the electron velocity approached the velocity of acoustic shear waves.

The electron mobilities measured both through and across the film have been found to be in the range 10–40 cm^2 V^{-1} sec^{-1} and hole mobilities have been measured by Lichtensteiger et al.[133], who produced p-type CdS by sputtering in phosphine, to be 6–15 cm^2 V^{-1} sec^{-1}.

Takeuchi et al.[136] deposited cadmium sulphide films by R.F. sputtering in argon, and measured the photoconductivity and the photovoltaic effects in SnO_2–CdS–In structures. They found that the crystallinity of the sputtered films was superior to that of the target material, and the sputtered films had bulk resistivities of tens or hundreds of MΩcm. The photovoltage produced by their SnO_2–CdS–In devices was rather small, with a maximum of 0.1 volts.

They take this voltage to be the sum of the forward voltage produced by the SnO_2–CdS interface and a reverse voltage produced at the CdS–In interface, which they consider is of MIS form due to oxygen absorption on the cadmium sulphide surface prior to indium deposition.

Reactive sputtering of cadmium in Ar–H_2S has also been used to deposit cadmium sulphide films with some success[31], although in this case the crystallite orientation appeared to be less precise than that of thermally evaporated films[137].

4.2.3 Vapour deposition

The deposition of cadmium sulphide by vapour phase reaction is well established. Evaporation of cadmium sulphide produces cadmium and sulphur vapours which react at the substrate to give cadmium sulphide, so these techniques are, in a sense, vapour deposition techniques. It is only one step from hot wall evaporation to completely enclosing the source and substrate in a reaction chamber to give closed-space quasi-equilibrium vapour deposition. It is more usual to use a transport gas such as argon, nitrogen or hydrogen sulphide (or a mixture of these) within the enclosed reaction chamber, and these systems are commonly used to grow single crystals of cadmium sulphide or epitaxial thin films.

Transport from a low temperature source to the substrate can be achieved if chemical reactions are promoted which result in volatile products, e.g.

$$CdS(solid) + H_2(gas) \rightleftharpoons Cd(vapour) + H_2S \text{ gas}$$

$$CdS(solid) + I_2(vapour) \rightleftharpoons CdI_2 + \frac{1}{x} S_x(vapour)$$

Hydrogen transport can give quite large deposition rates, with the c-axis normal to the substrate[138,139], when depositing polycrystalline films on a glass substrate, but the adhesion is poor for thicknesses greater than 20μ. Iodine transport gives lower deposition rates, with better adhesion, but produces crystallites of random orientation[138,139].

4.2.4 Spraying techniques

Spraying of the cadmium sulphide layer would be the ideal process for producing low cost, large area solar cells if it could be developed to give cadmium sulphide films of the required properties.

The process involves the spraying of a solution of a cadmium salt and a sulpho-organic compound on to a heated substrate. The subsequent reaction on the hot substrate produces a cadmium sulphide film plus volatile products.

Most workers[140,141,142] have used cadmium chloride and thiourea, and it has been found that the crystallinity of the film depends on the cadmium salt which is used, and on the substrate temperature and material[140]. The crystallite orientation depends on the anion to cation ratio, and for stoichiometry, the c-axis is normal to the substrate[140].

This technique is being actively investigated in a number of countries and solar cells with efficiencies approaching 5% have been produced[143,144].

4.2.5 Heterogeneous conversion of cadmium

Thin films of cadmium sulphide can be produced by the sulphurisation of thin films of cadmium. This can be achieved by heating the cadmium film in hydrogen sulphide or sulphur vapour[145], but the reaction is slow.

The anodic sulphurisation of cadmium using Na_2S as the reactive electrolyte has been discussed[146] in connection with semiconductor/electrolyte junction solar cells, but is also under active investigation at the International Research and Development Co. Ltd. as a means of producing large area, low cost cadmium sulphide layers.

Lawrence[147] sputtered cadmium in air to deposit cadmium oxide which was then converted to the sulphide by heating in hydrogen sulphide.

4.2.6 Sintered and screen-printed layers

These layers are formed by mixing cadmium sulphide powder with cadmium chloride in a slurry, and then firing. This pre-sintered material is then ball-milled to a fine powder and suspended in a slurry using either water or an organic solvent, to which a binding agent is often added. This slurry can be applied to a substrate by spraying or screen printing[148] amongst other methods. The slurry is then dried and fired at about 500°C–600°C.

The cadmium chloride in the slurry acts as a flux for the crystallisation of cadmium sulphide. The cadmium chloride melts at 568°C and cadmium sulphide dissolves in the molten cadmium chloride, so the sintering process causes growth of the powder particles by fusion. The chlorine acts as a donor in cadmium sulphide, but the concentration may be controlled by the sintering temperature, since cadmium chloride is volatile above 400°C and completely volatilises at 600°C. The sintered films resemble single crystals in many of their properties and have been used very successfully in making solar cells[31,149,150,151].

4.3 Formation of the Copper Sulphide–Cadmium Sulphide Junction

Many different methods of forming the junction have been used since 1954 and these have been reviewed by Stanley[31].

The vast majority of cadmium sulphide–copper sulphide solar cells are now formed by a chemiplating technique first proposed by Cusano[39] in 1962 and subsequently developed mainly at the Clevite Corporation, S.A.T. and the Institute of Energy Conversion.

4.3.1 Chemiplating processes

The most usual chemiplating technique used for producing the junction is similar to the one devised by Cusano[39] and developed at the Clevite Corporation[152]. The manufacture of these cells by the thermal evaporation of cadmium sulphide on to metallised Kapton and formation of the copper sulphide layer by the chemiplating process described below is often referred to as the Clevite process.

The chemiplating takes place in a solution of cuprous chloride in water with a carefully controlled pH value. The replacement reaction

$$CdS + 2CuCl \rightleftharpoons Cu_2S + CdCl_2 \quad (52)$$

takes place, giving a layer of insoluble Cu_2S on the surface of the cadmium sulphide whilst the cadmium chloride goes into solution.

Thermodynamic considerations show that this reaction is highly favourable. The standard free energies for the formation of all the compounds in the reaction are well-known and one can re-write the reaction in the form:

$Cd(solid) + Cl_2(gas) = CdCl_2(solid)$	$\Delta G_{298} = -81.88$ kcal/mole
$2Cu(solid) + S(solid) = Cu_2S(solid)$	$\Delta G_{298} = -20.6$ kcal/mole
$CdS(solid) = Cd(solid) + S(solid)$	$\Delta G_{298} = +33.6$ kcal/mole
$2CuCl(solid) = 2Cu(solid) + Cl_2(gas)$	$\Delta G_{298} = +(2 \times 28.2)$ kcal/mole
$CdS(s) + 2CuCl(s) = Cu_2S(s) + CdCl_2(s)$	$\Delta G_{298} = -12.48$ kcal/mole
	$\Delta G_{298} = -1.4$ eV/molecule

The change of free energy with temperature can be found from

$$\left[\frac{\delta G}{\delta T}\right]_p = -\Delta S = -[\text{entropy of products} - \text{entropy of reactants}]$$

From standard tables it can be found that $\Delta S = 2.10^{-4}$ kcal mole^{-1} deg^{-1} and the free energy of reaction changes very little with temperature.

Similar calculations for the other halides of copper show that the free energy changes are lower

$CdS + 2CuBr = CdBr_2 + Cu_2S \quad \Delta G_{298} = -9.52$ kcal/mole
$\quad\quad\quad\quad\quad\quad\quad\quad\quad\quad\quad\quad\quad\Delta S = 0.0$ kcal mole^{-1} deg^{-1}
$CdS + 2CuI = CdI_2 + Cu_2S \quad \Delta G_{298} = -1.76$ kcal/mole
$\quad\quad\quad\quad\quad\quad\quad\quad\quad\quad\quad\quad\quad\Delta S = 5.9 \, 10^{-3}$ kcal mole^{-1} deg^{-1}

These thermodynamic data thus show that the reactions should all go to completion, but that the copper chloride should be superior to either the bromide or the iodide, as found experimentally by Shirland et al.[153].

Thermodynamic considerations can describe only equilibria and the rates of reaction are governed by other factors. The rate limiting factors are most likely to be the diffusion of copper through the cadmium sulphide and the diffusion of cadmium through the copper sulphide. This is particularly likely in the case of the "Clevite" type plating process[154] where the reaction proceeds rather quickly with the copper sulphide layer thickness increasing approximately linearly with time at a rate of up to one hundred monolayers per second. After a few seconds, the copper sulphide layer has reached a thickness of about 0.3 μm and the reaction is stopped by taking the device out of the plating bath and washing in distilled water. At this point the copper sulphide layer is highly disordered, and the photovoltaic response is small. A heat treatment is necessary to "form" the cells and in the Clevite process this is done by baking in air for two minutes at 250°C. The baking process causes a large reduction in the dark currents and an increase in photovoltage. The diode factor decreases at first, but then increases as the baking time increases, whilst the shunt resistance increases to a maximum and then decreases for longer baking times. These changes give rise to a significant improvement in power conversion efficiency during baking until a maximum efficiency is reached. For longer baking times the reduction in the photocurrent causes a reduction in efficiency.

These changes in junction characteristics must be the result of changes in copper sulphide stoichiometry and crystallinity and in the cadmium sulphide/copper sulphide interface.

The basic principles of the Clevite plating process can be summarised as:

(1) thermal evaporation of a cadmium sulphide layer;
(2) hydrochloric acid etch to reveal grain boundaries;
(3) plating in a bath containing cuprous chloride held at a pH of about 3.5 and a temperature of 90°C;
(4) heat treatment in air at 250°C for 2 minutes.

All chemical plating techniques have found that the etch in hydrochloric acid is necessary to achieve high photocurrent, and most groups use a plating bath of similar composition and conditions, although, as described below, other plating solutions are also used. The heat treatment step is the one with the widest divergence of procedure. Optimum baking times vary from the two minutes found by the Clevite group, one to two minutes found by I.R.D. Limited[155] and seven minutes found by Bube et al.[156], all at about 200°C in air. The Institute of Energy Conversion find that their cells have a maximum efficiency if oxygen is rigorously excluded from the junction formation process and the main heat treatment is carried out after lamination, although some heating in the presence of oxygen is necessary[157]. After lamination they heat the cells to about 130°C for many hours and find that the short circuit current and the shunt resistance improve considerably[158]. This heat treatment also has the effect of improving the alignment of the crystallites in the cadmium sulphide layer[159].

Reinhartz and Van Aershodt[160] have considered very carefully the thermodynamic of the replacement reaction. They investigated experimentally and theoretically the criteria for maintaining the highest possible ratio of cuprous to cupric ions and concluded that complexing the cuprous ions with ammonium radicals should ensure good thermodynamic stability. They investigated the characteristics of cells whose junctions had been produced by chemiplating in solutions of cuprous chloride and ammonium chloride as a function of the concentrations of cuprous chloride and ammonium chloride. They produced cells with open circuit voltages of about 0.5 volts and short circuit currents which were concentration and thickness dependent. The short circuit current for cells plated at 90°C in a bath of 2 g/litre of copper with 10% ammonium chloride at pH 3 for a time of about 10 minutes exceeded 20 mA/cm^2 for AM0 radiation. The fill-factors which they found were, however, only about 0.5–0.65, and the power conversion efficiencies were not as high as might be expected. They did not show the I/V curves of their cells, but in view of the high open circuit voltage, the low fill-factor can probably be attributed to high series resistance and/or low shunt resistance. Since their top contact was a pressure contacted gold grid, and no lamination was attempted, it might be expected that the fill-factor would be rather low.

It appears inherently likely that a slow plating process which maintains equilibrium throughout and does not suffer from concentration gradients within the solution should result in superior crystallinity and stoichiometry than the rather rapid Clevite type processes. However, it is also likely that dipping times of many minutes in solutions with a pH of 3 require very high quality cadmium sulphide films with packing densities close to unity if the shunt resistance is not to become unacceptably low.

Nakayama[150] has used a plating bath containing cupric sulphate for producing junctions on ceramic cadmium sulphide layers. The cadmium sulphide plate and a metallic copper plate are immersed in a solution of cupric sulphate, and a D.C. potential is established between the cadmium sulphide and copper plates. The replacement reaction is then

$$Cu^{++} + CdS \rightleftharpoons CuS + Cd^{++}$$
$$CuS + Cu^{++} + 2e \rightleftharpoons Cu_2S$$

Cells produced in this way show high power conversion efficiencies (6–9%) but degradation is a serious problem. The degradation in these cells occurs in short circuit conditions[161] rather than the usually observed open circuit conditions, and the degradation mechanisms are different. Degradation in cadmium sulphide–copper sulphide cells will be discussed in Section 4.6.

4.3.2 Evaporation techniques

A number of attempts have been made to produce the copper sulphide layer by evaporation[31]. Evaporation of cuprous sulphide by thermal evaporation is usually unsuccessful due to dissociation and flash evaporation, whilst more successful, suffers from the same problem. Sputtering of cuprous sulphide has been tried, using reactive sputtering techniques, and cells of efficiency up to 4% have been produced[162].

Most of the work on evaporated layers has involved the evaporation of cuprous chloride and a subsequent bake to produce the chemical exchange reaction shown in equation (52). This technique has been used by a number of workers[163,164,165,128] and has been described in detail by te Velde[166]. The process involves the evaporation of carefully purified cuprous chloride onto the base cadmium sulphide film to a thickness of a few thousand Ångstroms. The films are then baked in an inert or reducing atmosphere at about 180°C for a time of a few minutes. The unwanted cadmium chloride is removed from the surface by washing preferably in alcohol and the cell can then be gridded and encapsulated in the usual way. An air bake has been used by some workers to activate the cells[163,164] whilst te Velde[165,166] used an air bake subsequent to the bake in argon to improve the efficiency of the cell, which on devices formed from single crystal cadmium sulphide has exceeded 10% (te Velde, private communication). Michelson and Abbott[164] found that the efficiency of their cells improved if the cadmium sulphide was etched with hydrochloric acid before evaporation of the cuprous chloride. Casperd[128] has investigated the dry barrier process in detail and has found that the optimum baking times are dependent on the cadmium sulphide crystallinity and surface stoichiometry and on the stoichiometry and thickness of the cuprous chloride layer. Hill and Edwards[167] have shown that the baking process causes an increase in the sulphur concentration on the surface of the cadmium sulphide and that efficient dry barrier cells require a base cadmium sulphide film with a surface which is slightly deficient in sulphur. If the cadmium sulphide film has a sulphur rich surface layer, then etching of the cadmium sulphide before deposition of the cuprous chloride will improve the cell efficiency, whilst cadmium sulphide films with a slight surface deficiency of sulphur produce cells of higher efficiency if the etching step is omitted.

The details of the solid replacement reaction have not been worked out but te Velde[166] has shown that baking a 2000 Å cuprous chloride layer on the surface of a flat single crystal of cadmium sulphide results in a 1000 Å cuprous sulphide layer and a 1000 Å thick cadmium chloride layer, and the outer face of the cuprous sulphide is the same as the original cadmium sulphide surface, as shown in Fig. 19.

This suggests that the sulphur lattice is not affected by the baking process, to a first approximation; and only copper and cadmium diffuse. As shown in the next section, the cuprous sulphide lattice can be derived from the cadmium sulphide lattice by replacing each cadmium ion by two copper ions with only very minor modifications to the sulphur lattice. The result obtained by te Velde[166] shown in Fig. 19 is thus physically quite reasonable. It is, however, extremely difficult to analyse the process, since one must consider the diffusion of copper ions from the cuprous chloride diffusing first through a mixture of copper chloride and cadmium chloride, and then through a mixture of

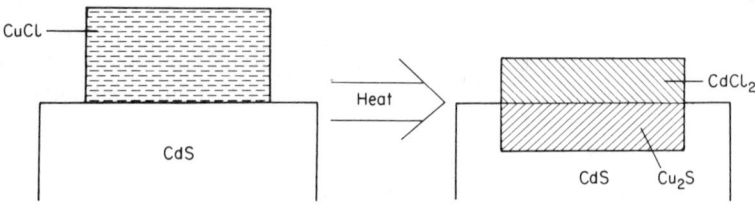

FIGURE 19. A representation of the dry-barrier process for thin cuprous chloride.

cadmium sulphide and copper sulphide, in approximate equilibrium with cadmium ions diffusing in the opposite direction. The much simpler process of the concentration dependent diffusion[168] of copper in cadmium sulphide does appear to offer some insight into the process, particularly on the effects of doping of the cadmium sulphide. The diffusion of copper in substitutional sites is likely to be small compared to the diffusion of interstitial copper, and the relationship between interstitial and substitutional copper can be written:

$$Cu_{int}^+ + V_{Cd}^- \rightleftharpoons Cu_{sub}^- + e^+ \quad (53)$$

Both the cadmium vacancy and the substitutional copper are acceptor centres and will thus be negatively charged in the n-type cadmium sulphide.

From the law of mass action the concentration C_s of copper substitutional is given by:

$$C_s = \frac{C_i C_v'}{Kp} \quad (54)$$

where C_i is the concentration of copper interstitial ions, C_v' is the equilibrium concentration of cadmium vacancies, K is the reaction constant and p is the hole density. If the hole density is due mainly to the substitutional copper acceptor centres, this leads to an effective diffusion coefficient for substitutional copper of

$$D_s = \frac{3KD_i}{C'_v} C_s \qquad (55)$$

where D_i is the diffusion coefficient of the interstitial copper ions. A diffusion coefficient which varied inversely as the vacancy concentration has been noted in copper sulphide by Okamoto and Kawai[169].

From equation (54), we can see that the concentration of copper substitutional ions depends on the hole concentration, and through this, is dependent on the doping of the cadmium sulphide. The diffusion coefficient D_s is thus also dependent on the doping level. If the cadmium sulphide is lightly doped, and hence the hole concentration is large, most of the copper ions will be interstitial and both the copper and cadmium ions will be mobile, as cadmium vacancies will not have been filled by copper ions. The inter-diffusion of copper and cadmium will then be rapid. High doping levels will increase the concentration of substitutional copper ions, reduce the cadmium vacancy concentration, and the mobility of both cadmium and copper ions will be reduced, thus leading to less rapid inter-diffusion. This may go some way to explaining the results of Egorova[170] and Palz et al.[171] who found that cells formed from doped cadmium sulphide needed longer baking times than those formed from undoped films, and the results of Buckley[172] who found that the chemiplating reaction rate decreased as the cadmium sulphide doping was increased.

Equation (55) suggests that the effective diffusion coefficient for substitutional copper will increase as the replacement reaction proceeds, but the initial assumption of copper diffusing in pure cadmium sulphide will no longer be valid since the copper ions must then diffuse through a mixture of copper sulphide and cadmium sulphide. In the final stages of the replacement reaction, the copper will be diffusing in cuprous sulphide, and the diffusion coefficient will then depend on the density of copper vacancies in the Cu_xS structure, and will decrease as the copper vacancy concentration decreases.

If cupric ions are present in the system, their reaction with cadmium vacancies could also produce copper substitutional acceptor centres via the reaction:

$$Cu_I^{++} + V_{Cd}^- \rightleftharpoons Cu_s^- + 2e^+ \qquad (56)$$

The concentration of substitutional copper centres from this reaction would be given by:

$$C_s = \frac{C_{I^{++}} C'_v}{K^1 p^2} \qquad (57)$$

where K^1 is the reaction constant for equation (56) and $C_{I^{++}}$ is the concentration of interstitial cupric ions. The dependence on the doping of the cadmium sulphide is now even stronger than for the cuprous ions. Since the presence of cupric ions is known to be highly detrimental to the efficiency and

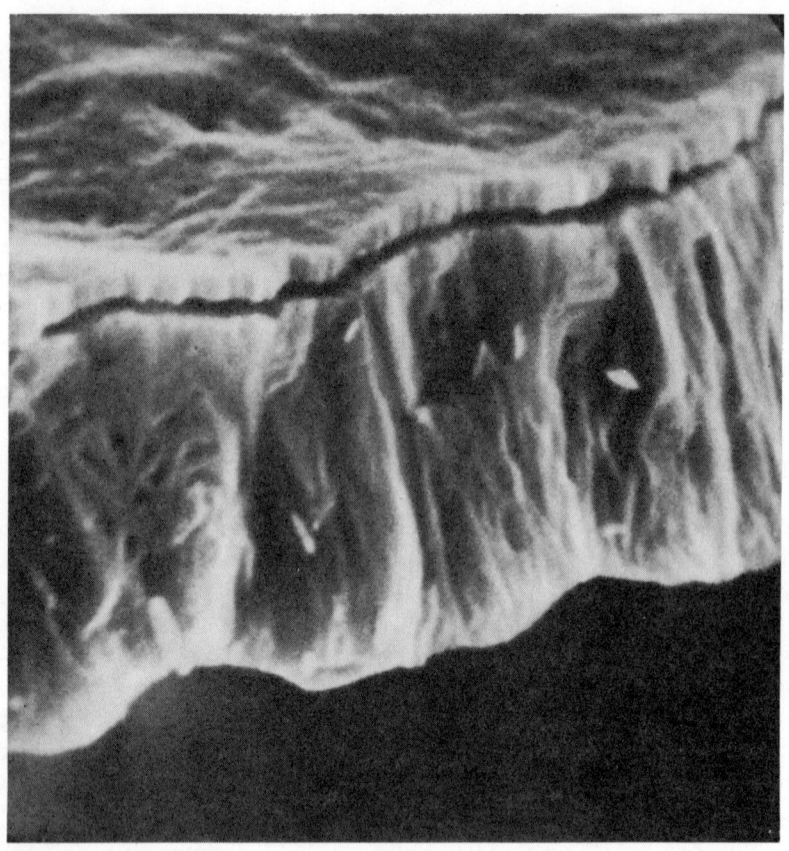

FIGURE 20. Micrograph of cuprous chloride deposited on cadmium sulphide. The columnar structure of the cuprous chloride can be seen.

stability of these cells, this may explain why doping of the cadmium sulphide has been found to improve these parameters considerably[173].

The type of dopant used can also have an effect on the concentration of substitutional copper ions. If the dopant substitutes for cadmium it may affect the concentration of cadmium vacancies and hence, through equations (54) and (57) the concentration of substitutional copper ions.

The formation of the junction on thin films of cadmium sulphide is complicated by the polycrystalline nature of the base film. If care is taken in the evaporation of the cuprous chloride, it can be deposited as a polycrystalline film, as shown in Fig. 20. There is thus likely to be preferential diffusion down the grain boundaries, and the cuprous sulphide thickness will be greater at the boundaries of the cadmium sulphide crystallites. This can be seen in Fig. 21 which shows the interface region of cadmium sulphide after the cuprous

(a) (b)

FIGURE 21. Micrographs of the surface of cadmium sulphide after the removal by KCN etch of junctions formed by (a) dry-barrier techniques on a 7 μm thick film and (b) chemiplating techniques on a 25 μm thick film.

sulphide has been etched from a completed cell. The penetration of the junction down grain boundaries is much smaller in the dry barrier process than in the chemiplating process, allowing the cadmium sulphide thickness to be reduced to between 5 μm and 10 μm without deleterious effect.

The thickness of the cuprous sulphide layer which can be produced by the solid state reaction is not dependent on the thickness of the cuprous chloride layer if this layer is more than about 4000 Å thick[128]. The thickness of cuprous sulphide produced by reacting a thick cuprous chloride layer is about 1500 Å to 2000 Å, leaving a sandwich structure shown in Fig. 22. The thickness of the cuprous sulphide is limited by the difficulty of cadmium diffusing through cuprous sulphide and cadmium chloride and the converse diffusion of copper ions. Furthermore, the more nearly stoichiometric and crystalline the cuprous sulphide layer, the smaller will be the diffusion coefficients of cadmium and copper, and thus the limiting thickness will decrease somewhat as the perfection of the cuprous sulphide layer improves.

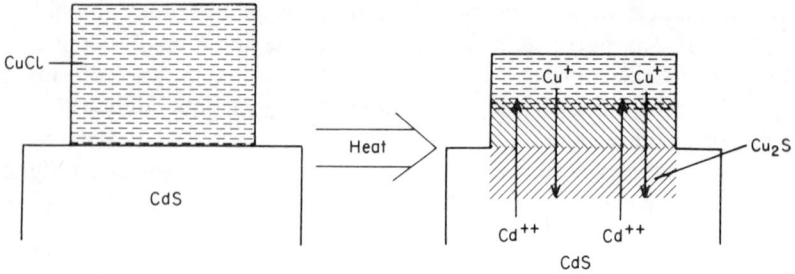

FIGURE 22. Representation of the dry-barrier formation process, showing the diffusion of cadmium ions from the CdS to the growing cadmium chloride face and the diffusion of cuprous ions from the cuprous chloride to the growing cuprous sulphide face.

4.4 Properties of Cuprous Sulphide

Cuprous sulphide is a semiconductor which can exist in a number of crystallographic phases. The phase diagram shown in Fig. 23 contains the copper sulphide phases between cupric sulphide CuS and cuprous sulphide Cu_2S. The composition region of interest in cadmium sulphide–copper sulphide solar cells extends from diginite ($Cu_{1.8}S$) to chalcocite (Cu_2S). It is convenient to denote the copper sulphide as Cu_xS and then use the value of x to label the phases.

The dignite phase occurs above $x = 1.77$ and has a pseudocubic structure[174] at room temperature whilst high diginite is cubic. The djurleite[175] phase occurs at $x = 1.96$ and is orthorhombic at room temperature, with a slow transformation at 93°C to a tetragonal form. An expansion of the phase diagram in the important composition region between $1.96 < x < 2.000$ is shown in Fig. 24. Chalcocite is often considered to be orthorhombic, but Evans[176] has shown that pure chalcocite is monoclinic, but with cell dimensions very similar to those of the orthorhombic cell postulated by Buerger and Buerger[177] i.e.

$$a_{OR} = b_{MC} : b_{OR} = 2a_{MC} \sin \beta_{MC} : c_{OR} = c_{MC}$$

Evans[176] found that the sulphur lattice has the close-packed hexagonal structure found by earlier workers[174], and points out that the structure of the high temperature phase, hexagonal chalcocite, is compatible with the space group $P2_1/c$ which he assigns to pure chalcocite below the phase transition temperature. The pseudo-orthorhombic aspects of most chalcocite crystals are due to fine scale twinning on the (100) plane and the pseudo-orthorhombic β'

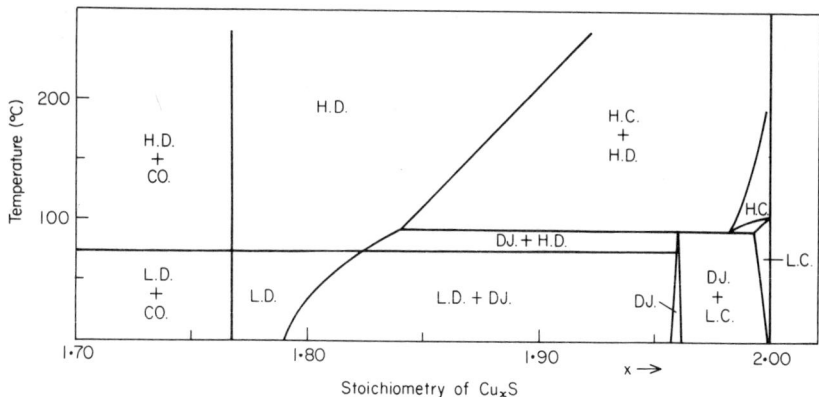

FIGURE 23. Part of the phase diagram of copper sulphide (after Roseboom[174]). CO = covalite; H.D. = high diginite; L.D. = low diginite; DJ = djurleite; H.C. = high chalcocite; L.C. = low chalcocite.

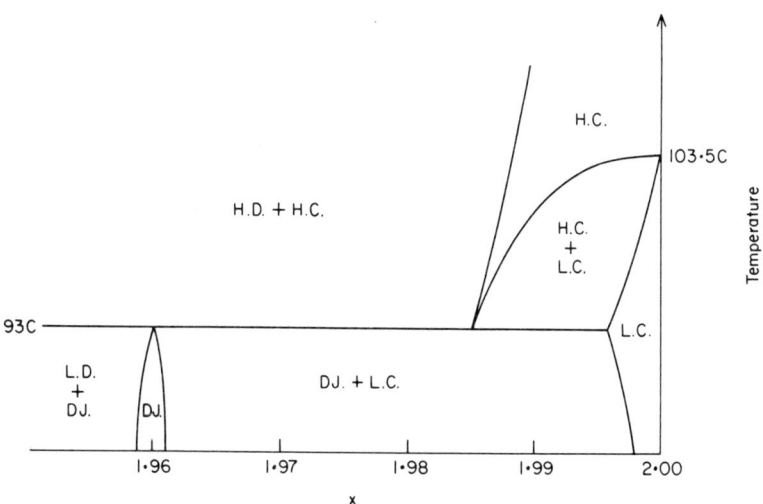

FIGURE 24. Detailed view of the phase diagram of copper sulphide.

angle of $90.08°$[176]. The copper atoms in the monoclinic structure are arranged in triangular co-ordination with the sulphur atoms, although some distortion of the regular array occurs with the Cu-S bond length varying from 2.21 Å to 2.89 Å with an average of 2.33 Å.

The high temperature phase of chalcocite has been studied by Buerger and Wuench[178] who found that the sulphur atoms were arranged in a close-packed hexagonal structure, but the copper atoms were in disordered sites as shown in Fig. 25. The cadmium sulphide–copper sulphide junctions are almost always formed at temperatures well above the phase transition

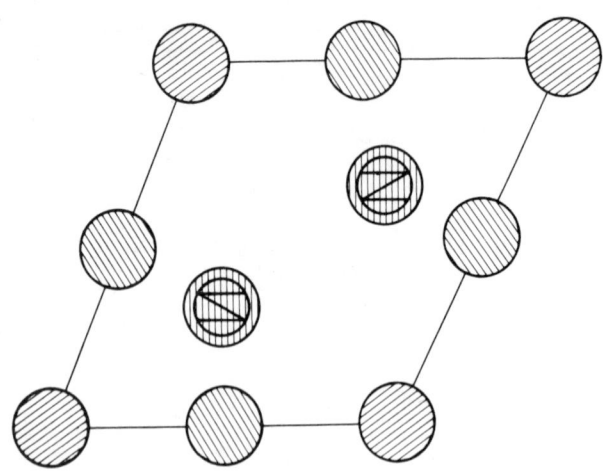

FIGURE 25. Representation of the structure of high chalcocite.

- ⊘ Sulphur at $\frac{3}{4}$
- ⊖ Sulphur at $\frac{1}{4}$
- ◕ Copper at $\frac{1}{4}, \frac{3}{4}$ 0.19 copper/site
- ◕ Copper at $0, \frac{1}{2}$, 0.62 copper/site
- ◉ Copper at $\pm\frac{1}{3}, \frac{2}{3}, 0.578$ $\pm\frac{1}{3}, \frac{2}{3}, 0.922$ } 0.41 copper/site

temperature of 103.5°C, so the junction is formed between cadmium sulphide and hexagonal chalcocite. As the formed solar cells cool below 103.5°C the chalcocite will change to the monoclinic phase, most probably with a considerable degree of twinning. Even though the phase transition is rapid[179] it seems likely that some disorder in the copper sites will remain unless the cells can be annealed in some way.

Some djurleite is inevitably present in the copper sulphide layer. In the chemiplating process Boer[180] has suggested that djurleite forms initially on the cadmium sulphide as an interlayer before chalcocite starts to grow and this would be promoted by the presence of cupric ions in solution. Cook et al.[179] completely converted single crystals of cadmium sulphide to copper sulphide by chemiplating and observed small amounts of djurleite amongst the chalcocite. They also observed that the relative amount of djurleite increased after the copper sulphide crystals were heated at 90°C in argon for 88 hours. Hill and Edwards[167] have postulated the existence of a layer of djurleite in cells produced by dry-barrier techniques.

Djurleite[175] is the phase which occurs at the composition $Cu_{1.96}S$[174] and has orthorhombic crystal structure[179,181]. The local atomic arrangement in djurleite is quite well matched to that in chalcocite[179], as can be seen in Table 2, and the composition range $1.96 < x < 1.998$ consists of a mixture of chalcocite and djurleite. For copper sulphide in the composition range $1.96 > x > 1.8$ i.e. between djurleite and diginite, the structure appears to be hexagonal[179], with the local atomic arrangement still reasonably well matched to chalcocite and cadmium sulphide. Table 2 shows that the c-axes of low chalcocite and cadmium sulphide match to within $\frac{1}{2}\%$ but the a- and b-axes mis-match by about 5%. This causes stress during the formation of the junction, which is relieved by the formation of cracks[179]. These can play an important role in the degradation of the solar cell since they promote the migration and agglomeration of copper atoms.

The crystallographic relationships between copper sulphide and cadmium sulphide were graphically illustrated by te Velde[182] who observed the growth of the different phases of copper sulphide in an operational cell heated through the various transition temperatures. He showed that the cadmium sulphide and chalcocite c- and a-axes coincided and observed the three possible arrangements of the chalcocite structure corresponding to the three a-axes of cadmium sulphide.

The optical properties of copper sulphide have been measured by a number of workers, and much of the data have been discussed by Stanley[31]. The most careful and complete measurements of the optical constants of copper sulphide single crystals have been made by Mulder[183,184,185] and summarised by Dielman[186]. Mulder[183] prepared single crystals of copper sulphide of various compositions by the complete conversion of very high quality single crystal platelets of cadmium sulphide. He then measured the optical constants along the various crystallographic directions, assuming the chalcocite to be orthorhombic. The optical constants measured by Mulder[183] for copper sulphide of various compositions are shown in Fig. 26. Mulder's experimental points have been omitted for clarity, and it can be clearly seen that the absorbance in chalcocite is highly anisotropic. Furthermore chalcocite is the only material which has a large absorbance in the photon energy region between 1.2 and 1.5 eV, and this occurs only when the light is incident along the c-axis. The absorption coefficients derived from these

Table 2

Hexagonal CdS	Hexagonal Cu$_2$S	Chalcocite Orthorhombic	Chalcocite Monoclinic	Djurleite Orthorhombic	Hexagonal Cu$_x$S 1.96 > x > 1.80	Hexagonal Cu$_{1.91}$S	Tetragonal Cu$_{1.96}$S	Diginite Pseudocubic	Cubic CdS
$a/\sqrt{3}$ 4.1367	a 3.961	a 11.848	a 15.246	a 15.71	a 15.475	a 11.355		a 5.56	a 5.839
		$a/3$ 3.949	$b/3$ 3.961	$a/4$ 3.93	$a/4$ 3.869	$a/3$ 3.785	a 4.008	$a/\sqrt{2}$ 3.93	$a/\sqrt{2}$ 4.129
		b 27.330	b 11.884	b 13.56	$\sqrt{3}a$ 26.803	$\sqrt{3}a$ 19.667			
$\sqrt{3}a$ 7.1650	$\sqrt{3}a$ 6.861	$b/3$ 6.832	$2a\sin\beta/4$ 6.794	$b/2$ 6.78	$\sqrt{3}a/4$ 6.701	$\sqrt{3}a/3$ 6.556	$3\langle 104\rangle$ 6.81	$\sqrt{3}a/\sqrt{2}$ 6.81	$\sqrt{3}a/\sqrt{2}$ 7.151
		c 13.497	c 13.494	c 26.84	c 13.356	c 13.506	c 11.268		
c 6.7161	c 6.722	$c/2$ 6.748	$c/2$ 6.747	$c/4$ 6.71	$c/2$ 6.678	$c/2$ 6.753	$2\langle \bar{1}02\rangle$ 6.532	$2a/\sqrt{3}$ 6.42	$2a/\sqrt{3}$ 6.742

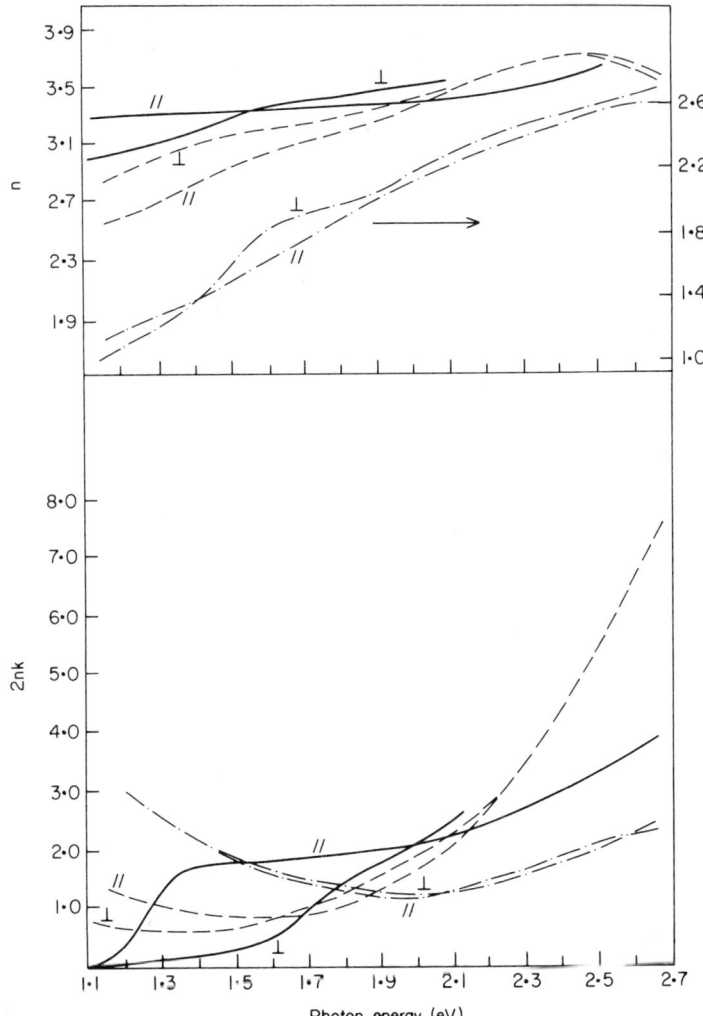

FIGURE 26. The optical constants of —·—·— diginite; ———— djurleite and ——— chalcocite for light perpendicular and parallel to the c-axis.

measured optical constants are shown in Fig. 27 together with the absorption coefficients measured by other workers.

Mulder[183] fitted the experimental measurements of optical constants to expressions describing both direct and indirect band-to-band transitions. This

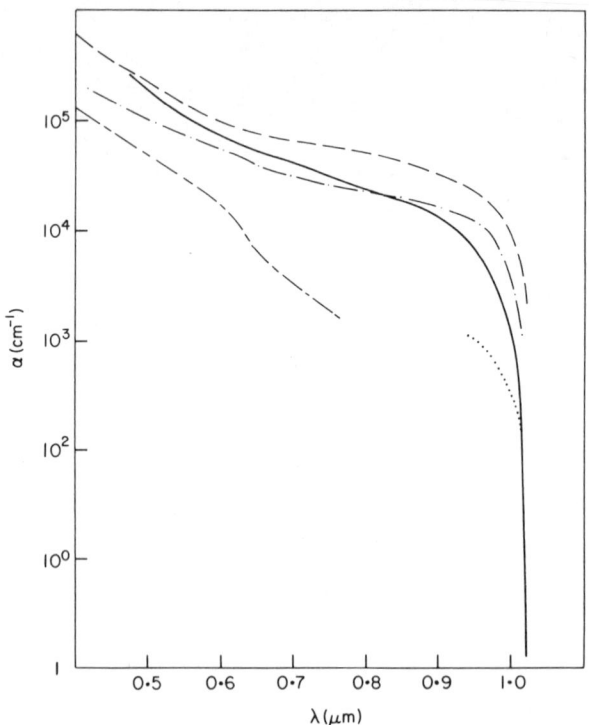

FIGURE 27. The absorption coefficient of cuprous sulphide as measured by ——— Mulder[183]; ——— Eisenman[297]; —·—·— Shiozawa et al.[152]; ————— Sorokin et al.[298] and .... Marshall and Mitra[299].

led him to propose[185] an electron band picture of chalcocite in which there are two valence bands, with one having its maximum at $k = 0$, being lower in energy than a valence band with a maximum at $k \neq 0$. Transitions between the conduction band with a minimum at $k = 0$ and the higher energy valence band give rise to the initial absorption between 1.2 eV and 1.5 eV, whilst the increase in absorption above 1.5 eV is due to the onset of direct transitions. Mulder[185] also measured the effect on the optical constants of heating the chalcocite from room temperature to above the phase transition temperature and found that the absorption between 1.2 eV and 1.5 eV was hardly changed whilst the absorption increased by a significant amount for photon energies above 1.5 eV. Mulder[185] interpreted these results in terms of his proposed band scheme and suggested that the $k \neq 0$ valence band is due mainly to the sulphur sub-lattice which changes very little with the change in phase whilst the $k = 0$

valence band contained a large contribution from the copper sub-lattice. The copper sub-lattice is much more disordered in the high temperature phase, and one might thus expect an increase in the energy of this valence band, giving rise to the increased absorption which he observed. Mulder[185] also proposed that this band-structure exists in djurleite and diginite and explained their different optical constants as being due to a Moss–Burstein shift (see Section 2). As the copper deficiency increases, the hole density increases, and the Fermi Level will fall. Djurleite is assumed to be degenerate, with the Fermi level within the higher ($k \neq 0$) valence band, but above the lower ($k = 0$) valence band. The indirect transitions would thus show a larger energy gap whilst the direct transitions would be unaltered, in agreement with Mulder's[183] data. As the copper deficit is further increased, the Fermi level will drop until it enters the $k = 0$ valence band. When this occurs, the direct transitions will be affected and the effective direct energy gap will increase with increasing copper deficit, which is again in agreement with Mulder's[183] observations. The free carrier absorption in all the copper sulphide samples was found[185] to increase in approximate proportion to the copper deficit as might be expected.

The absorption coefficient of 4.10^4 cm^{-1} seems rather high for an indirect transition, and Mulder's[185] interpretation has been challenged by Loferski[187] who considered that the transition is more likely to be direct, and pointed out that a valence band with a maximum $k \neq 0$ is unknown amongst the well-understood semiconductors. The small temperature dependence of the optical absorption in the 1.2 eV to 1.5 eV region found by Mulder[185] between room temperature and 100°C seems more typical of direct than of indirect transitions, particularly when compared with the rather larger temperature variations found for the optical constants at photon energies above 1.5 eV. Loferski[187] has found that chalcocite is cathodoluminescent at low temperatures with the luminescence peaking at 0.966 μm (1.28 eV). The djurleite and diginite phases were not observed to produce any cathodoluminescence[187] Glew and Bryant[188] have studied the luminescence from actual cells at 77 K excited by 10 kcV electrons, and found three peaks, at 0.78 μm, 0.885 μm and 0.98 μm. The peak at 0.98 μm increased in intensity after ion implantation of copper whilst the 0.78 μm peak decreased in intensity. Oxygen implantation increased the intensity of the peak at 0.78 μm but decreased the intensity of the peaks at 0.98 μm and 0.885 μm. Glew and Bryant[188] identify the peak at 0.98 μm as being due to chalcocite, in agreement with Loferski and Shewchun[187] and suggests that the 0.78 μm peak is due to djurleite. The peak at 0.885 μm was tentatively identified as coming from the sulphur $3p \to$ copper $4s$ transition. As pointed out by Loferski and Shewchun, djurleite seems to be degenerate and should thus have a very small recombination lifetime, and it is surprising that luminescence is observed from this phase with an intensity similar to that from chalcocite. It seems more likely that the 0.78 μm emission is due to oxygen centres at the junction to which the implanted oxygen could easily diffuse after the 10 minute 200°C anneal which Glew and Bryant used after implantation, which could be excited by 10 keV electrons whose range is greater than the thickness of the copper sulphide layer.

It can be seen from the discussion above, that the optical properties of copper sulphide are now quite well known, but their interpretation in terms of a band-structure is open to question.

The electrical properties of copper sulphide have been measured by a number of workers. Copper sulphide is a p-type semiconductor in which copper vacancies act as acceptor centres. Measurements of resistivity, Hall mobility and carrier concentrations are listed by Stanley[31]. There seems to be general agreement amongst the measurement values and it is found that the chalcocite phase has a resistivity between about $10^{-2}\,\Omega\,cm$ to above $1\,\Omega\,cm$ as the carrier concentration varies from $10^{19}\,cm^{-3}$ to $10^{15}\,cm^{-3}$ with decreasing copper deficit. The hole mobility seems to be less dependent on the stiochiometry[189] and values[31] of $1-3\,cm^2\,V^{-1}\,sec^{-1}$ have been measured for copper sulphide films and up to $25\,cm^2\,V^{-1}\,sec^{-1}$ for single crystals. The electrical conductivity of copper sulphide increases approximately linearly with copper deficit[169,189] as might be expected. Djurleite thus has a resistivity of a few $m\Omega\,cm$, whilst diginite has a resistivity about an order of magnitude lower. Windawi[190] has measured the carrier concentration of a nominally chalcocite film produced by chemical conversion from cadmium sulphide and found a value of $10^{21}\,cm^{-3}$ which results in the copper sulphide being degenerate with the Fermi level 0.2 eV inside the valence band. This is confirmed by the observation by Bougnot et al.[189] of a semiconductor to semi-metal transition at a composition of about $Cu_{1.95}S$.

The electron diffusion length has been measured in chalcocite, djurleite and diginite. Gill and Bube[191] formed a thick copper sulphide layer on flat single crystal cadmium sulphide. The sample was then angle-lapped with an angle of a few degrees and the photocurrent due to a small light spot was measured as the light spot was scanned across the bevelled junction region. The minority carrier diffusion length was estimated from these measurements to be between 0.1 μm and 0.4 μm, depending on the sample. Mulder[192] has derived the diffusion length in single crystal copper sulphide from measurements of the spectral response using a small single crystal of cadmium sulphide as a probe. Mulder[192] found that the diffusion length was about 350 Å in the direction perpendicular to the c-axis and about 950Å in a direction $35°$ off the c-axis[186], and that the diffusion length of djurleite and diginite were below 50 Å[186], as might be expected for degenerate semiconducting materials. Partain et al.[193] have used the electron beam induced current technique in a scanning electron microscope to estimate the minority carrier diffusion length in a chemiplated solar cell. The non-uniformity and roughness of the junction in a solar cell make other techniques extremely difficult. Partain et al.[193] found that the diffusion length in their copper sulphide was between 0.1 and 0.56 μm and in their cadmium sulphide films the diffusion length was between 0.1 and 0.31 μm which seems rather short. Their technique measured the diffusion length parallel to the c-axis of the copper sulphide and thus confirms the view[186] that the diffusion length in chalcocite is much longer for diffusion parallel to the c-axis than for diffusion perpendicular to the c-axis.

The effective mass of holes in copper sulphide increases slightly as the copper deficit increases, going from about $1.5\,m_0$ for chalcocite to about $2\,m_0$ for diginite[189].

The electrical properties of copper sulphide are quite temperature dependent. The variation of resistivity, hole density and Hall coefficient with temperature has been measured by Bougnot et al.[189,194] from about 5 K to 900 K for chalcocite, djurleite and diginite. Their results show that copper sulphide with a copper deficit of more than about 0.05 mole is degenerate in agreement with the optical data of Mulder[185]. In chalcocite the Fermi level rises with temperature and is about 50 meV above the valence band edge at room temperature. A number of investigations have been made of the variation of resistivity with temperature when copper sulphide is heated through the phase transition temperatures. As can be seen from Fig. 28 there is a clear change in resistivity at the phase transition temperatures which provides an extremely useful means of detecting the existence of the phases.

Doping of copper sulphide can have some very useful effects. Okamoto and Kawai[169] found that doping of diginite by indium caused the electrical and structural properties to change to those of chalcocite. This has been confirmed by Bougnot et al.[194] who found that dopants such as cadmium and zinc also produced this effect. This result is of importance since the junction formation process often leads to some cadmium doping of the copper sulphide layer and this can be seen to be possibly beneficial.

4.5 Analysis of the Cadmium Sulphide–Copper Sulphide Junction

For the first ten years or so after the cadmium sulphide cell had been introduced by Reynolds[28,118], it was generally thought that the cell operated as some kind of metal-semiconductor junction. Cusano[39] first suggested that the copper might be in the form of copper sulphide and evidence from the Philips Laboratories[195] and the Clevite Corp.[196] confirmed this suggestion. Shiozawa et al.[196] considered the cells to be p–i–n junctions with a strongly p-type layer of cuprous sulphide with a thickness of about 0.3 μm, an approximately intrinsic layer of cadmium sulphide compensated by copper, and an n-type layer of cadmium sulphide. Both te Velde[195] and Shiozawa et al.[196] proposed that the cuprous sulphide layer was responsible for most of

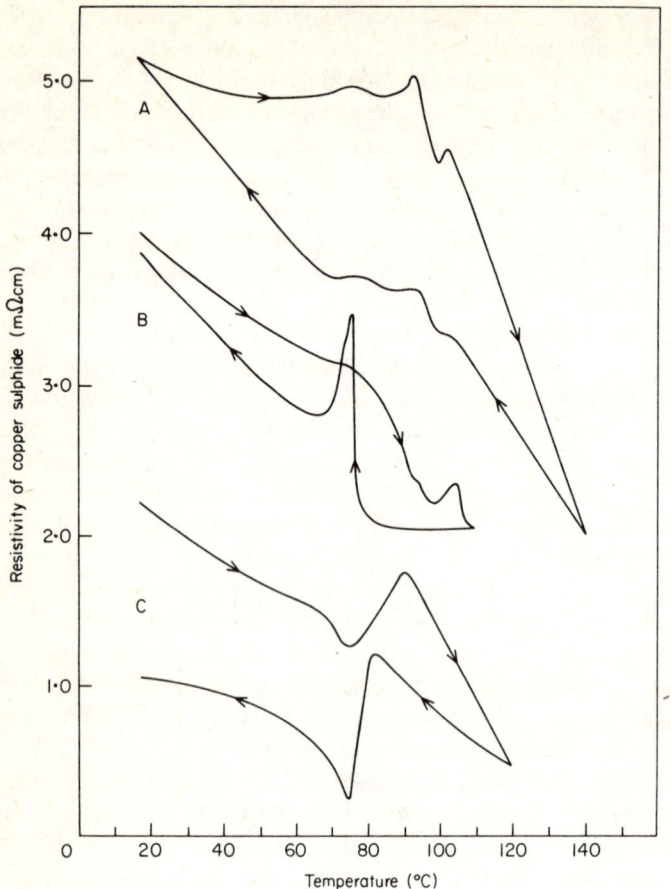

FIGURE 28. Resistivity of cuprous sulphide as a function of temperature on dry-barrier cells (A) $\eta > 4\%$, (B) $1 < \eta < 4\%$, (C) $\eta < 1\%$.

the optical absorption, and the cell was basically an anisotype heterojunction. The intrinsic region of cadmium sulphide was proposed by Shiozawa et al.[196] partly to explain some of the properties of the cells and partly because their measurements of copper diffusion in cadmium sulphide suggested that the cadmium sulphide close to the junction would be quite strongly compensated, by copper acceptor centres. This intrinsic region, about 1 μm thick, would be photoconductive and would cause an illumination dependent series resistance, and would affect the spectral response and the response time of the cell.

Van Aerschodt et al.[197] disputed the existence of the photoconductive intrinsic region and suggested that the barrier height and the occupancy of the interface states could be illumination dependent.

Bube and co-workers[191,198] considered that the heterojunction had a conduction band spike and that tunnelling through this spike and recombination via interface states played a major role in determining the current transport. Later work by Fahrendruch and Bube[199] also emphasised the dominance of tunnelling–recombination currents but concluded that the conduction band discontinuity at the junction was negative, i.e. a notch rather than a spike. They confirmed the suggestions by van Aerschodt et al.[197] that a photoconductive layer was not present in the cells and that the probability of recombination via the interface states is illumination dependent.

Te Velde[166] has analysed the heterojunction in the cases of either a notch or a spike in the conduction bands at the junction. He assumed that tunnelling does not contribute to the currents across the junction but that recombination occurred at the junction by unspecified processes. For the case of a notch in the conduction bands, te Velde[166] found that the electron current injected into the copper sulphide is small compared to the recombination current at the junction, but the transport of photogenerated carriers across the junction into the cadmium sulphide can be very efficient if the interface recombination velocity is small compared to the velocity of electrons at the top of the barrier. He also showed that the open-circuit voltage would decrease as the notch in the conduction bands at the interface increased and that the optimum condition would be achieved if the conduction bands were continuous.

When a spike in the conduction bands exists at the interface, te Velde[166] showed that the photocurrent collection efficiency decreased as the spike height increased, whilst the open-circuit voltage was independent of spike height. In this case also, therefore, the optimum condition is when the conduction bands were continuous.

The model proposed by te Velde[166] assumed that the injected currents were thermally activated over the barriers at the junction. The temperature and bias dependence of the nett current would be altered if tunnelling were to be introduced into the model, but the broad conclusions of the importance of conduction band continuity and small recombination velocity at the junction would still apply.

Pfisterer et al.[200,201] have analysed the results on their cells assuming a model similar to that proposed by Gill and Bube[191]. Their results are consistent with a spike of height 0.27 eV in the conduction bands at the interface and depletion layer widths of about 10 Å in copper sulphide and about 100 Å in cadmium sulphide. They also consider that copper diffuses into the cadmium sulphide, giving some degree of compensation. Illumination of the junction changes the occupancy of the copper acceptor levels. Infrared light will decrease the positive space charge at the junction whilst light of energy above about 1.4 eV will cause an increase in the positive space charge. The width of the spike is thus dependent on both the intensity and spectral content of the illumination.

Martinuzzi et al.[73,202] have investigated the dark conduction mechanisms in cells formed by chemiplating cadmium sulphide in both single crystal[202] and thin film[73] form. The dark current/voltage curves of both types of cells were measured over a wide temperature range, and fitted to generally empirical expressions. Their results were in good agreement with the tunnelling expressions [equations (31) and (32)] discussed earlier for temperatures below 300 K. For reverse dark currents Martinuzzi et al. found that about 5000 tunnelling steps were needed in thin film[73] cells whilst about 230 tunnelling steps were used in single crystal[202] cells. For a typical tunnelling step length of a few Ångstroms, these data would suggest that total tunnelling distances were about 1 μm in thin film cells and about 0.1 μm in single crystal cells.

The forward currents found both for single crystal and thin film cells could not be represented only by multi-step tunnelling-recombination mechanisms. At low forward bias and room temperature, thermal activation was required, with the activation energy of 0.65 eV being identified as the average barrier height for current flow by tunnelling through interface states in the single crystal cell[202]. For the thin film cells at low temperature and low forward bias a tunnelling process with between 50 and 80 steps was dominant, with thermally activated processes becoming more important as the temperature and forward bias increased[73].

Martinuzzi et al.[202] concluded from their measurements on single crystal cells that the diffusion potential was 0.82 eV and that the junction had a notch in the conduction bands of depth 0.14 eV resulting in a valence band discontinuity of 1.34 eV. They found that the Fermi level was 0.15 eV below the conduction band in the bulk cadmium sulphide and 0.12 eV below the conduction band in the depletion region, suggesting that only slight compensation occurred in these single crystal cells.

Boer[157,203–206], Rothwarf[207,208] and Rothwarf and Boer[55] have considered the operation of cadmium sulphide–copper sulphide cells in detail. Rothwarf[207] has considered the current mechanisms in these heterojunctions and has concluded that the simple thermal excitation model gives far too low a current density, and that the most probable mechanism at room temperature near the maximum power point is thermally activated tunnelling of electrons from the conduction band of cadmium sulphide into interface states with subsequent recombination taking place close to the interface. Non-activated tunnelling of electrons and the tunnelling of holes from the valence band of the cuprous sulphide were regarded as important only at low voltage, in agreement with Martinuzzi et al.[73]. Thermally activated tunnelling-recombination currents depend critically on the interface recombination velocity, which can be written as the product of the interface state density, capture cross-section and electron thermal velocity. Assuming that the interface states are due to edge dislocations, the lattice mis-match of cadmium sulphide and chalcocite leads to an estimated interface state density of about 5.10^{13} cm^{-2} with a capture cross-section of 10^{-14} to 10^{-15} cm^2. The electron thermal velocity is about 10^7 cm sec^{-1}. The forward current for thermal activated tunnelling-recombination can be written

$$J_F = es_I N_C \exp\left(\frac{e\phi_a}{kT}\right) \exp\left(\frac{eV}{kT}\right) \qquad (58)$$

where s_I is the interface recombination velocity, $N_C = 2(2\pi m_e kTh^{-2})^{3/2}$ for CdS $\sim 2.10^{18}$ cm^{-3} at room temperature and $e\phi_a$ is the activation energy, which for cells at the Institute of Energy Conversion is about 0.9 eV. This expression for forward current leads to an open-circuit voltage which also depends on the activation energy and the interface recombination velocity

$$V_{OC} = \phi_a + \frac{kT}{e} \ln\left(\frac{J_L}{es_I N_C}\right)$$

Rothwarf[207,208] has calculated the photogenerated current to be expected for various cell geometries, taking into account the recombination at grain boundaries. The collected current depends significantly on the properties of the cuprous sulphide layer such as thickness, electron diffusion length and outer surface recombination velocity.

Boer[157,203–206] has extended Rothwarf's model by considering in detail the fields and electron density distribution around the depletion layers. Boer makes the point that an illuminated junction is effectively reverse biassed at voltages below the open-circuit voltage. As the voltage across the junction is reduced below the open-circuit voltage, the junction becomes more "reverse" biassed, and the reverse current increases. The rapid transport across the junction of the electrons photogenerated in the copper sulphide causes the electron density at the interface to fall to a low level, and this limits the reverse current. The increase in junction fields which accompany the increasing "reverse" bias can result either in a rapid increase in tunnelling current from copper sulphide to cadmium sulphide or in the formation of high field domains in the depletion layer of the cadmium sulphide. Boer's[157] calculations of the electron density, field and potential distributions are shown in Fig. 29. The probability of formation of high field domains depends on the properties of the copper-doped depletion region of the cadmium sulphide. If these domains do form, then the electron density remains constant over the domain region and the electric field is limited to the domain field, as shown in Fig. 29. Increasing

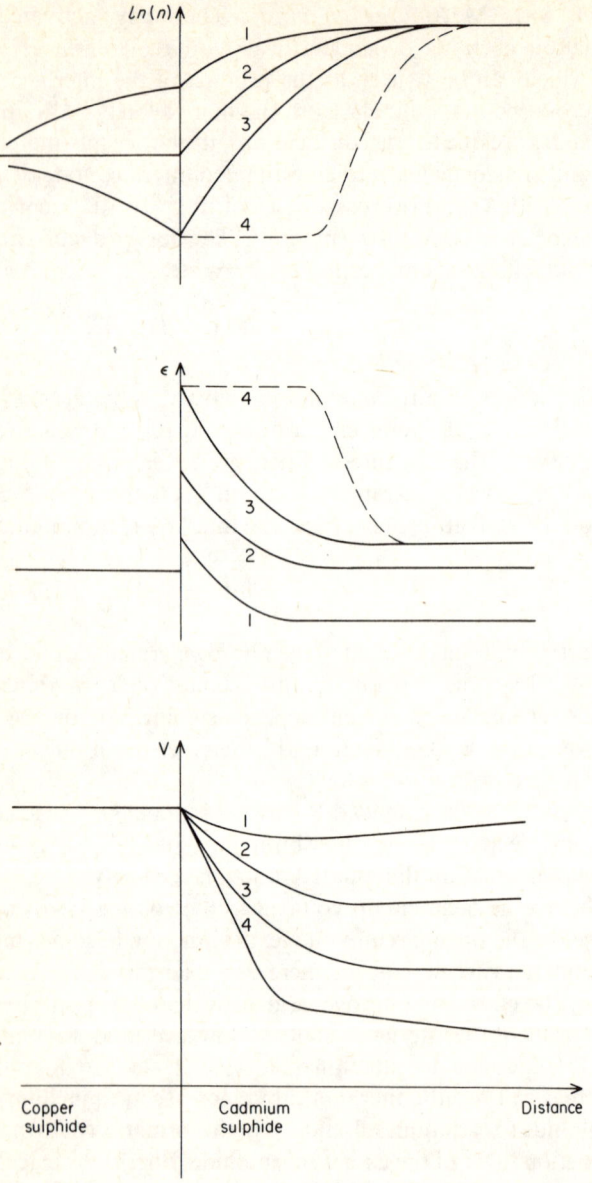

FIGURE 29. The variations at the interface of electron concentration, electric field and potential under various bias conditions according to Boer[157].

(1) $V > V_{OC}$ (2) $V = V_{OC}$ (3) $V \approx V_m$ (4) $V \sim 0$.

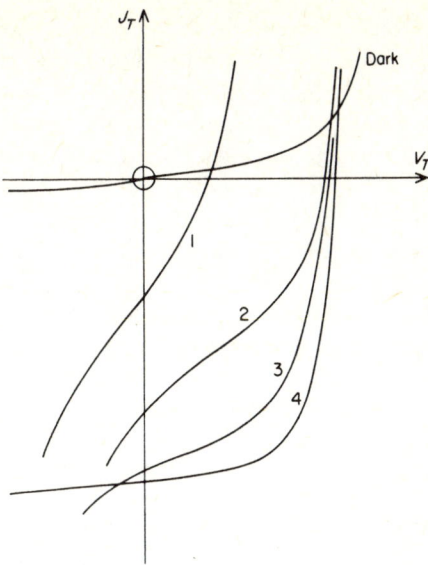

FIGURE 30. The effect of heat treatment after encapsulation on cells produced by the Institute of Energy Conversion[205]. (1) Before heat treatment; heat treatment at 130°C for (2) 6 hours, (3) 9 hours, (4) 12 hours.

the "reverse" bias then increases the width of the domain but does not alter the electron density or the field. The reverse current thus saturates and gives a current/voltage characteristic with a high fill-factor and a large value of dV_T/dI_T [see equation (39)]. If high field domains do not form in the cadmium sulphide depletion region, the fields at the junction will increase with increasing "reverse" bias until reverse breakdown occurs due to tunnelling. The reverse current will then increase rapidly with increasing "reverse" bias and this will result in a current/voltage characteristic with a low fill-factor and a small slope dV_T/dI_T which would, on the analysis given in Section 2.3 be ascribed to a low shunt resistance. The development of current saturation as the copper doped region in cadmium sulphide is extended by heat treatment is shown in Fig. 30. If Boer's[157,203-206] explanation is correct the change in short-circuit current with heat treatment can now be understood quite well, since the tunnelling currents in the untreated samples will be limited by the small diffusion lengths of electrons in the copper sulphide layer. It is not clear from Boer's analysis why the tunnelling currents through the junction in the absence of high field domains should not be larger than the saturation current if the copper sulphide layer has a long diffusion length. Boer[157] has also shown that the recombination current at the junction has little effect on the observed short-circuit current, but this interface recombination reduces the open-circuit voltage, in agreement with Rothwarf's[207] conclusion.

As discussed in Section 2.2, Gandham (private communication) has calculated the current which would be found in a planar junction, taking account of the optical interference effects which must exist in layers whose optical thickness is of the same order as the wavelength of the incident light. Gandham used the optical data of Mulder[183] for the chalcocite layer and also took account of the reflections from the substrate using the optical constants from standard tables for the substrate metals. His results are shown in Fig. 31 and, compared with those of Rothwarf[207], it can be seen that when optical

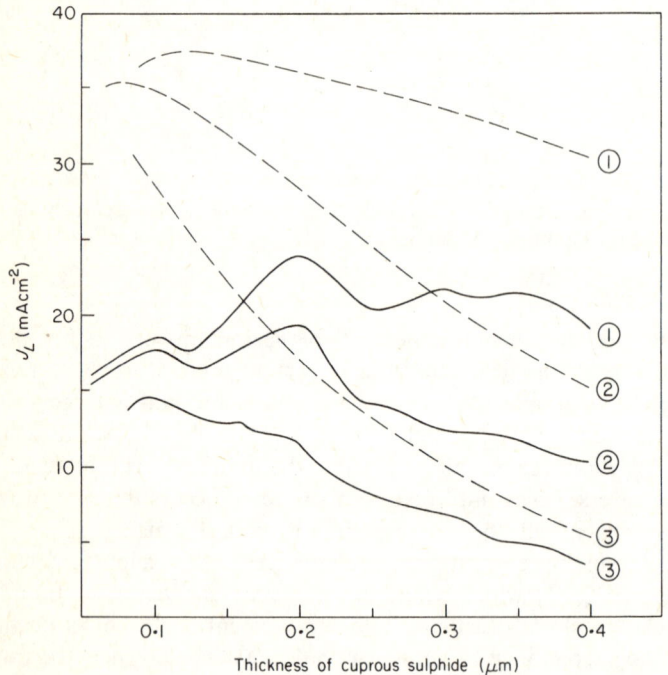

FIGURE 31. Variation of photogenerated current with cuprous sulphide layer thickness for the cases: ——— Reflectivity of the cell as a whole equal to zero. ——— Layers assumed to be optically plane and flat with the appropriate reflectivity. ① $L = 0.5\,\mu m$ ② $L = 0.2\,\mu m$ ③ $L = 0.1\,\mu m$
Surface recombination velocity = $10^4\,\mathrm{cm\,sec^{-1}}$.

interference is taken into account, the optimum copper sulphide layer thickness is smaller, and the expected maximum current is larger than when these effects are ignored. Part of the difference in magnitude of the currents calculated by Gandham and Rothwarf[207] is due to the recombination at crystallite boundaries which Rothwarf[207] included in his calculation but which was ignored by Gandham. The current through the junction depends critically on the diffusion length of electrons in the copper sulphide and the outer surface recombination velocity, and the current for a range of values of diffusion length and recombination velocity is shown in Fig. 32. These results suggest that the diffusion length should be greater than about 0.4 μm and the surface recombination velocity should be less than 10^4 cm sec^{-1} if efficient

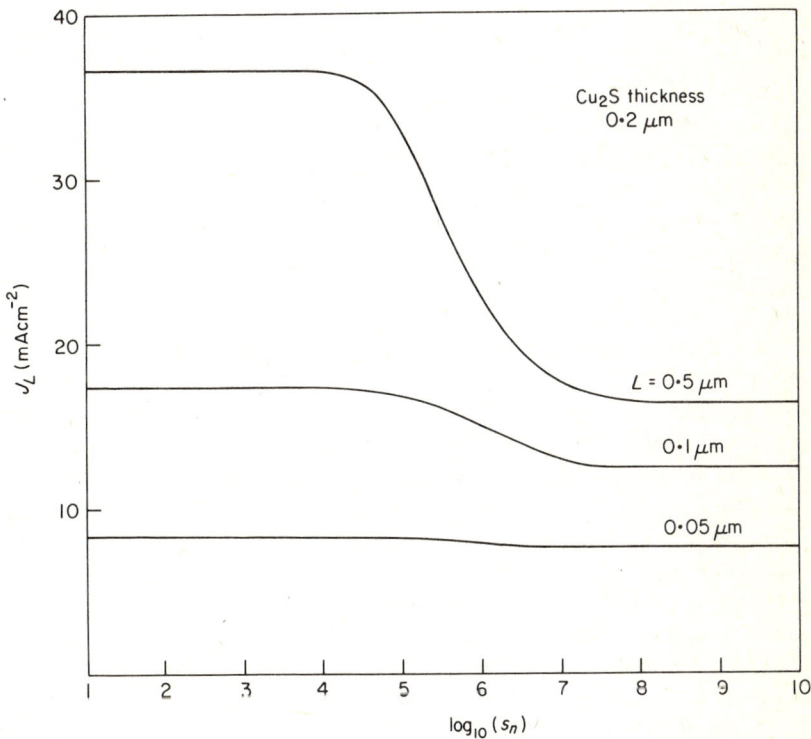

FIGURE 32. The photogenerated current as a function of the outer surface recombination velocity of a cuprous sulphide layer of thickness 0.2 μm and the given values of diffusion length.

current collection is to be achieved, which is in broad agreement with the conclusions of Boer[157,206] and Rothwarf[207].

Boer[203] has discussed the question of whether the junction between copper sulphide and cadmium sulphide is abrupt or graded. He pointed out that the solubility limit of copper in cadmium sulphide is about 100 ppm[209] and that Auger depth profiling shows negligible cadmium concentration in chemiplated copper sulphide layers. Boer[203] also suggested that the absence of any evidence for a mixed copper sulphide–cadmium sulphide phase supported the conclusion that the junction is abrupt.

Sullivan's[209] results on the solubility limit of copper in cadmium sulphide were obtained by dipping grains of single crystal cadmium sulphide in a chemiplating bath to give a surface layer of copper sulphide. These grains were heated to between 250°C and 500°C until a uniform copper distribution was obtained throughout this grain, and the average copper concentration in the grain was then measured after removal of the remaining copper sulphide layer. The solubility of copper in single crystal cadmium sulphide is thus being measured and this will depend on the availability of cadmium vacancies, which appears[209] to be dependent on the dislocation density. The dislocation density near the interface in thin film copper sulphide–cadmium sulphide solar cells must be quite large in order to accommodate the lattice mis-match, and one might thus expect that the copper solubility near the interface in these thin film cells would be much larger than that found in single crystal cadmium sulphide. Furthermore, the dislocation density will decrease at greater distances from the interface, and the copper solubility should also decrease. These qualitative arguments would suggest that a copper gradient should exist, and this seems to be confirmed by depth profiling data. The composition against depth curves measured by Meakin[210] and Casperd[128] are shown in Fig. 33. These data appear to show clear evidence for a composition gradient in both cadmium and copper near the junction, but the uncertainties in depth profiling must be taken into account. The uncertainties in Auger depth profiling using ion beam etching techniques depend to some extent on the ion energy used, but are usually between 10% and 15% of the depth when more than 0.1 μm has been removed. The uncertainties in depth arise because of the surface roughness induced by the ion etching process, and the Auger spectra are thus produced from a range of depths. The apparent gradients observed around the interface by Meakin[102] fall within the uncertainties in the depth measurement, and the copper concentration 0.1 μm from the interface is probably due, at least in part, to preferential diffusion[211] of copper down the c-axis direction of the cadmium sulphide in regions of less-than-perfect crystallinity such as dislocations etc. so these data are somewhat inconclusive. Van Aerschodt et al.[197] measured the variation of composition with position for very thick layers of copper sulphide plated on to single crystal cadmium

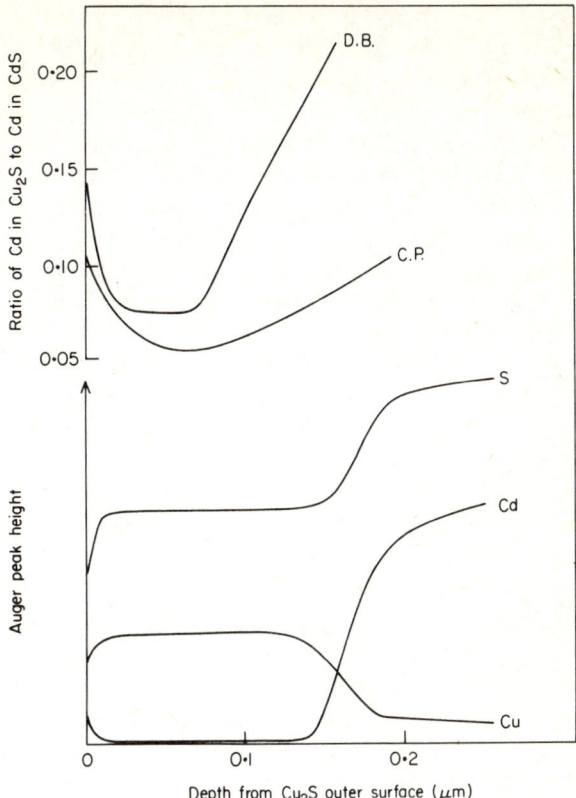

FIGURE 33. The upper curves show the variation in cadmium concentration in a dry-barrier (D.B.) and chemiplated (C.P.) cell as measured by Rutherford backscattering[128]. The lower curves show composition profiles of sulphur, cadmium and copper measured by Auger depth profiling[210].

sulphide, by scanning an electron beam across the edge of the junction and measuring the intensity of the characteristic x-rays produced. They observed an apparent gradient in the junction for both cadmium and copper but the graded region was of the same size as the diameter for their electron beam. There were graded regions on either side of the junction where copper concentrations decreased exponentially from a few percent to negligible values over a distance of about 20 μm, whilst cadmium concentrations in the copper sulphide decreased by a similar amount over 10 μm. Van Aerschodt et al.[197]

concluded that the thin film cells would have a significant and position dependent concentration of cadmium throughout the copper sulphide layer and of copper in most of the cadmium sulphide film. The results of Rutherford back-scattering measurements on the cadmium concentration in Casperd's[128] dry-barrier cells (Fig. 33) show that cadmium is present in the copper sulphide layer in these cells and does have a significant composition gradient. Meakin's[210] results on plated single crystals of cadmium sulphide suggest that cadmium is quite efficiently leached out of the copper sulphide layer whilst the junction is being formed in the plating bath. This difference between the two junction formation processes might be expected on the basis of the discussion in Section 4.3.

Brandhorst[212] has also measured a cadmium gradient extending for about 10 μm into a thick copper sulphide layer plated on to single crystal cadmium sulphide, in agreement with Van Aerschodt et al.[197]. Brandhorst[212] also measured the capacitance/voltage characteristics of the junction and concluded that the nett donor concentration was constant in the cadmium sulphide for a distance of about 0.2 μm and then increased abruptly by about two orders of magnitude over 50 Å–100 Å. Since cadmium vacancies also act as acceptor centres, these measurements do not preclude a decrease in cadmium concentration from approximately bulk cadmium sulphide level to a lower level over about 0.2 μm with an equal but opposite variation in copper concentration. Brandhorst[212] concluded from his measurements on operational cells that the copper sulphide close to the junction was heavily doped with cadmium, giving a hole concentration of about 10^{16} cm^{-3} with the Fermi level 0.2 eV above the valence band, whilst the majority of the copper sulphide had a hole concentration of about 10^{20} cm^{-3}.

The rather scant evidence available does suggest that the interface region between copper sulphide and cadmium sulphide extends over many lattice spacings and is probably not abrupt. There may well be a quite rapid change of composition with distance in the immediate vicinity of the junction, but there are probably more extensive regions in which the compositions are changing more slowly. In view of the method of formation of these junctions via the replacement reaction, this seems to be inherently more likely than a change from cuprous sulphide to cadmium sulphide taking place over only one or two lattice spacings.

Cheung et al.[213] have considered the effect of a grading in composition on the energy-bands near the junction. They have shown that a notch or spike in the conduction bands will be reduced in magnitude to a value ΔE_{CG} from its magnitude ΔE_{CA} in an abrupt junction when the composition is graded in the n-type material over a distance L where

$$\Delta E_{CG} = \Delta E_{CA} - [2eN_d(V_D-V)/\epsilon]^{1/2}L \qquad (59)$$

N_d is the donor concentration (more exactly it should be $N_d - N_a$) V_D is th diffusion potential, V the applied bias and ϵ the dielectric constant. Taking typical values for cadmium sulphide of $N_d \sim 10^{21}\,\text{m}^{-3}$, $V_D = 0.85V$ and $\varepsilon = \epsilon_0 \epsilon_r = 8.8 \cdot 10^{11}\,\text{Fm}^{-1}$, one finds at zero bias that the barrier is reduced from say 0.35 eV to 0.19 eV if the composition is graded over a length of 1000 Å. For the rather larger depletion layer electron concentrations of $10^{16}\,\text{cm}^{-3}$ found by Brandhorst[212] or Gill and Bube[191] the barrier effectively disappears if the composition is graded over 600 Å, whilst for the values of $5.10^{17}\,\text{cm}^{-3}$ observed by Luquet et al.[214], a gradient over only about 60 Å would lead to the disappearance of any conduction band discontinuity.

From these examples, it can be seen that the conduction band discontinuity depends strongly on the compositional variations around the junction and thus on the details of junction fabrication and post-formation treatments. It is quite probable therefore that the different results obtained by workers using different procedures reflect genuine differences in the junctions which are formed. It is also clear that Boer's[206] assumption that no discontinuity exists in the conduction bands at the junction is quite reasonable given a modest composition grading length. The form of junction which seems most likely is one which is graded over a distance which is short compared to the minority carrier diffusion length, but long compared to the effective radius of donor or acceptor states. The majority of workers find that the capacitance/voltage characteristics are those suggestive of an abrupt junction, although there are results on ceramic cells suggestive of a graded junction (see Section 4.6).

4.6 The Characteristics of Copper Sulphide–Cadmium Sulphide Solar Cells

The dark current/voltage characteristics of these cells have been measured by a number of groups and it has usually been found that two mechanisms are in operation.

Martinuzzi et al.[202] found that the forward current density in junctions formed on single crystal cadmium sulphide could be expressed by the equation

$$J = A^* T^2 \exp\left(\frac{e\phi_b}{kT}\right) \exp\left(\frac{eV}{AkT}\right) \quad \text{for} \quad T > 300\,\text{K} \qquad (60)$$

where A^* is the effective Richardson constant and the barrier height $\phi_b = 1.08 \pm 0.08$ volts. The diode factor A was found to be 1.2.

For thin film solar cells, Martinuzzi and Mallem[73] found that the forward currents were described by a tunnelling relationship of the form

$$J = C \exp(\beta V + \gamma T) \quad \text{for} \quad V > 0.3\,\text{volts} \qquad (61)$$

where the constants $\beta \simeq 15\,\text{V}^{-1}$ and $\gamma \simeq 2.5 \cdot 10^{-2}\,\text{K}^{-1}$.

Luquet et al.[214] found that the forward dark currents in the cells produced by S.A.T. were compatible with two independent tunnelling–recombination mechanisms and could be expressed by a relationship of the form

$$I_{FD} = 0.35 \exp[-11(1-1.8\cdot 10^{-3}\,T)]\,(e^{11.5V}-1)$$
$$+ 0.45[-44(1-1.8\cdot 10^{-3}\,T)]\,e^{36V}$$

At a temperature of 300 K, this gives forward current densities of

$$J_{FD} = 5.5\cdot 10^{-4}(e^{11.5V}-1) + 5.3\cdot 10^{-7}\,e^{36V} \quad \text{mA cm}^{-2}$$

Luquet et al.[214] also investigated the forward dark current densities of doped cells produced by S.A.T. and at 300 K found a relationship

$$J_{FD} = 2.10^{-2}(e^{8V}-1) + 2.5\cdot 10^{-6}\,e^{32V} \quad \text{mA cm}^{-2}$$

The forward current densities from the doped cells are smaller than those from undoped cells for a bias of around 0.5 V, and this should result in higher open-circuit voltages in the doped cells.

A semilogarithmic graph of the dark forward current/voltage characteristic has been found by most investigators to show two distinct slopes, as shown in Fig. 34. The cross-over point from a lower slope to a higher slope occurs at larger bias voltages for lower temperatures, but is usually about 0.2 to 0.3 volts at room temperature.

The current/voltage characteristics of cells under illumination are almost always poor until the cells have been subjected to a heat treatment. The effect of heat treatment on the characteristics of a cell seems to be fairly similar for all production procedures, and typical variations are shown in Fig. 30. There are major variations in the time interval represented by the stages indicated on Fig. 30, which can be minutes[152] or a few hours[205] for chemiplated cells and minutes[166] to tens of minutes[128] for dry barrier cells. The interpretation of these changes has been discussed in the previous section, and is not yet universally agreed.

Hadley and Philips[215] have analysed the current/voltage characteristics under illumination of cells produced at the Institute of Energy Conversion. They point out that equivalent circuits of the type shown in Fig. 12 will give valid results only if the ratio of the series resistance and shunt resistance is much less than unity. They have shown that the short-circuit currents in their cells are linear functions of light intensity over four orders of magnitude variation in light intensity, and that the open-circuit voltage varies linearly with the logarithm of the light intensity over three orders of magnitude of light intensity for high efficiency cells. Cells of lower efficiency showed a break in linearity with the diode factor changing from unity for high intensity to two at low intensity. The diode factor for high efficiency cells was found to be 1.00 ± 0.01 over the whole intensity range from 0.1 mW cm^{-2} to 100 mW cm^{-2}.

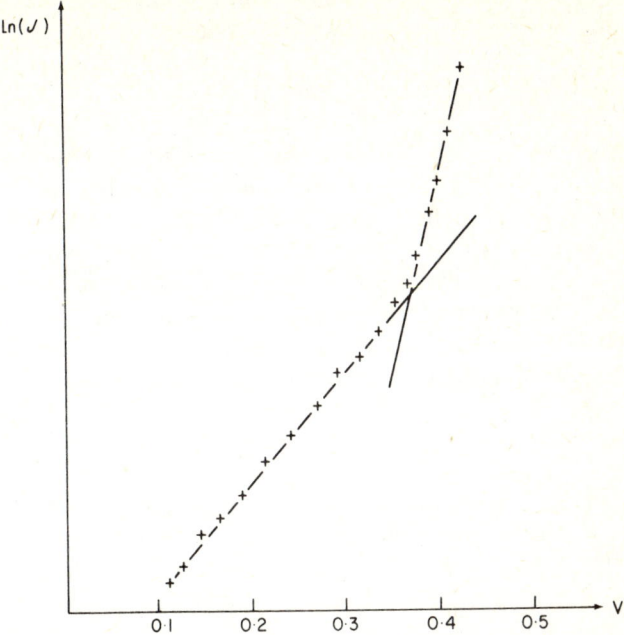

FIGURE 34. Logarithm of forward dark current against forward bias.

The forward currents could be described by a relationship of the form

$$J = J_{00} \exp\left(-\frac{e\phi}{kT}\right)\left[\exp\left(\frac{eV}{kT}\right) - 1\right]$$

where

$$J_{00} \approx 10^5 \quad \text{cm}^{-2} \quad \text{and} \quad \phi \simeq 0.9 \text{ volts}$$

For cells of low efficiency, a second diode mechanism became operative with a diode factor of two and $J_{00L} \sim 0.6 \text{ A cm}^{-2}$ and $\phi \sim 0.4 \text{ V}$.

Typical values of series resistance found on the I.E.C. cells are about 20 mΩ although this can be as high as a few tenths of an ohm on lower efficiency cells. The shunt resistances which they found were typically 500 Ω. The fill-factors of their cells have reached as high as 74%, and Hadley and Philips[215] report roof-top efficiencies between 5% and 6% for their better cells.

Bryant and Glew[84] have analysed the characteristics of cells produced by the International Research and Development Co. Limited in terms of an equivalent circuit similar to that shown in Fig. 12. The current/voltage characteristics were expressed by the relationship

$$J_T = J_{OR}\left[\exp\left\{\frac{e}{AkT}(V_T - J_T R_s)\right\} - 1\right] - J_L + \frac{V_T - J_T R_s}{R_{SH}} \qquad (62)$$

and the measured characteristics were fitted to this equation for a wide range of light intensities. The fitting procedure produced values of J_{OR}, J_L, A, R_s and R_{SH} for the range of light intensities used, and the variations of these parameters with light intensity are shown in Fig. 35 together with the power conversion efficiency. The variations shown in Fig. 35 for these parameters are fairly typical of their cells, although the magnitudes are somewhat different for different cells. The "series resistance" would be expected to become negative at low light intensities, as discussed earlier in Section 2.3, if

$$\left[\lambda' J_d(V_{OC}) + \frac{1}{R_{SH}}\right]^{-1} > \left(\frac{dV_T}{dJ_T}\right)_{J_T=0}$$

but Bryant and Glew appear to be the only investigators to have observed such an effect. The pre-exponential factor I_{OR} can be seen from Fig. 35 to be about 2.10^{-8} A cm^{-2} for 100 mW cm^{-2} illumination intensity, which is significantly higher than the figure of 10^{-10} A cm^{-2} quoted by Rothwarf[207] for the cells produced at the Institute of Energy Conversion, and this is reflected in the lower open-circuit voltage and fill-factor of the I.R.D. cells compared to those from the I.E.C.

Bogus and Mattes[216] found that the characteristics of their cells could be improved quite dramatically by treatment after junction formation. They evaporated a very thin film of copper on to the cuprous sulphide layer and then heat treated the cell in air for about 30 minutes at 180°C. The effects of copper deposition and subsequent heat treatment in the current/voltage characteristics are shown in Fig. 36. The short-circuit currents increased from about 24 mA cm^{-2} to about 33 mA cm^{-2} whilst the open-circuit voltage increased to about 0.52 volts and the power conversion efficiency increased to about 8%. A notable feature of this treatment was that even cells of initially low efficiency improved to a final efficiency of about 8%. The effect of the copper treatment can be explained in terms of a reduction in the copper vacancy concentration in the cuprous sulphide layer. This explanation seems to be confirmed by the large increase in resistivity of the cuprous sulphide and reduction in free-carrier absorption after treatment. Bogus and Mattes[216] also found that post-junction-formation treatment in a hydrogen glow-discharge had an effect on the cell characteristics similar to the copper deposition. The explanation of the effect of hydrogen glow-discharge treatment is not so straightforward as for the copper deposition, although te Velde[166] has postulated that hydrogen treatment of his dry-barrier cells reduced the amount of oxygen in the cuprous

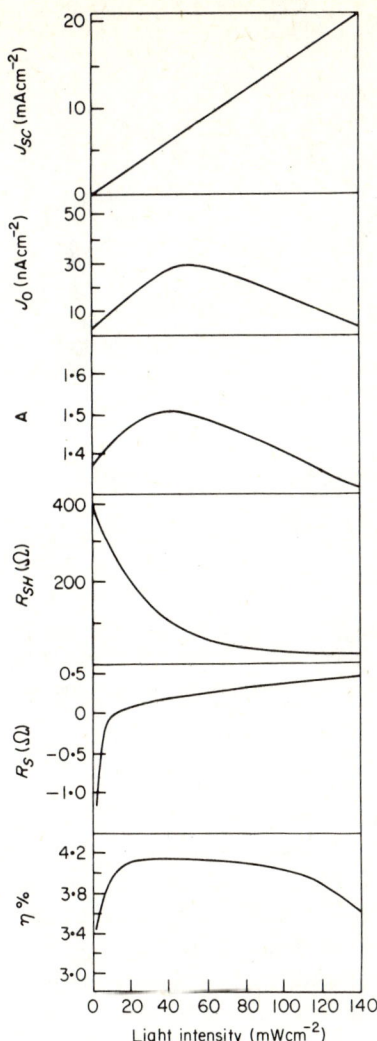

FIGURE 35. Variation of the equivalent circuit parameters of a solar cell with illumination intensity, as measured by Bryant and Glew[84].

sulphide layer, thus altering the shape of the electron energy levels at the junction. The copper deposition treatment[216] has been attempted by other groups[128,200,96] with much less success, showing once again how dependent

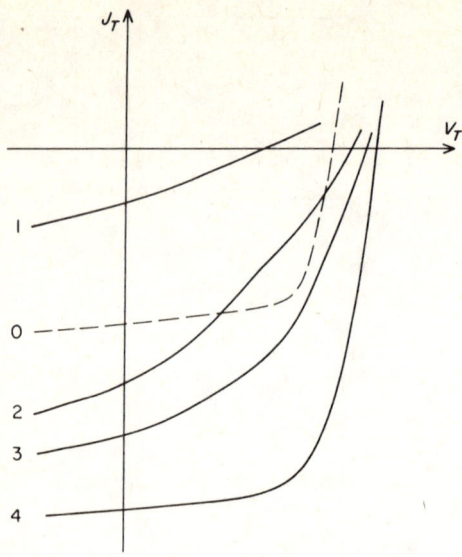

FIGURE 36. Current/voltage characteristics of a cell subject to copper treatment after junction formation (the Bogus Effect)[216]. (0) Before copper deposition. (1) After copper deposition. (2) → (4) Increasing heat treatment times.

on the details of the production process are the properties of cadmium sulphide–copper sulphide cells.

The final processes of gridding and encapsulation can also have a significant effect on the current/voltage characteristics. The encapsulant is either transparent plastic or glass and is usually glued to the outer surface of the copper sulphide with epoxy. The encapsulant thus acts as an antireflection coating and increases the intensity of radiation entering the cuprous sulphide. In view of the effect of high surface recombination velocity on photocurrent, some passivation of the outer surface of the cuprous sulphide would be highly beneficial. Very little has been published on this aspect of manufacture, although Boer[157] has suggested that the formation of a surface layer of copper oxide might give some degree of passivation. The development of satisfactory counter-electrodes or grids has been reviewed by Stanley[31]. The usual form of counter-electrode consists of a grid of fine (0.001 inch) wires with between five and one hundred wires to the inch (LPI), usually made of gold-flashed copper and glued to the cuprous sulphide with gold-filled epoxy. The work of Wolf[86] and Handy[87] can be used to estimate the optimum spacing of the wires, and this has been calculated in the author's laboratories to be about 3 mm for typical values of the cuprous sulphide parameters. Boer et al.[217] have measured the current/voltage characteristics of cells with counter-electrodes having grid spacings of 60×10 LPI, 30×10 LPI, 20×10 LPI, 10×10 LPI and 5×5 LPI on 7.5×7.5 cm cells. The short-circuit current increased as the grid spacing increased due to the increasing optical transparency of the grid. The series resistance was about the same for the first four grids but increased significantly for the 5×5 LPI grid. The optimum spacing would thus appear to be about 0.1 inches, in good agreement with the 3 mm calculated in the author's laboratories.

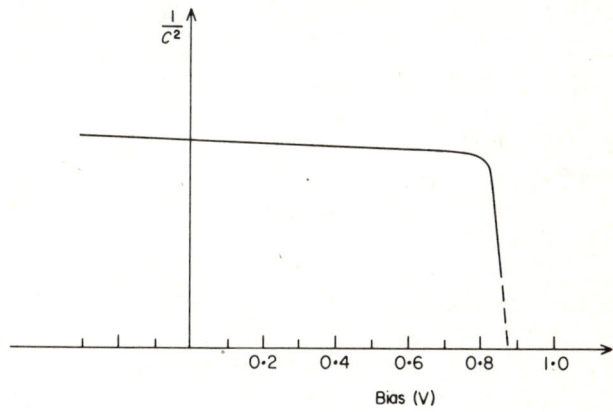

FIGURE 37. The variation of capacitance with bias.

The capitacitance/voltage characteristics have been investigated by a number of groups, and enable estimates to be made of the diffusion potential and the carrier densities near the junction. A typical characteristic for a cell made from evaporated cadmium sulphide is shown in Fig. 37. It is interesting to note that linear variations are usually found for the square of the capacitance, which is a relationship appropriate to a junction in which the conductivity "type" changes over a distance which is small compared to the depletion layer width of the ideal abrupt junction. As discussed in Section 4.5 the change in conductivity type over a short distance does not necessarily preclude the change in composition over longer distances.

If conventional junction theory is applied, the intersect on the voltage axis is the diffusion potential of the junction. The value of diffusion potential measured in this way for a particular cell is found to correlate well with the value estimated from the current/voltage characteristics. The intercept usually occurs at a bias between 0.8 and 1.0 V and some spread in the measured values is found for cells produced within one group as well as between cells produced by different groups. Luquet et al.[214] have found an intercept at about 0.5 V which agreed with the value of diffusion potential derived from the current/voltage characteristics of their cells. The ceramic cells produced by Nakayama[150] did not show a capacitance/voltage characteristic similar to that in Fig. 37 but had a single linear relationship between $1/C^2$ and bias for cells before heat treatment, and after heat treatment at 100°C for 30 minutes. After heat treatment at 150°C for 20 minutes, a single linear relationship was found between $1/C^3$ and bias. The intercept on the bias axis was 0.6 V for the untreated cells, 0.7 V for cells held at 100°C for 30 minutes and 0.5 V for cells held at 150°C for 20 minutes. These cells, therefore, had diffusion potentials similar to that found by Luquet et al.[214] but developed a graded junction after heat treatment at 150°C. Heat treatment at 300°C caused the formation of an intrinsic photoconducting layer and a junction capacitance effectively independent of bias.

The current/voltage characteristics of Nakayama's[150] cells showed excellent current saturation at low forward bias even though the capacitance/voltage curve did not exhibit the change in slope at about 0.4 V shown in Fig. 37. This casts some doubt on Boer's[157] suggestions that this change in slope is due to the onset of formation of high field domains, and the necessity of domain formation for producing current saturation and good fill-factors. The discontinuity in slope is usually ascribed to the existence within the cadmium sulphide of a wide layer which is compensated with copper which has diffused from the cuprous sulphide[212]. The measured carrier concentrations vary by about an order of magnitude for cells produced by different groups, but typically it is found that electron concentrations are about 10^{17}–10^{18} cm^{-3} in the bulk cadmium sulphide and about 10^{15}–10^{16} cm^{-3} in the copper compensated region which is usually between $\frac{1}{4}\mu$m and $\frac{1}{2}\mu$m thick. The hole concentrations in the copper sulphide layer are usually between 10^{19}–10^{20} cm^{-3}, and, as discussed earlier, there is some evidence[128,212,217] for a cadmium compensated layer close to the junction in which hole concentrations are two to three orders of magnitude less.

The spectral response of copper sulphide–cadmium sulphide cells depends to some extent on the production procedure, but it is found for all efficient cells that the response begins at the optical absorption edge of chalcocite. The response increases with increasing photon energy because of the increase in the product of absorption coefficient and thickness of the chalcocite layer (see Fig. 27). For photon energies above 2–2.5 eV, the response tends to fall since carriers are being generated close to the outer surface of the cuprous sulphide, and surface recombination reduces the density of photogenerated carriers which can reach the junction. The detailed shape of the spectral response curve is dependent on the cuprous sulphide thickness, electron diffusion length and surface recombination velocity, the arrangement of the electron energy bands in the junction region and the density and distribution of recombination centres and compensating centres, and to a lesser extent on the properties of the cadmium sulphide layer. For cells with a different geometrical arrangement than usual, i.e. cells in which the light does not enter through the cuprous sulphide layer, but enters through the cadmium sulphide, the spectral response is more dependent on the properties of the cadmium sulphide films, and the response to photons of energy higher than the cadmium sulphide energy-gap can be enhanced[218]. The spectral response of the usual "front wall" cells, where the light enters through the cuprous sulphide, is affected by the optical reflectivity of the substrate on which the cadmium sulphide is deposited[128,208]. If photons which are not absorbed on their first pass through the cell can be reflected by the back-surface contact, the total photogenerated current is increased[128], and since the longer wavelength photons are absorbed less strongly on the first pass than the short wavelength photons, this tends to enhance the red response[218]. A typical spectral response curve is shown in Fig. 38 and three regions can be seen. For wavelengths between

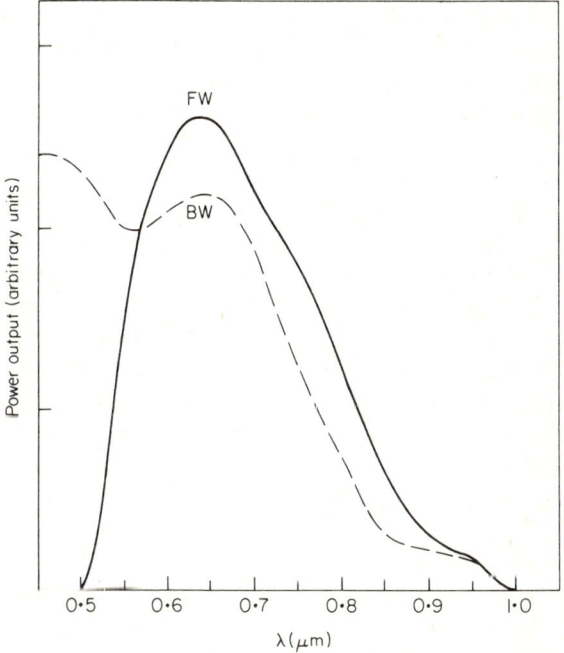

FIGURE 38. Spectral response curves of CdS–Cu$_2$S cells with light incident on Cu$_2$S (F.W.) or incident on CdS (B.W.). The spectral responses of different cells are different in many details but have the same basic shape.

0.65 µm and 1 µm the response increases due to the increasing absorption coefficient of cuprous sulphide. For wavelengths just below 0.65 µm, a dip in response is often observed, and this is probably due to the photoexcitation of centres at the interface. The response increases again for wavelengths about 0.55 µm when carriers photogenerated in the cadmium sulphide also contribute to the photocurrent. The final decrease occurs as higher energy photons are absorbed close to the surfaces of the cuprous sulphide and cadmium sulphide.

4.7 Degradation of Power Conversion Efficiency

The decrease in power conversion efficiency of cadmium sulphide–copper sulphide cells after use for some period of time has been a major problem[31]. These cells are far more sensitive to attack by atmospheric constituents such as water vapour and oxygen than single crystal silicon cells.

The effect of moisture on the properties of cadmium sulphide–copper sulphide solar cells has been investigated by Spakowski et al.[219]. They found that the degradation rate increased with relative humidities above about 40% but was very small at humidities less than 40%. The degradation occurred mainly in short-circuit which decreased faster than the open-circuit voltage by factors of four or five. The series resistance increased by up to one order of magnitude whilst the pre-exponential factor increased by up to three orders of magnitude after exposure to high humidity for two months. The diode factor also increased on exposure to water vapour, and the spectral response altered, mainly due to a loss of red response.

Efficient encapsulation of cells can prevent water vapour reaching the outer surface of cuprous sulphide and thus avoid this form of degradation. "Aclar" or "Mylar" film is often used as an encapsulant with the film stuck into the cell using an epoxy adhesive. Some moisture can penetrate these encapsulants giving lifetimes of only a few years in a normal atmosphere at room temperature, and glass has been used to form a more impervious seal[143]. The lifetime of a cell, i.e. the time required for the power conversion efficiency to decrease to 1/e of its original value, depends quite strongly on temperature and the illumination to which a cell is subjected. Sayed and Partain[220] have shown that the lifetime can be expressed in the form

$$\text{Lifetime} = A_L \exp(E_a/kT)$$

where A_L is a constant and E_a is an activation energy which depends on the ambient gas in contact with the cell outer surface, i.e. with the encapsulant, which in their case was probably Mylar. They found that the activation energy was 0.885 eV for nitrogen ambient, 0.580 eV for argon ambient and 0.397 eV for a dry air ambient. The predicted lifetime of the cells in argon or nitrogen ambient with cyclic illumination was over 30 years at a constant temperature of 25°C and 3 years at a constant temperature of 50°C[220]. With cyclic variations in temperature corresponding to day/night variations, they predicted a lifetime of over 20 years in a nitrogen ambient[220].

The mechanisms which cause the decrease in power conversion efficiency have been considered in detail by Besson et al.[143]. They considered that the degradation was the result of three mechanisms—oxidation processes, electronic processes and electrochemical processes.

The oxidation process was the result of oxidising agents such as moisture or oxygen penetrating the encapsulation and oxidising the cuprous sulphide, thus reducing its copper concentration and hence the short-circuit current. They reduced this process to negligible proportions by covering their cells with a thin sheet of glass[143].

The electronic degradation occurs during illumination and is reversible when the cell is left in the dark. Besson et al.[143] considered that this process is due to the filling of traps close to the interface, thus changing the charge density distribution near the interface, with a resultant change in the height of the barrier at the heterojunction. This type of process increases the diode factor and the pre-exponential factor and reduces the open-circuit voltage. The magnitude of the electronic degradation can be reduced by decreasing the density of traps at the interface, and Besson et al.[143] showed that it could be reduced considerably by judicious but unspecified doping of the cadmium sulphide.

The electrochemical degradation is a result of the decomposition of cuprous sulphide and subsequent diffusion of the copper ions into and through the cadmium sulphide[143,221,222]. There is a critical E.M.F. below which decomposition of cuprous sulphide will not occur, and this E.M.F. varies from zero for $Cu_{2.000}S$ to 0.269 V for $Cu_{1.995}S$[143]. In order that free copper be formed from this decomposition, a crystallisation overvoltage must be applied[143,222], which allows $Cu_{1.997}S$ to be subjected to an E.M.F. of 0.290 V before decomposition occurs. It is usually found in practice that no significant decomposition occurs for voltages less than about 0.33 V due to reaction kinetic considerations. Since the through-film resistance of the cadmium sulphide is two or three orders of magnitude larger than that of the cuprous sulphide, no electrochemical degradation can occur even when the full open-circuit voltage is applied across the cell if the cuprous sulphide is contained in a well-defined layer. A voltage greater than the decomposition potential can exist across the cuprous sulphide only if it penetrates deeply into the cadmium sulphide layer. Besson et al.[143] consider that this penetration is likely to occur either during the junction formation process, or in a completed cell by diffusion down grain boundaries, aided by the electric fields in the cell. If a spike of cuprous sulphide reached through the cadmium sulphide to the back contact, the shunt resistance would decrease leading mainly to a decrease in fill-factor. However, the full cell voltage would now fall across this spike, and if the cell voltage exceeded the decomposition potential, free copper would be plated out giving rise to a short-circuit. If the cell voltage did not exceed the decomposition potential, no free copper, and hence no short-circuits would occur, but a progressive reduction in shunt resistance would be observed as more and more spikes of cuprous sulphide penetrated through to the back contact. If diffusion of cuprous sulphide to the back contact can be prevented, then the electrochemical degradation process can be eliminated. Careful control of the junction formation is needed to prevent undue penetration down grain boundaries into the cadmium sulphide, the stoichiometry of the cuprous sulphide should be high so as to reduce the ionic mobility and the cadmium sulphide layer should be produced with large crystallite size and high packing density, and doped with complexes which block cuprous ion diffusion[223]. After taking these precautions, Besson et al.[143] found that their cells degraded very little after accelerated lifetime tests equivalent to nine years of actual use.

The degradation mechanism in ceramic cells also appears to be the result of copper diffusion[161]. Matsumoto et al.[161] found that ceramic cells degraded in the short-circuit mode but not in the open-circuit mode, and attributed this to migration of copper ions within the cuprous sulphide layer towards their collecting electrodes. However, the electrodes which they used were of silver, and electrodes of this material have been found to give rise to degradation[152].

The effects of radiation damage on cadmium sulphide–copper sulphide solar cells has been reviewed by Stanley[31] whilst, more recently, Bryant and Glew[224] have investigated the effect of positive ion bombardment on the cell characteristics. They found that the range of the positive ions determined the nature of the degradation observed. Ions which penetrated to the junction region caused an increase in the pre-exponential factor, by up to four orders of magnitude for a fluence of about 10^{16} protons, and hence a reduction in open-circuit voltage to about half its original value. The same fluence of protons produced only a 10% change in short-circuit current. Irradiation by xenon ions of 50 keV energy whose range is shorter than the cuprous sulphide layer thickness, reduced the short-circuit current to about half its original value after a fluence of about 2.10^{16} ions, whilst the open-circuit voltage changed by a few percent. As a general result, Bryant and Glew[224] concluded that if the incident irradiating particles deposit most of their energy within the cuprous sulphide layer, the major degradation will occur in the short-circuit current, and the reduction in short-circuit current increases as the radiation fluence increases. If the irradiating particles deposit their energy in the junction region, the major degradation occurs in the pre-exponential factor and the diode factor, and hence the open-circuit voltage, and again the degradation increases as the fluence is increased. Irradiation by electrons has been investigated by

Glew and Bryant[225] and they have shown that the degradation in short-circuit current is due mainly to the reduction in transparency of the encapsulants. In non-encapsulated cells, they found an increase in the pre-exponential factor by one or two orders of magnitude after a fluence of 2.10^{17} electrons of energy above 200 keV, and this caused a reduction in the open-circuit voltage. As might be expected, lower energy electrons (100 keV) caused greater degradation because of their shorter range.

It can be seen from the above discussion that cadmium sulphide–copper sulphide thin film solar cells have good resistance to radiation damage for space uses, and that the lifetime under terrestrial conditions can be expected to exceed twenty years if the procedures suggested by Besson et al.[143] are adopted.

5. Other Thin Film Solar Cells

Any semiconductor junction will produce some electrical power when illuminated, but we are concerned in this review only with those junctions which can convert sunlight into electricity with reasonable efficiency. We have seen in Section 2 that the energy-gap for optimum conversion efficiency by a p–n homojunction is about 1.2 eV to 1.4 eV, and the materials suitable for such junctions range from silicon with an indirect gap of 1.1 eV to aluminium antimonide with an indirect gap of 1.6 eV. Within this range are indium phosphide with a direct gap of 1.27 eV, gallium arsenide with a direct gap of 1.43 eV and cadmium telluride with a direct gap of 1.44 eV, as well as cuprous sulphide, and some of the less familiar ternary compounds such as copper indium selenide[226]. Thin film homojunctions have been formed from silicon gallium arsenide and cadmium telluride with some success.

Schottky barrier cells have been investigated fairly extensively, and the MIS structures have shown considerable promise. Since the open-circuit voltage of an MIS junction depends far more on the interface than on the energy-gap of the semiconductor, it is likely that the optimum energy-gap for this type of solar cell will be somewhat lower than that for a homojunction. The most successful MIS cells so far have been formed on silicon single crystals, but work on thin film versions of these cells is proceeding.

Many semiconductors have been combined to form heterojunctions, but it is clear from the work of de Vos[33] that the optimum junction consists of a material of energy-gap 1.1 eV and another material with an energy-gap greater than about 1.3 eV. Thin film heterojunctions, apart from copper sulphide–cadmium sulphide, which have been investigated include InP–CdS, InCuSe$_2$–CdS, GaAs–ZnSe and CdTe–CdS. The first three of these systems have well-matched lattice constants, and should thus have fairly small interface state densities, which we have seen in the last section are very important.

Thus all three types of junction have been produced in thin film form using materials with energy-gaps close to the optimum, and these will be discussed in more detail below.

5.1 Thin Film Homojunctions

5.1.1 Cadmium telluride homojunctions

The deposition of thin films of cadmium telluride has been quite extensively studied[88]. The phase equilibria[227] and free energies[228] of formation have been studied and films have been deposited by thermal evaporation[88], electron beam evaporation[94] and vapour transport[39,229]. Epitaxial films have been produced on silicon[230] and germanium[231] substrates.

The first homojunction solar cells seem to have been produced by Vodakov et al.[232] who produced p–n homojunction cells with a semitransparent top contact. The efficiency of their cells in sunlight was 4%. Their cells were developed until the open-circuit voltage reached 0.75 V, the short-circuit current was 9.8 mA cm^{-2} and the fill-factor was 0.63 giving an efficiency of 6% in light of intensity 77.2 mW cm^{-2}[233].

A theoretical investigation into the performance of cadmium telluride homojunctions has been carried out by Bell et al.[234]. They have shown that the open circuit voltage and power conversion efficiency are critically dependent on the minority carrier lifetimes which must be greater than 0.1 μsec for an efficient cell. The optimum dopant concentrations were calculated to be about 10^{16} cm^{-3} and the junction depth should be less than 1 μm. Bell et al.[234] considered that it would be possible to produce cadmium telluride with dopant concentrations of about 10^{16} cm^{-3} and lifetimes of about 1 μsec if the density of trapping states in the middle of the band-gap could be reduced sufficiently. From their work on cadmium telluride gamma detectors, they identify the deep traps as doubly ionised cadmium vacancies, and suggest that these can be transformed by complexing with halogens into relatively shallow traps. They measured the diffusion length of carriers using EBIC techniques and found that the diffusion length and carrier lifetime did increase with halogen dopant concentration.

Most of the work done on cadmium telluride cells have used a nominally heterojunction structure, and these will be discussed in Section 5.3.

5.1.2 Silicon homojunctions

Thin films of silicon are usually produced by vapour phase epitaxial growth on crystalline substrates. The use of crystalline substrates rules this technique out of consideration in this article, but an interesting development involves epitaxial growth of silicon onto edge-fed ribbon-grown silicon sheets.

The EFG ribbon has the possibility of being significantly cheaper than cut single-crystal plates and, being rectangular, can fill over 90% of the area of an array as opposed to about 70% for circular discs. D'Aiello et al.[235] have produced p^+/p/graded n/n^+ structures with open-circuit voltages of 0.636 V, and short-circuit currents of about 25 mA cm^{-2} at incident intensities of 97.5 mW cm^{-2}. The fill-factor was about 80% leading to efficiencies of over 12%.

Chu and co-workers have attempted to produce silicon homojunctions on cheaper substrates. The substrates investigated have included metallurgical grade silicon[236,237,238], graphite[236,237,239] and coated steel[237,240]. The coated steel substrates produced poor cells[237], but the other two substrates have given cells with efficiencies of about 4%. The silicon layers were produced by thermal decomposition of trichlorosilane in a conventional reactor furnace and the dopants were introduced into the gas flow during growth since this produced cells of higher efficiency than conventional diffusion techniques[236]. Using an n^+/p^+ structure on re-crystallised metallurgical grade silicon, Chu[236] has produced a 5 cm^2 area cell with an open-circuit voltage of 0.58 V, short-circuit current of 17 mA cm^{-2} and a power conversion efficiency of 5.2% at AM0. The major problem would appear to be the high degree of recombination in the junction region, since the diode factor is two and the pre-exponential factor is about 10^{-7} A cm^{-2}.

The characterisation of polycrystalline silicon deposited on cheap substrates has been studied by Miyazaki et al.[241] whilst Fang et al.[242] have produced polycrystalline silicon films on aluminium by electron beam evaporation. Aluminium appears to be a very suitable substrate for the growth of silicon films, since apart from its character as an acceptor in silicon, a silicon/aluminium eutectic is formed at temperatures above 577°C and this can significantly increase the crystallinity of silicon films[243].

5.2 Schottky Barrier Cells

5.2.1 Silicon Schottky barrier cells

A considerable amount of work has been done on the performance of silicon Schottky barrier solar cells, some of which was discussed in Section 2. The majority of this work has used single crystal slices of silicon so as to be able to characterise the properties of the barrier, particularly the effects of an insulating interfacial layer. Very little work appears to have been done on forming Schottky barriers on polycrystalline films, possibly because of the difficulty of avoiding short circuits.

The formation of Schottky barriers on amorphous films of silicon is a subject of growing interest and promise[244]. Amorphous silicon films can be produced by the evaporation of silicon on to substrates at temperatures below about 300°C, but are produced most cheaply in dense, pin-hole free layers by the plasma decomposition of silane. Wronski et al.[244] used stainless steel substrates with a thin layer of n-type amorphous silicon[245] to give good ohmic contact to the steel, and a 1 μm thick undoped layer of amorphous silicon on top of the n-type layer. Metal contacts were then evaporated on to the outer surface of the amorphous silicon and the current/voltage characteristics were measured. The diodes had nearly ideal characteristics, with diode factors close to unity and pre-exponential terms of about 10^{-12} A cm^{-2}. The barrier heights produced by the metals investigated were higher than those found on single crystal silicon, giving the potential of higher open-circuit voltages. The series resistance of these diodes decreased as the temperature increased, with an activation energy of 0.57 eV, similar to the activation energy for resistivity of amorphous silicon[246].

Amorphous silicon, as produced by glow-discharge decomposition of silane, is an n-type semiconductor with a room temperature resistivity between 10^{11} Ω cm[247] and 10^9 Ω cm[245] with a mobility in the extended electron states of about 10 cm^2 V^{-1} sec^{-1}[24]. It can be doped[245] n-type by flowing phosphine with the silane up to conductivities of about 10^{-2} $(\Omega$ cm$)^{-1}$ and can be doped p-type using diborane to conductivities of 10^{-3} $(\Omega$ cm$)^{-1}$. Amorphous silicon has an absorption coefficient which is up to an order of magnitude greater than that of single crystal silicon at around 2 eV[248,249].

Spear et al.[250] and Carlson and Wronski[249] have produced thin film p–n homojunctions using amorphous silicon. The diode characteristics showed a diode factor of about two due to recombination from the localised tail states[250]. Carlson and Wronski[249] investigated the photovoltaic efficiency of p–i–n structures on indium–tin–oxide coated glass. Short-circuit currents of over 10 mA cm^{-2} were observed in AM1 sunlight with open-circuit voltages of 0.6 V. The highest open-circuit voltage observed was 0.79 V (AM1 illumination) but the fill-factors were poor, series resistance limited to about 40%, so the maximum efficiency observed was 2.4%.

The Schottky barrier amorphous silicon solar cell appears to be a very promising candidate for large area, low cost devices, if the limitations caused by the high series resistance can be overcome.

5.2.2 Gallium arsenide Schottky barrier cells

Gallium arsenide has been used by Stirn and Yeh[251] for Schottky barrier solar cells using processes which could be adapted to the use of polycrystalline thin films[251]. They used tellurium-doped GaAs$_x$P$_{1-x}$ ($0 < x < 0.48$) films grown by vapour phase epitaxy on to single crystal gallium arsenide substrates. The surface of the GaAs–P is oxidised, and then a semitransparent gold top contact is evaporated on. The effect of oxidation is to increase the open-circuit voltage whilst leaving the short-circuit current and fill-factor almost unaffected, leading to increases in efficiency of the order of 50% (up to about 15%). The techniques by which this process could be adapted for the production of large area devices were not discussed by Stirn and Yeh[251] but vapour deposition on to polycrystalline gallium arsenide substrates is a possibility.

5.3 Thin Film Heterojunction Cells

5.3.1 Cadmium telluride–copper telluride

The first cadmium telluride–copper telluride solar cell was produced by Cusano[39]. Cusano[39,252] produced cadmium telluride films by a vapour reaction between cadmium plus cadmium iodide and tellurium. These films were then dipped in a cuprous ion solution to produce a replacement reaction directly analogous to that for copper sulphide formation [equation (52)].

Copper telluride is a semiconductor whose properties are much less well-known than copper sulphide. Copper telluride occurs naturally in the form of Rickardite (Cu_3Te_2)[253] and Weissite (approx. Cu_2Te)[253]. The crystal structure of Cu_2Te has been studied by Nowotny[254] who found a hexagonal unit cell with $c = 7.27$ Å and $a = 4.23$ Å. For tellurium rich samples ($\sim Cu_{1.95}Te$) Nowotny[254] measured values of $c = 7.24$ Å and $a = 4.19$ Å. The atom-to-atom distances in Cu_2Te were estimated to be Cu–Cu $= 2.33$ Å, Te–Te $= 2.82$ Å and Cu–Te $= 2.67$ Å. It seems therefore that cuprous telluride has phases which are closely analogous to those of cuprous sulphide, and that the cell dimensions are a poorer match to those of cadmium telluride than is the case for cuprous sulphide and cadmium sulphide.

Cusano[39] produced very thin layers of cuprous telluride (50–100 Å) and found that these were p-type with sheet resistances of the order of $100\,\Omega\,\square^{-1}$, suggesting a bulk resistivity of about $10^{-4}\,\Omega\,cm$, corresponding to a degenerate material analogous to diginite. The structure of Cusano's cells was[39,252] (1) base electrode, often of cadmium sulphide, (2) n-type cadmium telluride, with a dopant gradient to produce an upper surface of compensated or even p-type cadmium telluride, (3) p-type cuprous telluride and (4) pressure contacted gold or nickel grids.

The spectral response of the cells was that of cadmium telluride, suggesting that cuprous telluride is opto-electronically inactive and may simply have provided a low resistance current-collecting layer to a cadmium telluride p–n homojunction. The diode characteristics were far from ideal with a diode factor of about 2.7 and a rectification ratio of only a few hundred. The photovoltaic response of the films was good, with open-circuit voltages of about 0.7 V for single crystal cadmium telluride and about 0.5 V for thin film cells. The short-circuit current was higher in the thin film cells, approaching 20 mA cm^{-2} AM1, than for single crystal cells, but the fill-factors were much lower, 0.45 to 0.6 compared to 0.7 for single crystal cells. The efficiencies of Cusano's cells at about 80 mW cm^{-2} illumination reached 6% for thin film cells and 7.5% for single crystal cells[39]. It should be noted however, that these efficiencies are calculated using the illumination incident only on the non-gridded area of the cells[39]. For comparison with the results on other cells, in which the radiation incident over the whole cell area is used, at an intensity of 100 mW cm^{-2}, these efficiencies should be reduced to about 4.5% for thin film cells and about 6% for single crystal cells. Later development of the thin film cells[252] produced small (1.8 cm^2) cells with efficiencies of 6% and large (50 cm^2) cells with efficiencies of 5%, both for approximately AM2 incident radiation.

Bernard et al.[255] have produced copper telluride–cadmium telluride cells by flash evaporation of copper telluride on to either single crystal cadmium telluride or evaporated cadmium telluride films. The properties of the copper telluride were investigated for different tellurium to copper ratios using data previously obtained[256,257]. The junctions formed were found to be non-abrupt with high recombination and Bernard et al.[255] concluded that the cell was probably a cadmium telluride p–n homojunction with the copper telluride acting as a semitransparent conducting electrode. The photovoltaic properties of the cells were limited by the high series resistance which gave low fill-factors. The maximum efficiency obtained by Bernard et al.[255] was 5% at an incident intensity of 62 mW cm^{-2}.

Justi et al.[258] have prepared thin films of cadmium telluride by co-evaporation of cadmium and tellurium. Cuprous telluride was formed in a thin layer on the top surface of the film by a process similar to that of Cusano[39] and grids of various materials were investigated for their effect on power conversion efficiency. Open-circuit voltages of over 0.6 V were observed provided that the cadmium telluride film thickness was greater than 20 μm. Rhenium plated copper grids were found to give the best efficiencies, up to 4.1% under unspecified conditions, but the fill-factors were seriously reduced by the high series resistance and a rather low shunt resistance. All the thin film cadmium telluride–copper telluride cells investigated so far[259] seem to suffer from low fill-factors due to high series resistance, low shunt resistance and excessive recombination near the junction[234].

5.3.2 Cadmium sulphide–cadmium telluride solar cells

Bonnet and Rabenhorst[41,260] have investigated the properties of thin film cells of cadmium sulphide–cadmium telluride with a view to producing graded band-gap solar cells which in principle give the highest possible efficiency[42,261]. Films of p-type cadmium telluride were produced by high temperature gas-phase deposition on molybdenum substrates, and cadmium sulphide was subsequently deposited on the cadmium telluride by thermal evaporation, and an indium top contact was evaporated on to the cadmium sulphide. Very poor contact was found between the p-type cadmium telluride and the molybdenum back contact, but this was improved by heavy doping of the cadmium telluride adjacent to the contact. Open-circuit voltages of over 500 mV were found with short-circuit currents up to 15 mA cm^{-2}, but again with low fill-factors of about 45%, leading to efficiencies of between 5% and 6% for 50 mW cm^{-2} illumination. The cells were exceedingly stable, being able to withstand temperatures of 200°C for hours in vacuum and have worked under water for over one hour without degradation[260].

Martinuzzi[262] has produced cadmium sulphide–cadmium telluride heterojunctions by evaporating cadmium sulphide on to single crystal platelets of cadmium telluride. The diode characteristics were rather poor, with a diode factor between two and three and a pre-exponential factor of about 5.10^{-7} A cm^{-2}. The spectral response of these cells was rather more uniform between 0.8 and 0.5 μm than the thin film cells of Bonnet and Rabenhorst[260] but the efficiency was lower, having the low fill-factor which seems typical of cells involving cadmium telluride.

Hill and co-workers[94,128,263,264] have investigated the opto-electronic properties of cadmium sulphide–telluride alloys. They found that the energy-gap E_A of the alloy CdS_xTe_{1-x} could be expressed as $E_A = 1.5 - 1.1x + 2.0x^2$. The energy-gap of the alloy thus varies from about 1.3 eV for $x = 0.2$ to about 1.5 eV for $x = 0.5$. For alloys with $x \gtrsim 0.5$ films can be produced by thermal or electron-beam evaporation with resistivities, mobilities and diffusion lengths similar to those of good quality cadmium sulphide films[264], and it may be possible to replace cadmium telluride in heterojunction devices by $CdS_{0.5}Te_{0.5}$ with notable reductions in series resistance.

Work on heterojunctions of cadmium telluride with a variety of other materials has been reported[35,265] and some work has been done on their usefulness as solar cells[265].

5.3.3 Indium phosphide and ternary indium compound cells

Indium phosphide and copper indium selenide should form better solar cells in heterojunction with cadmium sulphide than the more usual cadmium sulphide–copper sulphide cell. The higher ultimate efficiency is due partly to the fact that indium phosphide has an energy-gap rather closer to the optimum[53] than that of copper sulphide, and partly to their much better transport properties. Practical efficiencies should be higher also because of the better match of lattice parameters. The mis-match between the lattice parameters of cadmium sulphide and indium phosphide is only 0.36%, whilst for cadmium sulphide and copper indium selenide the mis-match is 1.16%.

Junctions formed between cadmium sulphide and these indium compounds should thus have much lower interface state densities, and open-circuit voltages and fill-factors close to the theoretical limits might be achievable.

Shay et al.[266] have evaporated cadmium sulphide on to single crystals of indium phosphide. They quoted a power conversion efficiency of 12.5% for these single crystal cells at 53 mW cm^{-2} isolation and estimated an efficiency of 9.8% for AM0 isolation. These efficiencies are calculated from the light incident only on the non-gridded areas, and are thus not directly comparable with efficiencies quoted for other cells. In large area cells, grid transparency is about 90%–95%, thus reducing the short-circuit current, whilst the fill-factor is reduced by 5%–10% due to the inevitable increase in series resistance and decrease in shunt resistance compared to small cells. The AM0 efficiency of a large area cell with the same junction properties as the small cells of Shay et al.[266] would thus be 8.5%–9%, about the same as the cadmium sulphide–copper sulphide thin film cells produced by Bogus and Mattes[216].

Bachmann et al.[267] have prepared thin films of indium phosphide on carbon and molybdenum substrate by chemical vapour deposition. Cadmium sulphide was evaporated on to the indium phosphide and small areas (about 0.5 mm^2) were found to exhibit quite good photovoltaic properties. For an illumination of 93 mW cm^{-2} an open-circuit voltage of 0.4 V, short-circuit current of 90 μA from an illuminated area of 0.52 mm^2, and a fill-factor of 0.31 were measured. The efficiency was thus 2.3%, and this was increased to 2.8% by the addition of an antireflection coating. The cells were extremely stable showing negligible degradation in efficiency after heating in air at about 400°C.

Kazmerski et al.[268] have prepared thin film solar cells of copper indium selenide and cadmium sulphide. The copper indium selenide was evaporated on to a gold-metallised glass substrate and cadmium sulphide was then evaporated on top without breaking vacuum. The top contacts were fingers of aluminum. With a calibrated light source of intensity 100 mW cm^{-2} an efficiency of 5.7% was observed with a short-circuit current of over 20 mA cm^{-2}. The open-circuit voltage of 0.44 V was rather lower than that of 0.55 V measured on the single crystal cell of Shay et al.[266] which had an "active-area-only" efficiency of 12.0% for AM1 isolation. Kazmerski et al.[268] also investigated copper indium selenide on cadmium sulphide, i.e. with the light incident on the copper indium selenide side of the junction. These cells were less efficient than the alternative arrangement, as would be expected[53], giving a short-circuit current of 16 mA cm^{-2}, an open-circuit voltage of 0.4 V and an efficiency of 4.4% at 100 mW cm^{-2} illumination intensity. Copper indium selenide/cadmium sulphide cells seem to be much more resistant to degradation than cuprous sulphide–cadmium sulphide cells although much more work on lifetime needs to be done.

Copper indium sulphide has also been considered as a suitable material for use in solar cells since it has a direct gap of 1.55 eV and can be evaporated to form either n- or p-type films. Shewchun et al.[269] have also produced films of copper indium sulphide by the sulphurisation of copper–indium films. Copper indium sulphide has a lattice mis-match with cadmium sulphide of 5.6%, significantly larger than that of cuprous sulphide, so it is unlikely that heterojunctions of copper indium sulphide with cadmium sulphide will be as good as those using indium phosphide or copper indium selenide.

Considering the limited amount of work which has been done on heterojunctions of cadmium sulphide with these indium compounds, the achievements to date (end of 1976) are impressive and encouraging.

5.4 Semiconductor–Electrolyte Junction Solar Cells

A junction between a semiconductor and an electrolyte can behave in a manner similar to a Schottky barrier. The electrolyte must be capable of both oxidation and reduction reactions, i.e. it must contain a redox couple. These cells originate from the work of Becquerel[1] and are variously described as Becquerel effect cells[270], photogalvanic cells[271,272] or electrochemical solar cells[273,274].

The interface between a semiconductor and an electrolyte has been discussed in some detail by Gerischer[3] and the reactions which can occur in these cells have been considered by Williams[274]. Anderson and Chai[270] have shown that short-circuit currents of about 30 mA cm^{-2} could in principle be produced by AM1 illumination of a photogalvanic cell using a semiconductor with an energy gap of about 1.3 eV, and Gerischer[4] has shown that an open-circuit voltage of up to $(E_g - 0.5)$ eV could be generated. Even with quite modest fill-factors therefore, these cells are potentially of useful efficiency.

Gerischer[4] has investigated the variation of open-circuit voltage and short-circuit current with light intensity for single crystal electrodes of cadmium sulphide, cadmium selenide and gallium phosphide. Both cadmium sulphide and gallium phosphide produced open-circuit voltages of about 1.25 V at 100 mW cm^{-2} illumination and short-circuit currents of over 1 mA cm^{-2}. The open-circuit voltage of the cadmium sulphide cell varied logarithmically with light intensity, whilst the short-circuit current varied linearly with light intensity up to about 10 mW cm^{-2} when saturation occurred. Recently Gerischer and Gobrecht[273] have produced a photogalvanic cell using single crystal cadmium sulphide and a redox electrolyte of potassium ferrocyanide/potassium ferricyanide which has given open-circuit voltages of 1 V, short-circuit currents of about 6 mA cm^{-2} and an efficiency of 9.4% under 40 mW cm^{-2} illumination from a xenon lamp. This is estimated to be equivalent to an efficiency of 5.5% to solar radiation of the same intensity.

The characteristics of these cells degraded rapidly due to photodecomposition of the cadmium sulphide which produced a layer of sulphur at the interface.

Anderson and Chai[270] have investigated the properties of a photogalvanic cell using a cadmium sulphide single crystal electrode and a solution of potassium chloride in water as the electrolyte. They used light of about 400 nm wavelength at very low intensities (about 1 mW cm^{-2}), and found conversion efficiencies of over 4%, even though the fill-factors were about 20%.

So far no work on thin film semiconductor electrodes appears to have been published, but work in the author's laboratory has shown that 30 μm thick films of cadmium sulphide deposited with a high packing density on molybdenum substrates can produce open-circuit voltages of over 1 V with short-circuit currents of about 2 mA cm^{-2} under 100 mW cm^{-2} illumination from a tungsten lamp. The short-circuit current decreased over a period of a minute or so after the illumination commenced and became stable at between a half and a quarter of the initial 2 mA cm^{-2}. This seemed to be due to some form of depletion process since stirring of the electrolyte increased the short-circuit current to above 2 mA cm^{-2}. The electrolyte used was identical to that used by Gerischer[4,273] and the effects of pH on the cell characteristics and degradation were also the same as found by Gerischer[4,273]. The spectral response of the cell was that of the cadmium sulphide electrode, also in agreement with the results of Gerischer and Gobrecht[273]. It would seem from these results that the single crystal electrodes used so far could be replaced by thin films provided that these are prepared carefully, although there is evidence[275] that surface irregularities on the electrode affect the properties of the semiconductor/electrolyte interface.

The development of photogalvanic cells into useful cells depends on the possibility of preventing their rapid degradation due to photodecomposition of the semiconductor electrode. An attempt to counter the degradation of cadmium sulphide electrodes has been made[276] by using an aqueous Na_2S–Na_2S_x redox couple. The reduction reaction at the semiconductor anode for this couple is much more probable than the photodecomposition reaction, and any sulphur which is produced on the surface of the cadmium sulphide is dissolved. The degradation problem is alleviated at the expense of a large reduction in open-circuit voltage and hence in cell efficiency. A great deal of work needs to be done on optimising the electrolyte for these cells, and it would be helpful if semiconductor materials with energy-gaps in the 1.0 eV to 1.6 eV range were used rather than cadmium sulphide.

One very attractive use of these cells, if they could be developed for long term use, would be as combined solar thermal and solar electric generators. Most of the energy absorbed from the solar radiation appears as heat, and the electrolyte could be pumped through a heat exchanger, thus keeping the cell cool and providing a heat output. If the fluid pump were driven by the electrical output of the cells, an interesting self-controlled system could be devised, in that pumping would occur only when the solar insolation was large enough to provide the necessary pumping power, and hence large enough to have raised the temperature of the electrolyte sufficiently to provide a useful heat output.

6. Future Uses of Thin Film Solar Cells

"... behold the blessed vision of the sun, no longer pouring his energies unrequited into space, but by means of photo-electric cells and thermo-piles, these powers gathered into electric storehouses, to the total extinction of steam engines, and the utter repression of smoke."

It is intended in this final section to discuss the extent to which Appleyard's[12] far-sighted vision might become reality.

The advent of the so-called energy crisis has led to a widespread realisation of the importance of energy and its production and conservation, and has initiated a search for sources of energy which will not be depleted. Solar energy, in one form or another, has supplied all of mankind's energy requirements in the past, and, in principle at least, could continue to do so for the foreseeable future. The total power available from one square kilometre of solar cells in bright sunlight is about 100 MW, so a small fraction of the area of the Sahara desert could supply all the electrical power requirements of Europe, Africa and the Middle East. It can be seen therefore that solar cells have the capability of supplying the total electrical energy needs of the world, but this simple picture ignores a number of problems. Some of these are inherent in the nature of solar energy, whilst some are technological and thus capable of technological solutions.

Two aspects inherent to solar energy create problems—its variability and its low power density. When the sun is directly overhead, in a clear sky, each square metre of horizontal surface receives about 1 kW. This power is, of course, reduced to zero during the hours of darkness. The power is also reduced at other latitudes by additional atmospheric absorption and scattering, and for surfaces which are not normal to the sunlight, the intensity is reduced according to the cosine law. For any given geographical location, the solar insolation has both daily and seasonal variations, and the power received at any particular time is a less useful parameter than the total energy received in the course of one year. Even this total energy received per year varies from year to year due to variable weather conditions. It is usual to quote the mean energy received within a standard period (i.e. a day, month or a year) with a cumulative frequency distribution showing the probability that the energy received during any particular period will differ from the mean by a given amount. The mean daily insolation varies from about 22 MJ m^{-2} in tropical areas to about 6 MJ m^{-2} in latitudes of 70° or so. The seasonal variation in average energy received per day is quite small in tropical latitudes, but can be more than a factor of ten in northern Europe, and this gives rise to the major problem in the large scale use of solar cells in northern latitudes.

A design study of the possibility of using solar cells to provide all the electricity used in a house near Newcastle upon Tyne has shown that it would now be possible to construct a system to do this, but the energy storage requirements are prohibitive[277]. The electrical power used over a three year period was taken for a fairly typical house in which power for space and water heating and cooking was provided by gas. The data for solar insolation was obtained from the Meteorological Office station at Eskdalemuir, and the energy flow calculations were similar to those described by Biran and Erlicki[278]. Using the available south-facing area, an array with a power conversion efficiency of 13% would be required to provide the necessary energy. Using nickel–cadmium battery storage, 120 tons of batteries would be required with a volume, including access and ventilation, of about 360 m^3 and a cost at 1976 prices of £680,000, although some discount for quantity would probably be available. The cost of the solar array would be £75,000 at 1976 prices and the array efficiency of 13% is very high but is now possible with silicon single crystal cells of 18% individual efficiency. This design study is instructive for a number of reasons:

(1) it shows that it is possible in principle to provide the electrical needs of a house from solar energy even in a region where the year average insolation is only 8 MJ m^{-2} day^{-1};
(2) the study shows that the cost of solar cells must be reduced at least by a factor of 100 before they could be considered economically viable for this type of application;
(3) the study also shows that large seasonal variations (1.3 MJ m^{-2} day^{-1} to 16.8 MJ m^{-2} day^{-1} in this region) which are out of phase with the power usage variations lead to energy storage requirements which can not be provided economically with any technology which is commercially available at the moment.

The major problem in the use of solar cells for this type of application is thus seen to be in the storage capacity required to meet the power needs in winter when the insolation is small. In geographical locations where only minor seasonal variations in insolation occur, then the energy storage needs to be sufficient only to obviate the effects of diurnal variations in insolation and could be two to three orders of magnitude less than in this study.

This design study considered the electrical power requirements of individual houses. If the total electrical power consumed by a community is considered, it is found that the power required during working hours is significantly larger than that required during the night. It is thus possible to use solar cells to supply a large part of this extra daytime load[279]. The close phase matching of solar insolation variations with variations in power requirements leads to a system with small energy storage and efficient use of the available energy. This load shaving scheme could considerably reduce the required "stand-by" generating capacity needed by the electricity generating authorities, but it would be feasible only in localities where the seasonal variations in insolation are small.

Both of these studies have considered the generation and use of electricity in competition with the usual large electricity generating and distribution networks. The requirements of an array of cadmium sulphide–copper sulphide cells capable of providing electricity at a cost which is competitive with present generating stations have been considered by DeMeo[280]. He found that conversion efficiencies greater than 12% were required if costs of $800 per kW_{rated} were to be met, whilst 10% arrays costing $0.70 ft^{-2} could be used if costs of $1200 per kW_{rated} were permissible. These figures refer to arrays in high insolation areas and a cell lifetime of at least twenty years and are independent of the material of which the cell is composed.

It is now clear how the low power density and variability of solar energy give rise to technological problems relating to the efficiency required of an array of solar cells and the permissible cost of such an array. We have seen in the previous sections that no thin film cell has yet been constructed with an efficiency of 10% the highest efficiency to far being 8.5% at AM0 for cadmium sulphide–copper sulphide cells[105]. Array efficiencies greater than 12% are possible with silicon single crystal cells, but the cost of these cells is more than two orders of magnitude higher than the $1 ft^{-2} needed for competitive electricity production.

Hunt and co-workers[281,282,283] have considered the production process of silicon cells and identified the high energy and hence high cost processes. Even with present techniques of producing semiconductor grade silicon, modifications in the wafer cutting techniques could reduce the energy pay-back time from 12 years to 3 years. Low energy processes for the production of semiconductor grade silicon could further reduce the pay-back time to 4 months[281], and the cell costs would be reduced by a similar factor of 30 or so, whilst the array costs would be reduced by a rather smaller factor. The improvement of the yields in the various production stages would also have a dramatic effect on the cost of the cells and with improvements in the gridding, encapsulation and array fabrication processes, could give rise to a "learning curve" leading to a further reduction in cost by a factor of ten[284]. New methods of producing silicon plates suitable for use as solar cells are being developed[285], and may well lead to major reductions in the cost of silicon cells. It is possible, at least in principle, that arrays of silicon single crystal cells could just about meet the requirements set out by DeMeo[280], but the capital investment necessary would only occur if a very large market existed and, unfortunately, the market is unlikely to exist before the product is available.

The competition faced by thin film solar cells is therefore severe. One of the major advantages of thin film cells is that the production processes are at present far from optimum, and the cost reductions possible in moving from the present batch process to a production line process should be quite large. Silicon production processes, on the other hand, have already been subject to a considerable degree of optimisation, and the improvements discussed above are likely to be technically difficult and very expensive. Many of the production techniques for thin film solar cells are much less capital intensive than the silicon cell production processes and this could be a major factor in determining the cost per watt of cells in the period when usage of cells is growing but is not widespread. The ability to produce cells using relatively low technology would also be a major advantage to those countries in the Third World which would benefit most from the ability to produce small amounts (10 kW or so) of power independently of conventional electricity generating stations. These countries do not have an existing electricity generating and supply network, but they usually do have a high average solar insolation and small seasonal variations compared to the more northerly industrialised countries. It would seem sensible for these countries to consider the use of self-powered workshops with the power required being generated from solar cells on the roof of the workshop with a minimum of storage to smooth out load fluctuations. Since the economic criterion in this case is that it should be cheaper to provide the electricity from solar cells than to provide it by building conventional power stations plus a distribution network, the allowable cost/watt of the cells would be higher than that considered by DeMeo[280]. The ready availability of electrical power in all regions of these countries could transform their rural economies and could play a large part in easing the catastrophic population pressures on their major cities. The development of relatively low technology production processes would allow the cells to be produced within these Third World countries, providing employment for their scientifically trained personnel who at the moment often have to leave their countries to find useful employment.

In the high latitude industrialised countries, the major market for thin film solar cells which avoids the need for major storage facilities would seem to be in the leisure industries, where the cells would be in competition with storage batteries or petrol generators in providing small amounts of electrical power in situations where mains electricity is not available. In applications of this type, lifetimes of five to ten years and a cost per watt which is a factor of ten or so lower than the present figure would lead to substantial market penetration. The building of a market for solar cells would be greatly eased in this case also if the production processes were not highly capital intensive, and thin film cells would thus seem to be preferable to single crystal cells.

There are also many specialised markets such as active light sensor units for which the ability to produce thin film cells in geometrically complex forms makes them much more suitable than single crystal cells. For these markets the cost per watt is relatively unimportant, but the stability of the output may be crucial.

We have seen above that Appleyard's[12] vision is unlikely to become reality in the northerly industrialised regions until electrical storage can be increased by a factor of at least ten in power density and reduced by about three orders of magnitude in cost compared to nickel–cadmium batteries. There are large-scale uses possible for solar cells which do not require this dramatic improvement in storage: load sharing in regions with small seasonal variations in insolation, and the provision of power for workshops etc. in rural areas of the non-industrialised countries. For any large-scale use of solar cells, a number of requirements are apparent:

(1) the array of cells should have an efficiency of greater than 10%;
(2) the efficiency should be maintained above about 10% over a period of the order of twenty years;
(3) the cost of the array should be less than £0.50 per peak watt;
(4) the materials of which the cells are made should be available in sufficient quantities that production is not limited by depletion of resources.

The work at the Institute of Energy Conversion has shown that cuprous sulphide–cadmium sulphide cells can be produced with an efficiency of nearly 7% averaged over all the cells in one production run, with 83% of the cells produced having an efficiency above 4.4% up to a maximum efficiency of 7.64%[286]. These figures show the considerable improvements in process uniformity which have been made in the past ten years since the average efficiency of the Clevite cells was about $3\frac{1}{2}$%[286]. The efficiency of the cuprous sulphide–cadmium sulphide cell can probably be increased to about 10%[287] and the addition of zinc sulphide to the cadmium sulphide giving a better match of lattice constant and electron affinity can increase the open-circuit voltage to 0.7 V[288] and should allow arrays of these cells to be built with efficiencies in excess of 10%. The discussion in Section 4 showed that the lifetime of cuprous sulphide–cadmium sulphide cells can exceed twenty years if the cells are properly encapsulated, although their resistance is not yet adequate to the temperatures which might easily be reached if only natural convective cooling is employed in conditions of high insolation. The probable cost of these cells when manufactured in large quantities has been estimated to be between $0.20 per peak watt[289,290] and $0.10 per peak watt[291].

Cuprous sulphide–cadmium sulphide solar cells thus have a very good chance of fulfilling the first three requirements, but the possibility of depleting the cadmium resources is not negligible, and imposes a limitation on the thickness of the cadmium sulphide layer. Cadmium is produced as a by-product of zinc mining with a present production capacity throughout the world of about 20000 tonnes/annum[292]. Known reserves of zinc total about 6.10^8 tonnes[293,294] with an average zinc/cadmium ratio approximately equal to the ratio of their relative crustal abundances[294], leading to an estimate of about 2.10^6 tonnes of cadmium in known reserves, whilst future discoveries and technical advances enabling poorer deposits to be worked might increase this figure to 20.10^6 tonnes[293,294]. A similar estimate for the total reserves of cadmium has been given by Jorden[295]. A 10% efficient cell producing 1 kW under AM1 illumination would contain 10 ml $\equiv$ 4.8 g of cadmium sulphide for each micron thickness of the cadmium sulphide layer. A layer 10 μm thick would thus contain about $\frac{1}{2}$ kg of cadmium sulphide of which $7/9 \sim 400$ g would be cadmium, so one tonne of cadmium would be needed to produce 2.5 MW under AM1. An annual programme of 10^{11} m^2 of cells giving 10^{10} peak watts would require 4000 tonnes/year of cadmium which could be accommodated in the present world productive capacity[292], and would thus not disrupt the existing supply or price structure. After twenty years, the lifetime of the cells, this would permit the production of peak power equivalent to 200 nuclear power stations. In order to conserve cadmium reserves and avoid pollution, re-cycling of cadmium within the production processes and of old cells would be essential. A major enlargement of the production programme would probably cause difficulty of supply since cadmium production is related to zinc production and existing uses of cadmium would come into competition with solar cell requirements. Any attempt to use nickel/cadmium batteries for storage in such a production programme would of course quickly deplete the existing reserves. The programme described would not be of major value to the power requirements of the northerly industrialised nations unless a new storage method can be devised, but it could have a major impact on the rural economies of the underdeveloped countries.

The other types of thin film solar cell are in a much earlier stage of development than the cuprous sulphide–cadmium sulphide cell, although the copper indium selenide–cadmium sulphide cell looks very promising. The efficiency of these cells has now reached 6.7% and it is hoped that 8% efficient cells will be produced by 1979 (Kazmerski, private communication). Copper indium selenide–cadmium sulphide cells have degradation properties which are much superior to those of the cuprous sulphide–cadmium sulphide cell, and could possibly be produced at a similar cost. A large production programme of these cells would, however, have to take into account not only the availability of cadmium, but also that of indium and selenium which have only half the natural abundance of cadmium. This constraint is eased somewhat by the fact that the copper indium selenide layer can be much thinner than 10 μm and it may well be that the availability of cadmium is the major constraint. Similar considerations would apply to the indium phosphide–cadmium sulphide cell, but cells using tellurium would face severe problems of supply.

From the consideration of available resources, silicon is the obvious material to use, since depletion would be negligible over timescales measured in millenia even if the world electricity requirements were to be met entirely by solar cells. The development of a low energy route to silane and improvements in the production and the efficiency of amorphous silicon cells would thus provide the basis of solar electricity generation far into the future.

7. Conclusions

This article has discussed the production of thin film solar cells and the physical principles on which they operate. It has shown how many of the problems which have bedevilled cuprous sulphide–cadmium sulphide cells have now been overcome, and ways of overcoming the remaining problems can be seen in principle. These cells are likely to be commercially available in the near future[296] and their cost should fall dramatically as production increases.

Other types of thin film solar cell are not so well-developed although they have properties which may allow them to play an important role in the future. The use of photogalvanic cells for combined solar thermal/photovoltaic applications seems highly appropriate, whilst the rapid progress being made in copper indium selenide–cadmium sulphide cells combined with their degradation resistance could lead to their being an important alternative to cuprous sulphide–cadmium sulphide cells. For the longer term, silicon would seem to be the most useful material for thin film solar cells, since there is no question of resource depletion.

The major obstacle to the fulfilment of Appleyard's[12] vision in the high latitude industrialised countries has been seen to be the lack of suitable methods of electrical storage, but thin film solar cells with relatively low technology production techniques could have a dramatic impact on the rural economies of the Third World, and this could serve as a major goal for thin film solar cell development.

There is therefore good reason for believing that thin film solar cells will have a very important role to play in the future and will be of such public benefit that the cost and effort expended on their development will be more than justified.

References

1. E. Bequerel, *Compt. Rend.*, **9**, 561, (1839).
2. J. Dewar, *Proc. Roy. Soc.*, **27**, 354, (1878).
3. H. Gerischer, "Advances in Electrochemistry and Electrochemical Engineering", Vol. 1, ed., P. Delahey, John Wiley, New York, (1961).
4. H. Gerischer, *Electroan. Chem. Interfac. Chem.*, **58**, 275, (1975).
5. H. Gerischer, *Ber. Bunsenges Physik. Chem.*, **80**, 328, (1976).
6. W. Smith, *J. Soc. Telegraph Engineers*, **2**, 31, (1873).
7. R. Sabine, *Phil. Mag.*, **5**, 401, (1878).
8. W. G. Adams and R. E. Day, *Roy. Soc. Proc.*, **24**, 163, (1876); *Phil. Mag.*, **1**, 295, (1877).
9. Ch. E. Fritts, *Am. J. Science*, **26**, 465, (1883); *La Lumiere Electrique*, **15**, 226, (1885).
10. G. M. Minchin, *Phil. Mag.*, **35**, 354, (1893).
11. G. M. Minchin, *Nature*, **42**, 80, (1890).
12. R. Appleyard, *Telegraphic J. and Electrical Review*, **28**, 124, (1891).
13. Munk af Rosenchold, Pogg. A, *Annalen der Physik und Chemie Poggendorff Wiedemann, Leipsig*, **34**, 437, (1835).
14. F. Braun, *Ann. Phys. Chem.*, **153**, 556, (1874).
15. A. Schuster, *Phil. Mag.*, **48**, 251, (1874).
16. J. A. Fleming, "Principals of Electrical Wave Telegraphy", Longmans, London, (1908).
17. K. Hallwachs, *Phys. Zschr.*, **5**, 489, (1904).
18. L. O. Grondahl, *Phys. Rev.*, **27**, 813, (1926).
19. B. Lange, *Phys. Zschr.*, **31**, 139, (1930) loc. cit. pp. 916 and 964.
20. H. K. Henisch, "Metal Rectifiers", Clarendon Press, Oxford, (1949).
21. G. Barraz and E. Virasoro, *Anales Inst. Inv. Cient. Technol.*, **12/13**, 119, (1943), (Sante Fe, Argentina).
22. C. G. Fink and J. S. MacKay, *Trans Electrochem. Soc.*, **77**, 299, (1940).
23. F. M. Smits, *IEEE Trans. Electron Dev.*, **ED-23**, 640, (1976).
24. D. M. Chapin, C. S. Fullerand G. L. Pearson, *J. Appl. Phys.*, **25**, 676, (1954).
25. M. B. Prince, *J. Appl. Phys.*, **26**, 534, (1955).
26. J. J. Loferski, *J. Appl. Phys.*, **27**, 777, (1956).
27. P. Rappaport and J. J. Loferski, Proc. 11th Annual Battery Rand. D Conf., p. 96, U.S. Army Signals Eng. Lab., Fort Monmouth, N.J., U.S.A., (1957).
28. D. C. Reynolds, G. Leies, L. L. Antes and R. E. Marburger, *Phys. Rev.*, **96**, 533, (1954).
29. D. A. Hammond and F. A. Shirland, Proc. Electron Compt. Conf., p. 98, Philadelphia, Pa., U.S.A., (1959).
30. F. A. Shirland, A. R. L. Tech. Report, 60–293, Harshaw Chem. Co., (1960).

31. A. G. Stanley, *Appl. Sol. St. Sci.*, **5**, 251, (1976).
32. A. Carlson, Research in Semi-conductor Films, WADC–TR–56–52AD97494, (1956).
33. F. A. Shirland and J. R. Heitanen, 5th PVSC, (1965), Section IIC–3, Ref. 38.
34. F. A. Shirland and F. Augustine, 5th PVSC, (1965), Section IIC–4, Ref. 38.
35. Proc. Int. Conf. Phys. and Chem. Semicond. Heterojunctions and Layers Structures, Akademia Kiado, Budapest, (1971).
36. "Solar Cells" Proc. Int. Colloq., Toulouse, 6–10 July, (1970). Gordon and Breach, London, (1971).
37. H. J. Hovel, "Solar Cells", Vol. 11 of Semiconductors and Semimetals, eds., R. K. Willardson and A. C. Beer, Academic Press, London and New York, (1975).
38. Proceedings of 1st to 12th Photovoltaic Specialists Conference Reports, published by IEEE, New York.
39. D. A. Cusano, *Sol. St. Electron.*, **6**, 217, (1963).
40. D. Bonnet, M. Selders and H. Rabenhorst, *Festkorperprobleme* **XVI**, 293, (1976).
41. D. Bonnet and H. Rabenhorst, p. 155, Ref. 36.
42. J. Tauc, *Rev. Mod. Phys.*, **29**, 308, (1957).
43. L. W. James and R. L. Moon, *Appl. Phys. Lett.*, **26**, 467, (1975).
44. R. A. Smith, "Semi-conductors", Cambridge University Press, (1959).
45. M. P. Thakaekora, *Solar Energy*, **18**, 309, (1976).
46. R. A. Smith, "Wave Mechanics of Crystalline Solids", Chapman and Hall, London, (1969).
47. T. S. Moss, G. J. Burrel and B. Ellis, "Semi-conductor Opto-Electronics", Butterworth, London, (1973).
48. P. T. Landsberg, *Sol. St. Electron.*, **18**, 1043, (1976).
49. W. Shockley and H. J. Queisser, *J. Appl. Phys.*, **32**, 510, (1961).
50. N. A. Leontovich, *Sov. Phys. Uspekhi*, **17**, 963, (1975).
51. J. J. Loferski, *Proc. IEEE*, **51**, 667, (1963).
52. J. J. Wysocki and P. Rappaport, *J. Appl. Phys.*, **31**, 571, (1960).
53. A. de Vos, *Energy Conversion*, **16**, 67, (1976).
54. International Workshop on CdS solar cells and other abrupt heterojunctions, University of Delaware, Newark, Del U.S.A., May 1975, NSF–RANN AER75–15858.
55. A. Rothwarf and K. W. Boer, *Prog. Sol. St. Chemistry*, **10**, 72, (1975).
56. A. G. Milnes and D. L. Feucht, "Heterojunctions and Metal-Semi-conductor Junctions", Academic Press, London and New York, (1972).
57. B. Ellis and T. S. Moss, *Sol. St. Electron.*, **13**, 1, (1970).
58. M. Wolfe, *Proc. IEEE*, **51**, 674, (1963).
59. S. C. Tsaur, A. G. Milnes, R. Sahai and D. L. Feucht, "Symposium on GaAs" Boulder, Colorado, p. 156, Inst. of Physics, London, (1972).
60. J. G. Fossom, Sandia Laboratories Energy Report, SLA74–0273, June (1974).
61. M. P. Godlewski, C. R. Baraona and H. W. Brandhorst, 10th PVSC, (1973), p. 212, Ref. 38.
62. C. T. Sah, R. N. Noyce and W. Shockley, *Proc. Instn. Radio Eng.*, **45**, 1228, (1957).
63. D. A. Evans and P.T. Landsberg, *Sol. St. Electron.*, **6**, 169, (1963).

64. S. C. Choo, *Sol. St. Electron.*, **11**, 1069, (1968).
65. R. L. Anderson, *Sol. St. Electron.*, **5**, 341, (1962).
66. U. Dolega, *Z. Naturforsch*, **18**, 653, (1963).
67. C. J. M. Van Opdorp Thesis, Technische Hogeschool Eindhoven, Netherlands, (1969). (See also Ref. 35, p. 58.)
68. R. H. Rediker, S. Stopek and J. H. R. Ward, *Sol. St. Electron.*, **7**, 621, (1964).
69. P. C. Newman, *Electronics Letters*, **1**, 265 (1965).
70. A. R. Riben and D. L. Feucht, *Sol. St. Electron.*, **9**, 1055, (1966). *Int. J. Electronics*, **20**, 583, (1966).
71. R. L. Anderson, p. 55, Vol. 2, Ref. 35.
72. B. L. Sharma and R. K. Purohit, "Semi-conductor Heterojunctions", Pergamon Press, Oxford, (1974).
73. S. Martinuzzi and O. Mallem, *Phys. Stat. Sol.*, (a), **16**, 339, (1973).
74. W. Franz, Hdb. der Phys., Vol. XVIII, p. 155, Springer-Verlag, Berlin, (1956).
75. S. Kurtin, T. C. McGill and C. A. Mead, *Phys. Rev. Letts.*, **22**, 1433, (1969).
76. R. F. McQuat and D. L. Pulfrey, *J. Appl. Phys.*, **47**, 2113, (1976).
77. C. A. Mead, *Sol. St. Electron.*, **9**, 1023, (1966).
78. L. C. Olsen and R. C. Bohara 11th PVSC, (1975) p. 381, Ref. 38. Y. M. Yeh and R. J. Stirn, 11th PVSC, (1975) p. 391, Ref. 38.
79. W. A. Anderson and R. A. Milano, *Proc. IEEE*, **64**, 206, (1975). D. R. Lillington and W. G. Townsend, *Appl. Phys. Lett.*, **27**, 978, (1976).
80. H. C. Card and E. H. Rhoderick, *J. Phys. D.*, **4**, 1589, (1971).
81. H. C. Card and E. S. Yang, *Appl. Phys. Lett.*, **29**, 57, (1976).
82. S. J. Fonash, *J. Appl. Phys.*, **46**, 1286, (1975); **47**, 3597, (1976).
83a. J. Shewchun, M. Green and F. King, *Sol. St. Electron.*, **17**, 563, (1974).
83b. M. Green, F. King and J. Shewchun, *Sol. St. Electron.*, **17**, 551, (1974).
84. F. J. Bryant, and R. W. Glew, *Energy Conversion*, **14**, 129, (1975).
85. R. J. Stirn, 9th PVSC, (1972), p. 72, Ref. 38.
86. M. Wolf, *Proc. IRE*, **48**, 1246, (1960).
87. R. J. Handy, *Sol. St. Electron.*, **10**, 765, (1967).
88. D. B. Holt, *Thin Solid Films*, **24**, 1, (1974).
89. B. Joyce, *Rep. Prog. Phys.*, (GB), **37**, 363, (1974).
90. B. K. Chakraverty, *H. Phys. Chem. Solids*, **28**, 2401, 2413, (1967).
91. D. L. Kirk and M. S. Raven, *J. Phys. D.*, **9**, 2015, (1976).
92. R. Hill and D. Richardson, *Thin Solid Films*, **15**, 303, (1973).
93. S. Ogura, *Thin Solid Films*, **30**, 3, (1975).
94. R. Hill and D. Richardson, *Thin Solid Films*, **18**, 25, (1973).
95. P. S. Vincett, W. A. Barlow and G. G. Roberts, *Nature*, **255**, 542, (1975). P. S. Vincett, G. G. Roberts and W. A. Barlow, "Physics in Industry", Proc. Int. Conf., Dublin, March, (1976). Pergamon Press, Oxford, (1976).
96. Institute of Energy Conversion, University of Delaware, U.S.A. Report NSF/INN/SE/GI/-34872/PR/73/4.
97. C. Munakata, *Jap. J. Appl. Phys.*, **5**, 1251, (1966).
98. J. Dresner and F. V. Shallcross, *J. Appl. Phys.*, **34**, 2390, (1963).
99. K. Shima, *Jap. J. Appl. Phys.*, **15**, 983, (1976).
100. S. B. Hussain, *Thin Solid Films*, **23**, S21, (1974).

101. R. B. Hall, p. 284, Ref. 54.
102. D. Bednarczyk and A. Wegrzyn, *Thin Solid Films*, **36**, 165, (1976).
103. V. Snejdar and J. Jerhot, *Thin Solid Films*, **36**, 427, (1976).
104. R. L. Petritz, *Phys. Rev.*, **104**, 1508, (1956).
105. A. Waxman, V. E. Henrich, F. V. Shallcross, N. Borkan and P. K. Weimer, *J. Appl. Phys.*, **36**, 168, (1965).
106. Z. T. Kuznicki, *Thin Solid Films*, **33**, 349, (1976).
107. J. C. Anderson, *Thin Solid Films*, **37**, 127, (1976).
108. D. S. H. Chan and A. E. Hill, *Thin Solid Films*, **35**, 337, (1976).
109. M. Le Contellac and J. Richard, *Thin Solid Films*, **36**, 151, (1976).
110. H. Berger, *Phys. Stat. Sol. (a)*, **1**, 739, (1961).
111. A. Many, Y. Goldstein, and N. B. Grover, "Semi-conductor Surfaces", North Holland, Amsterdam, (1965).
112. K. L. Chopra, "Thin Film Phenomena", McGraw-Hill, New York, (1969).
113. K. Matsuura and I. Isurumi, *J. Phys. Soc. Japan*, **36**, 1543, (1975).
114. N. A. Komons, Office of Aerospace Research, U.S. Air Force, (1964).
115. L. D. Massie, Space/Aeronautics, 60, (1964).
116. F. A. Shirland, *Adv. Energy Conversion*, **6**, 201, (1966).
117. D. M. Perkins, *Adv. Energy Conversion*, **7**, 265, (1968).
118. D. C. Reynolds and G. M. Leies, *Elec. Eng.*, **73**, 734, (1954).
119. G. A. Somorjai, Proc. Int. Symp. on Evaporation of Condensation Solids, Dayton, Ohio, (1962).
120. G. A. Somorjai and D. W. Jepson, *J. Chem. Phys.*, **41**, 1389, (1964).
121. G. A. Somorjai and N. R. Stemple, *J. Appl. Phys.*, **35**, 3398, (1964).
122. J. S. Browder and S. S. Ballard, *Appl. Optics*, **11**, 841, (1972).
123. J. Philips, p. 475, Ref. 54.
124. A. Vecht, W. Grindle, and R. Mears, *J. Appl. Phys.*, **36**, 2935, (1966).
125. A. Vecht, W. Grindle, and R. Mears, *J. Appl. Phys.*, **37**, 3321, (1966).
126. C. Laermans, L. Michiels and A. De Bock, *Thin Solid Films*, **15**, 317, (1973).
127. M. H. Christman, K. A. Jones and K. H. Olsen, *J. Appl. Phys.*, **45**, 4295, (1974).
128. A. N. Casperd, Ph.D. Thesis, Newcastle upon Tyne Polytechnic, (1977).
129. J. P. David, S. Martinuzzi, F. Cabone-Brouty, J. P. Sorbier, J. M. Mathieu, J. M. Roman and J. F. Bretzner, p. 81, Ref. 36.
130. I. Lagnado and M. Lichtensteiger, *J. Vac. Sci. Tech.*, **7**, 318, (1970).
131. D. B. Frazer and H. Melchior, *J. Appl. Phys.*, **43**, 3120, (1972).
132. W. H. Leighton, *J. Appl. Phys.*, **44**, 5011, (1973).
133. M. Lichtensteiger, I. Lagnado and H. C. Gatos, *Appl. Phys. Lett.*, **15**, 418, (1969).
134. S. Durand, P. Bugnet, J. Deforges and G. Batailler, *Thin Solid Films*, **11**, 237, (1972).
135. C. E. Weitzel and L. K. Monteith, *Surface Sci.*, **40**, 555, (1973).
136. M. Takeuchi, Y. Sakagawa and H. Nagasaka, *Thin Solid Films*, **33**, 89, (1976).
137. J. Dresner and F. V. Shallcross, *Sol. St. Electron.*, **5**, 205, (1962).
138. F. H. Nicoll, *J. Electrochem. Soc.*, **110**, 1165, (1963).
139. J. J. Hegyi, *Extended Abstracts Electron. Div. Electrochem. Soc.*, **13**, Abstract 150, (1964).
140. R. R. Chamberlin and J. S. Skarman, *J. Electrochem. Soc.*, **113**, 86, (1966).

141. F. B. Micheletti and P. Mark, *Appl. Phys. Lett.*, **10**, 136, (1967).
142. C. Wu, R. S. Feigelson and R. H. Bube, *J. Appl. Phys.*, **43**, 756, (1972).
143. J. Besson, T. Nguyen Duy, A. Gautier, W. Palz, C. Martin and J. Vedel, 11th PVSC, (1975), p. 468, Ref. 38.
144. J. F. Jorden, 11th PVSC, (1975), p. 508, Ref. 38.
145. C. Weissmantel, G. Hecht, J. Herbenger and J. Tuphorn, *Proc. 2nd Colloq, Thin Films*, (1967), p. 399.
146. B. Millar and A. Heller, *Nature*, **262**, 680, (1976).
147. R. Lawrence, *Brit. J. Appl. Phys.*, **10**, 298, (1959).
148. S. Vojdani, A. Sharifnai and M. Doroudian, *Electron. Lett.*, **9**, 128, (1973).
149. L. R. Shiozawa, F. Augustine and W. R. Cook, Prog. Contract F33615-68-C-1732, Clevite Corp., (1969).
150. N. Nakayama, *Jap. J. Appl. Phys.*, **8**, 450, (1969).
151. E. Konstantinova and S. Kanev, *J. Appl. Phys.*, **42**, 5851, (1971).
152. L. R. Shiozawa, F. Augustine, G. A. Sullivan, J. M. Smith and W. R. Cook, 5224 Clevite Corp., Final Report, (1969), Contract AF33 (615).
153. F. A. Shirland, J. R. Hietanen, F. Augustine and W. K. Bower, Final Report, NAS3-6461, Clevite Corp., (1965).
154. F. A. Shirland, p. 465, Ref. 54.
155. L. Clark, R. W. Gale, K. Moore and R. J. Mytton, IRD Report 72/30, Project 10273, July (1972).
154. R. H. Bube, W. Gill and P. Lindquist, Prog. Rep. 1, Grant NGR-05-020-214, Stanford University, (1967).
157. K. W. Boer, *Phys. Rev. B.*, **13**, 5373, (1976).
158. K. W. Boer, IEC Tech. Report, NSF/RANN/AER72-03478A04/TR75/5, University of Delaware, (1975).
159. R. B. Hall, p. 284, Ref. 54.
160. K. K. Reinhartz and A. van Aerschodt, p. 95, Ref. 36.
161. H. Matsumoto, N. Nakayama, K. Yamaguchi and S. Ikoegami, *Jap. J. Appl. Phys.*, **15**, 1849, (1976).
162. E. J. Hsieh, D. Miller, K. W. Vindelov and T. G. Brown, p. 301, Ref. 54.
163. L. Clark, R. Gale, K. Moore, R. J. Mytton and R. S. Pinder, p. 241, Ref. 36.
164. R. A. Mickelsen and D. D. Abbott, NASA Report CR-120812, Contract No. NAS3-13232, also p. 484, Ref. 54.
165. T. S. te Velde, 8th PVSC, (1970), p. 372, Ref. 38.
166. T. S. te Velde, *Sol. St. Electron.*, **16**, 1305, (1973); *Energy Conversion*, **14**, 111, (1975).
167. R. Hill and I. A. S. Edwards, *Vacuum*, **27**, 277, (1977).
168. B. Tuck and A. Hooper, *J. Phys. D.*, **8**, 1806, (1975).
169. K. Okamoto and S. Kawai, *Jap. J. Appl. Phys.*, **12**, 1130, (1973).
170. I. V. Egorova, *Sov. Phys. Semicond.*, **2**, 266, (1968).
171. W. Palz, G. Cohen-Solal, J. Veldel, J. Fremy, T. N. Duy and J. Valerio, 7th PVSC (1968), p. 54, Ref. 38.
172. R. W. Buckley, Ph.D. Thesis, University of Durham, (1973).
173. W. Palz, J. Besson, T. N. Duy and J. Vedel, 10th PVSC, (1973), p. 69, Ref. 38.
174. E. H. Roseboom, *Econ. Geol.*, **61**, 641, (1966).

175. S. Djurle, *Acta Chem. Scand.*, **12**, 1415, (1958).
176. H. T. Evans, *Nature (Phys. Sci.)*, **232**, 69, (1971).
177. J. J. Buerger and N. W. Buerger, *Am. Minerologist*, **29**, 55, (1944).
178. M. J. Buerger and B. J. Wuench, *Science*, **141**, 276, (1963).
179. W. R. Cook, L. Shiozawa and F. Augustine, *J. Appl. Phys.*, **41**, 3058, (1970).
180. K. W. Boer, NASA–CR–126975, Grant NGR–08–001–028, University of Delaware, (1971).
181. H. Takeda, J. D. H. Donnay, E. H. Roseboom and D. E. Appleman, *Z. Krist.*, **125**, 404, (1967).
182. T. S. te Velde, *Philips Res. Reports*, **28**, 573, (1973).
183. B. J. Mulder, *Phys. Stat. Sol. (a)*, **13**, 79, (1972).
184. B. J. Mulder, *Phys. Stat. Sol. (a)*, **15**, 409, (1973).
185. B. J. Mulder, *Phys. Stat. Sol. (a)*, **18**, 633, (1973).
186. J. Dielman, p. 92, Ref. 54.
187. J. J. Loferski and J. Shewchun, p. 318, Ref. 54.
188. R. W. Glew and F. J. Bryant, *Thin Solid Films*, **29**, 269, (1975).
189. J. Bougnot, F. Guastavino, S. Couve-Duchemin and M. Savelli, p. 327, Ref. 54
190. H. M. Windawi, p. 177, Ref. 54.
191. W. D. Gill and R. H. Bube, *J. Appl. Phys.*, **41**, 1946, (1970).
192. B. J. Mulder, *Phys. Stat. Sol. (a)*, **13**, 569, (1972).
193. L. D. Partain, J. J. Oakes and I. G. Greenfield, p. 346, Ref. 54.
194. J. Bougnot, F. Guastavino, G. M. Mousalli and M. Savelli, p. 337, Ref. 54.
195. T. S. te Velde, Agard Conf. Proc., Agard-CP-21, p. 927, (1967).
196. L. R. Shozawa, G. A. Sullivan and F. Augustine, 8th PVSC, (1968) p. 39, Ref. 38.
197. A. E. van Aerschodt, J. J. Capart, K. H. David, M. Fabbriocotti, K. H. Heffels, J. J. Loferski and K. K. Reinhartz, *IEEE Trans. Electron Devices*, **ED-18**, 471, (1971).
198. P. F. Lindquist and R. H. Bube, *J. Appl. Phys.*, **43**, 2839, (1972).
199. A. L. Fahrenbruch and R. H. Bube, *J. Appl. Phys.*, **45**, 1264, (1974).
200. F. Pfisterer, G. H. Hewig and W. H. Bloss, 11th PVSC, (1975), p. 460, Ref. 38.
201. H. W. Schock, G. Bilger, W. H. Bloss, G. H. Hewig and F. Pfisterer, *Vacuum*, **27**, 281, (1977).
202. S. Martinuzzi, O. Mallen and T. Cabot, *Phys. Stat. Sol. (a)*, **36**, 227, (1976).
203. K. W. Boer, p. 159, Ref. 54.
204. K. W. Boer, p. 194, Ref. 54.
205. K. W. Boer, *Festkorperprobleme*, **XVI**, 315, (1976).
206. K. W. Boer, 12th PVSC, (1976), p. 475, Ref. 38.
207. A. Rothwarf, p. 9, Ref. 54.
208. A. Rothwarf, p. 167, Ref. 54.
209. G. A. Sullivan, *Phys. Rev.*, **184**, 796, (1969).
210. J. D. Meakin, p. 84, Ref. 54.
211. J. A. Borders, *J. Electrochem Soc.; Sol. St. Sci. Tech.*, **123**, 37, (1976).
212. H. W. Brandhorst, NASA Report 120–33–01–09–22, Dec. 12 (1968), 7th PVSC (1968), p. 33.
213. D. T. Cheung, S. Y. Chiang, and G. L. Pearson, *Sol. St. Electron.*, **18**, 263, (1975).
214. H. Luquet, L. Szepessy, J. Bougnot, M. Savelli and F. Guastavino, 11th PVSC, (1975), p. 445, Ref. 38.

216. K. Bogus and S. Mattes, 9th PVSC, (1972), p. 106, Ref. 38.
217. K. W. Boer, G. E. Birchinall, I. Greenfield, H. C. Hadley, T. L. Lu, L. Partain, J. E. Philips, J. Schutz and W. F. Tseng, 11th PVSC, (1975), p. 77, Ref. 38.
218. L. C. Burton, T. Hench, G. Storti and G. Haake, *J. Electrochem. Soc.; Sol. St. Sci. Tech.*, **123**, 1741, (1976).
219. A. E. Spakowski, F. L. Acampora and R. E. Hart, NASA Report TND-3663, (1966).
220. M. M. Sayed and L. D. Partain, *Electronics Lett*, **10**, 163, (1974).
221. H. J. Mathieu, K. K. Reinhartz and H. Rickert, 11th PVSC, (1975), p. 93, Ref. 38.
222. H. Rickert, C. Wedde, W. Palz, J. Vedel and T. Nguyen Duy, 11th PVSC, (1975), p. 439, Ref. 38.
223. M. Aven and R. E. Halstead, *Phys. Rev. (A)*, **137**, 228, (1965).
224. F. J. Bryant and R. W. Glew, *Radiation Effects*, **28**, 103, (1976).
225. R. W. Glew and F. J. Bryant, *Radiation Effects*, **29**, 83, (1976).
226. K. J. Bachman, E. Buehler, J. L. Shay and S. Wagner, *Z. Phys. Chem.*, **98**, 365, (1975).
227. M. R. Lorenz, *J. Phys. Chem. Solids*, **23**, 939, (1962).
228. R. F. Brebrick and A. J. Strauss, *J. Phys. Chem. Solids*, **25**, 1441, (1964).
229. C. Piaget, *Rev. Phys. Appl.*, **201**, (1966).
230. D. B. Holt and M. I. Abdalla, *Phys. Stat. Sol. (a)*, **26**, 507, (1974).
231. P. R. Kamadjiev, L. K. Mladjov and I. S. Vassilev, *Krystal und Technik*, **9**, 1249, (1974).
232. Yu A. Vodakov, G. A. Lomakina, G. P. Naumov and Yu A. Maslakovets, *Sov. Phys. Sol. St.*, **2**, 1, 11, (1961).
233. G. P. Naumov and O. V. Nicholaeva, *Sov. Phys. Sol. St.*, **3**, 2718, (1962).
234. R. O. Bell, H. B. Serreze and F. V. Wald, 11th PVSC, (1975), p. 497, Ref. 38.
235. R. V. D'Aiello, P. H. Robinson and H. Kressel, *Appl. Phys. Lett.*, **28**, 231, (1976).
236. T. L. Chu, Report NSF/RANN/SE/AER/73–07843/PR/75/4.
237. T. L. Chu, J. C. Lien, H. C. Mollenkopf, S. S. Chu and K. W. Heizer, *Solar Energy*, **17**, 229, (1975).
238. T. L. Chu and K. N. Singh, *Sol. St. Electron.*, **19**, 837, (1976).
239. T. L. Chu, K. Y. Duh, H. C. Mollenkopf and S. S. Chu, Electrochemical Society Spring Meeting, Washington D.C., U.S.A., 2–7 May, (1976), Princeton, N.J., U.S.A. Electrochemical Society, (1976), p. 565.
240. T. L. Chu, H. C. Mollenkopf and S. S. Chu, *J. Electrochem. Soc.*, **122**, 1681, (1975).
241. T. Mayazaki, T. Warabisako, N. Nakamura, T. Saitah, S. Minagawa and T. Tokuyama, Ref. 239, p. 593.
242. P. H. Fang, L. Ephrath and W. B. Nowak, *Appl. Phys. Lett.*, **25**, 583, (1974).
243. T. Saito and Y. Seki, *Appl. Phys. Lett.*, **29**, 600, (1976).
244. C. R. Wronski, D. E. Carlson and R. E. Daniel, *Appl. Phys. Lett.*, **29**, 602, (1976).
245. W. E. Spear and P. G. Le Comber, *Sol. St. Commun.*, **17**, 1193, (1975).
246. P. G. Le Comber, A. Maden and W. E. Spear, *J. Non-Cryst. Sol.*, **11**, 219, (1972).
247. P. G. Le Comber and W. E. Spear, *Phys. Rev. Lett.*, **25**, 509, (1970).

248. R. J. Loveland, W. E. Spear and A. Al-Sharbaty, *J. Non-Cryst. Sol.*, **13**, 55, (1973/74).
249. D. E. Carlson and C. R. Wronksky, *Appl. Phys. Lett.*, **28**, 671, (1976).
250. W. E. Spear, P. G. Le Comber, S. Kinmond and M. H. Brodsky, *Appl. Phys. Lett.*, **28**, 105, (1976).
251. R. J. Stirn and Y. C. M. Yeh, *Appl. Phys. Lett.*, **27**, 95, (1975).
252. D. A. Cusano, *Rev. Phys. Appl.*, **1**, 195, (1966).
253. S. A. Foreman and M. A. Peacock, *Am. Minerologist*, **35**, 441, (1949).
254. H. Nowotny, *Metallforsch.*, **1**, 40, (1946).
255. J. Bernard, R. Lacon, C. Paparoditis and M. Rodot, *Rev. Phys. Appl.*, **1**, 211, (1966).
256. V. C. Kieu and H. Rodot, *C. R. Acad. Sci.*, **260**, 1908, (1965).
257. C. Paparoditis, C. Stella, D. Darmagna and J. Bernard, Comm. Coll. Internat. Phys. Couches Minces, Clausthal-Gottingen, (1965).
258. E. W. Justi, G. Schneider and J. Seredynski, *Energy Conv.*, **13**, 53, (1973).
259. J. Lebrun, 8th PVSC, (1970), p. 33, Ref. 38.
260. D. Bonnet and H. Rabenhorst, 10th PVSC, (1971), p. 129, Ref. 38.
261. Y. Marfaing and J. Chevallier, *IEEE Trans. Electron. Dev.*, **ED-18**, 465, (1971).
262. S. Martinuzzi, *Phys. Stat. Sol. (a)*, **34**, K21, (1976).
263. R. Hill and A. N. Casperd, *Sol. St. Commun.*, **17**, 735, (1975).
264. R. Hill and A. N. Casperd, p. 265, Ref. 54.
265. A. L. Fahrenbruch, F. Buch, K. Mitchell and R. H. Bube, 11th PVSC, (1975). Ref. 38, p. 490.
266. J. L. Shay, S. Wagner, K. J. Bachmann, E. Buehler and H. M. Kasper, 11th PVSC, (1975), p. 503, Ref. 38.
267. K. J. Bachmann, E. Buehler, J. L. Shay and S. Wagner, *Appl. Phys. Lett.*, **29**, 121, (1976).
268. L. L. Kazmerski, F. R. White and G. K. Morgan, *Appl. Phys. Lett.*, **29**, 268, (1976).
269. J. Shewchun, J. J. Loferski, A. Wold, R. Arnott, E. A. De Meo, R. Beaulieu, C. C. Wu and H. L. Hwang, 11th PVSC, (1975), p. 482, Ref. 38.
270. W. W. Anderson and Y. G. Chai, *Energy Conv.*, **15**, 85 (1975).
271. R. Gomer, *Electrochimica Acta*, **20**, 13, (1975).
272. W. J. Albery and M. D. Archer, *Electrochimica Acta*, **21**, 1155, (1976).
273. H. Gerischer and J. Gobrecht, *Ber. Bunsenges. Physik. Chem.*, **80**, 328, (1976).
274. R. Williams, *J. Vac. Sci. Technol.*, **13**, 12, (1976).
275. E. C. Dutoit, R. L. Van Meirhaeghe, F. Cardon and W. P. Gomes, *Ber. Bunsenges. Physik. Chem.*, **79**, 1206, (1975).
276. B. Miller and A. Heller, *Nature*, **262**, 680, (1976).
277. R. Hill and R. Heyworth, Photovoltaic Specialists Workshop, NPL, London, (1977).
278. D. Biran and M. S. Erlicki, *Solar Energy*, **17**, 325, (1975).
279. K. W. Boer, *J. Environ. Sci. (U.S.A.)*, Jan/Feb, 8, (1974).
280. A. E. De Meo, Final Report, Research Project W575-21 E.P.R.I. and p. 109, Ref. 54.
281. L. P. Hunt, 11th PVSC, (1975), p. 259, Ref. 38.
282. L. P. Hunt, 12th PVSC, (1976), p. 347, Ref. 38.

283. L. P. Hunt, V. D. Dosaj, J. R. McCormick and L. D. Crossman, 12th PVSC, (1976), p. 125, Ref. 38.
284. M. Wolf, *Energy Conversion*, **14**, 49, (1975); P. E. Glaser, in "Sunlight to Electricity", ed., J. A. Merrigan, M.I.T. Press, (1975).
285. J. A. Zoutendyk, p. 34 "Sharing the Sun", Joint Conf. Am. Section I.S.E.S. and Sol. Energy Soc. Canada, Aug 15–20, 1976. Winnipeg, Canada. Am. Section I.S.E.S. Florida, U.S.A., 1976.
286. J. D. Meakin, B. Baron, K. W. Boer, L. Burton, W. Deveney, H. Hadley, J. Philips, A. Rothwarf, G. Storty and W. Tseng, p. 113, Ref. 285.
287. A. Rothwarf, Int. Conf. Solar Electricity, Toulouse, (1976). NSF/RANN/AER72-03478 A04/TR76/1.
288. L. C. Burton and T. L. Hench, *Appl. Phys. Lett.*, **29**, 613, (1976).
289. K. W. Boer and J. Olsen, 10th PVSC, (1973), p. 253, Ref. 38.
290. T. P. Brady, p. 499, Ref. 54.
291. J. F. Jorden, 11th PVSC, (1975), p. 508, Ref. 38.
292. "Metal Statistics (1965–1975)", p. 49, 63rd Edition. Metallgesellschaft, Frankfurt am Main, W. Germany, (1976).
293. U.S. Geological Survey, Professional Paper 820, p. 706.
294. W. von Engelhart, Resources Policy, June (1975), p. 186.
295. J. F. Jorden, Int. Conf. Photovoltaic Power Generation, p. 221, Hamburg, W. Germany, Sept. 25–27, 1974. Deutch Gesellschaft fur Luft und Rannfahrt e V Koln (1975).
296. S. F. DiZio, p. 108, Ref. 285.
297. L. Eisenman, *Ann. Physik.* (*Leipsig*), **10**, 129, (1952).
298. G. P. Sorokin, Yu. M. Papshev and P. T. Oush, *Sov. Phys. Sol. St.*, **7**, 1810, (1966).
299. R. Marshall and S. S. Mitra, *J. Appl. Phys.*, **36**, 3882, (1965).

Chapter 11†

Magnetic Thin Films and Devices

B. K. Middleton

*Department of Electrical and Electronic Engineering,
Manchester Polytechnic, Chester Street, Manchester, England*

1. Introduction 604
2. The Properties of Uniaxial Magnetic Thin Films 605
   2.1 Uniaxial Anisotropy 605
   2.2 Single Domain Behaviour 609
   2.3 Reversal Processes in Films 615
   2.4 Domain Walls 621
   2.5 Magnetisation Dispersion and Ripple 625
   2.6 High Speed Switching 628
3. High Speed Memory Applications 630
   3.1 Store Operation 631
   3.2 Advanced Planar and Cylindrical Storage Element Design 638
4. Applications in Magnetic Recording 641
   4.1 The Record-Replay Process 642
   4.2 Recording Properties of Thin Films 649
   4.3 High Coercivity Films 652
       4.3.1 Evaporated films 654
       4.3.2 Films produced by electroless chemical deposition 656
       4.3.3 Electrodeposited films 661
       4.3.4 Application to devices 661
   4.4 Thin Film Recording Heads 662
       4.4.1 Thin film inductive heads 663
       4.4.2 Thin film magnetoresistive heads 667
5. Domain Storage Devices 669
   5.1 Bubble Domains 669
       5.1.1 Bubble domain materials 674
       5.1.2 Domain propagation and devices 676
   5.2 Domain Wall Motion Devices 678
6. Optical Beam Storage Devices 683
7. Concluding Remarks 686
   References 687

† For a list of symbols used in this chapter and their definitions see p. xix.

1. Introduction

In 1955 Blois[1] demonstrated that thin ferromagnetic films could be prepared by vacuum evaporation which had uniaxial anisotropy and exhibited rectangular hysteresis loops with low switching fields. The subsequent suggestion that these properties made thin films ideal for application to computer storage devices, evoked a response from many commercial and academic organisations who committed themselves to the development of high speed magnetic film computer stores. The first operational store was announced in 1959[2] and since then many working stores have been developed and marketed. Despite the fact that thin films offered potential for the development of devices with performances much superior to those of the familiar core stores, the many new properties and problems that were uncovered in the ensuing study made their potential difficult, if at all possible, to realise in full. Nevertheless, successfully operating thin film stores in the form of the so-called plated wire memories have been available for a number of years.

Other areas of application for high coercivity thin films include disc and tape recording. Associated with these has been the development of thin film recording heads which themselves use low coercivity films of the type used in the thin film store. Many other applications have been studied and these include shift register and logic devices using low coercivity materials, and high anisotropy materials for bubble domain, optical, and electron beam stores.

In this chapter it is proposed firstly to give a brief résumé of the properties of thin ferromagnetic films in as much as they are important to their application to memory devices. Soft magnetic films are discussed and their behaviour compared with that expected of materials exhibiting uniaxial anisotropy. The application of the films to computer memories is then described and particular attention is paid to the properties which make them suitable for such applications, and also to the properties which make such applications difficult to achieve. The requirements of a magnetic recording system are then considered and it is shown that hard magnetic materials are needed. The production and properties of such films are then studied in the light of the requirements previously established. Magnetic thin film recording heads are described and their potential discussed. The potential application of mobile domains, of which bubbles are the foremost example, to a new range of devices is then presented. The properties required of the new types of materials needed and the progress towards achieving them is summarised. Finally, the optical beam storage and many other applications which have been considered for magnetic films are briefly reviewed.

This work is concerned largely with thin films of nickel iron in approximate proportion 81:19 and of the order of 1000 Å thickness. The high coercivity films are usually of cobalt or cobalt phosphorus of approximately the same thickness. These parameters should be borne in mind as the "norm" and variations from these will be clearly pointed out. Some of the work reviewed here, particularly that on soft magnetic materials has been the subject of earlier reviews[3-8], but a particular feature of this work will be the attention paid to hard magnetic materials and their applications. A complete review of all the material published on thin magnetic films is not possible in the space available but references included should be sufficient in number to provide a nucleus for further study.

2. The Properties of Uniaxial Magnetic Thin Films

Concern in this section is not with the reasons for the occurrence of ferromagnetism in thin films but with the arrangement of the magnetisation therein. The emphasis of the discussion is therefore on the domain structures and their behaviour under the influence of applied fields. The reasons why domains occur in bulk ferromagnetic specimens are well known[9] (see Section 2.2) and the factors which affect the shape of hysteresis loops widely appreciated. However the fact that the material is now in the form of a thin film introduces some important new aspects of behaviour.

The properties of thin films centre around the presence of the uniaxial anisotropy induced in their formative stages. Therefore the chapter proceeds with a study of uniaxial anisotropy, followed by sections dealing with its consequences to the reversal behaviour in firstly ideal and then in real magnetic films. The importance of domain walls and anisotropy dispersion will emerge and will be treated in separate sections. Finally the switching, or magnetisation reversal, speeds will be discussed before going on to device applications.

It has already been stated that 81:19 NiFe Permalloy films of the order of 1000 Å thickness are the subject of the study but it should also be emphasised that the nominal film composition and thickness are not the only factors determining the observed magnetic properties. Invariably the film deposition parameters are critical and their role in establishing magnetic properties will be discussed at the appropriate stages, but no effort will be made to exhaustively catalogue the variation of film properties with production conditions.

2.1 Uniaxial Anisotropy

Thin magnetic films are often produced by evaporation in vacuum of the appropriate metals and condensation of the vapour onto heated substrates. During deposition a magnetic field is usually applied in a direction in the film plane as shown in Fig. 1. Provided the field is sufficiently large an easy axis of magnetisation will be induced in the films and will be coincident with the field direction.

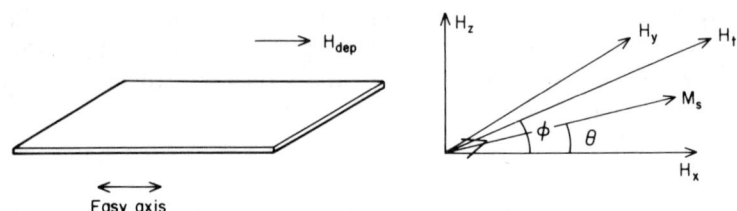

FIGURE 1. Thin film with easy axis of magnetisation along the x direction. Applied field and magnetisation directions defined for reversal in the film plane. H_{dep} represents the direction of the applied field during deposition of the film.

The meaning of the term "easy axis" is that in the absence of applied field and other effects the magnetisation vector naturally tends to point along this axis. Mathematically this situation is adequately represented by the introduction of an anisotropy energy density term E_k in the free energy such that

$$E_k = K_u \sin^2 \theta \qquad (1)$$

where θ is the angle between the magnetisation vector and the easy axis, and K_u is the induced uniaxial anisotropy constant, for which a typical value might be $1500\,\mathrm{ergs/cm^3}$. Clearly there are two stable states, corresponding to the energy being minimised by $\theta = 0$ or π, with an energy maximum between them opposing attempts to reverse the direction of magnetisation from one state to the other.

This anisotropy is sometimes called field induced but a simple consideration will show that a better title would be magnetisation induced anisotropy. The field of 100 or so Oersteds applied during film deposition is sufficiently large to saturate the film in the direction of the field. However the total induction in the film is made up of field and magnetisation contributions and for Permalloy the field contributes only 1% of the total. Therefore it can be said that the anisotropy is magnetisation induced and the field only serves to line up the magnetisation so as to produce a common easy axis for the entire film.

Equation (1) has been shown to be a reasonable representation of the induced uniaxial anisotropy energy density since hysteresis loops of the experimentally observed shape can be predicted using it, as can torque curves which are a more direct measure of the anisotropy[10]. However it should be emphasised that K_u represents an induced anisotropy which is completely distinct from the magnetocrystalline anisotropy which is also present in the crystallites that make up the film structure. The composition of the films is usually chosen to be around 81:19 NiFe at which point the magnetostriction is zero[3] thus reducing the sensitivity of the film properties to applied stress. At this composition the magnetocrystalline anisotropy constant has a finite but small value[11] and thus provides extra contributions, which are biaxial in nature, to the total anisotropy energy of the film.

The origins of induced anisotropy have been studied for a number of years but a satisfactory understanding of them has not yet emerged. Two possible mechanisms of anisotropy, in films deposited at normal incidence, have received particular attention and been shown to be capable of explaining much of the observed data although there still remains a sufficiency of unexplained results to lead one to the conclusion that a number of mechanisms contribute to the anisotropy[12]. In films deposited at angles other than normal incidence different geometrical situations pertain and consequently there are other contributions to the anisotropy[13]. This makes either decisive or revealing experiments difficult to design and renders it improbable that a simple overall description will be achieved.

The first mechanism to be discussed is that of directional ordering[14-16]. This occurs in alloys and is associated with the arrangements of atoms therein. Neighbouring pairs of like atoms embedded in a matrix of atoms of a different type and oriented differently to the magnetisation direction, have different energies. In a film containing a random array of such pairs of atoms there would be no variation of total film energy with magnetisation direction. However, during the deposition of films, atoms tend to take up the lowest available energy configurations by arranging themselves appropriately with respect to the applied magnetic field and therefore magnetisation directions. This being done, the magnetisation has been used to define a direction at which the film energy is a minimum. This is the easy direction and any subsequent deviation of the magnetisation from this direction causes increases in the film energy. Detailed analysis of this mechanism has predicted an energy, with angular dependence given by equation (1), and magnitude which reduces to zero in materials containing atoms of a single type. However, induced anisotropy is known to occur in such materials as iron and nickel and therefore this mechanism is not capable of explaining all the available experimental data.

The second mechanism involves the contributions arising as a result of the films being magnetostrictive. This implies that there are strains induced in the films when they are magnetised in particular directions[17,18]. This being so it should also be the case when the films were formed on their substrates when it would be expected that the strain would be frozen into the film in that one orientation since the film-substrate adhesion forces are very strong. The later application of a magnetic field in a different direction would require a strain to be set up in the film as the magnetisation rotates from its initial direction but since the substrate is rigid a stress is produced which leads to energy being stored in the film. This energy depends on the orientation of the magnetisation and is therefore a form of anisotropy energy. The energies from this source and that arising from directional ordering have been calculated by Robinson[17,18] and the total compared with experimental results. Robinson found it necessary to assume a value for the average magnetostriction constant of the film material while West[19] pointed out that the calculation required a correct averaging of the magnetoelastic energy over the polycrystalline sample. The resulting predictions of anisotropy constant as a function of composition of NiFe alloys are compared[8] with experimental results of Robinson[17,18], Takahashi et al.[20] and Takahashi[21] in Fig. 2. Within the scatter of the experimental points (only the average curves are reproduced in the figure) the agreement between experiment and theory is reasonable. Despite this, there is

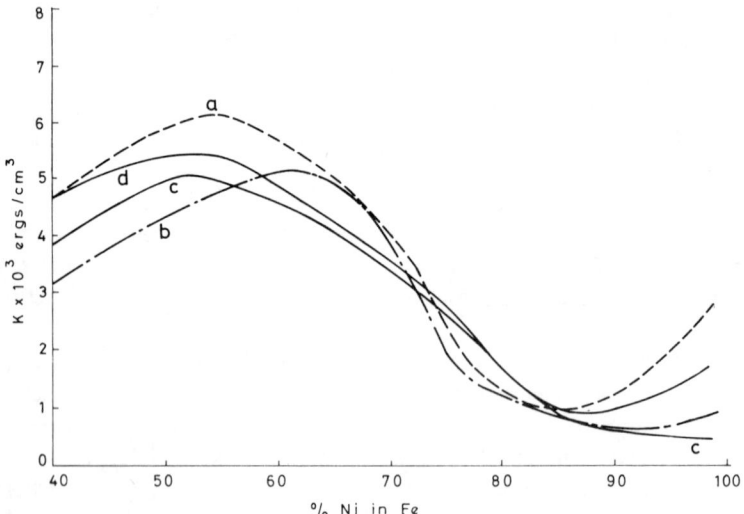

FIGURE 2. Induced uniaxial anisotropy constant as a function of film composition. (a) Measurements of Robinson, substrate temperature 513 K, (b) measurements of Takahashi, substrate temperature 573 K, (c) calculated values from directional ordering and magnetostriction theories, substrate temperature 600 K, (d) similar calculation for 500 K. [K. D. Leaver, *Contemp. Phys.* **9**, 475, (1968).]

evidence that impurities and defects play an important role in determining the anisotropy in thin films[12]. Quantitative predictions in this area are not yet available and more work is needed before a completely satisfactory description of the origins of induced anisotropy can be given.

In addition to the induced planar easy axis which has, so far, been the subject of discussion, there is substantial evidence that another form of induced anisotropy also exists. This is usually called "perpendicular anisotropy" and manifests itself as an induced easy axis normal to the film plane. Evidence seems to point to magnetostrictive and microscopic shape effects within the film as being responsible for its occurrence[22]. Anisotropy constants are often large and of the order of 10^6 ergs/cm^3 but their effect is somewhat diminished in Permalloy by the overriding shape anisotropy of the film. Nevertheless it has been shown to be responsible for the appearance of so-called "stripe domain" structures, which consist of long narrow domains having components of magnetisation normal and parallel to the film plane[23,24]. Recently interest in perpendicular anisotropy has been revived by the observation of bubble domains (see Section 5) in amorphous alloys[25]. In bubbles the magnetisation is oriented normal to the film plane and their occurrence specifically requires the presence of a perpendicular anisotropy. Its origins in amorphous films have not yet been determined although microscopic shape effects and pair ordering have been mentioned[25].

2.2 Single Domain Behaviour

From a consideration of the various contributions to the total magnetic energy of a bulk ferromagnetic specimen it can easily be demonstrated that the former can be reduced below that corresponding to saturation by the formation of a magnetic domain structure. This is possible because magnetostatic energy is reduced by the creation of magnetic domains. During the latter process the anisotropy and exchange energies are increased by the formation of domain walls. The final configuration is decided at the point where a further creation of domains would lead to greater increases in wall energy than it would decreases in magnetostatic energy[26]. In this process there is no guarantee that the particular minimum energy domain structure achieved is the one with the lowest energy. Therefore the formation of metastable as well as stable states is always possible.

Similar, but modified, considerations apply to thin films and it is important to draw some conclusions about their stable domain structures at this stage. Figure 3a shows the section through a thin film saturated along its easy axis. It is clear that, due to the demagnetising fields that will exist in the film plane as a result of poles formed at the end of the specimen, there will be a finite magnetostatic energy associated with this structure. Could this be reduced by

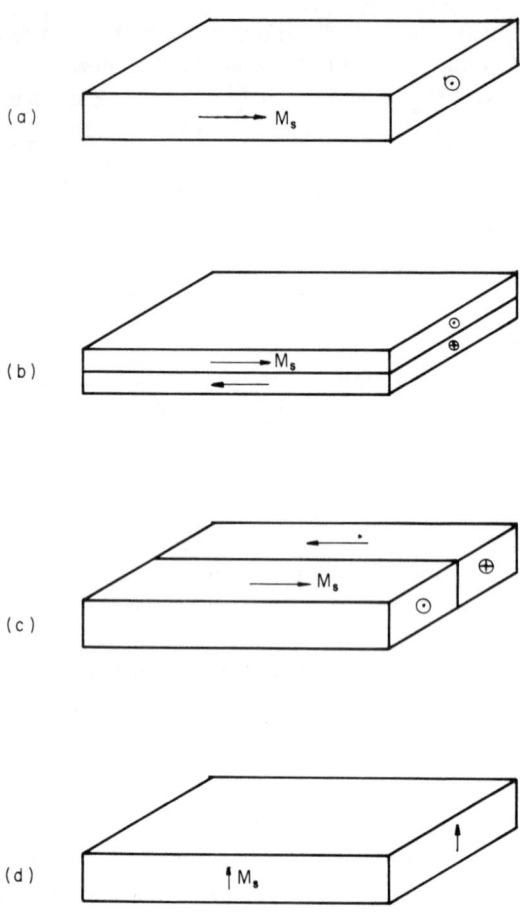

FIGURE 3. Possible thin film domain structures. (a) Film saturated along the easy axis. (b) Two-domain structure with plane of wall parallel to plane of film. (c) Two-domain structure with plane of wall normal to plane of film. (d) Film saturated in a direction perpendicular to the easy axis and the plane of the film.

the formation of domains? This would necessarily require the formation of domain walls and with the structure shown in Fig. 3b a wall would need to be formed in the film plane. For Permalloy, the width of a domain wall would approach 10 000 Å whereas films of most interest are of the order of 1000 Å thick and so it seems unlikely that such a structure can exist. It is therefore concluded that there will be little variation of magnetisation direction through the thickness of the film. An alternative structure is shown in Fig. 3c where the film is split into domains by walls whose planes are perpendicular to the plane of the film. This type of structure is found in practice and is clearly stable. The application of a field in the film plane would produce the single domain structure shown in Fig. 3a and the transference from the multidomain to the single domain state would be by a process of the motion of the domain walls so as to favour the growth of those domains with magnetisation direction parallel to the applied field. This mode of reversal contrasts with that of coherent rotation of magnetisation without the formations of domains: the latter will be discussed shortly. The only other simple domain structure needing mention at this stage is that shown in Fig. 3d where the magnetisation is directed normal to the film plane. Clearly this structure is associated with a high anisotropy energy, since the film is magnetised in a hard direction, and also a large magnetostatic energy as a result of magnetisation components normal to the film plane. Therefore components of magnetisation normal to the film plane are not expected in such simple structures.

The reversal of magnetisation by coherent rotation in elongated fine particles was first discussed by Stoner and Wohlfarth[27] and the theory presented has been adapted and applied to reversal in thin films[28-31]. Following this line, suppose the film shown in Fig. 1 had previously been magnetised to saturation by a high field applied along the positive easy axis. A field H_t is now applied in a direction making an angle ϕ with the easy axis, and consequently the magnetisation vector makes an angle θ with the easy axis. The total energy per unit volume of film is then made up of anisotropy and field energy contributions and is given by

$$E = -M_s H_t \cos(\phi - \theta) + K \sin^2 \theta \qquad (2)$$

where, from here onwards, K replaces K_u.

The exchange energy contribution is zero since the magnetisation is uniform. If the field has components H_x and H_y, as shown, equation (2) can be replaced by

$$E = -M_s H_x \cos \theta - M_s H_y \sin \theta + K \sin^2 \theta \qquad (3)$$

Differentiation of the energy with respect to θ gives an equation which defines the angle θ for the minimum energy. Therefore

$$\frac{\partial E}{\partial \theta} = M_s H_x \sin \theta - M_s H_y \cos \theta + 2K \sin \theta \cos \theta = 0 \qquad (4)$$

For any given H_x and H_y the corresponding values of θ which satisfy the above equation can be found. The same equation is sometimes called a torque equation since it simply states that the torques on any spin must sum to zero if that orientation is to be stable. However having solved equation (4) for θ, the second differential of E with respect to θ is needed to ascertain whether the solutions obtained correspond to stable or unstable states. Differentiation of equation (4) with respect to θ yields

$$\frac{\partial^2 E}{\partial \theta^2} = M_s H_x \cos\theta + M_s H_y \sin\theta + 2K(\cos^2\theta - \sin^2\theta) \qquad (5)$$

If the right hand side of equation (5) is positive then the energy is a minimum and if it is negative in value there is a maximum.

As a first example consider a field applied along the easy axis alone so that $H_y = 0$. Then equation (4) becomes

$$\frac{\partial E}{\partial \theta} = M_s H_x \sin\theta + 2K \sin\theta \cos\theta = 0 \qquad (6)$$

and has solutions

$$\sin\theta = 0 \quad \text{i.e.} \quad \theta = 0 \text{ or } \pi \qquad (7a)$$

$$H_x = -\frac{2K}{M_s}\cos\theta \qquad (7b)$$

Substitution from equation (7) into equation (5) with $H_y = 0$ gives a form of the second differential which can only have negative values and therefore represent unstable states, except at $\theta = 0$ or π when its value is zero. Corresponding to the other solutions of $\theta = 0$ or π the second differential is given by

$$\frac{\partial^2 E}{\partial \theta^2} = M_s H_x + 2K \qquad \theta = 0 \qquad (8a)$$

$$= -M_s H_x + 2K \qquad \theta = \pi \qquad (8b)$$

Clearly $\theta = 0$ is stable when $H_x > -2K/M_s$ and $\theta = \pi$ when $H_x < 2K/M_s$ corresponding respectively to components of magnetisation along the easy direction of M_s and $-M_s$ respectively. These conditions constitute part of the complete hysteresis cycle shown in Fig. 4 where it can be seen that there are only two stable states and the critical fields for discontinuous change from one to the other are $H_x = \pm H_k$ where $H_k = 2K/M_s$ is the anisotropy field.

The other case of particular interest corresponds to the field being applied in the hard direction, i.e. when $H_x = 0$. Equation (4) then becomes

$$\frac{\partial E}{\partial \theta} = -M_s H_y \cos\theta + 2K \sin\theta \cos\theta = 0 \qquad (9)$$

which has solutions given by

$$\cos\theta = 0 \quad \text{or} \quad \theta = \pi/2 \qquad (10a)$$

$$M_s H_y = 2K \sin\theta \qquad (10b)$$

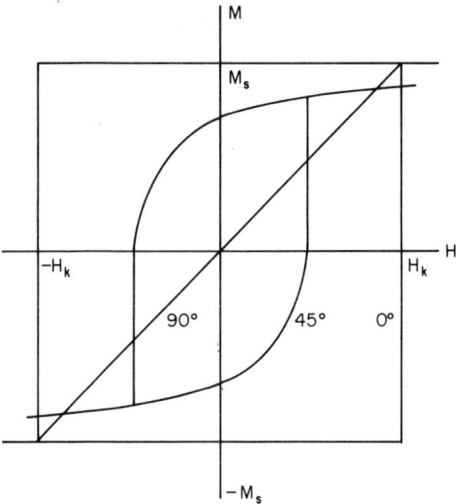

FIGURE 4. Predicted hysteresis loops for fields applied at angles of 0°, 45° and 90° to the easy axis.

The second differential of equation (9) can be shown for the second solution to be given by

$$\frac{\partial^2 E}{\partial \theta^2} = 2K \cos^2 \theta \qquad (11)$$

which is positive and so represents stable states. Therefore since the y component of magnetisation is given by $M_s \sin \theta$ it can easily be seen that

$$M_y = M_s \frac{H_y}{H_k} \qquad H_y \leqslant H_k \qquad (12)$$

Using the first solution, equation (10a), the second differential can be shown to be

$$\frac{\partial^2 E}{\partial \theta^2} = M_s H_y - 2K \qquad (13)$$

from which it can be seen that the state $\theta = \pi/2$ is stable for

$$H_y \geqslant H_k$$

This means that the hysteresis loop for hard axis reversal is as shown in Fig. 4. A linear dependence of magnetisation on field caused by a gradual rotation of magnetisation away from the easy axis is observed for field values up to H_k. For higher field values the magnetisation vector remains in the hard direction.

For other applied field directions the procedure for finding the magnetisation direction is in principle the same as for the simple cases already studied. However equation (4) does not have simple solutions and for a fixed field must be solved numerically for θ. The solutions must then be substituted into equation (5) to test their stability. Such calculations have been commonplace and the resultant hysteresis loop corresponding to a field applied at 45° to the easy axis is also shown in Fig. 4.

It is now convenient to recall that there are critical fields of $H = \pm 2K/M_s$ for reversal in the easy and hard directions. There are also critical fields at which there is a discontinuous change of magnetisation for reversal at 45° to the easy axis, and it is not difficult to appreciate that there will be critical field values for all applied field directions. These occur when a field changes so that the original orientation of magnetisation corresponding to an energy minimum then corresponds to a maximum. The original orientation is no longer stable and therefore reversal of magnetisation takes place. This instability is defined by the second differential of the energy changing its sign from plus to minus and so the relevant condition for the instabilities is

$$\frac{\partial^2 E}{\partial \theta^2} = 0 \qquad (14)$$

Elimination of θ from the corresponding equations (4) and (5) leads to the formula

$$H_x^{2/3} + H_y^{2/3} = H_k^{2/3} \qquad (15)$$

which is the locus of the critical fields for discontinuous magnetisation reversal. This is plotted in Fig. 5. Not only is it used for predicting critical fields but it can be used as the basis for a graphical construction which allows magnetisation directions to be quickly and easily determined thus obviating the need for the tedious and often repetitive work of solving the relevant equations.

The construction to which reference has just been made can be achieved as follows. Equation (4) applied for all positions of equilibrium and for a chosen value of θ represents a straight line relationship between H_x and H_y. Such a line is drawn in Fig. 5 for $\theta = 30°$ and the direction of the line pointing away from the easy axis is also the direction of the magnetisation. This, and any other similar lines for different values of θ, eventually meet the critical curve and are tangential to it. At any point on the line the corresponding field components can be read from the axes and it is known that at this field the magnetization will make an angle of 30° with the easy axis. Now consider any point (e.g. A) within the critical curve. It can be seen that two lines can be drawn through this point, tangential to the critical curve, and therefore two possible orientations of magnetisation exist. For a point B outside the critical curve only one line

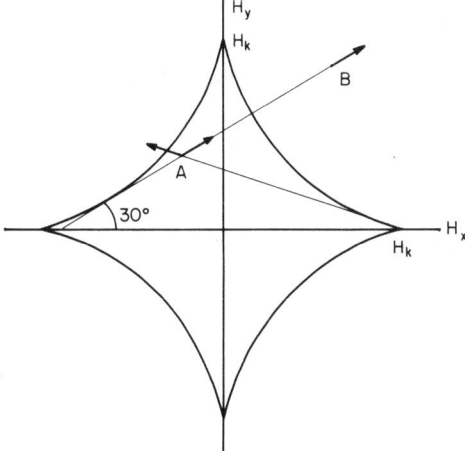

FIGURE 5. Critical curve for discontinuous magnetisation reversal. Straight lines refer to a geometrical construction described in the text.

can be drawn and so only one magnetisation direction is stable. These facts can be used to find the magnetisation direction(s) for given applied fields by adopting the following procedure. Draw in the line(s) which run from the point of interest towards the easy axis and which is(are) tangential with the critical curve. Within the critical curve there will be two possible lines while outside the critical curve there will be only one line. The direction of this(these) line(s) represents the only possible equilibrium direction(s) of the magnetisation. By measuring the gradient of the line(s) the direction(s) can readily be found.

Having developed the theory of reversal for coherent rotation it is opportune to recall that domain structures were quoted as being stable and that reversal could take place by wall motion. The occurrence of rotation and wall motion processes is discussed in the next section which is concerned with the behaviour of real as opposed to ideal films.

2.3 Reversal Processes in Films

There have been many studies of the reversal processes taking place in thin films and several experimental techniques have been used. These involve hysteresis measurements[30], Kerr magneto-optic observations[32], Bitter techniques[31,33-37], and Lorentz microscopy[38], which have all contributed in some measure to the overall understanding of the reversal process that now prevails. Each technique has particular advantages but in this discussion all the data are compounded and a composite picture of the general features of the reversal process is given. In giving such a description it should be emphasised that films produced by different workers under different conditions have properties nominally the same but differing in important detail which affects the mode of reversal.

The hysteresis loops observed for fields applied along the easy direction are generally rectangular[30]. However it is found that the coercivity is usually less than the theoretical value of $2K/M_s$ corresponding to coherent rotation and is a function of film thickness. Further, no flux can be detected in a direction perpendicular to the easy axis during reversal and so coherent rotation is ruled out as a possibility. Observation by both Kerr effect[32] and Bitter techniques[33-37] reveal that reversal takes place by the motion of roughly parallel domain walls which are nucleated, or are already present, at the edge of the films (see Fig. 6a) where demagnetising fields are highest. When the field necessary for wall motion is reached the walls pass through the entire film thus reversing the direction of magnetisation. Reversal is often achieved by the movement of a small number of domains. Sometimes the domains originate from particular spots on the film with which imperfections can be identified.

In some films the observed values of coercive force are greater than $2K/M_s$ and these are often referred to as inverted films[33]. In these, reversal begins by the formation of domains, as shown in Fig. 6b, which have their long axes perpendicular to the easy axis[31,33,37,39]. As the reversing field is increased the magnetisation in the domains forms a larger angle with the easy axis. The consequent domain structure is associated with a high energy which inhibits the further rotation of magnetisation. Reversal eventually takes place when walls move from the edges of the film as in the normal films already discussed.

The splitting or "rippling" of the magnetisation so that it rotates in opposite directions is a general feature of the reversal behaviour of thin magnetic films. If the film was uniform in properties with a field applied near to the easy direction the magnetisation would rotate coherently away from the easy axis. However, in practice, films are rarely uniform and there are variations in the direction of the easy axis from place to place so that when the field is applied along the mean easy direction some of the magnetisation will be affected by the local easy direction, which will cause it to rotate in one sense, while some of the magnetisation will be similarly influenced to rotate in the opposite sense Consequently the local variation of the easy axis of the film has affected the reversal fields of the film and the domain structures involved. This aspect of behaviour appears in many forms during the following considerations of reversal processes.

Domains formed under the influence of local variations of the direction of the easy axis are found to have a long axis perpendicular to the direction of mean magnetisation and to be narrow in a direction parallel to the direction of

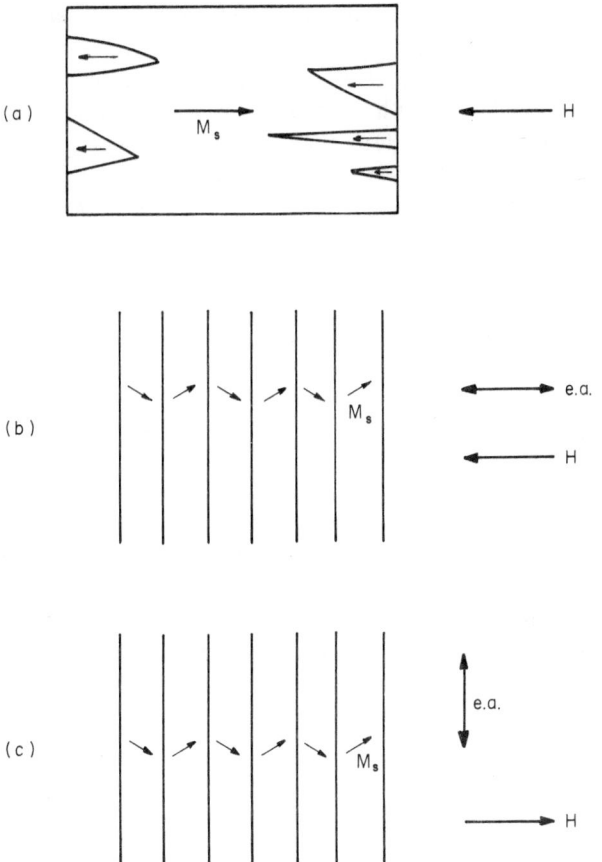

FIGURE 6. Reversal of magnetisation for: (a) field applied along easy axis causing domains to grow from film edges. (b) Field applied along easy axis of inverted film, showing the splitting into narrow domains. (c) Field applied along hard axis and being reduced from a large value. Splitting of film into domains is depicted.

the mean magnetisation. Such properties can be predicted from the theory of magnetisation ripple to be discussed in a later section. However it will be said at this stage that it is stray magnetic fields which give rise to a strong coupling in the transverse direction, and thus a long wavelength of magnetisation variations, while the coupling in the direction of the magnetisation is weak and connected with the exchange energy and is therefore associated with short wavelength variations of magnetisation direction. This explains the structures shown in Fig. 6b.

Now consider reversal taking place as a result of fields applied in directions making a small angle with the easy axis. Reversal takes place more or less as if the field were applied along the easy direction and is accomplished by the motion of domain walls. However if the field is applied at larger angles to the easy axis as shown in Fig. 7 then reversal takes place by a different process. First of all there is some rotation of magnetisation away from the easy axis towards the field direction. However as the applied field is increased towards a value defined by the critical curve, where a discontinuous rotation of magnetisation would be expected to take place, a stage is reached which is marked by the appearance of domains[3,31-37]. This domain structure persists until a further critical field value H_w is reached when reversal is completed by the movement of the domain walls already present or by domain walls originating from the edges of the film. Middelhoek[31] and Methfessel et al.[37] report that the domains are formed by a partial rotation process which is illustrated in Fig. 7. This shows first of all the magnetisation rotating away from the easy axis towards the field direction. However in non-uniform films it has been shown that when there are deviations of magnetisation from the mean magnetisation direction the resultant domains have their long axes perpendicular to the mean magnetisation direction. When reversal begins the magnetization of some of the domains rotates towards the field direction as shown in Fig. 7c. The structure so formed is associated with a high energy and the larger field H_w is required before reversal is completed by domain wall motion (Fig. 7d). The presence of a high energy can easily be demonstrated by removing the applied field at the intermediate stage before it has been increased to the value H_w whereupon the domain structure changes significantly to reduce its energy as shown in Fig. 7c. This mode (partial rotation) of domain formation is different from that observed by Smith[34], Smith and Harte[35,36], and Cohen[40] which was termed labyrinth switching. In the latter case domains are nucleated at the film edges and propagate through the film by extension of the domain tip. The propagation is accomplished by regions of film just beyond the tip of the existing domains being influenced in such a way that they are able to reverse their direction of magnetisation so as to extend the size of the domain.

Now consider fields applied in the hard direction. The hysteresis loops are observed to be similar to those predicted by the simple theory outlined earlier except for the fact that they are not straight lines but are slightly open and have a finite coercivity associated with them[30]. While this may seem a minor deviation from the theory, the absence of a flux component in the easy direction can usually be detected during the reversal process indicating that the magnetisation in the film is not rotating as a coherent unit. Observations of

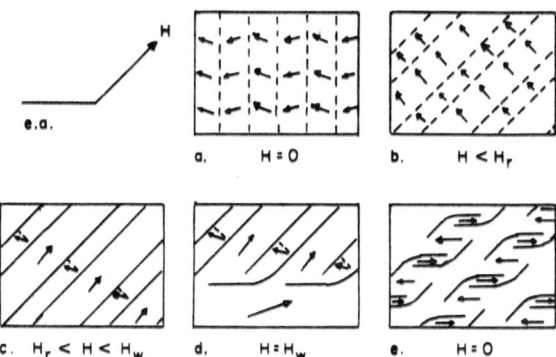

FIGURE 7. Magnetisation distribution at various stages of the reversal process: (a) before the application of the reversing field. (b) $H < H_r$: rotation of magnetisation away from the easy axis. (c) $H > H_r$: partial rotation of magnetisation. (d) $H = H_w$: wall motion. (e) H reduced to zero after application of a peak field not exceeding H_w. [S. Methfessel, S. Middelhoek and H. Thomas, *J. Appl. Phys.* **32**, 1959, (1961).]

the domain structure during reversal give satisfactory explanations for this situation. When a film previously saturated in either sense of the easy direction is subjected to a hard direction field the magnetisation rotates coherently until it is aligned with the hard axis and saturation has been achieved. On gradual removal of the field domains are seen to be formed lying with their long axes parallel to the easy direction[31,32,41]. This process is somewhat reminiscent of that found to take place during reversal in the easy direction. Films have already been shown to be made up of local regions with easy axes which vary in direction and consequently when the film is saturated in the mean hard direction some parts of the film will, when the field is removed, find it easier to rotate in the clockwise direction while the remaining portions of the film relax in the opposite direction as shown in Fig. 6c. A small field applied along one or other sense of the easy axis at the instant that the saturating field was removed would have caused the magnetisation of the whole film to tip to one sense or other of the easy direction. This property is used in the applications to data storage devices discussed later.

All the results are summarised in Fig. 8 where the appropriate threshold fields are drawn on a theoretical switching curve. The curve labelled H_w is the wall motion threshold field and where it cuts the easy axis it is approximately equal to the coercivity. Where this field runs inside the critical curve reversal

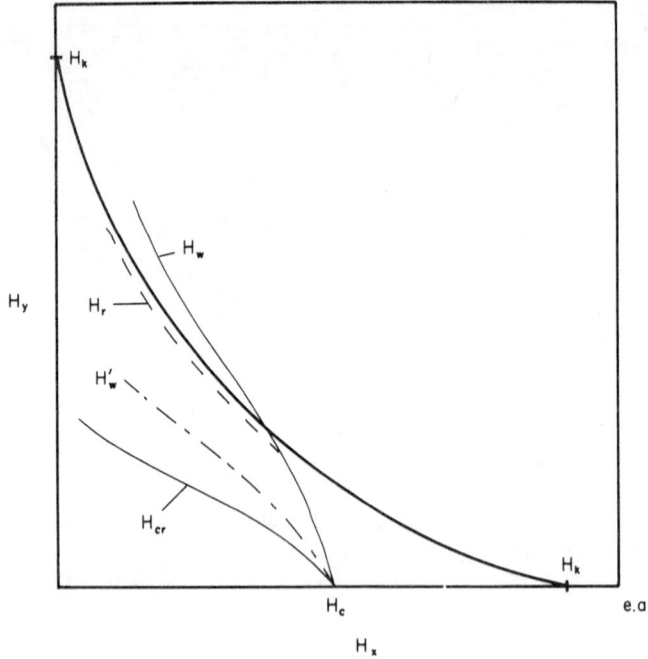

FIGURE 8. Theoretical critical switching curve compared with observed switching thresholds for wall motion H_w and H'_w, partial rotation H_r, and creep H_{cr}.

takes place completely by wall motion but where the line is outside the critical curve reversal begins by partial rotation at field values H_r which are near to the critical curve and reversal is only completed by wall motion. When fields are near to the hard axis reversal is almost totally by rotation of magnetisation. Smith and Harte[35,36] also observed that in certain thin films the reversal process may be by wall motion when fields are applied at large angles to the easy axis since the wall motion threshold does not cross the critical curve.

Another means of magnetisation reversal, called domain wall creeping, has also been observed. This takes place when a domain wall is subject to a D.C. applied field directed along the easy direction while at the same time a pulsed or A.C. field is applied in the hard direction. When the total field is greater than the critical field values already marked in Fig. 8 reversal will take place by the mechanisms already discussed. However if the field is reduced below those critical values wall motion still occurs. This is not a continuous process since the motion consists of discrete displacements of the wall corresponding to the application of each hard direction pulse. Middelhoek[31,42,43] studied the relationship between wall structure and creep and came to the conclusion that creep was associated with changes of domain wall structure induced by the applied magnetic fields. Creeping in Bloch walls was shown to be associated with Bloch–Néel transitions and in thinner films, containing cross tie walls, with the movement of Bloch lines, as was confirmed elsewhere[44]. In pure Néel walls creep was observed but the threshold fields were little different from the normal wall motion thresholds. Creep in multipolar Néel walls has been shown to be connected with the movements of sections of alternate polarity[45]. The details of wall structure to which reference has just been made will be described in Section 2.4.

Creep has been observed as a result of unipolar and bipolar hard axis pulses[46] and also in the absence of easy direction applied fields[47]. The function of the easy axis field is usually to ensure a nett wall displacement in a particular direction[47]. It is thought that many factors contribute but from this brief account it can be seen that it is associated with changes occurring in the structure, and therefore stray fields and energies, of the walls subjected to hard direction pulsed fields. It appears that the variation of energy etc. during the application of the field pulses makes alternative wall positions energetically favourable thus causing the wall to move. Many different mechanisms have been postulated to account for creep[44,48-52] but the validity of these is perhaps less important than the observation that higher coercivity films show less creep. In Bloch walls the creep threshold has been shown to increase linearly with wall motion threshold[46] and therefore control of coercivity is one means of eliminating creep.

This mechanism of creep is not to be confused with another observed by Olmen and Mitchell[53] which was also termed creep. The latter observation was that domain walls could be caused to move very slowly under the influence of D.C. fields alone which are lower than the threshold values given in Fig. 8. It was later shown that such motion had its origin in thermal effects[54].

2.4 Domain Walls

The domain wall structures that can be observed in thin films are substantially different from those present in bulk media. The reason for this becomes apparent upon examining the situation shown in Fig. 9a. This represents a section through a magnetic film where the magnetisation in the classical Bloch wall is rotating out of the plane of the film and about an axis normal to the plane of the wall. It can be seen that the poles formed on the top and bottom surfaces of the film will give rise to high demagnetising fields and therefore high magnetostatic energy for this type of wall. When the film is thin this energy is very large and consequently Néel[55] predicted that the energy of the film could be reduced by the magnetisation rotating in the film plane as

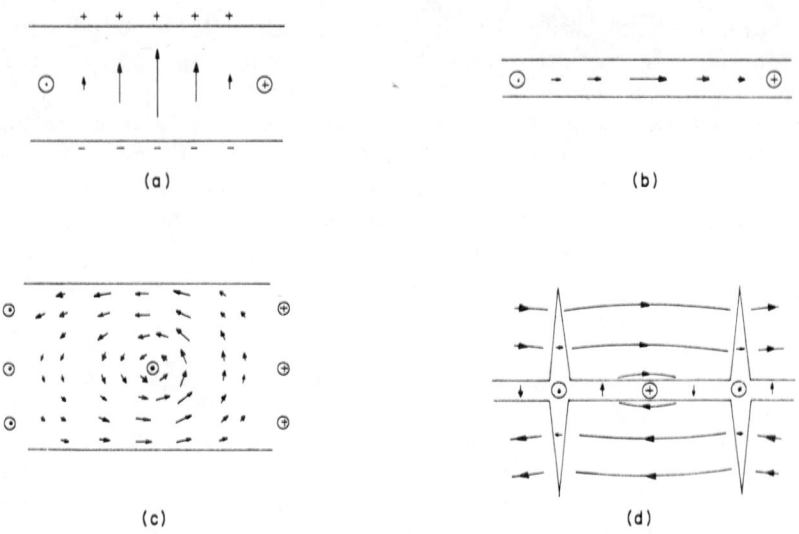

FIGURE 9. Cross sectional view of domain wall structures running normal to the plane of the page: (a) simple Bloch wall, (b) Néel wall, (c) two-dimensional Bloch wall, (d) is a plan view of a cross tie wall. It is made up of sections of Néel wall of alternating polarity separated by Bloch lines, where the magnetisation tips out of the film plane. The "limbs" protruding from the main wall are of the Néel type and are known as cross ties.

shown in Fig. 9b. This type of wall is known as the Néel wall and like other domain wall structures, to be discussed, is a consequence of the attempts of the magnetisation to reduce its magnetostatic energy. Bloch walls, which will later be shown to have a more complicated structure than indicated by Fig. 9a, and Néel walls have been observed in thin films as has another type of wall known as the cross tie wall[56] whose structure is shown in Fig. 9d. Methfessel et al.[57] and Middelhoek[31] have investigated the structure of domain walls as a function of film thickness and it has been demonstrated that Bloch walls were stable in films thicker than about 900 Å, cross tie walls in the range 250 to 900 Å, while in films less than about 250 Å thick Néel walls were stable.

Many workers have calculated the energy of these domain walls and their efforts have been reviewed earlier[58-61]. The calculation of Holz and Hubert[62] gives the lowest value so far obtained for the Néel wall energy and this is shown in Fig. 10. The computer calculations of Brown and Labonte[63] gave the energy of a Bloch wall and this is also plotted in Fig. 10. Although this minimisation of the energy gave the lowest energy structure for the case where

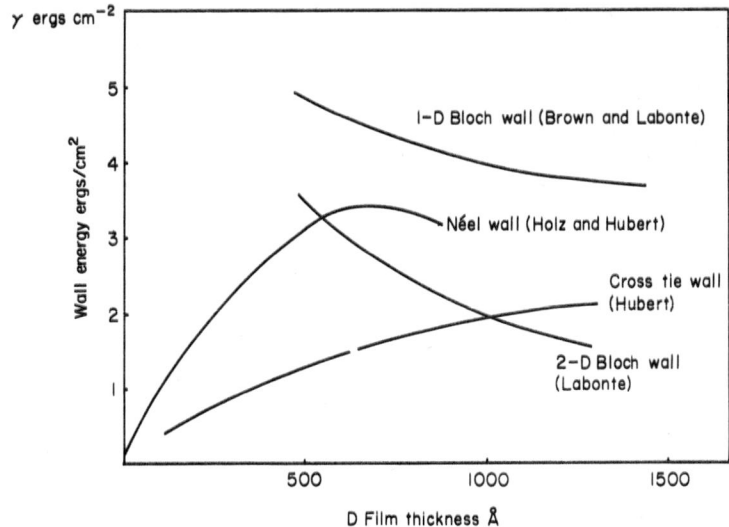

FIGURE 10. Calculated energy per unit area of domain wall plotted as a function of film thickness assuming, for Permalloy, $M_s = 800$ gauss, $K = 1000$ ergs cm^{-3}, exchange constant $A = 10^{-6}$ ergs cm^{-1}.

magnetisation tips out of the film plane as shown in Fig. 9a it was subsequently shown that much lower energies could be predicted if magnetisation variations through the thickness of the film were allowed in the calculations. Labonte[64] thereby obtained the minimum energy structure shown in Fig. 9c for which the corresponding energy is plotted in Fig. 10. Examination of this structure shows that it is a move towards a nearly stray field free structure by the formation of an almost divergence-free magnetisation distribution[65,66]. Experimental evidence for this is demonstrated simply from the earlier mentioned investigation into the structure of domain walls where it was found that Bloch walls did not attract Bitter colloid as strongly as other wall structures which are associated with higher stray fields.

The cross tie wall remains a difficult problem for the theoretician and its energy has not been satisfactorily calculated although a number of efforts have been made[60]. Figure 10 shows the estimate made by Hubert[66] which was obtained by summing the energies of the constituent Néel walls of a structure which was an approximation to that shown in Fig. 9d.

The structures of domain walls are also altered by the presence of magnetic fields. This is partly because the magnetisation in the domains either side of the domain walls rotate towards the direction of the applied field so that the walls are no longer of the 180° variety, and partly because the internal structures of domain walls are also affected. Investigations of the structure of walls in films of differing thicknesses, which were subject to hard direction applied fields, have been made[58,61] and the results summarised in Fig. 11 where the regions of stability are shown as a function of wall angle and film thickness. The curves

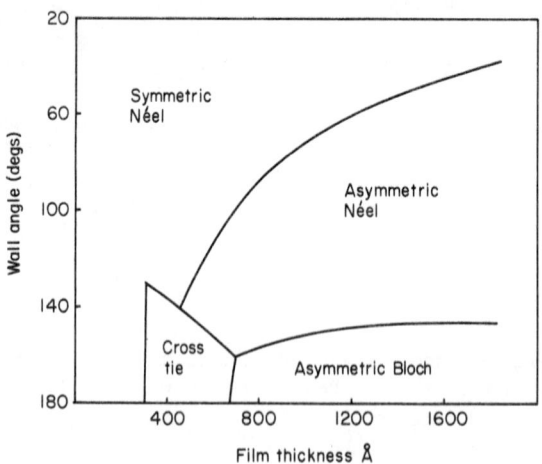

FIGURE 11. Stability of walls of various angles as a function of film thickness.

are the theoretical predictions of Hubert[61] which were found to be supported by observations made using Lorentz microscopy, in films of thickness greater than 400 Å, while the extension of the curves to thinner films is now cautiously made in the light of the obs…vations of Middelhoek[58]. It is clear that Néel walls are stable for all wall angles in thin films while in thicker films cross tie walls eventually become unstable when the wall angle is reduced. The instability of the cross tie wall is preceded by substantial changes in structure as the magnitude of the applied field is increased. This takes the form of movement of the Bloch lines in the walls, so that the sections of Néel wall with magnetisation in the same sense as the applied field grow at the expense of those oppositely magnetised, and a tilting of the cross ties so that they no longer form a right angle with the Néel sections of the main wall[31,57]. The significance of these changes to the reversal mode known as creep is discussed in the section on magnetisation reversal. In thicker films Bloch walls are stable only for small hard direction fields and give way to an asymmetric form of Néel wall before changing finally to the symmetric Néel wall structure at high fields.

Attempts to explain the wall transitions in terms of simple wall structures[57,67] which were modifications of the Bloch wall structure of Fig. 9a and the Néel wall structure of Fig. 9b resulted in a prediction that the Bloch wall eventually became unstable and was replaced by the Néel wall. However the more realistic structure of Fig. 9c, when subjected to a field in the hard direction, changes to a form of Néel wall which does not have the symmetric structure shown in Fig. 9b but has a distinctly asymmetric structure which was initially predicted[65,66] as a possible stray field free structure but which is not stable at wall angles close to 180°. Further increases of applied field are necessary before the symmetric Néel wall structure similar to that of Fig. 9b is obtained[61].

2.5 Magnetisation Dispersion and Ripple

The term dispersion is taken to mean all deviations of magnetisation from the mean magnetisation direction within a previously saturated film. Deviations have been observed, by the technique of Lorentz microscopy[38], which are microscopic in scale and known as magnetisation ripple. A micrograph showing magnetisation ripple is reproduced in Fig. 12 while Fig. 13a shows, schematically, a possible interpretation in terms of the deviations of magnetisation direction in a form known as longitudinal ripple. It is probably apparent that in making macroscopic measurements of dispersion there will be contributions from the ripple properties of the films but that there may also be contributions from variations which take place on a larger scale. Therefore observations of ripple will not give complete information about the dispersion and the behaviour of films unless larger scale variations have been avoided.

Magnetisation dispersion, as discussed above and observed in experimental situations, is concerned with deviations of magnetisation from an average direction and must be contrasted with its causes which are usually discussed in terms of dispersion of local or effective local anisotropies. The origin of these anisotropies will be discussed shortly. The connection between anisotropy dispersion and magnetisation dispersion is complicated and is the subject of ripple theories which have been introduced and tested over a number of years[68]. Measurements of magnetisation dispersion have been used as a measure of the uniaxial nature of the films and their suitability for storage device applications[69]. Many possible measures of dispersion exist but one such quantity, quoted only by way of example, is α_{90} which can simply be defined by the angle which a field greater than H_k must deviate from the hard direction so that when it is reduced to zero in magnitude the resultant component of magnetisation along the easy axis is 90% of the saturation value. Dispersion in NiFe alloys has been measured as a function of composition and shown to be a minimum near to where the magnetostriction goes to zero[34]. In

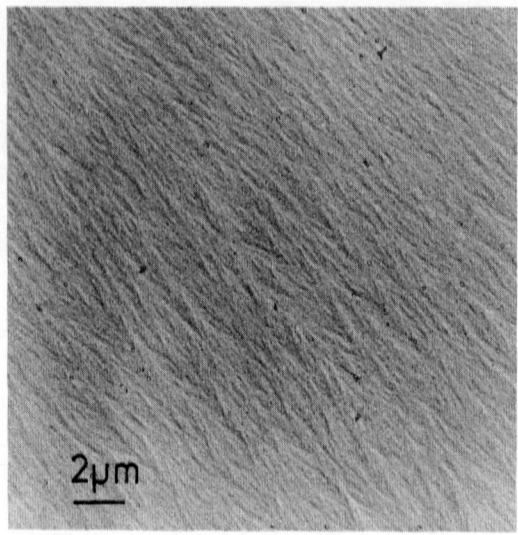

FIGURE 12. Micrograph showing magnetisation ripple observed using Lorentz electron microscopal techniques. (Micrograph provided by P. J. Grundy.)

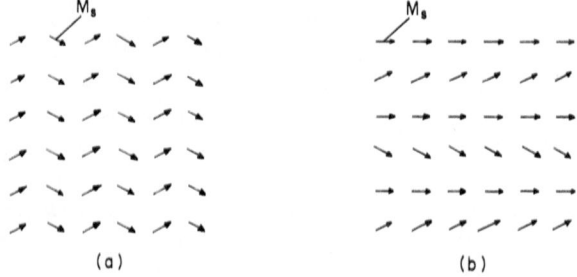

FIGURE 13. Schematic representation of (a) longitudinal and (b) transverse magnetisation ripple.

NiFeCo alloys, in which the nickel and iron contents have been adjusted for zero magnetostriction, the dispersion has been shown to go through a minimum as the percentage of cobalt is increased[70]. Varying the composition in this way is known to produce films in which the anisotropy goes through zero and near this point the minimum in dispersion is observed. It would therefore appear that dispersion is connected with the presence of both anisotropic and magnetoelastic effects; the latter not surprisingly when very large stresses have been observed in thin films[71,72]. A measure of confirmation comes from the work of Uchiyama et al.[73] who measured dispersion as a function of composition and temperature and were able to explain the results in terms of ripple theories only if magnetocrystalline and magnetoelastic sources were assumed to contribute to the random anisotropies. However composition alone is not sufficient to specify the dispersion and factors such as magnetic field applied during deposition[74], substrate temperature during deposition[75], and film thickness[76,77] contribute either in connection with the anisotropy and magnetoelastic effects or in some way not yet ascertained.

It has already been mentioned that dispersion measurements would be a reflection of the microscopic observations of ripple were it not for the large wavelength deviations of the easy axis[78]. Longer wavelength variations have been shown to result in larger values of dispersion over the entire film than is observed in small areas[79] and unless care is taken to avoid them it is not possible to use dispersion measurements to test ripple theories.

Magnetisation ripple was first observed by Fuller and Hale[38] using Lorentz electron microscopy which was later used to establish a connection between grain diameter and ripple wavelength[80] and to show that ripple amplitude is a minimum approximately at the composition[81,82], at which the crystalline anisotropy is zero. This seems to be in agreement with the observation of Uchiyama et al.[73] if one considers that the unsupported films in the electron microscope would probably not have the stress that is experienced by films attached to substrates. Despite the connection between dispersion, crystalline anisotropy and magnetostriction it is not by any means certain that these are the only possible reasons of its occurrence. Other factors such as chemical inhomogeneity[83] and surface roughness[84,85] may also contribute.

So much for the origin of ripple; the consequences of its occurrence have been discussed in terms of the reversal properties of films in a previous section. Middelhoek[31] was the first to calculate the properties of ripple and he showed that magnetostatic effects were responsible for the large attenuation and therefore long wavelength of ripple observed in the transverse direction (Fig. 13b) while exchange effects attenuated the amplitude of the longitudinal ripple (Fig. 13a) but to a lesser degree allowing shorter wavelengths to be observed. This prediction has not been altered by the more detailed and sophisticated theories of Hoffman[86-89], Rother[90] and Harte[91] except that more than one characteristic wavelength is needed to describe the properties of longitudinal ripple. The ripple theories have also been used to provide predictions about the nature of initial susceptibility which has been used as a vehicle for testing the theories. Experimental results seem to have provided some measure of confirmation[92-94] of the theoretical work of Hoffman. Also the splitting of the film into domains after saturation in the hard direction as well as domain wall coercivity for easy axis reversal can now be predicted as well as verified in practice[95,96], thus bringing the ripple theories nearer to success in predicting the overall behaviour of real thin films.

2.6 High Speed Switching

In the section dealing with the reversal processes occurring in thin films the domain structures were observed in either static or quasi-static situations. In applying the films to storage devices they will be subject to pulsed fields and the dynamic behaviour then becomes relevant.

In our earlier studies three modes of magnetisation reversal were identified; domain wall motion, partial rotation or labyrinth switching, and coherent rotation. On the basis of these observations it could be anticipated that at least three types of pulse switching will occur[3]. Evidence for such behaviour was obtained by Olson and Pohm[41] who made measurements of the switching time of films subjected to pulse fields applied along the easy direction while D.C. fields were present along the hard direction. The switching time was defined as the time between the application of the field pulse and the falling of the voltage in the pick-up coil to 10% of its maximum value. The relationship between applied pulse amplitude and the reciprocal of switching time is shown in Fig. 14 where three regions of the curves can be identified. At very low fields, just above the switching threshold, reversal in the easy direction takes place by wall motion with switching times of the order of a few microseconds. For this and other directions of applied field there is a change of slope of the graphs at higher applied field values and therefore higher switching speeds are observed. Extrapolation of the curve to the field axis gives an intercept for this mode of intermediate speed switching which has been shown to be close to that predicted for coherent rotation by the critical curve of Fig. 8. However it was shown that in practice the critical curve usually marks the onset of various types of incoherent reversal process and therefore this region of switching is thought to be connected with this mode of switching. The corresponding switching times vary roughly in the range 50 nsec to 1 μsec. For higher applied fields reversal takes place at much higher speeds (less than 20 nsec) and this is assumed to be by rotational reversal. The general trend observable from these results is that the higher the applied fields and the larger the angle between the easy direction and the field the greater the proportion of the flux that undergoes reversal by rotation. The advent of low and high speed switching processes in these types of experiments has been confirmed by Dietrich *et al.*[97] who examined voltages picked up due to flux components along the easy and hard directions to differentiate between reversal by rotation and wall motion. When large pulse fields were applied along the easy direction, switching times of less than 20 nsec were obtained.

Kryder and Humphrey[98] have also measured switching times, as a function of applied field, and found them to fit a three part curve similar to that shown in Fig. 14. They were also able to make observations on the "dynamic" domain structures occurring during reversal using a Kerr magneto-optic

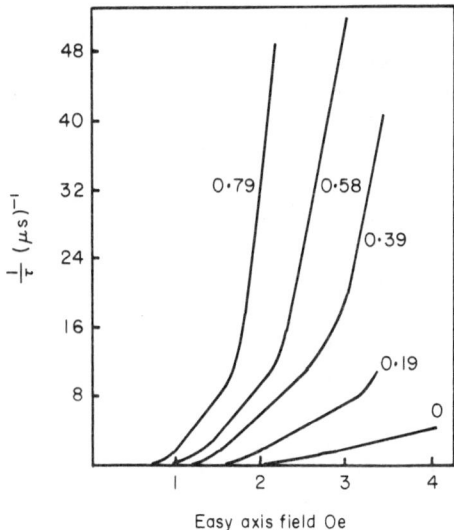

FIGURE 14. Reciprocal of switching time as a function of magnitude of pulsed easy axis field: the points are omitted and only the lines drawn. The field magnitudes quoted (in Oe) are those of the bias field applied in the hard direction. [C. D. Olsen and A. V. Pohm, *J. Appl. Phys.* **29**, 274, (1958).]

camera[99] with exposure times of around 10 nsec. These show that in the absence of a hard axis bias field and with field pulses, of varying magnitude, applied along the easy direction, different types of reversal take place. For fields varying from just above the coercivity, H_c, to the nucleation value H_n, marginally above H_k, reversal takes place by the movement of broad diffuse boundaries oriented perpendicular to, but moving in a direction parallel to, the easy axis. These are distinct from normal domain walls and represent a broad front of reversing magnetisation moving across the film[100]. At fields larger than H_n the whole film reverses simultaneously at high speed, while at low fields, very near H_c, diffuse boundaries predominate but some normal longitudinal domain walls (parallel to the easy axis) are produced and move in a direction perpendicular to the easy axis. The three modes of reversal are associated with the three parts of the switching curves. With D.C. bias fields, applied in the hard directions, and total fields less than the critical switching threshold, diffuse boundaries propagate at an angle to the easy axis, while at higher fields a type of non-coherent reversal is detected by the formation of a striped domain pattern. These observations appear to be at some variance with the interpretations given to the earlier results, particularly in respect of the diffuse boundaries, although this does not necessarily invalidate the latter which were carried out on different films. However the lesson seems to be that the use of static switching characteristics, particularly in respect of such properties as ripple[98], to interpret dynamic observations, should be made with caution.

Pulsed fields have also been applied along the hard direction[101] of films where it has already been shown that coherent or nearly coherent reversal can take place. Fields of less than 1 nsec rise time were applied and it was observed from the signal picked up from easy axis components of magnetisation that switching times of less than 1 nsec could be obtained.

The fact that domain wall or boundary motion is a slow process compared with coherent rotation is easily explained by the fact that the former is associated with the sequential rotation of the spins while in coherent rotation all the spins rotate in unison. The intermediate switching processes are obviously somewhere between these two extremes with an increased proportion of the film rotating with a higher degree of coherence than in the slow process of wall motion.

In summary it can be said that the ramifications of the above observations are obvious for high speed applications of magnetic films. High speed switching takes place due to rotational processes which are induced by fields which have large components in the hard direction.

3. High Speed Memory Applications

For many years the random access memory requirements of computers have been met, at least in part, by the core store. The property that makes the core suitable for information storage is that it possesses a rectangular hysteresis loop and can be magnetised to remanence in either of two senses. If one of these states corresponds to a binary one and the other to a binary nought and the core can be switched quickly from one state to the other then a digital storage element is obtained. The cores are magnetised circumferentially and reversal between the states is achieved by wall motion which is a process taking of the order of tenths of a microsecond. In thin films it has been shown that reversal of magnetisation can take place either by wall motion or by coherent rotation of magnetisation which can take place in times of the order of nanoseconds. Therefore a storage device dependent on the rotational switching speeds of thin films could operate very rapidly. In addition the ability to deposit fairly large areas in one evaporation has led to the belief that a low cost thin film store could ultimately be developed. These and other factors are now considered in as much as they are important to the design and development of thin film storage devices.

To understand the process of information storage in a thin film consider the situation in Fig. 1 when $\theta = 0$. The film is then magnetised in the positive direction of the easy axis and fields applied parallel to the easy axis, but of opposite sense, would cause magnetisation reversal to take place by a process of wall motion which has been shown to be a relatively slow process and so this mode of switching is not used in fast switching devices. However, consider the situation where the magnetic field is applied in a direction perpendicular to the easy axis. In this case the magnetisation would rotate towards the hard axis and if the field were sufficiently large the magnetisation would align itself with the hard axis. Removal of the field would then result in the magnetisation splitting up into finely spaced domains parallel to the easy axis and with alternate domains magnetised in opposite senses. However, if a small additional field had been applied along one or other sense of the easy axis it would have ensured that the magnetisation tipped into that sense of the easy axis. Thus by choosing the sense of the small applied field it is possible to select the sense of magnetisation when the hard direction field is removed. Therefore two fields are required to achieve storage by fast rotational reversal of magnetisation.

3.1 Store Operation

When writing onto planar thin films the above operation may be achieved, in a practical storage situation, as follows. The fields are applied to the chosen storage locations by strip line conductors positioned just above the film and aligned parallel and perpendicular to the easy axis (Fig. 15). The line running parallel to the easy axis is called the word line while the line parallel to the hard direction is called the digit line. A current passed down the word line produces a field component in the film plane directed along the hard direction and consequently the magnetisation is rotated into the hard direction. A current down the digit line produced a field component in the plane of the film directed along the easy axis. When the word field is removed the magnetisation is tipped by the digit field towards the appropriate direction of the easy axis and when the digit field is removed the magnetisation falls to that sense of the easy direction. The sequence of current pulses needed to produce the storage action is also shown in Fig. 15. Since the storage action is achieved only where the two

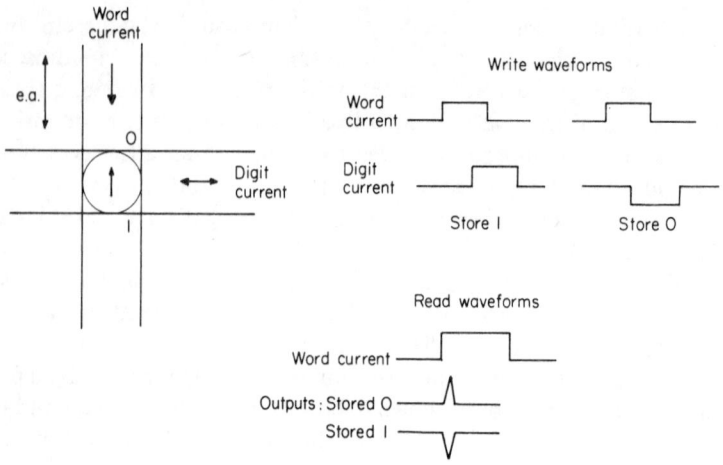

FIGURE 15. Single storage element and corresponding write and read waveforms.

fields coincide it can be seen that only the magnetisation under the intersection of two strip lines is used for the storage of the information.

A storage matrix is shown in Fig. 16 to be formed by the intersection of many parallel sets of word and digit lines. This is called a word organised store where the large current passed down the word line causes the magnetisation under all the strip line to be rotated to the hard direction, and where writing of information into the bit locations takes place coincidently with the simultaneous passage of currents down all of the digit lines. Therefore all of the bits in a word are recorded at one time.

Reading of information is achieved as follows. A current pulse is passed down the word line and voltages are induced in the digit lines, now used as sense lines. Alternately a second line may be sandwiched between the digit line and the film and used for this purpose. Consider firstly what happens if a one has been stored in the film (Fig. 15). The magnetisation under the digit line changes from a negative value to zero as the magnetisation rotates to the hard direction whereas if a nought had been stored in the location then as the magnetisation rotated to the hard direction the component of magnetisation under the strip line would change from a positive value to zero. Therefore the changes of sense of magnetisation are opposite and so will be the fluxes linking the digit lines and therefore the induced voltages. On removal of the word field the magnetisation would split into a multidomain structure, as discussed earlier, but if the easy axis were rotated a few degrees clockwise the magnetisation would fall back to the nought direction[102].

Magnetic Thin Films and Devices

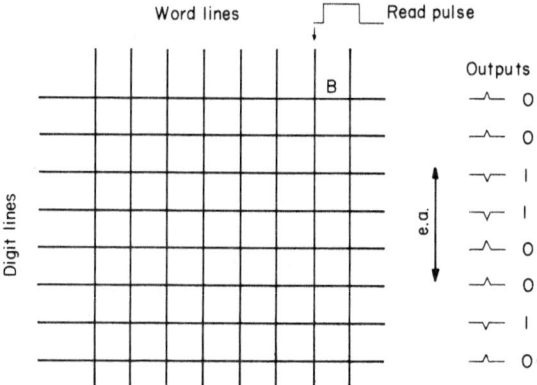

FIGURE 16. 8 × 8 word organised storage matrix. Readout of word B illustrated.

The working of the store depends on the current margins that are available during its operation. In an ideal film it can be seen that provided the word field is large enough to rotate the magnetisation to the hard direction, the digit field need but be finite for it to select in which direction the magnetisation will tip. The maximum digit field that could be used would be that which would cause the magnetisation almost to reverse if it were applied to a bit in the absence of a word field. Therefore in an ideal film its minimum value would be zero and its maximum H_k. The minimum digit field that would be used in practical circumstances would be that required to overcome the effect of dispersion of the easy axis and tip all the magnetisation to the chosen easy direction, while the maximum digit field should be less than the wall motion threshold field H_w. Writing can be achieved with word fields less than H_k but since it is desirable to use the same drive electronics for the record and replay processes the word field is usually chosen to be greater than H_k. Bradley[102] analysed the margins in the fields if the easy direction deviates from the direction of the word line and has shown that wide margins still exist for fairly high angular deviations of the easy axis. Therefore on the basis of simple analysis an operating store is feasible if H_w is not small and the angular deviation of easy axes is not large.

It is now appropriate to consider how well actual films, whose general properties have been discussed earlier, in some detail, match up to the requirements of a successfully operating store. In most cases the thin films have been deposited by evaporation onto glass substrates which, because of their smoothness, aid the production of films with reproducible properties[103].

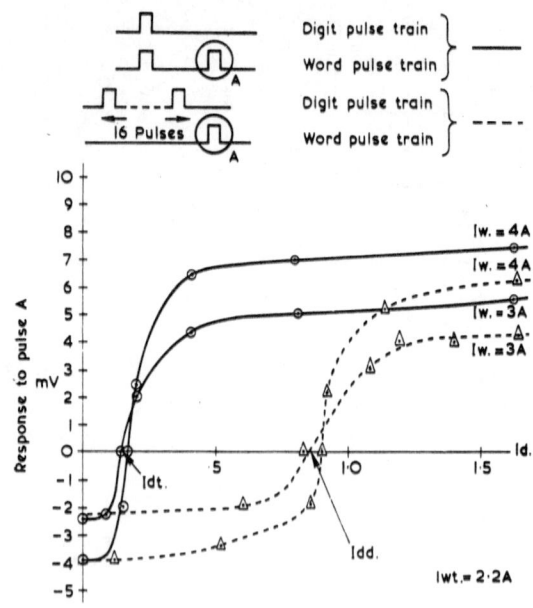

FIGURE 17. Output voltage from a single storage cell as a function of digit current. [E. M. Bradley, *J. Brit. I.R.E.* **20**, 765, (1960).]

Electrodeposited films[104] have also been prepared but metallic substrates are necessary and these are difficult to produce with surfaces as smooth as that of glass. Nevertheless with suitable values of H_k and H_w of a few Oersteds and angular dispersions of a few degrees being produced by both evaporation[102,103,105] and electrodeposition[104,106], of thin films of 81:19 NiFe it may be anticipated that good storage performance may be achieved.

Now consider the performance of a single storage cell. Suppose that, during the write operation, the word field is applied at an angle of five degrees to the hard axis and that its value is chosen to be above the minimum value needed for writing. When only the word field pulse, of suitable polarity, is applied a nought is written in the film. However, as the digit field pulse, with polarity appropriate to the recording of a one, is increased the tendency is for more of the flux to be rotated into the one state. Therefore the output voltage read, immediately after the write operation is firstly reduced to zero, and then increased in the opposite sense as the one state is written[102]. Figure 17 shows this variation and that it is affected by the magnitude of the word current. A second set of points, taken when the write word field was not applied, shows the stability of the film under the disturbing action of the digit field on its own. The variation of the output shows that a recorded nought is fairly stable under the influence of disturbing fields until the digit fields are sufficient to switch the film in the easy direction. Clearly the limits between which the store can be operated successfully are fixed by the critical currents I_{dt} and I_{dd}, marked in Fig. 17 and sometimes called the digit threshold and the digit disturb threshold currents respectively. In an ideal uniaxial film these curves should be flat except for the discontinuous changes at I_{dt} and I_{dd}. In addition a word threshold current I_{wt} can be defined as that word field applied in conjunction with a digit field greater than I_{dt} which will switch half of the flux from the nought state to the one state.

The variation of the three critical currents with variations of the angle between the word field and the hard direction are shown in Fig. 18. The first point to note is that there is always a reasonable operating margin between the critical digit currents and therefore successful storage operation on these considerations is possible. The variation of these parameters is more or less as would be expected from simple analysis except for the fact that the fields required to produce the switching are higher than would be expected from the hysteresis properties of the entire film. However, hysteresis measurements made on small areas of film gave values for H_k which then allowed closer fit between theory and experiment.

Not only has it been observed that the rotational behaviour of small areas of film are different from that of the film as a whole but simple experiments with discrete etched storage elements have shown that the latter need higher fields to facilitate successful storage operation[102,103]. The rotation of magnetisation under the influence of non-uniform strip line fields and in discrete film elements have therefore been studied[107-112] in view of their obvious importance. Deviations from the behaviour expected of large area films subjected to uniform fields arise because non-uniform magnetisation distributions and film edges introduce demagnetising fields which modify film behaviour. The effect is complicated but it is observed generally that higher fields are needed than is apparent from considering the anisotropy field alone. The increase in field that is necessary is related to the non-uniformity of the applied field and the geometry of the storage elements. The situation is further complicated by the presence of a conducting substrate when pulsed fields are applied. In these circumstances eddy currents induced in the substrate can cause a further increase in the effective anisotropy fields[113].

In practical stores it is necessary to make the storage elements as small as possible and to pack them close together to enable high storage densities to be obtained. Many new problems arise which begin to erode the operating margins which earlier made the construction of the store attractive. With the elements being made smaller the output voltages are correspondingly reduced.

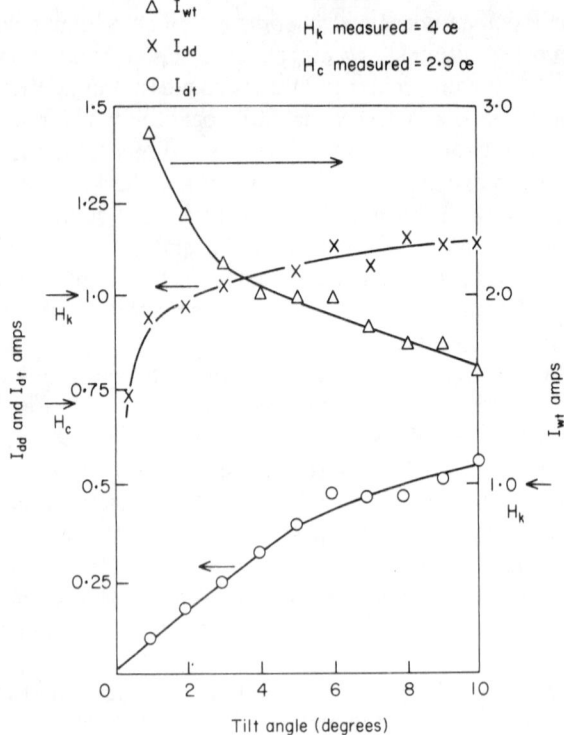

FIGURE 18. Variation of current threshold parameters with angle between word conductor and the easy axis. [E. M. Bradley, *J. Brit. I.R.E.* **20**, 765, (1960).]

This occurs when output voltages are of the order of mV and their reduction aggravates an already difficult signal to noise problem. Further difficulties arise as the storage elements are packed closer together and interactions increase causing more loss of output.

Firstly consider the effect of reducing strip line widths on current operating margins. Making the word line width smaller only decreases the values of word current needed and so does not affect the margins. Changing the width of the digit lines has little effect on the observed values of I_{dt} but reduces the observed values of I_{dd} and consequently there is a loss of operating margins. Interaction between elements has also been estimated by the effect it has on the output derived from the element that is suffering the interaction. The method used by Bradley[102] for estimating firstly digit interactions is to record one and nought into neighbouring digits of the same word and then read out. The variation of the output from the nought state with variations in digit line spacing gives a measure of the interaction. Similarly in estimating the effect of word interaction a one and nought were recorded into the same digits of neighbouring words and the variation of the amplitude of the nought output was taken as a measure of the interaction. In this way the loss of output due to reduced storage element size and interactions could be observed. Consequently, as a compromise between deterioration of operation and gain of storage density, an optimum store design was chosen. Word and digit drive strip widths of 1 mm were chosen while the centres of the word lines were spaced 1.2 mm apart. The digit lines were spaced 2.4 mm; the overall effect being a packing density of 230 bits per square inch for this early type of store design.

The problem of interactions is complex and is manifested most strongly in the reduction of output suffered by affected elements as a result of the passage of a large number of pulses[114]. Measurements of the effects of interactions usually involve variations on the theme of writing into the storage location under study and then writing a large number of times into neighbouring locations before reading from the original[112]. Using 10^6 "disturb" pulses Bonyhard[112] found word interactions to be strongly affected by film thickness resulting in a total loss of information in film thicker than 1400 Å. The reversal of magnetisation under pulsed fields, such as those which could be produced by one element on another, has been discussed earlier and is known as domain wall creeping. These observations may be consistent with results also quoted earlier which stated that wall creeping is strongly correlated with film coercivity[103] since this is a decreasing function of increasing film thickness. Discrete elements were found to suffer less from interactions but were still also strongly affected[112,115]. Therefore by careful choice of film thickness and coercivity, discrete as opposed to continuous films, by shaping the elements and tapering their edges to avoid the nucleation of domain walls, it is possible to reduce the susceptibility of films to creep[103].

Interactions are also dependent on strip line dimensions and spacings. Word interaction effects have been shown to be most severe[112], since word currents are large, and related to the interaction fields arising from the strip lines themselves rather than the storage elements. This is in contrast to the smaller effects of digit interactions which are thought to be related to fields arising from the recorded elements[112].

The threshold fields for domain wall creeping have been shown to be much lower than those of wall motion and coherent rotation. In addition it is difficult to control, let alone eliminate, the occurrence of domain walls in storage elements and therefore storage design has developed in a way which has reflected the need to produce elements less susceptible to creep. Usually this has involved either production of films with creep resistant properties or the design of storage elements which do not produce stray fields, which are the cause of creep, capable of interfering with neighbouring elements.

3.2 Advanced Planar and Cylindrical Storage Element Design

Some attempts have been made to reduce the effects of creep by depositing films in multilayer structures consisting of layers of magnetic material separated by very thin films of non-ferromagnetic materials. The properties of such composites are controlled by the properties of the magnetic layers, the nature of their interactions and the properties of the intermediate layers. However, it has been observed that creep still occurs in such films[116] but to a lesser extent and can be reduced by varying the thicknesses of the intermediate layers so as to give films of high coercivity which are creep resistant and potentially suitable for storage application[117,118]. Despite a measure of success with this approach the real success has been achieved by depositing films in geometries which minimise the stray magnetic fields that cause creep.

The problems caused by stray fields are all too apparent from the foregoing discussions. If systems of magnetic flux closure could be produced then some of the problems highlighted could possibly be reduced. Systems have been suggested[119-122] which employ pairs of films, an arrangement of which is shown schematically in Fig. 19. It can be easily shown that the sequence of pulses already discussed for storage on a single film would leave the films in this structure magnetised in opposite senses of the easy direction. Storage is then achieved in two films and there is a partial flux closure pattern when the films are magnetised not only along the easy direction but also when along the hard direction during readout. With demagnetising effects reduced, the thickness of the films and therefore the output can be increased as can packing density, according to one proposal[123], to 4400 bits/cm^2 of storage plane.

Other configurations based on this structure have been made with a view to achieving non-destructive readout (NDRO) of the stored information[124,125]. These are again based on a pair of coupled films one of which has high anisotropy and coercivity and is used to store the information while the other is of low anisotropy and coercivity and is used for the readout of information. When a readout pulse is applied to this combination the current level is chosen such that the field is sufficient to rotate the magnetisation only of the readout element into the hard direction. However, when the readout pulse is removed the magnetic field from the storage element is strong enough to rotate the magnetisation in such a sense as to regain flux closure and so the original information is restored to that element. Therefore non-destructive readout has been achieved. A second variation of this scheme is used when high speed switching is involved and this can also be appreciated from a study of Fig. 19. If the lower film is sandwiched between the readout line and the ground plane and a short duration pulse is sent down the storage line, then the same film is

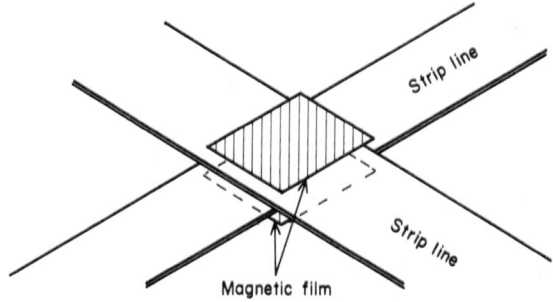

FIGURE 19. A storage location having partial flux closure as a result of films, above and below the strip line cross over, being magnetised in opposite directions.

subjected to the strip line field and the field of the image current in the ground plane[126] and its magnetisation is rotated towards the hard direction. The upper film is not subjected to the same nett field since it experiences the difference between the fields from the strip and the image current and is also shielded by eddy currents in the other conductor. When the pulse ends, the effect of the stray field from the upper element, and the eddy currents, push the magnetisation in the lower element back to the original direction. Clearly, pulse durations must be short compared with the decay time of eddy currents and are of the order of a few nanoseconds. When writing information into the films, pulses in both of the lines are present and the pulse duration may be increased.

Most success in the application of films to devices has been achieved in plated wire memories. A typical example of an application is shown in Fig. 20, due to Danylchuk et al.[127], which employs approximately 80:20 NiFe films electroplated onto BeCu wire of diameter 125 µm. The easy axis of the magnetic material is arranged to be circumferential to the wire and in the storage situation flux closure is achieved. This allows films of the order of 1 µm thickness to be used thus giving rise to higher outputs than is achieved with the open flux structure of the flat films. Other advantages are that the wire itself is used for the drive and sense conductors thus ensuring that lower current levels are required and that closer coupling between the film and pick up line is achieved[128]. The store itself consists of a parallel arrangement of plated wires enclosed, as illustrated, by strip conductors which form the word lines. In the storage operation a pulse is passed down the word line and the magnetisation is turned into the hard direction which is parallel to the wire axis. A pulse passed along the wire, i.e. the digit line, decides which way the magnetisation tips and therefore the information stored.

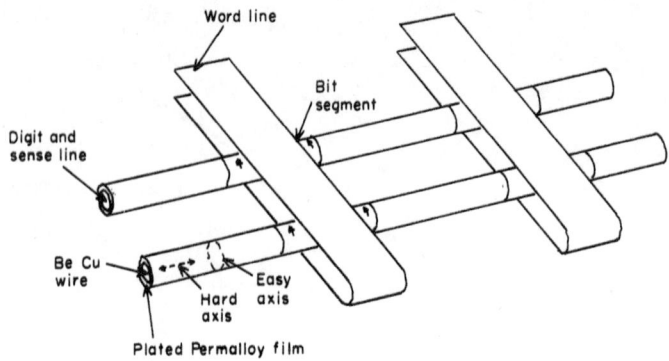

FIGURE 20. Part of a plated wire store. [I. Danylchuk, A. J. Perneski and M. W. Sagal, *Proc. Intermag Conf.*, Paper 5-4, (1964).]

From this brief description it is evident that the plated wire store incorporates uniaxial magnetic films and operates in a "coincident current" fashion analogous to the planer film store despite the fact that the magnetic film and one conductor now have cylindrical geometry. This means that while the basic philosophy of storage operation developed in connection with planar films applies with negligible modification to cylindrical films the many disadvantages associated with "open flux" planar structures have been substantially alleviated by the cylindrical film geometry. With the latter the harmful role of stray fields in producing bit interactions have been diminished and as a consequence the limitations to performance, which were so crippling to many planar film memories, are such that viable plated wire stores can be built and successfully operated.

Electrodeposition has been shown to be of advantage in that films with highly reproducible properties can be obtained which are nominally the same as those produced by evaporation[104,106]. In addition the process for plating wire is convenient for automation[127-129]; this being particularly important if commercial success is to be obtained. Further, wide operating margins have been demonstrated on films produced by different vendors thereby demonstrating the design stability of the original store structure[130]. Multilayer films have also been tested[131] and found to be satisfactory in their storage performance although a greater spread of film properties was observed. Planar packing densities of 1333 bits/in^2 and potential for development to 5000 bits/in^2 have been reported[132].

Plated wire memories have been marketed by a number of companies[133], for normal commercial use, and have also been shown to be particularly suited to military and space applications[134]. Improvements in store performance have been predicted and are said to require thinner films plated onto smaller diameter wires thereby allowing higher packing densities and lower drive currents[135,136]. To accompany the latter improved production control and testing systems are necessary[137]. The possibility of NDRO operation has also been receiving attention[138,139]. However, in predicting future applications for plated wire stores it should be borne in mind firstly that over the years magnetic film storage technology has never fulfilled its early promise to completely replace core store technology and secondly that in fast stores there is a growing movement away from magnetic storage devices towards semiconductor storage devices.

4. Applications in Magnetic Recording

As in the thin film store information is recorded by selecting the sense of the magnetisation in a recording medium. However, in this case, the magnetised regions are not discrete spots but are selected regions of a continuous medium which taked the form of a magnetic recording tape, the surface of a magnetic drum, or the surface of a disc. Thin films have obtained application mostly in disc and drum machines and in this discussion reference will only be made to the medium and the possible applications will be taken as understood. The underlying principles are broadly the same in all the applications and it is only in detail that they differ as far as the magnetics are concerned.

The essential parts of a magnetic recording system are shown in Fig. 21 and consist of a magnetic medium which moves past an erase, a record, and a replay head. In some systems the heads may be separate items, as shown, or combined in one read–write head assembly. In addition the erase head may be assembled with the record head or simply be omitted from the system altogether in which case the erase function is achieved by the overwriting of information already recorded.

The record-replay process takes place as follows. When a current is passed through the coil of the record head it magnetises the core material in the direction of the field produced by the head coil. A high flux density occurs in the region of the gap of the head and there is a fringing field in the front of the head gap. It is this field which penetrates the medium and causes recording to take place. As the medium passes the head gap it will be magnetised to one sense or other depending on the sense of the record current. It can be seen in Fig. 21 how the recording of a region of magnetisation is the same sense as the direction of medium motion. Shortly before the instant represented by the diagram the current in the head coil was of the opposite sense and the magnetisation was being recorded in the opposite direction. It can therefore be seen that a change in the direction of the record current caused a change of

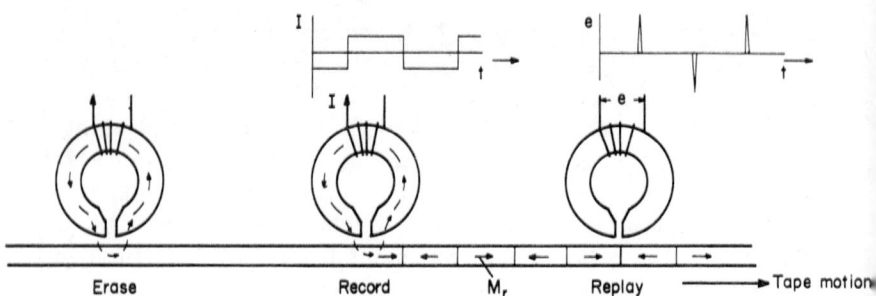

FIGURE 21. Simplified diagram of a digital recording system.

sense of recorded magnetisation in the medium. Such a reversal of magnetisation or transition is fundamental to digital magnetic recording and information is often represented by the polarity of flux reversals rather than by the sense of magnetisation, although the interpretation of the magnetisation in terms of ones and noughts depends on the particular coding system used. Once these transitions are recorded they are transported by the medium and eventually pass the replay head where a flux is induced in the head coil. The output voltage is the rate of change of flux in the head coil and for an ideal head this is the rate of change of flux below the centre of the head gap. Therefore the replay voltage will appear as a series of pulses.

If the recording medium were very thin and had a high coercivity the magnetisation transitions would be infinitely narrow. The output would then consist of a series of voltage spikes if the replay head were ideal. In such a case it would be possible to record an infinite number of transitions per unit length of medium and the output voltages would be distinguishable at this density since they would be voltage spikes. However in reality losses are introduced by many factors, transitions are never infinitely sharp and the output voltages are always pulses of finite width. There is therefore a limit to the density at which transitions can be recorded on the tape before resolution is lost. The factors limiting the transition and pulse widths are the subject of the next section.

4.1 The Record-Replay Process

There are four main parts of the record-replay cycle which need to be considered before an adequate description of the whole process can be given. These are: (1) the record process where the magnetisation orients itself under the influence of the record head field, (2) the relaxation or self-demagnetisation process where the recorded transition moves away from the record head and settles down to a new equilibrium width, (3) the remagnetisation process where the transition comes into the close proximity of the replay head and there is a further readjustment of the magnetisation, and (4) the replay process where an output voltage is generated in the replay head. Each of these processes will now be discussed in turn in this and the next section and particular attention will be paid to the requirements of the recording medium and the record and replay heads.

A precise description of the above cycle involves the computation and manipulation of demagnetising fields at each of the stages listed above. This introduces complexity which is further increased when the calculations have to be done in a dynamic situation in which the tape is moving and the record current is continuously changing. A number of workers following on from Iwasaki and Suzuki[140] have produced computer programmes capable of analysing such situations and giving good agreement with experimental results[141-144]. The advantage of the computer programmes is their precision but this is only obtained at a price which, in this case, amounts to the fact that it is often difficult to see trends in what is happening because there are usually a large number of variables being considered and large quantities of information need to be plotted out in order that trends can properly be studied. With this fact in mind other works have attempted to analyse the situation, without recourse to numerical techniques, using simplifying but plausible approximations. The first to do this, for the processes described earlier, were Williams and Comstock[145] followed by Maller and Middleton[146] while the later work of van Herk and Wesseling[147] is a first attempt to put this approach on a more rigorous foundation. However, the first two works mentioned gave simple formulae which show the effects of the relevant parameters at the various stages of the process.

The results of the simplified theories have been compared with the predictions of computer simulations by Tjaden and Tercic[144] and it was found that many of the predictions of these theories were qualitatively correct and in some cases quantitatively quite accurate. Therefore the results described here are those produced by Maller and Middleton[146] along with comments at appropriate stages indicating the limitations to their validity.

The component of the record head field in the plane of the medium, the x direction, was calculated by Karlqvist[148] to be

$$H_h = \frac{H_0}{\pi} \left[\arctan \frac{x}{y} - \arctan \frac{2g+x}{y} \right] \quad (16)$$

where H_0 is the field along the surface of the head nearest the tape, $2g$ the gap length and y the distance from the pole faces as shown in Fig. 22. The variation of H_h with x is shown in Fig. 22 for various values (y/g) where it can clearly be seen that the field gradient gets smaller the greater the distance from the record head being considered. The hysteresis loop system used is shown in Fig. 23

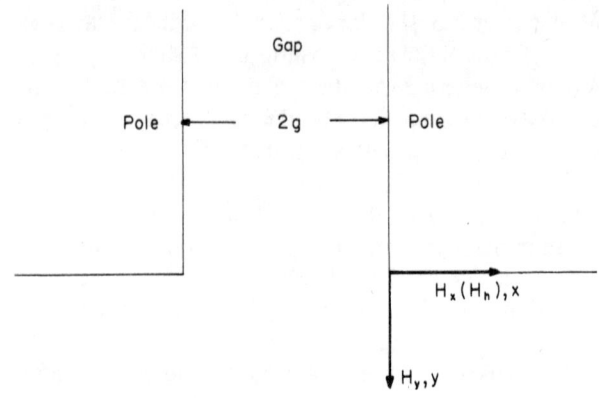

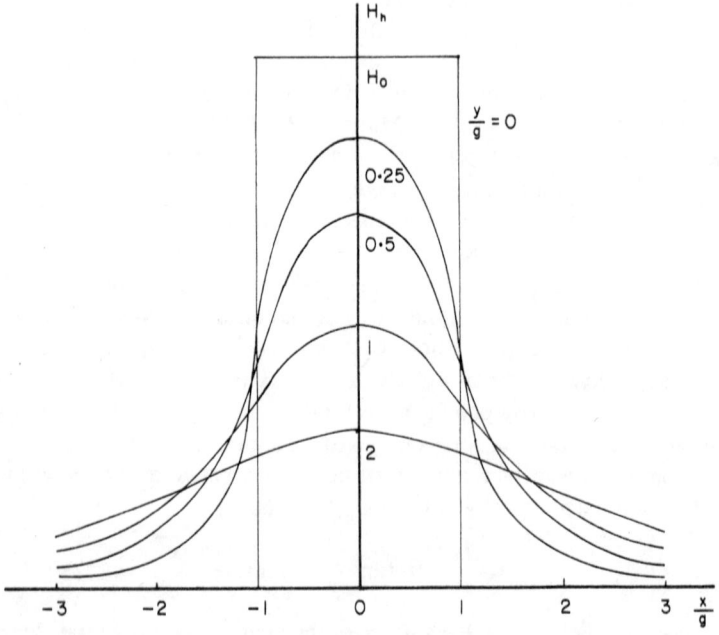

FIGURE 22. Field components parallel to the plane of the recording medium, predicted by Karlqvist equation, from a head gap of dimensions shown.

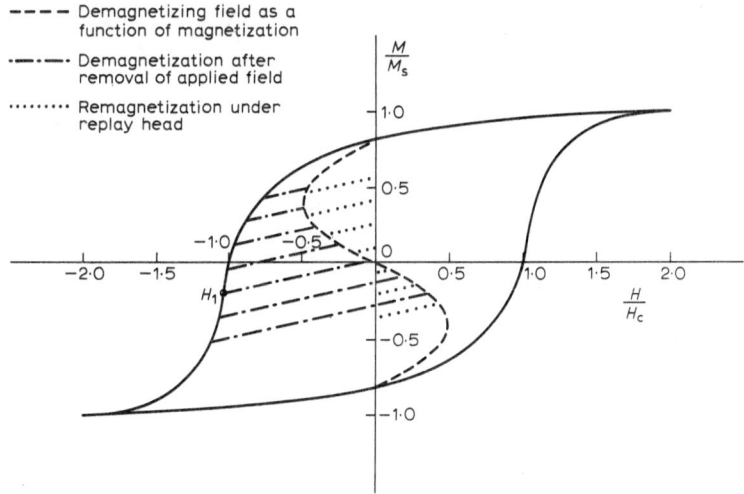

FIGURE 23. Assumed hysteresis loops according to equation (17). [V. A. J. Maller and B. K. Middleton, *The Radio and Electronic Engineer* **44**, 281, (1974).]

where the major hysteresis loop is assumed to be of the form

$$M(H) = \frac{2M_s}{\pi} \arctan\left\{\frac{H \pm H_c}{H_c} \tan(\pi s/2)\right\} \quad (17)$$

where H_c is the coercivity, M_s the saturation magnetisation and s the squareness M_r/M_s. The shape of this curve has been shown to be a good approximation to the experimentally obtained curves in the second quadrant, which is the region of most interest for these calculations.

Now suppose the tape had been previously erased by a D.C. field such that its magnetisation was at the positive remanent value. Then the sudden application of a negative field by the record head as described in equation (16) leads to a magnetisation distribution with properties which can be demonstrated by substitution for H_h into equation (17). It can be seen that there will be finite values of dM_x/dx for positive and negative values of x. However, as the tape moves, the distribution created at the trailing gap of the recording head will move with the medium towards the replay head. The distribution recorded at the leading gap of the head will move in the tape and be located always under the leading edge of the gap. Therefore it is only the distribution produced at the trailing edge which will be of significance to the information recorded in the tape.

The magnetisation gradient, or recorded pulse, can easily be seen to be given by

$$\frac{dM_x}{dx} = \frac{dM}{dH} \cdot \frac{dH}{dx} \qquad (18)$$

The field H experienced by the tape is made up of head field and any demagnetising field that arises from the tape, i.e.

$$H = H_h + H_d \qquad (19)$$

The demagnetising field H_d is calculated[149] on the assumption that the recorded magnetisation is of the form

$$M_x = \frac{2}{\pi} M_r \arctan \frac{x}{a} \qquad (20)$$

where a is a parameter related to the transition width and which is to be calculated. The centre of the recorded transition is thus assumed to be under the edge of the gap at $x = 0$ and at the field H_1 marked on the hysteresis loop so that after remagnetisation it still corresponds to the centre of the transition. Making the additional assumption that recording is achieved by the largest head field gradient the minimum transition width parameter for the record process a_1 can be shown, in a film medium of thickness D, to be

$$a_1 = \left(\frac{\pi s y}{8}\right)\sin(\pi s) + \sqrt{\pi y}\left(\frac{2M_r D}{H_c}\right)^{1/2}\sin(\pi s/2) \qquad (21)$$

where y is the record head to tape separation.

A similar calculation can be carried out for the relaxation process which takes place along the demagnetisation parts of the hysteresis loop as the transition moves away from the influence of the record head field and the magnetisation readjusts itself to be solely under the influence of the demagnetising field. In this case the relevant magnetisation curve is assumed to be linear and given by

$$M_2 - M_1 = \chi(H_2 - H_1) \qquad (22)$$

where the subscripts refer to the final and initial states of magnetisation and field, and χ is the slope of the remagnetisation curve. The minor loops are assumed to have slopes equal to that of the major hysteresis loop at $H = 0$. The latter approximation is thought to be reasonable and is found necessary since detailed formal knowledge of minor loops is not available. The transition width after relaxation a_2 can then be shown to be given by

$$a_2 = \left(\frac{a_1}{2V}\right) + \left[\left(\frac{a_1}{2V}\right)^2 + \frac{2\pi\chi D a_1}{V}\right]^{1/2} \quad V = \sin^2(\pi s/2) \qquad (23)$$

Finally, the remagnetisation process takes place along the same hysteresis curve and for simplicity of manipulation it is assumed that since the replay head is not subject to fields sufficient to cause saturation its permeability will be very high and so the image fields induced in the replay head cancel the self demagnetising fields in the tape and remagnetisation is complete. In this case the expression for the transition width at replay is

$$a_3 = \frac{a_2^2}{a_2 + 2\pi\chi D} \qquad (24)$$

Substitution from equation (23) into (24) and simplification leads to the following expression for the replayed transition width

$$a_3 = a_1 \operatorname{cosec}(\pi s/2) \qquad (25)$$

The above theory has been derived for the situation where a fairly wide transition has been assumed to be recorded. However, cases can exist where the recorded transition is very narrow and the above considerations give way to others. Suppose a narrow transition is recorded and it is approximately of the shape given by equation (20). When this transition moves away from the record head the maximum value of demagnetising field experienced by the tape can be calculated[149]. Evaluation of its maximum value may reveal that it exceeds the coercive force of the material. What this means is that for, say, positive values of magnetisation there would be fields which are more negative than the coercive force. Inspection of the hysteresis loop in Fig. 23 shows that this situation can never exist for a stable magnetisation distribution. For the positive magnetisation values to be stable the transition width would need to increase to a value sufficient to make the magnitude of the demagnetising field less than the coercivity. This can easily be shown to correspond to

$$a_d > \frac{2M_r D}{H_c} \qquad (26)$$

When hysteresis loops are not rectangular, then for any given magnetisation the corresponding value of demagnetising field must always be less than the corresponding value allowed by the major hysteresis loop of Fig. 23. Values of a_d for this situation can easily be computed and the results are

$$a_d = \frac{2M_r D}{H_c} f(\theta) \qquad (27a)$$

where to a reasonable approximation

$$f(\theta) = 1 + (\pi - 1)(1 - s) \qquad (27b)$$

When this transition passes under the replay head it would also undergo the process of remagnetisation as was described by equation (24). Values have been computed for the final transition width and for squareness values of $s > \frac{1}{2}$ it was found that to a good approximation

$$a_f \approx \frac{2M_r D}{H_c} \tag{28}$$

Emphasis is now placed on the question of which of the mechanisms, relaxation or self demagnetisation, limits the particular situation under consideration? If a narrow transition is recorded then self demagnetising fields will widen the transition as explained and self demagnetisation considerations will be the ones limiting the transition width. However if the written transition is wider such that when it moves away from the record head it is wider than that expected from simple self demagnetisation considerations, equation (27), it will remain unaltered and the transition is said to be recording-limited. Overall, these considerations boil down to the fact that the limiting process is the one which gives rise to the widest transition.

Figure 24 shows a graph of a/D against M_r/H_c, for a value of y/D of 2, calculated from equations (21), (23) and (25). Note that for small values of M_r/H_c a narrow transition is present in response to the record head and that this widens to a value a_2 when it moves away from the record head. When the transition moves under the replay head it becomes somewhat narrower once again (i.e. a_3). Note also that according to equation (21) a/D is approximately dependent on the square root of M_r/H_c and this fact is not substantially altered by the succeeding processes. The essential behaviour is therefore established at the record process and remains throughout. Figure 24 also shows the variation of transition width according to equations (27) and (28). It can be clearly seen that self demagnetisation is the controlling factor at larger values of M_r/H_c. In these circumstances the transition width does not depend on the record process but only on the properties of the recording medium.

The computer analysis of Tjaden and Tercic[144] also confirmed the square root and linear dependencies described earlier but the shapes of the transitions were found to diverge somewhat from the simple form given in equation (20). Clearly the latter approximation has not caused too much error in the calculation of the transition widths and so the above results can be taken as being satisfactory for qualitative and quantitative considerations provided a high degree of precision is not required. Consequently it is valid to say that, as far as keeping the transition widths to a minimum is concerned, it is necessary that the ratio $(M_r D/H_c)$ be kept as small as possible.

In the replay process the output voltage developed in the replay-head. windings depends on the rate of change of flux directly under the replay head gap caused by the movement of the medium. Clearly the output is then related to the transition width which has just been calculated and consequently predictions can be made concerning the relationship between recording

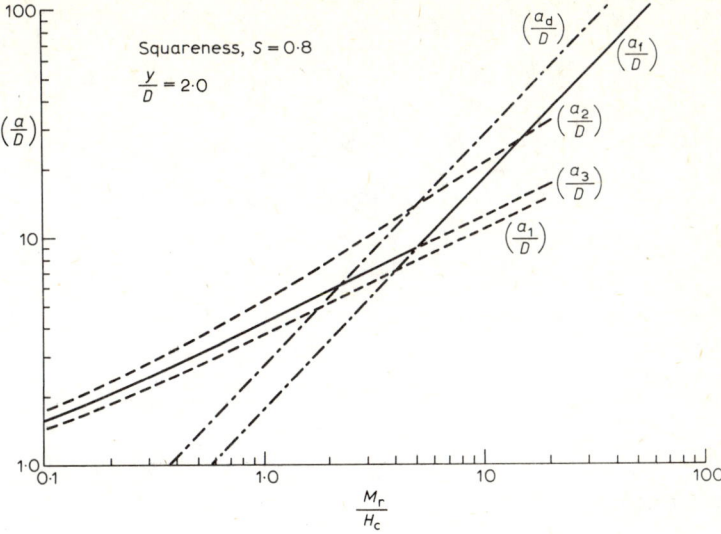

FIGURE 24. Graphs of reduced transition widths (a/D) as a function of M_r/H_c. [V. A. J. Maller and B. K. Middleton, *The Radio and Electronic Engineer* **44**, 281, (1974).]

properties and the magnetic properties of the medium. This task is tackled in the next section.

4.2 Recording Properties of Thin Films

Attempts have been made to observe directly the form of the transitions recorded on thin films. The information gained is valuable in that it gives a direct check on the theories which predict transition widths via the parameter a of equations (25) and (27). Observations by Lorentz microscopy[150–152] have shown that the recorded transition shape in thin films of coercivity less than 400 Oe is not independent of the dimension across the recorded track but that there is some superstructure, commonly called sawtooth structure, as shown diagrammatically in Fig. 25. Clearly this feature is not predicted by the original assumption [equation (19)] but measurements of the amplitude of the saw tooth structure reveal[152], despite the scatter of experimental results, an approximate correlation with $D/H_c^{1.5}$. This is qualitatively similar in its dependence on the variables to the prediction of equation (28) for the variation of transition width assuming self demagnetisation to be the forming process. In similar fashion agreement has been obtained between computed transition

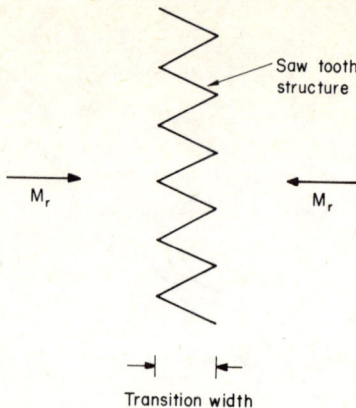

FIGURE 25. The sawtooth structure observed within the recorded transitions of evaporated films.

widths and observed sawtooth amplitudes despite the additional complexity of the latter[150].

As a direct test of the recording properties of thin films it is necessary to study the output from a recording head as a result of the passage of a recorded transition. Under the assumption that the transition is of the form given by equation (20) the output voltage in the replay head can be shown to be given by (putting $y = d$, the head to medium separation)[149]

$$e \propto M_r D \left[\arctan\left(\frac{g+x}{d+a}\right) + \arctan\left(\frac{g-x}{d+a}\right) \right] \quad (29)$$

which has amplitude

$$e \propto 2M_r D \arctan\left(\frac{g}{d+a}\right) \quad (30)$$

and pulse width at 50% of peak amplitude given by

$$p_{50} = 2[(d+a)^2 + g^2]^{1/2} \quad (31)$$

The pulse width at 25% of peak amplitude can also be calculated[149] and it has been determined in some of the experimental works but most emphasis is usually put on p_{50}. It is clear that, at large values of a, the pulse width is proportional to a and is therefore a measure of the transition width. However if a is small then the effects of g and d on pulse width also need to be taken into account.

The predicted dependence of p_{25} and p_{50} respectively on a, where a is linear with $M_r D/H_c$ according to self demagnetisation theory, has been confirmed experimentally for evaporated cobalt[153] and for cobalt chromium films[154] where the transitions were a factor of two greater than anticipated. In a variation on this approach it is possible to use observed pulse widths[153] along with the appropriate values of g and d and extricate values for a from the corresponding pulse width formulae[146]. Then graphs of a/D against M_r/H_c can be plotted and the variation of one with the other more easily established. For evaporated cobalt the linear relationship is confirmed in Fig. 26 while

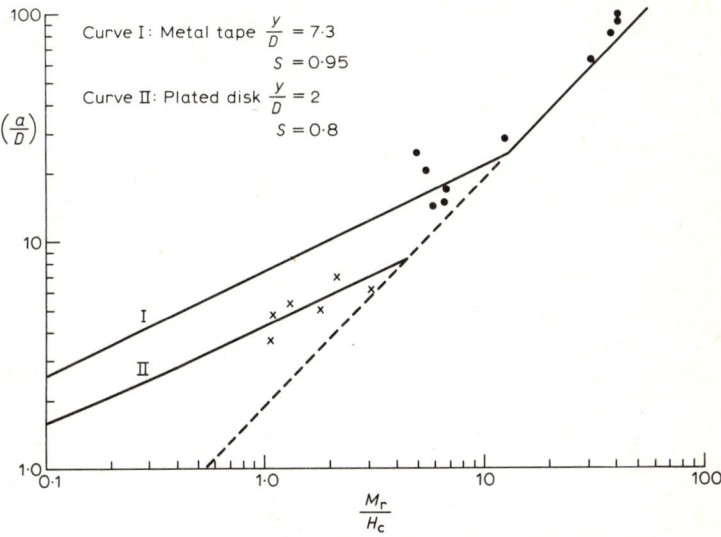

FIGURE 26. Experimental values of (a/D), for different M_r/H_c, obtained from evaporated cobalt film tapes and electroless plated discs. [V. A. J. Maller and B. K. Middleton, *The Radio and Electronic Engineer* **44**, 281, (1974).]

other points on the same graph, which were obtained for chemically deposited films, show a square root relationship corresponding to recording losses[146]. Therefore two possible regimes of operation are confirmed.

In addition to the above there has been a considerable number of works which have been concerned with the correlation between the recording and magnetic properties of chemically deposited films. The work of Speliotis, Morrison, and Judge[155,156] shows a square root relationship between pulse width and the film parameters D and H_c, with some confusion as to the role of M_r. However, it can be seen that although the square root relationship was found for most results when either film thickness or remanence was large, something nearer a linear dependence was observed. Therefore it would seem that these works confirm to some extent the existence of both regimes of behaviour, although the square root relationship does predominate suggesting losses at the writing stage. However, two pitfalls can be highlighted when drawing conclusions about the recording properties of thin films. Firstly, although the transition width may nominally be given by equation (28), the fact that the transition may not be of the arctangent form could lead to pulse widths which depend in a different way on the transition width[144,147,157]. Secondly, as shown by equation (31), there are head losses involved in the formulae which need to be taken into account when making correlations[158]. Nevertheless the results do show the trends predicted by the theory although a more rigorous approach reveals a smooth, rather than abrupt, variation of the relationship between a/D and M_r/H_c from a square root form to something nearer linear as M_r/H_c increases[143].

Pulse amplitudes have also been observed experimentally and a simple empirical relationship with magnetic properties obtained. Noting that for recording losses a is proportional to $\sqrt{M_r D/H_c}$ and that using equation (30) with $g, d \ll a$ the pulse amplitude should be proportional to $\sqrt{M_r D H_c}$, which is the observed relationship[155].

Variation of pulse amplitude with packing density has been investigated and correlations between packing density and recording properties made[155]. The fall of output with increasing packing density is shown in Fig. 27 for a number of recording media and it has been confirmed that the packing density at 6 dB reduction of output is inversely proportional to the pulse width p_{50}. Therefore isolated pulse width measurements can be used as an indicator of the packing density obtainable in thin film recording media.

The results therefore confirm that $M_r D/H_c$ is a prime parameter determining the recording properties of a thin film, thus demonstrating the need for high coercivity films, as a means of keeping $M_r D/H_c$ small, although other techniques such as partial penetration recording on thick media have also been used with great success. Nevertheless high coercivity thin films have been used on disc and drum machines in computer storage applications.

4.3 High Coercivity Films

Before discussing high coercivity films in detail it is appropriate to examine some of the reasons for the occurrence of high coercivity in magnetic materials. As an example consider a small spherical magnetic particle and suppose a magnetic field is applied along its easy axis in such a sense as to cause its magnetisation to reverse. Recalling that for uniaxial anisotropy the field at which rotational reversal takes place is $2K/M$, where for cobalt K would be the first order anisotropy constant K_1 which has value[159] $4.1 \times 10^6 \,\text{ergs cm}^{-3}$

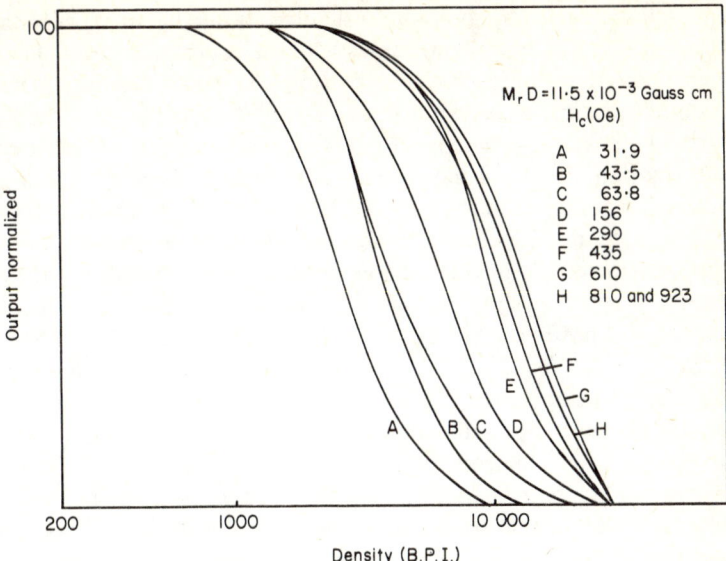

FIGURE 27. Fall of output with increasing packing density for a number of chemically deposited films (individual points omitted). [D. E. Speliotis, J. R. Morrison and J. S. Judge, *I.E.E.E. Trans. on Magnetics* **MAG-1**, 348, (1965).]

and taking $M_s = 1440$ gauss a coercivity of 5700 Oe is obtained. Clearly the rotational reversal of magnetisation against such a large anisotropy energy has resulted in a high coercivity.

Alternatively if the particle had zero magnetocrystalline anisotropy but the shape anisotropy of a prolate spheroid it would have energy arising from magnetostatic effects given by[160]

$$E = \tfrac{1}{2}(N_a - N_b)M_s^2 \sin^2\theta \qquad (32)$$

where N_a and N_b are the demagnetising factors in the direction of the semiaxes a and b. Equation (32) can be seen to be of the same form as the anisotropy energy equation (1). Hence the corresponding coercivity for a field applied along the easy axis would be[161]

$$H_c = (N_a - N_b)M_s \qquad (33)$$

and for a very long particle of cobalt for which $N_a = 2\pi$ and $N_b = 0$ the coercivity would be 9050 Oe. Clearly the point to be noted from these two examples is that reversal by rotation is usually associated with high coercivity in contrast to reversal by wall motion which usually occurs at low fields.

Reversal does not always take place by simple coherent rotation, there are more complicated processes known as curling[162], buckling[162] and fanning[163]. These are associated with lower critical reversal fields than is coherent rotation, and they may be further reduced if the fields are applied along directions other than the easy direction and if particles are so closely spaced as to interact. Experimental observations confirm coercivities lower than those predicted by $2K_1/M_s$ and equation (33) and reveal that they are dependent on the particle dimensions[164]. Interactions between particles are complicated and have been shown to have no effect on coercivity when it is caused by magnetocrystalline anisotropy but to have a large effect if it is controlled by shape anisotropy[165].

Overall, it may be concluded that fine particle behaviour is complicated and that coercivities will be high and depend on the nature of the anisotropies and inter-particle interactions involved. Therefore in discussing the properties of high coercivity films attention is paid to the nature of the reversal processes involved, be it rotation or wall motion, and the structure of the films, be they continuous or discontinuous, the latter often being associated with particle like behaviour.

4.3.1 Evaporated films

Hard magnetic films have been produced by evaporation techniques. Evaporation at normal incidence onto unheated substrates has been reported to give rather low coercivities but these may be increased to around 350 Oe, at the expense of reducing the magnetisation, by increasing the pressure during evaporation and reducing the deposition rate[153]. Lazzari et al.[166] optimised the substrate temperature during deposition and the thickness of the cobalt films to obtain maximum coercivity and then deposited a multilayer structure consisting of films of optimum thickness separated by thin layers of chromium. Coercivities of 600 Oe were obtained and were shown to be suitable for magnetic recording purposes[154]. The very high coercivities that have been obtained by evaporation resulted from oblique incidence deposition[167,168]. The variation of coercivity with angle of incidence is shown in Fig. 28. The results of Speliotis et al.[168] are interesting in that if the maximum values of coercivity are compared then they are seen to be in the ratio of their saturation magnetisations. This would seem to suggest, according to equation (33) that the high coercivity is obtained as a result of shape anisotropy factors in the film. Cohen[169] when studying the anisotropy of films produced at grazing incidence obtained a micrograph showing two interesting anisotropic effects. Firstly that the crystallites are elongated by virtue of the fact that the incident beam tends to cause the particles to become elongated and secondly that they

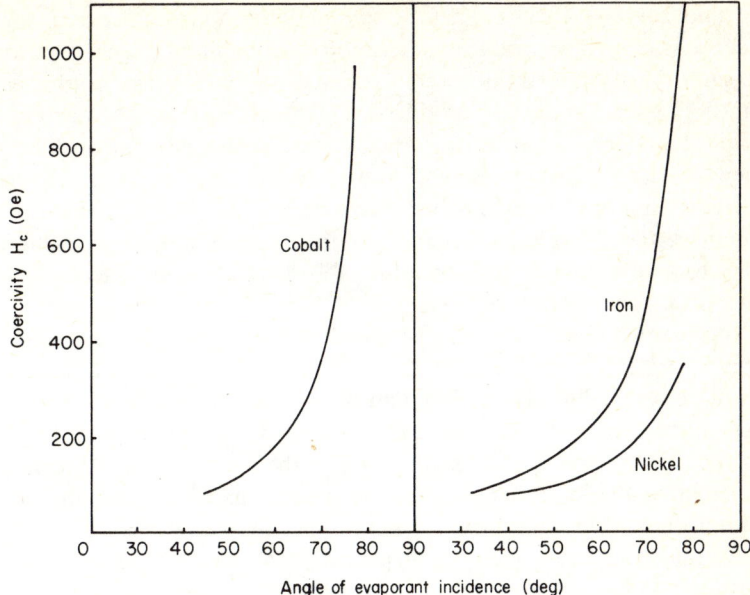

FIGURE 28. Variation of coercivity with angle of incidence (i.e. angle between film normal and direction of motion of depositing atoms) of evaporated metal atoms during film deposition (individual points omitted). [D. E. Speliotis, G. Bate, J. K. Alstad and J. R. Morrison, *J. Appl. Phys.* **36**, 972, (1965).]

become organised into chains along an axis perpendicular to the direction of incidence of the beam. The particle elongation effects have also been confirmed, but not with such clarity elsewhere[167,168]. These observations would seem consistent with the conclusions concerning the magnitudes of the coercivities and the possible occurrence of shape anisotropy controlled reversal.

Attempts have been made using a variety of techniques to observe the mechanisms by which the reversal of magnetisation is achieved. Domain structures have most easily been observed in films produced at normal incidence and which have low values of coercive force. The structures observed in demagnetised films are typical of those found in isotropic Permalloy films in that domain walls and pronounced magnetisation ripple are clearly observable[168,170]. In films with coercivities of the order of 150 Oe the reversal process is more typical of thin films with uniaxial anisotropy described earlier.

This is not surprising if it is remembered that deposition at other than normal incidence is known to produce high anisotropy in films. All the symptoms of reversal by partial rotation were observed: removal of a field causing saturation in the hard axis produced a parallel arrangement of magnetic domains along the easy axis, while fields in the easy direction introduced walls somewhat reminiscent of those observed in easy axis reversal. In higher coercivity films the domain structures were not resolvable using Lorentz techniques, and it was found necessary to investigate the structure using Hall probes to map out the field distribution above the films and interpret this in terms of magnetisation reversal processes. This was done initially on electrodeposited films[171] but the procedure was applied with equal success to evaporated films. The field plots gave results which indicated that reversal took place by the formation of domains of reverse magnetisation at various points in the films and that these domains grew in length such that their long axes were perpendicular to the direction of the applied field. Reversal was completed by the formation of many such domains which grew in size and eventually covered the whole of the film.

The domain structure and reversal properties of cobalt chromium thin films have been successfully observed using Lorentz microscopy[151]. Domains were seen to persist in thin films with coercivities as high as 900 Oe although the scale of the variation was much finer and the patterns more difficult to interpret. The reversal behaviour of films of coercivity of around 400 Oe was also observed to consist of the formation of reverse domains which grew by the movement of the tips so eventually reversing the magnetisation of the entire film.

4.3.2 *Films produced by electroless chemical deposition*

Chemical or electroless deposition (discussed in Chapter 2) can be successfully applied to the deposition of metallic cobalt, iron or nickel from aqueous solution onto a catalytic substrate in the presence of a reducing agent[172]. The process once started continues since the freshly deposited film acts as the catalyst for the further deposition of metal. First used by Brenner and Ridell[173], it has had wide application to the deposition of magnetic films. There have been substantial variations[174] in the solution compositions from those originally suggested and new baths have been introduced which have properties more suitable for the deposition of magnetic films. Fisher and Chilton[175], for example, used a bath containing sodium hypophosphite as the reducing agent, cobalt chloride, citric acid and sodium lauryl sulphate. This solution is typical and variations usually involve different reducing agents, chlorides instead of sulphates of the metal, and the use of other additives which control the concentration of the free metal ions in the solutions. The solutions are usually alkaline with pH values in the range 7–9 and temperatures of the order 60–80°C are commonly used. Since there has been a wide variety of bath compositions, it is only reasonable to expect a variation in the results obtained since the film properties are sensitive to the deposition parameters in general. Therefore the following discussions will concentrate on the nature of the magnetic properties observed and the reasons for their occurrence rather than list in catalogue fashion all the results obtained.

Coercivity is usually found to decrease monotonically with increasing film thickness[175], with the magnitude involved strongly dependent on the hypophosphite concentrations of the plating bath. Clearly the latter is a major factor in deciding the film coercivity, as is the solution pH[176] (see Fig. 29).

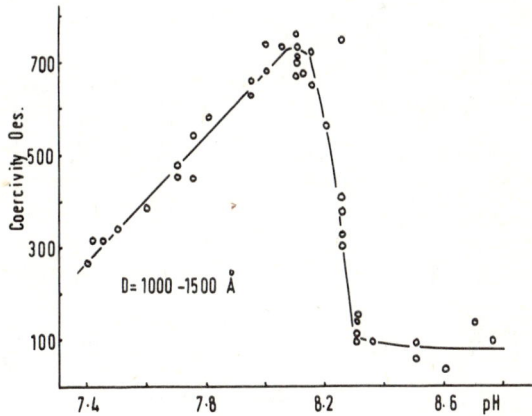

FIGURE 29. Coercivity as a function of pH of the hypophosphite solution used in the chemical deposition of cobalt phosphorus films of 1000 to 1500 Å thickness. [M. Aspland, G. A. Jones and B. K. Middleton, *I.E.E.E. Trans. on Magnetics* **MAG-5**, 314, (1969).]

Increasing pH has been shown to cause an increase in the phosphorus content of the films[177], as have variations in such factors as bath composition and hypophosphite concentration, and its presence in films prepared in these baths, and in other baths containing hypophosphite, is frequently postulated to be the cause of the high coercivity[174,178]. This applies, broadly speaking, to films of cobalt and cobalt nickel where the effect of increasing the nickel content[174,179] is shown in Fig. 30. It can be clearly seen that there is a maximum in the coercivity obtained at a composition of 80:20 Co:Ni, which

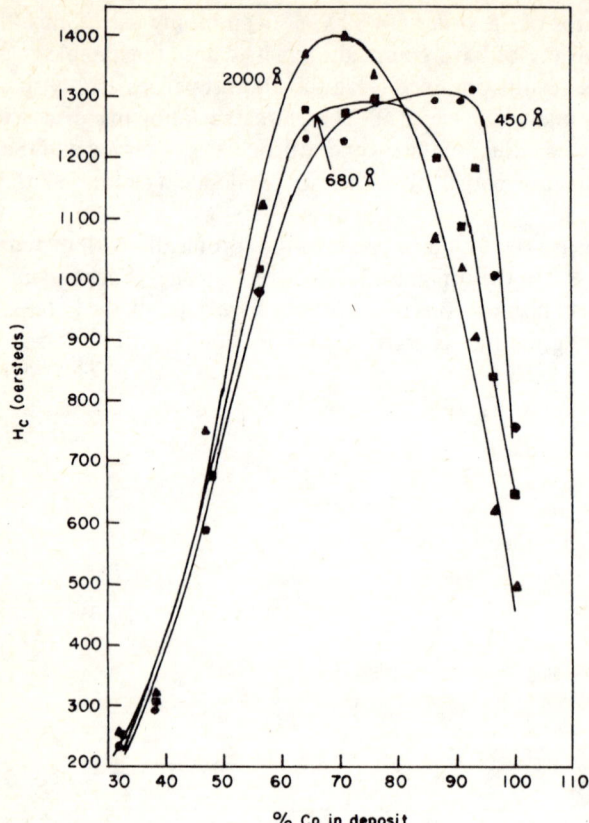

FIGURE 30. Variation of coercivity of chemically deposited Co–Ni films with composition for different film thicknesses. [J. S. Judge, J. R. Morrison and D. E. Speliotis, *J. Appl. Phys.* **36**, 948, (1965).]

coincides with maximum mixing of f.c.c. and h.c.p. phases. The variation of the induced magnetic anisotropy constant in Co:Ni alloys is almost identical[180] over the same composition range.

Structural investigations either by x-ray or electron diffraction techniques invariably show the films to be of the hexagonal close packed phase with substantially a random orientation of crystallites less than 1000 Å in diameter[178,181]. There are occasional reports of additional quantities of the cubic phase being present and also that there may be some degree of orientation[181]. Further, almost all of the investigations report, or predict, that the films contain some phosphorus but there is no evidence that this is in a crystalline form. Indeed the general conclusion is that the phosphorus is present either on its own in an amorphous form or as an amorphous cobalt phosphorus compound[174]. The presence of the phosphorus has been verified by annealing the films whereupon the diffraction of crystalline cobalt phosphorus compounds have resulted. Further, it is argued that with the values commonly quoted for the percentage by weight of phosphorus, it must occupy a substantial proportion by volume. Since phosphorus is insoluble in cobalt and it is not present as crystalline cobalt phosphorus, it seems that it must be in an amorphous form. Such evidence was provided by Jones and Middleton[182] who showed that in films with high coercivity there are regions or channels between the crystallites which allow electron beams to penetrate more easily than do the metallic cobalt crystallites. These boundaries between the crystallites were assumed to contain phosphorus in a form which does not give rise to diffraction patterns. With increasing coercivity the boundaries become more pronounced (Fig. 31) and eventually the entire films become substantially amorphous[176]. It could be argued that as the percentage of phosphorus increases the boundaries between the crystallites grow to accommodate it. However, a stage is reached when the phosphorus content is so high that during the growth process it is not possible for it to all be expelled to the surface of the grains and it is retained more uniformly in the film as a whole thus resulting in the amorphous phase.

Observations of the reversal processes and the domain structure of chemically deposited films has proved difficult[176], but by taking care it has been shown that the structure of high coercivity films can be observed using Lorentz electron microscopy[183]. Films with coercivity of the order of 200 Oe were shown to have domain structures during reversal typical of those of Permalloy but as the coercivity is increased the detail on the domain patterns becomes finer. However, domain patterns were still observed in films with coercivities of up to 600 Oe and what appeared to be domain walls were still in evidence during the reversal process. In the highest coercivity film the observed reversal process was not seen to be accompanied by any process which could be interpreted as domain wall motion. Other work by Bate *et al.*[170] and Green[184] indicated the formation of domains which grew laterally eventually covering the whole film. The latter workers employed techniques which do not have the same resolution as Lorentz microscopy and so it is difficult to make comparisons between the two sets of observations.

The above results now need to be discussed in the context of making deductions about the origins of high coercivity. Firstly high coercivities have been obtained and these are often associated with rotational reversal-type processes rather than wall motion. Secondly there is substantial evidence from structural observations that particle-like situations are achieved since the

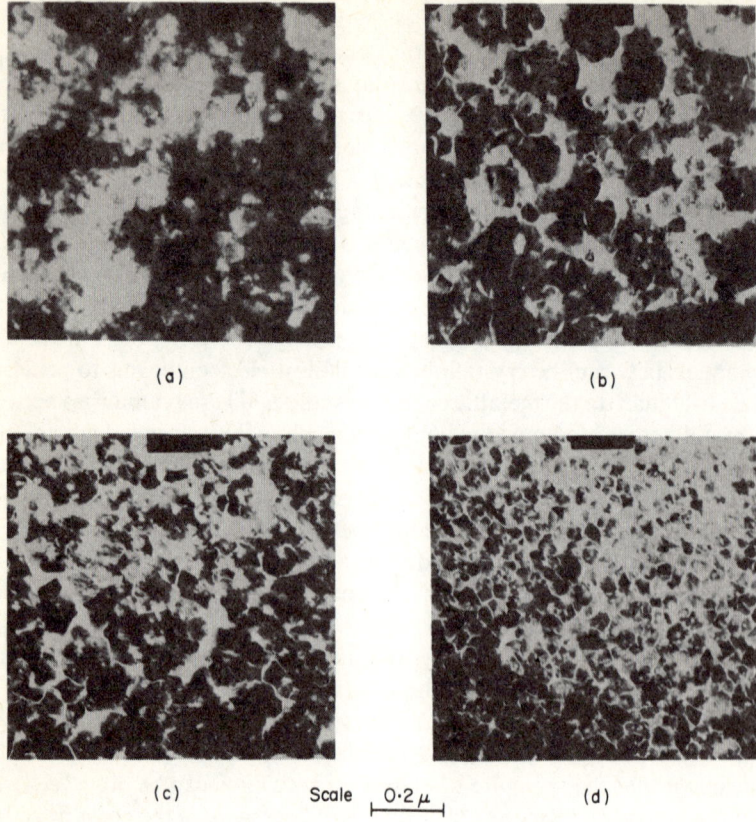

FIGURE 31. Electron micrographs of chemically deposited films from baths with pH values (see Fig. 29) of (a) 7.4, (b) 7.45, (c) 7.75 and (d) 8.05.

grains in the films are somewhat isolated. Thirdly that as coercivity increases the scale of the domain structures decreases until for very high coercivities no wall-like domain motions are observed, i.e. particle-like behaviour has been obtained. The latter may be due to increasing isolation of the domains and decreasing interactions. Fourthly films produced by chemical deposition tend to have squareness ratios (M_r/M_s) typically of the order of 0.7 to 0.9 when it is known that a randomly oriented array of non-interacting uniaxial particles would have a squareness ratio[185] of 0.5. Interactions can cause squarenesses to be increased[186] and the evidence for these comes in the form of the domain-like structures that have already been described. These observations conspire to give credibility to the proposition that high coercivity films of cobalt–phosphorus can be considered as approximating to an array of interacting particle-like structures exhibiting high coercivities and remanence ratios.

4.3.3 Electrodeposited films

Electrodeposited films have also been studied but not in the same detail as have chemically deposited films. Typically, the baths contain sodium hypophosphate, metals in the form of the chloride or sulphate, ammonium chloride, and ingredients to maintain the pH of the solution in the acid region[187]. Sallo and co-workers[188–191] have investigated the properties of electroplated cobalt and cobalt nickel thin films and found them to contain small amounts of phosphorus which profoundly influence their magnetic properties. Coercivities of up to approximately 2000 Oe have been achieved and their values attributed to the superstructure of the films. "Lamellar" and "rod"-like structures have been observed and careful use of Bitter patterns has indicated that these details may correspond to independent magnetic particles. Diffraction patterns have shown that the films are of hexagonal cobalt with the c-axes randomly oriented. A semiparticulate model for these films has been proposed in line with the discussions given above for chemically deposited films.

There is little evidence to suggest how the reversal of magnetization takes place in these films. The Bitter patterns taken by Sallo et al.[188–191] suggest that the superstructure observed corresponds to single domain particles while other observations[192] suggest that the reversal process is very similar to that taking place in evaporated and chemically deposited films. These results would also seem to give some credence to the prediction of particle behaviour in electrodeposited as well as the chemically deposited films.

4.3.4 Applications to devices

It has now been demonstrated that thin films can be produced with the required high coercive force, predicted as necessary for high density recording, and that very high density recording can be achieved with such films. The question remains "what prevents the films being produced and used very extensively?" In this chapter consideration has, so far, only been given to the magnetic properties of the films while in any application it is necessary that they have a very high resistance to wear. The answer to the above question is that unfortunately the wear resistance is rather low and either protective coatings or non-contact recording have to be considered[193]. In either case the large head to tape separation introduces a loss of performance, predicted by equation (31), and the potential value of high coercivity media is somewhat reduced. Further, it has proved difficult to produce uniform properties over the length required for magnetic tape applications and so application has been limited to disc and drum machines in computer systems where non-contact recording is used and the full value of the high density potential of such media has not yet and indeed may never be fully exploited.

4.4 Thin Film Recording Heads

Currently available conventional magnetic recording heads can be produced to satisfy a wide range of requirements in magnetic recording. However, the situation has been reached where narrow gap heads for high track density applications, for example in disc machines, need to be made and positioned to an extremely high degree of precision. It has, for some time, been realised that these applications could well benefit from the use of integrated circuit manufacturing techniques applicable to the manufacture of thin film integrated recording heads. Besides the advantages in mass production that integration of heads might have, the improved performance that has been demonstrated in terms of resolution could eventually lead to wider areas of application.

Firstly, consider the properties needed in a recording head to make it suitable for high density applications. To obtain high fields it is necessary to have a head material with a high saturation magnetisation and, further, these high fields should require only small magnetising currents. This means that a high permeability material must be used for their construction, but it must be borne in mind that the effective permeability of the head is determined not only by that of the head material but also by its geometry. In the latter respect, the gap length and depth (i.e. the dimension perpendicular to the recording medium and track width) are important. In a conventional head the effect of these parameters on its field can be approximately represented by simple formulae but in a thin film head their effect is more complicated. In addition to the field being large enough to record in the highest coercivity material used it has been shown that a high field gradient is needed for highest record resolution. Therefore in discussing heads, attention will be paid to the fields that they produce in the tape and the corresponding head field gradients.

To demonstrate the requirements of the replay head it is most convenient to assume that the replay process is linear and apply the well-known reciprocity formula to determine the output voltage[194]. Therefore if it is assumed that the magnetisation in the tape lies in the film plane and that components normal to the plane are negligible, then the output voltage is given by

$$e(\bar{x}) = K \int_{d}^{d+D} dy \frac{\partial}{\partial x} M_x(x - \bar{x}, y) H_h(x, y) \, dx \qquad (34)$$

where H_h is the field component that the head would produce, in the plane of the tape, if it were energised. If recording takes place so efficiently that a very narrow transition is written then the output predicted by equation (34) can be reduced to

$$e(\bar{x}) \propto H_h(\bar{x}, d) \qquad D \ll d \qquad (35)$$

and so the output pulse shape is a reflection of the head field shape. Therefore in making a head which is capable of producing high fields a high output is also assured. Further, in making the head field gradients large the pulse width derived from equation (35) will be narrow. Therefore in producing the head structure with suitable recording performance a satisfactory replay head could also be produced, although for each type there will be some minor individual optimisation.

4.4.1 Thin film inductive heads

As far as the thin film heads are concerned the simplest proposed structures are those shown[195] in Figs 32a and 32b. These consist of strip conductors which

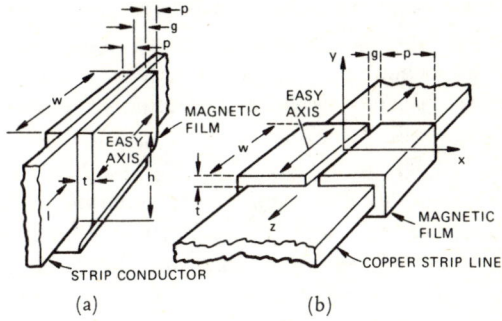

FIGURE 32. Single turn thin film recording heads often called (a) vertical and (b) horizontal. [E. P. Valstyn and L. F. Shew, *I.E.E.E. Trans. on Magnetics*, **MAG-9**, 317, (1973).]

have been coated with thin films of magnetic material and parts of the magnetic material have been removed to produce magnetic gaps. In case (a) the width of the conductor defines the gap width while in case (b) the width of the removed section of film is the definitive quantity. The easy axis of the film is arranged to lie along a direction parallel to the length of the strip. This means that in the absence of current the circumferential component of magnetic field is zero and the magnetisation will lie in the easy direction. For finite values of current there will be circumferential components of magnetic field and also, therefore, components of magnetisation in the same direction. Consequently where the magnetisation component is normal to the gap faces there will be magnetic poles and therefore a magnetic field produced.

Since the magnetisation induced in the hard direction of a thin film by a field applied in the same direction is proportional to the magnitude of the field, it has a constant permeability, and so the head is in effect a smaller version of currently used recording heads. However, the permeability in this case is smaller than that of the conventional head and the thickness of the material is greatly reduced, leading in both counts to the need to drive the heads into saturation using large currents. Valstyn and Kosy[196] have calculated the field distribution and magnitude expected from the horizontal head and the results are shown in Fig. 33. The field produced by a conventional head is also shown and it is clear that the thin film structure is capable of producing the fields needed for recording on magnetic media and also gives rise to higher field gradients than are available with conventional heads. Recording at high densities was reported by Kaske et al.[197] although the same head was not used for recovering the information. Although the horizontal head has been shown to be capable of recording at high densities, and has the advantage of being fabricated by photolithographic techniques, which are commonly used in integrated circuit technology, it suffers from the obvious disadvantages that virtually the whole of the head is exposed to wear as a result of abrasion from the tape. Therefore despite promising properties most attention has been paid to the vertical record head structures.

In the vertical structure the lapping and grinding techniques used on conventional recording heads are required for smoothing the front edge of the substrate which is parallel to the gap dimension. The wear is then limited by the nature of the substrate which may be glass or any other more suitable material. With the single turn head it has been shown that recording can be achieved on media of 600 Oe coercivity and that replay can also be obtained using the same heads[195]. It was shown that the pulse shapes observed are qualitatively similar to those predicted by Potter et al.[198] in that there is a small tail region in which the voltage goes negative, as shown in Fig. 34. It would be expected that if narrower pulses could be obtained higher packing densities should be achieved and this was indeed found to be the case. This observation is in agreement with the prediction of Potter et al.[198] that most of the improvement would be due to read head performance rather than to any improvement offered by these heads when used for writing[199].

The two disadvantages of these heads are that when used in the record mode they require rather high currents and that when they are used for replay the signal levels are small. Both of these aspects of behaviour can be improved by increasing the number of turns. A number of structures have been proposed[200] but the design favoured by Lazzari and Melnick[201] has received most attention and has been developed to an advanced stage. Its structure is shown

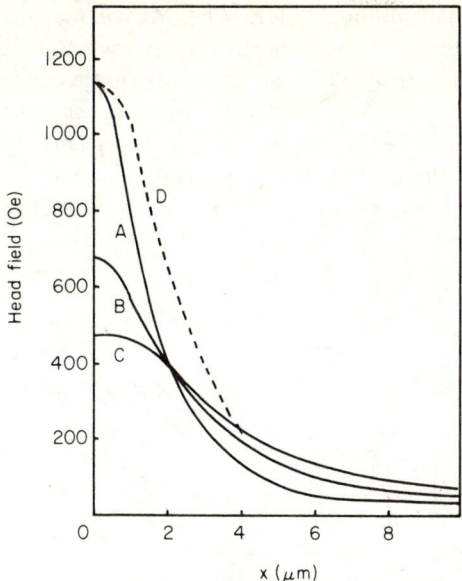

FIGURE 33. Variation of x component of head field with distance from the centre plane of the head gap. For A $y = 1\,\mu\text{m}$, B $y = 2\,\mu\text{m}$ and C $y = 3\,\mu\text{m}$. D is for conventional head with $y = 1\,\mu\text{m}$. [E. P. Valstyn and D. W. Kosy, *I.E.E.E. Trans. on Magnetics*, **MAG-5**, 442, (1969).]

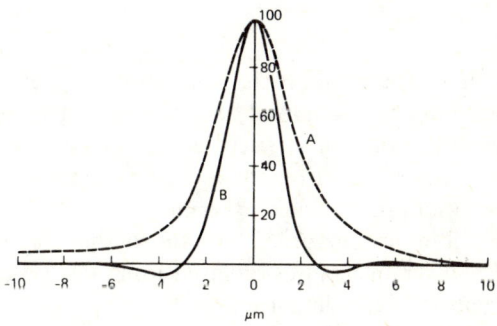

FIGURE 34. Output pulse shape produced by vertical thin film head structure B after recording on a thin film of 600 Oe coercivity. The pulse produced by a conventional ferrite head is also shown, A. Amplitudes have been normalised. [E. P. Valstyn and L. F. Shew, *I.E.E.E. Trans. on Magnetics*, **MAG-9**, 317, (1973).]

in Fig. 35 with five turns although this may be varied as may other details of the design. The magnetic films are multilayer structures which, in the absence of an applied field, are magnetised in opposite directions to the easy axis. This has an advantage when track widths are small, because the demagnetising fields are reduced and there are fewer problems arising from high magnetostatic energy. The head operates in much the same way as does the single turn head except that it has many turns and therefore requires a lower record current and gives a

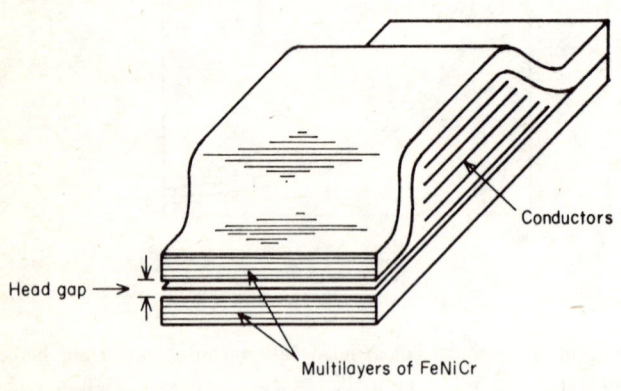

FIGURE 35. Multiturn head structure used by Lazzari and Melnick, [J. P. Lazzari and I. Melnick, *I.E.E.E. Trans. on Magnetics*, **MAG-6**, 601, (1970).]

higher output voltage. In addition the multilayer magnetic film structure allows the magnetisation to adopt a different distribution from that of a single thin film. The field distribution from such a head has been calculated[202] and shown to give fields which differ from those of a conventional head and which are in agreement with those obtained experimentally[203]. Making use of the predicted field distributions Lazzari[200] has shown that narrower transitions can be recorded and slimmer pulses replayed using this type of thin film as opposed to conventional recording heads.

Many variations of head design and applications in tape and disc recording have been proposed. In the latter case Chynoweth and Kayser[204] have described a working disc file with a hundred heads per inch of track and discussed the possibility of systems operating with much higher track densities.

4.4.2 Thin film magnetoresistive heads

As an alternative to the thin film head already described, Hunt[205] suggested the use of the magnetoresistive head structure shown in Fig. 36a. This consists of a thin film of Permalloy, nickel–cobalt, or similar alloy, which has an easy axis parallel to the z direction. The film is of approximate thickness 2000 Å, although much thinner films have been used, and therefore fields in the x direction have little effect on its magnetisation orientation. However, y components of field arising from the recording medium cause the magnetisation direction in the head to alter and make an angle θ, say, with the easy axis. A current is passed through the head and since the resistivity ρ of the film material is given by[205]

$$\rho = \rho_0 + \Delta\rho \cos^2 \theta \qquad (36)$$

where ρ_0 is the isotropic resistivity and $\Delta\rho$ the magnetoresistivity, the resistance of the metal changes in response to the field from the recording medium. Since $\sin \theta = H_y/H'_k$, where H'_k is an effective anisotropy field of the film made up of an actual anisotropy field and a self demagnetising field term, the resistance is proportional to H_y^2 but if a bias field H_0 is applied in addition to the signal field H_y the first order changes of resistance become proportional to H_y. The field H_0 may be thought of as a linearising field.

The response of magnetoresistive heads to recorded sine waves[205] and pulses[205–207] have been studied theoretically and the pulse width arising from a recorded transition of the shape given approximately by equation (20) is[207]

$$p_{50} = 2\sqrt{(d+a)(d+a+D+h)} \qquad (37)$$

where h is the height of the film. For practical values of h, of the order of 10 μm, the calculated pulses are wider than those expected from conventional heads with typical gap lengths. To overcome this poor resolution Potter[208] suggested the use of a shielded magnetoresistive head structure as shown in Fig. 36b. The function of the shields is to protect the magnetoresistive element from the field of the recording medium until the magnetisation transition is directly under the head. As a result the observed pulse widths are given, for a thin magnetoresistive element, by[207–209]

$$p_{50} = 2\sqrt{(d+a)^2 + (g/2)^2} \qquad (38)$$

which can be seen to be similar to that of a conventional head except that the term g has been replaced by $g/2$ and therefore the output pulses are narrower. This lower pulse width indicates that a higher packing density would be achieved with such a head and this has been confirmed by experimental results[208].

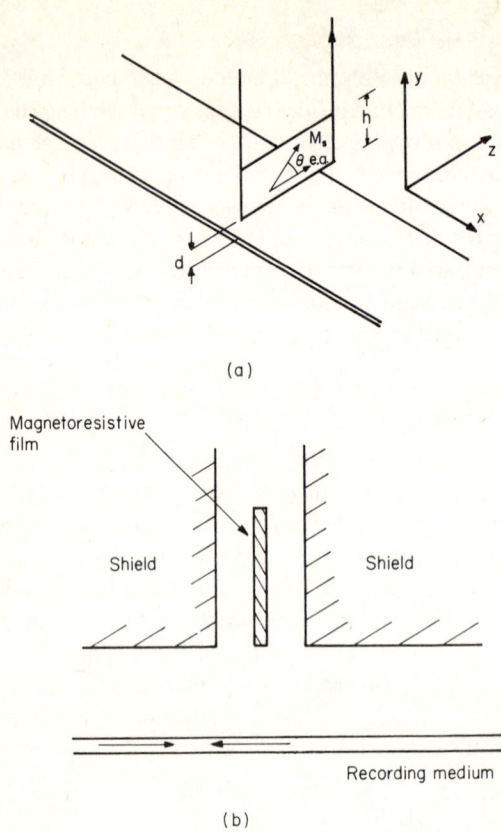

FIGURE 36. (a) Vertical magnetoresistive head structure proposed by Hunt. (b) Cross section of shielded magnetoresistive head structure proposed by Potter.

Magnetoresistive heads have been shown to give higher signal levels than conventional heads[208]. However, they suffer from the disadvantage that they can only be used in the read mode but a recent proposal has been made for a combined thin film record-magnetoresistive replay head combination[210]. Such a structure has been shown to be capable of recording at high densities and able to read out the information written.

Many problems remain to be tackled but it appears that substantial progress is being made in the quest for an integrated thin film head technology and confident predictions are being made as to their viability and successful commercial future.

5. Domain Storage Devices

A number of moving domain storage devices are currently under investigation because of their potential application in high capacity digital storage systems. These can be split into two groups, the first being bubble domain devices[211-214], which involve domains magnetised in directions normal to the storage plane, while the second, termed domain wall motion devices, involves domains magnetised in the film plane. In both cases the domains are moved through the media, to the required locations, by appropriately applied fields, to complete the storage operation. This process can be achieved electronically and therefore there is no motion of the recording medium as in disc and tape file machines. Compact, electronically accessible, high capacity storage devices have special attractions for particular storage applications such as has been predicted for bubbles in telephone switching systems. Perhaps overshadowing this requirement is the potential breakthrough into the computer storage market either as a disc replacement or as a consequence of a redistribution of the disposition of the storage hierarchy of current computing systems.

This section proceeds with a study of bubble domains followed by a brief review of other types of domain wall motion devices.

5.1 Bubble Domains

By contrast with other applications of thin magnetic films bubbles are structures arising from the orientation of magnetisation normal to the film plane as opposed to in the film plane. Consider Fig. 37a which shows a magnetic film with uniaxial anisotropy and easy axis normal to the film plane. If the magnetisation makes an angle θ with the easy axis the sum of anisotropy and magnetostatic energies per unit volume of film is given by

$$E = K \sin^2 \theta + 2\pi M_s^2 \cos^2 \theta \qquad (39)$$

It can easily be shown that the energy is minimised when $\theta = 0, \pi$ if $K > 2\pi M_s^2$ otherwise $\theta = \pi/2, 3\pi/2$. Therefore the condition for a film to be saturated in a direction normal to its plane is

$$K > 2\pi M_s^2, \quad H_k > 4\pi M_s \text{ or } q = K/2\pi M_s^2 > 1 \qquad (40)$$

where q is a quality factor of the material.

In practice the domain structure illustrated in Fig. 37b is found to have a lower energy[215,216] and this consists of strip domains magnetised alternately in opposite senses of the easy axis. The application of a field as shown in Fig. 37(c,d,e) causes the movement of the domain walls so that the domains with magnetisation of the same sense as the applied field grow at the expense of the domains magnetised in the opposite sense until a new minimum energy structure is reached. With increasing field alternate domains continue to

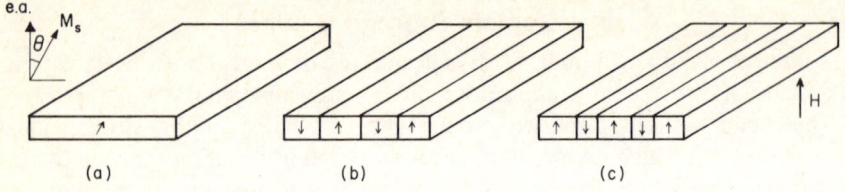
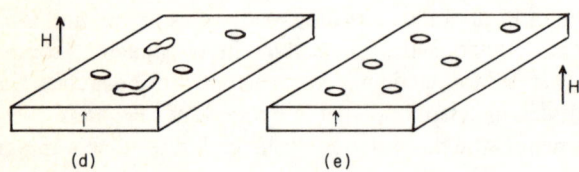

FIGURE 37. Domain structures in a thin film with easy axis normal to its plane: (a) saturated film with magnetisation making an angle θ with the easy axis: used in the text for defining conditions under which magnetisation directions normal to the film plane are stable, (b) demagnetised film with strip domains, (c) strip domains subjected to an applied field, (d) the instability of strip domains in a high field, (e) stable bubble domains.

narrow until a stage is reached where strip domains are no longer stable. This is usually marked by the length of the domains reducing rapidly[215,217] so that small domains of circular cross section is all that is left. The resulting cylindrical domains are known as bubbles[217]. Further increases in applied field eventually result in the collapse of the bubbles and therefore the saturation of the film.

Calculations of the spacing of strip domains as a function of applied field have been completed and shown to be in good agreement with observations[216]. More detailed calculations of the comparative energies of bubble and strip lattices illustrated in Fig. 38 have also been completed[216,218,219]. These show that the energies of bubble and strip lattices are very close together and that in low fields, as shown in Fig. 39, strip domains are stable. However, as the applied field is increased to a value H_1 it is the bubble lattice that assumes the lowest energy and is therefore favoured. Bubble lattices are not usually formed immediately the field H_1 is reached since energy barriers between different domain structures need to be overcome before changes can take place. As the applied field is increased to a value H_2 then, for lowest

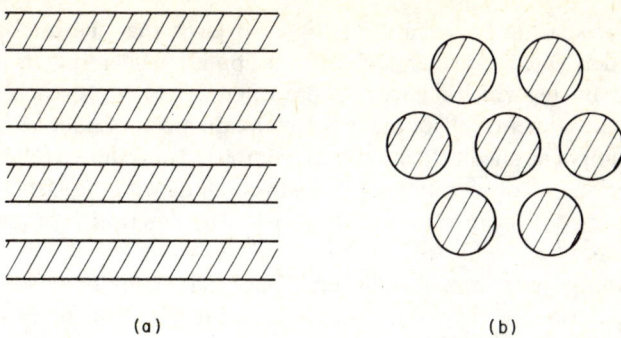

FIGURE 38. Plan view of sections of (a) strip and (b) bubble domain lattices where the hatched areas are magnetised in opposition to the rest of the film.

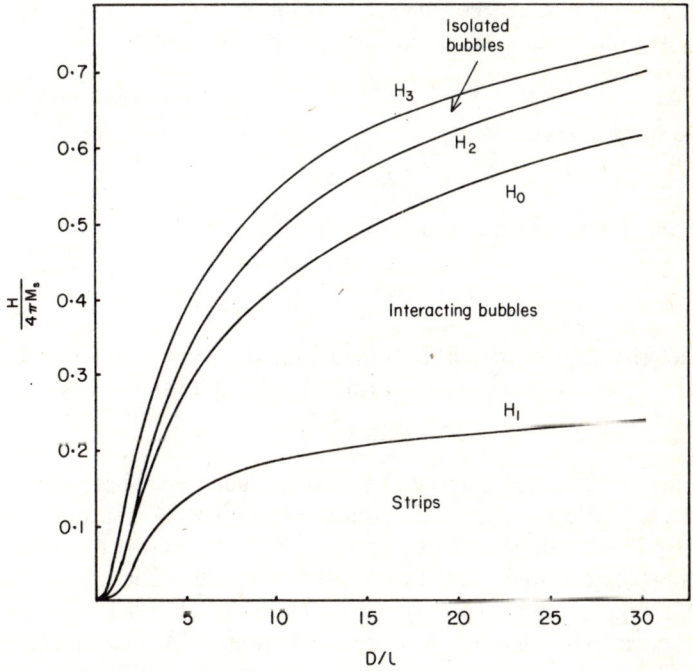

FIGURE 39. Reduced critical field values as a function of reduced film thickness. H_0 is the run out field and H_3 the collapse field.

energy, the spacing between domains is so large that the bubbles can be considered isolated. Further increases in applied field result in bubbles of reduced diameter which finally collapse when a field H_3 is reached. Experimental measurements[220] of the spacing and diameter of bubbles in a lattice, as a function of applied field, have been completed and shown to be in accord with experimental observations. Taking these along with the earlier results on strip domains[215] it can be seen that a satisfactory description of simple lattice structures has been achieved.

The stability and suitability of isolated bubble domains for devices were first described by Bobeck[217] but later investigated theoretically in great detail by Thiele[221-224]. In the latter works the ranges of fields and diameters over which bubbles are stable was established. The collapse field has already been shown in Fig. 39 while the isolated bubble, subject to a decreasing field, becomes elliptically unstable and runs out into a strip domain at a field shown as H_0 also in Fig. 39. The variation of the bubble diameters at the collapse and runout fields are shown as a function of film thickness in Fig. 40. Taking account of the stability data and considering the operating requirements of devices[222,224] the preferred material requirements can be established. Consequently the preferred film thickness has been chosen as

$$D = 4l$$

at which a bubble has diameter

$$2R = 8l$$

where, if γ is the energy per unit area of domain wall,

$$l = \frac{\gamma}{4\pi M_s^2}$$

is a characteristic length of bubble domain materials. The optimum bias field occurs midway between collapse and runout fields at a value of

$$H = 0.28 \times 4\pi M_s$$

The structure of the domain wall surrounding a bubble can have a profound effect upon its static and dynamic characteristics. Figure 41a shows a bubble bounded by a simple Bloch wall in which the spins are always perpendicular to a radius and rotate, with distance through the wall, in the same sense about the radius axis. Figure 41b shows a bubble in which the sense of rotation in the wall is not always the same and there are two sections of the wall with opposite polarity. The locations at which the sense of the wall changes were termed Bloch lines by Grundy et al.[225,226] when they were first observed in thin cobalt foils. They have also been observed in amorphous films[227] which are currently being applied to bubble devices. The significance of these structures

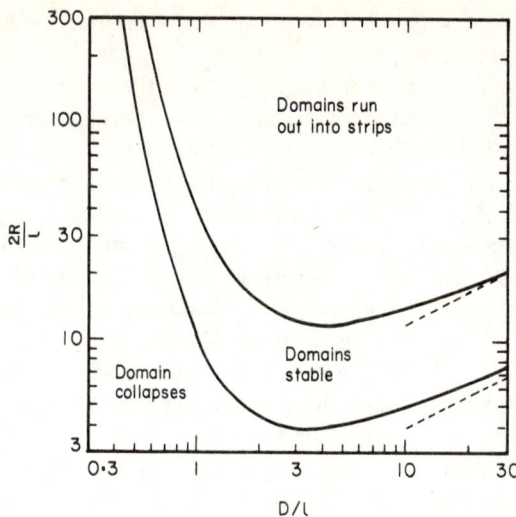

FIGURE 40. Reduced bubble diameter $2R$, at the collapse and runout fields, as a function of reduced film thickness. [A. A. Thiele, *Bell Syst. Tech. J.* **50**, 725, (1971).]

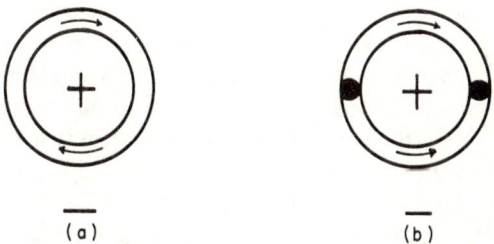

(a)   (b)

FIGURE 41. Plan view of a bubble showing, in exaggerated scale, the direction of magnetisation at the middle of its wall. In (a) the spins in the wall always rotate in the same sense about a bubble radius as axis, whereas in (b), the sense of rotation varies with position along the wall. The dark areas are termed Bloch lines where the sense of the magnetisation rotation changes. The magnetisation direction, on a radius through the middle of the Bloch line, rotates in a plane containing the radius and the film normal.

is that they are thought to be the cause of the "misbehaviour" of certain types of bubbles known as hard bubbles.

Hard bubbles are those which do not behave in the same way as normal bubbles for which the simple theory is known to apply. In static observations they are noted by their different sizes[228], shapes[229], and collapse fields[228,229] from normal bubbles. Their dynamic characteristics are marked by the fact that while they do move in the direction of an applied magnetic field gradient they also have a transverse velocity component[228,229] as well as a significantly reduced mobility[230]. These various properties are being explained on the basis of models which contain various numbers of Bloch lines of this type and also of a type which lie in the film plane[229]. However, these problems can be avoided by employing various means of hard bubble suppression. Multilayer films have been shown to be less susceptible to hard bubbles[231] as have low q films which have been shown to be almost hard bubble free[232]. However, the implantation of ions[233] into the storage medium has been shown to result in films free of hard bubbles and to be the simplest and most reliable method of hard bubble suppression.

Notwithstanding the problems of hard bubbles, a high wall velocity, and therefore mobility, is determined by the intrinsic properties of the materials. Highest mobilities have been obtained with garnet and amorphous films which are currently being most actively investigated.

5.1.1 Bubble domain materials

The types of materials that have been tried for use in bubble domain devices are summarised in the chart[234,235] shown in Fig. 42. Shaded areas indicate the approximate ranges of values of H_k and $4\pi M_s$ that have been observed and lines are drawn giving an estimate of the bubble domain sizes expected on the assumptions already given. Clearly, as predicted earlier, all the most promising materials lie above the line $H_k = 4\pi M_s$. The crystalline materials that have been studied[235] are orthoferrites [which have the composition $(R$ or $Y)FeO_3$, where R is a rare earth element], hexagonal ferrites ($MFe_{12}O_{19}$, where M is a metal), and garnets [$(R$ or $Y)_3Fe_5O_{12}$]. Referring to Fig. 42 it can be seen that in orthoferrites the bubbles tend to be too large for high packing density applications while in hexagonal ferrites the domain mobility has been shown to be too low[235] for high speed applications. Most interest is therefore being focused on garnet materials.

Epitaxial garnet films of high quality have been produced by liquid phase epitaxy[236-238] and by chemical vapour deposition[239] methods. The first has proved itself most suitable, and garnets themselves most adaptable to the production of different magnetic properties since different elements can be incorporated into their structure[235,236]. The garnet materials are, crystallographically, cubic but only slight deviations of the structure from cubicity

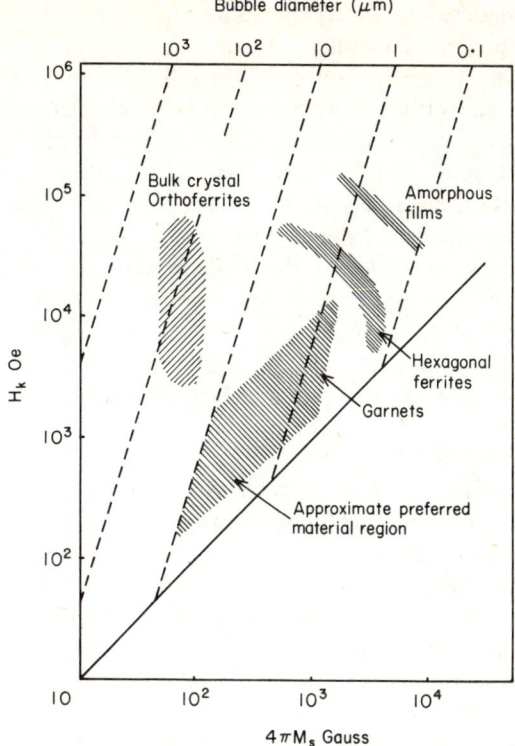

FIGURE 42. Material status chart, first produced by Gianola et al.[234] but later updated by Chang[235], showing approximate range of material properties that have been investigated. The region of preferred properties is also shown.

during the growth process are needed to introduce the necessary uniaxial anisotropy[239], sometimes called growth induced anisotropy. Therefore uniaxial garnet films with suitable magnetic properties can be produced, thereby staking their claim as candidates for inclusion in future bubble domain memory systems.

Amorphous magnetic films have been prepared using sputtering techniques and shown to be capable of supporting bubble domains[25,240]. Films of GdCo and GdFe were shown to exhibit the required perpendicular anisotropy and to support bubbles with diameters in the micron region. As a test of their suitability to devices, some of the films were also used in shift register systems[25]. However, inspection of Fig. 42 shows that these binary alloys suffer from rather high values of magnetisation which can result in rather low q values and also impose a need for high bias fields in device applications. In the search for improved properties ternary alloys have been investigated. By altering the type and concentration of the third element added to the GdCo system such alloys as GdCoAu[241], GdCoCu[241], GdCoMo[241-243], GdCoCr[244], and GdCoNi[245] have been produced. Of these GdCoMo, with properties near to those of garnets (see Fig. 42), appears to be the most suitable for further investigation[243]. Film reproducibility and uniformity remain problems to be tackled while on the fundamental side only a limited understanding of the mechanism of the perpendicular anisotropy exists[243] although pair ordering and shape effects may be contributors[25]. Nevertheless their potential for application has been demonstrated by the inclusion of GdCoMo films in storage devices with a high performance[246].

5.1.2 Domain propagation and devices

The movement of domains can be achieved in many ways but two, in particular, have greatest potential for application in storage devices. The first involves the use of conductors deposited on the surface of the bubble material[247] as the simple arrangement shown in Fig. 43 will be used to demonstrate. Consider a bubble initially positioned under the hair pin conductor numbered 1. When a current is passed along the similarly shaped conductor marked 2 it produces a field with components directed downwards into the film, i.e. opposite in sense to the applied bias field. Therefore the total applied field below the centre of the hair pin is less than in position 1 and so the bubble moves to position 2 to lower its energy. Therefore a variation of field, or more precisely a field gradient, is necessary to cause a bubble to move. Similarly a current along conductor 3 would, when no current flows in conductor 2, cause the bubble to move to position 3. Using more complicated variations on this basic circuit, shift register magnetic stores[247] have been proposed and investigated, although other more ingenious arrangements of conductors are possible[248].

The second and most important method of moving domains is shown in Fig. 44. This consists of a T and I bar structure[249], typically of NiFe Permalloy, overlaying a film or chip of bubble material. The thickness of the propagating structure is typically of the order of a few thousands of Ångstroms and is used to secure propagation of the bubbles as follows. A magnetic field, rotating in the film plane, is applied and adjusted to be sufficiently large to magnetise the overlay structure, as shown in Fig. 44a. The limbs of the T and I bars have such a large shape anisotropy that they can easily be magnetised only in the direction parallel to the length of their limbs and not in a direction perpendicular. Therefore when the circulating field is in the direction shown in Fig. 44a, the vertical limb of T and I bars are magnetised to saturation

Magnetic Thin Films and Devices 677

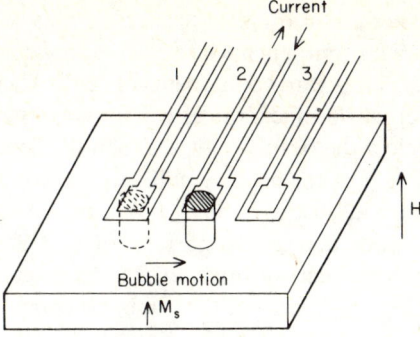

FIGURE 43. A simple conductor arrangement, for moving bubble domains.

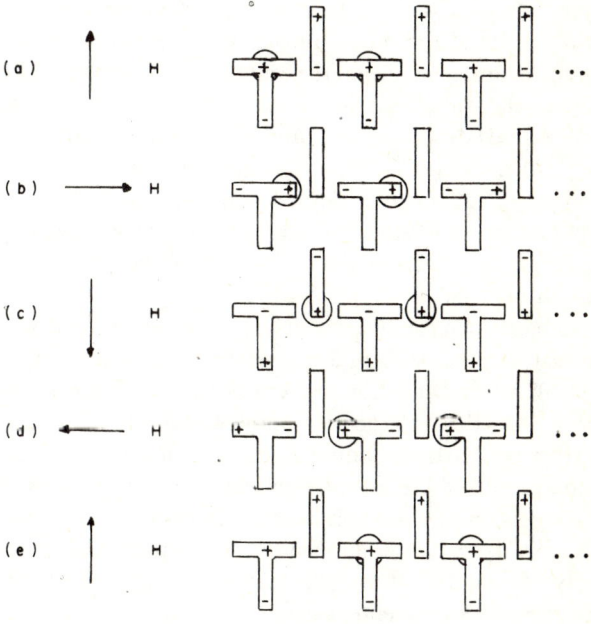

FIGURE 44. The T and I bar system of propagating domains. The arrows represent the direction of the externally applied rotating magnetic field and the + and − signs the induced poles in the Permalloy pattern. [A. J. Perneski, *I.E.E.E. Trans. on Magnetics*, **MAG-5**, 554, (1969).]

vertically and the poles induced are shown by the plus and minus signs. The positive poles cause field components directed downwards into the film while the negative poles cause upward components of fields. Therefore in agreement with the earlier observations, the bubbles will move to the positions of lowest energy which are below the positive poles. When the field has rotated to the position shown in Fig. 44b, the positive poles are seen to be positioned to the right of the original and so the bubbles move to the right. The following sequences of field position depicted by parts (c), (d) and (e) of Fig. 44 show that the rotating field always causes positive poles to be produced immediately to the right of the bubble and therefore the bubble is compelled to move to the right. Consequently, using a rotating field and a suitable Permalloy overlay pattern it is possible to produce shift register type action. Other types of overlay have been developed and are known as Y bars[250] and chevron patterns[251].

High capacity storage devices which use bubbles are invariably of the major–minor loop variety[213,252–254], as shown in Fig. 45. In this system bubbles are created at a generator[255] and, when required, introduced at a transfer gate[256] into the major loop. Transfer into the minor loops for storage takes place at the other transfer gates. Readout is achieved by the bubbles being transferred from the minor loops into the major loops so that they pass the bubble sensor. Clearly some sort of counting system is needed to keep track of all the bits of particular words when they are distributed into the minor loops of the store. Sensing can be achieved using inductive pickup, Hall, magneto-optic or magnetoresistive effects[257,258]. Of these the last mentioned seems to have the most promise since Permalloy is used in the magnetoresistive elements and is therefore easily fabricated into the shifting structure.

Memories have been proposed which have capacities as large as 15.10^6 bits, bit data rates up to the MHz region, and packing densities of information of the order of 10^6 bits/in^2 and more. Potential exists for improvement in all these figures particularly through the development of the new amorphous materials and the further application of integrated circuit technology. Therefore it is to be anticipated that bubble technology will advance at a steady rate so that eventual entry of high performance bubble systems into the data storage market is highly likely. Other applications exist, particularly in logic[235], but undoubtedly the future of bubble technology hinges on the success or otherwise of the work on memory devices.

5.2 Domain Wall Motion Devices

Domain wall motion devices, employing films with axes in their plane, have been known and investigated for many years. An early design of a thin film shift register, proposed by Broadbent[259], consisted of a thin uniaxial

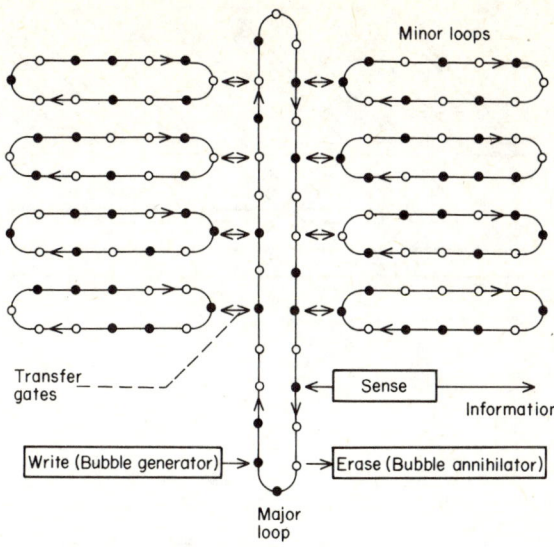

FIGURE 45. Schematic representation of a major/minor loop bubble store.

magnetic film situated adjacent to a conductor array as shown in Fig. 46a and in cross section in Fig. 46b. Suppose a domain of reverse magnetisation has been created in the otherwise saturated film by the passage of a current pulse of sufficient magnitude along the write conductor. If a pulse train of the form shown in Fig. 46b is passed along the conductors then the domain can be made to move as will now be described. When the first current pulse flows down the strip line nearest the film the field pattern shown in (2) is produced. The sense of the fields are such that the leading edge of the domain grows while the trailing edge of the domain diminishes. The domain is therefore effectively shifted by a distance equal to half of the width of one of the conductors. The next pulse down the second conductor pattern produces the field distribution shown in (3) and by similar reasoning the domain will move to the new position shown. The remaining pulses of the sequence ensure that the domain is moved a similar distance by each pulse so that when the train is complete the domain has reached such a position whereby repetition of the complete train will continue the motion of the domain in the same direction. All the domains in the register are moved by the application of the same pulses and so all the information is moved serially as required for shift register devices.

The tolerances in driving fields are defined on the low side by the fields needing to be in excess of the wall motion threshold field by a sufficient amount

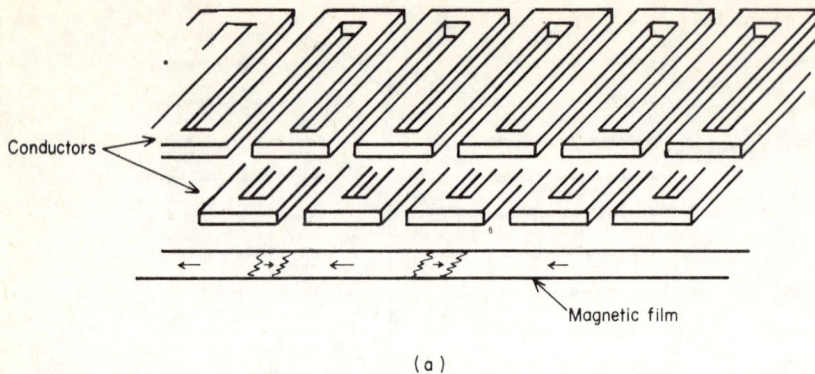

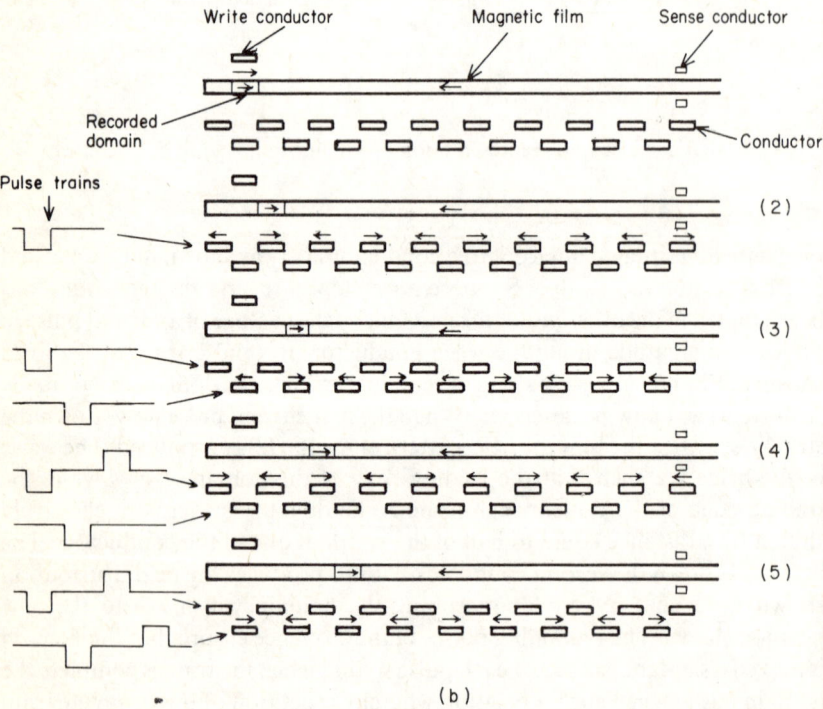

FIGURE 46. Thin film shift register proposed by Broadbent. (a) Exploded view of conductor and film pattern. (b) Cross section view of conductors and film showing field and magnetisation directions.

to give the walls the necessary velocity to sustain a high data rate of operation, whereas on the high side the fields must be less than those needed to cause the spontaneous reversal of magnetisation that would result in new and unwanted domains being formed. Therefore the tolerances for shift register operation in a perfect film lie within the range $H_w < H < H_k$. Tickle[260] suggested the use of evaporated NiFeCo films in which the addition of cobalt causes increases in the value of H_k; the relative proportions of nickel and iron are adjusted for zero magnetostriction of the resultant alloy. The value of H_w decreases with increasing film thickness and therefore by using an appropriate film thickness and composition, wide operating tolerances could hopefully be achieved.

In operation the appearance of spurious domains during the shifting cycle imposes a limit on performance. These domains can occur at the edges of the magnetic film but can be avoided by ensuring that during the deposition process the edges of the film are tapered. However, the most serious cause of their occurrence is that the trailing edges of domains are not always shifted and remanant nucleii are left behind. These form larger domains during subsequent shifting cycles and therefore result in errors of operation. Tickle[260] suggested the use of a D.C. bias field of the order of one Oersted applied to the entire film to ensure the erasure of the trailing edges of the domains. However, it was later shown that a register operating a 0.5 MHz bit rate with an error rate of 1 in 10^5 had such small tolerances that its further development and application were not contemplated[261].

Despite the apparently limited success with the form of shift register already described other types have been developed and operated with satisfactory tolerances thereby giving potential for wide application to commercial devices. Spain[262] used the structure shown in Fig. 47 in which the sawtooth pattern is that of a low coercivity NiFeCo channel situated in a background film of high coercivity. The difference in coercivity is achieved by evaporating the NiFeCo onto a substrate which had previously been coated with aluminium in the areas where high coercivity is required. The NiFeCo film on the rough aluminium surface is of high coercivity while that on the glass substrate is of low coercivity. In this way a wall motion threshold field of the order of 1–2 Oe is obtained in the low coercivity regions, when film thicknesses of 1500 Å are used, while the coercivity in the other regions is of the order of 25 Oe.

The shifting of domains is achieved as follows. If a domain is present as shown in Fig. 47 and a field H is applied, as shown in (b), the domain will expand along the low coercivity channel nearest to the field direction. The subsequent application of a field, shown in (c), causes the trailing edge of the domain in the upper region of the channel to be erased leaving only a small domain which is subject to a smaller local reverse field as a result of the field contributed by a current carrying conductor running underneath the sub-

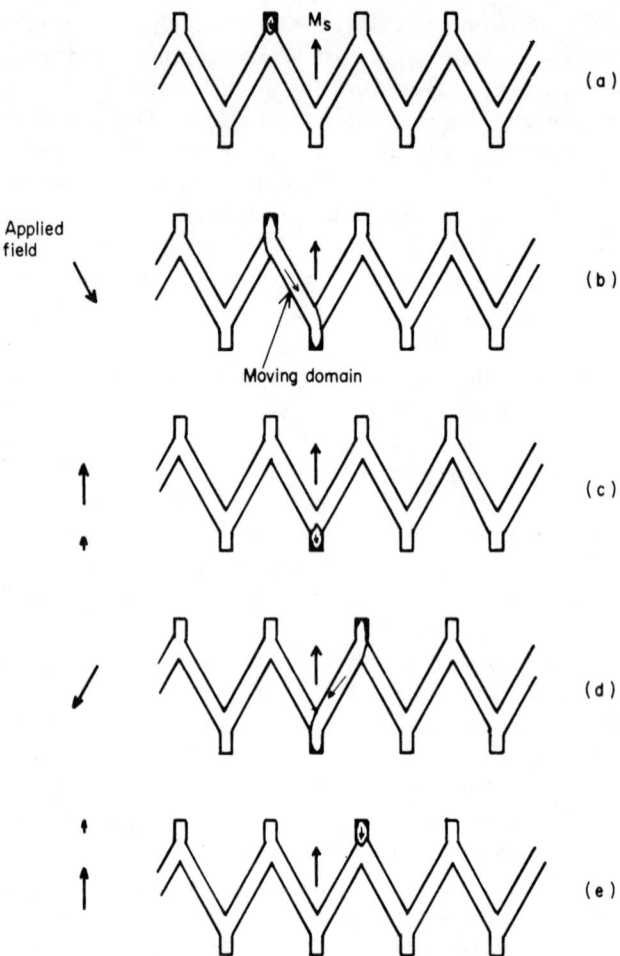

FIGURE 47. Thin film shift register, proposed by Spain, showing sequence of shifting fields. [R. J. Spain, *I.E.E.E. Trans. on Magnetics*, **MAG-2**, 347, (1966).]

strate, in a horizontal direction in (c). A similar shifting field cycle but with the fields shown in (d) and (e) moves the pulse to the position shown in (e). A repetition of the field patterns produces further propagation of the data. Similar considerations apply to this type of shift register as did to that described earlier as regards the choice of film composition and thickness, but there are additional conditions related to the different combinations of applied field and channel orientation with respect to the applied field[262-264]. Readout can be achieved inductively by passing the domain under a pickup loop or by utilising Hall or magnetoresistive effects for higher signal outputs[265].

This and other types of shift register design have been built and tested[263] at data rates of 500 kHz with typically 2000 bits per square inch of substrate and channel widths of 0.002 inches. Further, prototype memories of around 2 Mbit capacity, with multiloop organisation and employing multilayer films, have been built and operated satisfactorily[266,267], while a 24 Mbit memory was under construction[267] with the potential of being available as an alternative to disc-type storage systems.

Domain tip (DOT) motion has potential application not only to shift register type memories but also to other types such as "push down list" memories[263,268], buffer memories[263], and also to logic[262,269] and other devices such as counters[263].

In contrast to the above devices, which employ moving domains, there have been attempts to utilise domain walls[270,271] and more recently the detailed structure within a cross tie wall[272] for the storage function. In the latter case the movement of Bloch lines within the main wall takes place at high speed and gives potential for the development of high data rate shift register systems, although many problems can be envisaged when working on such a small scale.

6. Optical Beam Storage Devices

The possibility of recording information on magnetic thin films using electron beams was demonstrated some time ago[273] but it is only in recent years that there have been significant attempts to exploit this potential. Generally the electron beam method of storage has given way to optical beam techniques in view of the availability and performance characteristics of lasers[274]. Therefore not only does the mode of recording require a suitable thin film technology but it also requires the development of suitable optical technology to accompany it. As an illustration of this point, the essential components of a system are shown diagrammatically if Fig. 48. It can be seen that the recording medium is but a small part of a complex system and it has been pointed out in a recent review[274] that the optical technology of such a system has not yet been developed to a stage to make it economically competitive with existing mass memory technologies. Notwithstanding these difficulties there are many types of media, magnetic and non-magnetic, that could be used for recording either directly in bits or in holographic form. There are further complications related to the fact that certain materials can only be used for permanent recording while others are potentially suitable for record and replay functions. The

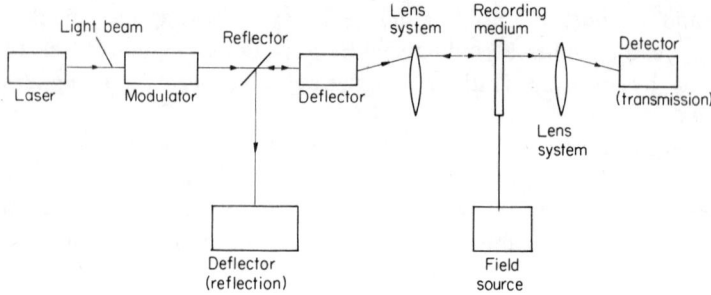

FIGURE 48. Block diagram representation of the elements of a laser beam store.

magnetic thin film media are in the latter group[275] but it is in the other areas that most progress has been made. Nevertheless a longer term potential application of thin magnetic films exists and will now be discussed.

The system shown in Fig. 48 consists of a laser from which light is directed via a modulator, deflector, and lens system onto the recording medium causing a local heating where information can be stored, sometimes with the aid of a magnetic field, as will be discussed shortly. Readout of information can be achieved using magneto-optic effects. For example, by allowing a beam to pass through the film, and utilising the Faraday effect, the state of the magnetisation in the film could be detected by the part of the system marked transmission, whereas reflected light can be utilised and detected by the system marked reflection.

The first type of recording to be described is termed Curie point writing[273] for which thin films with easy axes normal to the film plane are used (see Fig. 49). A suitable material is MnBi, which can be prepared with the required orientation by evaporation onto mica or glass substrates[276]. A system has been built[272] using a 15.2 cm diameter glass disc coated with such an MnBi film, of thickness 400–500 Å, having a rectangular hysteresis loop and a coercivity of 1500 Oe. The disc is rotated, as in a conventional magnetic disc machine, and light is reflected by a mechanical deflector onto the appropriate tracks. Laser light pulses of 0.1 μsec duration, when focused onto small areas of film are sufficient to raise the temperature of the film locally to above the Curie temperature of 360°C. When the temperature of the spot exceeds this value it has no net magnetisation (Fig. 49b), but there is stray field within it which, when the spot cools, causes the reintroduced magnetisation to be of the same sense. Therefore writing is achieved without the application of a field, but when

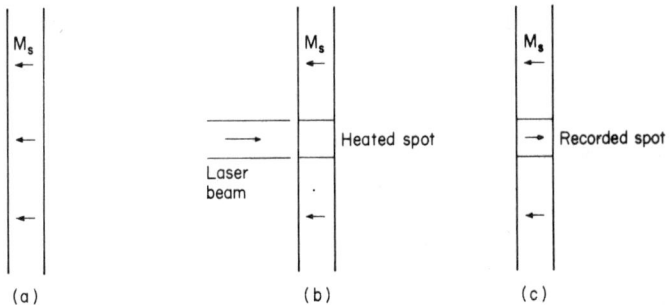

FIGURE 49. Thin film at various stages of the Curie point writing process. (a) At ambient temperature prior to recording. (b) Heated to above the Curie temperature by a light pulse. (c) Cooled after removal of incident light.

erasure of a written magnetisation is necessary a field of 700 Oe is required. Since this field cannot be limited in its area of application to the size of a single spot, it is applied over an area covering many spots. Therefore its magnitude must be insufficient for it to cause interference with other recorded spots and areas of film when it is applied coincidentally with a laser pulse. A disc system of this type could be operated with 10^4 tracks and with 10^5 bits per track giving a total capacity of 10^9 bits. Reliability of the storage medium has been demonstrated by the recording being repeated up to 10^7 times without significant deterioration of the signal[278].

Another system based on the rotating disc but using EuO or Fe-doped EuO has also been proposed[279-281]. EuO was chosen because of its large magneto-optic effects and high thermosensitivity although a cryogenic system is needed since the required properties are only obtained in the temperature range 77 to 150 K. The films were coated on a three inch diameter glass disc with storage locations being assessed with a rotating mirror deflector. The magnetisation in EuO lies in the plane of the film and a magnetic field is needed for write and erase functions. Fields of $0.5H_c$ were needed to effect recording on spots of 5 μm diameter in films of the order of 3300 Å thickness. Readout is achieved using the longitudinal Faraday effect with a beam directed at about 20° to the film surface. The light source is a GaAs semiconductor laser which has a high conversion efficiency thereby keeping power consumption low[282,283].

A further mode of recording was demonstrated by Treves et al.[284] using films of cobalt phosphorus produced by electrodeposition from hypophosphite solution[285]. Such films have coercivities which are strongly dependent on temperature as a result of the first order anisotropy constant of cobalt being a strong function of temperature. Films used had coercivities of 200 Oe which were reduced by a factor of about 3 at 150°C. Therefore by heating a spot with a laser pulse to 150°C and applying a field with a ferrite recording head, positioned on the opposite side of the substrate, recording could be achieved. The field was sufficiently large to record on spots down to 3.5 μm in diameter but not large enough to affect other areas of film. Readout was achieved using Kerr magneto-optic effects.

Thin films of material such as GdIG[286], which is ferromagnetic, have also been used in devices where use is made of the occurrence of a phenomenon known as the compensation temperature. The latter occurs when the two sublattice magnetisations, which are oriented in opposition, cancel and the overall magnetisation is zero. Since coercivity varies inversely with magnetisation it is large at the compensation temperature but decreases rapidly with increasing or decreasing temperatures so that reversal of magnetisation can be easily stimulated. Therefore by using a material, such as GdIG, kept at or near its compensation temperature and heating local regions with a laser beam, the application of a magnetic field will be sufficient to reverse the region of low coercivity. Therefore recording is observed.

One aspect of current activity is concerned with the production of materials which require a low temperature excursion during writing, have low switching fields, and large magneto-optical effects. In addition the films must be uniform in properties and be able to sustain repeated temperature cycling. Amongst the materials that have been given attention, in addition to those already mentioned, are $CrTe$[287], $MnAlGe$[288], $MnGaGe$[288], $MnAs$[289], $GdCo$[288,290], $PtCo$[288], partially substituted $GdIG$[291], cobalt ferrites[292], and iron garnet[293]. However, with available materials it has already been possible to demonstrate the potential[274,294] for storage densities as high as 10^8 bits per square inch, system capacities as high as 10^{11} bits, data rates in the MHz region, and access times varying from microseconds to milliseconds, depending on the particular system envisaged.

7. Concluding Remarks

In this brief survey of magnetic films the greatest attention has been focused on their applications in digital storage devices for computers. There have been other less important areas of potential application and these include analogue storage devices[295,296], ferroacoustic memories[297,298], logic[299-301], display panels[297,302,303], magnetometers or sensors[304-307], and parametrons[308-310].

The properties of thin films and their digital storage applications have been reviewed in a way which reflects the evolution and shift of interests over the years. For example the bulk of published works have been connected with soft

Permalloy films and their use in fast random access storage devices. Although a measure of success was obtained with the development of plated wire memories the thin film store never completely replaced the core store in its role in the computer and it now appears that it will be semiconductor storage devices that achieve this aim. Emphasis has therefore shifted towards the large capacity disc machines and their potential replacements. Hard magnetic films have been used as disc coatings although there has been considerable development of thin oxide media as an alternative and the latter have been widely used. In the context of the disc system thin film recording heads have potential high density applications. However, it is in the area of disc replacement that most current debate is concentrated. Bubbles[214] and thin film shift registers have been predicted to be the eventual replacements although there is considerable weight of opinion that disc systems will maintain current predominance for many years to come[311]. Current knowledge seems to indicate that bubble storage will probably be an order of magnitude more expensive than disc storage but will offer two orders of magnitude improvement in performance[312]. Therefore application will not be decided simply on the technical performance of bubble systems but on cost performance "trade offs" between alternative system structures. Consequently no predictions are made about future large capacity digital storage devices, although the potential of magnetic film devices is emphasised.

References

1. M. S. Blois, *J. Appl. Phys.*, **26**, 975, (1955).
2. J. I. Raffel, *J. Appl. Phys.*, **30**, 60S, (1959).
3. D. O. Smith, In "Magnetism", Eds., G. T. Rado and H. Suhl, Vol. 2, 465, Academic Press, New York and London, (1963).
4. E. W. Pugh and T. O. Mohr, In "Thin Films" (American Society for Metals), Chapman and Hall, (1964).
5. M. Prutton, "Thin Ferromagnetic Films", Butterworths, London, (1964).
6. R. F. Soohoo, "Magnetic Thin Films", Harper and Row, London, (1965).
7. C. E. Fuller, In "The Electrical Properties and Applications of Thin Films", ed., R. A. Coombe, Pitman, (1967).
8. K. D. Leaver, *Contemp. Phys.*, **9**, 475, (1968).
9. D. J. Craik and R. S. Tebble, "Ferromagnetism and Ferromagnetic Domains", North Holland, Amsterdam, (1965).
10. W. D. Doyle, *J. Appl. Phys.*, **33**, 1769, (1962).
11. E. L. Boyd, *I.B.M. J. Res. and Dev.*, **4**, 116, (1960).
12. M. Prutton, *Trans. Ninth Nat. Vac. Symp. Am. Vac. Soc.*, 69, (1962).
13. D. O. Smith, M. S. Cohen and G. P. Weiss, *J. Appl. Phys.*, **31**, 1755, (1960).
14. L. Néel, *Compt. Rend.*, **237**, 1468, (1953).
15. L. Néel, *J. Phys. Rad.*, **15**, 225, (1954).
16. S. Taniguchi, *Sc. Repts. Res. Inst., Tohuku Univ.*, **A7**, 269, (1955).

17. G. Robinson, *Proc. Symp. on Elec. and Mag. Props. of Thin Metallic Layers*, Louvain, Belgium, 140, (1961).
18. G. Robinson, *Proc. Int. Conf. Mag. and Cryst.*, Kyoto, Japan (*Phys. Soc. Japan*), **1**, 558, (1962).
19. F. G. West, *J. Appl. Phys.*, **35**, 1827, (1964).
20. M. Takahashi, D. Watanabe, T. Kono and S. Ogawa, *J. Phys. Soc. Japan*, **15**, 1351, (1960).
21. M. Takahashi, *J. Appl. Phys.*, **33**, 1101S, (1962).
22. H. Fujiwara and Y. Sugita, *I.E.E.E. Trans. on Magnetics*, **MAG-4**, 22, (1968).
23. R. J. Spain, *Appl. Phys. Letts.*, **3**, 208, (1963).
24. N. Saito, H. Fujiwara and Y. Sugita, *J. Phys. Soc. Japan*, **17**, 1116, (1964).
25. P. Chaudhari, J. J. Cuomo and R. J. Gambino, *I.B.M. J. Res. and Dev.*, **17**, 66, (1973).
26. R. S. Tebble, "Magnetic Domains", Methuen, London, (1969).
27. E. C. Stoner and E. P. Wohlfarth, *Phil. Trans. Roy. Soc.*, **A240**, 599, (1948).
28. S. Slonczewski, I.B.M. Rept. RM 003 111 224, (1956), cited in Ref. 5.
29. D. O. Smith, *J. Appl. Phys.*, **29**, 264, (1958).
30. E. M. Bradley and M. Prutton, *J. Electronics and Control*, **6**, 81, (1959).
31. S. Middelhoek, Ph. D. Thesis, University of Amsterdam, (1961).
32. M. Prutton, *Brit. J. Appl. Phys.*, **11**, 335, (1960).
33. D. O. Smith, E. E. Huber, M. S. Cohen and G. P. Weiss, *J. Appl. Phys.*, **31**, 295S, (1960).
34. D. O. Smith, *J. Appl. Phys.*, **32**, 70S, (1961).
35. D. O. Smith and K. J. Harte, *J. Appl. Phys.*, **33**, 1399, (1962).
36. D. O. Smith and K. J. Harte, *J. Appl. Phys.*, **34**, 442, (1963).
37. S. Methfessel, S. Middelhoek and H. Thomas, *J. Appl. Phys.*, **32**, 1959, (1961).
38. H. W. Fuller and M. E. Hale, *J. Appl. Phys.*, **31**, 238, (1960).
39. H. W. Fuller and H. Rubinstein, *J. Appl. Phys.*, **30**, 84S, (1959).
40. M. S. Cohen, *J. Appl. Phys.*, **34**, 1941, (1963).
41. C. D. Olsen and A. V. Pohm, *J. Appl. Phys.*, **29**, 274, (1958).
42. S. Middelhoek, *I.B.M. J. Res. and Dev.*, **6**, 140, (1962).
43. S. Middelhoek, *Z. Angew. Phys.*, **14**, 191, (1962).
44. E. Fuchs and H. Pfisterer, Proc. Fifth. Int. Congress for Electron Microscopy, Academic Press, New York and London, (1962).
45. T. H. Beeforth and P. J. Hulyer, *Nature*, **199**, 793, (1963).
46. D. S. Lo and M. M. Hanson, *I.E.E.E. Trans. on Magnetics*, **MAG-5**, 115, (1969).
47. T. T. Kusuda, H. C. Bourne and D. S. Bartran, *I.E.E.E. Trans. on Magnetics*, **MAG-7**, 165, (1971).
48. A. Green, K. D. Leaver and M. Prutton, *J. Appl. Phys.*, **35**, 812, (1964).
49. A. L. Olson and E. J. Torok, *J. Appl. Phys.*, **37**, 1297, (1966).
50. S. Middelhoek and D. Wild, *I.B.M. J. Res. and Dev.*, **2**, 93, (1967).
51. W. Kayser, *I.E.E.E. Trans. on Magnetics*, **MAG-3**, 141, (1967).
52. W. Kayser, A. V. Pohm and R. L. Samuels, *I.E.E.E. Trans. on Magnetics*, **MAG-5**, 236, (1969).
53. R. W. Olman and E. N. Mitchell, *J. Appl. Phys.*, **30**, 258S, (1959).
54. E. P. Wohlfarth, *J. Electronics and Control*, **10**, 33, (1961).

55. L. Néel, *C.R. Acad. Sci. Paris*, **241**, 533, (1955).
56. E. E. Huber, D. O. Smith and J. B. Goodenough, *J. Appl. Phys.*, **29**, 294, (1958).
57. S. Methfessel, S. Middelhoek and H. Thomas, *I.B.M. J. Res. and Dev.*, **4**, 96, (1960).
58. S. Middelhoek, *J. Appl. Phys.*, **34**, 1054, (1963).
59. A. Aharoni, *J. Phys.*, **32**, 966, (1971).
60. G. A. Jones and B. K. Middleton, *Int. J. Mag.*, **6**, 1, (1974).
61. A. Hubert, *I.E.E.E. Trans. on Magnetics*, **MAG-11**, 1285, (1975).
62. A. Holz and A. Hubert, *Z. Angew. Phys.*, **26**, 145, (1969).
63. W. F. Brown, Jnr. and A. E. Labonte, *J. Appl. Phys.*, **36**, 1380, (1965).
64. A. E. Labonte, *J. Appl. Phys.*, **40**, 2450, (1969).
65. A. Hubert, *Phys. Stat. Sol.*, **32**, 519, (1969).
66. A. Hubert, *Phys. Stat. Sol.*, **38**, 699, (1970).
67. E. J. Torok, A. L. Olson and H. N. Oredson, *J. Appl. Phys.*, **36**, 1394, (1965).
68. K. D. Leaver, *Thin Solid Films*, **2**, 149, (1968).
69. T. S. Crowther, *J. Appl. Phys.*, **34**, 580, (1963).
70. I. W. Wolf, *J. Appl. Phys.*, **33**, 1152S, (1962).
71. M. Prutton, *Nature*, **193**, 565, (1962).
72. G. P. Weiss and D. O. Smith, *J. Appl. Phys.*, **33**, 1166S, (1962).
73. S. Uchiyama, T. Fujii, M. Masuda and Y. Sakaki, *Jap. J. Appl. Phys.*, **6**, 512, (1967).
74. C. H. Tolman and P. E. Oberg, *Proc. Intermag Conf.*, *Am. Inst. of Elec. Eng.*, Paper 12-3, (1963).
75. J. H. Engelman and A. J. Hardwick, *Trans. Ninth Nat. Vac. Symp.*, *Am. Vac. Soc.*, 100, (1963).
76. H. Clow, *Nature*, **191**, 996, (1961).
77. K. D. Leaver, *Nature*, **196**, 158, (1962).
78. J. Eckardt, *Z. Angew. Phys.*, **17**, 202, (1964).
79. S. Fujii, S. Uchiyama, E. Yamada and Y. Sakaki, *Jap. J. Appl. Phys.*, **6**, 1, (1967).
80. A. Baltz and W. D. Doyle, *J. Appl. Phys.*, **35**, 1814, (1964).
81. A. Baltz, *Proc. Int. Conf. Magnetism*, Nottingham, 845, (1964).
82. T. Suzuki and C. H. Wilts, *J. Appl. Phys.*, **39**, 1151, (1968).
83. W. D. Doyle and M. Prutton, *J. Appl. Phys.*, **34**, 1077, (1963).
84. H. N. Oredson and E. J. Torok, *J. Appl. Phys.*, **36**, 950, (1965).
85. R. D. Fisher and H. E. Haber, *App. Phys. Letts.*, **2**, 11, (1963).
86. H. Hoffman, *Phys. Kondens. Materie*, **2**, 32, (1964).
87. H. Hoffman, *J. Appl. Phys.*, **35**, 1790, (1964).
88. H. Hoffman, *I.E.E.E. Trans. on Magnetics*, **MAG-2**, 566, (1966).
89. H. Hoffman, *I.E.E.E. Trans. on Magnetics*, **MAG-4**, 32, (1968).
90. H. Rother, *Z. Phys.*, **179**, 229, (1964).
91. K. J. Harte, *J. Appl. Phys.*, **39**, 1503, (1968).
92. K. D. Leaver, *J. Appl. Phys.*, **39**, 1157, (1968).
93. C. S. Comstock, A. C. Sharp, R. L. Samuels and A. V. Pohm, *I.E.E.E. Trans. on Magnetics*, **MAG-4**, 39, (1968).
94. T. Fujii, S. Uchiyama, S. Tsunashima and Y. Sakaki, *I.E.E.E. Trans. on Magnetics*, **MAG-5**, 223, (1969).

95. H. Hoffman and M. Okon, *Z. Angew. Phys.*, **21**, 406, (1966).
96. H. Hoffman, *I.E.E.E. Trans. on Magnetics*, **MAG-9**, 17, (1973).
97. W. Dietrich, W. E. Proebster and P. Wolf, *I.B.M. J. Res. and Dev.*, **4**, 189, (1960).
98. M. H. Kryder and F. B. Humphrey, *J. Appl. Phys.*, **41**, 1130, (1970).
99. M. H. Kryder and F. B. Humphrey, *Rev. Sci. Inst.*, **40**, 829, (1969).
100. M. H. Kryder and F. B. Humphrey, *I.E.E.E. Trans. on Magnetics*, **MAG-7**, 725, (1971).
101. B. R. Hearn, *J. Electronics and Control*, **16**, 33, (1964).
102. E. M. Bradley, *J. Brit. I.R.E.*, **20**, 765, (1960).
103. R. M. Pickard, J. A. Turner, J. K. Birtwistle and G. R. Hoffman, *J. Phys. D.*, **1**, 1685, (1968).
104. I. W. Wolf, *J. Electrochem. Soc.*, **108**, 959, (1962).
105. E. M. Bradley, *J. Appl. Phys.*, **33**, 1501, (1962).
106. R. D. Fisher, H. E. Austin and H. E. Haber, *J. Electrochem. Soc.*, **111**, 39, (1964).
107. K. D. Leaver and M. Prutton, *J. Appl. Phys.*, **33**, 1095, (1962).
108. K. D. Leaver and M. Prutton, *J. Electronics and Control*, **15**, 173, (1963).
109. P. I. Bonyhard and I. C. Buckingham, *I.E.E.E. Trans. on Magnetics*, **MAG-1**, 258, (1965).
110. D. B. Dove and T. R. Long, *I.E.E.E. Trans. on Magnetics*, **MAG-2**, 194, (1966).
111. P. C. Arnett, *I.E.E.E. Trans. on Magnetics*, **MAG-7**, 171, (1971).
112. P. I. Bonyhard, *J. Appl. Phys.*, **35**, 764, (1964).
113. P. I. Bonyhard, *J. Electronics and Control*, **16**, 339, (1964).
114. J. B. James, B. J. Steptoe and A. A. Kaposi, *Proc. I.R.E. Western Electronic Show and Convention*, (1962).
115. P. Mossman, *Proc. I.E.E.*, **111**, 1411, (1964).
116. K. U. Stein, *Z. Angew. Phys.*, **21**, 400, (1966).
117. K. U. Stein and E. Feldtkeller, *I.E.E.E. Trans. on Magnetics*, **MAG-2**, 184, (1966).
118. E. Feldtkeller, *J. Appl. Phys.*, **39**, 1181, (1968).
119. E. E. Bittman, *Proc. Intermag. Conf.*, Paper 9-1, (1963).
120. T. S. Crowther, *Proc. Intermag. Conf.*, Paper 5-7, (1964).
121. H. Chang, *J. Appl. Phys.*, **38**, 1203, (1967).
122. A. V. Pohm, J-M. Wang, W. Schnasse and T. A. Smay, *I.E.E.E. Trans. on Magnetics*, **MAG-5**, 408, (1969).
123. P. C. Arnett, J. F. Fresia, C. Lin, K. D. Przekurat and C. H. Stapper, Jnr., *I.E.E.E. Trans. on Magnetics*, **MAG-9**, 31, (1973).
124. L. J. Oakland and T. D. Rossing, *J. Appl. Phys.*, **30**, 54S, (1959).
125. R. J. Petschauer and R. D. Turnquist, *Proc. West. Joint Comp. Conf.*, **19**, 411, (1961).
126. F. R. Janisch, *I.E.E.E. Trans. on Magnetics*, **MAG-1**, 266, (1965).
127. I. Danylchuk, A. J. Perneski and M. W. Sagal, *Proc. Intermag. Conf.*, Paper 5-4, (1964).
128. T. R. Long, *J. Appl. Phys.*, **31**, 123S, (1960).
129. M. W. Sagal, *J. Electrochem. Soc.*, **112**, 1, (1965).
130. F. E. Luborsky and B. J. Drummond, *I.E.E.E. Trans. on Magnetics*, **MAG-10**, 78, (1974).

131. F. E. Luborsky, B. J. Drummond, R. O. McCary, W. Fahy and W. Sheerin, *I.E.E.E. Trans. on Magnetics*, **MAG-10**, 560, (1974).
132. G. A. Fedde, *I.E.E.E. Trans. on Magnetics*, **MAG-8**, 620, (1972).
133. J. S. Mathias and G. A. Fedde, *I.E.E.E. Trans. on Magnetics*, **MAG-5**, 728, (1969).
134. W. A. England, *I.E.E.E. Trans. on Magnetics*, **MAG-6**, 528, (1970).
135. J. P. McCallister, *I.E.E.E. Trans. on Magnetics*, **MAG-6**, 525, (1970).
136. A. C. M. Chen, W. D. Barber, F. E. Luborsky, *I.E.E.E. Trans. on Magnetics*, **MAG-7**, 494, (1971).
137. R. Girard, G. Grunberg, B. Lorang and G. Nicolas, *I.E.E.E. Trans. on Magnetics*, **MAG-5**, 501, (1969).
138. F. E. Luborsky, R. E. Skoda and W. D. Barber, *J. Appl. Phys.*, **42**, 1428, (1971).
139. F. E. Luborsky and W. D. Barber, *I.E.E.E. Trans. on Magnetics*, **MAG-7**, 490, (1971).
140. S. Iwasaki and T. Suzuki, *I.E.E.E. Trans. on Magnetics*, **MAG-4**, 269, (1968).
141. N. Curland and D. E. Speliotis, *I.E.E.E. Trans. on Magnetics*, **MAG-6**, 640, (1970).
142. N. Curland and D. E. Speliotis, *I.E.E.E. Trans. on Magnetics*, **MAG-7**, 538, (1971).
143. D. E. Speliotis, *Ann. N.Y. Acad. Sci.*, **189**, 21, (1972).
144. D. L. A. Tjaden and E. J. Tercic, *Philips Res. Repts.*, **30**, 120, (1975).
145. M. L. Williams and R. L. Comstock, *Proc. AIP Conf.*, No. 5, 738, (1972).
146. V. A. J. Maller and B. K. Middleton, *The Radio and Electronic Engineer*, **44**, 281, (1974).
147. A. van Herk and P. Wesseling, *I.E.E.E. Trans. on Magnetics*, **MAG-10**, 761, (1974).
148. O. Karlqvist, *Trans. Roy. Inst. Stockholm*, **86**, 3, (1954).
149. P. I. Bonyhard, A. V. Davies and B. K. Middleton, *I.E.E.E. Trans. on Magnetics*, **MAG-2**, 1, (1966).
150. N. Curland and D. E. Speliotis, *J. Appl. Phys.*, **41**, 1099, (1970).
151. J. Daval and D. Randet, *I.E.E.E. Trans. on Magnetics*, **MAG-6**, 768, (1970).
152. D. D. Dressler and J. H. Judy, *I.E.E.E. Trans. on Magnetics*, **MAG-10**, 674, (1974).
153. A. V. Davies, B. K. Middleton and A. C. Tickle, *I.E.E.E. Trans. on Magnetics*, **MAG-1**, 344, (1965).
154. J. P. Lazzari, I. Melnick and D. Randet, *I.E.E.E. Trans. on Magnetics*, **MAG-5**, 955, (1969).
155. D. E. Speliotis, J. R. Morrison and J. S. Judge, *I.E.E.E. Trans. on Magnetics*, **MAG-1**, 348, (1965).
156. D. E. Speliotis, J. R. Morrison and J. S. Judge, *I.E.E.E. Trans. on Magnetics*, **MAG-2**, 208, (1966).
157. D. L. A. Tjaden, *I.E.E.E. Trans. on Magnetics*, **MAG-7**, 544, (1971).
158. P. I. Bonyhard, B. K. Middleton and A. V. Davies, *I.E.E.E. Trans. on Magnetics*, **MAG-1**, 423, (1965).
159. C. Kittel, *Rev. Mod. Phys.*, **21**, 541, (1949).
160. E. C. Stoner and E. P. Wohlfarth, *Phil. Trans. Roy. Soc.*, **A240**, 599, (1948).
161. E. P. Wohlfarth. In "Magnetism", eds., G. T. Rado and H. Suhl, Vol. 1, 351, Academic Press, New York and London, (1963).
162. E. H. Frei, S. Shtrikman and D. Treves, *Phys. Rev.*, **106**, 446, (1957).
163. I. S. Jacobs and C. P. Bean, *Phys. Rev.*, **100**, 1060, (1955).

164. F. E. Luborsky, *J. Appl. Phys.*, **32**, 171S, (1961).
165. F. E. Luborsky, *J. Appl. Phys.*, **37**, 1091, (1966).
166. J. P. Lazzari, I. Melnick and D. Randet, *I.E.E.E. Trans. on Magnetics*, **MAG-3**, 205, (1967).
167. W. J. Schuele, *J. Appl. Phys.*, **35**, 2558, (1964).
168. D. E. Speliotis, G. Bate, J. K. Alstad and J. R. Morrison, *J. Appl. Phys.*, **36**, 972, (1965).
169. M. S. Cohen, *J. Appl. Phys.*, **31**, 1755, (1960).
170. G. Bate, D. E. Speliotis, J. K. Alstad and J. R. Morrison, *Proc. Int. Conf. on Magnetism*, Nottingham, 816, (1964).
171. G. Bate and D. E. Speliotis, *J. Appl. Phys.*, **34**, 1073, (1963).
172. D. E. Speliotis, *J. Appl. Phys.*, **38**, 1207, (1967).
173. A. Bremner and G. Riddell, *Proc. Am. Electroplaters Soc.*, **34**, 156, (1947).
174. D. E. Speliotis, J. S. Judge and J. R. Morrison, *J. Appl. Phys.*, **37**, 1158, (1964).
175. R. D. Fisher and W. H. Chilton, *J. Electrochem. Soc.*, **109**, 485, (1962).
176. M. Aspland, G. A. Jones and B. K. Middleton, *I.E.E.E. Trans. on Magnetics*, **MAG-5**, 314, (1969).
177. B. K. Middleton, Ph.D. Thesis, University of Salford, (1970).
178. J. S. Judge, J. R. Morrison and D. E. Speliotis, *J. Electrochem. Soc.*, **113**, 547, (1966).
179. J. S. Judge, J. R. Morrison and D. E. Speliotis, *J. Appl. Phys.*, **36**, 948, (1965).
180. M. Takahashi and T. Kono, *J. Phys. Soc. Japan*, **15**, 936, (1960).
181. R. D. Fisher and D. E. Koopman, *J. Electrochem. Soc.*, **111**, 263, (1964).
182. G. A. Jones and B. K. Middleton, *J. Mat. Sci.*, **3**, 519, (1968).
183. G. A. Jones and A. Farnsworth, *Phys. Stat. Sol.*, **15**, 545, (1973).
184. A. Green, *J. Sci. Inst.*, **1**, 671, (1968).
185. E. P. Wohlfarth and D. G. Tonge, *Phil. Mag.*, **2**, 1333, (1957).
186. N. Bertram and A. K. Bhatia, *I.E.E.E. Trans. on Magnetics*, **MAG-9**, 127, (1973).
187. T. H. Bonn and D. C. Wendell, U.S. Patent 2 644 787, (1953).
188. J. S. Sallo and K. H. Olson, *J. Appl. Phys.*, **32**, 203S, (1961).
189. J. S. Sallo and J. M. Carr, *J. Electrochem. Soc.*, **109**, 1040, (1962).
190. J. S. Sallo and J. M. Carr, *J. Appl. Phys.*, **33**, 1316, (1962).
191. J. S. Sallo and J. M. Carr, *J. Appl. Phys.*, **34**, 1309, (1963).
192. G. Bate and D. E. Speliotis, *J. Appl. Phys.*, **35**, 972, (1974).
193. J. R. Morrison, *Electrochem. Tech.*, **6**, 419, (1968).
194. L. G. Sebestyen, "Digital Magnetic Tape Recording For Computer Applications", Chapman and Hall, London, (1973).
195. E. P. Valstyn and L. F. Shew, *I.E.E.E. Trans. on Magnetics*, **MAG-9**, 317, (1973).
196. E. P. Valstyn and D. W. Kosy, *I.E.E.E. Trans. on Magnetics*, **MAG-5**, 442, (1969).
197. A. D. Kaske, P. E. Oberg, M. C. Paul and G. F. Sauter, *I.E.E.E. Trans. on Magnetics*, **MAG-7**, 675, (1971).
198. R. I. Potter, R. J. Schmulian and K. Hartman, *I.E.E.E. Trans. on Magnetics*, **MAG-7**, 689, (1971).
199. R. I. Potter and R. J. Schmulian, *I.E.E.E. Trans. on Magnetics*, **MAG-7**, 873, (1971).
200. J. P. Lazzari, 19th AIP Conf. MMM., Boston, (1973).

201. J. P. Lazzari and I. Melnick, *I.E.E.E. Trans. on Magnetics*, **MAG-6**, 601, (1970).
202. D. Augier and J. P. Lazzari, *I.E.E.E. Trans. on Magnetics*, **MAG-7**, 679, (1971).
203. J. P. Lazzari, *I.E.E.E. Trans. on Magnetics*, **MAG-9**, 322, (1973).
204. W. Chynoweth and W. Kayser, *AIP Conf. Proc.*, No. 24, 534, (1975).
205. R. P. Hunt, *I.E.E.E. Trans. on Magnetics*, **MAG-7**, 150, (1971).
206. R. L. Anderson, C. H. Bajorek and D. A. Thompson, *AIP Conf. Proc.*, No. 10, 1445, (1972).
207. A. V. Davies and B. K. Middleton, *I.E.E.E. Trans. on Magnetics*, **MAG-11**, 1689, (1975).
208. R. I. Potter, *I.E.E.E. Trans. on Magnetics*, **MAG-10**, 502, (1974).
209. R. W. Cole, R. I. Potter, C. C. Lin, K. L. Deckert and E. P. Valstyn, *IBM J. Res. and Dev.*, **18**, 551, (1974).
210. C. H. Bajorek, S. Krongelb, L. T. Romankiw and D. A. Thompson, *AIP Conf. Proc.*, No. 24, 548, (1975).
211. T. D. O'Dell, "Magnetic Bubbles," Macmillan, London, (1974).
212. A. H. Bobeck and E. Della Torre, "Magnetic Bubbles," North Holland, Amsterdam, (1975).
213. A. H. Bobeck and H. E. D. Scovil, *Scientific American*, **224**, 78, (1971).
214. A. H. Bobeck, P. I. Bonyhard and J. E. Geusic, *Proc. I.E.E.E.*, **63**, 1176, (1975).
215. C. Kooy and U. Enz, *Philips Res. Repts.*, **15**, 7, (1960).
216. J. A. Cape and G. W. Lehman, *J. Appl. Phys.*, **42**, 5732, (1971).
217. A. H. Bobeck, *Bell Syst. Tech. J.*, **46**, 1901, (1967).
218. W. F. Druyvesteyn and J. W. F. Dorleijn, *Philips Res. Repts.*, **26**, 11, (1971).
219. F. A. De Jonge and W. F. Druyvesteyn, *Festkorpeprobleme*, **XII**, 531, (1972).
220. F. A. De Jonge and W. F. Druyvesteyn, *AIP Conf. Proc.*, No. 5, 130, (1972).
221. A. A. Thiele, *Bell Syst. Tech. J.*, **48**, 3287, (1969).
222. A. A. Thiele, *J. Appl. Phys.*, **41**, 1139, (1970).
223. A. A. Thiele, A. H. Bobeck, E. Della Torre and U. F. Gianola, *Bell Syst. Tech. J.*, **50**, 711, (1971).
224. A. A. Thiele, *Bell Syst. Tech. J.*, **50**, 725, (1971).
225. P. J. Grundy, R. S. Tebble and D. C. Hothersall, *J. Phys. D: Appl. Phys.*, **4**, 174, (1971).
226. P. J. Grundy, D. C. Hothersall, G. A. Jones, B. K. Middleton and R. S. Tebble, *Phys. Stat. Sol.*, **A9**, 79, (1972).
227. P. Chaudhari and S. R. Herd, *IBM J. Res. and Dev.*, **20**, 102, (1976).
228. W. J. Tabor, A. H. Bobeck, G. P. Vella-Coleiro and A. Rosencwaig, *Bell Syst. Tech. J.*, **51**, 1427, (1972).
229. J. C. Slonczewski, A. P. Malozemoff and O. Voegeli, *AIP Conf. Proc.*, No. 10, 458, (1973).
230. G. P. Vella-Coleiro, *AIP Conf. Proc.*, No. 10, 424, (1973).
231. A. H. Bobeck, S. L. Blank and H. J. Levinstein, *Bell Syst. Tech. J.*, **51**, 1431, (1972).
232. A. B. Smith, M. Kestigian and W. R. Bekebrede, *AIP Conf. Proc.*, No. 18, 167, (1973).
233. R. Wolfe and J. C. North, *Bell Syst. Tech. J.*, **51**, 1436, (1972).
234. U. F. Gianola, D. H. Smith, A. A. Thiele and L. G. Van Uitert, *I.E.E.E. Trans. on Magnetics*, **MAG-5**, 558, (1969).

235. H. Chang, "Magnetic Bubble Technology: Integrated Circuit Magnetics for Digital Storage and Processing", I.E.E.E. Press, New York, (1975).
236. L. J. Varnerin, *I.E.E.E. Trans. on Magnetics*, **MAG-7**, 404, (1971).
237. L. K. Shick, J. W. Nielsen, A. H. Bobeck, A. J. Kurtzig, P. C. Michaelis and J. P. Reekstin, *App. Phys. Letts.*, **18**, 89, (1971).
238. H. J. Levinstein, S. Licht, R. W. Landorf and S. L. Blank, *App. Phys. Letts.*, **19**, 486, (1971).
239. A. H. Bobeck, E. G. Spencer, L. G. Van Uitert, S. Abraham, R. L. Barns, W. H. Grodkiewicz, R. C. Sherwood, R. H. Schmidt, D. H. Smith and E. M. Walters, *App. Phys. Letts.*, **17**, 131, (1970).
240. P. Chaudhari, J. J. Cuomo and R. J. Gambino, *App. Phys. Letts.*, **22**, 337, (1973).
241. P. Chaudhari and S. R. Herd, *IBM J. Res. and Dev.*, **20**, 102, (1976).
242. R. Hasegawa, *J. Appl. Phys.*, **46**, 5263, (1975).
243. C. H. Bajorek and R. J. Kobiliska, *IBM J. Res. and Dev.*, **20**, 271, (1976).
244. J. W. Schneider, *IBM J. Res. and Dev.*, **19**, 587, (1975).
245. R. Hasegawa and R. C. Taylor, *J. Appl. Phys.*, **46**, 3606, (1975).
246. M. H. Kryder, K. Y. Ahn and J. V. Powers, *I.E.E.E. Trans. on Magnetics*, **MAG-11**, 1145, (1975).
247. A. H. Bobeck, R. F. Fisher, A. J. Perneski, J. P. Remeika and L. G. Van Uitert, *I.E.E.E. Trans. on Magnetics*, **MAG-5**, 544, (1969).
248. J. A. Copeland, J. P. Elward, W. A. Johnson and J. R. Ruch, *J. Appl. Phys.*, **42**, 1266, (1971).
249. A. J. Perneski, *I.E.E.E. Trans. on Magnetics*, **MAG-5**, 554, (1969).
250. I. Danylchuk, *J. Appl. Phys.*, **42**, 1358, (1971).
251. A. H. Bobeck, R. F. Fisher and J. L. Smith, *AIP Conf. Proc.*, No. 5, 45, (1972).
252. P. I. Bonyhard, I. Danylchuk, D. E. Kish and J. L. Smith, *I.E.E.E. Trans. on Magnetics*, **MAG-6**, 447, (1970).
253. P. C. Michaelis and I. Danylchuk, *I.E.E.E. Trans. on Magnetics*, **MAG-7**, 737, (1971).
254. P. C. Michaelis and P. I. Bonyhard, *I.E.E.E. Trans. on Magnetics*, **MAG-9**, 436, (1973).
255. T. J. Nelson, Y. S. Chen and J. E. Geusic, *I.E.E.E. Trans. on Magnetics*, **MAG-9**, 289, (1973).
256. J. L. Smith, D. E. Kish and P. I. Bonyhard, *I.E.E.E. Trans. on Magnetics*, **MAG-9**, 285, (1973).
257. W. Strauss, *J. Appl. Phys.*, **42**, 1251, (1971).
258. G. S. Almasi, *I.E.E.E. Trans. on Magnetics*, **MAG-9**, 663, (1973).
259. K. D. Broadbent, *I.R.E. Trans.*, **EC-9**, 321, (1960).
260. A. C. Tickle, *J. Appl. Phys.*, **35**, 768, (1964).
261. A. C. Tickle, *I.E.R.E. Conf. Proc.*, No. 7, Paper No. 8, (1966).
262. R. J. Spain, *I.E.E.E. Trans. on Magnetics*, **MAG-2**, 347, (1966).
263. R. J. Spain and M. Marino, *I.E.E.E. Trans. on Magnetics*, **MAG-6**, 451, (1970).
264. R. J. Spain, *I.E.E.E. Trans. on Magnetics*, **MAG-3**, 334, (1967).
265. R. J. Spain, H. I. Jauvtis and D. M. Franklin, *I.E.E.E. Trans. on Magnetics*, **MAG-7**, 365, (1971).
266. R. J. Spain and H. I. Jauvtis, *Proc. Intermag Conf.*, Paper 6-4, (1974).

267. C. Batterel and M. Hanaut, *I.E.R.E. Conf. Proc.*, No. 35, 63, (1976).
268. R. J. Spain, M. J. Marino and H. I. Jauvtis, *Spring Joint Comp. Conf. AFIPS Proc.*, **30**, 491, (1967).
269. H. I. Jauvtis and R. J. Spain, *I.E.E.E. Trans. on Magnetics*, **MAG-5**, 537, (1969).
270. D. O. Smith, *I.R.E. Trans.*, **EC-10**, 708, (1961).
271. J. M. Ballantyne, *J. Appl. Phys.*, **33**, 1067S, (1962).
272. L. J. Schwee, H. R. Irons and W. E. Anderson, *I.E.E.E. Trans. on Magnetics*, **MAG-10**, 564, (1974).
273. L. Mayer, *J. Appl. Phys.*, **29**, 1003, (1958).
274. D. Chen and J. D. Zook, *Proc. I.E.E.E.*, **63**, 1207, (1975).
275. C. Chen, *Appl. Opt.*, **13**, 767, (1974).
276. D. Chen, *J. Appl. Phys.*, **42**, 3625, (1971).
277. R. L. Aagard, T. C. Lee and D. Chen, *Appl. Opt.*, **11**, 2133, (1972).
278. R. L. Aagard, F. M. Schmit, T. S. Lui and D. Chen, *I.E.E.E. Trans. on Magnetics*, **MAG-9**, 463, (1973).
279. G. Y. Fan and J. H. Greiner, *J. Appl. Phys.*, **39**, 1216, (1968).
280. K. Y. Ahn, *App. Phys. Letts.*, **17**, 347, (1970).
281. G. Y. Fan and J. H. Greiner, *J. Appl. Phys.*, **41**, 1401, (1970).
282. A. M. Patlach, *IBM J. Res. and Dev.*, **16**, 313, (1972).
283. B. R. Brown, *IBM J. Res. and Dev.*, **16**, 19, (1972).
284. D. Treves, R. P. Hunt and B. Dickey, *J. Appl. Phys.*, **40**, 972, (1969).
285. D. Treves, I. W. Wolf and N. Ballard, *J. Appl. Phys.*, **40**, 976, (1969).
286. R. E. MacDonald and J. W. Beck, *J. Appl. Phys.*, **40**, 1429, (1969).
287. R. L. Comstock and P. H. Lissberger, *J. Appl. Phys.*, **41**, 1397, (1970).
288. B. R. Brown, *Appl. Opt.*, **13**, 761, (1974).
289. A. M. Stoffel and J. Schneuder, *J. Appl. Phys.*, **41**, 1405, (1970).
290. S. Matsushita, K. Sunago and Y. Sakurai, *I.E.E.E. Trans. on Magnetics*, **MAG-11**, 1109, (1975).
291. J-P. Krumme and H. J. Schmitt, *I.E.E.E. Trans. on Magnetics*, **MAG-11**, 1097, (1975).
292. R. Ahrenkiel and T. Coburn, *I.E.E.E. Trans. on Magnetics*, **MAG-11**, 1103, (1975).
293. J. M. Robertson, P. L. Larsen and P. F. Bongers, *I.E.E.E. Trans. on Magnetics*, **MAG-11**, 1112, (1975).
294. G. Y. Fan, *I.E.E.E. Trans. on Magnetics*, **MAG-7**, 590, (1971).
295. M. C. Paul and P. E. Oberg, *J. Appl. Phys.*, **39**, 4783, (1968).
296. S. Koniki, S. Matsushita and Y. Sakurai, *I.E.E.E. Trans. on Magnetics*, **MAG-7**, 179, (1971).
297. R. M. Hornreich, H. Rubinstein and R. J. Spain, *I.E.E.E. Trans. on Magnetics*, **MAG-7**, 29, (1971).
298. H. Rubinstein, R. M. Hornreich, J. Teixeira and E. Cohler, *I.E.E.E. Trans. on Magnetics*, **MAG-6**, 475, (1970).
299. D. W. Dick and D. F. Wayne, *I.E.E.E. Trans. on Magnetics*, **MAG-2**, 343, (1966).
300. J. G. Edwards, *Proc. I.E.E.*, **112**, 40, (1965).
301. I. Williams, *Electronics*, **36**, 62, (1963).
302. R. J. Spain and H. W. Fuller, *J. Appl. Phys.*, **39**, 953, (1968).

303. S. Sugatani, S. Konishi and Y. Sakurai, *I.E.E.E. Trans. on Magnetics*, **MAG-5**, 464, (1969).
304. S. Oshima, T. Watanabe and T. Fukui, *I.E.E.E. Trans. on Magnetics*, **MAG-7**, 436, (1971).
305. P. E. Gise and R. B. Yarborough, *I.E.E.E. Trans. on Magnetics*, **MAG-11**, 1403, (1975).
306. H. R. Irons and L. J. Schwee, *I.E.E.E. Trans. on Magnetics*, **MAG-8**, 524, (1972).
307. C. J. Bader and C. S. DeRenzi, *I.E.E.E. Trans. on Magnetics*, **MAG-10**, 524, (1974).
308. A. R. Johnson, *Proc. Intermag. Conf.*, Paper 11-4, (1965).
309. R. F. Schauer, R. M. Stewart Jnr., A. V. Pohm and A. A. Reid, *I.R.E. Trans.*, **EC-9**, 315, (1960).
310. C. P. Wang, *I.E.E.E. Trans. in Control and Electronics*, **83**, 813, (1964).
311. K. E. Haughton, *Proc. I.E.E.E.*, **63**, 1148, (1975).
312. C. D. Mee, *I.E.E.E. Trans. on Magnetics*, **MAG-12**, 1, (1976).

Chapter 12†

Pyroelectric and Ferroelectric Thin Film Devices

J. C. Burfoot

*Physics Department, Queen Mary College,
London University, London, England*

1. Introduction. 697
2. Ferroelectric Thin Film Fabrication and Properties 698
3. Theories of Spontaneous Polarisation 703
4. Pyroelectrics. 704
   4.1 Measurement. 704
   4.2 Parameters 706
   4.3 Thermal Imaging 718
   4.4 Materials 721
5. Detectors and Imaging Devices 723
   5.1 Commercial Pyroelectric Detectors 723
   5.2 Comparison with Other Detectors 726
6. Thin Film Aspects. 728
   6.1 Theories and Measurement 728
      6.1.1 Vibrations and noise 728
      6.1.2 Polarisation and switching 730
   6.2 Non-homogeneity of Temperature 733
   6.3 Films 733
7. Other Devices 735
   References. 739

1. Introduction

Ferroelectrics and pyroelectrics are dielectrics whose crystal symmetry belongs to one of the ten polar classes. Not all pyroelectrics are ferroelectric. Prominent applications of ferroelectrics at present are in pyroelectric detection, including thermometry and thermal imaging, and a number of optical applications—displays, memories, optical modulators. In some cases, advan-

† For a list of symbols used in this chapter and their definitions see p. xxi.

tages in using the thin film form are shared also by thinned crystal specimens or thinned ceramics; the thinning is achieved by etching or polishing techniques. Commercially, the largest use of ferroelectrics is still in piezoelectric transducers—since ferroelectrics make very good piezoelectrics—and in miniature capacitors. The author suspects that in thin films the pyroelectric detection is the most active of these fields, and consequently he has devoted most of his space to that topic. As the selection is arbitrary and personal, a brief statement will be given next regarding the content and some aspects which have been excluded.

Thinned bulk specimens are not included, both to save space, and because they seemed not to be appropriate in this book, even though they may be of order ten microns thick. After a section on fabrication and a short discussion of fundamentals, Section 4 outlines the principles of pyroelectric detectors without special reference to thin films, and Section 5 briefly mentions and compares some commercial devices. Section 6 then surveys the same ground in its thin film aspects, covering stability, IR effects, dependence on thickness, depolarisation and other difficulties, and the effects of thermal gradients in the specimen. Section 7 concludes with applications as capacitors, piezoelectrics, and a number of optical devices.

2. Ferroelectric Thin Film Fabrication and Properties

Thin dielectric films in general are disappointingly reluctant to exhibit properties similar to those of bulk crystal materials, and this is true of ferroelectric films in particular. There are many thin film fabrication methods, but even more sets of property values; prediction of properties is virtually impossible. However a thin film sometimes approximates the behaviour of bulk polycrystalline ("ceramic") material, presumably because the thin film structure often consists of microcrystallites (grains) embedded in some sort of matrix.

As an example of this, for barium titanate, we found[1] that flash evaporation ("discrete evaporation") from rhenium at 10 to 20 Å per minute onto NaF substrate at 500°C to 580°C gave well-orientated films from 200 Å thick up to 2 microns and beyond, when carried out in oxygen at 5×10^{-5} torr; the crystallite sizes were up to 5000 Å. Double electron-diffraction spots from twinned regions corresponded to the 1% tetragonality characteristic of the bulk crystal. Annealing spoils the orientation except in the thinnest specimens (up to 1500 Å) and when carried out in vacuum.

Another indication of the similarity to polycrystalline properties is that the switching hysteresis loops are quite well formed, and in respect of the non-linear i-V characteristic these films behaved remarkably much as a single crystal might have done. It seems that, for example, space-charge layers at the

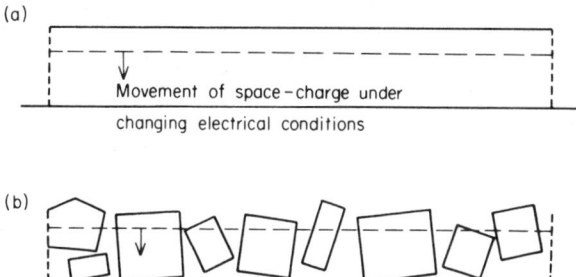

FIGURE 1. Illustrating the suggestion that space-charge movements may be not very different in a thin film (Fig. 1b) from those in a single crystal (Fig. 1a).

surface, under applied voltage, may be regarded as moving coherently in neighbouring grains (Fig. 1), while remaining parallel to the surface, just as they might in an ideal single-crystal film[2]. A domain also (see later) may encompass, and move coherently in, several grains. The i-V parameters have been studied also by Mesnard et al.[3], who reported the variation of conductivity with temperature in thin films of barium titanate to be like that in a ceramic.

There are many other methods of production of thin film ferroelectric specimens other than those referred to above, even to simple melting[4] though such films will be perhaps as much as 100 micron in thickness. But it is difficult to make thin films which exhibit full ferroelectric behaviour. Ferroelectrics and pyroelectrics are in principle dielectrics whose crystal symmetry belongs to one of the ten polar classes[5], though the simplicity of that concept must later be modified for thin films. Both ferroelectrics and pyroelectrics possess an electric moment P_s called "spontaneous", meaning that it does not have to be maintained by an external applied electric field (or by an external applied stress). Pyroelectricity is the variation of P_s with temperature and the pyroelectric coefficient is defined as:

$$P = -\frac{dP_s}{dT} \quad (1)$$

Speaking only of primary pyroelectricity—we shall not discuss secondary pyroelectricity—ferroelectrics comprise that subgroup of the pyroelectrics in which P_s can be reversed (or, in some few cases, merely reoriented) by application, and subsequent removal, of a suitable electric field; that subgroup is not distinguishable by any consideration of symmetry. Symmetry is affected by the presence of domains, and that in turn is affected by poling; domains and poling will be discussed later.

In discussing pyroelectric devices, it is not necessary to insist on ferroelectricity, and indicators other than a hysteresis loop may be acceptable. In fact, it may often be impossible to display a loop in a ferroelectric, due to a high electrical conductivity σ'. The existence of a sharp peak in permittivity ε as a function of temperature T is often quoted as a reliable indication of the existence of a spontaneous electric moment but this is not necessarily correct since such a peak can arise from many other causes. The peak occurs at the ferroelectric transition temperature T_c (Fig. 2), but in thin films there has often been no sharp peak and one has come to accept—as successful—even a fairly mild hump (see Fig. 3). In some cases this may be because the film is in high tensile stress, even as high as $3 \times 10^8 \, \text{N m}^{-2}$[7]. Certainly it is well known that the sharp peak of a crystal material degenerates when the material is made in a ceramic form. In some cases the degeneration is possibly due to the depolarising effects of the crystallite boundaries, which reduce the polarisability. There will usually be a spread of T_c values among the crystallites, due to the differing stresses from grain to grain; this will of course show as a flattening of the peak. There is a requirement for minimum total surface energy which may lower T_c in thinner films[8], but Fig. 3 is just one example showing that a precise overall value of T_c does not exist.

While it has sometimes been possible in bulk ceramics to calculate the intergrain stress effects from a model, the complexion of our thin films will certainly make this impossible. We face the fact that our present knowledge of thin film structures is not sufficient to allow us to predict the effects of a given change in grain structure. Direct measurement of the pyroelectric coefficient p remains necessary, and must be made over a range of temperatures to obtain $p(T)$. When we have also determined the permittivity at these temperatures, $\varepsilon(T)$, it will be possible to give a value to an appropriate figure of merit (see later) for the material in its thin film form, but for comparison with other possible materials, the properties of the substrate available for each particular thin film are influences at least as important, as is the nature of the mountings.

A few fabrication methods will be mentioned as an indication of the possibilities although these films were not grown for device application. The flash evaporation of barium titanate has been mentioned already[1,2]. In 1972 Mesnard et al.[3] grew barium titanate thin films on quartz at 500°C to 600°C; barium titanate films have also been grown by reactive cathodic sputtering of titanium in 50/50 oxygen/argon atmosphere at 10^{-1} torr[9] to give a TiO_2 film on a substrate at 300°C to 500°C. This was followed by evaporation of a calculated amount of BaO from a platinum filament at 1400°C; the composite was then annealed at 1100°C. The thicknesses d were up to one micron. On iridium substrates, these films gave both an ε peak and a hysteresis loop. Earlier attempts had not often given ferroelectric films because of a lack of stoichiometry caused by the fractionation of the barium titanate which

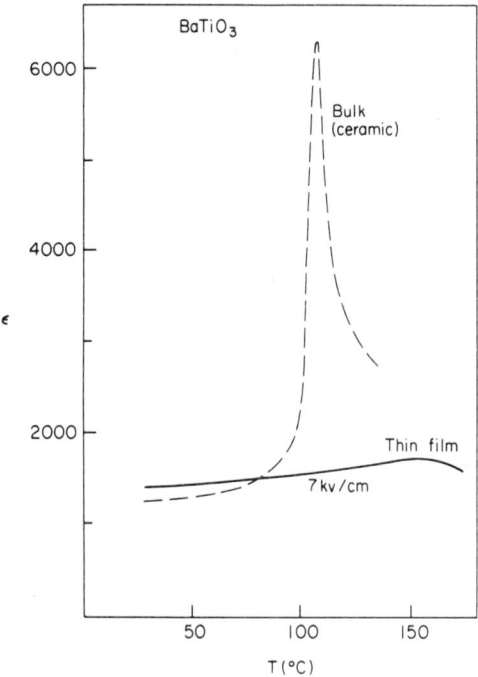

FIGURE 2. Permittivity ε of barium titanate thin film (after[6]), with the corresponding curve of a bulk ceramic specimen for comparison.

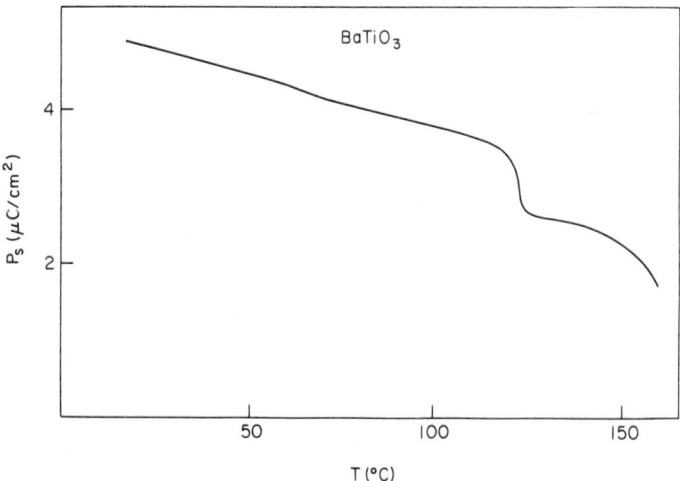

FIGURE 3. Apparent P_s values from the hysteresis loop of a barium titanate thin film[3].

evaporates from the source as two oxides at different rates; flash evaporation may be used to counter this problem. Green[10], and Feldman[11] evaporated the two oxides simultaneously from a tungsten boat. Brown[7] flash evaporated barium titanate to form thin films 0.05 to 0.7 micron thick; these were subsequently annealed in an oxygen atmosphere. A thermodynamic model was developed which accounted for observed hysteresis loop distortions, the flattened $\varepsilon(T)$ peak and a second-order transition. He found an amorphous structure on Pyrex substrates but observed epitaxial growth on sapphire and silicon substrates at about 900°C. The loop was badly distorted by substrate clamping of the film, but its vertical dimension gave a P_s value about $0.12\ \mu C\ cm^{-2}$, by comparison with a bulk value perhaps 200 times larger. $\varepsilon(T)$ had a low broad maximum around 85°C; the value of T_c is 120°C in bulk crystalline material. Absence of anisotropy of refractive index n showed that the polar axis (c) was perpendicular to the surfaces in Brown's films.

Rather thicker films of 10 microns or greater, of ferroelectric ceramics in materials similar to lead zirconate titanate ("PZT") have been made by a technological process which has been called "doctor blading"[12]. Very large areas of film have been made, having well-shaped hysteresis loops and with good uniformity, for use in a matrix electroluminescent display (Section 7).

In 1971, in a search for display devices, Wu[13] R.F.-sputtered bismuth titanate epitaxially on MgO. Segments of the larger 90° domains were detached, annealed and poled, and gave a loop comparable to that in the bulk but with a higher coercivity.

Lawson[14], and Wieder and Collins[15], deposited 9 µm thin films of indium antimonide on glass, and used zone crystallisation to obtain the required structure. Their interest was specifically in the semiconducting properties.

Ling et al.[16-18] grew InSb thin films by flash evaporation, again with a particular interest in the conductivity. They found breakdown under fields of $10^4\ V\ cm^{-1}$.

Triglycine sulphate (TGS) and selenate are well-known pyroelectric materials. TGS/TGSe mixtures have been prepared[19] as 10 to 40 micron thick films by pressing fine powders, the powders being produced either by freeze-drying aqueous solutions, or by settling out from a suspension in acetone. These specimens had an A.C. resistivity similar to that of the single-crystal form, and D.C. resistivity of only two orders less. They exhibited P values of approximately half of the single-crystal values and ε values of about one third of the single-crystal values. In some cases the value of ε is considerably reduced, which is an advantage in pyroelectric detection. (See Section 4.1.)

An early study of the ferroelectric transition known as "I to III" in evaporated KNO_3 layers, gave the thickness dependences shown in Fig. 4.

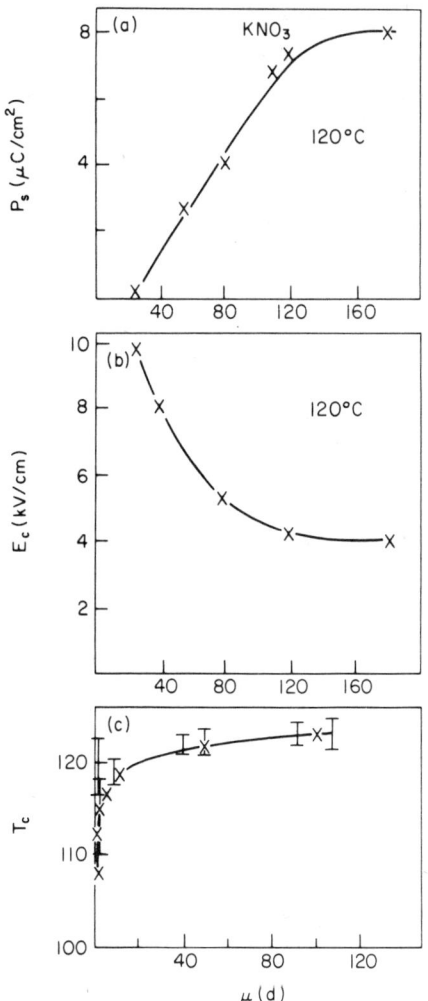

FIGURE 4. Properties of KNO_3 thin film versus thickness[4] at 120°C. (This author— J.C.B.—has replotted Fig. 4c from the logarithmic plot given in ref. 4.)

3. Theories of Spontaneous Polarisation

Sections 3 and 4 give brief indications of basic concepts developed for bulk materials, which will then be further discussed in the thin film context. We may consider a non-polar "prototype" configuration related to the existence of P_s in the polar structure. The existence of P_s is the result of a co-operative phenomenon (at the microscopic level) which causes distortion of the prototype configuration. In terms of lattice vibrations, it has become common to hypothesise an unusually low frequency ω_0 for one or more transverse optic modes of vibration, in the prototype structure. These modes have a dynamic configuration which is closely related to the static crystal configuration of the polarised state. This frequency, ω_0, decreases toward zero as the temperature of the prototype decreases (the mode is called the "soft" mode) and "becomes zero near T_c", that phrase representing, in the harmonic model, what in the real non-harmonic material is the onset of instability in the mode. The simple lattice-vibration theories are harmonic, so they are not able to describe that process, but we suppose that the mode configuration falls over into the static configuration of the polar phase. Progress has been made with more elaborate, anharmonic, theories to fill in the detail.

If we wish to measure the macroscopic polarisation P directly—for example as a means of determining the microscopic dipole moments—we find we are obliged to measure in fact only its changes, under various circumstances, and the best methods do so in a time short compared with the time it takes mobile charge to move through the specimen or through some attached circuit. For although P is defined as the dipole moment per unit *volume*, its effects are indistinguishable from those of a charge q per unit *area* of a plane across the polar direction; $q = \text{div } P$. If in fact for any reason some other charge (q' per unit area) arises in addition, the div P charge will be apparently increased or decreased by q' decreases it, the effect is referred to as "compensation" and this creates a partial "depolarisation". For example, q' may be a mobile charge which may be moving relatively slowly if it has to leak through the pyroelectric material, with time-constant $\varepsilon\rho'$ (where ρ' is the resistivity, $\rho' = 1/\sigma'$), but quickly if it has only to move through an external circuit, as for example a circuit provided to measure a current, as in the next section.

There may be ferroelectric domains within the crystal, and crystallites within any ceramic form. To measure P_s, the material must have been electrically "poled" by a suitable applied electric field so that each crystal or crystallite has become a single domain and so that the polar directions of each crystallite have been aligned in the directions nearest to the poling direction; otherwise it is possible for P_s of the various crystallites and domains to sum to an effective overall P_s value which is zero. We recognise that when a specimen is left standing, P_s may become apparently zero through compensation.

4. Pyroelectrics

4.1 Measurement

When the temperature of a pyroelectric is changed, a pyroelectric current i_p may be derived from a pair of electrodes suitably disposed on its surfaces. The pyroelectric coefficient p may be measured as:

$$p = -\frac{dP}{dT} = \frac{-dP/dt}{dT/dt} = \frac{-i_p/A}{r}$$

when the temperature is increased at a constant rate:

$$r = \frac{dT}{dt} \qquad (2)$$

usually by means of an incident light beam or infrared beam. (A constant rate is rarely achieved, especially as it is usually convenient[20] to use a chopped beam.) Thus the pyroelectric current is given by:

$$-i_p = Apr \qquad (3)$$

Here A is the projected area of the electrodes perpendicular to P_s in the material whose temperature is being changed. A usually refers to the electrode area, or the heated area if that is smaller (Fig. 5). In most cases, we assume that the temperature of the whole specimen is changing.

At earlier dates a common method for measurement of the pyroelectric coefficient was to refer to a knowledge of the variation of P with field E, $P(E)$, rather than to $P(T)$. That is, a hysteresis loop was measured (for example, by the Sawyer–Tower circuit, Fig. 6, or some modification of it), the remanent polarisation P_s was read off the P axis, and the temperature was adjusted before repeating this measurement of P_s. But this technique is only applicable to a pyroelectric which is ferroelectric; and it depends on the naive assumption that the remanent polarisation at a steady temperature, when switched through a hysteresis loop, is necessarily the same as in the very different circumstances of the Chynoweth experiment. These circumstances are more typical of those likely to be experienced by the material in a practical pyroelectric device. A third method—the discharge method—consists of measuring the charge available, q_1, from electrodes of area A on the specimen, when its temperature is increased rapidly from T_1 to well above T_c. q_1 should be equal to AP_{T_1} and the measurement can be repeated using different temperatures $T_1, T_2, \ldots$. Then:

$$p = \frac{1}{A} \mathop{\mathrm{Lt}}_{\Delta \to 0} \frac{q_1 - q_2}{\Delta}$$

where $\Delta \equiv T_1 - T_2$.

All methods are subject to error in the presence of other charge, due to depolarisation or disordering effects, or "embedded" charge (electric effect[23]).

The pyroelectric effect can be used for thermometry, since the pyroelectric current i_p results from a change of temperature, or for detection of a beam able

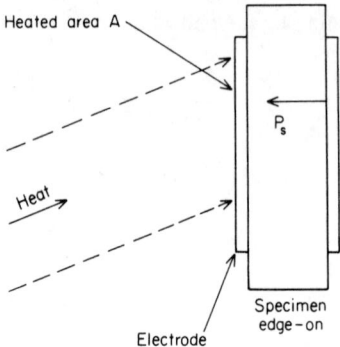

FIGURE 5. Schematic diagram of an incident beam falling on an electroded pyroelectric target. Only the face-electroded case is considered here; the edge-electroded case[21,22] is not suitable for thin films.

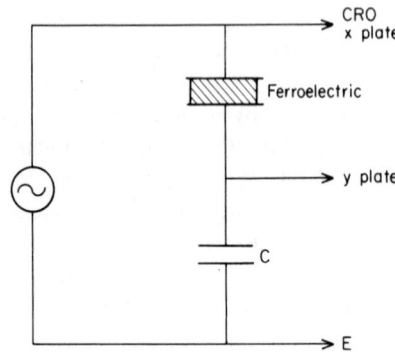

FIGURE 6. Sawyer–Tower circuit for display of the P–E loop of a ferroelectric.

to cause a temperature change in the pyroelectric specimen, or for thermal imaging; see Table 1.

4.2 Parameters

In Figure 5, if all the radiant energy W is absorbed by the pyroelectric target, then:

$$W = rC_T \qquad (4)$$

Table 1
Applications of the Pyroelectric Effect

Measurement of the temperature of a body in contact with the specimen (thermometry)

Measurement or detection of the heating due to a radiant heat beam falling on the target
$\left\{\begin{array}{l}\text{Intensity of IR beam in a spectrometer (broadband detector)}\\ \text{Form of a fast transient, as in a laser pulse}\\ \text{Integrated charge from a shock-pulse signal (total energy detector)}\end{array}\right.$

Measurement or detection of changes in radiant energy across a scene (thermal imaging)

---

where the specimen's thermal capacity is

$$C_T = cAd\rho \tag{5}$$

It has here been assumed that the target has a uniform thickness d. In equation (5), c is the specific heat and ρ is the density. Use of equation (3) then gives:

$$i_p = \frac{pAW}{C_T} \tag{6}$$

which will be quoted in the form of responsivity i_p/W, for example in microamps per watt.

$$\mathscr{R}_i = \frac{pA}{C_T} \tag{7}$$

More commonly, the electrodes on the target specimen will feed into a high-impedance amplifier, and the voltage responsivity (volts per watt, V_p/W) will be quoted.

$$\mathscr{R}_v = \frac{pAR}{C_T} \tag{8}$$

The circuit resistance R is a parallel combination of the amplifier input and the "leakage" of the target itself. We have here used equations (4) and (6) and $V_p = Ri_p$. Notice that equations (6, 7, 8) do not imply that responsivity is necessarily proportional to the area A; for example, if equation (5) holds, those expressions are independent of A, but proportional to $1/d$, and this is the reason for using thin film sensors.

This simple account has taken into consideration little more than the pyroelectric coefficient p and the device geometry. To be realistic we must consider the effect, on this theory, of thermal losses, of electric circuit impedances, of a modulation of the incident beam, of noise from various sources, and of non-homogeneous heating through the thickness of and laterally across the target. Figure 7 represents the simple account; in Fig. 8

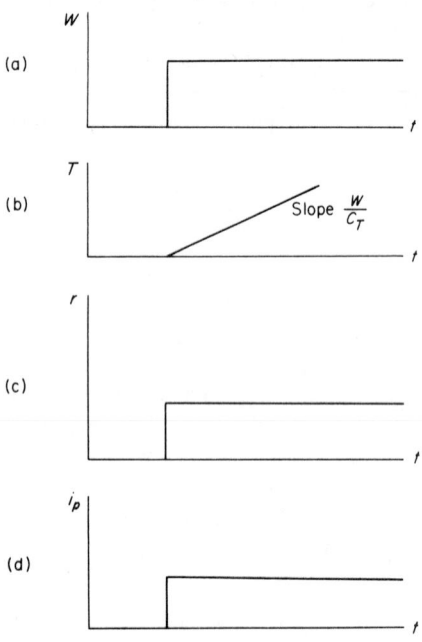

FIGURE 7. The simplest model—no thermal losses and no modulation. In Fig. 7d it has been assumed that the temperature change is never sufficient to make the variation of p with T significant (see Fig. 10).

there is represented also a simple form of thermal loss and a squarewave modulation, produced by a chopper in the incident beam. We have here assumed as an approximation that all thermal loss contributions to the environment are proportional to the temperature θ of excess over ambient:

$$\theta = T - T_a \qquad (9)$$

Even for Stefan radiation, this is reasonable if $\theta \ll T_a$. We have lumped all

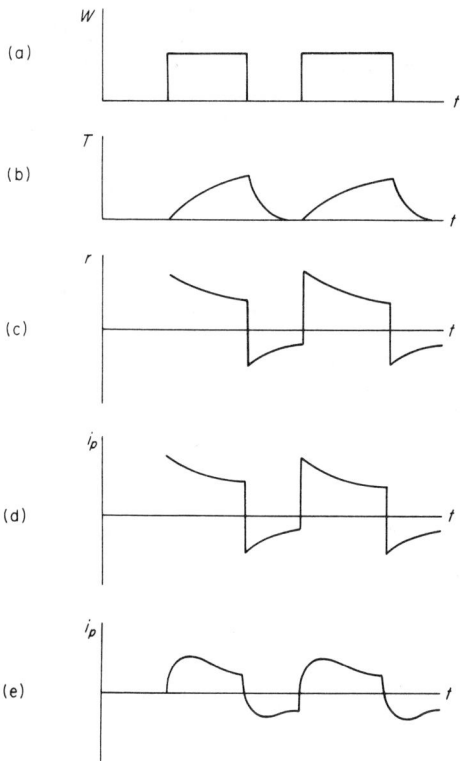

FIGURE 8. Simple model—including loss via a "thermal link" (see text) and "square" modulation; compare Fig. 7. Figure 8b has been drawn to correspond to a time after reaching a steady state under the pulsed heat beam. For Fig. 8e, see equation (12).

contributions together into the simple concept of a thermal link as in Fig. 9. Its thermal resistance is R_T. In Fig. 8b the initial rate of rise of temperature is equal to that in Fig. 7b; the rise is exponential:

$$\theta = WR_T(1 - e^{-t/\tau_T}) \qquad (10)$$

with thermal time-constant given by:

$$\tau_T = R_T C_T \qquad (11)$$

Figure 8e shows a more realistic form of the pyroelectric current which includes the effect of the electronic time-constant of the circuit, viz.:

$$\tau_e = RC \qquad (12)$$

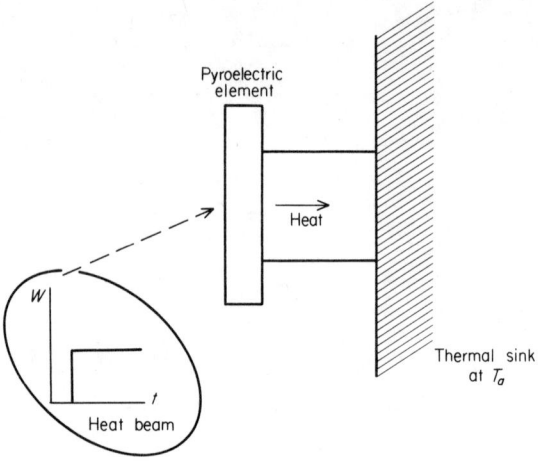

FIGURE 9. The concept of the thermal link, which lumps together thermal output from the pyroelectric target to the surroundings and to the mounting (at temperature T_a). C_T is given by equation (5).

Here C is the sum of the capacity of the target and of the amplifier input, and R is the circuit resistance including the effect of the specimen leakage in parallel. For large temperature excursions θ, it will also be necessary to recognise that the pyroelectric coefficient, p, in equation (6), is not a constant, as assumed when constructing Figs 7 or 8 but is a function of temperature T, as indicated in Fig. 10.

With sinusoidal modulation of the heat beam at angular frequency ω,

$$W = \hat{W}e^{j\omega t} \tag{13}$$

the solution of the temperature equation

$$eW = C_T \frac{d\theta}{dt} + \frac{\theta}{R_T} \tag{14}$$

is

$$\frac{\theta}{eWR_T} = \frac{e^{j\omega t}}{1+j\omega\tau_T} \tag{15}$$

Here the multiplier e (e is less than unity) has been introduced to allow for

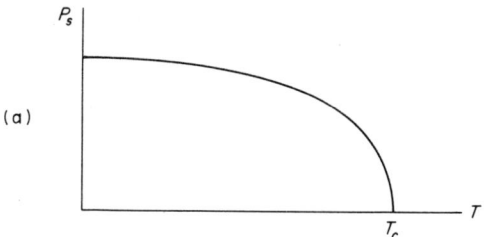

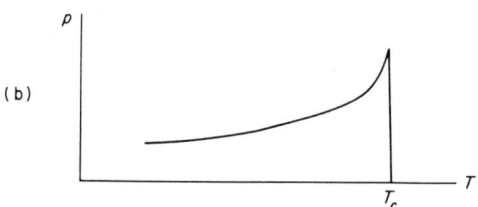

FIGURE 10. The relationship of polarisation P_s and pyroelectric coefficient p, as functions of temperature. (No thermal hysteresis is shown.)

absorption of less than the whole incident energy. The pyroelectric current which results is:

$$i_p = j\omega p A\theta \qquad (16)$$

with θ given by equation (15). Thus at low modulation frequency, $\omega \ll 1/\tau_T$, i_p is proportional to the frequency ω and has $\pi/2$ phase lead on the modulation of temperature; it is in phase with the energy modulation (see Fig. 11).

At high frequency, $\omega \gg 1/\tau_T$, i_p is independent of ω, and in phase with the temperature. In terms of magnitudes, the responsivity is:

$$\frac{i_p}{\dot{W}} = \frac{\omega p A R_T}{1+\omega^2\tau_T^2} \qquad (17)$$

When utilising the voltage signal, as is most commonly done, the equivalent circuit gives responsivity equal to:

$$\frac{\hat{V}_p}{e\hat{W}} = \frac{\omega p A R_T R}{\sqrt{1+\omega^2\tau_T^2}\sqrt{1+\omega^2\tau_e^2}} \qquad (18)$$

With a typical bulk target, τ_T may be of order one second; τ_e will often be less, but it is likely that the modulation frequency used will be appreciably larger than both $1/\tau_T$ and $1/\tau_e$. Then the high-frequency limit of equation (18) is

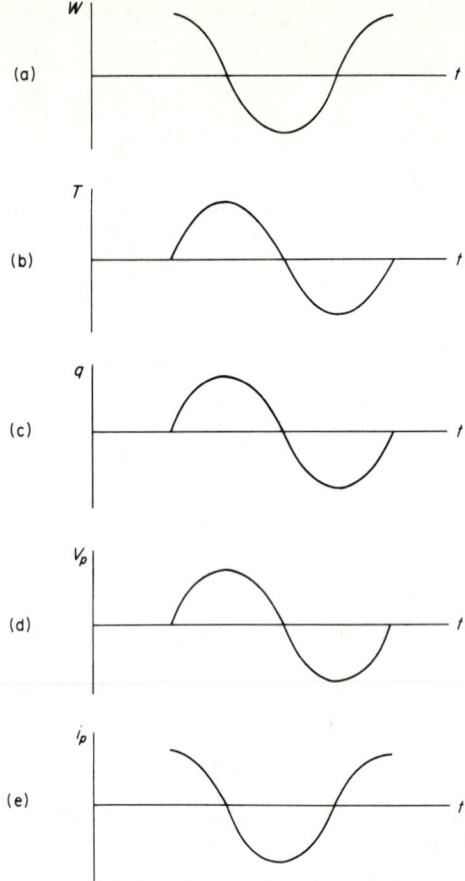

FIGURE 11. Phase relations of incident energy, target temperature, pyroelectric charge q, and pyroelectric voltage and current (highly schematic).

appropriate, that is:

$$\frac{\hat{V}_p}{\hat{W}} = \frac{pA}{\omega C_T C} \tag{19}$$

At low modulation frequencies, both equation (17) and equation (18) show responsivity proportional to frequency ω—a straight line with slope unity on a log–log plot of $\mathscr{R}$ versus ω. Table 2 and Fig. 12 indicate the expectations at other modulation frequencies.

Table 2
Form of Equations (17), (18)

	Low frequency		High frequency
$\mathscr{R}_v/pA$	$\omega R_T R$	$R/C_T{}^a$; R_T/C	$1/\omega C_T C$
$\mathscr{R}_i/pA$	ωR_T		$1/C_T$

^{a} If τ_T, τ_e are well separated, there is a range of moderate frequencies where the expression is valid:

$$\begin{cases} R/C_T & \text{for the case } \tau_T \gg \tau_e \\ R_T/C & \text{for the case } \tau_T \ll \tau_e \end{cases}$$

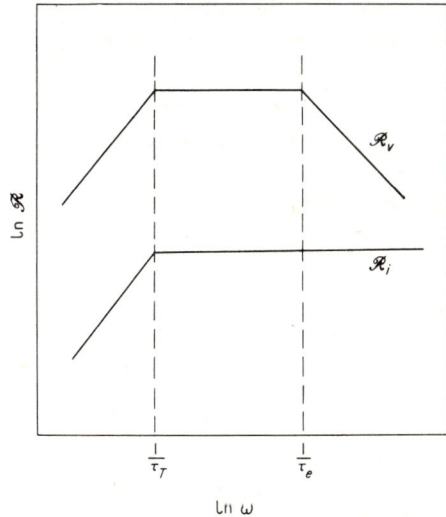

FIGURE 12. Responsivities versus modulation frequency, on a log–log plot (schematic); corresponds to Table 2.

Detectivity of a device is limited by noise from a number of sources. The "noise equivalent power" (NEP) is defined as that incident power which would give the observed noise output signal V_n, such that NEP is $P_n = V_n/(\mathscr{R}_v \sqrt{B})$ where B is the bandwidth within which the noise is measured. It is conventional to take $B = 1$ Hz.

The detectivity D is the reciprocal of the limiting NEP

$$D = \frac{1}{P_n} \tag{20}$$

Detectivity is measured in reciprocal watts, or more precisely, $Hz^{1/2} W^{-1}$.

(If it is known with certainty that the limiting noise is proportional to $\sqrt{A}$, an "area normalised" detectivity is quoted instead, $D^* = \sqrt{A}/P_n$.)

For all detectors, a standard value for comparison of detectivity is that of an "ideal" detector, which is a detector limited only by noise from temperature fluctuations at their minimum, viz. when the thermal link R_T of the detector to the ambient is entirely radiative. At that limit, R_T is given by:

$$\frac{1}{R_T^{\text{limit}}} = 4\sigma e A \theta^3 \tag{21}$$

where σ is Stefan's constant and e the emissivity. We write W_n for noise power in general, and the noise-power squared corresponding to that radiative limit is:

$$W_R^2 = \frac{4kT^2B}{R_T^{\text{limit}}} \tag{22}$$

In that limit, the noise-voltage squared is:

$$\overline{V_n^{2\text{limit}}} = \frac{\mathcal{R}_v^2 W_R^2}{e^2} \tag{23}$$

and the corresponding NEP is:

$$P_n^{\text{limit}} = \frac{V_n}{\mathcal{R}_v \sqrt{B}} \tag{24}$$

(where $V_n \equiv \sqrt{\overline{V_n^2}}$) and is therefore given by:

$$P_n^{\text{limit}} = \frac{W_R}{e\sqrt{B}} \tag{25}$$

with W_R given by the square root of equation (22). The limit is shown in Fig. 20.

The parallel expressions from temperature noise and Johnson noise (Nyquist noise due to voltage fluctuations in any resistor) are given in Table 3.

If amplifier noise sources have to be taken into account, it was once the practice to separate out the grid and anode noise sources, but it has become conventional to separate out those sources which are independent of the circuit, and therefore measurable when the amplifier input is shorted. These are called voltage noise; the remainder are current noise.

Whether it is source 2, 3, 4, or 5, in Table 3, which is the predominant one

Table 3
Noise Power for Various Noise Sources

		W_n
1^a	Radiation noise (ideal detector limit)	$\sqrt{4kT^2/R_T}^{\text{limit}}$
2	Temperature noise	$\sqrt{4kT^2/R_T}$
3	Johnson noise	$\sqrt{4kT\dfrac{c\rho}{p\sqrt{\sigma}}}\sqrt{Ad}$
4	Voltage noise	$\omega\dfrac{c\rho\varepsilon}{p}A\Delta V$
5	Current noise	$\dfrac{c\rho}{p}d\Delta i$

^{a} row 1 is *included* in row 2.

depends on several parameters, such as the detector area desired, the modulation frequency found convenient, and the amplifier characteristics. Therefore it is not useful for us to concentrate attention on one of these sources of noise and neglect the others. Of these many variants, we mention here only one which has certainly been of significance during the development of pyroelectric detectors—viz. the use of modulation frequencies much exceeding $1/\tau_T$ and $1/\tau_e$. This practice has arisen because, although the signal voltage, which depends on $\mathcal{R}_v$ given by equation (19), falls off as $1/\omega$ under those circumstances, the noise voltage also falls off, and quite often in a rather similar fashion. In any case, frequency dependent compensating amplifiers are sometimes used.

From each noise power W_n in Table 3, as in equation (25), the corresponding NEP P_n is given by:

$$P_n^2 = \frac{W_n^2}{e^2 B} \qquad (26)$$

In row 4 of Table 3 the voltage fluctuations ΔV must be measured in the apparatus actually to be used, and at the chosen modulation frequency ω; similarly for the current fluctuations Δ_i in row 5 of the table. Some measurements given by Putley[25] are shown in Fig. 13. For our purposes in this chapter, the dependences on the thicknesses d are the important results, and they are discussed in Section 6.

The dependences on area, A, have been considered in many studies. Figure

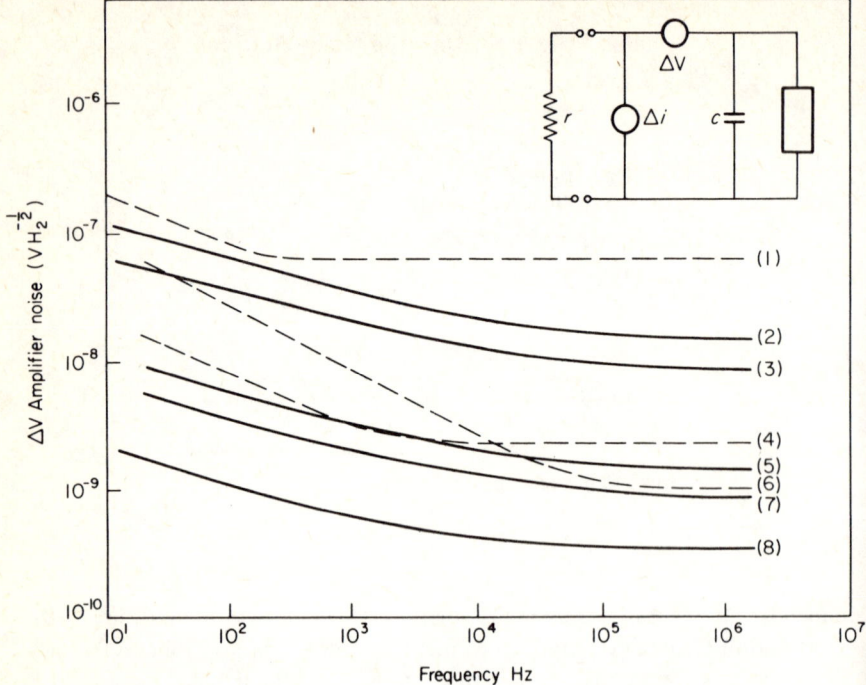

FIGURE 13. Noise characteristics of amplifiers. The frequency dependence of the equivalent current and voltage generators is shown for the XE 5886 miniature electrometer, triode and pentode connected, and for the BFW 11 FET[24].

14 shows data relevant to the circuit named. In this case the conclusion is that the voltage noise is most significant for large areas and high frequency. Johnson noise could only be reduced to a negligible level at thicknesses less than one micron. (If a substrate were then necessary, temperature noise would increase.)

To compare available materials, a figure of merit more useful than p may be needed. We will now mention some figures of merit which have been used.

From equation (18) we see that $\mathscr{R}_v$ is proportional to ωpA multiplied by a factor from which the frequency complication can be separated† only at high

†The implied assumptions, that permittivity ε and resistivity ρ' values are independent of frequency, are not very realistic, because there is usually some dispersion.

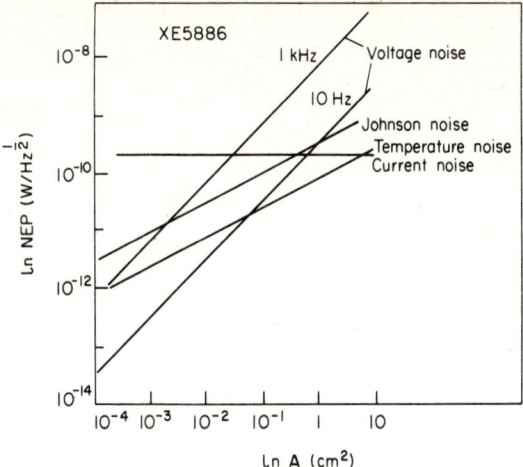

FIGURE 14. NEP as a function of A, calculated for a 10 micron thickness[24] for a circuit of Fig. 13, and for "a material similar to lithium sulphate", at several frequencies. The detector has a high impedance, which keeps Johnson noise low.

frequency, where we have:

$$\mathscr{R}_v \propto p \times \frac{1}{C_T C} \qquad (27a)$$

or at low frequency, where:

$$\mathscr{R}_v \propto p \times R_T R \qquad (27b)$$

The material is used in an electrical and thermal circuit, and a figure of merit for the *material* will be useful only in cases where circuit parameters such as C, R, C_T, R_T are determined by the specimen rather than by the amplifier connected across it. Then $R_T \propto 1/K$ (K is the thermal conductivity), $C_T \propto c\rho$, and $C \propto \varepsilon$, $R \propto \rho'$ if $1/R$ and C much exceed the amplifier input values. The simple high-frequency figure of merit would then be, from equation (27a),

$$F = \frac{p}{c\rho\varepsilon} \qquad (28a)$$

Because of the dispersion referred to in the footnote, it is sometimes thought better to use:

$$F = \frac{p}{c\rho\sqrt{\varepsilon}} \qquad (28b)$$

The latter figure is also rather less dependent on temperature. A figure of merit can also be used to represent the signal-to-noise ratio rather than the responsivity of the detector. Such a modified figure of merit can be selected only in cases where a particular noise source can be considered in isolation. For example, if thermal noise originating in the specimen is the limiting factor, it can be shown to give a noise voltage proportional to:

$$\frac{\sqrt{\varepsilon}}{\tau_e} \times \frac{R}{\sqrt{1+\omega^2\tau_e^2}} \qquad (29)$$

(The first factor is sometimes quoted as $\sqrt{\omega}\sqrt{\varepsilon\tan\delta}$ since $\omega\varepsilon'' = 1/\rho'$, where ρ' is the resistivity.) Then at high frequency $V_n \propto 1/(\varepsilon\sqrt{\rho'})$. That would lead to signal-to-noise proportional to $(p/c\rho\varepsilon) \times \varepsilon\sqrt{\rho'}$, where ρ is the density.

$$\frac{\mathcal{R}_v}{V_n} \propto \frac{p\sqrt{\rho'}}{c\rho} \qquad (30)$$

or some dispersive modification of this. Figures of merit for the thin film pyroelectric vidicon will be discussed in Section 6.

4.3 Thermal Imaging

Thermal imaging devices for military or anti-intruder surveillance and similar needs exist in which a thermal image of the scene is formed, by suitable optical components, on a pyroelectric target. These devices use a crystal or ceramic target. Recently, bulk targets have been used with a thickness of the order of ten microns, but here we shall exclude discussion of these and direct our attention (in Section 6) specifically to thin film targets. The principles already discussed for single-element detectors can be applied; we are now concerned essentially with an extension from the single-element target to an array of targets (or a line of targets). That is, one can use a two-dimensional array of single-element detectors (see Fig. 15a), instead of a two-dimensional scan of the object to be imaged, onto a single detector element (Fig. 15b). Compromises between these two extremes are also known[26]. Problems of uniformity and matching arise with multi-element arrays, and these have been given much attention.

A different technique has also been extensively used. In this method, the pyroelectric charge, generated by changes of temperature at each target element, is not read as a pyroelectric current i_p, but rather by means of an

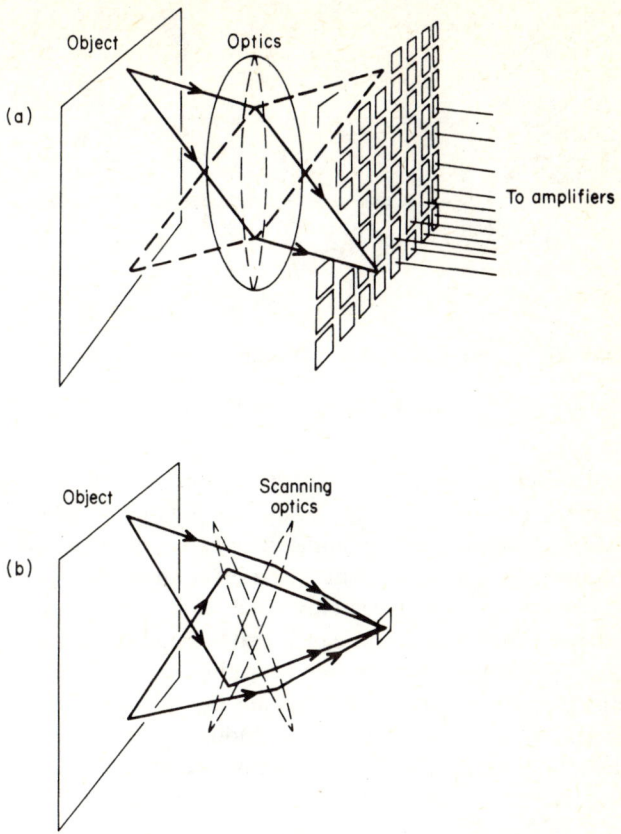

FIGURE 15. Schematic diagrams of two possibilities in thermal imaging. Figure 15a shows an array of detectors, and fixed optics, while Fig. 15b shows a single detector element, and scanning optics.

electron beam which discharges the target elements. It is of course not necessary that the target should consist of discrete elements. This device is called the "pyroelectric vidicon", by analogy with the classical vidicon for imaging by means of the photoelectric effect; the pyroelectric and photoelectric effects are quite different but the configurations of the cameras are very similar (see Fig. 16). By comparison with photoconductive imaging devices, the pyrovidicons are small and cheap and need no refrigeration.

Since the pyroelectric signal depends on a time differential, the incoming beam must be modulated by chopping, or a simple "panning" motion of the

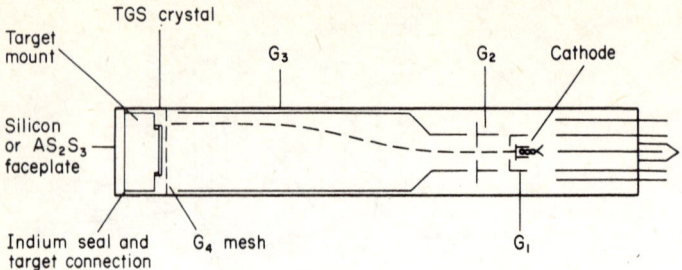

FIGURE 16. Outline diagram of English Electric Valve Co.'s pyroelectric vidicon, from[26].

camera over the scene will suffice. The latter possibility is one reason why the pyroelectric camera can be cheap, but it implies the existence of thermal gradients in the target. The simple theory given above assumes a single value of the temperature θ throughout the target at any instant but it can be extended to allow for the thermal currents which flow in the target when these thermal gradients exist[27,28]. Because of lateral thermal gradients, one of the principal limitations in performance has been due to lateral heat diffusion, and the optimum panning rate, which depends on the thermal diffusivity of the target material, is that rate which is fast enough to prevent lateral diffusion from spoiling the spatial resolution of the image, but which is also slow enough to avoid thermal lag effects from becoming prominent. At low spatial frequencies, of the order of ten lines per millimetre or less, at the target, temperature resolutions are of the order of 1/10 degree. Figure 17 shows some values given by Watton et al.[29]. (We may compare 1 line pair per millimetre to 36 T.V. lines per picture height.)

Thermal currents may also flow longitudinally—that is, through the thickness of the target. As a mathematical model, we may consider a thin target linked to a semi-infinite heat-sink, and if the geometry is such that a one-dimensional analysis is valid, we may consider the pyroelectric voltage generated in an elemental thin layer, and integrate the function over the target thickness[30].

$$V_p = \frac{p}{\varepsilon} \int_0^d T(x)\,dx$$

A simplified version has been given by Van der Ziel and Liu[31].

The pyroelectric vidicon must be electronically "stabilised", and if this is done by redistribution of secondary electrons between the target and the mesh shown in Fig. 16, it is called "anode-potential-stabilisation". In "cathode-

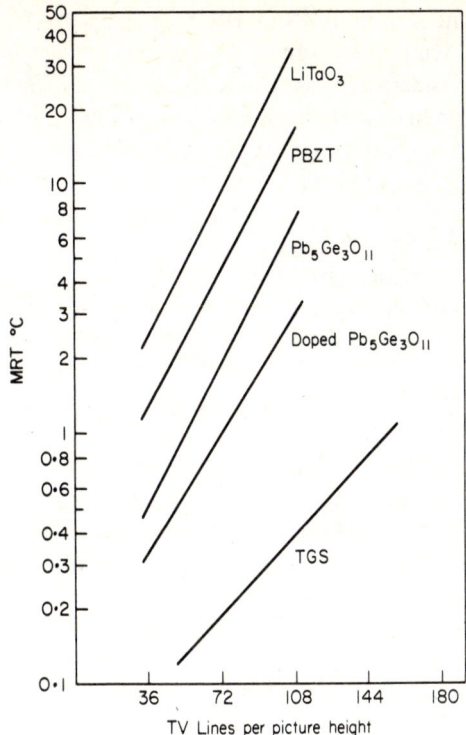

FIGURE 17. Temperature resolution as a function of spatial resolution[29]. (Crown copyright. Reproduced by permission of the Controller of H.M. Stationery Office.)

potential-stabilised" operation, the temperature emission at the target is the principal parameter characterising the electron beam behaviour (see for example[26]).

To remove the positive–negative nature of the pyroelectric signal it is necessary to "lift" the signal by provision of a "pedestal"; the reader may refer to ref. 32 for details of various possible techniques. Noise can arise from the pedestal, depending on the method of its provision, and may be one of the limitations to a good signal-to-noise ratio. The potential advantages of thin film targets are discussed in Section 6.

4.4 Materials

One prominent characteristic of an available pyroelectric material is its value of transition temperature T_c. Above the temperature T_c, the pyroelectric coefficient p is in principle zero (see Fig. 10). A value for T_c which is not too far above the working temperature gives a high value of p, but this may not necessarily be the desideratum, as it may imply a need for excessively fine thermal stability, both in order to have stable operation, and also to avoid risk of accidental depolarisation by an excursion above T_c. A T_c value near the working temperature may also carry with it a high value of the electrical permittivity ε, because both p and ε have peak values around T_c; a high value of p is offset if it is accompanied by high ε.

It was the depolarisation problem which tended to oust crystalline triglycine sulphate (TGS), the classical material for pyroelectric detectors, until modified materials (based on TGS) with locked-in polarisation were found. These modified materials utilised the tendency of TGS to have an internal bias (see Fig. 18), which may be to some extent dependent on the method of crystal growth. The modified material, LATGS, has a L-alanine locking in the polarisation; its values are listed in Table 4. It has low noise and is cheap. We saw that several figures of merit, characterising materials, have been in use, because in general differing uses call for different compromises. Table 4 shows some values of p and of figures of merit. A very complete tabulation, with sources, has been given by Lang[42].

Table 4
Pyroelectric Tabulation

Ferroelectrics	p $(10^{-9} \text{C cm}^{-2} \text{deg}^{-1})$	ε	$10p/\varepsilon$	$p/\sqrt{\varepsilon}$
TGS	30	30	10	6
LATGS	37	23	16	8
$LiNbO_3$	4	32	1	0.7
$LiTaO_3$ ref. 35	19	45	4	3
BSN (33:67) refs. 36, 37	110	1800	0.6	3
$PaTiO_3$	20	160	0.1	2
PLZT (8, 67, 36)	300	5000	0.1	42
PVF_2 refs. 38, 39	3	11	3	1
GNO	5	10	5	1.6
$Bi_4Ti_3O_{12}$	12	140	0.8	1
GASH	14	6	23	6
$PbTiO_3$	40	~100	4	4
TGSe	300	1000	3	10

(*contd.*)

Table 4—contd.

Non-ferroelectrics	$p(10^{-9}\mathrm{C\,cm^{-2}\,deg^{-1}})$
LiSo$_4$	8
LiSoO$_4$	6
H$_4$IO$_3$	3
ZnS	0.03
Tourmaline	0.4
Tartaric acid	3

For abbreviations, see text. In the case of solid solutions, the proportions are given in brackets. The source of much of the data is Keve[34].

From Fig. 10, it is clear that if the transition has in addition a thermal *hysteresis* ΔT, attempts to use the high pyroelectric value near T_c will meet severe difficulties. In the case of some compositions related to PZT, it has been found worthwhile to reduce the hysteresis by suitable dopants, and it was reported[40] that if ΔT is less than 4 degrees, p values up to 70×10^{-9} C cm^{-2} deg^{-1} can be achieved with ε around 350. PZT (at 7% Ti) doped with uranium[41] gave 40×10^{-9} C cm^{-2} deg^{-1} and ε around 180.

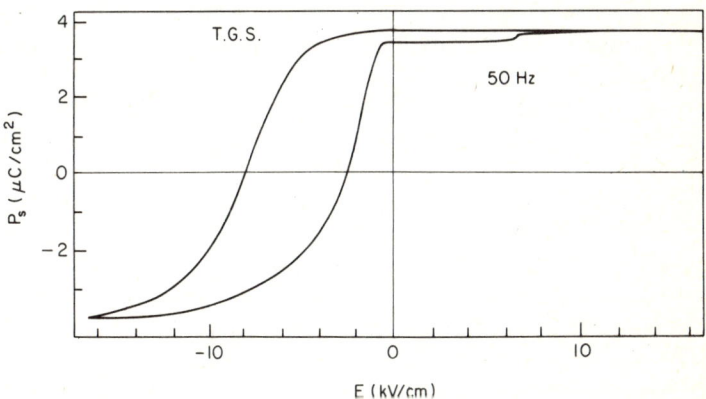

FIGURE 18. Hysteresis loop of TGS with L-alanine substitute[33].

5. Detectors and Imaging Devices

5.1 Commercial Pyroelectric Detectors

This section briefly surveys bulk forms of some of the devices which will be studied in thin film form in the next section. We should examine the wavelength range available, and any limitation of use due to properties other than the basic pyroelectric property, or due to the use of other materials in conjunction with the pyroelectric, such as windows, mountings, etc. We shall consider responsivity and speed of response, as well as the possible disadvantages of ancillary operations such as beam-chopping, or repoling.

Pyroelectric detectors are broadband, ("thermal") devices because the acceptable wavelengths λ are limited only by the ability of the pyroelectric material to absorb the energy of the beam. A pyroelectric detector was in use for microwaves in 1963[43]. Materials used in the half-decade 1970–75 included crystal barium titanate and triglycine sulphate (TGS), hot-pressed ceramic lead–zirconate titanates (PZT) and strontium–barium niobates. Pyroelectric detectors have a tendency to be microphonic because the above materials always have a very high piezoelectric coefficient. In some detectors, material used for windows has limited the usable wavelength values λ, but others have been mounted without windows and were therefore able to handle radiation from the UV out to beyond the far-IR. If the pyroelectric material itself is unable to absorb adequately at very short wavelengths, an absorbing coating a few microns thick may extend that end of the range, though the price paid in sacrificed speed is usually heavy. Also the full responsivity is usable only if speed can be sacrificed; a rise-time of 5 µsec, rather than 5 msec, is achieved by effectively reducing $\mathcal{R}_v$ three orders. Depoling due to the heating effect of the power in the beam itself may sometimes set an upper power limit and one manufacturer in 1972 quoted ½ watt.

With regard to responsivity and other characteristics available in makers' specifications, comparative quotations may not be reliable and we do not therefore present them here. The general characteristics may however be indicated by a few random examples. An early TGS detector with a target 6 mm across gave $220\,\mathrm{V W^{-1}}$ (or an order higher with a much smaller target) and less than 10 mV noise output from the amplifier—NEP $10^{-9}\,\mathrm{W\,Hz^{-1/2}}$. A detector manufactured in 1971 as a radiometer was quoted as having a resolution better than 1/100 degree (compare Fig. 17). One manufacturer's device had a rise-time of 2 nsec. In general, a shorter rise-time is to be expected from a pyroelectric device because it is one of the group of materials whose rise time depends on the *rate* of temperature change.

When modulation chopping is provided, its frequency varies considerably from maker to maker and the device referred to above had chopping available from $\omega = 1$ to 10^5 Hz. Some detectors may require repoling, especially if the chosen target material has low T_c. That is, the poling of the specimens may be lost if the working temperature is close to T_c. Repoling may require application of, for example, 10 V or 450 V and in some cases may also require the detector to be warmed simultaneously. Optics may be available with the device, and Fig. 19 shows those available with a Molectron P3 detector. Lang's excellent sourcebook[42] (page 135), includes references to other pyroelectric uses—UV and x-ray dosimeters, γ-ray dosimetry, measurement of fast atomic beams,

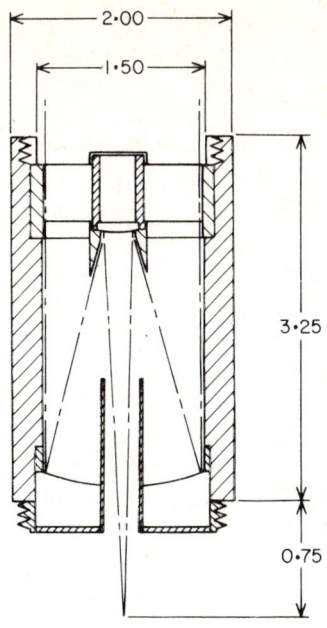

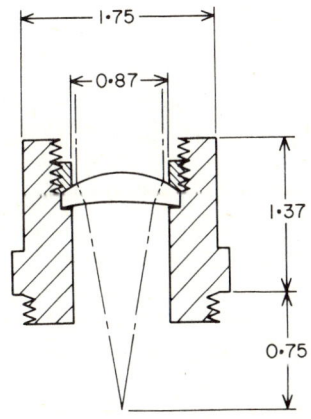
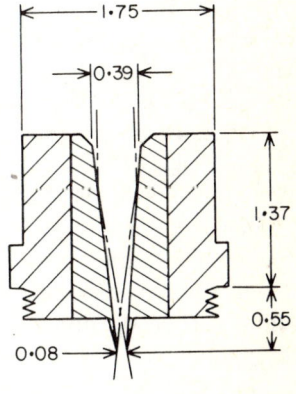

FIGURE 19. The optics available with a P3 pyroelectric detector (Molectron[44]).

large heat fluxes, and pyroelectric thermometers for very small temperature changes (6×10^{-6} degree). Thin film potentialities may exist in some of these areas.

5.2 Comparison with Other Detectors

At a time when the idea of the pyroelectric detector was new, a review of detector properties, with special emphasis on the far-IR, was given by Putley and Martin[45]. They classified their devices into photoconductive detectors, able to detect radiation of wavenumber as low as $1\,\text{cm}^{-1}$, crystal rectifying detectors able to detect radiation of wavenumber as low as $20\,\text{cm}^{-1}$, and "thermal" (broadband) detectors, such as bolometers, having in principle no limit. Some materials, for example germanium, may be usable in more than one class. These authors compared not only the spectral range of the detectors, but also (i) the responsivity, (ii) the detectivity, or its reciprocal the NEP (the "sensitivity"), and (iii) the response time, τ. Among room-temperature detectors, the Golay detector—a thermal detector—has long been very well known for use at wavenumbers above 5 or $10\,\text{cm}^{-1}$ with sensitivity $3 \times 10^{-10}\,\text{W}$. Below these values the crystal detector was used and, for example, the germanium crystal detector has a NEP $= 4 \times 10^{-9}\,\text{W Hz}^{-1/2}$ at $10\,\text{cm}^{-1}$. When greater sensitivity is needed, it was necessary to use cooled detectors, or less conveniently, to narrow the spectral range. For wavenumbers in the range 30 to $1\,\text{cm}^{-1}$, the cooled InSb detector was best, particularly when a rather slow response was acceptable; a germanium bolometer was particularly suitable for wavenumbers above $40\,\text{cm}^{-1}$. Even at the time of the above survey however the authors were able to comment that when the need is for a fast response (short τ), rather than a low NEP, then the pyroelectric detector may be used. Figure 20 summarises the situation nine years afterwards.

Photoconductive detectors can be sensitive *and* fast; and germanium as a photoconductive detector is good at $4\,\text{K}$. The NEP can be $10^{-11}\,\text{W Hz}^{-1/2}$ and τ about 1/10 microsec if used with a magnetic field to enhance its absorption at the higher frequencies. For pulsed signals, the thermal detectors, with high detectivity, may not be very suitable, as they are slow, with τ greater than one millisecond.

Cooled bolometers used with modulation frequency (ω) below $1/\tau$, have responsivity proportional to the temperature coefficient of resistance, α. A germanium bolometer is ten times better than a Golay device, and better than InSb unless rapid speed of response is needed. Superconducting bolometers have very high α, NEP $3 \times 10^{-12}\,\text{W Hz}^{-1/2}$, and τ about 1/20 sec, but they need to be controlled in temperature to perhaps 10^{-5} degree.

Commercial pyroelectric detectors have rise times about one microsecond, by comparison with the Golay's one millisecond, although much shorter time-constants have been claimed. The InSb detector does not respond to radiation

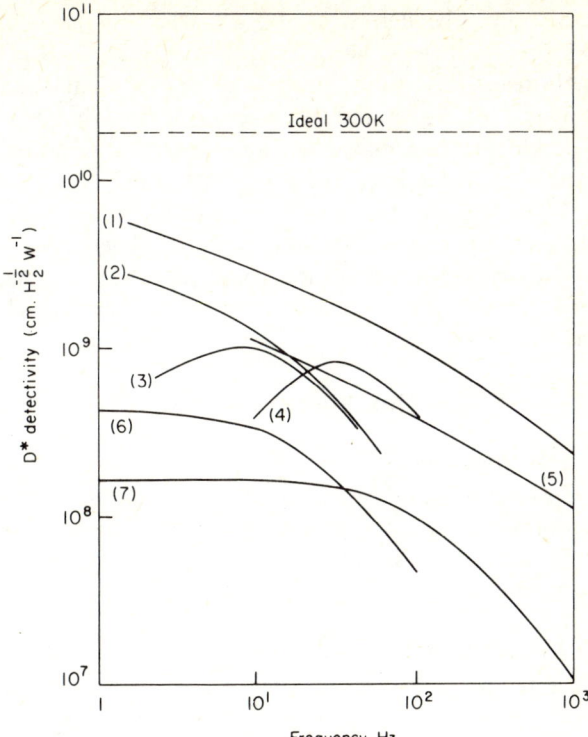

FIGURE 20. The detectivity (1/NEP) of a pyroelectric detector compared to those of a Golay cell, a thermistor bolometer, and room-temperature PbS and InSb detectors (areas given). For comparison, the detectivity of an ideal thermal detector at room temperature is also shown[24].

with wavelength λ exceeding about 7 microns, (wavenumber less than 1400 cm^{-1}), whereas the range of the thermistor bolometer extends to about 35 microns. There is no reason why the pyroelectric detector should not extend to far beyond 100 microns, for it is limited only by its absorbing ability. It has been useful as a detector for the pulsed far-IR from the CN maser at 337 microns. It is the only detector which can be easily used for that.

The pyroelectric detector was classified by Lancaster and Mytton[46] as an active capacitative bolometer, and these authors quoted a thin film Sb_2O_3 passive capacitative bolometer as being the equal of the pyroelectric detector, with the advantages of better reproducibility from element to element, and less subject to microphony. The detectivity D^* at 20 Hz was said to exceed $10^8 \text{ cm Hz}^{1/2} \text{ W}^{-1}$.

A pyromagnetic detector functions in a manner precisely analogous to the pyroelectric detector, and was found to make a fast broadband device[47]. Gadolinium and manganese germanide crystals gave good response even at temperatures rather different from their optimun. Low NEP should be attainable and the impedance of this detector, which is lower than that of pyroelectric detectors, might be a better match to the amplifiers. As an imaging device, it has already been pointed out that, by comparison with photoconductive imaging devices, the pyroelectric detector is small, cheap, and it needs no refrigeration.

6. Thin Film Aspects

6.1 Theories and Measurement

Four topics will be discussed in this section for specimens in thin film form—the effects of that form on theory of vibrations, noise in thin films, the variation of polarisation P_s with thickness, and some observations on hysteresis loops (P–E loops) in thin films.

6.1.1 Vibrations and noise

In the area of dielectric and pyroelectric properties, it is not yet possible to fit thin film behaviour to microscopic models with any degree of confidence. The soft-mode description mentioned in Section 3.1 traditionally was applied to "displacive" types of material (that is, those materials whose microscopic model has a single minimum in their diagram of potential against displacement) and descriptions in terms of microscopic potential "double-wells" were applied to order–disorder types. More recently that distinction has lost some of its clarity and we shall pay it little attention. In any case these models deal principally with reversible (switchable) materials, whereas we have pointed out that not all pyroelectrics are switchable. The soft-mode concept was developed for bulk crystals in which no surfaces or boundaries have to be considered. In the thin film specimens, not only are there two prominent parallel specimen boundaries, but usually also a considerable state of subdivision into crystallites (grains)—with sizes not only small but also variable. It is difficult to approximate realistically to the required modifications of that soft-mode description, for such complicated topographies.

The magnitude of the surface effects can be seen from work with KNO_3[48] in which a ferroelectric thin film has been shown to be still stable at room temperature, though the "bulk" form becomes unstable below 110°C. The effects of surfaces on the vibrations should show in the IR spectral properties, and also in the values of T_c and of ε. The effects to be expected have been calculated, for example⁽⁴⁹⁾, by assuming mode inter-relations based on the "random phase approximation", but the validity of this approach has not been tested adequately against experimental information.

The vibration frequencies are determined by the microscopic force-constants, so that instead of the lattice-vibration approach, direct use can be made of the force-constants, or the potential energy. That is, an approach equivalent to that above should be possible by calculating and comparing the potential energies of the two lattice configurations—viz. that configuration which is stable above T_c, and that configuration which is stable below. This type of calculation also has shown how the static configurations of atom layers at the surface are distorted. For a lattice of identical atoms[50] it was found in one case, for example, that layers of atoms parallel to the surface could be expected to distort by static displacements alternately outward and inward and that these distortions would become unstable at certain values of a force-constant ratio. These authors considered a crystal lattice fixed to a substrate at one face and concluded that a broad thin plate (such as a thin film) attached at a fixed surface, is more stable than one whose lateral dimensions are no greater than its thickness. Such calculations have obvious applicability to our problems, but they are rather too idealised. When more realistic models are available, it may become possible to predict values of pyroelectric coefficients p, and their dependence on the film geometry, from known force-constants.

There are also investigations of the effects on the IR lattice-vibration frequencies in cases where all three dimensions are restricted, as in powders. Much of this work has not been in the area of our interests here, but most thin film specimens do have configurational parameters not very different from those of powders, because the crystallites in the thin film do not usually grow much more, laterally, than they do in the thickness direction. When the particle size s is appreciably smaller than the appropriate mode wavelength Λ, the IR characteristics are almost entirely a function of surface modes determined by the depolarisation field. Both the mode frequency and its damping factor are affected. In cases where s is much smaller than Λ, the particle size ceases to control the spectra; only the particle shape then matters[51]. We pay little attention here to the displacive versus order–disorder distinction mentioned above, but it may be mentioned that the presence of surface affects tunnelling between the two wells in a double-well model (above), causing a distribution of tunnelling activation-energies.

Little work has been done on the thermodynamics of thin film ferroelectrics. Very early, a thermodynamic treatment of epitaxially grown barium titanate thin films, with tensile stresses, was given by Brown[7], based on ref (52). An analysis of KNO_3 thin films was given[4] in which the dielectric parameter T_{CW}, but not T_c, varied with the thickness d. When $\varepsilon(T)$, above T_c, has the

Curie–Weiss form, $\varepsilon \propto 1/(T - T_{CW})$, T_{CW} is defined by that equation. The proposed variation of T_{CW} was then $T_c - T_{CW} = 129 - (230/d)$, where the thickness d is given in microns. This was attributed to "hydrostatic" internal stresses in the thin film.

Now let us look at noise in thin films, first for pyroelectric detectors in general, and then for the rather complicated case of the pyroelectric vidicon. Table 3 allows us to discuss the dependence, on the thickness d, of the signal-to-noise ratio. The table shows that only Johnson noise and current noise will be smaller in a thin film, the noise power squared W_n^2 being proportional to d and to d^2 respectively. In fact, in the (rather unlikely) event that one of the noise sources in Table 3 can be considered in isolation, the table shows that $1/V_n$ must be proportional to d^m, with $m = \frac{1}{2}, -\frac{1}{2}, 0, -1$ respectively, for temperature noise, Johnson noise, voltage noise, and current noise. Then in the high frequency limit, equation (19), the signal-to-noise ratio is proportional to $d^m/C_T C$. Johnson noise-limited performance is then advantageously affected by using thin films, and current noise-limited performance is even more improved. This can be seen most readily by inspecting the special case when the circuit capacitance is dwarfed by the specimen capacitance, so that $C \propto d^{-1}$; in that case $C_T \propto d$, as in Fig. 22, leads to $C_T C$ being independent of d. (If temperature noise is the limit, no advantage will accrue from use of thin films, and similarly at low frequency.) The TGS/TGSe 40 micron films described in Section 2. ref. (20), having low ε and a good figure of merit, gave NEP of 5×10^{-10} W Hz$^{-1/2}$ from an area $\frac{1}{2} \times \frac{1}{2}$ mm^2.

The signal from a pyroelectric vidicon is proportional to p/d, but in a thin film target this might become significantly modified by pedestal noise. We must also bear in mind that the read-out efficiency of the device is proportional to $1/C$ unless C is much smaller than the ratio of frametime to electron beam resistance[32]. The pedestal noise is usually swamped by amplifier noise but it is proportional to the square root of the pedestal current, so $V_n^{(ped)} \propto (\varepsilon/d)^{1/2}$ in the target; if in a thin film d is small enough that that noise source becomes dominant, then p/d as an indicator of expected signal size must be divided by $(\varepsilon/d)^{1/2}$ so that the signal-to-noise ratio tends to be proportional to $p/(\varepsilon d)^{1/2}$. The proportionality of read-out efficiency to $1/C$ means that efficiency is proportional to d, which tends to cancel the thin film advantage. Watton states[33] that the resistance of a cold cathode gun together with ε of order 10 should enable efficient read-out of a 10 micron thick target.

6.1.2 Polarisation and switching

The symmetry class of a specimen should indicate whether it is polar or non-polar. It must be realised however, that for a thin specimen the class is not that of the corresponding bulk crystal. Ten of the 32 symmetry classes are polar and should indicate a pyroelectric crystal, but what ought to be considered is not the symmetry of the unbounded crystal but that of the specimen including its faces[54].

The direct macroscopic depolarising effect of the parallel surfaces perpendicular to P was calculated many years ago[55,56]. Its effect is that the spontaneous polarisation exhibited will be less in thin film specimens than in bulk specimens. At equilibrium it would result, in principle, in zero polarisation below a certain critical thickness—a few microns or less; a critical thickness of d around 200 Å was estimated for barium titanate[55].

That type of calculation deals with a single crystal in thin film format, which is rather far from being a realistic concept. Recent experimental studies[57] of thin film barium titanate less than $\frac{1}{2}$ micron thick have shown that the rate of variation of P_s with d changes markedly at d around 400 Å, and there is a similar effect when the diameter of ordered regions becomes less than 200 Å.

In a material which is not a perfect insulator, space-charge distribution is related to P_s distribution throughout the thickness. Recent computer calculations[58] of the charge distribution have shown that the donor concentration at the electrode critically affects the result. In barium titanate layers thinner than a micron, with A.C. applied, it has been shown[59] that high field and low permittivity occur in the surface layer, and that the effective overall ε should vary with d in a manner which has indeed been experimentally observed[60,61]. Würfel and Batra[62] have shown that the depolarising field due to the surfaces can never be fully cancelled by surface charge.

Polarisation may also vary in time ("ageing") and this occurs when the specimen is not in thermodynamic equilibrium. Two-micron to seven-micron reactively sputtered ferroelectric thin films of "PZBFN"

$$(Pb_{0.92}Bi_{0.07}La_{0.01}Fe_{0.405}Nb_{0.325}Zr_{0.27})O_3$$

with P_s about 10 to 20 $\mu C\, cm^{-2}$, under development for ferroelectric-photoconductive memory (Section 7), were found to decrease their switched polarisation spontaneously[63].

Dielectric losses have been studied; for example flash evaporated thin films of barium titanate have been reported[3] as having electrical conductivity like that of ceramics, and the variation with temperature indicates a small value of activation energy, around $\frac{1}{2}$ eV. But the loss variations with thickness are not known.

In principle the pyroelectric coefficient p can be derived from a study of P_s of the ferroelectric hysteresis loop, Fig. 21a, at a number of temperatures. But the loop is often distorted to a "footprint" (see Fig. 21b) because the material is not an ideal insulator, especially in a thin film specimen. Such a loop may not be easily distinguishable from that of a simple lossy dielectric (Fig. 21c). In any case, because of other effects, consideration of the *values* of the intercepts may not help reliably. For example, the abscissa value E must be deduced (from the

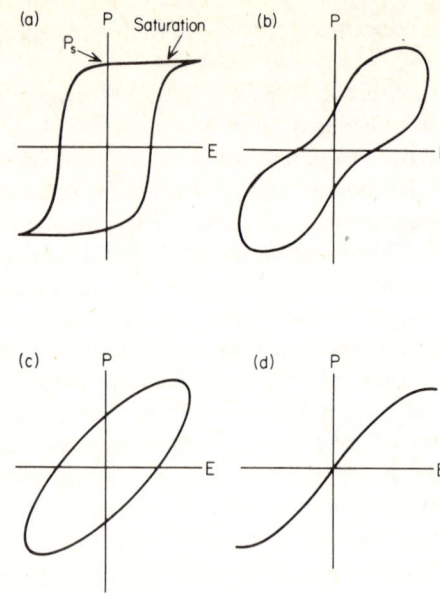

FIGURE 21. Some possible P–E plots (schematic). (a) Hysteresis loop of a ferroelectric. (b) The "footprint" which results from addition of some specimen conductivity. (c) The ellipse given by an ordinary lossy dielectric. (d) The essential dielectric non-linearity underlying Figs 21a and b.

voltage measured) with due allowance for porosity and granularity effects which may not be well known; and the effective thickness may be uncertain. The most significant characteristic of the hysteresis loop is its dielectric non-linearity, Fig. 21d (to be distinguished, of course, from the i-V non-linearity); even if the coercivity is very small, the non-linearity still shows.

In determining the pyroelectric coefficient p we may make direct measurement of polarisation at a number of temperatures. To go further and examine the figure of merit, measurement of the details of the peak, $\varepsilon(T)$, will also be needed—and we have mentioned already the reasons why the $\varepsilon(T)$ peak is usually degenerated to a mild hump in thin film specimens. Compensation by mobile charge (see Section 3) complicates these measurements, and compels us to make measurements in times appreciably less than the time-constant characterising mobile charge. In thin films, that time-constant may be uncomfortably small because the resistances are low and not sufficiently offset by the higher capacitance values of the thin films; their conductance is likely to be greater than $(1/\rho'_{bulk})(d/A)$, especially when d is very small.

6.2 Non-homogeneity of Temperature

The simple equations we have given for pyroelectric devices, based on uniform heating throughout the target, will be especially inappropriate for typical thin films, and improved equations may be very complicated. We can only indicate here their general nature. For example noise (Section 6.1) may be altered by change of specimen dimension, and we shall give a sketch of the effects to be expected which will suffice for general indications but is probably not reliable enough for predictive calculations.

The longitudinal non-homogeneity of heating is particularly relevant for thin films but difficult to calculate. For the incident radiation beam to be detected, if the absorption length for the radiation in the pyroelectric material, x_c, is appreciably less than the thickness d (or if an absorbent layer is made on the front face of the pyroelectric specimen) temperature changes which are directly due to the radiation will be all at the front face, and the temperatures of points in the body of the specimen will vary as a complicated function of depth and time and thermal conductivity and the modulation of the radiation. Such functions, calculated for x_c much less than d, would be more appropriate to thicker specimens than to those which concern us. On the other hand, if x_c is appreciably greater than d, a large part of the incident radiation will be lost but uniform heating may well be a reasonable approximation in predictive speculations; it will be necessary only to remember that the effective responsivity must fall by a factor x_c/d.

Figure 22a shows a choice of thermal element δx, whose parameters are $\delta C_T, \delta R_T$. It can be shown[24], combining such elements, that a sinusoidal variation at the front face is reduced to $1/e$ of its value at a depth of

$$x_c = \sqrt{\frac{2K}{\omega c \rho}} \qquad (31)$$

The typical values $c\rho = 1.5 \times 10^6 \, \text{J m}^{-3} \, \text{deg}^{-1}$; $K = 5 \times 10^{-2} \, \text{W m}^{-1} \, \text{deg}^{-1}$; $\omega = 2\pi 10^4$ (frequency 10 kHz) give $x_c = 3.3$ microns (and varying as $\omega^{-1/2}$ at modulation frequencies other than 10 kHz). This suggests that thin films may have thicknesses neither much greater, nor much smaller, than x_c, but in the very difficult intermediate region. The schematic diagram Fig. 22b does no more than indicate the modification which is called for. But qualitatively it may be seen that there are two effects, viz. a reduction in the effective d of the pyroelectric generator, and an increase in R_T.

6.3 Films

The film used for a pyroelectric detector does not have to be ferroelectric. Sputtered ZnO film has been used[64], 1/10 to 10 microns thick, because typically pyroelectric sensitivity varies inversely as d. It is easy to vary the

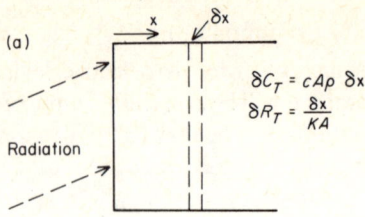

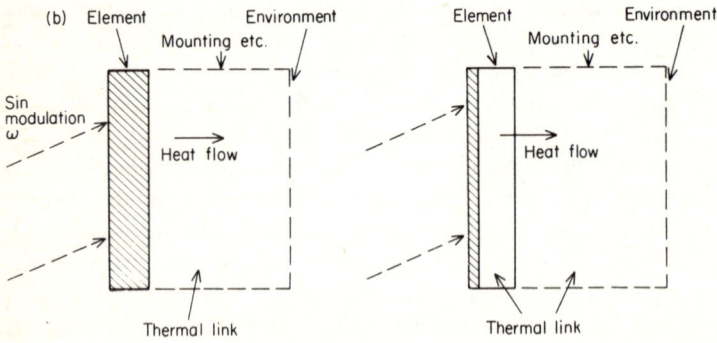

FIGURE 22. (a) The elements used to determine the effect, at various depths x, of a sinusoidal temperature variation at the front face of the target. (b) An indication of the back portion of a pyroelectric target functioning as part of the thermal link to the environment (schematic); compare Fig. 9. The shaded portion shows in each case the effective pyroelectric generator. The two cases illustrated represent a case in which the specimen is effectively uniformly heated, and, on the right, the difference caused by a decrease in x_c, such that $x_c \ll d$.

geometry of a sputtered film to suit convenience, and integration with semiconducting shift registers is being studied. A major current problem is the thermal conductivity of the substrate, if the film is not free-standing. There are also film stresses, originating in the substrate, which are thought to degenerate the response.

Commercially available polyvinylidene fluoride (PVF_2) film[65] has been used as a pyroelectric detector, after stretching and hot poling, which brings out polar properties. The NEP is high, but its cheapness and availability in large sizes, and its geometrical versatility, make it attractive. For example one may make a detector element with an area as large as $10 \, cm^2$, and the responsivity is uniform to 1% over that area. The structure of the material is not yet well understood; it is initially amorphous but may be made as much as half crystalline. When poled, the polarisation can be up to 3 µC cm$^{-2(66)}$. The detectivity, of order 10^8 Hz$^{1/2}$ W^{-1}, is available over a wide range of wavelengths and is still high at liquid nitrogen temperatures. The limiting noise is Johnson noise in the film. Current responsivity is as high as 3 µAW$^{-1(67)}$; $\mathscr{R}_v = 47$ VW$^{-1(38)}$. Peterson et al.[27] have given a model for this material which allows a polarisation value which is non-uniform through the thickness.

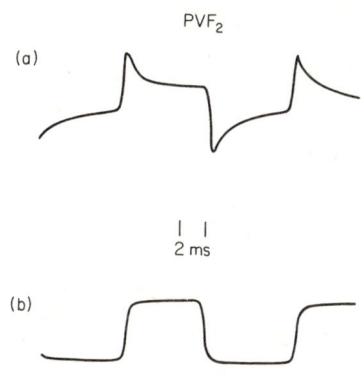

FIGURE 23. Pyroelectric signals from PVF$_2$ detector with CO$_2$ laser beam incident, chopped at 90 Hz. (a) $\tau_T = 2$ msec. (b) $\tau_T = 1$ sec (drawn from[68]).

Poled polyvinylidene fluoride film has been mounted[68] either in total contact with a heat sink which is also the back electrode, or in "drumhead" fashion with only its periphery in contact. The τ_T values were respectively 2 msec and 1 sec and the pyroelectric signals are shown in Fig. 23 when the incident radiation is chopped, with the on-time equal to $5\frac{1}{2}$ msec in each case. In the drumhead mounting, the rate of rise of temperature, $r = dT/dt$, is sensibly constant during the heat pulse. The front electrode was either highly absorbing gold-black or partially transmitting gold film.

The low electrical and thermal conductivities of PVF$_2$ film should make it function well in a pyroelectric vidicon.

7. Other Devices

Since ferroelectrics have "anomalously" high values of many physical properties they often become preferred materials for devices based on some

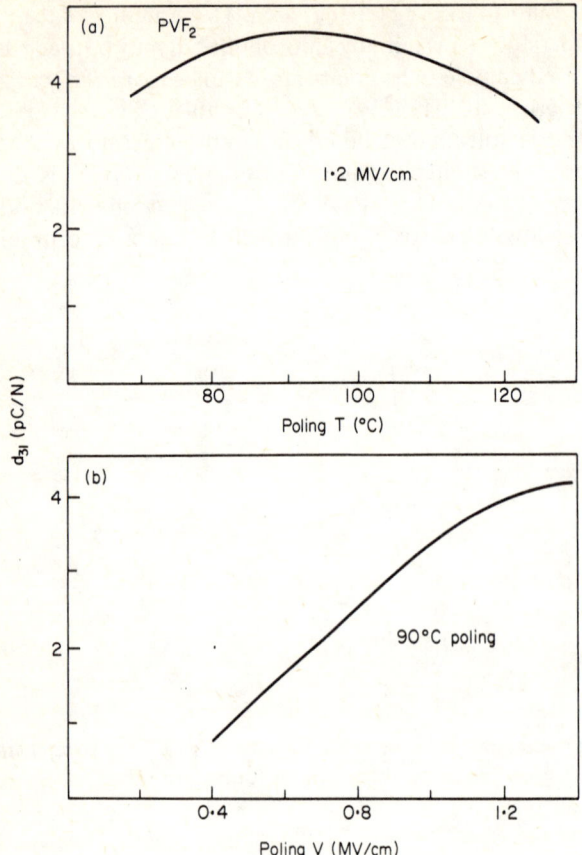

FIGURE 24. Effect of poling conditions on d_{31} of PVF_2 film (drawn from[70]). Effect of varying (a) poling temperature, (b) poling voltage.

property other than ferroelectricity. For example they have high permittivity and piezoelectric coefficients. In this section we discuss therefore capacitative and piezoelectric uses for thin film ferroelectrics, and similarly, applications for their high values of birefringence. In addition an adaptive transistor and optical waveguides are considered.

In 1967 barium titanate thin films less than one micron thick were being made, by R.F. sputtering and subsequent heat treatment. Much of the material was not tetragonal but cubic and the film was therefore only weakly ferroelectric[6] but nevertheless it had a high permittivity value, 1580, and tan δ

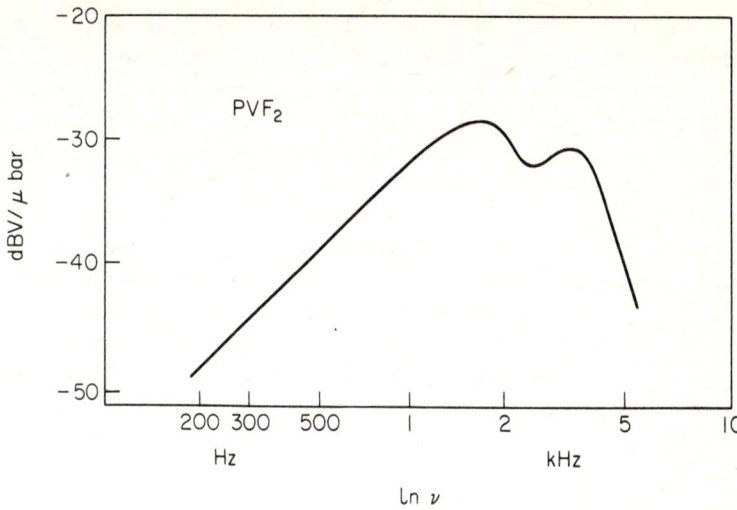

FIGURE 25. Frequency response of PVF_2 microphone (after[70]).

less than 0.04, useful for physically small capacitors. It is well known that the variations with temperature, especially near T_c, and the variations with voltage, may be a disadvantage in these uses. An early review of such films for small capacitors was given in 1969[69]. Bismuth and strontium titanate thin films were later found to be less lossy.

The piezoelectric properties may be exploited in transducers. The piezoelectric coefficient d_{31} of the PVF_2 is about $10 pC N^{-1}$ which is certainly usable, and it is commercially reproducible; the accompanying low elasticity is an advantage. (It is d_{31} which describes the transverse charge—that is, the charge separating through the film thickness—arising from stress applied along the film.) Piezoelectric effects in other plastics had not been permanent. The effects of the poling voltage, and the temperature, on d_{31} are shown in Fig. 24.

One example of the use of PVF_2 in a piezoelectric transducer is in a prototype microphone[70] compatible with P.O. phones, and with intelligibility and signal-to-noise ratio both superior to those of the carbon microphone. The microphone consists of two 25 micron films which are edge-clamped back-to-back. Ageing due to temperature and humidity occurs but the plots are satisfyingly flat (versus time) in the aged condition. Figure 25[70] shows the frequency response of the microphone, which extends into the ultrasound region.

Page composers, displays, and memories may also be made in thin film form. A page composer is a two-dimensional array of light gates, set up in some particular pattern by either electrical or optical means. The whole "page" may then be transferred in one parallel transfer into an optical store, or simply viewed as a display when backed by a suitable light source. Displays may often include also a memory function, though they are not usually required to be changeable as rapidly as a memory. Several thin film types were described in 1975 at the "Symposium on Applications of Ferroelectrics"[32]. An epitaxial layer, with thickness as low as one micron, of NiCl boracite on a Cr–Cl boracite substrate, made by chemical vapour deposition was described[71]. A thin film needs only a very small switching voltage, though in fact a film thicker than about seven microns gives optimum contrast ratio. The polarisation is perpendicular to the surface, and when it is reversed (switched) the optical indicatrix turns 90° around the P_s axis. The optical arrangement is "longitudinal", via transparent electrodes and crossed polars. The birefringence is high with Δn up to 0.03, and the thickness should be chosen to give half-wavelength phase difference. These promising films are to be further developed.

In 1971, sputtered bismuth titanate films on MgO gave birefringence values similar to those in the bulk[72]. Portions could be detached, annealed, and poled to a single-domain condition, and the polarisation was found also to be similar to that in the bulk, $50 \mu C \, cm^{-2}$, but with a rather high coercivity. Sputtered epitaxial bismuth titanate[73] and R. F.-sputtered barium titanate films[74] have been described. In use as a matrix-addressed display device, problems of crosstalk between adjacent elements, and of stability, remain to be tackled. In ref. (73) smaller or shorter switching pulses have been used to give partial switching, allowing a greyscale instead of only "black and white".

Ferroelectric-photoconductive devices have a two-layer construction with transparent electrodes. For example, a scanned light spot may be used to address the device, while a voltage V is applied to the whole sandwich. The applied field falls largely across the ferroelectric layer, and switches it, but only at the sites of those elements where the light spot incident on the photoconductor renders it conducting. At other places, V falls principally across the photoconductor. In some cases the ferroelectric element may be a thin film, in order to reduce the necessary switching voltage to a workable value. In 1972 cases were discussed[75] of films, a few microns thick, of PZBFN65[76] and of bismuth titanate[72], and also a one-micron sputtered ferroelectric film of a bismuth lanthanate derivative of PZT with 0.7 microns of CdSe as photoconductor[77]. This had a strip form for the top semitransparent electrode (about 2000 strips on a 1.27 cm crystal) which allowed $50 \mu sec$ switching. It had a discontinuous gold layer between the ferroelectric and the photoconductor to eliminate space-charge-layer effects[76,78]. "Unswitched" areas of the matrix would normally experience "disturb" voltages about 1/50 of the drive voltage and each would in fact switch partially, to the extent of about 1:1000, thus generating excessive disturb signals. But the device here described had two light beams, one being so arranged as to limit the drive voltage to one row only of the matrix; this substantially reduced disturb signals from areas not addressed.

A ferroelectric memory device previously made with bulk ferroelectrics was subject to instabilities, in time, of its semiconducting film. This is a metal-insulator-semiconductor transistor ("MIST"). It was recently proposed[79] that this device should be constructed with the insulator (gate) replaced by a three-micron ferroelectric film. Its memory and switching characteristics were described[80]. The ferroelectric polarisation controls the surface conductivity of the semiconductor at the interface and so changes the transitor characteristic. Source and drain islands are formed in the surface of bulk silicon by high temperature diffusion. The ferroelectric thin film, R.F.-sputtered bismuth titanate, is then deposited followed by a top metal electrode. Metal contact below the ferroelectric thin film, and on the islands, is made in two windows removed from the ferroelectric film over a portion of the islands. The source-to-drain conduction then differs depending on the state of the ferroelectric switching. The current, plotted against switching volts (gate volts) shows a hysteresis loop. Partial switching gives intermediate values of the source-drain current. P_s must be sufficiently large or the semiconductor carrier density sufficiently low. The device can be used as a variable resistor, the resistance values being preset or, "remembered". This configuration is the basis of a transistor with variable preset threshold voltage—an "adaptive" transistor. The device is fast and stable.

Finally, let us note an interesting amplitude modulator for light which has recently been described[81] and which uses two sputtered optical waveguides formed in the surface of bulk ferroelectric lithium niobate (by sputtering Nb_2O_5). The waveguides are operated as phase modulators for the light, in the legs of a Jamin interferometer, 10.8 volt giving a modulation of one radian in the component phase modulators. The phase modulation uses the electro-optic coefficient r_{33} of the lithium niobate; the propagation constant k of the waveguide is related to the refractive index n of the lithium niobate such that $dk/dn = 0.4$. The frequency characteristic is flat to 750 MHz.

References

1. J. R. Slack and J. C. Burfoot, *Thin Solid Films*, **6**, 233, (1970).
2. J. R. Slack and J. C. Burfoot, *J. Phys., C.* **4**, 898, (1971).
3. G. Mesnard, M. E. Treilleux and G. Metrat, *Thin Solid Films*, **10**, 21, (1972).
4. J. P. Nolta, N. W. Schubring and R. A. Dork, in "Ferroelectricity", ed., E. F. Weller, Elsevier, Amsterdam, (1967).
5. J. F. Nye, "Physical Properties of Crystals", Oxford University Press, Oxford, (1960).

6. R. Vu Huy Dat and C. Baumberger, *Phys. Stat. Sol.*, **22**, K67, (1967).
7. V. R. Brown, PhD dissertation, Univ. of Michigan, University Microfilms, Ann Arbor, Michigan, (1966).
8. N. T. Gladkich, *Thin Solid Films*, **16**, 257, (1973).
9. N. Schwartz, *Trans. 10th Nat. Vac. Symp.*, (1963).
10. J. P. Green, Tech. Mem. ESL-TM-105., M.I.T. Cambridge, (1961).
11. C. Feldman, *Rev. Sc. Inst.*, **26**, 463, (1955).
12. C. Wentworth and G. W. Taylor, *Am. Cer. Soc. Bull.*, **46**, 1186, (1967).
13. S. Y. Wu, *Proc. IEEE Conf. on Ferroelec.*, New York, (1971).
14. A. R. Lawson, *Thin Solid Films*, **12**, 291, (1972).
15. H. H. Wieder and D. A. Collins, *Thin Solid Films*, **20**, 201, (1974).
16. C. H. Ling, J. H. Fisher and J. C. Anderson, *Thin Solid Films*, **14**, 267, (1972).
17. C. H. Ling and J. C. Anderson, *Thin Solid Films*, **15**, 355, (1973).
18. C. H. Ling, *Thin Solid Films*, **16**, 199, (1973).
19. P. W. Whipps and K. L. Bye, *Ferroelectrics*, **7**, 183, (1974).
20. A. G. Chynoweth, *J. Appl. Phys.*, **27**, 78, (1956).
21. A. Hadni, in "Proc. Symp. Submm. Waves", ed., J. Fox, Polytechnic Press, Brooklyn, (1971).
22. A. Shaulov, A. Rosenthal and M. Simhony, *J. Appl. Phys.*, **43**, 4518, (1972).
23. B. Gross, "Charge Storage in Solid Dielectrics", Elsevier, Amsterdam, (1964).
24. E. H. Putley, *Semicond. & Semimet.*, **5**, 259, (1970).
25. E. H. Putley, *Semicond. & Semimet.*, **12**, (1976).
26. E. H. Putley, R. Watton and J. H. Ludlow, *Ferroelectrics*, **3**, 263, (1972).
27. R. L. Peterson, G. W. Day, P. M. Gruzensky and R. J. Phelan (jnr.), *J. Appl. Phys.*, **45**, 3296, (1974).
28. B. R. Holeman, *I. R. Phys.*, **12**, 125, (1972).
29. R. Watton, C. Smith and G. R. Jones, *Ferroelectrics*, **14**, 719, (1976).
30. H. Blackburn and H. C. Wright, *I. R. Phys.*, **10**, 191, (1970).
31. A. Van der Ziel and S. T. Liu, *J. Appl. Phys.*, **43**, 4260, (1972).
32. R. Watton, *Ferroelectrics*, **10**, 91, (1976).
33. E. T. Keve, K. L. Bye, P. W. Whipps and A. D. Annis, *Ferroelectrics*, **3**, 39, (1971).
34. E. T. Keve, At "Inst. Phys. 'Physics in Industry' Seminar: Ferroelectrics and Applications", Edinburgh, March (1972).
35. A. M. Glass and R. C. Abrams, *J. Appl. Phys.*, **41**, 4455, (1970).
36. A. M. Glass, *App. Phys. Letts.*, **13**, 147, (1968).
37. A. M. Glass, *J. Appl. Phys.*, **40**, 4699, (1969).
38. A. M. Glass, J. H. McFee and J. G. Bergman (jnr.), *J. Appl. Phys.*, **42**, 5219, (1971).
39. R. J. Phelan (jnr.), R. J. Mahler and A. R. Cook, *App. Phys. Letts.*, **19**, 337, (1971).
40. D. Appleby, S. G. Porter and F. W. Ainger, *Ferroelectrics*, **14**, 715, (1976).
41. B. Hardiman, C. P. Reeves and R. Zeyfang, *Ferroelectrics*, **12**, 163, (1976).
42. S. B. Lang, "Sourcebook of Pyroelectricity". Gordon and Breach, New York, (1974).
43. D. J. White and H. H. Wieder, *J. Appl. Phys.*, **34**, 2487, (1963).
44. Molectron Corporation, "The P3 Detector", (1972).

45. E. H. Putley and D. H. Martin, In "Spectroscopic Techniques", ed., D. H. Martin, North Holland, Amsterdam, (1967).
46. M. C. Lancaster and R. J. Mytton, *Thin Solid Films*, **13**, 243, (1972).
47. R. M. Walser, R. N. Bené and R. E. Caruthers, *I.E.E.E. Trans. in Electron Devices*, **ED 18**, 309, (1971).
48. J. P. Nolta and N. W. Schubring, *Phys. Rev. Letts.*, **9**, 285, (1962).
49. R. E. Nettleton, *J. Appl. Phys.*, **38**, 2775, (1967).
50. B. C. Clark, R. Herman, D. C. Gazis and R. F. Wallis, in "Ferroelectricity", ed., E. F. Weller, Elsevier, Amsterdam, (1967).
51. T. P. Martin and L. Genzel, *Phys. Rev.*, **B8**, 1630, (1973).
52. P. W. Forsbergh, *Phys. Rev.*, **93**, 686, (1954).
53. R. Watton, private correspondence, (1976).
54. R. Landauer, *J. Chem. Phys.*, **32**, 1784, (1960).
55. I. Ivanchik, *Sov. Sol. St.*, **3**, 3731, (1961), transl. p. 2705.
56. L. P. Kholodenko, *Sov. Sol. St.*, **5**, 897, (1963), transl. p. 600.
57. Yu Ya Tomashpolski and M. A. Sevostianov, *Ferroelectrics*, **13**, 415, (1976).
58. U. T. Höchli and H. Müller, *Ferroelectrics*, **13**, 399, (1976).
59. O. G. Vendik and I. G. Mironenko, *Ferroelectrics*, **9**, 45, (1975).
60. J. C. Burfoot and J. R. Slack, *J. Phys. Soc. Jap.*, **28**, Suppl. 417, (1970).
61. J. H. Pratt, *Proc. I.E.E.E.*, **59**, 1440, (1971).
62. P. Würfel and I. P. Batra, *Ferroelectrics*, **12**, 55, (1976).
63. B. S. Sharma, S. F. Vogel and P. I. Prentky, *Ferroelectrics*, **5**, 69, (1973).
64. C. B. Roundy and R. L. Byer, *Ferroelectrics*, **10**, 215, (1976).
65. Kureha Chemical Company, Nihonbashi, Horidomecho chuo ku, Tokyo, (1971).
66. G. W. Day, C. A. Hamilton, P. N. Gruzensky and R. J. Phelan (jnr.), *Ferroelectrics*, **10**, 99, (1976).
67. G. W. Day, C. A. Hamilton, R. L. Peterson, R. J. Phelan (jnr.) and L. O. Mullen, *App. Phys. Letts.*, **24**, 456, (1974).
68. J. H. McFee, J. G. Bergman and G. R. Crane, *Ferroelectrics*, **3**, 305, (1972).
69. A. E. Feuersanger, in "Thin Film Dielectrics", ed., F. Vratny, Electrochemical Society, New York, (1969).
70. G. M. Garner, In "Newsletter", Allen Clark Research Centre, Plessey Co. Ltd., (1975).
71. H. Schmid, In "Abstracts I.E.E.E. Symp. on Applic. of Ferroel.," (1975), (Paper 10.1).
72. S. Y. Wu, W. J. Takei, M. H. Francombe and S. E. Cummins, *Ferroelectrics*, **3**, 217, (1972).
73. S. Y. Wu, W. J. Takei and M. H. Francombe, *Ferroelectrics*, **10**, 209, (1976).
74. J. K. Park and W. W. Grannemann, *Ferroelectrics*, **10**, 217, (1976).
75. M. H. Francombe, *Ferroelectrics*, **3**, 199, (1972).
76. D. W. Chapman, In "Proc. I.E.E.E. Comp. Gp. Conf.", Washington, (1970).
77. D. W. Chapman and R. R. Mehta, *Ferroelectrics*, **3**, 101, (1972).
78. P. R. Mehta, *Ferroelectrics*, **4**, 5, (1972).
79. S. Y. Wu, *I.E.E.E. Trans. in Electron. Devices*, **ED21**, 499, (1974).
80. S. Y. Wu, *Ferroelectrics*, **11**, 379, (1976).
81. J. C. Webster and F. Zernike, *Ferroelectrics*, **10**, 249, (1976).

Chapter 13†

Superconducting Thin Film Devices

Gordon B. Donaldson
Department of Applied Physics
University of Strathclyde,
Glasgow, Scotland ‡

1. Introduction 744
2. Principles of Superconductivity 745
   2.1 Critical Properties. 745
   2.2 Magnetic Properties 747
   2.3 Fluxoid Quantisation 750
   2.4 The BCS Model 753
   2.5 Quasiparticles and the Energy Gap; Tunnelling 754
   2.6 The Superconducting Wavefunction 759
   2.7 Weak Links—the Josephson Effect 761
   2.8 Applying the Josephson Effects 769
   2.9 Summary. 769
3. Superconductivity in Thin Films. 770
   3.1 Transition Temperatures 770
   3.2 Penetration Depth and Coherence Length 771
   3.3 Energy Gap Anisotropy. 772
   3.4 Critical Magnetic Fields. 772
   3.5 Critical Currents; Ground Planes 775
   3.6 The Proximity Effect 776
4. Devices—General Considerations 777
   4.1 Fabrication 777
   4.2 Tunnelling Devices 779
   4.3 Quasibulk Application of Films 779
   4.4 Conclusion 780

† For a list of symbols used in this chapter and their definitions see p. xxii.
‡ Written during the author's tenure of a Fulbright-Hays Senior Scholarship, while on leave from the University of Lancaster, England, at the Department of Physics, University of California, Berkeley as a guest of the Materials and Molecular Research Division, Lawrence Berkeley Laboratory.

5. Zero Resistance and Magnetic Transition Devices. 780
   5.1 Transmission and Delay Lines 780
   5.2 The D.C. Transformer 781
6. Thermal Devices 784
   6.1 Thermometers and Bolometers 784
   6.2 Particle Detectors 791
7. Switching Devices and Amplifiers 791
   7.1 Cryotrons 792
   7.2 Giaever Tunnelling Switches 801
   7.3 Linear Amplifiers 802
8. Galvanometers and Fluxmeters: SQUIDs 802
   8.1 R.F. SQUIDs 803
   8.2 D.C. SQUIDs 807
   8.3 Application of SQUIDs 810
   8.4 R.F. versus D.C.; Point Contacts and Thin Films 814
9. Microwave and Infrared Energies and Frequencies 818
   9.1 Energy Gap Measurements and Phonon Spectroscopy 818
   9.2 Inelastic Tunnelling: Infrared and Raman Spectroscopy. . . . 820
   9.3 Phonon Generation and Detection 822
   9.4 Other Applications of the Giaever Effect 824
   9.5 The A.C. Josephson Effect 825
10. Fabrication of Tunnel Junctions and Weak Links. 831
    10.1 Tunnel Junctions 832
    10.2 Weak Links 834
    Acknowledgements 835
    Appendix 835
    References 836
    Conclusion Added in Proof 843
    References 843

1. Introduction

Superconductivity is first and foremost characterised by the property of zero resistance[1], and this phenomenon alone has led to some of its best known applications. For example[2] three Tesla superconducting magnets of several metres bore are in regular use, while industry is actively developing resistanceless generators and motors. On the small scale, too, there have been many applications of the perfect conductance property, as we shall see.

In all this, thin films have played their part, but it is also true that many thin film devices utilise much more subtle superconducting properties, such as the flux exclusion demonstrated by Meissner and Ochsenfeld[3] in 1933, the quantum mechanical phase coherence of the electrons forming the superconducting ground state, or the energy gap separating the ground state from its excitations. Before discussing devices themselves, therefore, we must review some experimental and theoretical principles of superconductivity (Section 2). In the space available, this review can hardly be comprehensive, but full accounts can be found in standard textbooks, which are available at all levels (see Appendix).

Special superconducting properties of thin films are treated in Section 3, and devices themselves in Sections 4–9. We concentrate largely on discussing the physical principles associated with each particular type of device and not on covering the many different versions of each which may be available. Fortunately, most topics have generated substantial reviews to which we can direct the reader for exhaustive details and full references.

Particular attention has been given to Josephson effect[4] devices. These are of growing importance in several scientific disciplines, and by their speed and sensitivity have rendered several earlier superconducting instruments defunct. So significant indeed are several recent Josephson effect applications, that we have occasionally felt it necessary to describe systems which have so far only been realised with bulk structures, even though thin film designs would be much more desirable. Interest is so intense that in such cases solutions to the film fabrication problems are likely in the near future.

In view of the material of Chapter 2, we will not normally discuss device preparation, except where the details of film deposition are crucial to the superconducting properties. However, we have included a short section (Section 10) on the preparation of the sandwich-type tunnel structures used in certain Josephson and other devices.

Following most current textbooks and papers, we have normally used mixed Gaussian cgs units, though we have tried not to be too pedantic about this and include practical (mks) units where it aids clarity.

2. Principles of Superconductivity

We first review those experimental and theoretical principles of superconductivity which are pertinent to devices. Initially we consider bulk properties, leaving special features of thin films to Section 3. Further details can be found in the texts listed in the Appendix.

2.1 Critical Properties

Superconductivity has now been observed in many thousand elements, alloys, and compounds[5]. In zero magnetic field the electrical resistance vanishes below a critical temperature (T_c) characteristic of the particular superconductor (see Table 1). The width of the transition (ΔT_c) (see Fig. 1), can be less than 10^{-4} K in well crystallised elemental specimens, but may rise to several tenths of a kelvin in less pure materials and can be even larger in films.

The resistance can be restored by application of a sufficiently large magnetic field. In the case of type I superconductors (see Section 2.2), the critical field at

Table 1
Bulk Critical Data for some Superconductors used in Thin Film Devices

	T_c Kelvin	$H_{cb}(T=0)$ gauss	$H_{c1}(T=0)$ gauss	$H_{c2}(T=0)$ gauss
Nb	9.3	1970	1800	3900
Pb	7.2	803	—	—
V	5.4	1400	800	2680
Ta	4.47	831	—	—
Sn	3.72	305	—	—
In	3.40	282	—	—
Al	1.18	105	—	—
Mo	0.92	94	—	—
Zn	0.88	55	—	—
Ti	0.39	100	—	—
Nn_3Sn	18	—	—	$\geqslant 200\,000$
NbN	15	—	—	$> 200\,000$
$Nb_{0.75}Ti_{0.25}$	9.9	—	300	90 500
PbIn	3.8	390	300	600

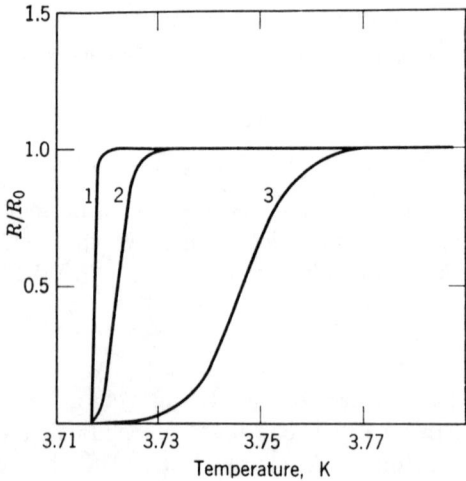

FIGURE 1. Resistance versus temperature curves for tin specimens: (1) pure single crystal, (2) pure polycrystal, and (3) less pure polycrystal. (After Newhouse, Fig. 1, see Appendix.)

temperature T obeys the empirical relation

$$H_c(T) = H_c(0)\{1-(T/T_c)^2\} \quad (1)$$

where $H_c(0)$ is typically 100–1000 gauss (Table 1). The transition to the normal state is first-order, involving the absorption of latent heat $T(H_c/4\pi)(dH_c/dT)$ per unit volume, except at $T = T_c$ and at $T = 0$, when it is second-order.

The resistance is also restored by the passage of a critical current sufficient to generate $H_c(T)$ at the surface of the material.

2.2 Magnetic Properties

In small applied fields a superconductor exhibits virtually perfect diamagnetism, excluding flux from all but a surface region, where persistent screening currents flow and there is a field penetration given closely by an expression like

$$H(x) = H(0)\exp(-x/\lambda) \quad (2)$$

The *penetration depth* λ, typically in the range 50–500 nm, is characteristic of the material, and varies with temperature as

$$\lambda(T) = \lambda(0)\{1-(T/T_c)^4\}^{-1/2} \quad (3)$$

This flux exclusion (Meissner effect)[3] does not depend on specimen history: the magnetisation is the same whether the specimen is cooled through the resistive transition in an applied field or the field is applied after previous cooling in zero field.

The usual magnetostatic laws yield a shape-dependent demagnetisation factor n, which takes account of field distortion around the specimen. In a uniform applied field H_a, the magnetisation is

$$M = -H_a/4\pi(1-n) \quad (4)$$

where $n = \frac{1}{3}$ for a sphere and 0 for a film parallel to the field. For a film of thickness d and typical transverse dimensions $R(\gg d)$ in a perpendicular field, n is nearly unity, and

$$M \simeq -4\pi H_a R/d \quad (5)$$

As H_a increases, the condensation energy is effectively reduced by the energy associated with screening currents and flux exclusion, and it eventually becomes energetically favourable for flux to penetrate the specimen. In this connection we must distinguish two types of behaviour, that of type I superconductors (essentially the superconducting elements except Nb and V), and that of type II materials (all others).

In type I specimens, the simplest behaviour is observed when $n = 0$, because as H_a reaches $H_c(T)$, flux penetrates uniformly and completely, and the normal

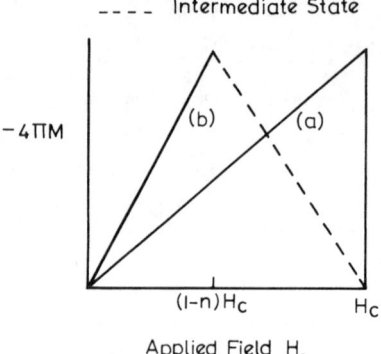

FIGURE 2. Magnetisation curve of type I superconductors: (a) demagnetisation factor $n = 0$; (b) $n \simeq \frac{1}{2}$.

resistance is fully restored (Fig. 2). For $n \neq 0$, the local field at the widest point of the specimen will reach $H_c(T)$ when

$$H_a = H'_c(T) = (1-n)H_c(T) \qquad (6)$$

For $H'_c < H_a < H_c$ the *intermediate state* is established, consisting of approximately laminar regions of superconducting material from which flux is excluded and of normal material in which $H \sim H_c$. The regions are typically microns in width, and as H_a increases, the normal ones grow at the expense of those which are superconducting, while M decreases until the material becomes fully normal at $H_a = H_c$.

Type II superconductors, on the other hand, remain in the Meissner state only up to a field

$$H_a = (1-n)H_{c1} \qquad (7)$$

where $H_{c1}(<H_c)$ is known as the *lower critical field*. Above this, flux penetrates (see Fig. 3), not in laminae, but in tubes (or vortices), each about λ in diameter. The vortices often form a symmetric lattice. Each vortex in this *mixed state* (or *vortex state*) contains the same flux—that is, one flux quantum (see Section 2.3) given by

$$\Phi_0 = hc/2e = 2.07 \times 10^{-7} \text{ gauss-cm}^2 = 2.07 \times 10^{-15} \text{ weber} \qquad (8)$$

Superconductivity persists, with zero resistance possible and with finite magnetisation (Fig. 4), up to an upper critical field H_{c2}, where

$$H_{c2}/H_c = \sqrt{2}\kappa \qquad (9)$$

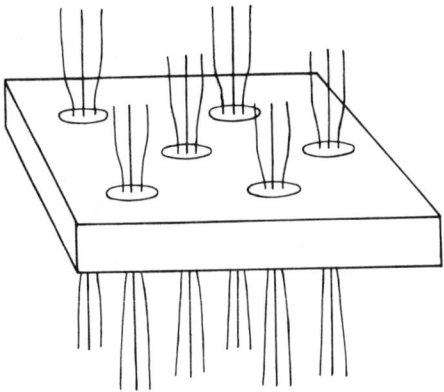

FIGURE 3. Penetration of flux by vortices in the mixed state.

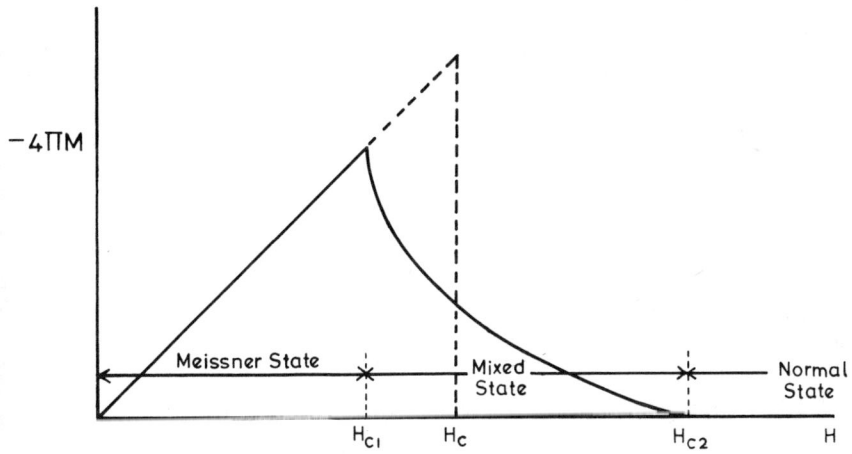

FIGURE 4. Magnetisation curve of ideal type II superconductor, with $n = 0$ (solid line). Dashed line shows type I material for comparison.

κ being a material-dependent parameter ranging from 0.8 to over 100. (A surface sheath of superconductivity can then persist up to $H_{c3} = 1.69\,H_{c2}$ in certain circumstances.) The ratio H_{c1}/H_c is also determined by κ, and

$$H_{c1}/H_c = (1/\sqrt{2}\kappa)\ln \kappa \quad \text{for} \quad \kappa \gg 1 \tag{10}$$

Ideally, both type I and type II magnetisation procedures are reversible. However, in impure material and in specimens with high defect densities, flux can be trapped, or *pinned*. Such pinning is important in preventing dissipation when current is passed through type II superconductors in the mixed state, and is vital in the technology of high field applications[6]. Here, it is sufficient to note that when there is a transport current density J_t in the superconductor, there is a Lorentz force f on each vortex given by

$$f = \frac{1}{c}(J_t \times \Phi_0) \tag{11}$$

where Φ_0 has the direction of the vortex. Alternatively, this may be written as a bulk force

$$F = \frac{1}{c}(J_t \times B) \tag{12}$$

where B is the internal flux. The vortex array remains pinned until F exceeds a critical value, when it starts to move. At velocity v there is a drag force

$$F_d = \eta v \tag{13}$$

where η is due to dissipation in the normal cores of the moving vortices.

When in motion, the vortex array induces, by Faraday's law, an electric field

$$E = \frac{1}{c}(B \times v) \tag{14}$$

Since E is parallel to J_t, it leads to losses commonly described as due to *flux flow resistance*. Clearly, provided the pinning force is great enough to prevent flux flow, the transport of current will be lossless.

2.3 Fluxoid Quantisation

An important experimental property of all superconductors is that of flux quantisation. As a starting point we may describe this approximately by the observation that whenever flux threads a hole in a multiply-connected superconductor (for example a ring), it is quantised in units of the *flux quantum* $\Phi_0 = hc/2e$ (in rationalised units $h/2e$). This is precisely stated as

$$\int_{\text{area defined by } K} B \cdot dS = n\Phi_0 \tag{15}$$

for any curve K lying wholly in the superconductor (see Fig. 5), and whose perimeter has no net current circulation around it.

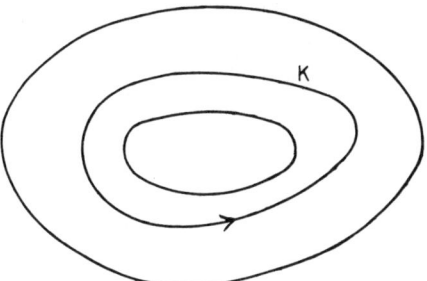

FIGURE 5. Integration path K in multiply-connected superconductor. Fluxoid is quantised within K, and if K is much further than λ from any surface, this is equivalent to quantisation of flux.

A more precise expression of the quantisation property, permitting us to drop the restriction on currents is

$$\oint_K \left(\frac{4\pi\lambda^2}{c} \mathbf{j}(r) + \mathbf{A}(\mathbf{r}) \right) \cdot d\mathbf{l} = \frac{nhc}{2e} = n\Phi_0 \qquad (16)$$

where $\mathbf{j}(\mathbf{r})$ and $\mathbf{A}(\mathbf{r})$ are the current density and the magnetic vector potential at point $\mathbf{r}$. The integral in equation (16) is known as the *fluxoid* of the curve K, and it is strictly this, and not flux, which is quantised. However, if K lies much further than λ from a surface or a vortex axis, the screening currents $\mathbf{j}(\mathbf{r})$ are negligible and equation (16) becomes

$$\oint_K \mathbf{A}(\mathbf{r}) \cdot d\mathbf{l} = n\Phi_0 \qquad (17)$$

Since

$$\mathbf{B} = \operatorname{curl} \mathbf{A} \qquad (18)$$

this transforms to equation (15), showing that flux quantisation is usually an adequate concept: unless the dimensions of the hole or other flux linkage region are comparable to λ, the error in not calculating the full fluxoid is negligible.

Equation (16) gives an insight into the nature of the superconducting condensation. Consider (following London[7]) γ resistanceless electrons per unit volume. Under an electric field, **E**, the acceleration is $\dot{\mathbf{v}}$, where

$$e\mathbf{E} = m\dot{\mathbf{v}} \tag{19}$$

Convert to current densities using

$$d\mathbf{j}/dt = \gamma e\dot{\mathbf{v}} \tag{20}$$

and applying Maxwell's equations (we may ignore $d\mathbf{E}/dt$)

$$\operatorname{curl} \mathbf{E} = (1/c)(d\mathbf{B}/dt) \quad \text{and} \quad \operatorname{curl} \mathbf{H} = 4\pi\mathbf{j}/c \tag{21}$$

we have

$$\operatorname{curl} \operatorname{curl} \mathbf{B} = (4\pi\gamma e^2/mc^2)\dot{\mathbf{B}} \tag{22}$$

The solution of this equation, although it does not give the Meissner effect directly, yields the penetration depth in the form

$$\lambda = (mc^2/4\pi\gamma e^2)^{1/2} \tag{23}$$

and if we insert this in the quantisation condition (equation 16) we have, with equation (20),

$$\int_K (2m\mathbf{v} + 2e\mathbf{A}/c) \cdot d\mathbf{l} = nh \tag{24}$$

The integrand in equation (2.24) is just the canonical quantum mechanical momentum **p** of a particle of charge $2e$ and mass $2m$, so that

$$\int_K \mathbf{p} \cdot d\mathbf{l} = nh \tag{25}$$

If, therefore, we imagine the superconducting carriers to be paired electrons, with some associated wavefunction ψ, equation (25) corresponds to the Bohr–Sommerfeld quantisation condition, and thus to the principle that the phase of the wavefunction is single valued, modulo 2π.

In fact, this treatment is over-simplified, especially in the value of λ derived, but the conclusion is essentially correct: in a superconductor, the current carriers are pairs of electrons condensed into a single quantum mechanical state, whose wavefunction is

$$\psi(\mathbf{r}) = |\psi(\mathbf{r})| e^{i\phi(\mathbf{r})} \tag{26}$$

where the phase $\phi(r)$ is given in the usual wave mechanical way by

$$\nabla\phi = \mathbf{p}/\hbar \tag{27}$$

and is coherent over macroscopic distances. This extensive phase coherence is of great importance in many aspects of superconductivity, and is central to the Josephson effect (Section 2.7).

2.4 The BCS Model

The definitive description of the superconducting state, first given by Bardeen, Cooper and Schrieffer (BCS)[8], is based on the principle of the Cooper pair[9]. Under a weak attractive interaction, caused by the exchange of virtual phonons emitted and absorbed as electrons polarise the positive ionic lattice, a pair of electrons forms a bound state in which their energy is reduced relative to the Fermi sea. The paired electrons optimally occupy states of equal and opposite wavevector and spin (represented by $(\mathbf{k}\uparrow, -\mathbf{k}\downarrow)$) and the virtual phonon exchange successively scatters them into similar states $(\mathbf{k}'\uparrow, -\mathbf{k}'\downarrow)$ and so on.

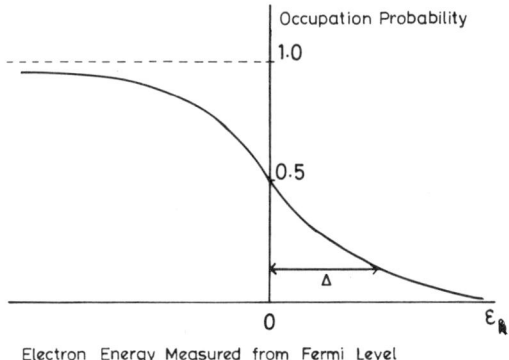

FIGURE 6. Occupancy of electron pair states in BCS ground state $(T = 0)$.

At $T = 0$, all electrons form Cooper pairs in principle, but the binding energy of any given pair depends on the number of empty states into which it may scatter. Since the energy which electrons exchange on scattering is typically $\leqslant \hbar\omega_D$ (ω_D being the Debye frequency) and hence $\ll \varepsilon_F$, the Fermi energy, the necessary empty states can only be made available by a slight smearing of the Fermi–Dirac occupancy function (Fig. 6) over an energy characterised by a parameter Δ. The overall kinetic energy is thus raised slightly relative to the normal state, but the resultant Cooper pairing condensation outweighs this. The functional form of the smearing which yields maximum net binding energy was calculated by BCS, both for $T = 0$, and also for finite temperatures when not all electrons are paired. They find that Δ is k dependent, being given by the integral equation

$$\Delta_{\mathbf{k}}(T) = -\tfrac{1}{2}\sum_{\mathbf{k}'}\frac{V_{\mathbf{k}\mathbf{k}'}\Delta_{\mathbf{k}'}(T)}{[\varepsilon_{\mathbf{k}'}^2+\Delta_{\mathbf{k}}^2(T)]^{1/2}}\tanh\frac{[\varepsilon_{\mathbf{k}'}^2+\Delta_{\mathbf{k}'}^2(T)]^{1/2}}{2k_BT} \quad (28)$$

where k_B is Boltzmann's constant, $\varepsilon_{\mathbf{k}'}$ is electron energy measured relative to the Fermi surface, and $V_{\mathbf{k}\mathbf{k}'}$ is the interaction which scatters pairs from $(\mathbf{k}\uparrow, -\mathbf{k}\downarrow)$ to $(\mathbf{k}'\uparrow, -\mathbf{k}'\downarrow)$. In fact,

$$V_{\mathbf{k}\mathbf{k}'} = V_s + V_{ph} \quad (29)$$

where V_s is a screened Coulomb term and V_{ph} reflects the strength of the electron-phonon coupling $\alpha^2(\omega)$ and the phonon density of states $F(\omega)$. The energy-dependent solutions of Δ contain details of $\alpha^2(\omega)F(\omega)$, which we will discuss further in Section 9. For most purposes, however, we may set

$$\begin{aligned}V_{\mathbf{k}\mathbf{k}'} &= V \quad \text{for} \quad |\varepsilon_{\mathbf{k}}|, |\varepsilon_{\mathbf{k}'}| \quad \text{both} \quad \leqslant \hbar\omega_D \\ &= 0, \quad \text{otherwise}\end{aligned} \quad (30)$$

This yields $\Delta(T)$ independent of energy and

$$\Delta(0) = 2\hbar\omega_D \exp[-1/N(0)V] \quad (31)$$

$N(0)$ being the density of states for one spin at the Fermi surface. The value of $\Delta(0)$ is typically a few meV, and as T increases $\Delta(T)$ decreases (see Fig. 7) reaching zero at T_c, where

$$\Delta(0) = 1.76 k_B T_c \quad (32)$$

The condensed state is highly coherent, as can be inferred by the spatial extent of a typical Cooper pair, which is hv_F/Δ, about 10^{-4} cm. Any given pair overlaps up to 10^6 others, and there is thus substantial momentum ordering between pairs as well as within pairs.

2.5 Quasiparticles and the Energy Gap; Tunnelling

An excitation from the BCS ground state is formed by breaking a pair $(\mathbf{k}\uparrow, -\mathbf{k}\downarrow)$ to form singly occupied states in (say) $\mathbf{k}'\uparrow$ and $-\mathbf{k}\downarrow$. The single occupation of these states blocks them off for scattering purposes and affects the binding energy of all remaining paired electrons. The net energy required is

$$E = E_{\mathbf{k}} + E_{\mathbf{k}'} = (\varepsilon_{\mathbf{k}}^2 + \Delta^2)^{1/2} + (\varepsilon_{\mathbf{k}'}^2 + \Delta^2)^{1/2} \quad (33)$$

Pair breaking thus requires a minimum energy of 2Δ, and can be said to produce two particle-like excitations (quasiparticles) of energy

$$E = (\varepsilon^2 + \Delta^2)^{1/2} \quad (34)$$

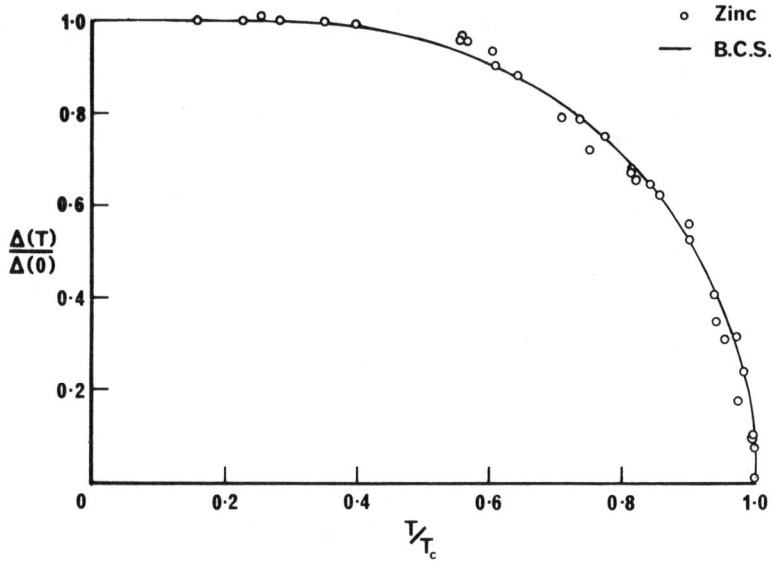

FIGURE 7. Variation of BCS energy gap with temperature, compared with experimental results for zinc (G. B. Donaldson, Proc. 10th Int. Conf. on Low Temp. Physics, 1966, p. 291).

These quasiparticles are long-lived, decaying by recombination with emission of a real phonon of energy 2Δ. Their density of states is

$$g(E) = g_N(\varepsilon)\rho(E) \qquad (35)$$

where $g_N(\varepsilon)$ is just the normal electron density of states, whose variation may normally be neglected over the range of a few times Δ, and $\rho(E)$ is the reduced quasiparticle density (Fig. 8) given in the BCS approximations by

$$\begin{aligned}\rho(E) &= |E|/(E^2 - \Delta^2)^{1/2} & E \geqslant \Delta \\ &= 0 & 0 < E < \Delta\end{aligned} \qquad (36)$$

Notice how the parameter Δ assumes the role of an energy gap in the quasiparticle spectrum. Some typical values of $\Delta(0)$ are given in Table 2.

At finite temperatures, the quasiparticles behave as Fermi–Dirac particles, so that their number is

$$N(E) = g(E)f(E) \qquad (37)$$

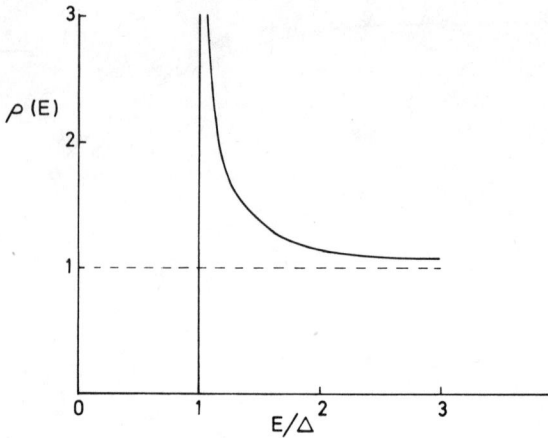

FIGURE 8. Variation of BCS reduced quasiparticle density of states with energy.

Table 2
Energy Gap Data for certain Superconductors (see Parks (Chap. III) or Solymar (Appendix) for Complete Reference List). Typical Measurement Uncertainties are ± 3%

	$\Delta(T=0)$ meV	$2\Delta(0)/kT_c$
Nb	1.52	3.8
Pb	1.33	4.3
V	0.79	3.4
Ta	0.69	3.6
Sn	0.58	3.6
In	0.53	3.6
Al	0.18	3.5
Zn	0.12	3.2

where

$$f(E) = (\exp(E/k_B T) + 1)^{-1} \tag{38}$$

is the Fermi function.

Too direct a comparison should not be made between these excitations and the picture of electrons crossing the energy gap in a semiconductor. For example, in the superconductor there is no valence band as such, and the quasiparticles do not have fixed quantised charges as with simple hole or electron excitations. However, the effects of equation (37) *can* be directly observed in the exponential decrease with temperature of the electronic specific heat, and of the attenuation of ultrasonic waves by quasiparticle scattering. An even more powerful demonstration of the effects of quasiparticles is seen in certain aspects of quantum mechanical tunnelling, a process in

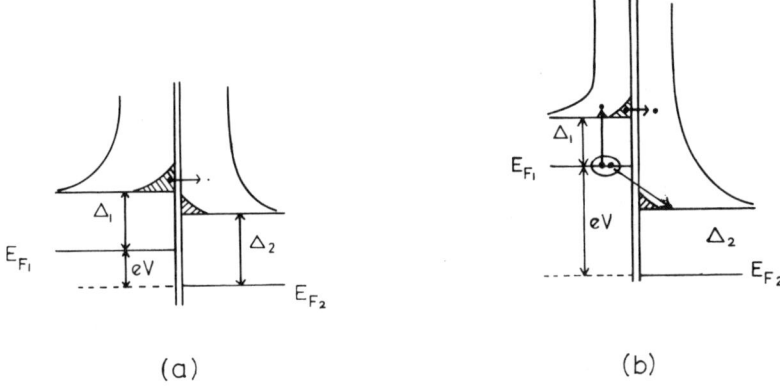

(a)   (b)

FIGURE 9. Simplified "semiconductor" model of tunnelling between non-identical superconductors ($T > 0$): (a) $(\Delta_2 - \Delta_1)/e < V < (\Delta_2 + \Delta_1)/e$, (b) $V > (\Delta_2 + \Delta_1)/e$.

which electrons pass between two metal electrodes separated by a thin barrier, typically about 20 Å thick. Two types of tunnel effect are possible with superconducting electrodes. The first involves coupling of the phases ϕ (equation 26) in the two electrodes, and gives rise to the Josephson effect[4]; however, since Josephson tunnelling is only one of a more general class of effects we will defer discussion of it to Section 2.7. The other, Giaever tunnelling[10], involves quasiparticles, and it can be shown, using models similar to the one in Fig. 9, that the net tunnelling current under a bias voltage V is

$$I = K \int_0^\infty \rho_1(E)\rho_2(E-eV)\{f(E-eV) - f(E)\}\, dE \quad (39)$$

where the subscripts refer to the respective electrodes, and K is the (ohmic) conductance of the junction at high bias or when the electrodes are normal. The $I-V$ characteristics depend in detail on whether one, two identical, or two different superconducting electrodes are used, but are (Fig. 10) highly non-linear, and reflect the strong variations in the densities of states. For example, if two identical superconductors are used (S-I-S tunnelling), I has a discontinuity

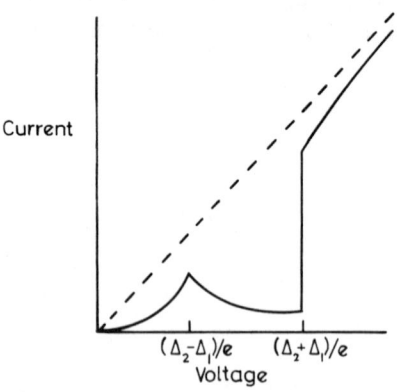

FIGURE 10. Current-voltage characteristic for model of Fig.9.

at $V = 2\Delta(T)/e$, and this structure has been of great value in the measurement of the energy gap in many superconductors. On the other hand, if one electrode is normal (S-I-M), we may apply equation (39) by writing $\rho_2(E) = 1$, and obtain the differential conductance in the form

$$(dI/dV) = -K \int_0^\infty \rho(E) f'(E - eV) dE \qquad (40)$$

Since $f'(E - eV)$ becomes a delta function at $T = 0$, we have

$$(dI/dV) \to K\rho(V) \quad \text{as} \quad T \to 0 \qquad (41)$$

so that at sufficiently low temperatures the differential conductance of an S-I-M junction is a direct measure of the reduced quasiparticle density of states in S. This result is valid with slight modifications even when $\rho(E)$ does not have the simple BCS behaviour.

2.6 The Superconducting Wavefunction

We saw in Section 2.3 that fluxoid quantisation was consistent with the principle of a macroscopic electronic wavefunction

$$\psi(\mathbf{r}) = |\psi(\mathbf{r})|e^{i\phi(\mathbf{r})} \quad (42)$$

In 1953, Ginzburg and Landau (GL)[11], in a theory which remains a powerful research tool to this day, combined the concept of $\psi(\mathbf{r})$ as a wavefunction with a treatment of it as an order parameter. They expressed the free energy of a superconductor in a magnetic field as a series expansion in ψ^2 and added a term to describe the kinetic energy of screening currents. This is proportional to p^2, where $\mathbf{p}$ is the momentum of the carriers. But, since ψ is a wavefunction, it is consistent to use the canonical momentum expression

$$p^2 = |\hbar\nabla\psi - (2e/c)\mathbf{A}\psi|^2 \quad (43)$$

the factor 2 appearing because the carriers are pairs. GL minimised the free energy with respect to variations in the field and in $\psi(r)$, for various boundary conditions such as superconductor-vacuum and superconductor-normal interfaces. A penetration depth

$$\lambda = [mc^2/16e^2|\psi|^2]^{1/2} \quad (44)$$

is found, which is essentially the London value (equation 23) with $|\psi|^2$ representing the superconducting pair density and e replaced by $2e$. However, a new length ξ, known as the *coherence length*, also emerges: this is a consequence of the $\nabla\psi$ term in equation (43), which imposes a free energy price on rapid spatial variations in ψ. This length, ranging from 100–15000 Å in various materials, is the minimum distance over which substantial changes in $|\psi|^2$ are possible. Gor'kov[12] showed that $|\psi|^2$ was in fact proportional to the gap parameter of Δ of BCS theory and that in pure material

$$\xi = \hbar v_F/\pi\Delta \quad (45)$$

$$= 0.18\, \hbar v_F/k_B T_c \quad (46)$$

which is just the size of a Cooper pair (Section 2.4).

The quantity κ of equation (9) also emerges from GL theory, and in a somewhat generalised way. It is given by

$$\kappa = \lambda/\xi \quad (47)$$

and its size determines the magnitude and sign of the wall energy of a superconducting-normal interface, and thus the behaviour of the specimen when it is penetrated by flux. For $\kappa < 1/\sqrt{2}$ the material is type I, while it is type II for $\kappa > 1/\sqrt{2}$.

Detailed GL solutions for the spatial variations of ψ and H show that an isolated vortex (Section 2.2) consists (Fig. 11) not only of a region $\sim \lambda$ in diameter over which fields and circulating currents are significant, but also of a smaller region (about ξ in radius) over which the energy gap (order parameter) is substantially reduced from its equilibrium value. A vortex is therefore roughly equivalent to a cylinder of normal metal ξ in radius.

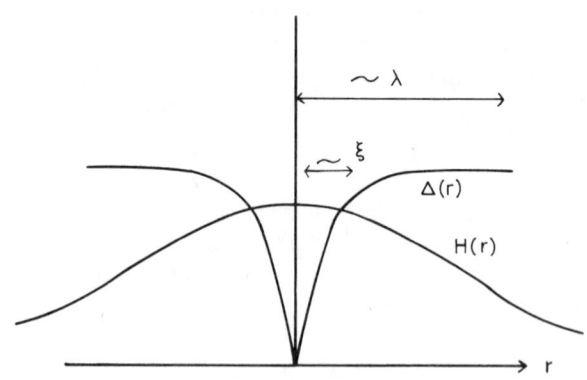

FIGURE 11. Local field and energy gap structure in an isolated vortex.

A substantial modification to GL theory is required, because it assumes the *local* electrodynamics implicit in equations (19) to (22). These lead, with equation (18), to

$$\mathbf{j}(\mathbf{r}) = -(ne^2/mc)\mathbf{A}(\mathbf{r}) \qquad (48)$$

whereas the strong correlation of condensed electrons over the range ξ in fact requires that $\mathbf{j}(\mathbf{r})$ be determined by values of $\mathbf{A}(\mathbf{r}')$ at points throughout the volume $|\mathbf{r}' - \mathbf{r}| < \xi$. Since a magnetic quantity like $\mathbf{A}$ can vary strongly on a scale of λ, it is not valid simply to set $\mathbf{A}(\mathbf{r}') = \mathbf{A}(\mathbf{r})$, at least in type I cases ($\xi > \lambda$). The necessary non-local theory was given by Pippard[13] who showed that the effect was to make the penetration depth larger than the London value λ_L (equation 23). He found

$$\lambda = \lambda_L (\tfrac{3}{4}\pi)^{1/6} (\xi_0/\lambda_L)^{1/3} \qquad (49)$$

where ξ_0 is the coherence length in the pure metal. When $\xi \ll \lambda$ (type II), local electrodynamics apply and the measured λ equals λ_L. Typical values of ξ and λ are shown in Table 3.

Table 3
Coherence Lengths, Penetration Depths, and Ginzburg-Landau Parameters of some Superconductors

	ξ/nm	λ/nm	κ
Al	1600	51	0.03
In	260	51	0.1
Sn	230	51	0.22
Pb	83	39	0.45
Ta	100	50	0.5
Nb	38	47	1.1
Nb$_3$Sn	3.1	75	25

Pippard's approach deals also with the case of impure superconductors of finite mean free path l. The results (slightly simplified) are

$$\text{Clean limit } l \gg \xi: \quad \frac{1}{\xi} = \frac{1}{\xi_0} + \frac{1}{l} \quad (50)$$

$$\lambda = \lambda_0 \quad (51)$$

$$\text{Dirty limit } l \ll \xi: \quad \xi = (\xi_0 l)^{1/2} \quad (52)$$

$$\lambda = \lambda_L (\xi_0/\xi)^{1/2} \quad (53)$$

We discuss the case of thin films in Section 3.2.

2.7 Weak Links—the Josephson Effect

If we rearrange the integrand in equation (24)

$$\mathbf{p} = 2m\mathbf{v} + 2e\mathbf{A}/c \quad (54)$$

substituting $\mathbf{j} = ne\mathbf{v}$ and $\nabla \phi = \mathbf{p}/\hbar$, we find an intimate connection between the supercurrent, the magnetic field and the phase of the wavefunction, valid for bulk materials at least:

$$\mathbf{j} = \frac{ne\hbar}{m} \{\nabla \phi - (2e/\hbar c)\mathbf{A}\} \quad (55)$$

This connection finds its most extensive application in the various Josephson[4,14] effects associated with the coherent properties of ϕ. We do not derive these here, and will simply state the results.

When two superconductors are weakly connected, by, for example, a tunnel junction, a small bridge of micron dimensions, or in a number of other ways (see Fig. 12), phase information can be exchanged by the leakage of a small number of Cooper pairs from one to the other. In the absence of currents and applied magnetic fields, a coupling energy given by

$$W = -W_0 \cos(\phi_1 - \phi_2) \tag{56}$$

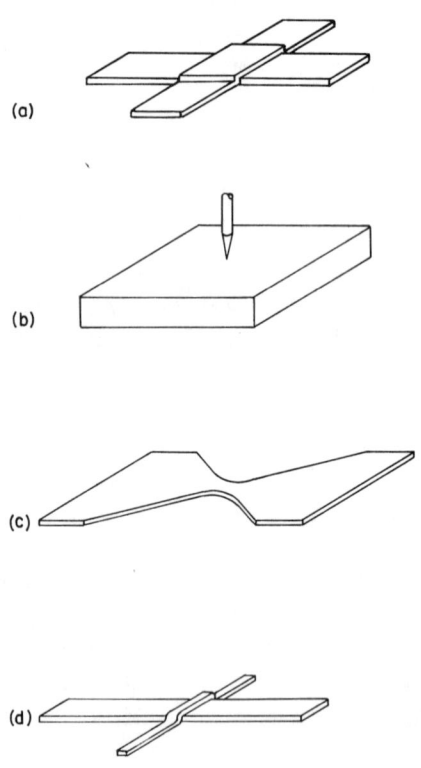

FIGURE 12. Josephson weak link structures: (a) tunnel junction, (b) point contact, (c) thin film microbridge, and (d) film with "weak" superconducting region caused by normal metal overlay. Not shown: S-N-S tunnel junction in which ~3000 Å of normal metal replaces the 20 Å oxide barrier. (From Ref. 67.)

tends to equalise the phases ϕ_1 and ϕ_2 on opposite sides of the link. E_0 depends on the nature and size of the link, but is typically about 1–10 eV for the *entire junction*. Provided the link is not exposed to external thermal or electrical radiation greater than W_0, the phases will be coupled and equal. In the presence of a magnetic field, we replace $(\phi_1 - \phi_2)$ by a gauge invariant phase difference†

$$\gamma = \phi_1 - \phi_2 - \frac{2e}{\hbar c}\int_1^2 \mathbf{A} \cdot d\mathbf{l} \qquad (57)$$

where the integral is taken from deep inside one superconductor where ϕ is constant ($\gg \lambda$ from the link) to a similar point in the other.

The transport properties of the link are governed by two further equations, often called the Josephson relations:

$$j = j_c \sin \gamma \qquad (58)$$

and

$$V = \frac{\hbar}{2e}\frac{d\gamma}{dt} \qquad (59)$$

where j is the supercurrent density at a point in the link (the integral in equation (57) being taken through that point), j_c again reflects the link properties, and V is the chemical potential difference (essentially the voltage) across the junction.

To see some of the effects covered by these simple equations, consider a narrow planar tunnel junction λ, across which the current density is uniform. Equation (58) can now be replaced by an equation for the total current

$$I = I_c \sin \gamma \qquad (60)$$

where I_c is typically 1 μA – 10 mA and given by[15]

$$I_c \leqslant \frac{\pi K \Delta(T)}{2e} \tanh \frac{\Delta(T)}{2k_B T} \qquad (61)$$

K being the conductance of the tunnel junction at high bias voltages.

If the junction is current biased at $I < I_c$ in the absence of external fields and with the self field of I being neglected, a static solution of equation (60) is

---

† The choice of $\mathbf{A}$ is arbitrary, in the sense that an alternative gauge, in which A is replaced by $\mathbf{A}' = \mathbf{A} + \nabla\chi$, χ being any scaler potential function will give the same value for the field, since curl $\mathbf{A}$ = curl $\mathbf{A}'$. Formula such as equation (57) are invariant under a change of χ, even though the individual terms are gauge dependent. See Tinkham (Appendix), p. 197 for further details.

possible, with

$$\gamma = \phi_1 - \phi_2 = \arcsin(I/I_c) \qquad (62)$$

Moreover, $V = 0$, (from equation 59), so that it is a *super*current which flows.

If I is made to exceed I_c, this solution is impossible: in the case of our tunnel junction the current has to be transported by the quasiparticle (Giaever) tunnelling process and the junction switches (Fig. 13) to a bias voltage $V = 2\Delta/e$. Hysteresis is observed as the system follows the quasiparticle tunnelling characteristic when I is returned to zero.

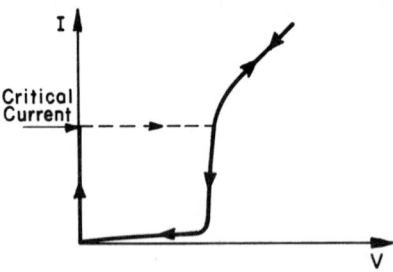

FIGURE 13. Current-voltage characteristic of unshunted Josephson tunnel junction.

Josephson effects persist in the voltage biased state, however: whenever V is finite, γ will be time-dependent, and equation (60) becomes, from the integration of equation (59)

$$I = I_c \sin 2\pi \nu t \qquad (63)$$

where

$$\nu = 2eV/h \qquad (64)$$

An A.C. supercurrent therefore flows in this case. Its amplitude is actually a function of voltage[16], peaking at $V = \Delta$ and decreasing above this. Since $2e/h = 484\,\text{MHz-}\mu\text{V}^{-1}$ and Δ is typically $\sim$meV, Josephson frequencies are usually in the GHz range.

More realistically, the link may be taken to be an ideal Josephson device shunted by a capacitance C and by its own normal conductance $G(V)$, which may be ohmic as in a microbridge, or have the more complex behaviour of a tunnel junction. We then have

$$I = I_c \sin\gamma + G(V)V + C(dV/dt) \tag{65}$$

and

$$V = (\hbar/2e)(d\gamma/dt) \tag{59}$$

No analytic solutions are possible here, except in special cases, but the following points emerge. First, for $I \leq I_c$, a zero voltage solution is possible. Secondly in the voltage biased state a supercurrent of fundamental frequency

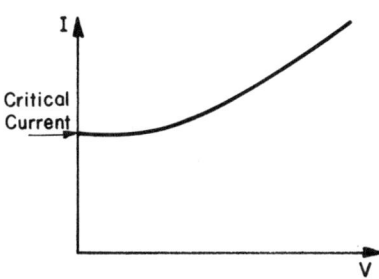

FIGURE 14. Current-voltage characteristic of Josephson point contact of shunted tunnel junction.

$v = 2e\bar{V}/h$ flows. Third, the A.C. supercurrent is strongly anharmonic and has a non-zero time average, so that the net D.C. current (Fig. 14) is thus larger than would be obtained from the shunt conductance alone. Stewart and McCumber[17] have shown that hysteresis is absent if

$$\beta_c = 2eI_c C/\hbar\{G(0)\}^2 < 1 \tag{66}$$

In an ideal tunnel junction $G(0) = 0$, so this formula gives the value of shunt resistance which must be used to eliminate any hysteresis.

Whereas the Josephson current equation (58) is valid only in the limit of weak coupling, and has to be replaced by a more complex relationship (still periodic in γ) in less ideal cases, equation (64) relating voltage to frequency has general validity. It is known to produce effects identical to better than 1 part in 10^8 in different metals[18] and is thought theoretically to be free of corrections to all orders of magnitude[19]. Note, though, that by voltage here we strictly mean electrochemical potential difference. This can depend on thermal and other effects, and must be measured by an instrument allowing electron flow; an electrostatic voltmeter does not do this and could conceivably introduce slight errors.

To see the effect of finite magnetic fields, consider a field B_z applied to the junction of Fig. 15. In a suitable gauge we have, from $\mathbf{B} = \operatorname{curl} \mathbf{A}$, the approximation

$$A_x = B_z y \tag{67}$$

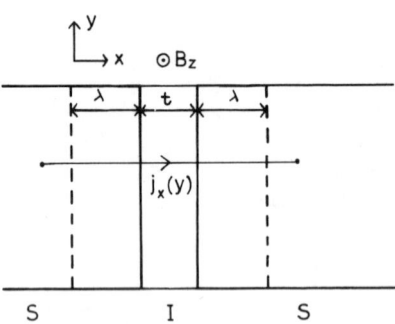

FIGURE 15. Model of *S-I-S* Josephson tunnel junction for discussion of critical current variation in perpendicular field B_z.

within the barrier (thickness t) and within λ on either side of it, and

$$A = 0 \tag{68}$$

elsewhere. From equations (57) and (58)

$$j_x(y) = j_c \sin\left\{\phi_1 - \phi_2 - \frac{2e}{hc}(t + 2\lambda) B_z y\right\} \tag{69}$$

To find $(\phi_1 - \phi_2)$ in the presence of a given total transport current I, we must integrate $j_x(y)$ from $-\tfrac{1}{2}l$ to $\tfrac{1}{2}l$. Note the algebraic similarity to the problem of single slit diffraction, with the term $(2e/hc)(t+2\lambda)B_z y$ providing the "path difference". The result is

$$I = I_c \frac{\sin\left\{(2e/hc)(t+2\lambda) B_z l\right\}}{(2e/hc)(t+2\lambda) B_z l} \sin(\phi_1 - \phi_2) \tag{70}$$

and the junction behaves as described earlier, except that I_c becomes field dependent, varying (see Fig. 16) as

$$I_c(B) = I_c \frac{\sin(\pi\Phi/\Phi_0)}{\pi\Phi/\Phi_0} \tag{71}$$

where $\Phi_0 = hc/2e$ and

$$\Phi = B_z(t+2\lambda) \tag{72}$$

is just the flux linking the junction area.

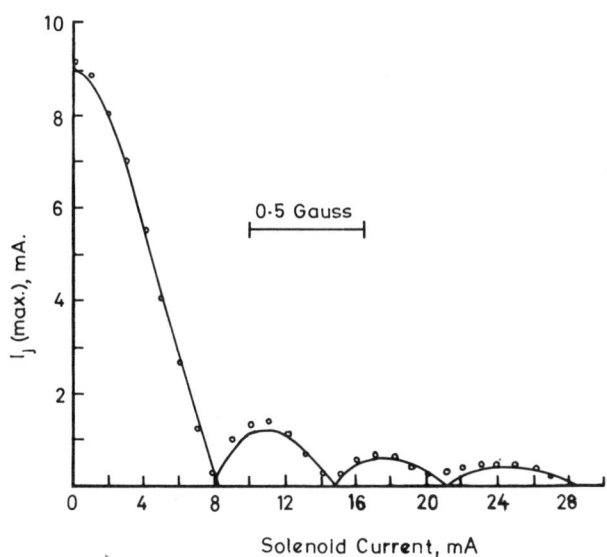

FIGURE 16. Dependence of maximum supercurrent upon flux through junction.

All this strictly applies only to "narrow" junctions, i.e. those for which the width $l \ll \lambda_J$, where $\lambda_J = \{c\Phi_0/8\pi^2 j_c(2\lambda+t)\}^{1/2} \simeq 0.1$ mm is known as the Josephson penetration depth. For $l \gtrsim \lambda_J$, the Meissner effect confines the flow of transport current to the junction edges, and solutions involving "vortices" in the junction area must be considered. We will return to this point in Section 7.1.2. Most practical applications, however, use "narrow" junctions.

Extending the "diffraction" principle, one might expect that with two narrow junctions in parallel in a ring, a Young's slits analogy would apply with

$$\{I_c(B)\}_{\text{ring}} = 2\{I_c(B)\}_{\text{single junction}} \sin(\pi\Phi_a/\Phi_0) \tag{73}$$

where Φ_a is the flux "applied" to the ring (the external field times the ring area).

Experimental results (Fig. 17) show that although equation (73) correctly describes the flux periodicity of I_c, the current does not modulate to zero as predicted. We can see why by considering the screening current induced in the ring by the applied field. This is

$$I_s = \Phi_d/L = (\Phi_a - \Phi_i)/L \tag{74}$$

where Φ_i is the flux inside the loop, and L the loop inductance. There are usually many possible solutions for Φ_d, but it can be shown[21] that for strong

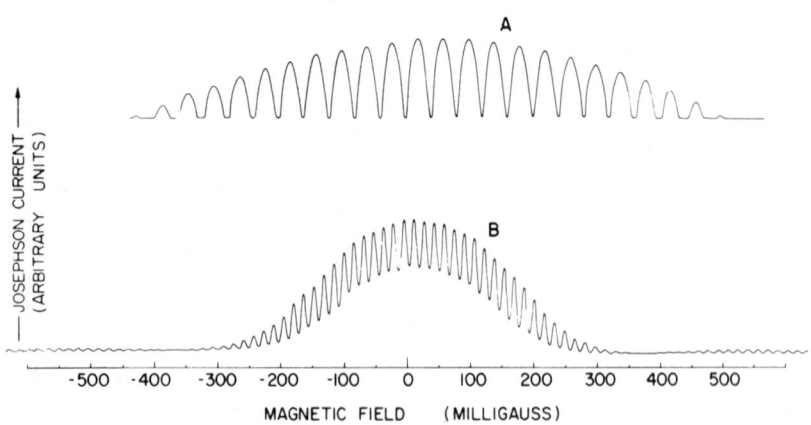

FIGURE 17. Maximum supercurrent as a function of field for two pairs (A and B) of junctions in parallel. The area enclosed by pair A is smaller than that of pair B[20].

screening (defined as $\Phi_0/L \ll I_c$), the most stable has $|\Phi_d| < \frac{1}{2}\Phi_0$. In such a case a ring fed symmetrically with a transport current I_t will have currents

$$\tfrac{1}{2}I_t \pm \Phi_d/L \tag{75}$$

in the respective junctions. The system goes normal when the current through either junction is I_c: the total transport current is then

$$2I_c - 2|\Phi_d|/L > 2I_c - \Phi_0/L \tag{76}$$

so that the maximum modulation, whatever the size of I_c, is

$$\Delta I_c = \Phi_0/L \tag{77}$$

in the strong screening limit. For weaker screening ($\Phi_0/L \gtrsim I_c$) it is more nearly possible to have $\Delta I_c \sim I_c$, but since I_c is now Φ_0/L or less, equation (77) is still approximately valid. In a typical double junction device (Section 8.2) $I_c \sim \Phi_0/L$ and $\Delta I_c \sim \Phi_0/2L$.

2.8 Applying the Josephson Effects

Three possible fields of application are revealed by the basic phenomena described in Section 2.7.

We first note that when I is made to exceed I_c in a tunnel junction, the rate at which the system switches to the $V \simeq 2\Delta/e$ state is ultimately limited by the frequency of plasma-like oscillations of the electrons within the barrier region. This can be shown to be

$$\omega_J = (8\pi e t j_c/\varepsilon \hbar)^{1/2} \sim 10^{13} \text{ sec}^{-1} \qquad (78)$$

ε being the dielectric constant of the tunnel barrier. We may thus contemplate switching times < 1psec, though as we shall see, capacitative time constants

$$\tau \sim 2C\Delta/I_c \sim 50 \text{ psec} \qquad (79)$$

tend to set the limit in practice. The hysteresis (Fig. 13) gives a bistable nature to the switch and suggests computer applications.

Secondly, the variation of I_c with period Φ_0, corresponding to 10^{-5} gauss in loop structures of area $\sim 1 \text{ mm}^2$, suggests sensitive magnetometers. Figure 17 even suggests the name—superconducting quantum interference device (SQUID).

Finally we have the possibility, raised by the Josephson A.C. effect, of precise voltage-controlled R.F. generators operating in the GHz range. Equations (63) and (64) also reveal the possibility of frequency to voltage converters, and of mixers and detectors, all making use of the extreme non-linearity of Josephson junctions[22].

We will discuss how these basic ideas have been developed into devices in Sections 7–9.

2.9 Summary

We have seen superconductivity phenomena at what are (superficially) three levels of sophistication. First are the "simple" properties of critical temperature, critical fields in type I and II cases, and zero resistance. Then follow "energy gap dependent" properties, relying on the finite energy separating the ground state from its excitations and on the Fermi–Dirac properties of the quasiparticles; we include here properties associated with the coherence length ξ. Finally there are properties arising from the macroscopic phase coherence of the ground state wavefunction, ranging from fluxoid quantisation to the Josephson effects.

Each of these phenomena has an application in one or more thin film devices.

3. Superconductivity in Thin Films

3.1 Transition Temperatures

Provided $d \geqslant \xi$, single crystal thin films behave like bulk material in the value and sharpness of T_c. Thickness dependencies of T_c appear for $d \leqslant \xi$, and T_c also differs slightly from bulk in most practical evaporated or sputtered films, which are usually granular, or polycrystalline at best. Impurities can also change T_c somewhat[23], and in particular ferromagnetic impurities (Mn, Fe, Gd, Co), whose strong localised spins can break the coupling of Cooper pairs, reduce T_c steeply, and destroy superconductivity if present at the 1% level[24]. Dramatic T_c increases are also possible: T_c can be increased by about × 800 in Be sputtered on to a 4.2 K substrate[25]. A complex literature (see Chopra[26]) shows that the influence of thickness, grain size, stress due to differential thermal contraction, and many other effects have been studied.

For the materials and film thicknesses ($d \gtrsim 300$ Å) used in devices, however, we may make a generalisation, which has only one or two exceptions, discussed below. Provided ferromagnetic impurities are kept to about 1 ppm or lower, the T_c of a film evaporated or sputtered in a conventional way on to a room temperature or hotter substrate will always be within 5% of the bulk value, and usually much closer. From the device point-of-view, such a change is usually unimportant. In the case of alloy films prepared by co-evaporation, substrate conditions must of course be suitable for any annealing which may be necessary to achieve correct stoichiometry.

Though T_c may change little, its width (ΔT_c) normally increases substantially in films. For example, the resistance-temperature curve may show a transition broadened to several K. In part this can be due to a small spread in T_c values caused by differences in thickness, lattice parameter, purity and strain across the film. Often, in fact, it may be necessary to mechanically trim the ultra-thin edges of films to eliminate resistanceless threads which can appear at temperatures well above the T_c of the film as a whole. A more fundamental cause of any ΔT_c increase may be due to the very small grain sizes (t) in polycrystalline or amorphous films. For $t \lesssim \xi$, one may have tiny superconducting volumes, each less than the size of a typical Cooper pair, weakly linked together. In such small volumes the co-operative nature of the superconducting ordering process is changed, and significant thermodynamic fluctuations of the order parameter become possible, so that precursors of superconductivity may be seen well above T_c. For example, evidence of the Meissner effect has been seen in granular indium[27] at temperatures up to 6 K ($T_c = 3.4$ K in bulk). Once again, however, we may make a generalisation: $\Delta T_c / T_c$ is usually less than 2% for device films prepared under conventional conditions.

One major exception to our rules is aluminium. Its affinity for oxygen is so great that small grained films ($t < 300$ Å) are common, even with evaporation pressures $\lesssim 10^{-5}$ torr[28]. T_c increases of 10% are hard to avoid, and 1.8 K is often obtained (bulk $T_c = 1.20$ K), with edges showing much greater increases. In fact lattice damaging by ion implantation can be shown to increase T_c to > 6.7 K[29]. Suitable edge trimming may reduce ΔT_c, but not the absolute change in T_c itself. Since devices are usually designed for operation in the ^{4}He range (1 K–4.2 K) this property has the perverse effect of reducing the usefulness of aluminium, not as a superconductor, but as an easily evaporated, readily oxidisable, *normal* metal. Above 3 K, its use is usually safe, though.

The other bad actors are niobium and to a lesser extent, tantalum. T_c is reduced by ~ 1 K per atomic percent by dissolved oxygen[30], which is easily absorbed during evaporation. In part this is due to the formation of complex oxides (some of which are superconducting), but it is also due to lattice paramater variations caused by interstitial inclusions, as can be demonstrated by sputtering in N_2 and A[31]. Without care, T_c may well fall below 4 K, and films will often remain normal at all temperatures. However, films sputtered in 10 microns of argon usually have T_c above 8 K, if no other impurities are released by the sputtering arrangements. By using substrates heated to 400°C, vacua better than 10^{-7} torr, and high electron-beam evaporation rates, transition temperatures close to bulk (9.46 K) can be obtained.

Alloy films, such as Nb_3Sn, V_3Si, and Nb_3AlGe have been difficult to prepare with T_c values close to bulk. Recent substantial improvements in high rate sputtering[32] and in computer-controlled UHV electron-beam co-evaporation techniques[33] have altered this picture and very satisfactory transition temperatures are now obtainable.

3.2 Penetration Depth and Coherence Length

The way in which the film thickness d reduces the bulk electron mean free path l to a smaller value l' is well understood. The equation

$$\frac{1}{l'} = \frac{1}{l} + \frac{1}{d} \qquad (80)$$

is a useful guide, though there are available much more precise formulae[34] which take the nature of surface scattering into account. It is often more accurate, however, to determine l' experimentally, either from the resistance ratio $R_{300K}/R_{4.2K}$ and the known room temperature mean free path of the pure metal, or from the low temperature resistivity and parameters such as valence and Fermi energy[35].

The effects of l' and d on λ and ξ are extremely complicated. In the dirty limit ($l' \ll \xi, d$) it is usually valid simply to replace l by l' in equation (52). The same substitution usually gives a reasonable guide for other purities. Precise calculations depend critically on the relative values of l', d, ξ_0, and λ_0, and have been done only for a few extreme limits, such as $d \gg \xi$[36]. Universal formulae are not available and once again particular cases are probably best dealt with by experimental methods. We may use the type II behaviour of films in perpendicular fields (Section 3.4.3) and measure the upper critical field $H_{c2\perp}$. Since this field corresponds to the point at which the system becomes completely normal, when the vortex separation is just equal to a core diameter ξ, we have

$$\xi = (\Phi_0/2\pi H_{c2\perp})^{1/2} \tag{81}$$

If the thermodynamic critical field (available from tables) is H_c, we also have $H_{c2} = \sqrt{2}\kappa$ and $H_c = \sqrt{2}(\xi/\lambda)H_c$ so

$$\lambda = (H_{c2\perp}\Phi_0/4\pi H_c^2)^{1/2} \tag{82}$$

3.3 Energy Gap Anistropy

Since the energy gap Δ is a function of wavenumber (equation 28), it can vary with direction relative to the Fermi surface. In a highly anisotropic material, such as niobium, $\pm 7\%$ variation with orientation may be observed[37]. If tunnelling experiments are performed on polycrystalline films, the junction is likely to cover many crystallites, and to sample the full range of energy gaps. Any singularities in the $I-V$ characteristic will be correspondingly smeared.

If the film is in the dirty limit ($l \ll \xi$) however, simple BCS theory is modified by a theory of Anderson[38]. The usual ($\mathbf{k}\uparrow$, $-\mathbf{k}\downarrow$) pairs are replaced by "pairs" involving linear combinations of states from all over the Fermi surface. A single energy gap then applies in all directions, and sharp tunnelling singularities are again observed.

3.4 Critical Magnetic Fields

The behaviour of films in magnetic fields is extremely complicated, essentially because so many parameters (including field, field orientation, H_c, T, ξ, λ, l, d, and even the film width w) are involved. It is possible here to give only the broadest outline of this behaviour. Much more detail is available in the reviews of Chopra[26], Burger and St. James[39], de Gennes[40], and Tinkham[41].

3.4.1 Thermodynamic artificial field

Since flux penetrates a specimen in the Meissner state to depth λ, the effective diamagnetism of any film with $d \sim \lambda$ will be much less than perfect. Simple magnetostatics[42] shows that for $d > \lambda$ the free energy per unit area of a film in field H_a is

$$F_H = \frac{1}{8\pi}(d-\lambda)H_a^2 \tag{83}$$

For $d \ll \lambda$,

$$F_H = \frac{\alpha}{12}\left(\frac{d}{\lambda}\right)^2 H_a^2 \tag{84}$$

where $\alpha \sim 1$ and depends on whether a local or other model is valid. The effective penetration depth, defined as the mean distance of field penetration $\int xH(x)\mathrm{d}x/\int H(x)\mathrm{d}x$ is

$$\lambda_d = \lambda \coth(d/\lambda) \tag{85}$$

The thermodynamic critical field H'_c is the value of H_a at which F_H equals the superconducting condensation energy $dH_c^2/8\pi$ per unit area. Thus,

$$H'_c/H_c \simeq 1 + \lambda/2d \quad \text{for} \quad d > \lambda \tag{86}$$

and

$$H'_c/H_c \sim \sqrt{12}(\lambda/d) \quad \text{for} \quad d \ll \lambda \tag{87}$$

3.4.2 Parallel critical field

When the field is parallel to the film, the nature of the transition to the normal state depends not only on κ (which is modified from the bulk value because the thickness d has reduced the mean free path), but also on d/λ. A detailed account has been given by Guyon et al.[43]. Briefly we may say that for $d/\lambda > \sqrt{5}$ and $\kappa < 1/\sqrt{2}$, films have a first order transition at H'_c, though superheating and surface superconductivity complexities may delay full development of the normal state when $d/1.8\lambda < \kappa < 1/\sqrt{2}$. For $\kappa > 1/\sqrt{2}$, and provided $d/\lambda > 1.8\kappa$, behaviour similar to that of bulk type II material is found, with an upper critical field $H'_{c2} = \sqrt{2}\kappa H_c$: H_{c2} is thus enhanced by the changes in κ caused by the reduction in mean free path. When films become thin enough to be comparable with ξ (roughly speaking, for $d/\lambda < \sqrt{5}$ when $\kappa < 1/\sqrt{2}$ and $d/\lambda < 1.8\kappa$ when $\kappa \geqslant 1/\sqrt{2}$), there is not enough thickness to sustain either an intermediate or vortex state. The energy gap drops smoothly to zero as the critical field H'_c is approached, and the transition to the normal state is second order.

3.4.3 Perpendicular critical fields

The demagnetising factor of a film (Section 2.2) in a perpendicular field leads to flux entry at fields

$$H_\perp \cong \left(\frac{d}{w}\right) H_c \tag{88}$$

for type I materials, and

$$H_{1\perp} \cong \left(\frac{d}{w}\right) H_{c1} \tag{89}$$

for type II materials. Here w is the width of the film. It is well established that all films, with whatever κ, enter the mixed state above H_1[44-46]. For $\kappa > 0.62$, this state persists up to $H_{c2} = \Phi_0/2\pi\xi^2$. For lower values of κ, the film remains in the mixed state up to H_{c2} provided κ exceeds a critical value, given by

$$\kappa_c^2 \gtrsim d/2\xi \gtrsim d(\pi H_{c2}/2\Phi_0)^{1/2} \tag{90}$$

If this condition is not satisfied, the simple mixed state is stable only up to a field given by

$$H = 4\Phi_0 \kappa^4/\pi d^2 \tag{91}$$

above which the film enters first a multiply-quantised vortex state and later perhaps even the intermediate state before going fully normal (Fig. 18).

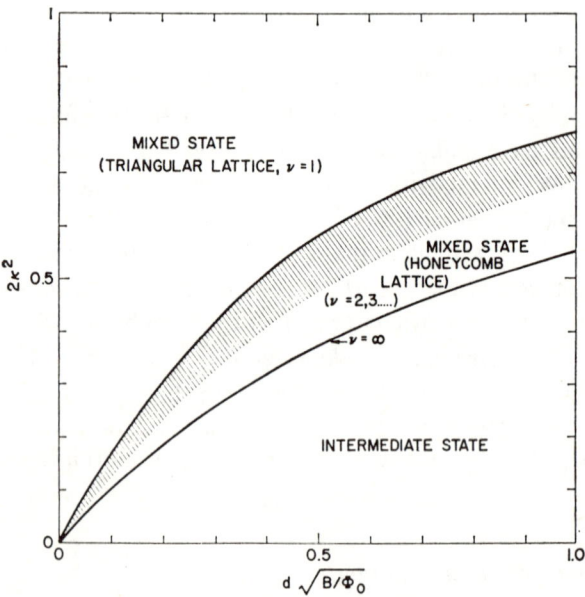

FIGURE 18. Magnetic phase diagram of thin films in perpendicular field B[45]. Here ν is the number of flux quanta per vortex.

From the device point-of-view, details of the upper perpendicular critical field are not very significant. It is $H_{1\perp}$ that matters most, for, as equation (88) shows, whenever a field has a transverse component $> (d/R)H_0$, there is a risk that vortices may enter the film. Once they do so there is a good chance that they will remain pinned after the field, which might be a transient pulse, is removed. For a 1 mm wide, 2000 Å thick lead film, $H_{1\perp} \sim 0.1$ gauss. This can represent a very real problem in Josephson effect devices, which could be sensitive to one flux quantum trapped in the wrong place.

3.5 Critical Currents; Ground Planes

A reasonable guide to the critical current of a bulk cylindrical wire (radius a) is given by Silsbee's Rule[47]

$$I_c = 2H_c a \qquad (92)$$

which is just the condition that the surface field due to the current is H_c. When the specimen is much thinner ($\sim \lambda$) the fact that current flows throughout the specimen must be taken into account. Provided the current density is uniform across the film, it is easy to show that the critical current density is[48]

$$j_c(T) = (\tfrac{2}{3})^{2/3} cH_0(T)/\lambda(T) \qquad (93)$$

A typical value for $j_c(0)$ is then $\sim 5 \times 10^7$ A-cm^{-2} (corresponding to $I_c = 100$ A for a 1 mm wide, 2000 Å thick, film.

For the usual geometry of a rectangular film, the current is not uniform, but is distributed so as to minimise the self-field component perpendicular to the film. The current density is highest along the edges. Calculations of current distribution near the critical condition have been attempted, but since edges are usually poorly defined, they are of little value. Empirically we may say that I_c is usually 20–100 times less than that expected from equation (93) with a uniform current. This still gives $I_c \sim 1$ A, and so is not usually a restriction in device work.

Uniform current density *can* be achieved by depositing the film on a cylindrical substrate of radius $\gg \lambda$ and length much greater than radius. A more common alternative is to deposit it on an insulated ground plane[49] (Fig. 19), which is an extended superconducting layer of thickness $\gg \lambda$. An image current, of polarity opposite to the film current, flows in the plane, ensuring that the net field is parallel to the film surface. The optimum current distribution then becomes the uniform one, the edges are of less significance, and critical currents are greatly increased.

The field due to the image current in the ground plane virtually cancels the primary field farther from the film edge than about twice the film-plane separation. This field screening implies that the inductance of the film-ground

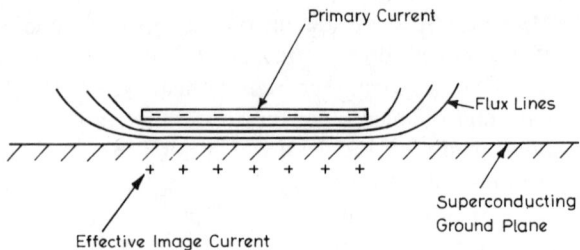

FIGURE 19. Field and current smoothing by a superconducting ground plane[49].

plane combination is about two orders of magnitude less than that of the isolated film, a reduction which also applies to the mutual inductance between the film and nearby circuit elements, unless these are deposited directly above the film. These properties make the ground plane very important in certain magnetic devices (Section 7).

If we assume that the field in the film-ground plane separation region is H, and that it is zero elsewhere, we have, by Amperes law

$$H = 4\pi I/w \qquad (94)$$

for a film of width w carrying current I.

3.6 The Proximity Effect

If a normal metal N and a superconductor S are in very good electrical contact, Cooper pairs can diffuse into N from a region in S within about ξ of the interface[50]. The thickness of a leakage region ($\sim 10^3$ Å) depends in detail on the mean free path in N, and on whether or not N is itself a superconductor with a critical temperature T_{cN} (lower than the operating temperature T). Throughout the leakage region, N essentially becomes superconducting, though it has a rather low critical current because its Cooper pair density is low.

One application of this is in S-N-S junctions, in which the usual 20 Å of a tunnel junction is replaced by 10000 Å of a normal metal, typically copper or a copper alloy. The normal metal acts as a weak link between the superconductors, with a typical normal resistance of 10^{-6} ohms, and $I_c \sim 20$ mA. The critical current is a strong function of temperature, and if N is a dirty alloy, it varies as [51]

$$I_c(T) = I_c(0)\exp(-T/T_0) \qquad (95)$$

with T_0 a constant, of order 0.1 K in the case of Cu/Al.

When an N film is superimposed on an S film of thickness $\sim \xi$, the composite structure has a single critical temperature T_{cNS}, less than that of the unperturbed superconductor T_{cS}. Figure 12(d) showed an application of this in which a narrow strip of normal metal overlays a superconducting film. When operated a little below T_{cNS}, the structure behaves as a weak link. In a development of this system[52] (Fig. 20) the main film consists of a composite structure. Microcircuit fabrication techniques are used to form a weaker region, of characteristic critical temperature T'_{cNS}, by increasing the $N:S$ thickness ratio over a very small length ($l \sim 1\,\mu$m) of the film. Since the

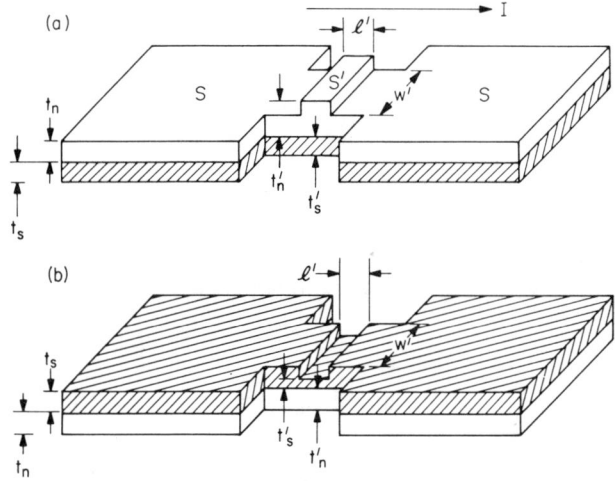

FIGURE 20. Proximity effect microbridges (s = superconductor, n = normal metal)[52].

proximity effect also operates along the film, the bridge region usually has a final critical temperature quite close to T_{cNS}. However, the bridge acts as a weak link, and its strength can be controlled by appropriately choosing l' and T'_{cNS} (i.e. the $N:S$ ratio). An example of I_c versus T is shown in Fig. 21.

4. Devices—General Considerations

4.1 Fabrication

The choice of thin films is not usually forced upon the designer of a superconducting device; instead he will be attracted to them by many of the simple properties that make films appealing for use at any temperature. For example, small mass, low cost and mechanical stability are all important

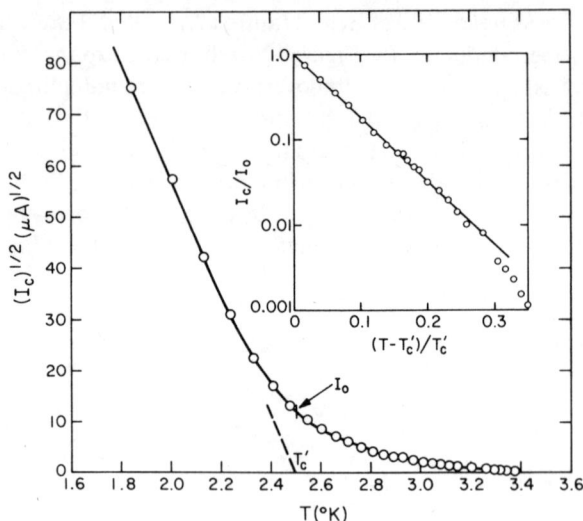

FIGURE 21. Variation of proximity effect bridge critical current with temperature[52].

considerations, some with enhanced significance when thermal contraction problems are considered.

Because of this, fabrication procedures do not usually need to be very sophisticated, and we may refer the reader to Chapter 2 for general outlines. Most devices can be produced if one has a 10^{-6} torr liquid nitrogen baffled diffusion pumped coating unit with resistively heated evaporation sources. UHV confers no easy advantages though it is a desirable facility if the ultimate in reproducibility is sought. It is also necessary if good T_c values are to be obtained from electron gun evaporation of Nb and Ta, though sputter guns are reasonable alternatives for refractory metals unless high width definition ($<100\,\mu$m) is needed. A quartz crystal thickness monitor is essential, but other desirable refinements, like mask-changing, co-evaporation of alloys, substrate heating, etc., are more optional.

Soda glass microscope slides are usually first choice as substrates, but practices vary on the amount of washing and other pretreatment that is thought necessary for good adhesion. Corning 7059 glass, whose contraction on cooling is close to that of many metals (and particularly suited to niobium) can help with differential shrinkage problems. Quartz and sapphire may be used when good thermal conductance is necessary, such as when a device is not to be directly immersed in liquid helium.

4.2 Tunnelling Devices

In contrast to the general rule in Section 4.1 about device design, one may have no choice but to employ thin films if tunnelling is involved. Tunnelling samples superconducting properties only within about a coherence length ($\xi \sim 10$–100 nm) of the barrier. For this reason, and also to prevent difficulties with field emission processes from points and to ensure that planar tunnelling theory is valid, it is important to have electrode surfaces which are free of local roughness on this scale or less. This condition is much more easily met by depositing a thin film, whose surface roughness will usually be proportional to $\sqrt{N}$ (N being the number of atomic layers in the film) than it is by preparing a bulk surface mechanically and chemically.

4.3 Quasibulk Application of Films

In considering possible applications of superconductors, it is often useful to recall that, in the Meissner state, all currents (whether transport or screening) are confined to within about λ (~ 100 nm) of an outside surface (Section 2.2). Provided the proposed device is not to be operated with flux penetrating it, it may then be perfectly acceptable to use a design in which all of the superconductor, apart from the surface layer, is removed and replaced by a quite different material, chosen for other properties such as thermal conductivity, mechanical strength or price. The usual means of achieving this is to coat the non-superconducting material with a superconducting film, usually $\leqslant 1\,\mu$ thick.

An example of this principle is one commonly found in interconnection use in cryostats, especially in stages operating below 1 K. Fine magnanin wire coated with a thin lead layer and covered with standard insulation is not only resistanceless, but also has negligible heat conductance, is malleable and can be soldered. A more substantial application is the chemical formation of a thin layer of Nb_3Sn on a niobium ribbon, which avoids the brittle problems of the bulk alloy while retaining the high T_c and H_{c2}. This has been extended to the computer-controlled electron-beam evaporation of two or more elements[33] to deposit high quality Nb_3Sn, V_3Al, $Nb_3(GeAl)$ and other films on thin ($\sim 60\,\mu m$) Hastelloy B ribbon at rates up to $5000\,\text{Å}\,\text{sec}^{-1}$. Continuous production of moving ribbon is possible, and the method may prove an economically competitive way of producing high field superconducting materials.

Another well developed use of surface superconducting layers has been in radio-frequency resonators and cavities[53] whose quality factor is

$$Q = 4\pi^2 G/Rc \qquad (96)$$

where G is essentially a geometrical form factor, and R the real part of the cavity surface impedance. For high energy nuclear particle accelerators, where large values of Q and low power dissipation are sought, superconductivity has an obvious attraction. In fact, one does not have $R = 0$ in such cases: instead

$$R = R_{qp} + R_{res} \qquad (97)$$

where R_{qp} depends on absorption by thermally excited quasiparticles and R_{res} depends largely on surface condition. Nevertheless, by comparison with $R \sim 7 \times 10^{-3}\,\Omega$ for copper at 4 K, values of $R_{qp} \sim 10^{-6}\,\Omega$ at 4.2 K and $10^{-9}\,\Omega$ at 1.8 K apply to lead and niobium, and R_{res} can be reduced to $\lesssim 10^{-7}\,\Omega$. Extensive work has therefore gone into developing superconducting cavities[53], and although niobium has only been used in bulk form, lead[54] has been used as a film electroplated onto a copper substrate. The two techniques are roughly competitive, with Q values of $>5 \times 10^{10}$ having been reached at 10 GHz, and $Q \sim 10^9$ at 100 MHz, even in the presence of 3 MV-m^{-1} accelerating fields. Nb$_3$Sn films have also been studied in the hope of being able to use them in high magnetic fields[55].

A final example of the use of films to simulate bulk material is in the application of the Meissner effect to magnetic screening[56], in which one places a specimen inside a superconducting container. It may often be adequate to use a brass container dipped in tin-lead solder, or else electroplated with lead. Care must be taken to ensure that the large demagnetising factor of films does not lead to flux trapping as the shield is cooled through the transition temperature. Moreover, it must not be exposed to transverse fields in excess of $H_\perp$, which can be very small for thin films (Section 3.4).

4.4 Conclusion

We now turn to the treatment of "true" thin-film devices, broadly following the development of Section 2. Thus Section 5 deals with magnetic transitions to the mixed or normal states (Section 2.2), Section 6 is devoted to thermally operated devices, and Sections 7 and 8 are mostly about the D.C. Josephson effect. The A.C. Josephson effect and other phenomena associated with infrared frequencies are discussed in Section 9. The classification is flexible, however, and we have treated all devices with the same function together, whatever their operating principle. For example, "old" cryotrons, which are switched by magnetically induced S–N transitions are treated, not in Section 5, but with their Josephson successors in Section 7.

5. Zero Resistance and Magnetic Transition Devices

5.1 Transmission and Delay Lines

Although there have been many applications of superconductivity to resonator devices[57], such as the lead-plated cavities mentioned in Section 4, few involve conventional thin film structures. An exception is based on work by Pippard[58] who showed that the penetration of magnetic field into a superconductor will slow a wave on a transmission line. For two plates separated by a layer of dielectric t thick, the phase velocity in the dominant mode[59] is

$$c' = c_d(1 + \{\lambda_1 + \lambda_2\}/t)^{-1/2} \qquad (98)$$

where c_d is the velocity of light in the dielectric, and λ_1 and λ_2 are the respective penetration depths (given by equation (85) if the plates are films with $d \ll \lambda$).

If the films are in the form of meander lines, consisting of series connected quarter-wave lines, c' is reduced even further by the increased length and by the interaction between adjacent lines. Values of $c'/c_s \simeq 10^{-2}$ are approachable. Passow et al.[60] have described structures involving superconducting films evaporated on Teflon, separated by SiO insulating films, and mounted between superconducting ground planes. By using various geometries they have produced non-dispersive 300 MHz delay lines with $Q \sim 8 \times 10^4$, and also dispersive lines suitable for pulse compression with input to output pulse length ratios of up to 1000. These are superior to acoustic delay lines in having low insertion loss and greater ease of fabrication. Passow et al.[60] also produced lines whose centre frequency could be tuned through 6% by using a thin film heater to vary λ (see equation 3) to which the surface inductance is proportional.

Superconducting coaxial lines with losses of $1\,\text{dB-km}^{-1}$ and $1\,\text{GHz}$ bandwidth have been made with information transmission in mind[61]. Thin films have not so far been involved.

5.2 The D.C. Transformer

The principle of the D.C. transformer, discovered by Giaever[62], and since investigated by him[63] and by others[64], is shown in Fig. 22. Two superconducting films are separated by a layer of insulator, too thick to permit tunnelling but usually less than 200 Å. A perpendicular magnetic field is applied, and because the films are ~ 1000 Å thick, they each enter the mixed state whatever material they may be prepared from (see Section 3.4). The vortices magnetically couple together through the insulating layer. If a current is passed through one film (the primary), there is a force on the vortex structure in that film, given (from equation 12) by

$$\mathbf{F}_p = \frac{1}{c}(\mathbf{J}_p \times \mathbf{B}) \qquad (99)$$

where $\mathbf{J}_p$ is the primary current density and $\mathbf{B}$ the flux. When the critical pinning force is exceeded, the primary array is set in motion across the film and provided the magnetic coupling with the vortex array in the secondary film is strong enough, it too will move, producing a D.C. secondary voltage V_s.

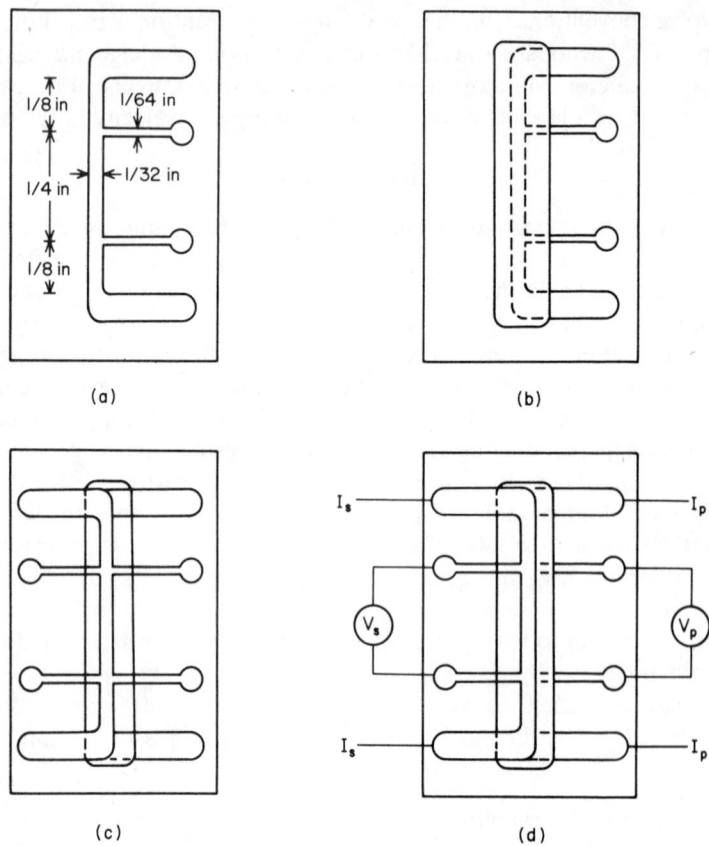

FIGURE 22. D.C. transformer fabrication[64]: (a) primary superconducting film, (b) SiO insulating layer, (c) secondary superconducting film, and (d) current and voltage connections.

At low voltages and currents the usual transformer laws apply: in particular $V_p = V_s$ for an open circuit secondary. However, with increasing J_p the system rapidly becomes very inefficient, with V_p breaking away and decreasing (see Fig. 23). The breakpoint occurs at lower values of J_p if the magnetic field or the insulator thickness is increased.

Clem[64] has recently given elegant theories of these phenomena which are in striking agreement with experiment[65]. For each film, there is an equation

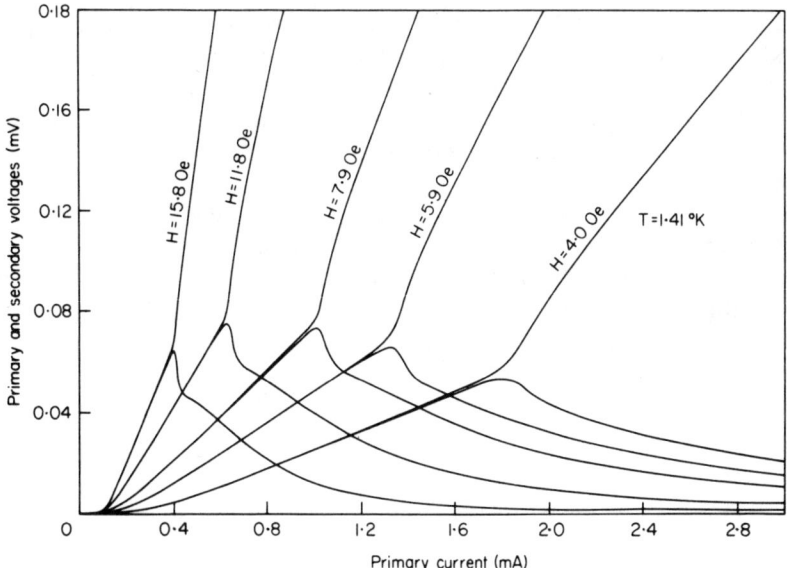

FIGURE 23. Primary and secondary voltages as a function of applied primary current for several magnetic field strengths (zero secondary current)[64].

of motion in which the viscous drag force (equation 13) is equated to the sum of forces due to the applied current, to pinning, and to the magnetic coupling with the other film. The last is represented by the periodic term

$$F_{mc} = F_m \sin[(2\pi/l)(x_p - x_s)] \qquad (100)$$

where l, the vortex spacing in the direction of flow, is given by

$$l = (2\Phi_0/B\sqrt{3}) \qquad (101)$$

and $(x_p - x_s)$ is the relative displacement of the two (triangular) vortex lattices. Since the viscous drag is related to the electric field generated and to the flux flow resistance R, the following equations emerge

$$V_p = \{(I_p \pm I_{pc}) - I_0 \sin 2\pi(x_p - x_s)/l\} R_p \qquad (102)$$
$$V_s = \{(I_s \pm I_{sc}) + I_0 \sin 2\pi(x_p - x_s)/l\} R_s \qquad (103)$$

where I_p, I_s are the transport currents, I_{pc}, I_{sc} are the measured depinning (flux flow) critical currents in the individual films, and I_0 is a constant proportional to F_m. The results of Fig. 23 emerge directly from these equations. The breakpoint corresponds to the coupling limit, which is reached when $(x_p - x_s) = \frac{1}{4}l$. Above this $(x_p - x_s)$ is time-dependent and there is dynamic coupling, whose time averaged value tends to zero as the average relative slip rate becomes even greater: V_s therefore decreases as I_p (and V_p) are further increased.

I_0, and hence F_m, can be deduced from experimental data, and are in good agreement with Clem's[64] theory of coupling. This takes account of the way in which the vortex field distribution, which is confined to a region of diameter $\sim \lambda$ within the superconducting films, spreads out within the insulator (thickness t). Clearly, the coupling will be small if the field has a chance to become essentially uniform within t, as it will do if the vortex spacing is made small relative to t, either by increasing t, or through increasing the vortex density by raising B. A simplified result of the theory is

$$F_m = \frac{3\Phi_0^2 d_p d_s}{32\pi^3 \lambda_p^2 \lambda_s^2} \tilde{F}_m \qquad (104)$$

where t is the insulator thickness, $\tilde{F}_m$ is a constant of order unity, d_p, d_s are the respective film thicknesses and λ_p, λ_s are penetration depths.

Ekin and Clem[66] have demonstrated experimentally that, as equation (104) predicts, F_m is much larger in thick film (~ 6000 Å) transformers than in thin film cases, so that much higher breakdown currents (~ 20 mA) are possible. With oxygen-doped granular aluminium films, in which the depinning currents are particularly small, power transfer efficiencies can exceed 80%. Significant applications of the transformer in cryogenic integrated circuit devices now seem feasible.

6. Thermal Devices

In this section we discuss several devices whose operation depends on the temperature dependence of some superconducting property. Since many of these devices are thermometers and bolometers, we also discuss here, for completeness, other such instruments which employ superconducting technology, although not in the temperature-sensing element.

6.1 Thermometers and Bolometers

A thin film bolometer consists basically of a film with a temperature-sensitive property (usually resistance). It is in good thermal contact with a substrate which is in turn a good absorber of radiation (usually infrared). The assembly is connected to a heat sink through a thermal conductance G.

Three terms contribute to the noise equivalent power (NEP) (also discussed in Chapters 9 and 12) of the system: first the Johnson noise in the thin film, secondly thermal fluctuations of the assembly as a whole and of the film relative to the substrate, and thirdly fluctuations in the black body background seen by the bolometer[67]. In low background applications the first and third terms can usually be made small, leaving

$$(\text{NEP})^2 = 4kT^2G \qquad (105)$$

for a film in perfect contact with its substrate.

The minimum detectable temperature rise δT_m is given by

$$(\text{NEP}) = G\delta T_m \qquad (106)$$

and a useful figure of merit is the specific detectivity

$$D^* = \sqrt{A}/\text{NEP} \qquad (107)$$

where A is the bolometer area. The time constant $\tau = C/G$, C being the heat capacity. For high D^* one seeks small G, but also a small τ (preferably much less than 1 sec), so C must be very small.

Thermometers, like bolometers, are based on a temperature-sensitive element, and each should have a small δT_m and C. They differ in that thermometers should be in good contact with their immediate surroundings, and G can be large so that the design conditions above can be relaxed. However, the choice of operating principle is more restricted for thermometers, because they must usually operate over a substantial temperature range and should have a thermometric parameter $\theta(T)$ which is a reasonable, even if empirical, function of T; a bolometer, by contrast, may be operated at a single temperature T_0, with excursions δT so small that the approximation $\delta T = \delta\theta/\theta'(T_0)$ is adequate even if $\theta(T)$ is poorly defined in general.

6.1.1 Transition edge devices

The small width (ΔT_c) of the resistive transition means that, for a film biased at constant current near $0.5 R_n$ (R_n being the normal resistance), small relative changes dT/T will produce much larger relative resistance changes dR/R_D. Since ΔT_c, like the normal state resistivity, varies widely and depends on evaporation conditions, a considerable range of dR/dT values are found. However, a useful guide is[68]

$$(T/R)(dR/dT) \approx 200 \qquad (108)$$

This property is of little use in thermometry, because operation is restricted to a single temperature, magnetic field modulation of T_c being unsatisfactory for films (Section 3.4.3). In bolometry, however, very satisfactory applications are possible. Typically a film (~ 2000 Å) is deposited on a sapphire substrate, the latter carrying a coating of, for example, bismuth to enhance absorption. The substrate is thermally isolated as far as is possible consistent with the need to make electrical connections. An interesting example of the principle of Section 4.3 here is the use of lead-coated nylon threads to support and connect to a bolometer[69].

Radiation, chopped at a few hertz, falls on the absorber, and changes in the resistance of the film are measured synchronously. If necessary, a heater on the substrate can be used to maintain the average temperature close to the transition midpoint[68].

Using a $38\,\Omega$ tin film Katz and Rose[70] obtained NEP values of $10^{-12}\,\text{W}/\sqrt{\text{Hz}}$. More recently Clarke et al.[71] have used a $6\,\Omega$ aluminium film, biased by 50 nA at the $1\,\Omega$ point, where $dR/dT = 15\,\Omega/\text{K}$, as the bolometer element. The film was deposited on a sapphire slice coated on the reverse with a Bi film matched in resistance to the substrate dielectric constant to give an infrared absorption efficiency of 50%. Resistance changes were measured by the effects on the loading of a $Q = 300$ resonant circuit. An NEP of $10^{-14}\,\text{W-Hz}^{-1/2}$ at 1 Hz was obtained, which is within 3 times the limit set by equation (105). Moreover though $G(5 \times 10^{-8}\,\text{W-K}^{-1})$ was small enough and A large enough to give an exceptionally high $D^*(4 \times 10^{13}\,\text{cm-W-Hz}^{-1/2})$, the value of C was kept down to the point at which $\tau = 0.13$ sec. The device should compete with the best semiconducting bolometers, particularly in high altitude astronomical work.

By pre-evaporating a thin metal layer, it appears to be possible to improve the thermal contact between a film and its substrate[72]. This can reduce the temperature fluctuations of the film from which $1/f$ voltage noise can arise. It seems possible that in this way NEP values closer to the limit of equation (105) may be reached.

6.1.2 Josephson devices—S-N-S structure

The critical current in a Josephson tunnel junction could possibly be used as a thermometric parameter (see equation 61), but no working devices appear to be in regular use. Clarke, Hoffer, and Richards[69], however, have used Pb/CuAl/Pb proximity sandwiches (Section 3.6) as highly successful bolometers (Fig. 24). The critical current is given by equation (95) with $T_0 = 0.1$ K, so

$$\partial I_c/\partial T \simeq -I_c/2\sqrt{TT_0} \qquad (109)$$

In practice the junction is current biased just above I_c, and the voltage modulation is then given by $\partial V/\partial I_c \simeq R_D$ where R_D is approximately the normal junction resistance.

With $I_c \simeq 20$ mA, and $R_D = \mu\Omega$, an NEP of 2×10^{-15} W/$\sqrt{\text{Hz}}$ was obtained with $\tau \sim 1$ second. The corresponding temperature resolution was 5×10^{-7} K/$\sqrt{\text{Hz}}$, only 2.5 times the Johnson noise limit.

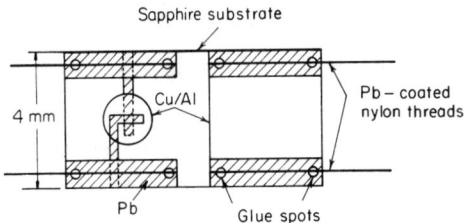

FIGURE 24. Configuration of superconducting-normal metal-superconductor (S-N-S) junction bolometers[69].

A disadvantage of the system is the very small output signal

$$\delta V \simeq (R_D I_c/2\sqrt{TT_0})\delta T \qquad (110)$$

This has to be measured using a SQUID (Section 8.3), thus adding to the cryogenic complexity.

Finally, we note that while tunnel junction critical currents are not used bolometrically, broad band detectors using non-thermal changes in I_c induced by the Josephson A.C. effect are practical. These are discussed in Section 9.5.2.

6.1.3 Giaever tunnelling devices

Several features of the Giaever tunnelling equation (equation 39), containing as it does the Fermi function

$$f(E) = (\exp E/kT + 1)^{-1} \qquad (111)$$

and the temperature-dependent density of states

$$\rho(E) = |E|/(E^2 - \Delta^2(T))^{1/2} \qquad (112)$$

for one or both electrodes, show potential for thermometry, as first observed by Giaever and his colleagues[73,74].

Superconductor-insulator-normal metal (S-I-N) junctions are easiest to apply, partly because computations of equation (39) are simpler and partly because these junctions are not subject to excess currents from secondary processes such as double particle tunnelling[75] and Josephson tunnelling. Manipulation of equation (39), and computation[76] of I and $\partial I/\partial V$ as functions of Δ/kT and eV/Δ show two particularly useful possibilities for absolute thermometry[77]:

1. For $\Delta/kT \geqslant 3$ (corresponding to $T \lesssim 0.6T_c$)

$$\sigma(0) = K^{-1}(\partial I/\partial V_{V=0} = (2\pi\Delta/kT)^{1/2}\exp(-\Delta/kT) \qquad (113)$$

so that provided Δ is known, or can be separately measured from the tunnel characteristic, T can be directly found from the zero bias conductance. For $\Delta/kT<3$, equation (113) is insufficiently accurate, but Δ/kT can still be obtained from tables.

2. For $eV<\Delta$, equation (39) yields, for S-I-N,

$$I = 2K\Delta \sum_{n=1}^{\infty} (-1)^{n+1} \sinh(neV/kT)H_1(n\Delta/kT) \qquad (114)$$

where H_1 is a modified Hankel function. For $\Delta/kT \gtrsim 3.5$ ($T \lesssim 0.4T_c$), this is adequately approximated, in the bias region $eV < \Delta - 1.5kT$, by

$$I = K\sqrt{2\pi\Delta kT}\exp(-\Delta/kT)\sinh(eV/kT) \qquad (115)$$

For $eV \gg kT$,

$$I = K\sqrt{2\pi\Delta kT}\exp(-\Delta/kT)\exp(eV/kT) \qquad (116)$$

Thus, T can be directly determined from the I–V characteristic without an intermediate measurement of Δ.

Two drawbacks mar the simplicity of these equations. First is the gap smearing typical in films (Section 3.3). The $\exp(-\Delta/kT)$ term in equation (113) gives undue weight to the smaller gaps, and this will probably not be compensated in the separate Δ measurement. Errors therefore appear in T. Second is the linear non-tunnelling background leakage conductance σ_L, which is typically about 10^{-3}–10^{-4} times K. At low temperatures this can dominate the tunnel current, but is a source of error even at higher T if the $\sigma(0)$ method is used.

In respect of method 2, each of these effects influences equation (116); the first, however, only modifies the terms containing Δ and preserves the $\exp(eV/kT)$ term, while the second adds a term linear in V which can in principle be separated out.

Donaldson and Band[78] have shown that in practice both methods can be used, in the helium temperature range and down to 0.3 K, to give absolute temperatures correct to within a few percent without any direct temperature calibration. (A "calibration" is, of course, contained in the voltage standard adopted and the text book value of k/e chosen.) They have also described a modulation system whose output displays,

$$I(dV/dI) = kT/e \tag{117}$$

directly on a digital voltmeter. Feedback loops can be easily added to subtract the effect of leakage conductances. Accuracies of order 1% in the absolute temperature have been achieved (Fig. 25). For $T \lesssim 0.2T_c$, however, the leakage

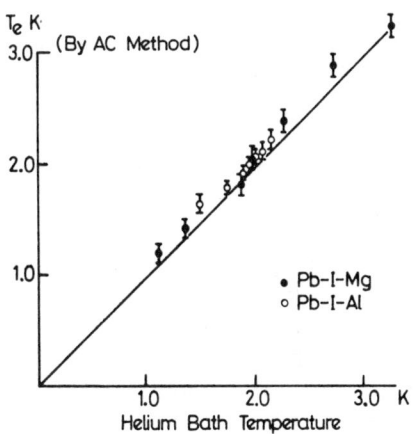

FIGURE 25. Temperatures obtained from S-B-N thermometer compared with helium bath temperatures[78].

current is too large for accurate compensation, and the thermometer is not usable.

Equation (116) has also been applied to bolometery[69], using a $Pb/Al_2O_3/Al$ junction. At $T = 1.5\,K$, where $\Delta/kT \gg 1$ and $eV/kT \gg 1$, the approximation

$$(\partial V/\partial T)I = -0.8\,\Delta/eT \tag{118}$$

is valid. With equipment similar to that used for S-N-S bolometery, $\delta T \sim 10^{-7}\,K/\sqrt{Hz}$ should be observable. In practice $\delta T \sim 10^{-5}\,K/\sqrt{Hz}$ was obtained. Improvements are possible, but leakage currents represent difficulties, and further efforts may not be profitable since transition edge and S-N-S devices appear to be adequate alternatives in this case.

6.1.4 Josephson effect noise thermometers

Temperature-dependent Johnson noise has been used as the thermometric property, with Josephson junctions as detectors, in two types of devices which are respectively voltage- and current-operated (V- and I-device)[79].

In the V-device, a Josephson junction is held at bias $V_B \ll \Delta$ by shunting it with a resistor $R \ll R_N$ (R_N is the normal resistance) and current biasing the combination just above I_c. The Josephson frequency $v_J = 2eV_B/h$ (equation 52) fluctuates because of the Johnson noise fluctuations in V_B. These are given, in a bandwidth Δf, by

$$\overline{V_n^2} = 4kRT\Delta f. \tag{119}$$

Frequency modulation theory shows that although V_n^2 is white noise, the v_J fluctuations have a finite bandwidth

$$\{\overline{(\Delta v_J)^2}\}^{1/2} = 4\pi e^2 RT/\hbar^2 \tag{120}$$

from which T can be determined absolutely if R is known.

This property has been exploited[80] using both Pb–I–Nb tunnel junctions and Nb–Nb point contacts. Typically $R \sim 10^{-5}\,\Omega$ and $V_B \sim 10^{-7}$ volts, making $v_J \sim 50$ MHz. This frequency is sampled digitally and from many successive samples, its variance (and hence the temperature) is determined by a dedicated computer. Agreement to within 1% with temperatures determined by ^{60}Co γ-ray anisotropy has been found[81].

In the I-device, a superconducting loop of inductance L is broken by a resistor R. The Johnson noise currents are then in Δf

$$\overline{I_n^2} = 4RkT\Delta f/(R^2 + 4\pi^2 L^2 f^2) \tag{121}$$

These generate a fluctuating flux $\Phi_n = LI_n$, which can be detected by coupling into a SQUID operated in the flux-locked loop mode (Section 8.1). Provided the operating bandwidth and other circuit parameters are known, T can be deduced from the appropriate $\{\overline{\Phi_n^2}\}^{1/2}$.

In practice[82], R, L, and the coupling mutual inductance (which modifies equation (121) slightly), must be carefully chosen to ensure that the Johnson noise dominates the intrinsic SQUID noise in a suitable bandwidth (typically 10^{-2} to 1 Hz). With an R.F. point contact SQUID, device noise temperatures of 0.05 mK have been achieved. The measured Johnson noise temperature was found to vary linearly with established scales to within 1%, and to agree absolutely with them to within the uncertainty of other circuit parameters ($\sim 10\%$). No thin film versions of this thermometer exist yet.

Wheatley and Webb[79] have reviewed the relative potentials of the two devices, and conclude that the I-device is superior from the point-of-view of thermal contact, and of the mass of the part which has to be cooled to ultra-low temperature. Claims of 10 μK and less have been made for the ultimate intrinsic noise temperature which might be achieved in each case, but this should be more easily reached with the I-device. With each thermometer increased sampling times become necessary at low temperatures when the variance is very small: at 1 mK a temperature measurement might require several thousand seconds.

6.2 Particle Detectors

Transition edge biased films and tunnel junctions have been applied to the detection of atomic and molecular particles of a wide range of energies using the effective temperature increase caused as a particle loses energy within the detector material.

An example of the first type[83] is the use of a 1 μm wide 500 Å thick Sn–In film held within 0.001 K of T_c to detect Ar, Ar^+, He, and He^+ particles of energies between 150 eV and 800 eV. The film is current biased, and the passage of a particle, by inducing the normal state across the whole width of the film, causes a voltage pulse to be developed, about 100 μV high and 0.1–1 msec long. There have been similar applications to molecular particles[84] and high energy α-particles[85].

Wood and White[86] used Sn–I–Sn tunnel junctions, voltage biased at $V < 2\Delta/e$ and $T \ll T_c$, as MeV α-particle detectors. In losing energy, the particle created excess quasiparticles, which were detected as a 5 μA, 400 nsec current pulse across the junction. Since each quasiparticle creation process involves an energy $\simeq 2\Delta$, this becomes the ultimate energy resolution. In practice, the improvement over Si semiconductor detectors was about ×5; however, tunnel junctions might ultimately be superior by the factor (Energy gap of semiconductor/$\Delta_{superconductor}$)$^{1/2} \sim 30$. Nevertheless, the difficulties of fabricating arrays of tunnel junctions (see Section 10.1) have so far prevented development of this as a significant nuclear physics technique.

7. Switching Devices and Amplifiers

In Sections 4 and 5 we examined a number of devices based on flux exclusion, or on magnetic transitions to the mixed or normal states. Within factors of order κ, and ignoring any demagnetisation factors, such devices are essentially sensitive to fields $\sim H_c$. In this section and the next, we are mostly concerned with devices based on the more subtle field–phase relations discussed in Section 2.7. In such devices, the critical quantity is a phase difference (typically $\gamma \sim \frac{1}{2}\pi$) induced by the presence of a flux quantum in some loop characterising the system. For device areas of order 1 mm^2, this implies field sensitivity of $\sim 10^{-5}$ gauss, many orders of magnitude below H_c; indeed, as we shall see, resolutions of $\sim 10^{-10}$ gauss are possible.

The "old" thin film cryotron (Section 7.1.1) is an S-N transition system. We include it here rather than in Section 5 for compactness, and because it serves as an introduction to its tunnelling successor, by which it has been completely superceded.

7.1 Cryotrons

Under appropriate current bias, both superconductors and superconducting weak links can be transferred from a zero voltage to a finite voltage state by a suitable magnetic field. There is obviously a potential here for switches with infinite on/off resistance, and we discuss such switches in this section.

7.1.1 Thin film cryotrons

In a thin film cryotron[87] (Fig. 26), a "gate" film, usually of tin, is held below T_c and current biased (I_g). A second "control" film of higher T_c (usually lead) is evaporated over the gate, either in a crossed film or an in-line configuration; the two films are electrically isolated, usually by a thin film of SiO. A control current I_c generates a field which induces the S-N transition in the gate film, so turning the normal resistance off and on. The entire structure is usually deposited on a superconducting ground plane (Section 3.5) to ensure uniform fields parallel to the films, and to reduce not only the film inductance (so improving switching speeds), but also the pick-up from neighbouring devices (allowing high component density).

The static gain of a cryotron is

$$G = I_{cc}/I_{gc} \tag{122}$$

where I_{gc} is the critical gate current in zero field, while I_{cc} is the control current which produces a critical field at the gate surface when $I_g = 0$. For films much thicker than the gate penetration depth, equation (94) applies, and if we rewrite equation (122) in terms of current densities and film widths W_g, W_c we obtain

$$G = K(W_g/W_c) \tag{123}$$

where $K \sim 1$. Thus, in a crossed film device, we may have $W_g \gg W_c$, so that there is current gain, and the gate current can be used, without amplification, as the control in a subsequent device. For in-line cryotons we must have $W_c \geqslant W_g$; however, the fields due to the gate and control currents are now either parallel or antiparallel, and reinforce or oppose each other, so that I_{cc} depends in an asymmetric way on the size and direction of I_g (see Fig. 27). In an appropriate bias region $\partial I_{cc}/\partial I_g$ can exceed unity. Thus, even though G itself is less than unity, the system can show dynamic gain.

In its applications, the cryotron is used, not so much to turn currents on and off as in conventional switches, but rather to steer them into alternative

Superconducting Thin Film Devices

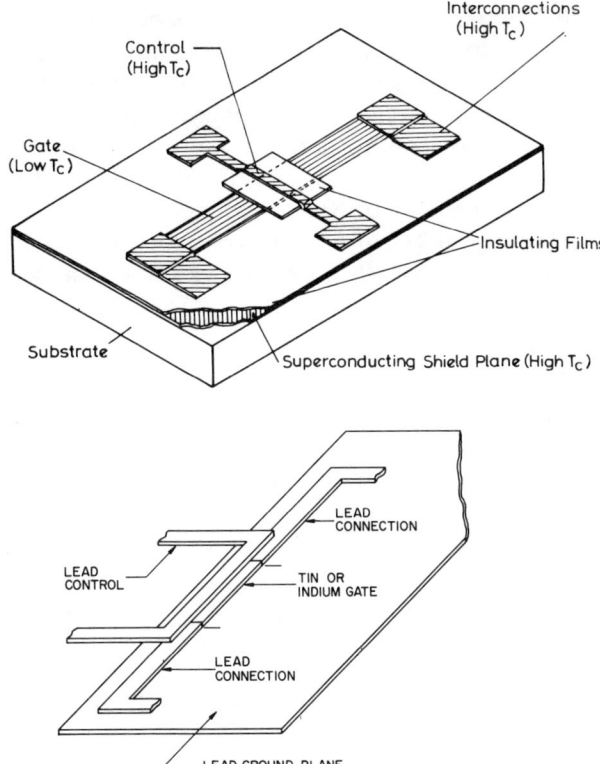

FIGURE 26. (a) Crossed film cryotron, (b) in-line cryotron.

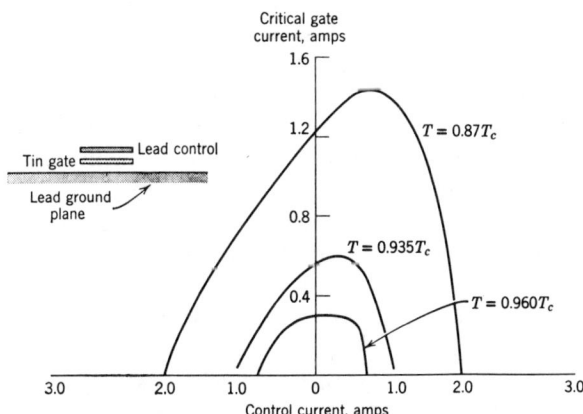

FIGURE 27. Variation of I_c and I_g for in-line cryotron.

superconducting paths. In Fig. 28 a simple logical element illustrates this point. The entire structure is superconductive, and current I_0 initially flows in arm A (having been established when B was held normal by its control). To operate the switch, a current pulse is applied to control A making gate A resistive; the voltage across it steers the current into B in a time $\tau \sim (L_A + L_B)/R_A$, where L_A, L_B are the arm inductances and R_A the resistance of gate A. The A pulse can now be turned off leaving I_0 flowing in B. The system is therefore a flip-flop, and simpler than its electronic cousins in not requiring any cross coupling between its arms. Its logical state can be determined by

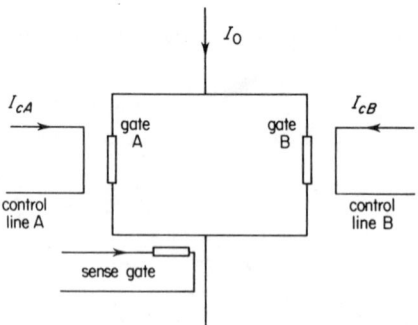

FIGURE 28. Principle of cryotron flip-flop.

using I_{gA} as the control current for a sense cryotron; the sense gate current will develop a voltage if I_{gA} is finite, but not if it is zero.

Storage is introduced if the external bias I_0 is now reduced to zero. Conservation of flux will then cause I_0 in B to be replaced by a persistent circulating current $I_0 L_B/(L_A + L_B)$. Had I_0 been in A, the circulating current would be $-I_0 L_A/(L_A + L_B)$. A logical "1" or "0" can thus be defined by the direction of the circulating current or by its size if L_A and L_B are made very different. In a complex array, the cell can be xy addressed by the appropriate choice of bias and gate lines. Readout is again possible with a sense cryotron.

Another storage system is the Crowe cell[88] (Fig. 29). Information is stored as a persistent current circulating round one or other of the holes on either side of the central bridge, whose resistance is controlled by xy addressing through insulated overlays.

Highly developed logic structures were produced on the basis of these principles[87]. For a time there were hopes that cryotrons, at densities of 10^5-10^6 cells/ft^2 might serve as computer memories with access times of < 1 nsec, determined essentially by the speed of the $N-S$ transition. However, phase boundary propagation speeds were found to limit switching to

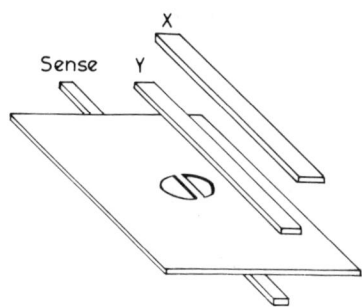

FIGURE 29. Principle of Crowe cell[88].

≳ 50 nsec. Moreover, the switching repetition rate could not exceed 10 MHz because of heat flow limitations imposed by the latent heat of transition. Ultimately competing computer technologies overtook the cryotron in switching speed, repetition rate and convenience of operation (room temperature rather than cryogenic). Even had they not, the simple cryotron would undoubtedly have been eclipsed by its tunnelling alternative, which we discuss in the next section.

Linear amplifiers, based on the high $\partial I_g/\partial I_c$ of both types of cryotron, are reviewed in Section 7.3.

7.1.2 Josephson tunnel junction devices

In the tunnelling cryotron[89] (Fig. 30), the gate element of the simple in-line cryotron has become a Josephson tunnel junction, nowadays usually Pb–In/I/Pb or Nb/I/Pb, biased at I_g, somewhat below the maximum critical current $I_J(H = 0)$. A current through a control film above, but insulated from, the junction, will reduce $I_J(H)$ to below I_g, and drive the system to a state with $V \sim (\Delta_1 + \Delta_2)/e$ (see Section 2.7). Ground planes are usually used. Logical structures such as flip-flops then become possibilities by analogy with those of the previous section.

Tunnelling systems have several advantages over earlier cryotron types:

(1) The output voltage, of order $(\Delta_1 + \Delta_2)/e$, or about 2.5 mV with Pb, is more easily detectable than the resistive state voltages typical of earlier cryotrons.

(2) The speed of the transition is ultimately governed by the Josephson plasma frequency v_s, which determines the characteristic decay time of transient fields in the junction cavity. This exceeds 10^{12} Hz, so sub-picosecond switching might be contemplated. In practice the junction capacity C dominates the switching, introducing a charging delay time $\sim C(\Delta_1 + \Delta_2)/eI_B$. Nevertheless, this can be made as small as a few picoseconds, and switching times below 29 psec (and probably of order 10 psec) have been observed.

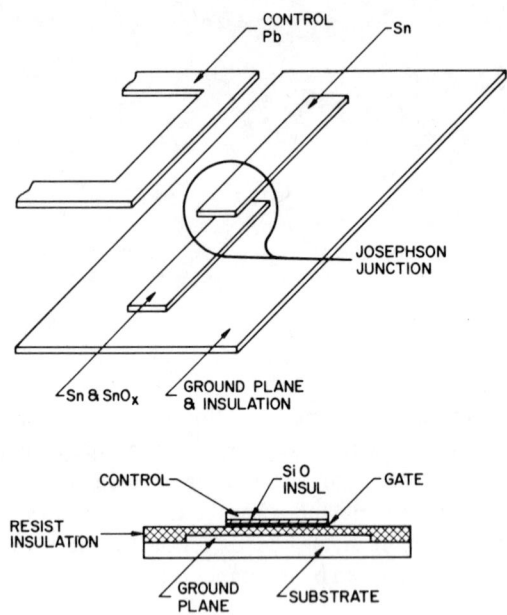

FIGURE 30. In-line tunnelling cryotron[89].

(3) Although neither type of cryotron flip-flop dissipates except while current is being steered, energy is involved during switching. For a basic cryotron this is typically 2×10^{-13} J. For a tunnel structure, by contrast, it is orders of magnitude smaller, being given by $\frac{1}{2}LI_g^2$, where L is the loop inductance: inserting $L \sim 10^{-10}$ H and $I_J \sim 4$ mA as typical, we have a dissipation of only 4×10^{-16} J.

As with in-line "old" cryotrons, the self field of I_g must be considered when calculating the critical control current I_{cc}. For junctions with $w/\lambda_J \gg 1$, where w is the junction width and λ_J the Josephson penetration depth, the junction critical current behaves approximately as[89]

$$I_J(H) = I_J(0)(1 - |H|/H_0) \tag{124}$$

where H_0 is a constant ~ 0.15 gauss; here

$$|H| = (\tfrac{1}{2}I_g + I_c)/w \tag{125}$$

Putting $I_g = I_J(H)$ and $I_c = I_{cc}$ gives a lopsided critical relation, in which high dynamic gain is again possible with suitable bias, so that the gate current of one junction can control another. This applies also[90,91] (Fig. 31) for

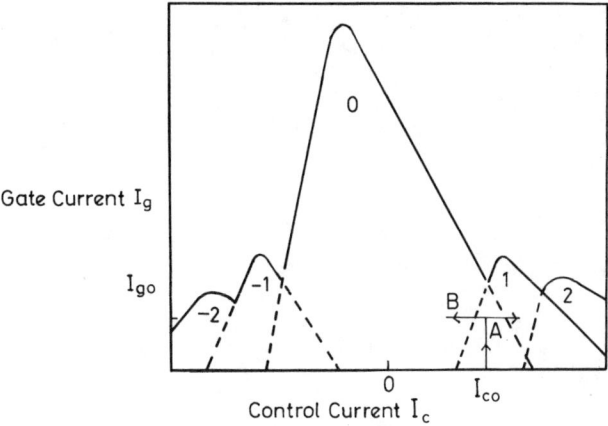

FIGURE 31. Schematic of variation of I_c with I_g for "wide" junction ($w/\lambda \sim 3$) (based on Refs. 90 and 97). The integers indicate number and direction of vortices in junction.

smaller values of w/λ_J; here, however, side lobes appear, corresponding to 1, 2, ... vortices in the junction area.

Loop structures. Most developments have used flip-flop memories. Damping is necessary to prevent overshoot as the current decays in the pulsed arm, but can be designed in, using the Stewart and McCumber[17] damping parameter (equation 66). The critical shunting condition[92] is $\beta_c = \pi L I_J/2\Phi_0$, where I_J is the maximum critical current of one junction. A recent example of a flip-flop cell is illustrated in Fig. 32. It is formally similar to Fig. 28, with a "1" or "0" being written into the system by simultaneously pulsing a word current (I_0) and one or other write current (I_{cA} or I_{cB}). When the pulses cease, a clockwise or anti-clockwise persistent current remains circulating in the loop. The state of the system can be read non-destructively by pulsing the sense line (I_s) together with I_0: if "1" is stored, the field due to these currents add, and the sense junction develops the gap voltage. The cell has strip widths down to 2 μm, and an area of 645 μm², comparable to that of static semiconductor memory cells. Typical switching currents are $I_0 = 4$ mA, $I_{cA} \sim I_{cB} \sim 15$ mA and I_{min} (the current below which a junction switches back to the zero voltage state) ~ 1 mA. Typical switching times are ~ 50 psec, but transients extend the total time for, say, the write operation to 150 psec. The energy dissipated during switching is about 10^{-16} J.

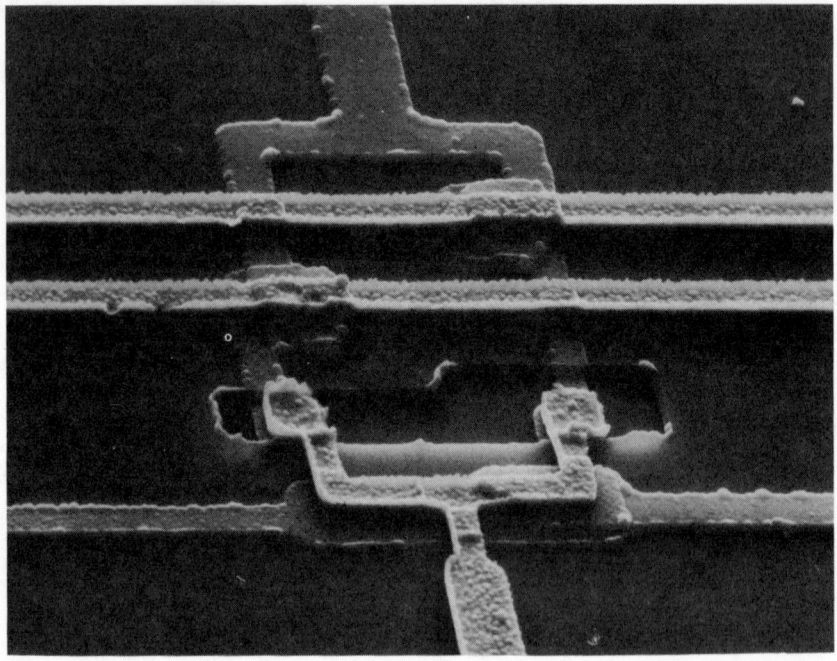

FIGURE 32. S.E.M. photograph of 645 μm^2 memory cell with 2 μm minimum line width[92].

Memory cells are inconvenient for use in logic operations (such as AND, OR, etc.), since they must be reset after each operation. Figure 33 shows a self-resetting device[93] which is more useful for such purposes. A load resistor R_L shunts a D.C.-biased junction whose state is determined by control lines. Provided $R_L \ll R_J$, most of the bias current will pass through R_L when the junction is resistive, and it can then act as the control current of the output sense junction. For AND operation the various current levels are chosen so that all control lines must be on to provide sufficient field to switch the junction; OR operates when the current levels are large enough to make any one input sufficient. When the controls are off, the junction resets automatically to the superconducting state. The inductance L of the shunt leads must be kept small to minimise the transient decay time L/R_L. Henkels et al. have observed characteristic times of 160 psec[94], but it has been estimated that with semiconductor barrier tunnel junctions (Section 10) on ground planes, switching times of ~50 psec and resetting times of 500 psec should be attainable.

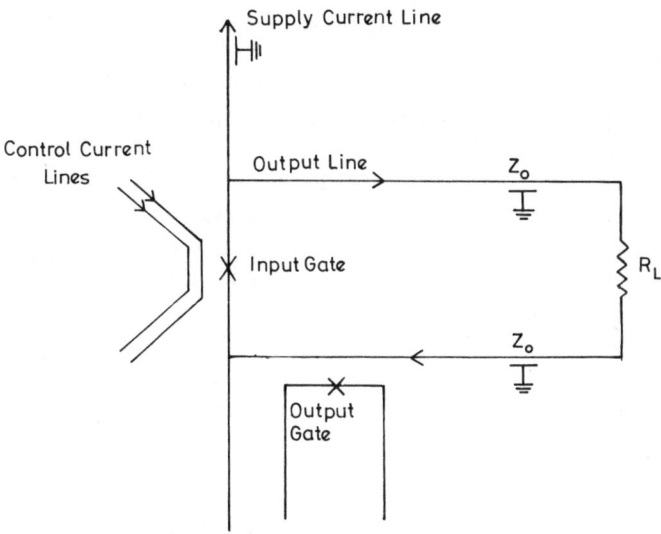

FIGURE 33. Self-resetting logical element[93].

Large scale integration of superconducting switching circuitry is now possible. For example, a 7 bit shift register clocking at more than 200 MHz has been described, and simpler systems operating at > 1 GHZ have been built[95]. Real prospects now seem to exist for computers with speeds 10–100 times faster than are currently available. They would use power levels 10^4 smaller than at present. The "Carnot" heat required to reject this energy at room temperature removes about × 600 of this gain, in a technical sense. From a design point of view, however, one is concerned with energy dissipated actually at the memory surface, and the full 10^4 factor is meaningful in this context. Processors with 30 Mbits in a 1 ft³ volume and requiring 1 W refrigeration capacity have been proposed[96].

Single flux quantum cryotrons. It is possible to use the flux in a junction itself, rather than in a loop, as the stored information. Gueret[97] has discussed the curves of Fig. 31 in detail, and has shown that the dotted extensions of the I_g–I_c lines define regions in which a particular number (N) of vortices can exist stably in the junction. In overlap areas bistable equilibrium is possible. Thus, if the bias conditions are changed to make the point (I_c, I_g) move from a bistable to a monostable region, a vortex may be forced to enter or leave the junction. If $N = 0$ initially, the sequence $(I_{co}, 0) \to (I_{co}, I_{go})$ followed by increasing I_{co} beyond A and then returning to $(I_{co}, 0)$ will cause a vortex to enter and remain stable within the system, storing a "1". It can be destroyed by crossing the other boundary at B. With $N = 1$ initially, there would have been no transition at A.

A word is thus written by simultaneously pulsing I_c and I_g. Readout is achieved by pulsing as though to store a "1": if there is a transition it will be because a "0" was stored, and energy will be released. This can be detected using a nearby junction biased just below I_J, which will be switched to the voltage state. Readout is thus destructive as both memory and sense junctions must be reset, but there is potential for a big increase in component density relative to double junction flip-flops, since junction areas as small as 0.22 mil^2 (55 μm^2) have been used.

The flux shuttle. Fulton *et al.*$^{(98)}$ have shown that a vortex can be moved around within a long junction. To see how this is possible, consider first a chain of identical junctions connected in loops of inductance L (Fig. 34), with an in-line control strip evaporated above each junction. The critical current I_c

but not more, because a larger number would require the persistent current in the loop to exceed I_c. Initially, let the loop $J3$-$J4$-$L3$ support a vortex (quantum), with a current $I_{\text{circ}} = \Phi_0/L$ circulating anti-clockwise as shown. If control 4 is now pulsed with a current $\sim 0.8 I_c$, junction $J4$ will be driven normal, and the voltage thus induced will tend to cancel the current in $L3$, while establishing current in $L4$. The effect is to transfer the vortex to the loop $J4$-$J5$-$L4$, where it remains when the control is turned off.

On the other hand, had the original anticlockwise current circulated in $J4$-$J5$-$L4$, or had both loops 3 and 4 contained an anticlockwise vortex so that there was no current in $J4$, an upwards current $\lesssim I_c$ in control 4 would not have made the total current in $J4$ greater than critical, and there would thus be no vortex transfer. It follows that the effect of sequentially pulsing the controls 1, 2, ... will be to appear to move all vortices one loop to the right as in a shift register.

Damping by shunt conductance is necessary to dissipate the vortex kinetic energy ($\sim \Phi_0^2/L$) transferred to the system by the control pulse. Failing this, the vortex may overshoot and subsequently oscillate.

A discrete structure such as this might have $I_c \sim 200\,\mu\text{A}$, $L \sim 10^{-11}$ H and a dissipation energy of $\sim 4 \times 10^{-19}$ J, and could have a transfer time of 40 psec, with a density of 10^6 bits-cm^{-2}. However, there is no reason, in principle, why the series of junctions in Fig. 34 should not be replaced by a single long one, in which the oxide barrier contains a series of vortex traps (such as magnetic impurities or local variations on oxide barrier thickness). These would act as magnetic potential wells, trapping the vortices between shift operations. Generalisation$^{(98)}$ of the Josephson equations (56, 58 and 59) to treat the fields within the barrier shows that vortices (or, if preferred, regions of spatially varying phase difference across the junction, each containing Φ_0) will travel in

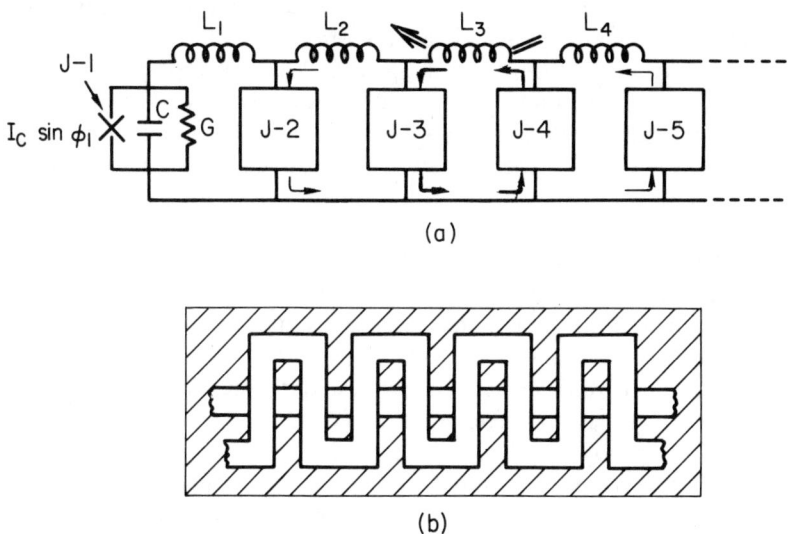

FIGURE 34. Discrete flux shuttle[98]. (a) Lumped circuit model. (b) Fabrication. In addition a control line overlays each junction.

such a barrier at $\bar{c} = \lambda_J \omega_J$, where λ_J is the Josephson penetration depth $\{c\Phi_0/8\pi^2 j_c(2\lambda + t)\}^{1/2}$ and ω_J is the Josephson plasma frequency $(8\pi e t j_c/\varepsilon\hbar)^{1/2}$ where λ is the penetration depth, t is the barrier width, ε is the barrier dielectric constant and j_c the critical current density. This gives $\bar{c} \sim 10^7$ m-sec^{-1} for a typical junction. Vortices will remain effectively discrete provided they are separated by $> 4\lambda_J \sim 2 \times 10^{-4}$ m, which thus gives the minimum separation of the traps. In operation, writing at the end of the strip would be by control line, and shifting would probably be carried out by an R.F. field of suitable wavelength, commensurate with the trap separation. Dissipation would be $\sim 2 \times 10^{-21}$ J per operation.

Preliminary experiments[98] on a 12 mm continuous junction show that vortices can indeed be trapped by a magnetic well—in this case a localised field about 3 mm long. They also show that, untrapped and undamped, a vortex will travel along the barrier at the expected velocity, reflecting from the ends as it reaches them.

7.2 Giaever Tunnelling Switches

For completeness, we recall that although quasiparticle tunnelling junctions do not pass supercurrents, S_1–I–S_2 types have a negative resistance region in the bias range $|\Delta_2 - \Delta_1|/e < V < (\Delta_1 + \Delta_2)/e$ (see Fig. 10). This raises the possibility of hysteretic switching action similar to that in a Josephson junction, and computer elements, along with other devices, were seriously considered when superconducting tunnelling was discovered[99]. Initial experiments[100] were promising, but Josephson cryotrons have since proved greatly superior in speed, besides having the advantage of zero stand-by power.

7.3 Linear Amplifiers

The application of superconductivity to the measurement of small voltages and other signals is now dominated by Josephson effect devices. (Section 8). However, several ingenious earlier uses of the basic N–S transition for such purposes are on record. They have been reviewed by Newhouse[101], and were usually based on the use of either thermal or magnetic modulation of resistance to chop a low level D.C. signal, rendering it suitable for narrow band A.C.-detection techniques. Sensitivities of 10^{-12} V were frequently attained, but time constants were a recurring problem: one is normally concerned with source impedances below $10^{-4}\,\Omega$ in these cases, and if the modulated resistance has a significant inductance, L/R values of several seconds may occur.

The crossed film grounded plane cryotron, with its tiny inductances (Section 7.1.1), proved a successful way of overcoming these problems. The cryotrons were operated in a biased small signal mode with high gain $(\partial I_g/\partial I_c)$, making cascading of stages possible. Ultimately[102] an eight stage amplifier with a current gain of 22000 and a noise current of 6.8×10^{-8} μA r.m.s. at 45 kHz was produced.

8. Galvanometers and Fluxmeters: SQUIDs

This section is devoted to superconducting quantum interference devices, or SQUIDs as they are universally known. A large literature already exists on the subject and our coverage can only be very brief: for further details and references, the reader is directed to one of the many books and review articles dealing with this topic[81,103–107].

SQUIDs are of two distinct types. First is the so-called R.F. version using a single weak link in an otherwise closed superconducting loop, while the second is the D.C. SQUID in which two roughly equal junctions are used in the loop. We will discuss their principles separately, but first make a general point. In each case the most pronounced quantum effects will be observed when the current induced in the loop by the application of one flux quantum is comparable to the critical current of a junction. Thus we expect to have

$$I_c \sim \Phi_0/L \qquad (126)$$

L being the loop inductance (see also Section 2.7). However, since to determine how the critical current varies, it is necessary for the SQUID to spend at least part of its time in the voltage biased condition, we must consider the effect of Johnson noise currents, which for wide band are given by

$$\overline{I_N^2} = kT/L \cdot \qquad (127)$$

Obviously, we wish $(\overline{I_N^2})^{1/2} \ll I_c$, to prevent spurious switching and the swamping of quantum effects. The most satisfactory combination is[104]

$$L < \Phi_0^2/4kT \qquad (128)$$

and

$$I_c \simeq \Phi_0/2L \qquad (129)$$

leading to typical values of $L \sim 10^{-9}$H and $I_c \sim 5\,\mu$A for a device operated in the liquid helium range. Note that the inductance of a single unscreened turn of radius r wound with wire of much smaller radius a is, in mks units

$$L \simeq 4\pi \times 10^{-7} r \ln(r/a) \qquad (130)$$

so that inductances as small as 10^{-9} H require loops of area about 1 mm^2, or if larger areas are to be used, substantial magnetic screening.

8.1 R.F. SQUIDs

Typical R.F. SQUID structures are illustrated in Fig. 35. In most cases the weak link has been a point contact bridging a space in a massive block of superconductor, usually niobium. The space may consist of one or two cylindrical holes, or a toroidal hole, these giving the necessary low inductance because of the screening effect of the surrounding material. Thin film structures, using tunnel junctions or other weak links in evaporated superconducting rings, have also been used (see Section 8.4).

No direct connection is made to the SQUID; instead (see Fig. 36) an R.F. magnetic flux is applied to the loop (or to a hole in a bulk structure) by a coil connected to a high Q tank current driven at its resonant frequency, usually 30–100 MHz. External unknown signals are also coupled into the SQUID magnetically.

The junction obeys equation (65)

$$I = I_c \sin\gamma + GV + C(dV/dt) \qquad (131)$$

where G includes any shunt conductance added to the link. Since I also circulates in the loop,

$$I = (\Phi_a - \Phi_i)/L \qquad (132)$$

where Φ_a is the total flux applied to the loop, and Φ_i is the internal flux.

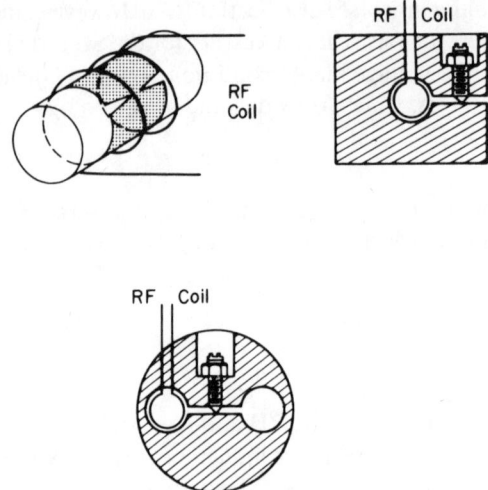

FIGURE 35. Typical R.F. SQUID structures[106].

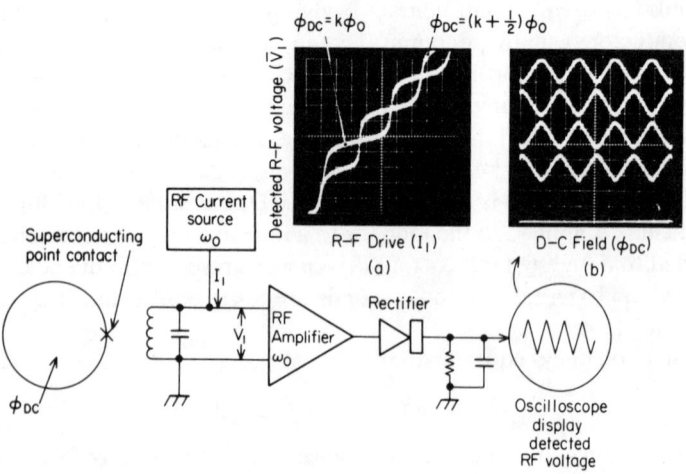

FIGURE 36. R.F. SQUID: simplified drive circuit with response[106].

The phase difference around the rest of the loop can be found by integrating equation (55), assuming that the path lies in the interior of the superconductor where $\mathbf{j} = 0$. Replacing $\int \mathbf{A}\cdot d\mathbf{l}$ by Φ_i, and using the condition that the total phase difference round the whole circuit must be $2n\pi$, where n is integral, we obtain

$$\gamma + 2\pi\Phi_i/\Phi_0 = 2n\pi \tag{133}$$

By Faraday's law,

$$V = L\, d\Phi_i/dt \tag{134}$$

and combining these equations gives

$$\Phi_a = \Phi_i + LI_c \sin 2\pi(n - \Phi_i/\Phi_0) + GL\, d\Phi_i/dt + LC\, d^2\Phi_i/dt^2 \tag{135}$$

for the relation between applied and internal flux. Quasistatic solutions of this equation, valid for frequencies $\omega \ll (LG)^{-1}$, are illustrated in Fig. 37, for the case $2\pi LI_c/\Phi_0 > 1$, which practical SQUIDs satisfy. For a given Φ_a, multiple solutions of Φ_i are possible, and on the stable regions (those of positive slope) the Φ_i solutions differ by integral multiples of Φ_0. If flux is forced to enter or leave the system, as when, for example, Φ_a is biased beyond a point such as A, at which $I = I_c$, the transition time is of order (LG); the damping is usually made large enough to prevent overshoot to stable states other than the nearest.

If the system is made to execute a loop such as $AA'B'B$, the dissipation, which is $1/L \times$ (area $AA'B'B$), is seen to be approximately $2\Phi_0 I_c - \Phi_0^2/L$, which is $\Phi_0 I_c$ if we assume $LI_c \sim \Phi_0$. This hysteresis loss is the key to the operation of the SQUID, as we now indicate following a very elementary approach by Zimmerman[106]. Consider first a total applied flux

$$\Phi_a(t) = \Phi_{DC} + \Phi_{RF} \sin \omega t \tag{136}$$

where $2\Phi_{RF}$ exceeds the width of a hysteresis loop ($2\Phi_0$), but is smaller than the full width of a quantum state ($2LI_c$). For $I_{DC} = l\Phi_0$ (l integral), $\Phi_a(t)$ will never reach any transition points and there will be no hysteretic loss; on the other hand, if $\Phi_{DC} = (l+\tfrac{1}{2})\Phi_0$, the system will continually describe loops, dissipating energy at a rate $2\pi\Phi_0 I_c/\omega$. Since this energy must be drawn from the tank circuit, its apparent Q will be correspondingly lowered. We may therefore expect the amplitude of the voltage across the tank circuit to vary periodically with Φ_a, the period being Φ_0.

In practice, the initial value of Φ_{RF} is made large enough to cause a transition whatever the value of Φ_{DC}. However, after a transition occurs and energy is drained from the tank circuit, it takes a time $\sim Q/\omega$ until Φ_{RF} has recovered sufficiently for another to happen. The peak detector output ($\bar{V}_1$) is thus a measure of the value of Φ_{RF} which will just cause a transition. Full analysis[81] shows that $\bar{V}_1$ is a continuous triangular function of Φ_{DC} (see Fig.

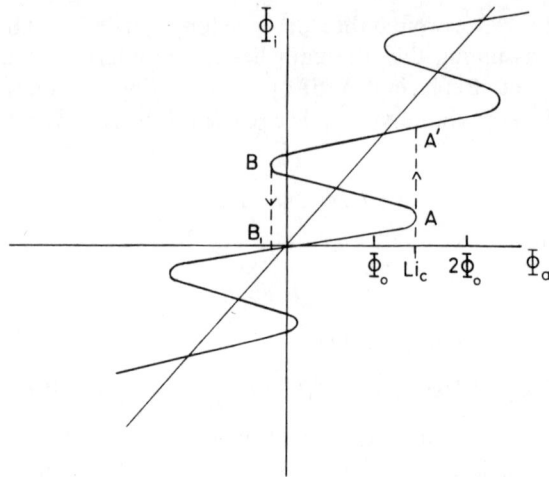

FIGURE 37. Relation between flux applied to R.F. SQUID loop and flux within loop, for $2\pi L I_c/\Phi_0 > 1$.

36) with period Φ_0 and

$$\frac{\partial V_T}{\partial \Phi_{DC}} = \pm \frac{\omega}{\alpha} \frac{L_T}{L} \qquad (137)$$

where L_T and L are the tank and SQUID circuit inductances respectively, and α the coefficient of coupling between them. The optimum value of α turns out to be $\sim Q^{-1/2}$.

In a typical system $L = 4 \times 10^{-10}$ H, $L_T = 1.4 \times 10^{-7}$ H, $Q \sim 200$, $\alpha = 0.2$, $\omega/2\pi = 30$ MHz, and V_T modulates with amplitude $\Delta V_T = 14\,\mu\text{V}$.

The electronics used to detect ΔV_T are fairly conventional. Individual flux quanta can be counted, but it is more usual to use the feedback flux nulling system illustrated in Fig. 38. A small audiofrequency flux $\Phi(\omega_a)$, is added to Φ_{DC} and Φ_{RF}, so modulating the output $\bar{V}_1$. After R.F. detection, a synchronous demodulator determines the ω_a, component of V_1 and its D.C. output is fed back to reduce this component to zero. This flux locks the system at one of the triangle points on the $\bar{V}_1 - \Phi_{DC}$ diagram (Fig. 36b) where V_1 will have components only at $2\omega_a$ and higher frequencies. The feedback current thus becomes a measure of the applied external flux Φ_{DC}. Dynamic ranges of $\pm 500\,\Phi_0$ are typical, and typical systems have a response up to 1–2 kHz.

The feedback system eliminates several low frequency drift effects, but there

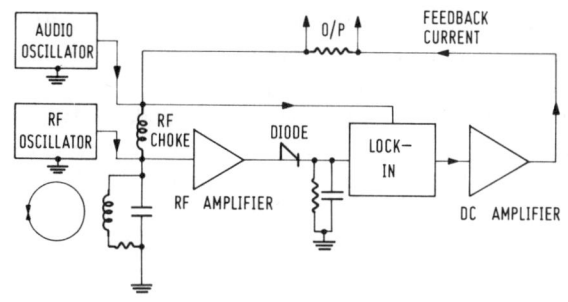

FIGURE 38. R.F. SQUID flux-locked feedback system[104].

remains inherent noise from the tank circuit and the electronics and from Johnson noise in the SQUID itself. Clarke[104] shows that for $LI_c = \Phi_0$, a full noise analysis[108] reduces to

$$(\overline{\Phi_N^2})^{1/2}/\Phi_0 = 3.1(B/v_{RF})^{1/2}(LkT/\Phi_0^2)^{2/3} \qquad (138)$$

where $\overline{\Phi_N^2}$ is the mean square flux noise in the SQUID in bandwidth B, and $v_{RF} = \omega_{RF}/2\pi$. For the example above $(\overline{\Phi_N^2})^{1/2} = 0.7 \times 10^{-4}/\sqrt{\text{Hz}}$. Amplifier noise degrades this figure somewhat, and the net signal resolution (unity $S:N$ ratio) of the system may be expected to be

$$\delta\Phi_N \sim 1 \times 10^{-4}\Phi_0/\sqrt{\text{Hz}} \qquad (139)$$

or about $2 \times 10^{-13} T/\sqrt{\text{Hz}}$ for a SQUID of area $1\,\text{mm}^2$. This resolution, which has been achieved, is about the best possible for a 30 MHz SQUID, but higher resolutions can be achieved by using higher R.F. frequencies together with improved electronics. Thus Clark and Jackel[109] have recently described a 450 MHz SQUID with $\delta\Phi_N = 3 \times 10^{-5}\Phi_0/\sqrt{\text{Hz}}$ while Gaertner[110] has achieved $\delta\Phi_N \sim 7 \times 10^{-6}\Phi_0/\sqrt{\text{Hz}}$. The upper frequency limit is set by the transition time constant $(GL)^{-1}$, and is 10–100 GHz. However, though SQUIDs have been operated in this region[111], noise performance has not been a consideration.

8.2 D.C. SQUIDs

The basic properties of double junction structures were outlined in Section 2.7. We recall that the critical current of a junction pair varies with a period corresponding to one flux quantum applied to the loop connecting them (equation 73), and that the modulation depth ΔI_c is certainly less than Φ_0/L. For a typical device loop (chosen to have $LI_c \sim \Phi_0$), a realistic estimate of ΔI_c is about $\Phi_0/2L$.

D.C. SQUIDs usually use a pair of point contacts or two thin film tunnel junctions each shunted by a conductance sufficient to prevent hysteresis (equation 66). The relevant Josephson equations for $I > I_c(B)$ have been solved for several special cases, and fuller solutions will shortly become available[112]. The general behaviour is shown schematically in Fig. 39. Note

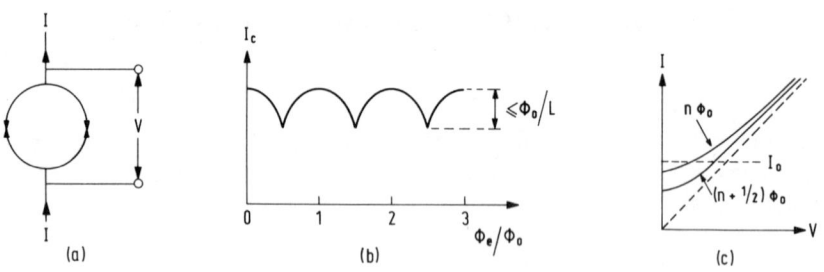

FIGURE 39. D.C. shunted junction SQUID[104]: (a) junction configuration, (b) oscillation in total critical current as a function of applied flux, (c) I–V curves for two values of applied flux, $n\Phi_0$ and $(n+\tfrac{1}{2})\Phi_0$. I_0 is the bias current.

that for $0 < V \lesssim \Delta/e$, the entire I–V characteristic changes with $I_c(B)$. If the system is biased at a constant current I_0 slightly greater than the minimum value of $I_c(B)$ the D.C. voltage will field modulate with period Φ_0 and a depth

$$\Delta V = (\Delta I_c)R' \approx \Phi_0 R/2L \qquad (140)$$

R' being the resistance at the operating point ($V \sim \tfrac{1}{2}\Delta$), which for simplicity we take equal to R, the normal resistance including the shunt. Once again digital operation is possible but, as with R.F. SQUIDs, a flux-locked loop arrangement is more common (Fig. 40). A small ($\ll \Phi_0$) audio-modulation at $v_0 (\sim 100\,\text{kHz})$ is added to the D.C. field to be measured, and the ΔV phase sensitively detected at v_0. The feedback current holds the system at $n\Phi_0$, or, for greater sensitivity, at a cusp point $(n+\tfrac{1}{2})\Phi_0$ on Fig. 39b.

For a typical case, with $L \sim 10^{-9}\,\text{H}$ (a SQUID area of $1\,\text{mm}^2$) and $R \sim 0.5\,\Omega$, the amplitude ΔV is of order $0.5\,\mu\text{V}$. There is no problem about detecting even the small fraction of this generated by the modulation flux, but

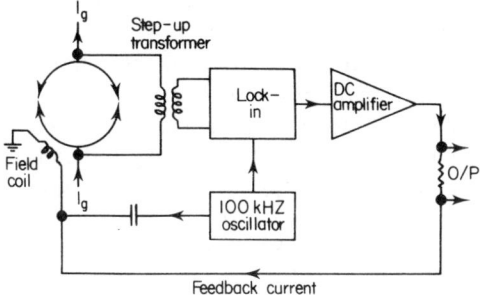

FIGURE 40. D.C. SQUID flux-locked loop circuit[104].

if the optimum noise performance is to be realised it is important to match the source impedance to the amplifiers. Clarke, Goubau and Ketchen[113,114] have done this using a high $Q(\sim 150)$ 100 kHz resonant circuit, wound with superconducting niobium, as a transformer. Johnson noise currents in the individual junctions produce two uncorrelated contributions to the SQUID noise—firstly voltage noise across the junction and secondly flux noise caused by induced currents circulating in the loop. These have been included, with the amplifier noise, together with a full discussion of feedback effects in a detailed calculation of the device noise[114]; an approximate result[113], for the flux noise in bandwidth B is

$$N = \frac{L\{(4k_B T R Q^2 + \overline{V^2})B\}^{1/2}}{\delta\Phi_N} \quad (141)$$

where $\overline{V^2}$ is the mean square amplifier noise, typically $2nV/\sqrt{Hz}$. Under optimum conditions the ideal noise at 4.2 K is $\delta\Phi_N = 3.2 \times 10^{-5}\Phi_0/\sqrt{Hz}$, and experimental results only 10% poorer than this were achieved[114]; at 1.8 K a value of $2.0 \times 10^{-5}\Phi_0/\sqrt{Hz}$ at 1 Hz was reached. The slewing rate was $2 \times 10^4\Phi_0\text{-sec}^{-1}$ and the dynamic range $\pm 3 \times 10^6$ for a 1 Hz band. A limitation of the system is its maximum bandwidth, which is restricted to about 2 kHz because of the 100 kHz used in the current driving the resonant circuit. Commercially available R.F. SQUIDs actually have very similar performance, but in special applications R.F. devices can be operated with bandwidths many orders of magnitude greater[109,111], because the bias frequency can extend up to many GHz.

Below about 0.1 Hz, SQUIDs (like other devices) show significant $1/f$ noise which degrade the above figures. We do not have space to review this here: a full discussion is given by Clarke et al.[114].

8.3 Application of SQUIDs

We defer for the moment a discussion of the relative merits of D.C. and R.F. SQUIDS, and of the programmes in developing thin film structures, and now turn to some of the ways in which SQUIDs have been applied, without distinguishing between types. For further details see the articles by Clarke[104,107] or by Giffard et al.[81], or the 1974 summary of progress by Kamper[115].

8.3.1 Magnetometry

With a basic resolution of $\leqslant 10^{-4}\Phi_0/\sqrt{\text{Hz}}$, corresponding to

$$\leqslant 2 \times 10^{-13} T/\sqrt{\text{Hz}}\, (2 \times 10^{-9}\, G/\sqrt{\text{Hz}})$$

for an area of $1\,\text{mm}^2$, SQUIDs have a sensitivity about three orders of magnitude superior to the flux gate magnetometer, the best available competitor. A major problem, however, is that in virtually any unscreened environment, whether laboratory or field, the variations in ambient field are relatively very large. The diurnal variation of the earth's field is about 10^{-4} gauss, and it is common to have mains-related fields (50 or 60 Hz and their harmonics) of 10^{-2} gauss and more. The earth's field itself, at ~ 1 gauss, can be sufficient to reduce or eliminate the critical current of a Josephson junction. Shielding is therefore usually necessary, with the cryostat inside, for example, one or more mu-metal shields; very often, too, the SQUID itself is mounted inside a superconducting shield (Section 4.3) which freezes the field at the value it had when the apparatus was first cooled.

Superconducting flux transformers (Fig. 41) have had an important role to play in SQUID magnetometry. For example, they make it possible to work with specimens too large to fit inside a SQUID, or which must be at a distance from the SQUID because they are exposed to high ambient fields. Schematically, a transformer consists of two loops (area A_1 and A_2, inductance L_1 and L_2 respectively) connected in series by superconducting leads whose inductance we assume negligible. The loop A_1 is exposed to the field to be measured (H_e) while A_2 is coupled to the SQUID with mutual inductance M. The supercurrent induced by H_e causes a flux

$$\Phi = MA_1 H_e/(L_1 + L_2) \qquad (142)$$

to be applied to the SQUID. Equation (142) suggests that a transformer, with suitably chosen parameters, might increase the basic field sensitivity $(\delta\Phi_N/A_{\text{SQUID}})$ of a SQUID without limit; in practice a factor of about 10 has been achieved. Careful discussions of coupling considerations[114,116] suggest that the figure of merit

$$L_2 \langle \delta\Phi_N \rangle^2/2M^2 \qquad (143)$$

is the most reasonable with which to categorise the performance of a SQUID-

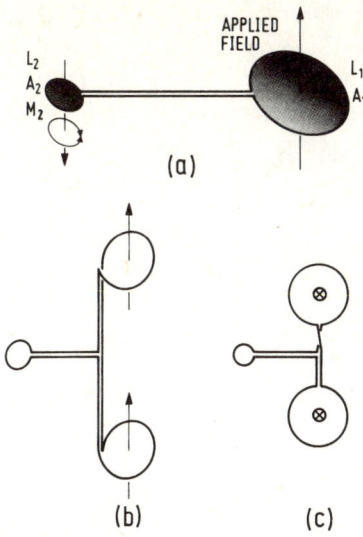

FIGURE 41. (a) Flux transformer, (b) and (c) gradiometers[104].

transformer combination in magnetometer and other applications. It corresponds to the energy resolution per Hz referred to the input coil coupled to the SQUID, and typical values are of order $(0.2–5) \times 10^{-29}$ J-Hz^{-1}[114].

Flux transformers with two equal opposed primary coils (see Fig. 41) act as gradiometers, because uniform applied fields generate no supercurrent in them, and only differences in flux linkage are observed. Suitable coil arrangements allow various gradient components, such as $\partial H_x/\partial x$, $\partial H_y/\partial z$, etc. to be measured; with more complex opposition arrangements[117], it is possible to discriminate against uniform gradients and so measure higher order spatial differentials of the field. Gradients are sometimes of interest in their own right, but the main advantage here is the possibility of detecting the magnetic effects of local dipole sources in the presence of large background fields, including mains frequency signals, which are more distantly generated, and usually more uniform. Accurately balancing the two pick-up coils can be a difficulty: usually a small piece of superconductor, positioned within one of the coils, is adjusted to slightly distort the field linkage in that coil. However, balance to 1 part in 10^6 is possible and gradient sensitivities of 3×10^{-12} gauss/cm-$\sqrt{\text{Hz}}$ have been obtained[118].

The best advertised use of SQUID magnetometry is probably its application to magnetocardiography (MCG) and other studies of physiological magnetic fields, by Cohen and others[119]. Some fields (for example those due to the alpha rhythm in the brain) are at the 10^{-8} gauss level and to be detectable require the best screening facilities. However, cardiac fields may rise to 10^{-6} gauss and are easily recorded with a simple unscreened gradiometer. MCG systems requiring helium refill only a few times per month are readily available, and the subject is of growing clinical interest[120]. For example foetal MCG traces are more easily distinguished from maternal traces than are the corresponding ECG measurements[121].

A full review of magnetometry applications would be impossible here. SQUIDs have been used in NMR[122], in millikelvin thermometry using the paramagnetic susceptibility of microgram specimens[82], in studying magnetism in tiny biochemical specimens[123], as detectors in gravity wave experiments[124], and in many other ways. Simply referring to four successive articles in a recent conference report, one finds SQUIDs being used to detect asbestos particles in the human lung[119]; tracking the magnetic equivalent of a small car over a 120 ft × 120 ft area with an accuracy of about 1 ft[118]; measuring the position of a levitated ball, and so, from the fluctuations in gravitational acceleration ($\Delta g/g \sim 10^{-11}$), analysing solid earth tides and other oscillations[125]; and appearing in a design study for an orbiting test of general relativity in which a gyroscopic precession of 6 arc-seconds/year will be detected with a resolution of 3.5×10^{-3} arc-seconds[126].

Geophysics is a particularly promising field which SQUID magnetometry has begun to invade. Passive studies of local geomagnetic fields may be useful in mineral and other surveys[127], and in elucidating the piezomagnetic precursors of seismic disturbances such as earthquakes[128]. Signals in various narrow bands have different significances[129]: Thus, the interaction of the solar wind and earth's magnetospheric plasma produces characteristic resonances in the band $(24\,h)^{-1}$ to $(1\,sec)^{-1}$, while Schumann resonances, caused by the transmission of fields due to lightning flashes around the world appear at 8 Hz, 14 Hz and higher. Moreover, active magnetometry, performed by studying the frequency dependent response at various peaks to fields generated by coils laid out on the earth's surface, yields useful information on the electromagnetic wave impedance at different depths[130]. The advent of SQUID magnetometers has made it possible to reduce, by several orders of magnitude, the coil currents necessary to generate measurable responses. It will now be much easier to make measurements in difficult places: for example, to have to generate and circulate 100 A at an arbitrary frequency in a 1 mile diameter coil in a desert location would be daunting, but there would be little difficulty if only 10 mA were required.

8.3.2 Ammeters and voltmeters

The magnetic field sensitivity of a SQUID can obviously be applied to the detection of currents, and in turn to measuring the fields which produce them.

As to ammeters, if we imagine the unknown current to be perfectly coupled to the SQUID by a single turn coil of inductance L (i.e. just the SQUID inductance), the change in I giving one SQUID period will be Φ_0/L. This is essentially the SQUID critical current, which is $\sim 1\,\mu A$. Even with a ten turn coil and a resolution of $10^{-4}\Phi_0/\sqrt{Hz}$, the current resolution would still only be $\sim 10^{-10}\,A/\sqrt{Hz}$, a figure which does not approach the capabilities of other current measuring instruments.

Nevertheless, SQUID ammeters have found a use for the specific purpose of using the truly zero D.C. input resistance which the instrument will have if the coupling coil is superconducting. This property makes it possible to design experiments in which the current to be measured is persistent. Thus D.C. bridges using superconducting inductors have employed SQUID ammeters to detect persistent out of balance current signals. Inductance changes of order $\Delta L \sim 10^{-13}\,\mu H$ should be resolvable. Such D.C. measurements not only eliminate high frequency capacitative and eddy current effects, but they permit one to distinguish between kinetic and magnetic inductance[131] and give information on the penetration depth. They can also be used to investigate departures from the simple $\sin\phi$ dependence of the Josephson current[132]. Persistent current ammeters have also been applied in the measurement of current ratios for standards purposes[133]: if n wires carrying a current I_1 in one direction and 1 wire carrying I_2 in the opposite direction are enclosed in a long superconducting tube, tightly drawn down against the wires, then the high magnetic coupling will cause a persistent current (nI_1-I_2) to flow on the inside of the tube and to return on the outside. This is transformer-coupled into the SQUID, and used as an error signal in a nulling arrangement to establish a $I_1:I_2$ ratio of $1:n$. Ratios of 100:1 have been measured to 1 part in 10^9 with a resolution of $5 \times 10^{-11}\,A$.

A typical SQUID voltmeter is shown in Fig. 42. The unknown voltage V_0, of source resistance R_0, is in series with a standard resistor R_S and a coil providing coupling to the SQUID (coil inductance L', mutual inductance $M = \alpha\sqrt{LL'}$)[104]. The SQUID operates in a null mode, with the current $V_0/(R_0+R_S)$ being cancelled by current feedback to R_S. The size of the feedback current thus determines V_0. For a loop current gain G, the input impedance is $\sim GR_S$; the gain also shortens the natural time constant $L'/(R_0+R_S)$ considerably.

A basic advantage of this system is that the inherent flux noise of the SQUID is substantially cancelled. Giffard et al.[81] show that provided the time constants are properly related to the measuring bandwidth B, the voltmeter noise figure is, for $R_0 \gg R_S$,

$$K = 1 + (\delta\Phi_N)^2 R_0/4k_B T M^2 \qquad (144)$$

(Noise figure is the ratio of the total voltmeter mean square noise to that of the

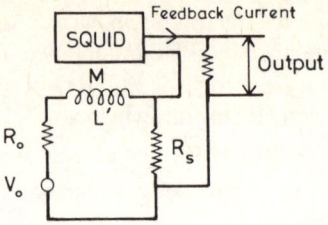

FIGURE 42. SQUID voltmeter.

Johnson noise in the source resistor itself.) The noise temperature T_N, defined as the value of T for which $K = 2$, is thus

$$T_N = (\delta\Phi_N)^2 R_0/4k_B\alpha^2 L'L \qquad (145)$$

Inserting typical values ($\delta\Phi_N = 10^{-4}\Phi_0$, $\alpha = 1$, $L \sim 10^{-9}$ H) we have $T_N \sim 10^{-6}(R_0/L')$ so that for reasonable values of L'/R_0 (say 1–100 seconds, bearing in mind that G reduces the time constant essentially to L/GR_0), T_N is far below helium temperatures. That is to say, the system noise is dominated by the Johnson noise in the resistor, for which

$$\begin{aligned}(\overline{V_N^2})^{1/2} &= (4kTR_0)^{1/2}/\sqrt{\text{Hz}} \\ &= 10^{-11} \text{ volts/(ohm-Hz)}^{1/2} \text{ at } 4.2\,\text{K}\end{aligned} \qquad (146)$$

and this, rather than SQUID noise, determines the voltmeter sensitivity. We have already seen this property applied in the I-device Johnson noise thermometer discussed in Section 6.1.4, but for true voltmeter applications we can see why SQUIDs are normally thought of as suitable only for cryogenic environments and low source impedances. For a helium temperature $10^{-8}\,\Omega$ source, resolutions of $10^{-15}\,\text{V}/\sqrt{\text{Hz}}$ have been achieved, making the SQUID the most sensitive voltmeter yet produced, but even with very sophisticated electronics the system noise is lower than that of conventional voltmeters only for source impedances up to $1–10\,\Omega$.[134]

8.4 R.F. versus D.C.; Point Contacts and Thin Films

As equation (138) shows, R.F. SQUIDs can achieve flux resolutions of order $1 \times 10^{-4}\Phi_0/\sqrt{\text{Hz}}$ at 30 MHz and might reach about $6 \times 10^{-7}\Phi_0/\sqrt{\text{Hz}}$ at 100 GHz, the upper reasonable frequency limit. For D.C. SQUIDs, one might expect from equation (141) to reach $2.5 \times 10^{-6}\Phi_0/\sqrt{\text{Hz}}$ with a $5\,\Omega$ device.

From a resolution point of view, therefore, the two devices are competitive, and *a priori* one might expect most workers to avoid the inconvenience of having to include 30 MHz lines or 100 GHz waveguides in their cryostat design, by choosing D.C. SQUIDs, with their few "D.C." (usually 100 k Hz) connecting lines, whenever possible.

By the same token, thin films, which need no adjustment, should long since have ousted point contacts as basic SQUID elements. It would seem obvious, too, that the opportunity to develop instruments with all components on a single chip, or to produce higher order gradiometers and cascaded structures, should have led to intensive industrial activity paralleling that in cryotrons (Section 7.1.2).

Yet, at the time of writing, all commercially available SQUIDs are R.F. operated, and of these about one-half are point contact types[135]. Moreover the highest order gradiometer yet made is Zimmerman's 12-hole point contact bulk structure[117]. Clearly, it is of interest to enquire why this should be so.

As between point contacts and films, the key is in the 10^{-9} H inductance which we require of a SQUID loop. With point contacts in relatively long holes in bulk specimens this is easily achieved with reasonable dimensions. By contrast, as equation (130) showed, an evaporated ring of this inductance would require tiny dimensions, which call either for sophisticated evaporating techniques, or else for the extra stage of covering the junction area with a ground plane. Tiny loops also present more difficulty in respect of coupling coils. Thus, if only small numbers of SQUIDs are needed, thin film structures may cost more in time and facilities to make. Surprisingly, too, the all-niobium point contact structure is remarkably stable, and repeated use over months and years without adjustment is quite common.

The D.C. versus R.F. question is partly historical, since early D.C. SQUIDs had a poorer noise performance through being badly matched to their room temperature amplifiers. Largely, though, the preference for R.F. arises because of the difficulty of producing two weak links which start, and remain, more or less equal. This may seem paradoxical, in view of the computer researchers' success (Section 10) with just this problem. Once again, the explanation lies in the costs and time involved: these are on a scale not likely to appeal to workers who have R.F. SQUIDs available anyway.

However, it should not be felt that the present polarisation towards R.F. and point contacts is very extreme. In any case, the position is rapidly changing. As a matter of fact, much of the most significant physics done with SQUIDs has used a SLUG[136], an inexpensive form of bulk D.C. device. Likewise, at least one of the R.F. devices currently on sale uses a thin film[135]. The active element is a 300 Å metal layer evaporated on a quartz rod as a cylinder 7 mm long × $1\frac{1}{2}$ mm in diameter, with a narrow bridge 0.5 µm by 5 µm long scratched or etched in its circumference[105]. This solenoidal geometry provides the necessary low inductance. It has even been shown that R.F. SQUID characteristics can be obtained from a thinned cylindrical film of sputtered Nb without any localised link or bridge in it[137].

Other thin film R.F. SQUIDs have been produced. For example, Palmer and Decker[138] have described photoresist masking and etching techniques by which bridges 0.5×5 μm can be formed in 300 Å Ta and other metal films, and Notarys and Mercereau[52,139] have applied it to the fabrication of proximity effect weak links (see Section 3.4) in which an overlay of normal metal partly suppresses superconductivity in a narrow bridge. With this technique they have produced complete R.F. SQUID magnetometers (see Fig. 43) and other instruments by microcircuit techniques, with inductors, capacitors, and coupling coils all evaporated on the same chip. A 10 GHz thin film SQUID yielding $10^{-5}\Phi_0/\sqrt{Hz}$ has also been produced[140].

There is also a marked move towards D.C. in thin film SQUIDs. For

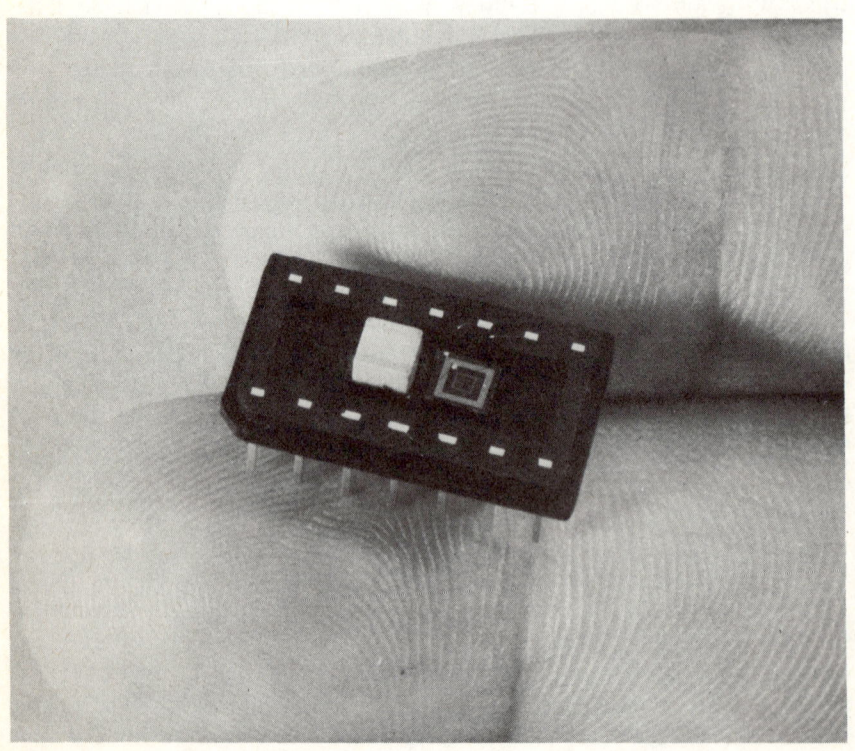

FIGURE 43. R.F. thin film SQUID magnetometer[139].

example, the Notarys–Mercereau proximity effect bridges were sufficiently similar and reliable to allow the formation of double junction galvometers by the microcircuit techniques[141]. The basic loop had an area of only 500 μm^2, and an inductance as small as 8×10^{-11} H. More recently, as techniques of producing reliable tunnel junctions on Pb, Pb/In, and Nb films with inexpensive equipment have been developed (see Section 10), this type of link

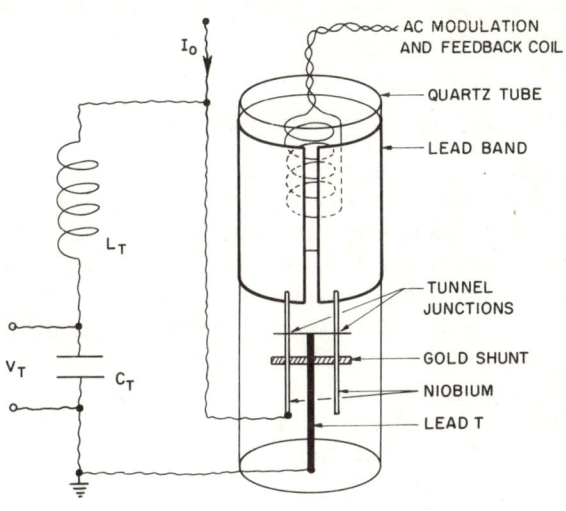

FIGURE 44. Cylindrical D.C. SQUID[113,114].

has begun to appear in reliable D.C. SQUIDs. Thus, Clarke, Goubau, and Ketchen[114] have developed a thin film device (Fig. 44) based on Nb–NbO$_x$–Pb tunnel junctions, with an inductance of 10^{-9} H. Ground planes are desirable, but can be dispensed with. The resolution is $3 \times 10^{-5} \Phi_0/\sqrt{Hz}$, the best yet obtained with D.C. SQUIDs. It is rugged and reliable, and provided the lead films are covered in a waterproof coating, such as a thin glue film, will stand thermal cycling indefinitely. An Nb–NbO$_x$–Nb version[142] of the device has proved almost indestructible whether by cycling or sheer bad mechanical handling[143]. Ketchen et al.[144] have recently developed this device, producing a $\partial H_x/\partial z$ gradiometer deposited entirely on one substrate with a resolution of $10^{-4}\Phi_0/\text{cm-Hz}^{1/2}$, corresponding to a gradient of

3×10^{-10} gauss/cm-Hz$^{1/2}$. Balancing in a uniform perpendicular field is carried out to 1 part in 10^5 by moving co-planar films, and the entire system rejects parallel fields (H_y, H_z) to better than $10^4:1$.

In the author's opinion D.C. thin film SQUIDs will have come fully into their own within a few years and should form the basis of a new generation of commercial devices except in cases where there is a need for the highest bandwidths coupled with minimum noise (Section 8.2). Tunnel junctions, whose critical currents are essentially temperature independent for $T \leqslant 0.6 T_c$, present the least problems of temperature control, and should be the most popular form of SQUID element.

9. Microwave and Infrared Energies and Frequencies

We now discuss applications of superconductive properties which are more frequency-selective than the broadband heating effects considered in Section 6. We deal with the incoherent excitation of quasiparticles above the energy gap, and also with the coherent Josephson A.C. effect. Although the latter does not specifically involve the gap, the A.C. supercurrent amplitudes decrease sharply for $eV > 2\Delta$[16], so that in both types of processes we are restricted to characteristic energies up to a few times Δ, that is up to ~ 20 meV (equivalent to frequencies up to $\sim 10^{13}$ Hz).

Many "devices" in this category employ Giaever tunnelling to study the basic physics of superconductivity itself. In a more general book such as this we have given little weight to these; a comprehensive account is given by Solymar (see Appendix).

9.1 Energy Gap Measurements and Phonon Spectroscopy

Giaever tunnelling was first applied to measuring superconducting energy gaps using the singularities in the I–V characteristics (Fig. 10). Later, quasiparticle densities of states were determined using direct dI/dV measurements on S-I-N junctions at $T \ll T_c$ (equation 41). These were extended to situations in which simple BCS densities of states would not be expected, such as proximity effect sandwiches, vortices in the mixed state, superconductors doped with magnetic impurities and others (see Solymar, appendix).

One discovery was that even in straightforward cases, the density of states shows differences of several percent from BCS behaviour (Fig. 45)[145]. The resolution of these anomalies was enhanced by second harmonic type (d^2I/dV^2) systems[146], and they were found to be particularly pronounced in materials, such as lead, for which there is strong coupling between electrons and the lattice vibration spectrum $F(\omega)$[147]. Moreover, it was found that the structure in (dI/dV) reflected singularities in $F(\omega)$ such as van Hove[148] critical points. Schrieffer and others[149] modified the BCS interaction to

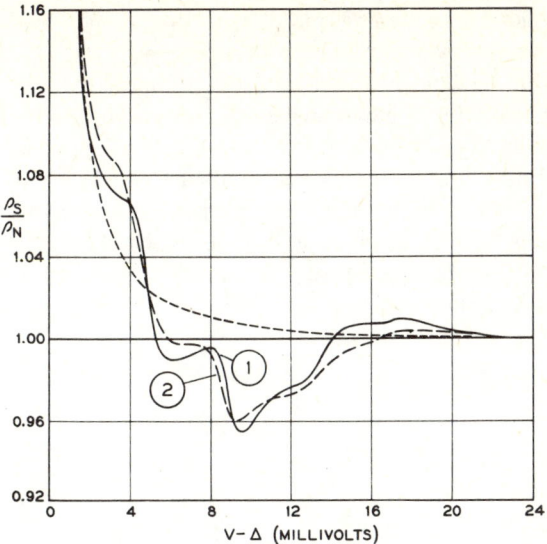

FIGURE 45. Tunnelling densities of states: BCS theory ① lead (theoretical) using simple model phonon spectrum, and ② lead (experimental).

include the effects of a screened Coulomb repulsion between electrons (expressed as a pseudopotential μ^*) and also an electron–phonon interaction $\alpha^2(\omega)F(\omega)$. They showed that the effect is to make the energy gap parameter complex

$$\Delta(\omega) = \Delta_1(\omega) + i\Delta_2(\omega) \qquad (147)$$

and that the density measured by (dI/dV) is

$$\rho(\omega) = \text{Re}\{|\omega|/\sqrt{\omega^2 - \Delta^2(\omega)}\} \qquad (148)$$

In principle, then, $\rho(\omega)$ can be used to determine μ^* and the function $\alpha^2(\omega)F(\omega)$. McMillan and Rowell[147] have described in great detail the experimental and theoretical procedures necessary to do this. Remarkable agreement with the values obtained by other techniques were found: in particular, there were excellent correlations between tunnelling and neutron diffraction results for $F(\omega)$. The phonon spectra of a wide range of metals has now been studied, and the method has recently been extended to normal metals such as Cu and Ag by using the proximity effect to induce superconductivity in them[150].

A principal use of tunnelling phonon spectroscopy has been in the verification of a theory by McMillan[151] of superconducting critical temperatures. For example, by determining μ^* and $\alpha^2 F$ in superconducting alloys of a particular structural type, one may make estimates of the maximum possible T_c of any alloy of that type. Details of the theory and its tests need not concern us here, but it should not be overlooked that substantial data of use outside superconductivity is now available[152,153], particularly on changes in $F(\omega)$ by phonon softening as one element is progressively alloyed with another (Fig. 46), and on various electronic quantities such as μ^*, and the electron–phonon coupling constant λ.

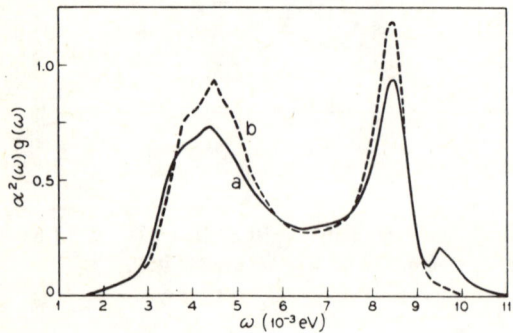

FIGURE 46. $\alpha^2(\omega)F(\omega)$ obtained by gap inversion[152]. (a) $Pb_{0.97}In_{0.03}$, (b) pure lead.

9.2 Inelastic Tunnelling: Infrared and Raman Spectroscopy

Inelastic tunnelling occurs if an electron crossing a tunnel barrier loses energy within the barrier by exciting some quantum process of characteristic energy E_s. At $T = 0$, when the most energetic electrons come from the Fermi surface in one electrode, the inelastic process requires bias voltages for which $eV \geqslant E_s$. If dI/dV, or d^2I/dV^2, is plotted against V, there will thus be structure in the traces at the onset points.

Tunnel barriers will not usually sustain voltages much greater than 1 V without electric field breakdown. The principle is thus useful for studying excitations in the 10 meV–100 meV range. Typical of these are vibrations in organic molecules, and the first work in this field, by Lambe and Jaklevic[154] demonstrated C–H stretch and other modes in organic molecules evaporated into the barrier after oxidation. Theoretical models were developed by these and other workers[155]. The technique has now been refined to the point where the resolution is many times that of conventional spectroscopy (see Fig. 47) being limited only by kT and modulation to about $7\,\text{cm}^{-1}$[156]. It has great sensitivity and microgram quantities deposited on a junction are easily detectable.

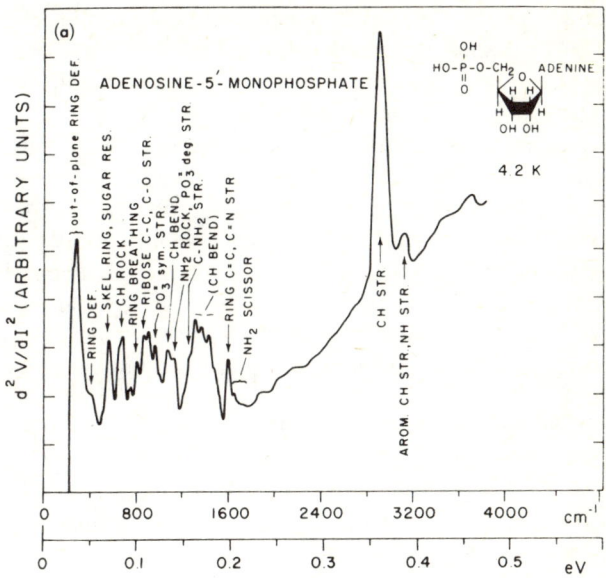

FIGURE 47. Tunnelling spectrum of adenosine-5'-monophosphate in triply distilled water[156].

Notice that although cryogenic temperatures are required to reduce kT smearing, it is not strictly necessary to use a superconducting film in either electrode. However, the logarithmic singularity in the superconducting density of states (Fig. 8) represents an even narrower energy probe than the $k_B T$ width of the Fermi tail in a normal metal, and so improves the resolution; an offsetting disadvantage can be the appearance of the phonon structure of Section 9.1 in the infrared spectrum.

It has recently been shown[156] that, as an alternative to evaporation, which may decompose large organic molecules, the dopant can be introduced in suitable quantities by dropping a dilute solution of it on to the oxidised metal surface. The excess is spun off at 3000 r/min. In this way molecules as large as DNA have been investigated, though they yield rather broad spectra. Molecular degradation of molecules by electron irradiation in an electron microscope has also been studied[157], and the method may have value in surface chemistry studies.

It is now clear that agreement between tunnel spectroscopy and other methods is best for internal modes of oscillation of organic molecules. (For example, within 1 meV for modes in a benzoate ion[156], but less good for oscillations of polar branches of a molecule (e.g. OH^- stretch vibrations).) The latter broaden and vary in position by up to 10%, in a way which depends on the nature of the oxide and of the final junction electrode. This appears possibly due to an image dipole effect in the final electrode, but might also be due to hydrogen bonding between the OH^- and the metal atoms.

Recently the method has been extended to the observation of electronic transitions in complex molecules including copper phthalocyanin and some laser dyes[158]. Conductance changes of a few percent at about 1 volt are observed. Unexpectedly, signals from certain optically forbidden singlet–triplet transitions were found to be as strong as permitted singlet–singlet transitions. No theoretical explanation has been given.

9.3 Phonon Generation and Detection

When electrons tunnel across a Giaever barrier (Section 2.5) biased at V, the new quasiparticles usually initially lose energy (Fig. 48) by relaxing to the gap edge in one or more stages, emitting phonons with a continuous energy (Ω) distribution from $\Omega = 0$ up to $\Omega = eV - 2\Delta$[159]. The decay process is completed, most probably from the gap edge but occasionally before relaxation is finished, by interaction with another quasiparticle to form a ground state pair, together with the emission of a recombination phonon ($\Omega \geqslant 2\Delta$).

Both the relaxation and the recombination spectra may be seriously modified before the phonons can traverse the superconductor and enter the substrate[160]. In strong coupling superconductors (Section 9.1), such as tin or lead, phonons with $\Omega > 2\Delta$ quickly cause a pair breaking process, and are reabsorbed generating new quasiparticles. The net result is a down conversion of most $\Omega > 2\Delta$ phonons, and a final output spectrum largely confined to the range $0 < \Omega \leqslant 2\Delta$ with a strong peak at 2Δ (corresponding to 290 GHz in Sn and 630 GHz in Pb). In aluminium however, the weak coupling allows the phonons to escape with a much less modified spectrum, which although peaking at $\Omega = 2\Delta$ extends to $\Omega = eV - 2\Delta$, where there is a discontinuity, beyond which a small tail extends to $\Omega = eV$. Phonon frequencies greater than $24\Delta_{Al}/h (\geqslant 1\,\text{THz})$ have been observed in this way.

Generation spectra similar to those of tunnel junctions can be obtained

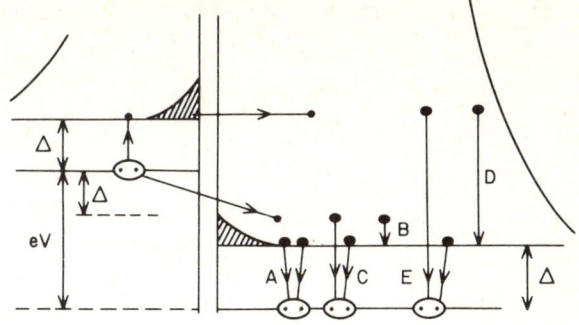

FIGURE 48. Quasiparticle decay, following S–I–S tunnelling processes (see Fig. 9b) with emission of phonons of energy Ω: A (recombination) $\Omega \geqslant 2\Delta$; B (relaxation) $\Omega \leqslant eV - 2\Delta$; C (recombination) and D (relaxation) continuous spectra with peak at $\Omega \sim eV$; E (recombination) $\Omega \geqslant 2\Delta + eV$.

from single films if the excess quasiparticle population is directly created by photon irradiation (laser, microwaves) or by black body phonons (normal metal overlaid heater). Dynes and Narayanamurti[161] have described a phonon fluorescence peak at 2Δ produced in this way.

Tunnel junctions may also be used for phonon detection, because for $\Omega \geqslant 2\Delta_s$ (the energy gap in the detecting superconductor) an incoming phonon will break pairs and so create excess quasiparticles, which can be observed as an excess current δI_s in the bias range $eV_s < 2\Delta_s$. Such detector junctions were first used to observe 2Δ phonons which had travelled down a quartz rod from a pulsed generator junction deposited at the opposite end. The frequency could be moderately tuned by magnetic field variation of Δ_g. Later Kinder[162] introduced frequency selectivity using the high energy relaxation spectrum obtainable from an aluminium junction. He took advantage of the high pass filtering property of the detector (and especially its sharp cut-off at $2\Delta_s$) and measured the synchronous response $(\partial I_s/\partial V_g)_{V_s}$ as a function of V_g, which gives (to first order) a direct plot of the phonon spectrum. For voltages such that $eV_g - 2\Delta_g \geqslant 2\Delta_s$, it turns out[163] that because of the discontinuity at $\Omega = eV_g - 2\Delta_g$, any structure in the response (best seen in $(\partial^2 I_s/\partial V_g^2)_{V_s}$) is due to phonons arriving at the detector in a very narrow band around $\Omega = eV_g - 2\Delta_g - 2\Delta_s$. In this way (Fig. 49) spectroscopic studies of phonons up to 1130 GHz in a bandwidth of 23 GHz have proved possible[164].

These generation and detection techniques have been applied to problems such as the analysis of heat flow at solid–helium interfaces[165], and to the

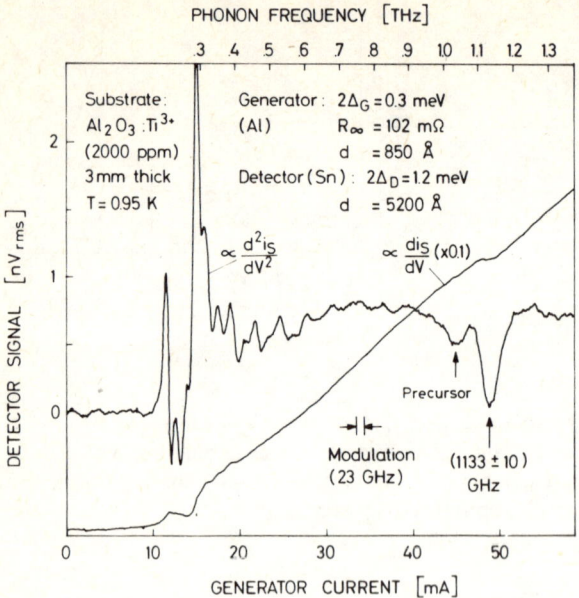

FIGURE 49. Phonon absorption in $Al_2O_3 = Ti^{3+}$ with Al generator and Sn detector tunnel junctions[164].

direct study of electron–phonon interactions in superconductors and normal metals[166]. By use of various pure and doped substrates (for example Al_2O_3, Si, MgO and alkali halides) for both generators and detectors, the propagation and dispersion of ultrasound in dielectrics has been investigated, and scattering studies performed on point defects, localised tunnelling states and isolated paramagnetic spins[163,164,167].

9.4 Other Applications of the Giaever Effect

9.4.1 Photon processes

Quasiparticle decay by photon, rather than phonon, emission is possible but is several orders of magnitude less probable. It is difficult to observe and of no device significance.

Excess tunnel currents due to pair breaking by photons and to other photon associated effects are, by contrast, easy to observe[168]. There have been no device developments, however.

9.4.2 Amplifiers and oscillators

As with switches (Section 7), some early progress was made with certain Giaever devices as electronic components for the 1–100 MHz range. Solymar (see appendix) has reviewed these, but once again progress in conventional devices and Josephson instruments have stopped their further development.

9.4.3 Detectors: the super-Schottky diode

A new type of Schottky diode, using quasiparticle tunnelling between a superconductor and a semiconductor[169], shows a curvature in its characteristic which is on a scale of meV, rather than the more usual eV. For detection and mixing this may compete favourably with Josephson devices (Section 9.5.2).

9.5 The A.C. Josephson Effect

As we said in Section 2.8, in the Josephson A.C. effect we have apparently the basis for coherent infrared radiation generators. For a junction biased at a D.C. voltage V_0, equations (63) and (64) give

$$I(t) = I_c \sin(2EV_0 t/\hbar) \tag{149}$$

where the current oscillates at 484 MHz per microvolt of bias. Moreover, a Josephson element is highly non-linear: if the applied voltage is

$$V = V_0 + V_1 \cos \omega_1 t \tag{150}$$

we have, applying equation (59)

$$I(t) = I_c \sin\left\{\frac{2e}{\hbar} \int V(t)\,dt\right\} \tag{151}$$

$$= I_c \sin\left\{\frac{2e}{\hbar}\left(V_0 t + \frac{V_1}{\omega_1}\sin(\omega_1 t + \phi_1)\right)\right\} \tag{152}$$

This is a frequency modulated signal which may be written in turn as

$$I(t) = I_c \sum_{-\infty}^{\infty} (-1)^n J_n\left(\frac{2eV_1}{\hbar\omega_1}\right) \sin\left[(\omega_0 - n\omega_1)t + \phi_0\right] \tag{153}$$

where J_n is a Bessel function of the first kind and $\omega_0 = 2eV/\hbar$. The large number of side bands characteristic of FM signals is evident, and the system is clearly an effective frequency converter and (for $\omega_0 = m\omega_1$) harmonic generator. Note also that when $\omega_0 = n\omega_1$, that is, for bias voltages $V_0 = n\hbar\omega/2e$, one side band has frequency zero. A singularity in the D.C. I–V characteristic therefore occurs, and is in practice (see Fig. 50) a current step rather than a spike[170], for various circuit reasons[171].

We now discuss some of the applications of these principles, notably to detection, mixing, and to some very interesting standards work. An immense amount of work has been done in this field and for fuller details, the reader should see the general review by Richards, Auracher, and van Duzer[22], a detailed review of detection devices by Richard[67], and the Proceedings of the 1973 Conference on the A.C. Josephson effect[172].

As a general point, we recall that I_c in equation (153) is not strictly a constant, but is frequency-dependent, rising slowly to a weak singularity (Riedel peak)[16] at the frequency $4\Delta/h$. (We assume identical superconductors.) This corresponds to $V_0 = 2\Delta$. At higher biases, $\hbar\omega$ is sufficient to break Cooper pairs and weaken the Josephson current, so I_c falls steadily. The upper frequency limit for reasonable Josephson effects is a few times $4\Delta/h$, corresponding to about 5×10^{15} Hz for a high T_c material such as Nb. A useful device figure of merit is RI_c, where R is the normal resistance and I_c refers to zero frequency. As equation (61) shows, the theoretical upper limit for this is $\pi\Delta/2e$.

Table 4
Typical Resistances and Capacitances of Josephson Elements

	R	C
Thin film microbridge	$0.2\,\Omega$	1 pf
Thin film tunnel junction	$100\,\Omega$	1000 pf
Point contact	$50\,\Omega$	100 pf

Microwave devices should preferably have impedances well matched to that of the outside world. Table 4 shows the difficulties for thin film devices here: microbridges tend to have resistances which are too small, and tunnel junctions to have capacitances which are short circuits at these frequencies. Films have therefore had less impact than point contacts in this field.

9.5.1 Generation

Direct emission of radiation has proved one of the disappointments of the Josephson effect. The problem is a power one: with $I_c \sim 10\text{--}100\,\mu\text{A}$ and $V \sim 1\text{--}5\,\text{meV}$, power levels within a device will be below a microwatt, and when impedance mismatches to free space are considered, the detectable power is of order 10^{-10} W. This is of no interest in the microwave region, although at millimetre and submillimetre wavelengths where there are no alternative tunable coherent sources, it is better than nothing. Single junction outputs up to 2×10^{-10} W at $18.0\,\text{cm}^{-1}$ have been observed[173].

Clark[174] has shown that a three-dimensional array of small lead spheres emits radiation as though from many junctions. The behaviour corresponds to a "super-radiant" state[175], in which M junctions self-synchronise by mutual radiative coupling with a cavity around them. The power density is then proportional to M^2. Thin film tunnel junction arrays can also show super-radiance and with 100 junctions 10^{-9} W has been obtained at 9 GHz[176]. Tunability of thin film arrays has also been demonstrated[177]. It is thus possible that with more extensive arrays power levels may be raised to the point at which competition with other microwave sources is possible.

9.5.2 Radiation detectors

There are several modes in which Josephson elements can serve as radiation detectors, and we now summarise the more promising of these, following the fuller treatments of Richards et al.[22], and Richards[67].

Square law detection. With $V_0 = \omega_0$ and $n = 0$ in equation (153), we have for the critical current I'_c in the presence of radiation

$$I'_c/I_c = J_0(2eV_1/\hbar\omega_1) \tag{154}$$

$$\approx (1 - e^2 V_1^2/\hbar^2 \omega_1^2) \tag{155}$$

so that square law detection of the applied signal is possible. It is usual to run the detector, with hysteresis eliminated by shunting if necessary, at a bias current just above I_c, though at a voltage $V_0 \ll \hbar\omega_1/e$. Chopping the applied radiation to vary I'_c then produces modulation in V_0 (similar to that obtained in the operation of a D.C. SQUID), which can be measured.

The technique is said to be "broadband" and "video", because for any input frequency the output is at zero (or chopper) frequency, and is associated with a particular D.C. bias (V_0). This contrasts with the linear detection systems described below.

Ulrich[177] has applied this method to observations of the Sun and Jupiter at 1 mm. He obtained a noise level equivalence to 73 flux units (1 flux unit $= 10^{-26}$ W m^{-2} Hz^{-1}) in a one second band, corresponding to a noise temperature of 0.25 K/Hz. He did not use thin films. However, Richards[67], while commenting that the transition edge bolometer (Section 6.1.1) might well prove better than the Josephson square law detector for such observations because of the ease of coupling to an incident field, suggests that arrays of Josephson thin film tunnel junctions might ultimately have an impedance high enough to make them useful in astronomy.

Narrow band square law regenerative detection, with bandwidth ~ 300 MHz is also possible, with NEP levels $\sim 10^{-14}$ W$/\sqrt{\text{Hz}}$ in a mode which uses the mutual interaction of a junction, a surrounding cavity and the incident radiation. There seem to have been no significant applications.

Linear detection and mixing. At the first of the steps predicted by equation (153), that is, where $\omega_0 = \omega_1$, the junction is effectively serving as a heterodyne detector with an internal oscillator, and producing a D.C. output which is just the step height. In the resistively shunted junction approximation[17] this is

$$\Delta I_1 \propto J_1(2eV_1/\hbar\omega_1) \qquad (156)$$

For small V_1 this is a linear function. Once again, the junction is current biased just above a step, and the applied signal is chopped: corresponding variations in ΔI_1 can thus be read out as voltage changes at the chopper frequency. Unfortunately, the incident signal must be strong enough to R.F.-synchronise the oscillations in the junction, otherwise the behaviour is more like the square law behaviour discussed above. The power necessary is $> 10^{-14}$ W, and too large to be of interest in small signal applications.

When an external local oscillator ($V_L \sin \omega_L t$) is available, however, its signal adds to $V_1 \sin \omega_1 t$ and equation (153) becomes

$$I = I_c \sum_{n,m=-\infty}^{\infty} (-1)^{m+n} J_n\left(\frac{2eV_1}{\hbar\omega_1}\right) J_m\left(\frac{2eV_L}{\hbar\omega_L}\right)$$
$$\times \sin(\omega_0 t - n\omega_1 t - m\omega_L t + \phi_0) \qquad (157)$$

and steps appear whenever $\omega_0 - n\omega_1 - m\omega_L = 0$, that is, at

$$V = (\hbar/2e)(n\omega_1 + m\omega_L) \qquad (158)$$

The necessary synchronisation is provided by the local oscillator, and this mode has provided the most promising of the Josephson effect detectors.

One application, to low noise mixers and receivers at frequencies $\ll \Delta/h$, uses the zero voltage critical current with $\omega_1 \sim \omega_L$ and $n = -m = 1$. V_L is made large enough to make $I'_c/I_c \sim 0.5$ (equation 147). The effect of V_1 ($\ll V_L$) is then to modulate V_L, and thus I'_c at the difference frequency ($\omega_1 - \omega_L$). Detection of the I'_c modulation then proceeds as for D.C. SQUIDs and the voltage output is coupled to an I.F. amplifier. A conversion gain of 1.3 at 36 GHz has been demonstrated[178] and with a local oscillator power of 10^{-9} W a mixer noise temperature of 53 K has been obtained; this compares well with figures of 10^{-3} W and 200 K, respectively, for a cooled Schottky diode. Indeed when certain heating difficulties have been overcome, the noise temperature T_N may approach the ultimate limit set by photon fluctuation noise in the receiver, which is $T_N = \hbar\omega_1/k_B$ and which corresponds to 14 K at 300 GHz. Operation at 891 GHz[179], and probably even at 32 THz[180], has been observed, but at much lower conversion efficiency.

Another use of the external local oscillator principle is in harmonic mixing for frequency comparison purposes. Substantial input power is available in such cases and secondary mixing processes make it possible to detect IF outputs at $n\omega_1 - m\omega_L$, independent of ω_0, at voltages near $V = 0$. Very high

$n:m$ ratios are possible: Thus, Blaney and Knight[181] have observed mixing between a 0.9 THz laser and the 825th harmonic of a 1 GHz crystal oscillator. The motivation was a new determination of the velocity of light[182], and the specific object to reduce the number of steps in a frequency comparison of the standard of time (atomic clock) and the standard of wavelength (krypton light).

Again, point contacts are usually used in these heterodyne detectors, but thin film arrays may ultimately prove useful, principally because they provide higher junction impedances.

Parametric amplifiers. A particularly promising application of thin film weak links and tunnel junctions uses the non-linear inductance of a Josephson element. Using the definition

$$V = \frac{d}{dt}[L(I) \cdot I] \tag{159}$$

one obtains for the inductance,

$$L(I) = \frac{h}{2eI_c}\left(\frac{\sin^{-1}(I/I_c)}{I/I_c}\right) \tag{160}$$

As with other non-linear reactance systems (e.g. Varactor), one can build this property into a parametric amplifier by adding a pump signal at ω_p to the signal input at ω_1. Amplifiers can be operated with external pumps, or by using internally general A.C. currents.

A successful parametric amplifier (the SUPARAMP)[183] has been made with a series array of eighty 1000 Å unbiased tin film microbridges 0.3 μm × 0.3 μm spaced 5 μm apart. The array is formed in a parallel plate thin film stripline, and is much shorter than the guide wavelength. Thin film tuning stubs and ground planes are also formed on the same substrate. The device operates at 10 GHz and has shown gain up to 16 dB, a bandwidth of 1.0 GHz, and a noise temperature of 26 K. This is competitive with varactor amplifiers of similar centre frequency and bandwidth.

The theory of the device has been given by Parrish *et al.*[184], and reviewed by Richards[67], who suggests that a large array of niobium (i.e. large Δ) junctions might make a very valuable amplifier up to, but probably not above, 100 GHz. Richards also discusses a number of point contact Josephson parametric amplifiers which are biased at a voltage V_0 such that the pump frequency is internally generated ($\omega_p = 2eV_0/h$).

9.5.3 Measurement of e/h; voltage standards
The relation

$$v = 2eV/h \tag{64}$$

for the frequency-voltage relation of the Josephson A.C. current clearly gives a basis for an accurate determination of e/h. After several years of work, the position has been reached that e/h is known to 0.12 ppm[185-187], and that the legal volt (in the U.S.A., at least), has been changed from a chemical to a Josephson standard[188].

The basis of this development is shown in Fig. 50. When irradiated with microwaves of frequency v, steps appear at biases

$$V_0 = V_n = nhv/e \tag{161}$$

The best steps are usually found in tunnel junctions, rather than point contacts, and there is, of course, always ample power available so that the

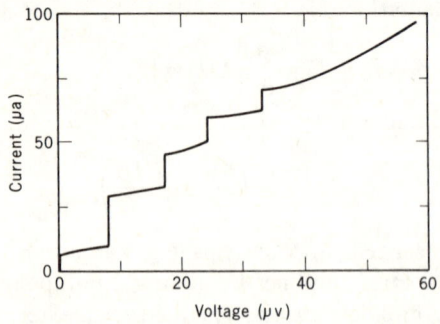

FIGURE 50. Current-voltage characteristic of tunnel junction irradiated with 4 GHz microwaves[195].

impedance mismatch is no problem. Initial work was directed to assessing the reliability of equation (64), and the results were very encouraging. First, Bloch[19] showed that the basic Josephson equation (59) can be derived directly using Faraday's law of induction and the principle that the free energy of a superconducting ring varies periodically with magnetic flux. The latter is essentially a basic result of quantum mechanics, and contains no approximations due to microscopic or other theories of superconductivity. Secondly, concern that the finite lifetime of Cooper pairs might have frequency pulling effects was eliminated by the realisation that all such interactions involved changes in the chemical potential: provided this, and not the simple electrostatic potential is used in equation (64) no corrections are needed[189]. On the experimental side, Clarke[18] used a SQUID voltmeter to detect any difference between the voltages of two junctions involving different metals irradiated with the same frequency and biased on the same order step. The null result showed the voltages to be equal (and the steps vertical) to 1 part in 10^8.

Likewise, using junctions irradiated with frequencies in the ratio 94:1 and biased on steps 1 and 94, respectively, it was shown that the voltage-frequency linearity is exact to better than 1 ppm[190].

The influence of external noise on the steps is also important. Theoretical and experimental treatments[191] show that while current noise reduces the height of a step and voltage noise broadens it (though in fact negligibly in a screened room), the step remains antisymmetric. Thus, the midpoint is quite unaffected by noise, and with this qualification equation (161) is completely accurate.

The major experiments of determining the applied frequency and the step voltages in terms of national standards were undertaken by groups at the University of Pennsylvania and NBS (U.S.A.)[185], and at NPL (U.K.) and NSL (Australia)[187]. Typically, they used $v \sim 10\,\text{GHz}$ and $n \sim 100$. The accuracy and agreement were so good that they were limited by the uncertainty (due to drift and other factors) of the intercomparison between the national standards and the maintained BIPM standards in Paris.

Two major developments have ensued. First is that the value of e/h determined was 38 ppm smaller than the result of the previous adjustment which was thought to be accurate to 10 ppm. Taylor et al. undertook a complete review[192] of fundamental constants, which in turn has led to a substantial reassessment of quantum electrodynamic theory and experiment, on which, through the fine structure constant α, many fundamental constant values depended. Various minor corrections were found to be necessary, and the disagreements are now essentially resolved[193].

On the other hand, the replacement of the chemical standard of voltage by the Josephson one is leading to much simpler intercomparisons of voltage. Great improvements have been made in the design of voltage dividers (necessary to compare the chemical volt $\sim 1\,\text{V}$ to the voltage of a Josephson step $\sim 10\,\text{mV}$), and it is now practical to pack into one cryostat this, a reference junction and a null detector—a SQUID naturally[194]!

An excellent elementary review of the fundamental constant work has been given by Clark[195].

Finally, there is considerable interest in the properties of series arrays of m thin film Josephson junctions of microbridges coupled in the superradiant state. These have been found[196] to show Josephson steps at frequencies v from 30 MHz–20 GHz, at voltages $V = nmhv/2e$. It may be possible to make direct standardisation at 1 V in this way also.

10. Fabrication of Tunnel Junctions and Weak Links

Only in the production of tunnel junction elements and certain weak link structures is the thin film superconductive device worker involved in techniques which are not of general use in the thin film world. It is therefore appropriate to devote a section to these. For further details see Solymar (see Appendix).

10.1 Tunnel Junctions

In principle, to form a tunnel junction one needs only to grow an oxide on a freshly deposited film and then over-deposit a counterelectrode. Manipulation of the oxidation parameters (pressure, temperature, time) should produce, by trial and error, the condition necessary for the right barrier thickness. The desired resistances are usually between 0.1 and 10 Ω, and since the low voltage resistance of a junction depends approximately exponentially on the barrier thickness[198] (which is normally about 20 Å), only a small range of thicknesses need be contemplated.

Only rarely is it this simple. Usually there are high failure rates, with pinholes producing shorts, which are sometimes superconducting with critical currents of many milliamps. Often junctions prepared simultaneously on the same substrate will have resistances differing by more than an order of magnitude, and different workers, even using nominally identical techniques on the same apparatus, will get wider variations. Even successful junctions will often fail (open circuit or short circuit), after a few thermal cycles, and some (usually those involving aluminium) will steadily increase in resistance if stored at room temperature, as trapped gases diffuse towards the tunnel barriers.

Moreover, there are problems of "leakage" currents. These are usually ohmic currents, observable as a spurious additional conductance (σ_L) at low T for $V < \Delta/e$, when the quasiparticle tunnelling current is nearly zero. Their origins have not been properly investigated, and though, in part, they can be due to trapping of "normal" vortex cores in a superconducting film[199] and reduced simply by careful magnetic procedures, there are certainly other contributory causes. These currents are not a serious problem in many Josephson applications, where one is shunting the junction anyway, though it is desirable to keep them small in tunnelling cryotrons (Section 7.1). However, for many Giaever applications, especially those using properties observable at $V < \Delta/e$ (e.g. thermometers) (Section 6.1), they are the ultimate limiting factor in sensitivity. McMillan and Rowell[147] suggest that to qualify as a "tunnel" junction, a device should have $\sigma_L \leqslant 10^{-3}$, and that results from measurements where σ_L was larger should be suspect.

Much effort has been devoted to forming good junctions on "hard" metals such as niobium, because their adhesion and other properties make them especially durable. However, progress has been slow, which has undoubtedly been responsible for much of the delay in tunnelling device development, particularly in the case of the double junction D.C. SQUID.

However, there have emerged many procedures, usually particular to a given type of junction, which give reliable long-lasting structures with tough uniform barriers. They broadly divide into two classes—those involving thermal oxidation and those involving a gaseous anodisation with a glow discharge. We will discuss examples of each.

A common procedure is to form a barrier on freshly evaporated aluminium. If the film is exposed to room temperature air for about 1 min at 760 torr and a counter-electrode of Sn or Pb deposited, one may expect a good success rate, with junction resistances in the 1–100 Ω range. The good success is presumably due to the ease of formation and toughness of Al_2O_3, but, even here, conditions in individual laboratories are critical. Presumably moisture content and other impurities in the local air are significant; certainly the cleanliness of the film itself is a factor, as can be demonstrated by evaporating the aluminium a variety of residual pressures[200].

Hansma[201] has recently described a thermal oxidation procedure for Nb. The films, 2000 Å thick, were sputtered through Al masks at > 20 Å sec, in $(1-5) \times 10^{-3}$ torr of argon, on to unheated microscope slides. They were then oxidised in air in a separate chamber for 10 min at temperatures between 100 and 160°C. The junctions were completed by over-depositing with Pb-6% In alloy, and finally protected from moisture (to which Pb is very sensitive), by spinning on a protective layer of KTFR photoresist. The ultimate 4.2 K tunnelling resistance was controlled by the oxidation temperature: thus, 100°C produced 20 Ω, while 140°C gave 1 kΩ.

Room temperature conductances were $10-10^3$ times higher than those at helium temperature, suggesting an unusually high thermally activated impurity conductivity in the barrier. However at 4K the junctions had excellent tunnelling characteristics with Josephson critical currents up to 70% of the theoretical maxima. A recent development extends this to Nb-barrier-Nb[142].

Techniques involving oxidation by activated molecules in low pressure glow discharge were first reported by Miles and Smith[202]. The most advanced version of this principle is that of R.F. sputter anodisation[203,204]. The unoxidised film is usually wider than the intended junction area, which is defined by a photoresist coating in which a 10^{-2} mm^2 window is opened. The area is chemically cleaned, to expose the metal, by sputter etch cleaning in argon; then, in the same chamber, oxidation proceeds using an R.F. discharge in $(2-50) \times 10^{-3}$ torr of oxygen for about 15 min and 10^2-10^{-1} W-cm^{-2} of R.F. power. Aluminium sputtering electrodes are used, and the final barrier thickness (resistance) is controlled by the oxidation pressure. The procedure is a dynamic equilibrium one, and at the final thickness there is a balance between the oxidation of the metal and the rate at which oxide is sputtered away. This results in excellent quality junctions, with resistances predictable to within × 3, and leakage conductances below 10^{-3}. Procedures have been published for Pb/In-barrier-Pb[203], Pb-barrier-Pb[204], and Nb-barrier-Nb[205]. D.C. glow discharge methods for Nb-barrier-Pb have also been described[206,207].

The surface physics of these procedures has been extensively studied[208] using ellipsometry to determine the thickness and other properties of the oxide as it forms. Good agreement with established models of tunnelling in barriers has been established, showing the very high quality of the oxide formed this way. The RF anodisation method is the basis of the very complex structures used in the logic circuits of Section 7.

Many alternatives to oxide barriers have been studied. The aim is to use a material with a small semiconductor energy gap. For the same tunnel resistance, a thicker barrier would be used, reducing the chance of pinholes, but also reducing the junction capacitance. Thus, carbon, cleaved $NbSe_2$, Te, CdS, and others[209] have been reported in barrier use, but seem unsuitable in various ways and have not been developed. A new possibility is Si[210]: by thinning wafers locally to ≤ 1000 Å and depositing films on each side, reasonable junctions with good Josephson characteristics are produced. However, the leakage conductance is high enough almost to eliminate hysteresis, and the junctions would be useful only in "shunted junction" applications, such as detection and mixing.

10.2 Weak Links

Thin film microbridges suitable for Josephson devices should have dimensions ~ 1 μm if they are to form weak enough links for satisfactory operation. Many techniques for preparing such small structures have been described, and they fall broadly into one group calling for mechanical scribing and another in which "optical" processes are used to render material suitable for etching. Typical of the former is the method of Gregers–Hansen et al.[211] who used a glass substrate with a semi-circular groove 0.5 μm wide etched in it. After a tin film had been evaporated, a sharp razor blade was drawn across it at right angles to the groove. This cuts a 0.2 μm wide space in the film except at the groove, which the blade does not enter. The result is a bridge 0.2 μm long and 0.5 μm wide, resistance 0.2–1 Ω at 4.2 K. The critical current is linear in $(T_c - T)$ and about 30 μA at 10 mK below T_c. Alternatives are to scribe the film with a diamond point or blade[212]. Amongst quasi-optical techniques, Adde et al.[213] have formed bridges 0.6 μm × 0.6 μm by using ion-bombardment beams to sputter away niobium from 500 Å thick films. Electron beam imaging has also been used to destroy polymer chains in a plastic film covering the metal: the affected plastic is then dissolved and unprotected metal etched away[214]. Such methods are particularly suitable for producing arrays for super-radiant generation.

We finally mention the techniques of Mercereau and his colleagues for the production of the proximity effect bridges discussed in Section 8.4. Conventional optical exposure on photo-resist is used to define patterns on films. However, the patterned metal is not then removed by etching: instead, it is converted to oxide by anodisation using boric acid as the electrolytes. A 10 sec voltage pulse is applied, and the maximum voltage determines the depth of metal converted, at the rate, for example, of 6 Å/volt for Nb and 8 Å/volt for Ta. By starting with a layered film structure (e.g. Ta on Nb), and anodising to different depths, one may make regions of Nb with different chosen thicknesses of Ta on top, or regions of Nb with only oxide on top. Thus, proximity bridges of different characteristics have been produced on single substrates, together with necessary interconnections, all from the same Nb/Ta deposition[215]. The technique has been applied to the production of integrated SQUID structures, and to arrays of 100 and more bridges exhibiting coherent radiation response.

Acknowledgements

The author is grateful to the University of Lancaster for leave, to the Fulbright–Hays Commission for a scholarship, and to the Department of Physics, University of California, Berkeley and the Materials and Molecular Research Division, Lawrence Berkeley Laboratory for hospitality and assistance. He is greatly indebted for useful discussion to J. Clarke, J. Ekins, T. Fulton, W. Goubau, E. Guyon, P. K. Hansma, G. Hawkins, M. Ketchen, H. Notarys, and P. L. Richards, and particularly thanks A. R. Long for a critical reading of the manuscript. He also thanks those authors who gave permission for the reproduction of their published diagrams and who often supplied original copies of their drawings as well.

Appendix

Books and articles for further reading on the physics of superconductivity and its applications.

J. P. Burger, "La Supraconductivité des Métaux, des Alliages, et des Films Minces", Masson, Paris, (1974).

K. L. Chopra, "Thin Film Phenomena", McGraw-Hill, New York, (1969).

W. D. Gregory, W. N. Mathews, and E. A. Edelsack (eds.) "The Science and Technology of Superconductivity", Plenum, New York, (1973).

R. E. Johnson, "Handbook of Thin Film Technology", eds., L. I. Maissel and R. Glang, McGraw-Hill, New York, (1970).

E. A. Lynton, "Superconductivity", 3rd Edition, Methuen, London, (1969).

V. L. Newhouse, "Applied Superconductivity", John Wiley and Sons, New York, (1964).

R. D. Parks (ed.) "Superconductivity", Marcel Dekker, New York, (1969).

A. C. Rose-Innes and E. H. Rhoderick, "Introduction to Superconductivity", Pergamon Press, New York, (1969).

L. Solymar, "Superconductive Tunnelling and Applications", Chapman and Hall, London, (1972).
M. Tinkham, "Introduction to Superconductivity", McGraw-Hill, New York, (1975).
Proc. I.E.E.E., Special Issue on "Applications of Superconductivity", **61**, January 1973.
Proc. Int. Conf. on Detection and Emission of Electromagnetic Radiation with Josephson functions. *Révue de Physique Applique*, **9**, 1–312, (1974).

See also successive Proceedings of the International Conferences on Low Temperature Physics and of the Applied Superconductivity Conferences.

References

1. H. Kamerlingh Onnes, *Leiden. Commun.* 122b, 124c, (1911).
2. For recent progress in applied physics and engineering aspects of superconductivity see *Proc. Appl. Superconductivity Conf.* 1974, *I.E.E.E. Trans. Magnetics*, **MAG-11**, (1975) and *Proc. 5th Int. Cryogenic Engineering Conference*, IPC Business Press (1975).
3. W. Meissner and R. Ochenfeld, *Naturwiss*, **21**, 787, (1933).
4. B. D. Josephson, *Phys. Letts.*, **1**, 251, (1962).
5. Properties of superconductors are listed in a periodically revised National Bureau of Standards publication; the current one is NBS Technical Note 825 (1974), supplementing Notes 724 (1972) and 482 (1969), ed., B. W. Roberts.
6. For a review, see D. Dew-Hughes, *Rep. Progr. Phys.*, **34**, 821, (1972).
7. F. London, "Superfluids", Vol. I, Wiley, New York, (1950).
8. Bardeen, L. N. Cooper and J. R. Schrieffer, *Phys. Rev.*, **104**, 1175, (1957).
9. L. N. Cooper, *Phys. Rev.*, **104**, 1189, (1956).
10. I. Giaever, *Phys. Rev. Letts.*, **5**, 147, and 464, (1960).
11. V. L. Ginzburg and L. D. Landau, *Zh. Eksperim. i. Teor. Fiz.*, **20**, 1064, (1950); see also "Collected Papers of L. D. Landau", ed., ter Haar, Pergamon Press, Oxford, (1967).
12. L. P. Gor'kov, *Zh. Eksperim. i. Teor. Fiz.*, **36**, 1918, **37**, 833, and 1407, (1959); *Soviet Phys. J.E.T.P.*, **9**, 1364, **10**, 593, and 998, (1960).
13. A. B. Pippard, *Proc. Roy. Soc.*, **A216**, 547, (1953).
14. B. D. Josephson, *Rev. Mod. Phys.*, **36**, 216, (1964); and "Superconductivity", ed., R. D. Parks, Marcel Dekker, New York, (1969).
15. V. Ambegaokar and A. Baratoff, *Phys. Rev. Letts.*, **10**, 486, (1963), erratum, **11**, 104, (1963).
16. E. Riedel, *Z. Naturforsch*, **19A**, 1634, (1964) and N. R. Werthamer, *Phys. Rev.*, **147**, 255, (1966).
17. D. E. McCumber, *J. Appl. Phys.*, **39**, 3113, (1968) and W. C. Stewart, *App. Phys. Letts.*, **12**, 277, (1968).
18. J. Clarke, *Phys. Rev. Letts*, **21**, 1566, (1968).
19. F. Bloch, *Phys. Rev.*, **B2**, 109, (1973).
20. R. C. Jaklevic, J. Lambe, J. E. Mercereau, and A. H. Silver, *Phys. Rev.*, **140**, A1628, (1965).

21. M. Tinkham, "Introduction to Superconductivity" (see Appendix), p. 215.
22. For a review, see P. L. Richards, F. Auracher, and T. van Duzer, *Proc. I.E.E.E.*, **61**, 36, (1973).
23. H. L. Caswell, *J. Appl. Phys.*, **32**, 105, 2641, (1961).
24. B. T. Matthias, H. Suhl, and E. Corenswit, *J. Phys. Chem. Solids*, **13**, 156, (1960).
25. W. Buckel and R. Hilsch, *Z. Phys.*, **131**, 420, (1952).
26. K. L. Chopra, (see Appendix).
27. J. P. Gollub, M. R. Beasley, R. Callarotti, and M. Tinkham, *Phys. Rev.*, **B7**, 3039, (1973).
28. B. Abeles, R. W. Cohen, and G. W. Cullen, *Phys. Rev. Letts.*, **17**, 632, (1966).
29. A. M. Lamoise, J. Chavmont, F. Meunier, and H. Bernas, *J. de Physique Letts.*, **36**, L271, (1975).
30. C. A. Neugebauer, *J. Appl. Phys.*, **35**, 3599, (1964).
31. G. Hawkins (private communication).
32. S. D. Dahlgren, M. Suenaga, and T. S. Luhman, *J. Appl. Phys.*, **45**, 5462, (1974). S. D. Dahlgren, *I.E.E.E. Trans. Magnetics*, **MAG-11**, 217, (1975).
33. A review is given by R. H. Hammond, *I.E.E.E. Trans. Magnetics*, **MAG-11**, 201, (1975).
34. See, for example, Newhouse (Appendix), p. 107.
35. See for example, C. Kittel, "Introduction to Solid State Physics", 4th edition, Wiley, London, (1976).
36. R. S. Thomson and A. Baratoff, *Phys. Rev.*, **167**, 361, (1968).
37. M. L. A. MacVicar, *Phys. Rev.* **B2**, 97, (1970).
38. P. W. Anderson, *Phys. Chem. Solids*, **11**, 26, (1959).
39. J. P. Burger and D. Saint-James, in "Superconductivity", ed., R. Parks, Marcel Dekker, New York, (1969).
40. P. G. de Gennes, "Superconductivity of Metals and Alloys", W. A. Benjamin, New York, (1966).
41. M. Tinkham, "Introduction to Superconductivity", McGraw-Hill, New York, (1975).
42. R. E. Johnson, in "Handbook of Thin Film Technology", eds., L. I. Maissel and R. Glang, McGraw-Hill, New York, (1970).
43. E. Guyon, F. Meunier and R. S. Thompson, *Phys. Rev.*, **156**, 452, (1967).
44. M. Tinkham, *Phys. Rev.*, **129**, 2413, (1963); D. E. Morris and M. Tinkham, *Phys. Rev.*, **134**, A1156, (1964).
45. G. Lasher, *Phys. Rev.*, **154**, 345, (1967).
46. G. J. Dolan and J. Silcox, *Phys. Rev. Letts.*, **30**, 603, (1973).
47. F. B. Silsbee, *Wash. Acad. Sci.*, **6**, 599, (1916).
48. J. Bardeen, *Rev. Mod. Phys.*, **34**, 667, (1962).
49. V. L. Newhouse, J. W. Bremer, and H. H. Edwards, *Proc. I.R.E.*, **48**, 1395, (1960). C. R. Smallman, A. E. Slade, and M. L. Cohen, *Proc. I.R.E.*, **48**, 1562, (1960).
50. The proximity effect is reviewed by G. Deutscher and P. G. de Gennes, in "Superconductivity", ed., R. Parks, Marcel Dekker, New York, (1969).
51. J. Clarke, *Proc. Roy. Soc. A*, **308**, 447, (1969).
52. H. A. Notarys and J. E. Mercereau, *J. Appl. Phys.*, **44**, 1821, (1973).

53. J. P. Turneaure, Proc. Appl. Superconductivity Conf. (1972), I.E.E.E., New York, (1972).
54. G. J. Dick (to be published).
55. B. Hillenbrand, H. Martens, M. Pfister, K. Schnitzke, and G. Ziegler, *I.E.E.E., Trans. Magnetics*, **MAG-11**, 420, (1975).
56. B. Cabrera and W. O. Hamilton, in "Science and Technology of Superconductivity", eds., W. D. Gregory, W. N. Matthews and E. A. Edelsack, Plenum Press, New York, (1973).
57. W. H. Hartwig, *Proc. I.E.E.E.*, **61**, 58, (1973).
58. A. B. Pippard, *Proc. Roy. Soc.*, **A216**, 547, (1953).
59. J. C. Swithart, *J. Appl. Phys.*, **32**, 461, (1961); P. V. Mason and R. W. Gould, *J. Appl. Phys.*, **40**, 2039, (1969).
60. C. A. Passow, V. L. Newhouse, and R. L. Gunshor, Proc. Appl. Superconductivity Conf. (1972), I.E.E.E., New York, (1972).
61. N. S. Nahman, *Proc. I.E.E.E.*, **61**, 76, (1973).
62. I. Giaever, *Phys. Rev. Letts.*, **15**, 825, (1965).
63. I. Giaever, *Phys. Rev. Letts.*, **16**, 460, (1966); R. Deltour and M. Tinkham, *Phys. Rev.*, **174**, 478, (1968).
64. J. R. Clem, *Phys. Rev.*, **B9**, 898, (1974); *ibid*, **B12**, 1742, (1975).
65. J. W. Ekin, B. Serin, and J. R. Clem, *Phys. Rev.*, **B9**, 912, (1974).
66. J. W. Ekin and J. R. Clem, *Phys. Rev.*, **B12**, 1753, (1975).
67. P. L. Richards, "Infrared Detectors", Semi-conductors and Semimetals, Vol. V, Willardson and Beer (Eds.), Academic Press, London and New York, (1977).
68. V. L. Newhouse, "Applied Superconductivity", John Wiley and Sons, New York, (1964).
69. J. Clarke, G. I. Hoffer, and P. L. Richards, *Rev. Phys. Appl.*, **9**, 69, (1974).
70. R. M. Katz and K. Rose, *Proc. I.E.E.E.*, **61**, 55, (1973).
71. J. Clarke, G. I. Hoffer, P. L. Richards, and N.-H. Yeh, Proc. 14th Int. Conf. on Low Temp Phys., North Holland–American Elsevier, (1975).
72. J. Clarke and T. Y. Hsiang, *I.E.E.E. Trans. Magnetics*, **MAG-11**, 845, (1975).
73. I. Giaever and K. Megerle, *Phys. Rev.*, **122**, 1101, (1961).
74. I. Giaever, H. R. Hart, and K. Megerle, *Phys. Rev.*, **126**, 941, (1962).
75. B. N. Taylor and E. Burstein, *Phys. Rev. Letts.*, **10**, 14, (1963); J. R. Schrieffer and J. W. Wilkins, *Phys. Rev. Letts.*, **10**, 17, (1963).
76. S. Bermon, Tech. Rept. 1, University of Illinois, Urbana, NSF Grant GP 1100, (1964).
77. G. B. Donaldson, *Revue Generale d'Electricité*, **79**, 828, (1970); G. B. Donaldson, to be published.
78. G. B. Donaldson and W. T. Band, Suppl. Bull. Int. Inst. of Refrigeration, Annex 1970-2, 27, (1970).
79. J. C. Wheatley and R. A. Webb, *Science*, **182**, 241, (1973).
80. R. A. Kamper and J. E. Zimmerman, *J. Appl. Phys.*, **42**, 132, (1971); R. J. Soulen and H. Marshak, Proc. 14th Int. Conf. on Low Temp. Physics, North Holland–American Elsevier, (1975).
81. R. P. Giffard, R. A. Webb, and J. C. Wheatley, *J. Low Temp. Phys.*, **6**, 533, (1972).
82. R. A. Webb, R. P. Giffard, and J. C. Wheatley, *J. Low Temp. Phys.*, **13**, 383, (1973).

83. J. A. Hoyle, R. R. Humphris, and J. W. Boring, *I.E.E.E. Trans. Magnetics*, **MAG-11**, 690, (1975).
84. M. Cavallini, G. Gallinaro, and G. Scoles, *Z. Naturforsch*, **24A**, 1850, (1969).
85. D. E. Spiel, R. W. Boom, and E. C. Crittenden, *App. Phys. Letts.*, **7**, 292, (1965).
86. G. H. Wood and B. L. White, *App. Phys. Letts.*, **15**, 237, (1969).
87. For reviews, see V. L. Newhouse, in "Superconductivity", ed., R. Parks, Marcel Dekker, New York, (1969); J. Matisoo, in "Science and Technology of Superconductivity", Plenum Press, New York, (1973).
88. J. W. Crowe, *I.B.M., J. Res. Dev.*, **1**, 295, (1957).
89. J. Matisoo, *Proc. I.E.E.E.*, **55**, 172, (1967).
90. S. Basaviah and R. F. Broom, *I.E.E.E. Trans. Magnetics*, **MAG-11**, 759, (1975).
91. H. H. Zappe, *App. Phys. Letts.*, **25**, 424, (1974).
92. R. F. Broom, W. Jutzi, and Th. O. Mohr, *I.E.E.E. Trans. Magnetics*, **MAG-11**, 755, (1975).
93. C. L. Huang and T. van Duzer, *App. Phys. Letts.*, **25**, 753, (1974).
94. W. H. Henkels, *I.E.E.E. Trans. Magnetics*, **MAG-10**, 860, (1974).
95. J. R. Matisoo, (to be published).
96. W. Anacker, *I.E.E.E. Trans. Magnetics*, **MAG-5**, 968, (1969).
97. P. Gueret, *App. Phys. Letts.*, **25**, 426, (1974); *I.E.E.E. Trans. Magnetics*, **MAG-11**, 751, (1975).
98. T. A. Fulton, R. C. Dynes, and P. W. Anderson, *Proc. I.E.E.E.*, **61**, 28, (1973); T. A. Fulton and R. C. Dynes, *Solid State Comm.*, **12**, 57, (1973).
99. I. Giaever, *Rev. Mod. Phys.*, **46**, 245, (1974).
100. I. Giaever and K. Megerle, *I.R.E. Trans.*, **ED-9**, 459, (1962).
101. V. L. Newhouse, in "Superconductivity", ed., R. Parks, Marcel Dekker, New York, (1969).
102. V. L. Newhouse and H. H. Edwards, *Proc. I.E.E.E.*, **52**, 1191, (1964).
103. See L. Solymar, "Superconductive Tunnelling and Applications", Chapman and Hall, London, (1972); M. Tinkham, "Introduction to Superconductivity", McGraw-Hill, New York, (1975) and O. V. Lounasmaa, "Experimental below 1° Kelvin", Academic Press, London and New York, (1974).
104. J. Clarke, *Proc. I.E.E.E.*, **61**, 8, (1973).
105. W. L. Goodman, V. W. Hesterman, L. H. Rorden, and W. S. Goree, *Proc. I.E.E.E.*, **61**, 20, (1973).
106. J. E. Zimmerman, *Cryogenics*, **12**, 19, (1972).
107. J. Clarke, *Science*, **184**, 1235, (1974).
108. J. Kurkijarvi, *Phys. Rev.*, **B6**, 832, (1972); *J. Appl. Phys.*, **44**, 3729, (1973). See also L. D. Jackel and R. A. Buhrman, *J. Low Temp. Phys.*, **19**, 201, (1975).
109. T. D. Clark and L. D. Jackel, *I.E.E.E., Trans. Magnetics*, **MAG-11**, 736, (1975).
110. M. R. Gaertnner, Paper 8-3, Intermag. Conference, Toronto, (1974).
111. R. A. Kamper, M. B. Simmonds, R. T. Adair, and C. A. Moer, Proc. Appl. Superconductivity Conf., 1972, I.E.E.E., New York, (1972).
112. A. Th. A. M. de Waele, and R. de Bruyn Oubuter, *Physica*, **41**, 225, (1969); J. Clarke and C. Tesche, *J. Low Temp. Phys.* (to be published).
113. J. Clarke, W. Goubau, and M. Ketchen, *App. Phys. Letts.*, **27**, 155, (1975).
114. J. Clarke, W. Goubau and M. Ketchen to be published in *J. Low Temp. Phys.*

115. R. A. Kamper, *I.E.E.E., Trans. Magnetics*, **MAG-11**, 141, (1975).
116. J. H. Claassen, *J. Appl. Phys.*, **46**, 2268, (1975).
117. J. E. Zimmerman, *J. Appl. Phys.*, **42**, 4483, (1971).
118. M. M. Wynn, C. P. Frahm, P. J. Carroll, R. H. Clark, J. Wellhoner, and M. J. Wynn, *I.E.E.E., Trans. Magnetics*, **MAG-11**, 701, (1975).
119. D. Cohen, *I.E.E.E., Trans. Magnetics*, **MAG-11**, 694, (1975); *Physics Today*, **28**, 35, (1975).
120. For example, D. Cohen and L. A. Kaufman, *Circulation Research*, **36**, 414, (1975). Other medical references are cited in Ref. 119.
121. J. Ahopelto, P. J. Karp, T. E. Katila, R. Lukander, and P. Mäkipää, Proc. 14th Int. Conf. on Low Temp. Physics, North Holland–American Elsevier, (1975).
122. D. J. Meredith, G. R. Pickett, and O. G. Symko, *J. Low Temp. Phys.*, **13**, 607, (1973).
123. H. E. Hoenig, R. H. Wang, G. R. Rossman and J. E. Mercereau, Proc. Appl. Superconductivity Conf., I.E.E.E., New York, 570, (1972).
124. S. P. Boughn, M. S. McAshan, R. C. Taber, W. M. Fairbank, and R. P. Giffard, Proc. 14th Int. Conf. on Low Temp. Phys., North Holland–American Elsevier, (1975).
125. J. M. Goodkind and R. J. Warburton, *I.E.E.E., Trans. Magnetics*, **MAG-11**, 708, (1975).
126. J. B. Hendricks, *I.E.E.E., Trans. Magnetics*, **MAG-11**, 712, (1975).
127. C. M. Swift, A Magnetotelluric Investigation of an Electrical Conductivity Anomaly in the S. W. United States, Ph.D. Thesis (unpublished, Geophysics Dept., M.I.T., Cambridge, Mass. (1967)).
128. M. J. S. Johnston and F. D. Stacey, *Nature*, **224**, 1289, (1969).
129. See, for example—Natural Electromagnetic Phenomena below 30 kHz, Proc. Adv. NATO Study Institute, eds., D. F. Bled and I. Estermann, Plenum, New York, (1964).
130. H. F. Morrison (private communication).
131. R. Meservey, P. M. Tedrow, and D. Paraskevopoulos, *J. Appl. Phys.*, **45**, 4601, (1974).
132. J. R. Waldram and J. M. Lumley, *Rev. Phys. Appl.*, **10**, (1974).
133. D. B. Sullivan and R. F. Dziuba, *Rev. Sci. Inst.*, **45**, 517, (1974); and *I.E.E.E. Trans. Magnetics*, **MAG-11**, 716, (1975).
134. A. Davidson, R. S. Newbower, and M. R. Beasley, *Rev. Sci. Inst.*, **45**, 838, (1974).
135. SQUID manufacturers include (TF = thin film type made): Canadian Thin Films, Lake City Way, Burnaby BC (TF); Cryogenic Consultants Ltd., The Vale, London W.3; Develco Inc. Logue Ave., Mountain View, California 94040 (TF); Instruments for Technology, Ltd., Kuusikallionkuja 3E, 02210, Tuonella, Finland (TF); SHE, Sorrento Valley Blvd., San Diego, California 92121. Superconducting Technology Inc., Independence Ave., Mountain View, California 94043.
136. J. Clarke, *Phil. Mag.*, **13**, 115, (1966).
137. J. M. Goodkind and J. M. Dundon, *Rev. Sci. Inst.*, **42**, 1264, (1971).
138. D. W. Palmer and S. K. Decker, *Rev. Sci. Inst.*, **44**, 1621, (1973).
139. J. E. Mercereau and H. A. Notarys, *J. Vac. Sci. Technol.*, **10**, 646, (1973).

140. J. M. Pierce, J. E. Opfer, and L. H. Rorden, *I.E.E.E. Trans. Magnetics*, **MAG-10**, 599, (1974).
141. S. K. Decker and J. E. Mercereau, *App. Phys. Letts.*, **23**, 347, (1973).
142. G. Hawkins and J. Clarke, to be published in *J. Appl. Phys.*
143. G. Hawkins (private communication).
144. M. Ketchen, J. Clarke, W. Goubau, and G. B. Donaldson, to be published in *J. Low Temp. Phys.*
145. J. M. Rowell, P. W. Anderson, and D. F. Thomas, *Phys. Rev. Letts.*, **10**, 334, (1963).
146. D. E. Thomas and J. M. Rowell, *Rev. Sci. Inst.*, **36**, 1301, (1965).
147. For a review, see W. L. McMillan and J. M. Rowell, in "Superconductivity", ed., R. Parks, Marcel Dekker, New York, (1969).
148. L. van Hove, *Phys. Rev.*, **89**, 1189, (1953).
149. J. R. Schrieffer, D. J. Scalapino, and J. W. Wilkins, *Phys. Rev. Letts.*, **10**, 336, (1963).
150. P. Chaikin and P. Hansma, to be published.
151. W. L. McMillan, *Phys. Rev.*, **167**, 331, (1968).
152. J. M. Rowell, W. L. McMillan and P. W. Anderson, *Phys. Rev. Letts.*, **14**, 633, (1965).
153. See, for example: R. C. Dynes and J. M. Rowell, *Phys. Rev. B.*, **11**, 1884, (1975). "A Tabulation of the Electron–Phonon Interaction in Superconducting Metals and Alloys, Part I", by J. M. Rowell, W. L. McMillan and R. C. Dynes is available from the authors (Bell Telephone Laboratories, Murray Hill, N. J.) Fuller data is to be published in J. Phys. Chem. Ref. Data.
154. J. Lambe and R. C. Jaklevic, *Phys. Rev.*, **165**, 821, (1968).
155. D. J. Scalapino and S. M. Marcus, *Phys. Rev. Letts.*, **18**, 459, (1967).
156. For a review, see M. G. Simonsen, R. V. Coleman and P. K. Hansma, *J. Chem. Phys.*, **61**, 3789, (1974); and P. K. Hansma, Proc. 14th Int. Conf. on Low Temp. Physics, North Holland–American Elsevier, New York, (1975).
157. P. K. Hansma and M. Parikh, *Science*, **188**, 1304, (1975).
158. S. de Cheveigné, A. Leger and J. Klein, Proc. 14th Int. Conf. on Low Temp. Physics, North Holland–American Elsevier, New York, (1975).
159. Phonon generation and detection by tunnel junctions is reviewed by A. H. Dayem, *J. de Physique, Supp.* **C4**, 15 (1972).
160. A. Rothwarf and B. N. Taylor, *Phys. Rev. Letts.*, **19**, 27, (1967). A. R. Long, *J. Phys. F.*, **3**, 2023, (1973). M. Welte, K. Lassman and W. Eisenmenger, *J. de Physique, Supp.*, **C4**, 25, (1972).
161. R. C. Dynes and V. Narayanamurti, *Phys. Rev.*, **B6**, 143, (1972).
162. H. Kinder, *Phys. Rev. Lett.*, **28**, 1564, (1972).
163. W. Forkel, to be published.
164. W. Forkel, M. Welte and W. Eisenmenger, *Phys. Rev. Letts.*, **31**, 215, (1973). W. Eisenmenger, in "Phonon Scattering in Solids", eds., L. J. Challis, V. W. Rampton and A. F. G. Wyatt, Plenum, London, (1976).
165. W. Dietsche and M. Kinder, in "Phonon Scattering in Solids", eds., L. J. Challis, V. W. Rampton and A. F. G. Wyatt, Plenum, London, (1976).
166. A. R. Long, *J. Phys. F.*, **3**, 2040, (1973).
167. R. Windheim and H. Kinder, in "Phonon Scattering in Solids", eds., L. J. Challis, V. W. Rampton and A. F. G. Wyatt, Plenum, London (1976).

168. P. K. Tien and J. Gordon, *Phys. Rev.*, **129**, 647, (1963).
169. A. Silver, *I.E.E.E. Trans. Magnetics*, **MAG-11**, 794, (1975); M. McColl, R. J. Pedersen, M. F. Bottjer, M. F. Millea, A. H. Silver, and F. L. Vernon, *App. Phys. Letts.*, **28**, 159, (1976).
170. S. Shapiro, *Phys. Rev. Letts.*, **11**, 80, (1963).
171. C. C. Grimes and S. Shapiro, *Phys. Rev.*, **169**, 397, (1968).
172. Proc. Int. Conf. on Detection and Emission of Electromagnetic Radiation with Josephson Junctions, *Rev. Phys. Appl.*, **9**, 1, (1974).
173. R. K. Elseley and A. J. Sievers, *Rev. Phys. Appl.*, **9**, 295, (1974).
174. T. D. Clark, *Rev. Phys. Appl.*, **9**, 207, (1974).
175. D. R. Tilley, *Phys. Letts.*, **33A**, 205, (1970).
176. T. F. Finnegan, J. Wilson, and J. Toots, *Rev. Phys. Appl.*, **9**, 200, (1974).
177. B. T. Ulrich, *Rev. Phys. Appl.*, **9**, 111, (1974).
178. J. Claasen, Y. Taur, and P. L. Richards, *App. Phys. Letts.*, **25**, 759, (1974).
179. T. G. Blaney, *Rev. Phys. Appl.*, **9**, 279, (1974).
180. D. G. McDonald, *App. Phys. Letts.*, **24**, 335, (1974).
181. T. G. Blaney and D. J. E. Knight, *J. Phys. D*, **7**, 1882, (1974).
182. T. G. Blaney, C. C. Bradley, G. T. Edwards, B. W. Jolliffe, D. J. E. Knight, and P. T. Woods, Conf. on Precision Electromagnetic Measurements, I.E.E.E., London, (1974).
183. M. J. Feldman, P. T. Parrish, and R. Y. Chiao, Proc. 14th Int. Conf. on Low Temp. Phys., North Holland–American Elsevier, New York, (1975).
184. P. T. Parrish, M. J. Feldman, and R. Y. Chiao, Proc. 14th Int. Conf. on Low Temp. Phys., North Holland–American Elsevier, New York, (1975).
185. W. H. Parker, D. N. Langenberg, A. Denenstein, and B. N. Taylor, *Phys. Rev.*, **177**, 639, (1969).
    A. Denenstein, T. F. Finnegan, D. N. Langenberg, and B. N. Taylor, *Phys. Rev.*, **B1**, 4500, (1970).
186. T. F. Finnegan, A. Denenstein, and D. N. Langenberg, *Phys. Rev.*, **B4**, 1487, (1971).
187. B. W. Petley and K. Morris, *Metrologia*, **6**, 46, (1970).
    I. K. Harvey, J. C. MacFarlane, and R. B. Frenkel, *Phys. Rev. Letts.*, **25**, 853, (1970).
188. B. F. Field, T. F. Finnegan, and J. Toots, *Metrologia*, **9**, 155, (1973).
189. D. E. McCumber, *Phys. Rev. Letts.*, **23**, 1228, (1969).
190. T. F. Finnegan, A. Denenstein, D. N. Langenberg, J. C. McMenamin, D. E. Novoseller, and L. Cheng, *Phys. Rev. Letts.*, **23**, 229, (1969).
191. V. E. Kose and D. B. Sullivan, *J. Appl. Phys.*, **41**, 169, (1970).
192. B. N. Taylor, W. H. Parker, and D. N. Langenberg, *Rev. Mod. Phys.*, **41**, 375, (1969).
193. T. Applequist and S. J. Brodsky, *Phys. Rev. Letts.*, **24**, 562, (1970).
194. R. F. Dziuba, B. F. Field, and T. F. Finnegan, *I.E.E.E. Trans. Inst. Meas.*, **IM-23**, 264, (1974).
195. J. Clarke, *Amer. J. Phys.*, **38**, 1070, (1970).
196. D. W. Palmer and J. E. Mercereau, *App. Phys. Letts.*, **25**, 467, (1974).
197. J. M. Eldridge and J. Matisoo, Proc. 12th Int. Conf. on Low Temp. Phys., Academic Press of Japan, Tokyo, (1971).

198. S. Basaviah, J. M. Eldridge, and J. Matisoo, *J. Appl. Phys.*, **45**, 457, (1974).
199. G. B. Donaldson, D. J. Brassington, and W. T. Band, *J. Phys. F—Metal Phys.*, **5**, 1726, (1975).
200. A. R. Long, private communication.
201. P. K. Hansma, *J. Appl. Phys.*, **45**, 1472, (1974).
202. J. L. Miles and P. M. Smith, *J. Electrochem. Soc.*, **110**, 1240, (1963).
203. J. H. Greiner, *J. Appl. Phys.*, **42**, 5151, (1971).
204. J. H. Greiner, *J. Appl. Phys.*, **45**, 32, (1974).
205. R. F. Broom, R. Jaggi, R. B. Laibowitz, and Th. O. Mohr, Proc. 14th Int. Conf. on Low Temp. Phys., North Holland–American Elsevier, New York, (1975).
206. L. O. Mullen and D. B. Sullivan, *J. Appl. Phys.*, **40**, 2115, (1969).
207. K. Schwitdal, *J. Appl. Phys.*, **43**, 202, (1972).
208. See, for example, J. M. Eldridge and D. W. Dong, *Surf. Sci.*, **40**, 512, (1973); J. M. Eldridge, *ibid*, 531.
209. Full details are given by P. Cardinne, J. Nordman, and M. Renard, *Rev. Phys. Appl.*, **9**, 167, (1974).
210. C. L. Huang and T. van Duzer, *I.E.E.E. Trans. Magnetics*, **MAG-11**, 766, (1975).
211. P. E. Gregers-Hansen, M. T. Levinsen, and G. Fog Pedersen, *J. Low Temp. Phys.*, **7**, 99, (1972).
212. J. E. Mooij, C. A. Gorter, and J. F. Noordam, *Rev. Phys. Appl.*, **9**, 173, (1974).
213. R. Adde, P. Crozat, S. Gourrier, G. Vernet, M. Bernheim, and D. Zenatti, *Rev. Phys. Appl.*, **9**, 179, (1974).
214. R. B. Laibowitz, J. M. Viggiano, and M. Hatzakis, *Rev. Phys. Appl.*, **9**, 165, (1974).
215. R. A. Kirschman, M. A. Notarys, and J. E. Mercereau, *I.E.E.E. Trans. Magnetics*, **MAG-11**, 778, (1975).

Conclusion Added in Proof

Since 1976 progress in superconductive thin film electronics and its applications has been more than ample to justify the optimism expressed in this chapter. Particularly exciting have been the developments in computer devices where all the formidable apparatus of microcircuit technology has been brought to bear on the problems, and in SQUID magnetometry which is notable for the rapidly widening number of scientific fields in which its unparalleled sensitivity is proving useful. The reader is referred to the proceedings of the most recent conference[1,2] for details, and to B. B. Schwartz and S. Foner (eds.), "Superconductor Applications: SQUID and Machines", NATO Advanced Study Institutes Series B, Vol. 21, Plenum Press, New York and London, (1977).

References

1. H. D. Hahlbohm and H. D. Lubbig (eds.), "SQUID: Superconducting Quantum Interference Devices and their Applications", Proc. Int. Conf. on Superconducting Quantum Devices, Berlin 1976. W. de Gruyter, Berlin, (1977).

2. B. S. Deaver, Jr., E. A. Edelsack, C. M. Falco and S. Wolf (eds.), Proc. Conf. on Future Trends in Superconductive Electronics, Charlottesville, Virginia, March 1978. To be published in Am. Inst. of Physics Conference Proceedings Series, (1978).

Subject Index

A

Absentee layer, 333
Absorption,
 in coatings, 365
 coefficient of cuprous sulphide, 550
 by cuprous sulphide, 529
 measurement of, 366
Absorption coefficient,
 as function of energy, 499
 of indirect gap semiconductors, 498
 Moss-Burstein shift, 500
 of semiconductors, 491
Absorption losses,
 in light guides, 380
Access time, 686
Acoustic shear waves, 533
Activation energy,
 in amorphous films, 274
 of cell degradation, 574
 electrostatic origins of, 166–167
 field dependence of, 169
 influence of permittivity, 179
 in mixed photoconductors, 462
 of photoconductors, 452
 size dependence of, 166–167
 of tantalum oxidation, 97
 for tunnelling via interface states, 556
Activators,
 in chemical reduction plating, 50
Addressing,
 of display panels, 479
Adsorption,
 effects on film properties, 186
Ageing,
 of film resistors, 98
 of optical coatings, 365
 of polarisation, 731
Alkali films, 78–79
Alloys,
 conduction in, 65

Aluminium,
 as visual reflector, 344
Aluminium antimonide,
 band gap of, 577
Aluminium oxide
 evaporation crucible, 30
 protective layers, 344
Ammeters,
 SQUID devices, 813–814
Amorphous semiconductors, 245–319
 acousto-optical properties of, 303
 applications of, 288
 bubble devices, 675
 interaction of radiation with, 300
 preparation of, 252
 solar cells, 580
Amplification factor,
 of photoconductors, 430
Amplifiers,
 superconducting, 791–802
Angle of deposition,
 effect on orientation, 529
Annealing,
 effect on conduction, 77
 of semiconductor films, 231
Anodisation, 42–44, 118–119
 gaseous, 39
 oxide formation, 118
 principles of, 42–44
Anthracene,
 photoconduction in, 463
Antimony trisulphide,
 spectral response of, 454
 in vidicon, 474
Antireflection coatings, 334–340
 infrared spectroscopy, 336
 use of magnesium fluoride, 335
 "moth's eye coating", 336
 performance of, 339
 reflectance of, 334

Subject Index

Arsenic triselenide,
  photoresponse of, 461
Asbestos particles,
  detection of, 812
Auger recombination, 507
Avalanche effects,
  in dielectrics, 43

B

Band bending, 210
  at crystallite boundaries, 523
  at interfaces, 141
Band diagram,
  of discontinuous film, 170
  of heterojunction, 509
  of ideal solar cell, 494
  of metal/dielectric contacts, 138
Band-pass filters, 349–361
Bandwidth,
  of Fabry Perot filters, 350
Barium stearate,
  monomolecular films, 49, 120
Barium titanate,
  ferroelectric effect in, 698
  preparation of, 115
  spontaneous polarisation of, 731
Barrier height,
  between grains, 442, 523
  dependence on illumination, 555
Barrier modulation, 225, 235
Barrier theory,
  of transistors, 211
Barrier trapping, 216
Beam splitters, 406–408
  effective index perturbation, 406
Becquerel cells, 585
Beer–Lambert law, 431
Beryllia,
  bias sputtering, 36
  as evaporation crucible, 30
Binding energy,
  of paired electrons, 753
Birefringence,
  in ferroelectrics, 738
Bismuth,
  p-type doping of cadmium sulphide, 454
Bismuth titanate,
  as ferroelectric element, 481
Bitter patterns, 615

Bloch walls,
  bubble domains, 673
  energy of, 623
Bolometers, 149
  cooled thermistor, 726
  superconducting, 785–787
Boltzmann's equation, 58
Borosilicate glass,
  light guides, 412
Bragg grating couplers, 394–399
  transmission and reflection types, 398
Breakdown, 118, 142–143
  of anodised films, 118
  self-healing, 143
Bubble, 609, 669–678
  domains, 609
  materials, 674–676
  propagation, 676
  sensing of, 678
  size of, 672
  stability of, 672
  storage elements, 669–674
Buckling,
  magnetisation reversal, 654

C

Cadmium selenide,
  photoconductor devices, 447
  transistors, 216
Cadmium sulphide,
  alloys of, 583
  evaporation of, 527
  photoconductor devices, 447
  selenium diffused light guides, 413
  solar cells, 526
  sputtering of, 532
  structure of, 520
  vapour deposition of, 534
Cadmium telluride,
  anomalous photovoltaic effect in, 523
  band gap of, 577
  homojunctions, 578
Capacitance,
  of crossovers, 106
Capacitance/voltage plots, 571
Capacitors, 118, 144–149
  pressure sensitivity of, 149
  production by anodisation, 118
  structure of, 147
  temperature sensitivity of, 149

Subject Index

Capture cross section,
  of interface traps, 222
  of recombination centres, 433
Carrier density,
  in bulk crystallites, 214
Catalysis, 50
Cathodoluminescence,
  in copper sulphide, 551
Ceramic solar cells,
  degradation of, 576
Cerenkov radiation, 410
Cermets,
  maximum thermometers, 163–206
  percolation conduction in, 193–194
  selective filters, 202
  sheet resistance of, 163–206
  stability of, 201
  strain gauges, 201
  temperature coefficient of, 195
  temperature coefficient of resistance, 153
  vidicons, 202
Chalcocite, 544
Chalcogenide glass,
  light guides, 413
Chemical reduction plating, 50–51
  electroless deposition, 50
Chemical vapour deposition,
  of cadmium sulphide, 449
Chemiplating,
  of cuprous sulphide, 536
Child–Langmuir law, 441
Cobalt,
  dispersion in, 626
Coercivity, 616, 652–662
  as function of film thickness, 657
  as function of temperature, 686
  origins of, 652
Coherence length,
  in thin films, 771–772
  of wave functions, 759
Colloids,
  deposition from, 51
Communications systems, 367
Compensation,
  in junction region, 554
Compensation temperature,
  in ferromagnets, 686
Computer simulation,
  of filter production, 364
Conductivity,
  of discontinuous films, 165–180

Conductivity—cont.
  of metals, 165–180
Conductors, 100–105
  materials, 101
  reliability of, 103
  requirements of, 101
  on semiconductors, 100
Connectivity of lattice, 183
Contacts, 135–142, 261–265
  accumulation layer, width of, 140
  to amorphous films, 261
  depletion layer, width of, 139
  to diectric films, 135
  diffusion of, 264
  effect on device lifetime, 264
  to photoconductors, 435
  to transistors, 237
Contamination,
  of evaporants, 27
  of substrates, 521
Continuity equations,
  of minority carriers, 502
Copper,
  doping of copper sulphide, 568
Copper indium sulphide,
  band gap of, 585
Copper sulphide,
  optimum thickness of, 561
  phase diagram of, 545
  solar cells, 526
Copper telluride,
  phases of, 581
Coupling, 382–402
  electron/phonon, 820
  layers, 343
  waveguides to optical circuits, 382
Creeping,
  of domain walls, 620
Critical currents,
  in superconductors, 775–776
Critical field,
  for magnetisation reversal, 612
Critical magnetic fields,
  in thin films, 772
Critical temperature,
  in superconductors, 745
Cross-tie walls,
  energy of, 623
Cryolite,
  refractive index of, 340
Cryotrons, 792–795

Croyotrons—*cont.*
  construction of, 793
Curie point,
  writing, 684
Curie–Weiss law, 730
Curling,
  magnetisation reversal, 654
Current density,
  in discontinuous films, 170
Current/voltage characteristics, 566
  of thin film transistors, 212

D

Damping,
  of cryotrons, 797
Dangling bonds, 207
Debye length, 222
  of grain boundary barrier, 216
Decay energy,
  of defects, 77
Decomposition, 46
Defects, 367
  in optical coatings, 367
  scattering by, 64
Degradation, 574–577
  of solar cell performance, 574
  of solar cells, 538
Demarcation level,
  of electron traps, 438
Dember field,
  origin of, 502
Depletion layer,
  at grain surfaces, 524
  in transistors, 210
  width of, 507
Depth profiling,
  by ion-etching, 562
Detectivity, 714
  of superconducting bolometers, 785
Dichroc beam splitters, 363–364
Dielectric films, 113–161
  continuity of, 122
  as crossovers, 150
  homogeneity of, 123
  mechanical stability of, 122
Dielectric loss, 127–130
  loss angle, 127
  polarisation effects, 129
Differential capacitance, 225

Diffusion,
  dependence of coefficient on doping, 541
  via grain boundaries, 543
  during junction formation, 540
Diffusion lengths,
  of minority carriers, 492
  in cadmium sulphide, 532
Diffusivity,
  of pyroelectric target, 720
Diginite, 544
Diode factor, 506
Discontinuous magnetisation reversal,
  critical field for, 615
Dislocation density,
  influence on lattice mis-match, 562
Dispersion,
  of magnetisation, 625–628
  of refractive index, 364
  of thin film guides, 404
Display devices, 477–482
  transistor addressed, 241
Disproportionation, 44
Dissociation, 527
  of cadmium sulphide, 527
Djurleite, 544
Domains, 609–615
  ferroelectrics, 704
  in magnetic films, 609
  wall motion of, 678–683
  wall stability of, 624
  wall width, 610
Doping,
  by gas discharge, 258
Drain current,
  dependence on illumination, 234
Drift,
  accelerated tests, 98
Dummy medium,
  design uses, 348
Dust particles,
  influence on laser performance, 366

E

Economics,
  of solar cell arrays, 588
Eddy currents,
  in conducting substrates, 635
Edge filters, 344–349
  use of eight wave layers, 345

Effective mass,
  of electrons, 497
Effective thickness,
  of light guides, 379
Efficiency,
  as function of energy gap, 500
  of solar cell, 496
Electrochemical degradation,
  of copper sulphide, 575
Electrodeposition,
  of magnetic films, 661
Electrofax, 475
Electroforming, 49, 154
  in discontinuous films, 188
Electroless deposition,
  of magnetic films, 656–661
Electroluminescence, 241
Electron affinity, 508
Electron beam heating, 31
Electron beam imaging, 834
Electron guns, 31
Electron pairing, 752
Electron transport, 58–87
Electro-optic devices, 415–421
  modulator efficiency, 417
Electrophotography, 475–477
  operational principles, 475
Electroplating, 44–47
  basic principles of, 47
  practical aspects of, 49
Electrotransport, 103
Encapsulation,
  of solar cells, 570
Energy,
  pay-back of, 589
  storage of, 588
Energy gap,
  anisotropy of, 772
  measurement of, 818–820
  in superconductors, 754–758
Enhancement theory,
  of transistors, 208
Equivalent circuit,
  of ideal solar cell, 493
  of practical solar cells, 513
Equivalent index, 340
  of zinc-sulphide/cryolite combination, 345
Etalon filters, 354–356
  construction of, 355
  temperature coefficient of, 355

Evanescent waves,
  power carried by, 378
Evaporation, 23–30, 115–116
  of dielectrics, 115
  of magnetic films, 654–656
  principles of, 26
  sources, 27
Exchange energy, 609
Excitons, 433
Extinction coefficient, 431

F

Fabry Perot filters, 350–358
  angular behaviour of, 356–358
  construction of, 351
  performance of, 353
  transmittance of, 350
Failure,
  mean time to, 104
Fanning,
  magnetisation reversal, 654
Faraday effect, 684
Fast states, 225–227
  effect on conductivity modulation, 225
Fermi distribution,
  of electrons, 168
Fermi surface, 84
Ferroelectric films, 697–739
  displays, 479
  fabrication of, 698–703
Ferroelectric memory, 739
Ferroelectric/photoconductive devices, 738
Field distribution,
  in light guides, 373
Field-effect mobility, 218
Filamentary growth,
  observation of, 284
Fill-factor,
  definition of, 495
Finesse,
  of Fabry Perot filters, 350
Flash evaporation, 29
Fluidised bed,
  coating, 120
Flux closure, 639
Flux exclusion, 747
Flux flow resistance, 750
Fluxmeters,
  superconducting, 802–818
Fluxoid, 751

Flux pinning, 750
Flux quantisation,
  magnitude of, 748
Flux shuttle,
  as storage element, 800
Flux transformers, 810
Forming, 273–275
  effect on carrier mobility, 273
  of metallic filaments, 273
  threshold voltage, 273
Forward current, 505–508
  in diodes, 505
Free carrier lifetime,
  in amorphous films, 276
Frenkel excitons, 432

G

Gain,
  of photoconductors, 436
Gallium arsenide, 421
  band gap of, 577
Galvanometers,
  superconducting, 802–818
Garnets,
  for bubble devices, 674
Gas detectors, 186
Gaseous anodisation, 39
Gate capacitance, 210
Gauge factor, 107
Generation rate,
  of carriers in heterojunctions, 504
Germanium,
  photoconductor devices, 446
  refractive index of, 335
  thermal detectors, 726
Giaever tunnelling, 787–791
  switches, 802
Glass absorption filters, 351
Glass formation, 252
Glow discharge, 35
Golay detector, 726
Gold,
  as light guide cladding, 381
  vapour pressure of, 26
Graded index films, 336
Grain boundaries, 211–213
  barrier model of, 213
  trapping, 211
Grating couplers, 391–393
Grating filters, 404–406

Grid collector,
  design of, 517
Gridding, 570
Growth rate,
  of gold films, 27
Guide thickness, 375

H

Hall coefficient,
  temperature dependence, 452
  thickness dependence, 451
Hall effect, 85
Heat treatment,
  optimisation, 539
  of solar cells, 537
Heterojunctions,
  solar cells, 581
High field conduction, 133–135
  in dielectrics, 133
  Poole Frenkel effect, 133
  Schottky effect, 133
High field domains, 557
High temperature films, 95
Holographic recording, 683
Hopping conduction,
  in amorphous materials, 172
  decay distance, 182
  in dielectrics, 132–133
  in organic photoconductors, 467
Hot wall techniques, 29, 530
Hydrogen discharge,
  treatment of copper sulphide, 568
Hysteresis, 613
  distortion of loop, 702, 731
  in ferroelectrics, 700
  of record/replay system, 645
  in superconductors, 766

I

Ideallity factor,
  of various efficiency solar cells, 567
Image forces, 174
Image intensifier, 479
Imaging systems, 304–307, 723–726
  using amorphous films, 304
Impurity band conduction, 183
Impurity scattering, 63
Indium,
  as cadmium sulphide dopant, 530
Indium antimonide,
  ferroelectric properties of, 702

Subject Index

Indium antimonide—*cont.*
  as infrared detector, 454–455
  transistors, 223
Indium copper selenide,
  in heterojunctions, 577
Indium phosphide,
  band gap of, 577
  heterojunction cells, 583
Indium/tin oxide,
  electrodes, 448
Induced transmission filters, 360–361
Inductive recording heads, 663–666
Inductors and capacitors, 105–107
Infrared detectors,
  based on amorphous films, 304
Infrared radiation generators, 825–826
Inhomogeneities,
  in refractive index, 365
Injection currents,
  in diodes, 506
Integrated optical circuits, 399–402
  interconnections to, 399
Interconnections,
  in displays, 478
Interface barrier height, 222
Interface recombination velocity, 557
Interface state density,
  influence on forward current, 510
Interface states, 220
Interfacial polarisation, 129
Interference effects,
  in solar cells, 504
Intrinsic photoconduction, 435
Inversion,
  in magnetic films, 616
Ion-beam sputtering, 37
Ion drift, 220
Ion-implantation,
  crystalline/amorphous transition, 257
  of metal films, 153
Ion-plating, 40
Island capillary interaction, 192, 195
Island separation,
  distribution of, 181

J

Josephson effect, 761–769
  alternating current, 825–826
  applications of, 769
Junction formation,
  by dry techniques, 539

Junction formation—*cont.*
  thermodynamics of, 538
Junction width,
  in solar cells, 565

K

Kerr effect, 615

L

Labyrinth switching,
  of domains, 618
Lamination,
  of solar cells, 537
Langmuir expression,
  of evaporation, 26
Langmuir films, 51
Large scale integration,
  of superconducting circuitry, 799
Laser beam evaporation, 32
Laser beam store, 684
Laser damage,
  influence of spot-size, 367
Laser-fibre coupling, 402
Laser mirrors, 366
Lead telluride,
  refractive index of, 353
Lead–tin telluride, 462
Lead zirconate titanate,
  ferroelectric properties of, 702
Leakage currents,
  in tunnel junctions, 832
Lifetime,
  of photogenerated carriers, 434
  of solar cells, 574
Light emitters, 188–189
Light guides, 367–422
  theory of, 370–376
Light valve, 480
Linear amplifiers,
  superconducting, 802
Liquid crystals, 478
Liquid phase epitaxy, 51
Liquid semiconductors,
  switching in, 269
Lithium ion bombardment,
  of fused quartz, 412
Lithium niobate,
  light amplitude modulator, 739
Logic elements,
  superconducting, 794
Lorentz microscopy, 615

Loss mechanisms,
  in light guides, 380
Luminescence,
  in displays, 477

M

Magnetic films, 603–696
  preparation of, 634
Magnetic modulation,
  of resistance, 802
Magnetic transition devices, 780–784
Magnetisation,
  induced anisotropy, 606
  reversal of, 611
  ripple of, 616
  of superconductors, 748
Magnetisation anisotropy,
  in evaporated films, 654
Magnetisation dispersion,
  measurement of, 625
Magnetisation gradient, 646
Magnetisation reversal,
  by domain wall creeping, 620
Magnetocardiography, 812
Magnetometry,
  using SQUIDs, 810–813
  superconducting, 769
Magneto-optic devices, 421
Magneto-resistance, 85–86
Magnetoresistive recording heads, 667
Magnetostriction, 607
Magnetron sputtering, 461
Manganese oxide,
  in electrolytic capacitors, 147
Mass action,
  law of, 540
Mass memory stores,
  using amorphous films, 299–303
  based on optical properties, 302
Materials characterisation, 258–261
  Auger electron spectroscopy, 260
  differential thermal analysis, 258
  electron spectroscopy for chemical analysis, 260
  electron spin resonance, 260
  thermo-gravimetric analysis, 258
  use of ultra-high vacuum, 259
  x-ray photoelectron spectroscopy, 260
Matrix-addressed displays, 738
Matthiessen's rule, 68

Mean free path, 79–82
  measurement of in thin films, 79
Membranes,
  electrical properties of, 470
Memories,
  plated wire, 639
Memory applications,
  of magnetic films, 630–641
Memory devices, 152, 282–285
  applications of, 288–292
  as data stores, 288
  fabrication of, 286
  instability of, 287
  materials, 283
  operation of, 282
  effect of pressure, 284
  in radiation environments, 289
  reliability of, 284
  telephone applications, 291
Metal-clad guides, 381
Microwave filters, 338
Military applications,
  of plated wire memories, 641
Mixed phase conduction, 191
Mixed photoconductors, 457–463
Mobility,
  of bubble domains, 674
  in cadmium sulphide, 530–531
  of ions in electrolyte, 44
  of minority carriers, 497
  reduction of by surface scattering, 452
Mobility edge,
  in mixed photoconductors, 460
Mode frequency,
  in pyroelectric detectors, 729
Mode numbers,
  in light guides, 375
Mode splitter, 407
Modulators, 418–421
  amplitude modulation, 420
  intensity modulators, 420
  phase modulation, 418
Moisture,
  effects on filter performance, 354, 365
Momentum ordering,
  of paired electrons, 754
Monte Carlo model,
  conduction simulation, 185
  of percolation conduction, 191
Moss-Burstein shift,
  in copper sulphide, 551

"Moth's eye coating",
  specular reflectance of, 336
Multiple cavity filters, 358–360

N

Néel walls,
  energy of, 623
Network conduction,
  in human body, 470
Network model,
  of polycrystalline films, 524
Neutron irradiation, 280
Nichrome films, 93–94
  preparation of, 93
  resistivity of, 93
  stability of, 94
Niobium pentoxide,
  light guides, 412
Nitriding, 46
Noise,
  in amplifiers, 716
  in D.C. SQUIDs, 807–809
  equivalent power, 713
  in R.F. SQUIDs, 807
  in superconducting bolometers, 785
  in thin film pyroelectrics, 730
  in transistors, 238
  from various sources, 715
Normalised conductivity, 177
Notation of optical filters, 334
Nucleation, 25

O

Olsen effect, 74
Open circuit voltage,
  as function of light intensity, 566
  in ideal solar cells, 494
Optical admittance, 327
Optical communications, 472
Optical devices, 321–427
Optical fibres, 382
Optical minus filters, 349
Optical properties,
  of copper sulphide, 548
Optical recording, 683–686
Optical thickness,
  control of, 364
Ordering,
  in magnetic films, 607
Organic photoconductors, 442, 463–470

Oscillators,
  for telephone systems, 92
Oxidation, 46
  of tantalum, 97
Oxide layers,
  in Schottky barriers, 512
Oxygen,
  effect on p-n junctions, 537
  inhibiting of photoconduction, 452

P

Packing density, 520
  in magnetic films, 652
Page composer,
  ferroelectric, 738
Parametric amplifiers, 829–831
Particle detectors, 791
Passivation,
  of discontinuous films, 186
Penetration depth,
  of flux in superconductors, 747
  in thin films, 771–772
Percolation conduction, 183–186
  critical probability, 183
Permittivity, 125–127, 701,
  of barium titanate films, 700
  of dielectric films, 125
  temperature coefficient of, 126
Phase factor,
  of harmonic wave, 326
Phase-grating couplers, 393–394
Phase modulators, 418
Phase transformation,
  laser induced, 301
Phonon fluorescence, 823
Phonons,
  generation and detection of, 822–824
  in indirect transitions, 498
Phonon spectroscopy, 818–820
Phonon spectrum, 819
Photoconductivity,
  theory of, 432–436
  in thin films, 429–485
Photoconductive devices,
  required properties of, 431
Photo-electric effect,
  in copper oxide, 489
Photo-electric threshold, 508
Photogalvanic cells, 586
Photogenerated carriers,
  density of, 437

Photogenerated current,
  in ideal solar cell, 493
Photographic effects,
  in amorphous films, 307
Photosynthesis, 482
Pin-holes, 529
Plasma frequency, 801
Plasma reactions, 39
Platinum
  alloy resistors, 95
  diffusion barrier, 102
Plutonium, 482
Polarisation, 126, 361–363
  headlight dazzle applications, 363
  locked in, 722
  in thin film pyroelectrics, 730–735
Polymer films, 52
Polymerisation, 45
  via glow discharge, 49
Polyvinylcarbazole,
  photoconduction in, 469
Polyvinyl chloride,
  deposition of, 52
Polyvinylidene fluoride,
  as pyroelectric detector, 734
Potential,
  due to impurities, 63
Potential barrier height, 170
Power flow, 376–380
  in light guides, 376
Poynting vector, 328
Prism input coupler, 384–391
  use of dust particles, 390
  efficiency of, 388
Prism output coupler, 383–384
  using evanescent waves, 383
Proton irradiation,
  of fused silica, 411
Proximity effect,
  microbridges, 776–777
Pyroelectric coefficient,
  definition of, 699
  measurement of, 705
Pyroelectric detectors, 723–726
  microphony in, 724
  specifications of, 724
Pyroelectric effect, 697–739
  applications of, 707
Pyroelectric materials,
  properties summary, 722

Q

Quantum efficiency,
  of pair generation, 492
Quasi Fermi level, 437
Quasiparticles, 754–758
  decay of, 823

R

Radiation,
  from amorphous films, 277
  damage, 576
  detectors, 470–472, 827–831
Radical distributions,
  in amorphous materials, 124
Radio frequency generators,
  superconducting, 769
Radio frequency SQUIDs, 803–807
  structure of, 804
Raman spectroscopy, 820–822
Reactive evaporation, 32
Reactive sputtering, 38
Recombination,
  in diodes, 507
  of excess minority carriers, 492
  in photoconductors, 433
Recording, 641–668
  computer analysis of, 643
  film heads, 662
  head density, 666
  losses, 652
  operation of, 642–648
  properties of magnetic films, 649
Rectification, 489
Reduction, 46
Reflectance and transmittance, 329
Reflectance zones,
  width of, 343
Reflection increasing coatings, 340–344
  scattering in, 341
Refractive index,
  imaginary and real parts of, 327
Refractory metals,
  as evaporation sources, 28
Registration, 239
Relaxation,
  of magnetic films, 642
Relaxation time, 60, 68
Reliability,
  of storage system, 685
Remanent polarisation, 705

Resistivity,
  of amorphous films, 255
  of cadmium sulphide, 530
  of copper sulphide, 554
  of dielectric films, 130
Resolving power,
  dependence on finesse, 350
Resonance condition,
  in light guides, 389
Response time, 436
Responsivity,
  definition of, 711
  thickness dependence of, 707
Resources,
  of cadmium, 592
R.F. sputtering, 38
R.F. supported sputtering, 37
Reverse saturation current,
  in p–n junctions, 492
Rhenium,
  resistors, 95
Ripple,
  in demagnetised films, 655
  of magnetisation, 625–628
  observation of, 627
Ripple elimination,
  in edge filters, 347
Roll-coating,
  of cadmium sulphide, 532

S

Sawtooth structure,
  in magnetic films, 650
Scattering,
  in coatings, 365
  grain boundary, 74
  losses, 380
  low angle, 74
  measurement of, 366
Schottky cells,
  analysis of, 512–513
  performance of, 579
Screen printing, 535
Second harmonic generation, 408–411
  efficiency of, 409
Selenium, 443–446
Semi-continuous films, 189–195
  conductivity of, 191
  critical exponent, 191
  critical probability, 190
  island/capillary interaction, 192

Semi-continuous films—*cont.*
  percolation conduction, 189
Sensitisers,
  in chemical reduction plating, 50
Series resistance,
  of dielectrics, 127
  negative values, 568
  origins of, 516
  effect on short circuit current, 514
  in transistors, 237
Sheet conductivity,
  of semiconductor layers, 210
Shift register, 680, 682
Short circuit current,
  as function of light intensity, 566
Shunt resistance,
  effect on open circuit voltage, 514
  origins of, 516
Silicon,
  band gap of, 577
  photoconductor devices, 446
  polycrystalline solar cells, 578
  refractive index of, 335
  solar cells, 489
Silicon oxide,
  protective layers, 344
  refractive index of, 353
Silver,
  reflectance of, 344
Sintering,
  of anodised tantalum, 148
Slow states, 219–225
  interfacial density of, 222
  ion drift, 220
Sodium tungstate,
  percolation conduction in, 193
Solar cells, 487–602
  historical development, 488–491
Solar spectrum,
  distribution of, 497
Solar thermal collectors, 202
Solubility,
  of copper in cadmium sulphide, 562
Solution deposition, 49–51, 120–122
  of monomolecular layers, 120
  of oxide films, 51
  of silicon dioxide, 120
Source/drain current,
  distribution of, 240
Space charge,
  in dielectrics, 142

Space charge—*cont.*
  in photoconductors, 441
Specific heat, 62
Spectral response,
  of semiconductors, 448
Specularity parameter, 71
  angular dependence, 73
  measurement of in thin films, 80–82
Spontaneous electric moment, 699
Spontaneous polarisation, 703–704
  as function of temperature, 701
  in thin films, 731
Spraying, 534
Sputtering, 33–41, 116–117
  basic principles of, 33
  of cadmium sulphide, 448
  of dielectrics, 116
  yield, 34, 255
Squareness,
  of chemically deposited films, 660
  of hysteresis loop, 648
SQUIDs
  direct current, 807–810
Stability,
  of film resistors, 96–100
Sticking coefficient, 519
Stop-bands,
  broadening of, 349
Storage,
  density of information, 686
  device design, 638–641
  devices, 632, 669–683
  matrix, 633
Strain gauges, 107–108, 187–188
Stress coefficient,
  of electron concentration, 525
Strip domains, 609, 670
Strip guides, 381–382
  radiation from, 382
Strip lines,
  interactions in, 636
Substrates,
  for cadmium sulphide, 528
  for solar cells, 519
  influence on stability, 98
Substrate temperature,
  influence on film properties, 520
Sulphurisation,
  of cadmium, 535
Superconducting devices, 777–780
  fabrication of, 777–778

Superconducting films, 744–835
  principles of, 745–769
Superconducting quantum interference devices,
SQUIDs, 802–818
Superconducting surface layers,
  applications of, 779
Superconductors,
  properties of, 746
Supercurrent,
  as function of applied field, 768
Surface band-bending, 218
Surface potential, 452
Surface recombination velocity,
  definition of, 502
  reduction of, 505
Surface roughness, 74
  influence on scattering losses, 365
  influence on specular reflection, 380
Surface scattering, 69
Surface space charge,
  penetration depth of, 215
Surface states, 509
Susceptibility,
  dependence on ripple, 627
Switching, 265–273, 628–630
  in amorphous films, 265
  of cryotrons, 797
  filament formation, 267
  forming, 269
  in magnetic films, 628
  memory switch, 266
  as function of pulse field, 629
  in superconductors, 791–801
  theories of, 278–281
  in thin film pyroelectrics, 730–735
  threshold switch, 266

T

Tantalum films, 88–93
  phases of, 89
  preparation of, 89
  resistivity of, 91
  stability of, 92
Tantalum oxide,
  light guides, 412
Tapered coupler, 376, 399
Temperature coefficient of resistance, 83–84
  effects of impurities, 84
  of thin metal films, 83

Temperature dependence,
  of film conductivity, 187
Temperature inhomogeneity,
  in thin film pyroelectrics, 733
Temperature resolution,
  of pyroelectric materials, 721
Terminations, 100
Tetracyanodiquinomethane,
  photoconduction in, 463
Thermal conductivity,
  of electrode materials, 277
Thermal detectors, 723–728
Thermal expansion,
  of substrates, 519
Thermal growth,
  of silicon dioxide, 41
Thermal imaging, 718–721
  on pyroelectric target, 718
Thermal loss,
  from pyroelectric specimens, 710
Thermal mis-match,
  effect on film stress, 525
Thermal scattering, 61
Thermionic emission,
  across crystallite boundaries, 523
  of electrons between islands, 165
Thermistors, 154, 187
Thermoelectric effect, 84–85
  curvature of Fermi surface, 84
  thin film thermocouples, 85
Thermometry, 705
Thin film cermets,
  percolation conduction in, 193
Thin film circuits, 149–150
  elements of, 149
Thin film lasers, 415, 421–422
  semiconductor injection devices, 421
Thin film lenses, 401
Third harmonic measurements, 131
Thorium fluoride, 342
Threshold devices, 275–278, 292–299
  as amplifiers, 295
  applications of, 292–299
  as display switches, 292
  as inductors, 297
  in logic circuits, 294
  ON state, 275
  in oscillator circuits, 296
  as strain gauges, 297
  as temperature sensors, 297
  thermal runaway, 275

Titanium,
  resistors, 94
Titanium dioxide,
  light guides, 412
Topography,
  of cadmium sulphide, 530
Torque curves, 607
Total reflection, 371
Transconductance, 235–237
Transducers,
  piezoelectric, 737
Transformer,
  superconducting D.C., 781–784
Transistors, 207–243
  cadmium-selenide, 227
  characteristics of, 232
  materials, 227
  matrix-addressed, 241
  production of, 228
  structure of, 150–152
  yield of, 231
Transition metal alloys,
  conduction in, 66
Transition metal oxides, 285–286
  switching in, 285
Transition temperature,
  of ferroelectrics, 700
  thickness dependence of, 770
  in thin films, 770–771
Transition time,
  between islands, 168
  of SQUIDs, 805
Transition width,
  of recorded data, 647
Transit time, 436
Transmission lines, 780–781
Trap density, 222–225
  effect of surface potential, 224
Trapping, 211
  at grain boundaries, 210
Trapping centres, 437
Trapping parameter,
  as function of gate voltage, 220
Triglycine sulphate,
  ferroelectric properties of, 702
Triode sputtering, 37
Tunnel junctions,
  fabrication of, 831–835
Tunnelling, 166–186, 179–180
  through conduction band spike, 555
  critical voltage for, 175

Tunnelling—*cont.*
  cryotrons, 795–801
  devices, 779
  in dielectric films, 135
  in diodes, 508
  in discontinuous films, 166
  inelastic, 820–822
  multi-step, 511
  optimum conduction path, 172
  spectrum of, 820–822
  influence of substrate, 179
  transition probability, 167
  transmission coefficient, 169
  influence of traps, 179
Two-dimensional optics, 402–404

U

Ultraviolet polymerisation, 45
Ultraviolet reflectance,
  of Fabry Perot filters, 352
Umklapp process, 74
Uniaxial anisotropy, 605–609
  as function of composition, 608
  field induced, 606
  growth induced, 675

V

Vapour phase growth, 44–46
Velocity of propagation,
  in light guides, 369
Vibration theories,
  of ferroelectricity, 704
Vidicon, 472–474
  construction of, 473
  pyroelectric, 720

Violanthrene,
  photoconduction in, 465
Voltmeters,
  SQUID devices, 813–814
Vortex pinning, 750

W

Wannier excitons, 433
  in organic photoconductors, 465
Wave function decay, 222
Waveguide effects,
  in amorphous films, 303
Waveguide materials, 411–415
Wear resistance,
  of magnetic films, 661

X

Xerographic effects, 307–309
  in amorphous films, 307
  radiography, 308
Xerography, 475

Z

Zinc oxide, 456–457
  in electrofax process, 456
  pyroelectric films, 733
  sputtered waveguides, 412
  waveguides, 410
Zinc selenide,
  in hetrojunctions, 577
  refractive index of, 336
Zinc sulphide,
  refractive index of, 335
  waveguides, 411